SKELETAL DEVELOPMENT AND REMODELING IN HEALTH, DISEASE, AND AGING

ANNALS OF THE NEW YORK ACADEMY OF SCIENCES
Volume 1068

SKELETAL DEVELOPMENT AND REMODELING IN HEALTH, DISEASE, AND AGING

Edited by
Mone Zaidi

Published by Blackwell Publishing on behalf of the New York Academy of Sciences
Boston, Massachusetts
2006

Library of Congress Cataloging-in-Publication Data

Skeletal development and remodeling in health, disease, and aging/edited by Mone Zaidi.
p. cm. – (Annals of the New York Academy of Sciences ; v. 1068)
Includes bibliographical references and index.
ISBN-13: 978-1-57331-583-8 (alk. paper)
ISBN-10: 1-57331-583-4 (alk. paper)
1. Bones–Growth–Congresses. 2. Bone remodeling–Congresses. I. Zaidi, Mone. II. Series.
[DNLM: 1. Bone Remodeling–physiology–Congresses. 2. Bone and Bones–physiology–Congresses. 3. Genetic Techniques–Congresses. 4. Osteoporosis–therapy–Congresses. W1 AN626YL v.1068 2006/WE 200 S62677 2006]
Q11.N5 vol. 1068
[QP88.2]
500 s–dc22
[612.8'6]

2006011996

The *Annals of the New York Academy of Sciences* (ISSN: 0077-8923 [print]; ISSN: 1749-6632 [online]) is published 28 times a year on behalf of the New York Academy of Sciences by Blackwell Publishing, with offices located at 350 Main Street, Malden, Massachusetts 02148 USA, PO Box 1354, Garsington Road, Oxford OX4 2DQ UK, and PO Box 378 Carlton South, 3053 Victoria Australia.

Information for subscribers: Subscription prices for 2006 are: Premium Institutional: \$3850.00 (US) and £2139.00 (Europe and Rest of World).
Customers in the UK should add VAT at 5%. Customers in the EU should also add VAT at 5% or provide a VAT registration number or evidence of entitlement to exemption. Customers in Canada should add 7% GST or provide evidence of entitlement to exemption. The Premium Institutional price also includes online access to full-text articles from 1997 to present, where available. For other pricing options or more information about online access to Blackwell Publishing journals, including access information and terms and conditions, please visit www.blackwellpublishing.com/nyas.

Membership information: Members may order copies of the *Annals* volumes directly from the Academy by visiting www.nyas.org/annals, emailing membership@nyas.org, faxing 212-888-2894, or calling 800-843-6927 (US only), or +1 212 838 0230, ext. 345 (International). For more information on becoming a member of the New York Academy of Sciences, please visit www.nyas.org/membership.

Journal Customer Services: For ordering information, claims, and any inquiry concerning your institutional subscription, please contact your nearest office:
UK: Email: customerservices@blackwellpublishing.com; Tel: +44 (0) 1865 778315; Fax +44 (0) 1865 471775
US: Email: customerservices@blackwellpublishing.com; Tel: +1 781 388 8599 or 1 800 835 6770 (Toll free in the USA); Fax: +1 781 388 8232
Asia: Email: customerservices@blackwellpublishing.com; Tel: +65 6511 8000; Fax: +61 3 8359 1120
Members: Claims and inquiries on member orders should be directed to the Academy at email: membership@nyas.org or Tel: +1 212 838 0230 (International) or 800-843-6927 (US only).

MIX
Paper from responsible sources
FSC www.fsc.org FSC® C013604

Mailing: The *Annals of the New York Academy of Sciences* are mailed Standard Rate. **Postmaster:** Send all address changes to *Annals of the New York Academy of Sciences*, Blackwell Publishing, Inc., Journals Subscription Department, 350 Main Street, Malden, MA 01248-5020. Mailing to rest of world by DHL Smart and Global Mail.

Annals are available to subscribers online at the New York Academy of Sciences and also at Blackwell Synergy. Visit www.annalsnyas.org or www.blackwell-synergy.com to search the articles and register for table of contents e-mail alerts. Access to full text and PDF downloads of *Annals* articles are available to nonmembers and subscribers on a pay-per-view basis at www.annalsnyas.org.

The paper used in this publication meets the minimum requirements of the National Standard for Information Sciences Permanence of Paper for Printed Library Materials, ANSI Z39.48_1984.

ISSN: 0077-8923 (print); 1749-6632 (online)
ISBN-10: 1-57331-583-4 (paper); ISBN-13: 978-1-57331-583-8 (paper)

A catalogue record for this title is available from the British Library.

ANNALS OF THE NEW YORK ACADEMY OF SCIENCES

Volume 1068
April 2006

SKELETAL DEVELOPMENT AND REMODELING IN HEALTH, DISEASE, AND AGING

Editor

MONE ZAIDI

This volume is the result of a conference entitled **Skeletal Development and Remodeling in Health, Disease, and Aging**, sponsored by the New York Academy of Sciences, and held on May 18–21, 2005 in New York City.

CONTENTS

Part II. Skeletal Remodeling

Part III. Mechanisms of Bone Loss

We gratefully acknowledge the support of:

National Institutes of Health

- National Institute on Aging
- National Institute of Arthritis and Musculoskeletal and Skin Diseases
- National Institute of Dental and Craniofacial Research

Corporate Co-sponsors

Alliance for Better Bone Health

- Procter & Gamble Pharmaceuticals
- Sanofi-Aventis

Novartis Pharmaceuticals
Roche Pharmaceuticals

Corporate Major Funder

Merck and Co.

Supporters

Amgen Inc.
Diane Wolf & Diane R. Wolf Foundation
Eli Lilly and Co.
Genzyme
Mitzi & Warren Eisenberg Family Foundation
Wyeth Pharmaceuticals

In Memorium: Gideon Rodan M.D., Ph.D.

Contributions to Bone Biology and Global Health

DWIGHT A. TOWLER AND JACK MARTIN

Internal Medicine, Bone and Mineral Diseases, Washington University Medical Center; and Barnes-Jewish Hospital, St. Louis, Missouri 63110, USA

Courtesy of Merck & Co., Inc.

Gideon Rodan died peacefully at home in Philadelphia a little more than 6 months after this Conference, at which he participated fully, chaired a session, and asked the penetrating questions he was known for over the years. His passing marks the loss of one, who as a scientist, teacher, and leader, did much to transform research into the physiology of bone by applying modern methods of biochemistry, and molecular and cell biology.

Born in Bucharest, Rumania in 1934, Gideon lost his father to the tyranny that covered Europe in the following years. He and his mother survived and migrated to Israel in 1950, where after completing his high school education, he studied first mathematics and then medicine at the Hebrew University. After completing his medical training and deciding upon research and an academic career he joined the Weitzman Institute as a staff scientist and completed his Ph.D. in 1970 on physicochemical aspects of mineral metabolism. Those years were productive ones, with the young Israeli scientist coming to be notice for his work on blood-bone calcium equilibrium and bone turnover, at a time when these research areas were sparsely populated. Much of that work was carried out together with young colleagues at the time—for example, Harell, Binderman, and Liberman, who ultimately remained in Israel to contribute to that country's

Address for correspondence: Dwight A. Towler, M.D., Ph.D., Barnes-Jewish Hospital, North Campus Box 8301, 660 South Euclid Ave., St. Louis, MO 63110. Voice: 314-454-7434; fax: 314-454-8434. e-mail: dtowler@im.wustl.edu

Ann. N.Y. Acad. Sci. 1068: xiii–xvi (2006). © 2006 New York Academy of Sciences.
doi: 10.1196/annals.1346.059

strength in bone research. The value of his early research became evident many times in later years, as he came to develop functional read-outs in bone of his basic research in biochemistry and molecular biology.

After the Weitzman appointment, his intention was to undertake a postdoctoral fellowship in the United States, but he was attracted to an assistant professorship at a new dental school at the University of Connecticut, Farmington. He rose rapidly through the ranks to head of the Department of Oral Biology in 1978, until his move to industry in 1985. Those were exciting years in Connecticut, it was a time when basic research in dental schools across the United States was at best patchy in distribution, but developments took place rapidly under the leadership of Gideon Rodan. His own research flourished in the 1970s, when with the constant collaboration of his wife, Sevgi, the group led the way in bringing cell biology to the study of bone. The Rodans developed a clonal rodent osteosarcoma cell line that for the first time allowed studies of the molecular regulation of adenylate cyclase in bone cells in response to parathyroid hormone (PTH), together with other aspects of cell signaling, particularly the role of calcium. A constant feature of Gideon's approach to basic research was his aim to apply it to function, and maybe this was what drove him several years later when he decided to move to industry. An example from that period was his work on differential cellular signaling responses being dependent upon mechanical and electrical stimuli to cells.

Gideon Rodan was never a slave to technology in his research—if you needed a method, you sought it out and established it. What was most important to him was the thinking that went into experiments, and what a pleasure it was to have scientific discussions with him. This was a very early stage of bone science, when to most of those interested, bone cells still seemed too difficult to study. Osteosarcoma cells and osteoblast-rich cultures from rodent bone that came from the work of Wong and Cohn, provided the means to investigate control mechanisms for the first time. These might have remained as "phenomenology" were it not for the Rodan approach, of putting such new information into context with what was known, for example, the electron microscopic evidence of Alan Boyde that bone lining cells responded to PTH by contracting. These and other data provided the means in the liberal discussions that could be enjoyed with Gideon Rodan, to lead to the concept of bone-resorbing factors using the osteoblast lineage in producing their effects on osteoclast formation and activation. This was regarded at the time almost as heresy but the Rodan view was always that as long as something was testable it was reasonable to propose.

The bone research scene at the University of Connecticut was an exciting one, with collaborations across the campus at the Health Center and the Department of Oral Biology at the focus of many of them. With Larry Raisz in Medicine there were effective research collaborations over many years. Gideon's early forays into prostaglandins in bone, where with typical thoroughness he prepared himself for that venture by studying blood platelets for a time, were influential in Larry Raisz's decision to study prostaglandins, a subject that he contributes to substantially to this day. There were a number of long-standing, important contributors to the research of the Rodan laboratory, a constant feature being, of course, Sevgi Rodan, who has led much of the research in the group for three decades. Bob Majeska was a key figure in studies of adenylate cyclase and alkaline phosphatase in the osteosarcoma cells. There were many Japanese Fellows over those years, one of whom, Masaki Noda, returned to Japan after 10 years of productive work and became a departmental head at Tokyo Medical and Dental College, with a large and successful research group. One of Gideon Rodan's greatest legacies is the long list of former students and postdocs who benefited so much from his teaching, training, and leadership, always effected with the gentleness of manner that so marked the man. His journal club discussion groups at his home in Connecticut provided a key educational asset that they all appreciated scientifically as much as they enjoyed them socially. The list of trainees surely comprises another Rodan "family," in addition to his wonderful Sevgi, Aylin, and Elan, and it is fitting indeed that the American Society of Bone and Mineral Research has named its mentorship award after Gideon Rodan, who set such an example to all.

Gideon's decision in 1985 to leave academia and take on leadership of the bone group at Merck came as a surprise to some of us. He seemed too interested in the unknown, maybe too purely scholarly. In the event he stayed that way, but also did all that was required of him by industry and more. Perhaps, as mentioned earlier, this decision was simply another manifestation of his wish to apply his research, as well as to enjoy it academically. Whatever the reason, he certainly made his mark in this way also. Setting out to search for drugs of benefit to bone, particularly osteoporosis, he approached this by establishing first-class research in bone biology, appointing excellent people, and seeming from the outset to have the ideal research position. Then with remarkable vision, perspicacity, and wisdom, he took an opportunity that came his way through Herbert Fleisch, to license a bisphosphonate that had been progressing slowly in Europe. The rest is history. This became alendronate, the first drug ever proven in a major clinical trial to prevent osteoporotic fractures, and thereby providing a stimulus that changed the bone field totally. The rest of industry applied itself eagerly to the search for anti-osteoporosis drugs, and substantial industry support came to research in the bone field. Gideon was the driver of the alendronate development, with his laboratory carrying out first-class work on bisphosphonate action, but he was also deeply involved in the planning of clinical studies and the translation of bisphosphonates to clinical medicine. There can be no doubt of the enormous impact of that development upon the field of research in bone.

—**JACK MARTIN**

Gideon was a vibrant, engaged, and wonderful man with a big heart and endless wisdom. Contemplative and analytical by nature, he was able to divine greater insight from data, drawing upon his unparalleled knowledge of the literature. He always sought the truth; his joy of science, the gleam in his eye, arose from divining novel insight, implication, and meaning—the novel insight, not just his novel insight. He relished the pleasure and privilege of being able to share it with you. His scientific curiosity and profound intellect were pure—completely absent the hubris that can afflict our best intentions. His unwavering commitment to the philosophy that "An observation is never a fact until repeated by another lab"—including those observations made within his own lab—attested to his absolute reverence for the truth. He once stated that he enjoyed drug discovery in part because "you can't fake it," his succinct way of stating how the robustness of drug development could vet and validate one's understanding of human biology. Yet he also truly saw pharmaceutical research as a calling—a civic and almost sacred duty to do one's best to improve the human condition. As Jack so well stated, Gideon was the mentor supreme.

Gideon often quoted the Talmud, saying it was our role to contribute to God's work, not complete it; he did his best to do both. Through his fostering of alendronate as a drug, he brought to the world the first effective pharmacotherapy for hip fracture and osteoporosis. He was committed to the fullest understanding of osteoclast cell physiology, and aggressively pursed all aspects of osteoclast formation, function, and pharmacological manipulation. It was deeply personal to him. However, his studies in bone formation were no less substantial. His work with Shun-ichi Harada and Sevgi provided pioneering insights into how osteogenic–angiogenic coupling by vascular endothelial growth factor was rate limiting in bone formation. His mechanistic studies with Marie-Helene Lafage-Proust on fluoride and alendronate actions in minipigs first highlighted how bone quality–viz., material properties arising from the integrated mineraline and matrix composite–was pharmacologically manipulable and thus an important therapeutic end point beyond bone mineral density.

Gideon spoke five languages, read voraciously, loved to write and debate—but not argue—the literature, and loved to sail. Before retiring, he arranged a "sabbatical" studying RNA processing at U. Penn—always the love of science and knowledge. Gideon was a great man, with a great intellect, beset with great suffering. But he never complained—always exuding his quiet warmth and grace.

Henry Wheeler Shaw, the 19th century humorist, stated "One of the rarest things that a man ever does, is to do the best he can." Mentor, scientist, husband, father, friend; in all that he did, Gideon Rodan was a rare man indeed.

—DWIGHT TOWLER

Editorial

Skeletal Biology and Skeletal Medicine

Just 20 years ago, we were optimizing methods to culture bone cells. Today, through the use of genetically manipulated mouse models, the application of genomics and proteomics to bone biology, and the systematic cataloging of gene mutations of human disease, we have seams of new information on how the skeleton develops and is thereafter remodeled. In addition, the identification of new obligatory molecules for both bone resorption and formation has led to a flurry of interest in new drug development. Skeletal biology and skeletal medicine has truly emerged as its own discipline that transcends specialties as diverse as developmental biology, cell and molecular physiology, mathematics, clinical biophysics, rheumatology, and orthopedic surgery, among many others.

The New York Academy of Sciences symposium on "Skeletal Development and Remodeling in Health Diseases and Aging" held at The Mount Sinai School of Medicine was an attempt to categorically disseminate this new information in an authoritative forum and more importantly, to foster areas of consensus that would trigger future investigation. Seven plenary lectures were interspersed with shorter state-of-the-art reviews. Opportunity also existed, on an invited basis, for investigators new to the field to present posters, four of which were selected for travel awards. It was indeed remarkable how each investigator took this opportunity to present not only published data, but also share yet unpublished information in a highly purposeful dialogue with their peers.

This volume consists of four key themes. The first theme focuses on skeletal development to encompass areas as broad ranging as the role of parathyroid hormone-related protein (PTHrP) signaling during chondrocyte maturation, the relevance of hypoxia to endochondral bone formation, effects of bone morphogenetic proteins and their antagonists during embryonic ossification and bone modeling, and the mechanisms of joint and articular cartilage formation. The second part consists of chapters that discuss mechanisms regulating bone remodeling, a process by which the skeleton is restructured throughout adult life. This section focuses on molecular signals for osteoclast and osteoblast development, function, and survival. The third part centers on novel insights into varied pathophysiologies that underlie bone loss as is seen in estrogen deficiency, steroid use, HIV, autoimmune diseases, and aging. Mechanisms of tumor osteolysis, particularly in multiple myeloma, are also discussed, as are pathways that link bone to blood vessels resulting in aberrant vascular calcification. In the last section, current and future therapies, both anabolic

Ann. N.Y. Acad. Sci. 1068: xvii–xviii (2006). © 2006 New York Academy of Sciences.
doi: 10.1196/annals.1346.060

and antiresorptive, are reviewed the following purposeful discussions of bone structure and bone strength. Additionally, the recently identified stem cell niche in bone marrow, the discovery that osteoblast precursors circulate, and the future role of stem cells in skeletal medicine have been discussed. This text is by no means complete, but I do hope that it would provide a valuable source of new information to senior scientists and young investigators alike.

However, while we gain some, we do unfortunately lose others. And, this year, we lost one of the most revered scientists in the bone field, Gideon A. Rodan, MD, PhD, who despite his protracted illness, chaired the most creative session of this meeting—Across Tissue Lines. Jack Martin and Dwight Towler have jointly reviewed the contributions of Gideon Rodan in an elegant tribute that precedes the chapters.

—MONE ZAIDI
Editor

ACKNOWLEDGMENTS

My sincere thanks are due to the Academy staff Dr. Rashid Shaikh (Director for Education), Dr. Shari Dermer (Program Manager), and Ms. Sherryl Usmani for their patience and perseverance, and for the thought and care they put into organizing this meeting. I am also grateful to Ms. Julia Keller (Blackwell Publishers) and Mr. Steve Bohall (NYAS) for editorial guidance.

This symposium would not have been possible without the generous support of the NIH institutes NIA, NIAMS, NIDCR and NIDDK, the corporate houses Proctor and Gamble Pharmaceuticals, Aventis Pharmaceuticals (Alliance for Better Bone Health), Roche Laboratories, Novartis Pharmaceuticals, Merck and Company, Amgen, Genzyme, Eli Lilly Pharmaceuticals, and Wyeth Pharmaceuticals, as well as private sponsors Mr. and Mrs. Warren Eisenberg, The Wolfe Foundation, and Ms. Diane Wolfe. These generous contributions are most appreciated.

This meeting would have also not come to fruition without the full assistance of my staff, most notably Ms. Mary Jo Sweeney, my worthy assistant, as well as my long-standing colleagues Drs. Baljit Moonga and Li Sun, and my M.D./Ph.D. student, Jameel Iqbal.

A debt of gratitude is owed to each and every member of the scientific community who participated so actively and made more than generous contributions to our knowledge base. I specifically thank my two co-chairs, Drs. Steven Teitelbaum and Solomon Epstein who helped put together the program.

Last, but not the least, I am grateful to my wife, Dr. Meenakshi Zaidi, our children, Neeha and Samir, and my father Dr. Sibte Zaidi, for their warm support through the months of this meeting's organization. I am dedicating this book to my late mother, Mrs. Qamar Zaidi, who was always my inspiring force.

PTHrP and Skeletal Development

HENRY M. KRONENBERG

Endocrine Unit, Massachusetts General Hospital and Harvard Medical School, Boston, Massachusetts 02114, USA

ABSTRACT: Parathyroid hormone-related protein (PTHrP) participates in the regulation of endochondral bone development. After the cartilage mold is established in fetal life, perichondrial cells and chondrocytes at the ends of the mold synthesize PTHrP. This ligand then acts on PTH/PTHrP receptors on chondrocytes. As chondrocytes go through a program of proliferation and then further differentiation into postmitotic, hypertrophic chondrocytes, PTHrP action keeps chondrocytes proliferating and delays their further differentiation. Indian hedgehog (Ihh) is synthesized by chondrocytes that have just stopped proliferating and is required for synthesis of PTHrP. The feedback loop between PTHrP and Ihh serves to regulate the pace of chondrocyte differentiation and the sites at which perichondrial cells first differentiate into osteoblasts. Activation of the PTH/PTHrP receptor leads to stimulation of both G_s and G_q family heterotrimeric G proteins. Genetic analyses demonstrate that G_s activation mediates the action of PTHrP to keep chondrocytes proliferating, while G_q activation opposes this action. Downstream from G_s activation, synthesis of the cyclin-cdk inhibitor, p57, is suppressed, thereby increasing the pool of proliferating chondrocytes. PTHrP's actions to delay chondrocyte differentiation are mediated by the phosphorylation of the transcription factor, SOX9, and by suppression of synthesis of mRNA encoding the transcription factor, Runx2. These pathways and undoubtedly others cooperate to regulate the pace of differentiation of growth plate chondrocytes in response to PTHrP.

KEYWORDS: parathyroid hormone-related protein; growth plate; chondrocytes; Indian hedgehog

INTRODUCTION

Investigators studying the hypercalcemia of malignancy discovered parathyroid hormone-related protein (PTHrP) as the cause of that syndrome, when the protein is secreted into the circulation.[1] That discovery led to a series of studies exploring the physiologic roles of PTHrP. PTHrP was shown to be synthesized in many types of cells, to activate the same receptor activated by PTH, and, to

Address for correspondence: Henry M. Kronenberg, Endocrine Unit, Massachusetts General Hospital, Thier 501, 50 Blossom St., Boston, MA 02114. Voice: 617-726-3966; fax: 617-726-7543. e-mail: hkronenberg@partners.org

Ann. N.Y. Acad. Sci. 1068: 1–13 (2006). © 2006 New York Academy of Sciences.
doi: 10.1196/annals.1346.002

exhibit a variety of actions ranging from relaxation of many smooth muscle beds to regulation of cellular proliferation and differentiation.[2] The importance of many of these actions of PTHrP has been established by analyses of the phenotypes of mice missing the *PTHrP* gene,[3] the *PTH/PTHrP receptor* gene,[4] and the *PTH* gene.[5] The mice missing PTHrP die at birth because all bones formed by endochondral bone formation develop improperly.[6,7] As a result, the rib cage is small and inappropriately mineralized, leading rapidly to death. Further studies have shown that PTHrP regulates the proliferation and differentiation of growth plate chondrocytes. These actions typify local actions of PTHrP in many organs. The similarities of the growth plates of *PTHrP* (–/–) mice and those of *PTH/PTHrP receptor* (–/–) mice bring genetic support for the idea that the PTH/PTHrP receptor mediates the actions of PTHrP on the growth plate. Here I will review the actions of PTHrP in the growth plate, the way that PTHrP signaling is coordinated with signaling in other pathways, and initial genetic studies that explore the intracellular mechanisms of PTHrP action.

ENDOCHONDRAL BONE FORMATION

The development of all bones begins with the formation of mesenchymal condensations. Groups of mesenchymal cells draw close together and initiate a characteristic genetic program. In a few bones, such as the flat bones of the skull, these condensations then differentiate directly into osteoblasts. In most bones, however, mesenchymal cells differentiate into chondrocytes and an adjacent perichondrium. The chondrocytes then direct the differentiation of perichondrial cells into osteoblasts, through a process called endochondral bone formation.[8] FIGURE 1 illustrates the steps in endochondral bone formation. After the formation of condensations (FIG. 1B), the cells enlarge and form round chondrocytes (FIG. 1C). These chondrocytes proliferate and secrete a matrix rich in collagen type II and aggrecan. In response to a still unknown signal, chondrocytes in the middle of the developing bone stop proliferating, enlarge (hypertrophy), and change their genetic program to secrete a matrix rich in collagen type X (FIG. 1D). These hypertrophic chondrocytes then direct the mineralization of the surrounding matrix, signal to adjacent perichondrial cells to direct their differentiation into osteoblasts, and also stimulate the invasion of blood vessels. With vascularization, preosteoblastic cells from the perichondrial region also invade the cartilage to begin formation of the first true bone inside the cartilage mold called the primary spongiosa (FIG. 1E,F). Osteoclasts, cells derived from the hematopoietic lineage, also enter the cartilage mold and, together with the cells of the osteoblastic lineage, digest the matrix that had been synthesized by the hypertrophic chondrocytes. These chondrocytes die through apoptosis at the border of the cartilage and primary spongiosa. At the ends of the cartilage mold, round chondrocytes continue to

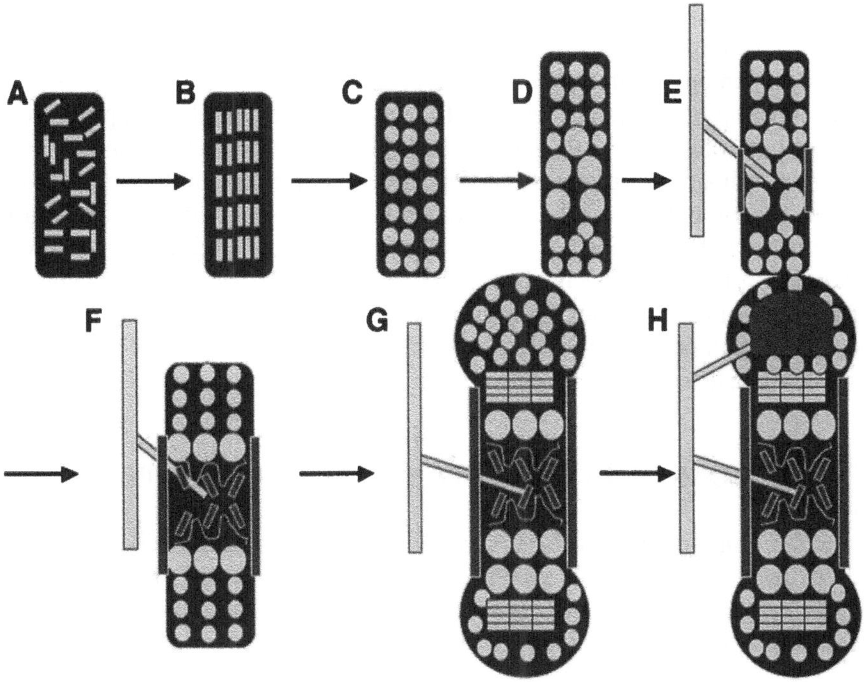

FIGURE 1. Endochondral bone formation. The sequential steps in endochondral bone formation are illustrated. (**A**) Mesenchymal cells. (**B**) Mesenchymal cells move close together to form a condensation. (**C**) Mesenchymal cells differentiate into chondrocytes. (**D**) Chondrocytes in the middle of the cartilage mold stop proliferating and become hypertrophic chondrocytes. (**E**) Hypertrophic chondrocytes induce vascularization and formation of a bone collar in the adjacent perichondrium. (**F**) Osteoblasts differentiate from cells brought into the cartilage mold with vascular invasion, forming the primary spongiosa. (**G**) Chondrocytes continue to proliferate, forming columns of flat chondrocytes. (**H**) Secondary ossification centers form.

proliferate. Those closest to the hypertrophic chondrocytes flatten out and form orderly columns of still proliferating chondrocytes (FIG. 1G). The cells closest to the hypertrophic chondrocytes stop proliferating (prehypertrophic cells) and subsequently hypertrophy. Bone lengthening is caused by the increase in chondrocyte number, the synthesis of chondrocyte matrix, and the substantial enlargement of chondrocytes that occurs during hypertrophy. Subsequently, secondary sites of ossification form. Chondrocytes in the center of the region of round proliferative chondrocytes stop dividing, hypertrophy, and attract a new, vascularized primary spongiosa. In this way, the sequence of endochondral bone formation is recapitulated (FIG. 1H). Chondrocytes between the primary and secondary ossification center form a true growth plate that persists as the engine of bone lengthening in post-natal life. Though the fetal growth region is not a true "plate" of cells, here I will refer to the fetal growth region as the growth plate, since this usage has become common. This overall process, thus, requires careful coordination of cell migration, proliferation, differentiation, and death among groups of cells of varying lineages. Not surprisingly, multiple signaling pathways are required to regulate these complicated processes. These pathways include those directed by fibroblast growth factors (FGFs), bone morphogenetic proteins (BMPs), Indian hedgehog (Ihh), C-type natriuretic protein (CNP), insulin-like growth factors (IGFs), wingless-int family members (wnts), ligands for nuclear receptors (retinoic acid, glucocorticoids, thyroid hormone, estrogens, androgens, and PTHrP). Here I will focus on PTHrP, but mention other pathways to emphasize that the actions of PTHrP can only be understood in the context of a complicated network of interacting signaling pathways.

ACTIONS OF PTHrP ON THE FETAL GROWTH PLATE

In fetal mouse bone PTHrP is synthesized by perichondrial cells and chondrocytes at the ends of the growing bones[9] (shown schematically in FIG. 2). Early in bone development, these cells appear to be the only cells in bone competent to synthesize PTHrP. Even when PTHrP production is stimulated,

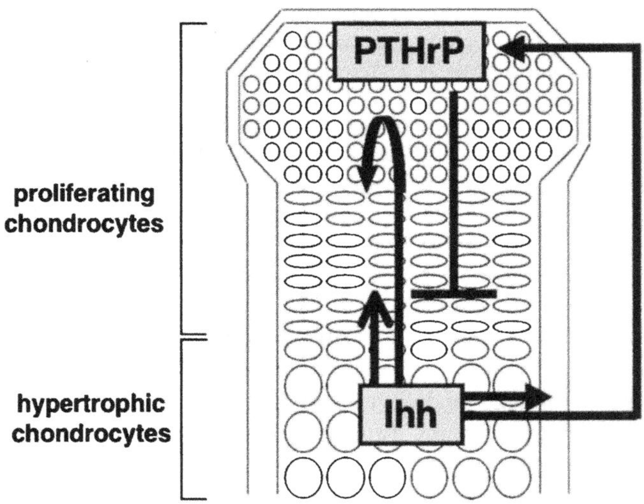

FIGURE 2. PTHrP-Ihh feedback loop. PTHrP is synthesized by chondrocytes and perichondrial cells at the ends of the developing bones. PTHrP acts on chondrocytes to keep the chondrocytes proliferating and to delay the differentiation of the chondrocytes into prehypertrophic and hypertrophic chondrocytes. After chondrocytes stop proliferating, they then synthesize (Ihh). Ihh acts to increase the synthesis of PTHrP, to accelerate the differentiation of round proliferative chondrocytes into flat proliferating chondrocytes, to increase the rate of proliferation of adjacent chondrocytes, and to direct perichondrial cells to differentiate into osteoblasts.

for example, by various genetic tricks that increase the synthesis of Ihh in fetal bone, only perichondrial cells and round chondrocytes at the ends of the bones increase their production of PTHrP.[10] PTHrP then diffuses away from its site of production and binds to PTH/PTHrP receptors on nearby chondrocytes.[7] Proliferating chondrocytes express low numbers of PTH/PTHrP receptors, but increase receptor expression dramatically as the cells stop proliferating. PTHrP acts on chondrocytes bearing PTH/PTHrP receptors to keep the chondrocytes proliferating and to delay their conversion into prehypertrophic and then hypertrophic chondrocytes. PTHrP also modestly increases the rate of chondrocyte proliferation, though this action is only detectable early during fetal development.[11]

The phenotypes of mice with a variety of mutant genes support this model of PTHrP action. Mice missing either the PTHrP or the PTH/PTHrP receptor genes undergo the normal endochondral sequence of proliferation and hypertrophy, except that columns of proliferating chondrocytes are much shorter than normal or even nonexistent (FIG. 3). Further, in chimeric mice with some chondrocytes missing the PTH/PTHrP receptor, the mutant chondrocytes become hypertrophic much sooner than their normal counterparts.[12] Conversely, mice overexpressing PTHrP in chondrocytes through a transgenic strategy are born with bones composed almost exclusively of proliferating chondrocytes, with a profound delay in the endochondral sequence.[13] Similar findings occur in mice expressing a constitutively active PTH/PTHrP receptor in

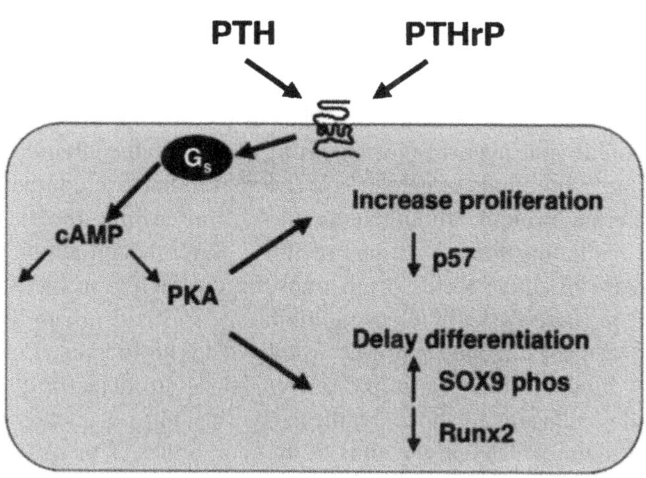

FIGURE 3. Intracellular signaling by the PTH/PTHrP receptor. Activation of the PTH/PTHrP receptor leads to activation of multiple G proteins. The activation of G_S leads to subsequent activation of adenylate cyclase and generation of cyclic AMP. Cyclic AMP has several actions, including the activation of protein kinase A (PKA). PKA then, by unknown mechanisms, leads to a fall in p57 and Runx2 mRNA and protein. PKA also phosphorylates SOX9, increasing the activity of SOX9.

chondrocytes.[14] Both of these transgenic constructs can prevent the neonatal death of *PTHrP* (−/−)[14,15]; this observation justifies the conclusion that the PTHrP (−/−) mice die at birth because of their skeletal abnormalities. Presumably, the rib cage, with its smaller circumference than normal, does not allow adequate respiration. Though no humans have been identified with absent PTHrP production, human fetuses with defective or absent PTH/PTHrP receptors (Blomstrand chondro-osteodystrophy) die *in utero* with skeletal abnormalities closely resembling those of the PTH/PTHrP receptor (−/−) mouse.[16,17] Furthermore, humans with Jansen chondro-osteodystrophy have point mutations in the gene encoding the PTH/PTHrP receptor that render the receptor active even in the absence of ligand.[18] Such people have growth abnormalities like those predicted from the actions of PTHrP in fetal mouse bone. These changes lead to delay in hypertrophic chondrocyte differentiation and resultant short stature. Thus, it is likely that the fetal mouse growth plate resembles that of the human in the way that PTHrP regulates chondrocyte proliferation and differentiation.

INTERACTIONS BETWEEN IHH AND PTHrP

Ihh is a major regulator of PTHrP synthesis in the growth plate. In the absence of Ihh, PTHrP mRNA levels are undetectable in fetal bone.[19] The chondrocytes in these bones exhibit accelerated differentiation, as would be in bones that lack PTHrP. In addition, the genetic introduction of constitutively active PTH/PTHrP receptors, using a transgenic construct that uses the collagen IIα (1) promoter to drive expression of these receptors, into Ihh (−/−) mice leads to reversal of the accelerated differentiation of chondrocytes.[11] The pathway used by Ihh to stimulate the production of PTHrP is unknown. Ihh is synthesized by prehypertrophic and hypertrophic chondrocytes just after chondrocytes stop proliferating.[20] Ihh binds to patched (ptc), a membrane protein that keeps another membrane protein, smoothened (smo) from reaching the plasma membrane and mediating further actions of Ihh.[21] Smoothened action then leads to an increase of active forms of gli transcription factors in the nucleus and a decrease in proteolytically cleaved, inhibitory forms of gli in the nucleus. Activation of the hedgehog pathway usually leads to increased synthesis of ptc mRNA. Such an increase in ptc mRNA is easy to detect in proliferating chondrocytes adjacent to those synthesizing Ihh, but not clearly evident in cells synthesizing PTHrP at the ends of the fetal bones. This observation and the considerable distance between Ihh-producing and PTHrP-producing cells raise the possibility that Ihh might not act directly on chondrocytes synthesizing PTHrP. Some data, for example, suggest that blockade of transforming growth factor (TGF)-β signaling can decrease PTHrP synthesis in response to Ihh.[22] Thus, the TGF-β signaling pathway may partly mediate stimulation of PTHrP synthesis by Ihh. Further, the movement of Ihh must be regulated in a complicated manner, given that the active form of Ihh is heavily lipid-modified, with a palmitate group at its amino terminus and a cholesterol moiety on its carboxy-terminus.[23] Whatever the mechanism whereby Ihh regulates PTHrP synthesis, the interactions of Ihh and PTHrP provide a powerful pathway for precisely regulating the proliferation and differentiation of chondrocytes in the growth plate (FIG. 2). PTHrP is produced by cells at the ends of bones[9] and acts on nearby chondrocytes bearing PTH/PTHrP receptors to keep them proliferating and delay their differentiation. Chondrocytes sufficiently far away from the source of PTHrP, however, stop proliferating and then and only then synthesize Ihh. This Ihh then, directly or indirectly, signals back to the end of the growth plate to stimulate the synthesis of more PTHrP. This feedback system, thus, determines the site at which chondrocytes stop proliferating and start making Ihh.

Such a feedback loop might be expected not only to determine the length of the zone of chondrocyte proliferation but might also contribute to the striking uniformity of chondrocyte morphology across the growth plate: the columns of proliferating chondrocytes quite uniformly end in a smooth line that separates the prehypertrophic and hypertrophic chondrocytes from the columns. In both the PTHrP (–/–) and the PTH/PTHrP receptor (–/–) mice, the early hypertrophic chondrocytes unevenly overlap with columnar chondrocytes.[6,24] This unevenness is an expected result of the absence of an effective feedback mechanism for determining the precise site at which proliferation stops. Presumably, in the normal setting, if, stochastically, a columnar chondrocyte stops proliferating sooner than otherwise expected, it will synthesize Ihh closer to the top of the growth plate than otherwise. This Ihh would be expected, then, to stimulate the production of an increased amount of PTHrP (because the Ihh-producing cell is closer to the top of the growth plate than average), and that PTHrP would act to keep nearby chondrocytes proliferating. This process would be expected to increase the uniformity of the sites at which the columns of proliferating chondrocytes across the growth plate choose to initiate the process of hypertrophy.

Ihh has a number of actions that are independent of PTHrP (FIG. 2). Ihh stimulates proliferation of chondrocytes and directs adjacent perichondrial cells to become osteoblasts.[25,26] Interestingly, this action of Ihh is not needed during intramembranous bone formation in the skull. In the absence of such signaling during endochondral bone formation, however, perichondrial cells can become chondrocytes. Further, Ihh stimulates the conversion of round proliferating chondrocytes to flat columnar chondrocytes.[10] Since the flat proliferating chondrocytes form columns that extend in the longitudinal axis of long bones, this action of Ihh serves to reinforce the asymmetry of bone growth in the long bones. Thus, together, PTHrP and Ihh regulate multiple steps in the development of the growth plate. Ihh stimulates the production of PTHrP and accelerates the conversion of round chondrocytes to flat columnar chondrocytes. PTHrP acts to keep chondrocytes proliferating and delays the production of Ihh and the further differentiation of chondrocytes. PTHrP and Ihh together, therefore, determine both the entry and the exit of chondrocytes from the pool of flat proliferating chondrocytes. Further, by determining the site of Ihh production, PTHrP thus also determines the site at which perichondrial cells first become true osteoblasts in the bone collar adjacent to the growth plate.

INTRACELLULAR PATHWAYS ACTIVATED BY PTHrP IN FETAL GROWTH PLATE CHONDROCYTES

Because the defects in *PTHrP* (–/–) mice and *PTH/PTHrP receptor* (–/–) mice so closely resemble each other, one is tempted to conclude that most and perhaps all of the morphologically obvious actions of PTHrP to regulate the growth plate are via activation of the PTH/PTHrP receptor. Having said that, PTHrP also clearly activates receptors distinct from the PTH/PTHrP receptor in other settings (such as the stimulation of placental calcium transport[27]) and also has actions directly within cell nuclei that have been documented *in vitro*.[28] Additional studies will be needed to determine the contributions of these other pathways to the actions of PTHrP on the growth plate. Binding of PTHrP to the PTH/PTHrP receptor stimulates multiple heterotrimeric G proteins,[29] including G_s, the G_q family, and $G_{12,13}$.[30] Activation of each G protein triggers a series of downstream events that culminate in obvious biological consequences. The relative stimulation of each of the distinct signaling pathways varies among cell types, as do the physiologic consequences of this activation. Furthermore, the activities of each intracellular pathway are influenced by pathways triggered by signals from adjacent cells and from matrix proteins, each distributed asymmetrically in the growth plate. Therefore, to make sure that we study the actions of PTHrP in a setting relevant to the *in vivo* setting, we have used genetic tools to identify the roles of activation of the G_s and G_q pathways by PTHrP in the fetal growth plate. To determine to role of G_s signaling, we took advantage of technology that allows the generation of chimeric mice with cellular contributions from genetically engineered embryonic stem (ES) cells and from normal hosts. Ung-il Chung mated mice heterozygous for ablation of the second exon of the gene encoding $G_s\alpha$ and derived ES cells homozygous for the $G_s\alpha$ ablation from resultant blastocysts.[31] He then produced chimeras that contained normal chondrocytes, as well as chondrocytes missing the gene encoding $G_s\alpha$. The chondrocytes missing $G_s\alpha$ stopped proliferating prematurely and became hypertrophic chondrocytes close to the top of the growth plate. In that way they closely resembled chondrocytes missing the PTH/PTHrP receptor generated using analogous chimeric mice.[12] From this similarity and other experiments we conclude that the PTH/PTHrP receptor uses G_s activation to keep chondrocytes proliferating. Experiments were subsequently performed using mice that have mutant PTH/PTHrP receptors that cannot activate the G_q family of G proteins but can activate G_s normally.[32] These mice were generated by taking advantage of information about a PTH/PTHrP receptor that exhibits such signaling after introduction of mutations in the sequence encoding the second intracellular loop of the receptor.[33] The mutant sequence was substituted for the normal sequence using homologous recombination to generate a "knock-in" mouse. By examining the bones of mice with normal G_s signaling but absent G_q signaling in response to activation of the PTH/PTHrP receptor, we could deduce actions of the G_q pathway in chondrocytes. These studies showed that activation of G_q by the PTH/PTHrP receptor mildly accelerates chondrocyte differentiation. Thus, activation of G_s and G_q by activation of the PTH/PTHrP receptor in chondrocytes leads to actions that oppose each other. The usefulness of this seemingly wasteful opposition of the two pathways is not certain. As in many other settings in which a stimulus activates opposing pathways, this complicated pattern may allow regulatory interactions with other pathways that are useful. Furthermore, the particular properties of the PTH/PTHrP receptor suggest another consequence of the activation of both G_s and G_q by the PTH/PTHrP receptor in chondrocytes. Since activation of the G_q pathway generally requires higher concentrations of PTHrP than does activation of the G_s pathway,[34] one might predict that, chondrocytes nearer to the source of PTHrP might respond more dramatically with G_q activation than chondrocytes further from the source of PTHrP. Thus, chondrocytes nearer to the top of the growth plate, where PTHrP is synthesized, would be expected to have G_q signaling oppose the actions of G_s. As the source of PTHrP moves further away from the proliferating chondrocytes, however, one would predict a more rapid loss of G_q activation than of G_s activation. The net effect would be to allow PTHrP to prolong proliferation more dramatically than otherwise expected, as the concentration of PTHrP falls. These contrasting responses of chondrocytes depending on their locations in the growth plate may help smooth the functional gradient of PTHrP action across the growth plate.

G_s activation leads to production of cyclic AMP and activation of protein kinase A (PKA). The actions of cyclic AMP and PKA that lead chondrocytes to continue proliferating and delay differentiation are incompletely understood. PTHrP and PKA are the major sources of phosphorylation of the chondrocyte transcription factor, SOX9.[35] Phosphorylated SOX9 activates target genes more efficiently than does unphosphorylated SOX9. Since SOX9 acts to slow the differentiation of chondrocytes, SOX9 phosphorylation probably contributes to the action of PTHrP to delay chondrocyte differentiation. PTHrP also decreases expression of the p57 gene in chondrocytes both in bone explants and *in vivo*, as deduced from measuring levels of p57 mRNA and protein in mice missing the *PTHrP* gene.[36] p57 is a member of the CIP/KIP family of inhibitors of cyclin-dependent kinases; this family also includes p21 and p27. Ablation of the p21 or p27 genes has little effect on the fetal growth plate, but ablation of the gene encoding p57 leads to an increase in proliferation and delay of differentiation of chondrocytes that partly resembles the phenotype of *PTHrP* (−/−) mice.[37] Strikingly, in double knockout mice missing both the *PTHrP* and the *p57* gene, many growth plates, including those of the ribs, sternum, and ulna, exhibit patterns of proliferation and differentiation much closer to normal than in the *PTHrP* (–/–) growth plates.[36] This result is consistent with the hypothesis that suppression of p57 synthesis is a major mechanism used by PTHrP to keep chondrocytes proliferating and delay their differentiation. Recent experiments demonstrate that PTHrP also decreases the production of the transcription factor, Runx2, in chondrocytes (but not in osteoblasts) in explants of fetal long bones.[38] In such explants, PTH administration rapidly leads to a fall in Runx2 mRNA and protein in chondrocytes. Since Runx2 is required for the hypertrophic differentiation of chondrocytes in most bones, the action of PTHrP to suppress Runx2 production probably contributes to the delay in the differentiation of chondrocytes caused by PTHrP. FIGURE 3 summarizes the actions of PTHrP just discussed. Undoubtedly, this figure is incomplete, since the multiple signaling cascades downstream from the PTH/PTHrP receptor must modify the actions of large numbers of pathways.

CONCLUSIONS

The focus here on the actions of PTHrP in the growth plate does not imply either that these actions are the most important actions of PTHrP or that PTHrP is the most important regulator in the growth plate. Certainly, multiple signaling systems are together required for proper coordination of bone development. The actions of PTHrP can only be understood in the context of these other signaling systems. Further, the presence of PTHrP and PTH/PTHrP receptors in virtually every organ at various times during development and adult life suggests that PTHrP signaling has actions in multiple physiologic settings. Whether hedgehog signaling regulates PTHrP signaling in other settings is currently unknown. It seems likely that actions of PTHrP to, for example, relax smooth muscle will use intracellular pathways that differ in important respects from those outlined here. Thus, while the growth plate may well serve as a useful model for the actions of PTHrP to regulate proliferation and differentiation in some settings, it seems likely that the varied sites of PTHrP action will reveal a broad array of cell-specific responses. The study of genetically altered mice will continue to provide powerful tools for establishing the physiologic relevance of proposed actions of PTHrP.

REFERENCES

1. SUVA, L.J., G.A. WINSLOW, R.E. WETTENHALL, *et al.* 1987. A parathyroid hormone-related protein implicated in malignant hypercalcemia: cloning and expression. Science **237:** 893–896.
2. STREWLER, G.J. 2000. The physiology of parathyroid hormone-related protein. N. Eng. J. Med. **342:** 177–185.

3. KARAPLIS, A.C., A. LUZ, J. GLOWACKI, *et al.* 1994. Lethal skeletal dysplasia from targeted disruption of the parathyroid hormone-related peptide gene. Genes Dev. **8:** 277–289.
4. LANSKE, B., A.C. KARAPLIS, K. LEE, *et al.* 1996. PTH/PTHrP receptor in early development and Indian hedgehog-regulated bone growth. Science **273:** 663–666.
5. MIAO, D., B. HE, A.C. KARAPLIS, *et al.* 2002. Parathyroid hormone is essential for normal fetal bone formation. J. Clin. Invest. **109:** 1173–1182.
6. AMIZUKA, N., H. WARSHAWSKY, J.E. HENDERSON, *et al.* 1994. Parathyroid hormone-related peptide-depleted mice show abnormal epiphyseal cartilage development and altered endochondral bone formation. J. Cell. Biol. **126:** 1611–1623.
7. LEE, K., B. LANSKE, A.C. KARAPLIS, *et al.* 1996. Parathyroid hormone-related peptide delays terminal differentiation of chondrocytes during endochondral bone development. Endocrinology **137:** 5109–5118.
8. KRONENBERG, H.M. 2003. Developmental regulation of the growth plate. Nature **423:** 332–336.
9. LEE, K., J.D. DEEDS & G.V. SEGRE. 1995. Expression of parathyroid hormone-related peptide and its receptor messenger ribonucleic acid during fetal development of rats. Endocrinology **136:** 453–463.
10. KOBAYASHI, T., D.W. SOEGIARTO, Y. YANG, *et al.* 2005. Indian hedgehog stimulates periarticular chondrocyte differentiation to regulate growth plate length independently of PTHrP. J. Clin. Invest. **115:** 1734–1742.
11. KARP, S.J., E. SCHIPANI, B. ST-JACQUES, *et al.* 2000. Indian hedgehog coordinates endochondral bone growth and morphogenesis via parathyroid hormone related-protein-dependent and -independent pathways. Development **127:** 543–548.
12. CHUNG, U.I., B. LANSKE, K. LEE, *et al.* 1998. The parathyroid hormone/parathyroid hormone-related peptide receptor coordinates endochondral bone development by directly controlling chondrocyte differentiation. Proc. Natl. Acad. Sci. USA **95:** 13030–13035.
13. WEIR, E.C., W.M. PHILBRICK, M. AMLING, *et al.* 1996. Targeted overexpression of parathyroid hormone-related peptide in chondrocytes causes chondrodysplasia and delayed endochondral bone formation. Proc. Natl. Acad. Sci. USA **93:** 10240–10245.
14. SCHIPANI, E., B. LANSKE, J. HUNZELMAN, *et al.* 1997. Targeted expression of constitutively active receptors for parathyroid hormone and parathyroid hormone-related peptide delays endochondral bone formation and rescues mice that lack parathyroid hormone-related peptide. Proc. Natl. Acad. Sci. USA **94:** 13689–13694.
15. PHILBRICK, W.M., B.E. DREYER, I.A. NAKCHBANDI, *et al.* 1998. Parathyroid hormone-related protein is required for tooth eruption. Proc. Natl. Acad. Sci. USA **95:** 11846–11851.
16. ZHANG, P., A.S. JOBERT, A. COUVINEAU, *et al.* 1998. A homozygous inactivating mutation in the parathyroid hormone/parathyroid hormone-related peptide receptor causing Blomstrand chondrodysplasia. J. Clin. Endocr. Metab. **83:** 3365–3368.
17. KARAPLIS, A.C., B. HE, M.T. NGUYEN, *et al.* 1998. Inactivating mutation in the human parathyroid hormone receptor type 1 gene in Blomstrand chondrodysplasia [see comments]. Endocrinology **139:** 5255–5258.

18. SCHIPANI, E., K. KRUSE, H. JUPPNER. 1995. A constitutively active mutant PTH-PTHrP receptor in Jansen-type metaphyseal chondrodysplasia. Science **268:** 98–100.
19. ST-JACQUES, B., M. HAMMERSCHMIDT & A.P. MCMAHON. 1999. Indian hedgehog signaling regulates proliferation and differentiation of chondrocytes and is essential for bone formation.[erratum appears in Genes Dev. 1999 Oct 1;13(19):2617]. Genes Dev. **13:** 2072–2086.
20. BITGOOD, M.J. & A.P. MCMAHON. 1995. *Hedgehog* and BMP genes are coexpressed at many diverse sites of cell-cell interaction in the mouse embryo. Dev. Biol. **172:** 126–138.
21. MCMAHON, A.P. 2000. More surprises in the hedgehog signaling pathway. Cell **100:** 185–188.
22. ALVAREZ, J., P. SOHN, X. ZENG, *et al.* 2002. TGFbeta2 mediates the effects of hedgehog on hypertrophic differentiation and PTHrP expression. Development **129:** 1913–1924.
23. INGHAM, P.W. 2001. Hedgehog signaling: a tale of two lipids. Science **294:** 1879–1881.
24. LANSKE, B., P. DIVIETI, C.S. KOVACS, *et al.* 1998. The parathyroid hormone (PTH)/PTH-related peptide receptor mediates actions of both ligands in murine bone. Endocrinology **139:** 5194–5204.
25. LONG, F., X.M. ZHANG, S. KARP, *et al.* 2001. Genetic manipulation of hedgehog signaling in the endochondral skeleton reveals a direct role in the regulation of chondrocyte proliferation. Development **128:** 5099–5108.
26. LONG, F., U.I. CHUNG, S. OHBA, *et al.* 2004. Ihh signaling is directly required for the osteoblast lineage in the endochondral skeleton. Development **131:** 1309–1318.
27. ABBAS, S.K., D.W. PICKARD, C.P. RODDA, *et al.* 1989. Stimulation of ovine placental calcium transport by purified natural and recombinant parathyroid hormone-related protein (PTHrP) preparations. Q. J. Exp. Physiol. **74:** 549–552.
28. HENDERSON, J.E., N. AMIZUKA, H. WARSHAWSKY, *et al.* 1995. Nuclear localization of parathyroid hormone-related peptide enhances survival of chondrocytes under conditions that promote apoptotic cell death. Mol. Cell. Endocrinol. **15:** 4064–4075.
29. ABOU-SAMRA, A.B., H. JÜPPNER, T. FORCE, *et al.* 1992. Expression cloning of a common receptor for parathyroid hormone and parathyroid hormone-related peptide from rat osteoblast-like cells: a single receptor stimulates intracellular accumulation of both cAMP and inositol triphosphates and increases intracellular free calcium. Proc. Natl. Acad. Sci. USA **89:** 2732–2736.
30. SINGH, A.T., A. GILCHRIST, T. VOYNO-YASENETSKAYA, *et al.* 2005. G alpha12/G alpha13 subunits of heterotrimeric G proteins mediate parathyroid hormone activation of phospholipase D in UMR-106 osteoblastic cells. Endocrinology **146:** 2171–2175.
31. BASTEPE, M., L.S. WEINSTEIN, N. OGATA, *et al.* 2004. Stimulatory G protein directly regulates hypertrophic differentiation of growth plate cartilage *in vivo*. Proc. Natl. Acad. Sci. USA **101:** 14794–14799.
32. GUO, J., U.I. CHUNG, H. KONDO, *et al.* 2002. The PTH/PTHrP receptor can delay chondrocyte hypertrophy in vivo without activating phospholipase C. Dev. Cell **3:** 183–194.
33. IIDA-KLEIN, A., J. GUO, M.T. DRAKE, *et al.* 1995. Structural requirements of PTH/PTHrP receptors for phospholipase C activation and regulation of phosphate uptake. Miner. Electrolyte. Metab. **21:** 177–179.

34. BRINGHURST, F.R., H. JÜPPNER, J. GUO, *et al.* 1993. Cloned, stably expressed parathyroid hormone (PTH)/PTH-related peptide receptors activate multiple messenger signals and biological responses in LLC-PK1 kidney cells. Endocrinology **132:** 2090–2098.
35. HUANG, W., U.I. CHUNG, H.M. KRONENBERG, *et al.* 2001. The chondrogenic transcription factor Sox9 is a target of signaling by the parathyroid hormone-related peptide in the growth plate of endochondral bones. Proc. Natl. Acad. Sci. USA **98:** 160–165.
36. MACLEAN, H.E., J. GUO, M.C. KNIGHT, *et al.* 2004. The cyclin-dependent kinase inhibitor p57(Kip2) mediates proliferative actions of PTHrP in chondrocytes. J. Clin. Invest. **113:** 1334–1343.
37. ZHANG, P., N.J. LIEGEOIS, C. WONG, *et al.* 1997. Altered cell differentiation and proliferation in mice lacking p57KIP2 indicates a role in Beckwith-Wiedemann syndrome. Nature **387:** 151–158.
38. GUO, J., U.I. CHUNG, A. SZCZEPANIK, *et al.* 2003. PTH/PTHrP receptor delays chondrocyte hypertrophy via both Runx2-dependent and independent pathways. J. Bone Miner. Res. **18:** Abst. 1144, p. S38.

Regulating Bone Growth and Development with Bone Morphogenetic Proteins

PHOEBE S. LEBOY

Department of Biochemistry, School of Dental Medicine, University of Pennsylvania, Philadelphia, Pennsylvania 19104-6030, USA

ABSTRACT: There are many gene products reported to promote osteoblast differentiation and thus increase bone formation, but only the transcription factor Runx2 and members of the bone morphogenetic protein (BMP) family of growth/differentiation factors have been shown to be absolute requirements for osteogenesis. Mice lacking the transcription factor Runx2 (also known as cbfa1) develop no bone. Similarly, osteoblast differentiation and bone formation is blocked when BMP signaling is suppressed by overexpression of noggin, a selective BMP antagonist. It is therefore not unexpected that several different mechanisms have evolved to regulate the effects of BMP-induced signaling. In this session we focus on the multiple ways in which cells can modulate BMP-induced osteogenesis and mechanisms by which BMP signaling can lead to transcriptional control of gene expression.

KEYWORDS: skeleton; bone growth; bone development; osteoblast; bone morphogenetic proteins; TGF-β

The term *bone growth* can encompass many events, including developmental establishment of skeletal elements, enlargement of bone primordia, and elongation of long bones. In this article we use the term to refer to actions leading to deposition of new bone via signaling molecules that stimulate osteoblast differentiation. The bone may be formed via endochondral or intramembranous pathways, and may be ectopic or at bony sites. The unifying characteristic is that this bone growth occurs when precursor mesenchymal cells or osteoprogenitors are stimulated with secreted molecules that are members of the TGF-β superfamily of growth and differentiation factors. Since there are many recent publications reviewing growth factor regulation of bone growth,[1–4] in this brief overview I will highlight some recent findings related to the specificity of action and signaling mechanisms used by these factors.

Address for correspondence: Phoebe S. Leboy, Department of Biochemistry, School of Dental Medicine, University of Pennsylvania, 240 South 40th Street, Philadelphia, PA 19104-6030. Voice: 215-898-8933; fax: 215-898-3695.
e-mail: phoebe@biochem.dental.upenn.edu

Ann. N.Y. Acad. Sci. 1068: 14–18 (2006).
doi: 10.1196/annals.1346.003

The major osteoblast-stimulating factors within the TGF-β family are found among both the bone morphogenetic proteins (BMPs) and the related growth and differentiation factors (GDF). The variety and evolutionary relationships of these proteins are described by Rosen later in this volume. BMPs were named for the ability to induce ectopic bone formation, and the name implies that they are capable of inducing the osteogenic differentiation of non-bone cells.[5,6] This is true for most of the best-studied BMPs, but not all. Furthermore, the name does not acknowledge the breadth of BMP actions. The observation that mouse development was blocked at the gastrula stage in the absence of either BMP2 or BMP4 was the first indication that BMPs are more appropriately characterized as "body morphogenetic proteins." They are now known to specify developmental cell fate during embryogenesis and determine the dorsoventral body axis. They are involved in the formation and patterning of the nervous system, heart, gut, and kidney as well as skeletal tissues and teeth.

With the discovery that BMPs affect development of many tissue types came the recognition that BMP actions in early embryos could not be explained simply as the presence or absence of ligands that can activate BMP receptors. Instead, the body morphogenetic actions of BMPs are now viewed as mediated by the formation and maintenance of gradients. For example, the current view of neural crest induction during gastrulation is that the neural plate forms at regions of low BMP levels, while intermediate BMP signaling leads to neural crest formation and high BMP signaling results in non-neural ectoderm.[7] One mechanism for developing such gradients is to have a BMP ligand produced and secreted at a discrete site, such that concentrations decrease with increasing distance from the source. Another mechanism is to form gradients of active ligand by gradually sequestering some of the ligand in an inactive form. Research in the past decade has revealed several secreted molecules, such as noggin, chordin, follistatin, and gremlin, which bind to BMPs, thereby inactivating them and creating a gradient of free BMP available for receptor binding. Studies of *Xenopus* development indicate that BMP antagonists derived from dorsal mesoderm diffuse in the ectoderm to create a gradient of BMP activity, establishing an intermediate level of BMP required for neural crest formation. Similarly, the low BMP levels leading to neural plate formation reflect high levels of BMP inhibitors, and zebrafish embryos lacking functional chordin have a reduced neural plate.[7]

Studies in postnatal animals indicate that the continued osteogenesis required for bone growth, bone remodeling, and bone repair in vertebrates is regulated by the availability of a subset of BMPs including BMP-2, -4, -6, -7, and -9. Like the systems operating in early embryonic development, the levels of functional BMPs can be controlled by the production of BMP-binding proteins that antagonize BMP action, such as noggin and chordin. Dr. Rosen's article reviews several additional ways in which levels of functional BMP can be regulated, including retention of the pro-region after BMP processing and the existence of compounds that can modify BMP binding to BMP receptors. However, regardless of the mechanism for regulating BMP availability, it is questionable whether gradients of active BMP similar to that seen in early morphogenesis play a significant role in defining localizing sites of osteogenesis. Many of the systems modulating BMP-induced osteogenesis do so because BMP-activated signaling triggers negative feedback mechanisms in the target cell. For example, exposure of mesenchymal and osteoblastic cells to BMP rapidly induces expression of the secreted BMP antagonist noggin as well as Smad6 and Smad7, the intracellular inhibitors of BMP signaling. It is logical that the major systems influencing BMP activity in bone formation are primarily regulated temporally rather than spatially, consistent with the very different roles of BMPs in early development and in postnatal maintenance of the skeleton. In embryogenesis, BMPs define embryo polarity and patterning, whereas in the maintenance of the skeleton they act to recruit osteoblast precursors to a defined location in response to a specific need.

Although BMPs have been intensively studied for 15 years, it is still not clear what controls whether a specific BMP will be osteoinductive. Curiously, BMP2 and BMP4, which are absolutely required for the embryo to develop beyond the gastrula stage, also can promote postnatal osteogenesis. BMP-2 and BMP-7 are the most extensively studied for their ability to promote osteogenesis *in vivo* and *in vitro*, but recent studies suggest that BMP-6 and BMP-9 may be more effective in promoting orthotopic bone formation.[8,9] In contrast, although BMP3 is included in the bone morphogenetic protein category on the basis of homology, it inhibits osteogenesis.[10] One way to provide specificity for BMPs would be to have a variety of receptors specialized for different functions. Functional BMP receptors are heterodimers, created by combinations of three type I and three type II receptors.[4] While multiple pairings could provide enough combinations to have specific receptors for each BMP, this does not appear to be the case, since both overexpression and knockout studies suggest that several BMPs can use one receptor pair and the absence of a receptor does not totally block signaling activated by specific BMPs.[11]

The number of intracellular pathways activated by BMPs is relatively limited, suggesting that this will not be a source of much specificity in BMP action. The major signaling pathway downstream of BMPs is receptor activation of Smad1, Smad5, or Smad8, although there is increasing evidence that p38 mitogen-activated kinase can function as an alternative pathway.[12,13] Activation of Smads occurs by serine phosphorylations at the C terminus, which permit binding to co-Smad4 and nuclear translocation of the Smad complex. The activated Smad complex is capable of binding to specific response elements on DNA and thus acting as a transcription factor. However, its DNA-binding affinity is relatively weak, and effective Smad transcriptional regulation appears to rely on the Smad complex dimerizing with other DNA-binding proteins (co-activators or co-modulators), which have higher affinity for their response element sequences on DNA. BMP-activated Smads,

TGF-β-activated Smads, and their common co-Smad, Smad4, all have DNA-binding regions that can recognize similar response elements. Initial studies examining the DNA sequences needed for Smad-binding identified a GTCT motif, a GCAT motif, and a GCCG motif. These seemed to work best as multiple copies and none of them appeared specific for individual activated Smads. Karaulanov *et al.*[14] have recently brought some clarity to the question of Smad-binding sites, characterizing the promotor regions of several BMP-responsive genes and demonstrating a pattern in which the sequence TGGCGCC (Bre7 motif) is located a short distance upstream of one or more SBE (Smad-binding element) motifs containing GTCTG. A more extensive genome scan for vertebrate BMP response elements, combined with functional promotor assays, has identified several dozen genes that are candidates for regulation by activated Smads.[15]

A central unanswered question in the field of BMP-induced osteogenesis is how BMP-activated signaling results in the expression of an osteoblast phenotype. Because BMP treatment (or overexpression of activated BMP receptors) ultimately results in elevated alkaline phosphatase expression as well as increased production of bone matrix proteins, it is frequently assumed that BMP-activated Smads directly regulate the expression of these genes. One popular theory is that BMP-activated Smads bind to activated Runx2 (cbfa1), the transcription factor required for bone formation, thereby forming a transcriptional complex that regulates the expression of key osteoblast genes. However, there is little evidence for this. In his presentation, Dr. Cao discusses one mechanism by which BMP-activated Smads might regulate osteoblast gene expression without DNA binding; activated Smad relieves repression of the osteopontin gene by binding to the repressor Hoxc8 and removing it from the osteopontin promoter.[16] An examination of genes identified as having Smad-responsive promoters, based on phylogenetic analysis combined with functional promoter assays, identified over 100 putative BMP target genes. None of them were osteoblast-specific genes, but a very high proportion of them were transcriptional regulators.[15]

It therefore seems increasingly likely that many reports showing BMP-mediated increases in osteoblast gene expression actually involve an indirect mechanism, in which the initial rapid downstream effect of BMP is the binding of activated Smads to the promoter of a gene encoding a transcriptional regulator or an activator of signaling intermediates. In this scenario, proteins coded by these genes would subsequently affect the expression of osteoblast genes such as alkaline phosphatase, osteopontin, and bone sialoprotein. From the viewpoint of achieving bone formation, this is a mechanistic distinction that might suggest additional therapies for modulating osteogenesis. From a semantic perspective, it would argue that bone morphogenetic proteins not only have a developmental role extending far beyond what the name would suggest, but they also have no direct effect on genes associated with an osteoblast phenotype.

REFERENCES

1. KOBAYASHI, T. & H.M. KRONENBERG. 2005. Minireview: transcriptional regulation in development of bone. Endocrinology **146:** 1012–1017.
2. PROVOT, S. & E. SCHIPANI. 2005. Molecular mechanisms of endochondral bone formation. Biochem. Biophys. Res. Commun. **328:** 658–665.
3. TEN DIJKE, P., O. KORSCHYNSKYI, G. VALDIMARSDOTTIR, *et al.* 2003. Controlling cell fate by BMP receptors. Mol. Cell. Endocrinol. **211:** 105–113.
4. WAN, M. & X. CAO. 2005. BMP signaling in skeletal development. Biochem. Biophys. Res. Commun. **328:** 651–657.
5. URIST, M.R. 1965. Bone: formation by autoinduction. Science **150:** 893–899.
6. WOZNEY, J.M., V. ROSEN, A.J. CELESTE, *et al.* 1988. Novel regulators of bone formation: molecular clones and activities. Science **242:** 1528–1534.
7. HUANG, X. & J.P. SAINT-JEANNET. 2004. Induction of the neural crest and the opportunities of life on the edge. Dev. Biol. **275:** 1–11.
8. CHENG, H.W., W. JIANG, F.M. PHILLIPS, *et al.* 2003. Osteogenic activity of the fourteen types of human bone morphogenetic proteins (BMPs). J. Bone Joint Surg. [Am.] **85A:** 1544–1552.
9. BLUNK, T., A.L. SIEMINSKI, B. APPEL, *et al.* 2003. Bone morphogenetic protein 9: a potent modulator of cartilage development *in vitro*. Growth Factors **21:** 71–77.
10. DALUISKI, A., T. ENGSTRAND, M.E. BAHAMONDE, *et al.* 2001. Bone morphogenetic protein-3 is a negative regulator of bone density. Nat. Genet. **27:** 84–88.
11. YU, P.B., H. BEPPU, N. KAWAI, *et al.* 2005. BMP type II receptor deletion reveals BMP ligand-specific gain of signaling in pulmonary artery smooth muscle cells. J. Biol. Chem. **280:** 24443–24450.
12. NOHE, A., S. HASSEL, M. EHRLICH, *et al.* 2002. The mode of bone morphogenetic protein (BMP) receptor oligomerization determines different BMP-2 signaling pathways. J. Biol. Chem. **277:** 5330–5338.
13. GUICHEUX, J., J. LEMONNIER, C. GHAYOR, *et al.* 2003. Activation of p38 mitogen-activated protein kinase and c-Jun-NH2-terminal kinase by BMP-2 and their implication in the stimulation of osteoblastic cell differentiation. J. Bone Miner. Res. **18:** 2060–2068.
14. KARAULANOV, E., W. KNOCHEL & C. NIEHRS. 2004. Transcriptional regulation of BMP4 synexpression in transgenic *Xenopus*. EMBO J. **23:** 844–856.
15. VON BUBNOFF, A. & K.W.Y. CHO. 2005. Phylogenetic footprinting. Dev. Biol. **281:** 210–226.
16. YANG, X.L., X.H. JI, X.M. SHI, *et al.* 2000. Smad1 domains interacting with hoxc-8 induce osteoblast differentiation. J. Biol. Chem. **275:** 1065–1072.

BMP and BMP Inhibitors in Bone

VICKI ROSEN

Department of Developmental Biology, Harvard School of Dental Medicine, Boston, Massachusetts 02115, USA

ABSTRACT: Bone morphogenetic proteins (BMPs) are signaling molecules that act locally on target cells to affect cell survival, proliferation, and differentiation. While first identified as bone-inducing agents, BMPs are now known to affect the formation and function of many organ systems. Here we focus only on the roles of BMPs in the skeleton. In the developing mouse embryo, BMPs direct skeletal patterning, chondrogenesis, and bone formation. In postnatal animals, BMPs are potent bone regeneration factors, affecting both the amount of new bone formed and the rate at which bone healing occurs. The amount of BMP available for signaling is tightly regulated in both the embryo and postnatally, and in the context of the skeleton, several structurally distinct BMP ligand antagonists have been shown to alter the ability of BMPs to bind to their receptors, blocking BMP activity in physiologically important circumstances. For example, noggin knockout mice display cartilage hyperplasia during skeletal development that results in the loss of joint formation (too much BMP activity), while mice that overexpress noggin in skeletal cells display severe osteopenia and bone fragility (too little BMP activity). Sclerostin, chordin, CTGF, follistatin, and gremlin are additional BMP antagonists that may act in the skeleton to regulate BMP availability. Another class of BMP inhibitors are the proteins that bind to BMP receptors but have no inherent signaling function and thus act as BMP receptor antagonists. To date, inhibin and BMP-3 have been identified as BMP receptor antagonists that can block BMP signaling in bone. Identification of BMP antagonists allows us to investigate their role in diseases that affect skeletal function, such as osteopenia and nonunion fracture, and may provide a novel therapeutic intervention point for treatment.

KEYWORDS: **BMP; bone; osteogenesis**

The remarkable regenerative capacity of bone has been a well-accepted phenomenon for over a century. Standard medical practices, such as casting broken limbs, cutting through skeletal structures in open heart and brain surgery, and using cadavers as a source of bone grafts rely on this intrinsic healing power.

Address for correspondence: Prof. Vicki Rosen, Ph.D., Department of Developmental Biology, REB 510, Harvard School of Dental Medicine, 188 Longwood Avenue, Boston, MA 02115. Voice: 617-432-5910; fax: 617-432-3246.
e-mail: Vicki_rosen@hsdm.harvard.edu

Ann. N.Y. Acad. Sci. 1068: 19–25 (2006).
doi: 10.1196/annals.1346.005

Only with the pioneering work of Dr. Marshall Urist in the 1960s did scientists and clinicians begin to explore what signals present in bone account for this regeneration. Urist was the first to report that protein extracts from demineralized bone matrix were able to induce bone formation at ectopic sites in rodents and that the process initiated by the implantation of these extracts closely resembled the cellular progression seen during endochondral bone formation and fracture healing. Urist called this bone-forming activity bone morphogenetic protein (BMP), and this seminal discovery marked the beginning of BMP research.[1] Major advances in our understanding of BMP biology were also made by Dr. Harri Reddi and his group at National Institutes of Health who determined that BMP activity is not species specific and demonstrated that the delivery matrix used to implant BMPs was a key regulator of its biological activity.[2,3] Reddi also began to dissect the cellular events initiated at the site of BMP implantation and it was this work that attracted many cell and molecular biologists to the BMP field.[4] Today, more than 15 years after the first BMP genes were identified, we know that BMPs are involved in the development of nearly all vertebrate organs and tissues.[5,6] We have also learned that very different organisms are built using similar regulatory molecules and developmental strategies, and that the same signaling pathways appear to underlie key steps in coral, sea urchin, fly, worm, frog, chick, mouse, and human development.[7] During these processes, BMPs act directly on target cells to affect cell survival, cell proliferation, and cell differentiation. Equally important, BMPs interact with other major signaling paths to shape organs and control their function.[8]

Individual BMPs, like other members of the transforming growth factor-β (TGF-β) superfamily, are transcribed as large precursor molecules with amino terminal signal sequences and prodomains of variable size and heterogeneity. These BMP precursors dimerize and are cleaved at RXXR consensus cleavage sites upon secretion from cells to release the mature carboxy terminal dimers of 110–140aa that comprise active BMP. The mature region of the BMP molecule contains seven absolutely conserved cysteines. The three-dimensional structure conferred by the location of six of the seven cysteines makes BMP molecules highly resistant to heat, denaturants, and extremes of pH. These unique properties were of fundamental importance in the initial discovery and subsequent purifications of native BMP activity from bone.[9]

When BMPs are secreted from cells, they have one of several fates: they may immediately exert their actions locally; they may be bound up by extracellular antagonists present at the site of BMP secretion; or, they may interact with extracellular matrix proteins that serve to sequester or enhance BMP activity by anchoring it to make it more available to target cells.[10–12] BMP receptors are serine-threonine kinases, classified as type I or type II based on sequence homology. To date, three type I BMP receptors have been identified; Alk3 binds only to BMPs, Alk6 binds to BMPs and anti-mullerian hormone (AMH), and Alk2 recognizes BMPs and activin. Type II receptors that exhibit BMP binding are BMP RII, a receptor that recognizes only BMPs, and Act RII and IIb that are shared by BMPs and activin.[13] Unlike TGF-β, both type I and type II BMP receptors can bind BMPs in the absence of the other receptor. Also in contrast to other TGF-β-like molecules, there is no direct connection between the extracellular domains of the individual receptors.[14,15] BMP ligand binding to type I receptor triggers the intracellular association of type I and type II receptors, allowing the constitutively phosphorylated type II receptor to phosphorylate the type I receptor. Once activated, type I receptors can recognize and phosphorylate pathway-specific receptor-regulated Smads (R-Smads) that reside in the cell cytoplasm. Phosphorylation of an R-Smad releases it from the receptor complex and allows it to associate with Smad 4, common to all TGF-β/BMP/activin/nodal signaling pathways. The R-Smad/Smad4 complex then translocates to the nucleus where it recognizes regulatory elements in target genes with the help of associated DNA-binding cofactors.[16] Recent structural analysis has shown that under physiological conditions Smad heterotrimers recruit DNA-binding transcription factors, co-activators, and co-repressors to a single genomic locus, providing a way in which many nuclear components can interact with Smads, producing both positive and negative regulation of gene expression.[17]

How the specificity of BMP-receptor signaling is determined remains to be established. Within the TGF-β gene superfamily, many ligands bind to the same receptors, and many receptors bind the same ligands, suggesting that ligand–receptor pairing is not sufficient to generate a specific signal. One current hypothesis is that biological output is determined by the R-Smad partners of each type I receptor.[18] The BMP type I receptors Alk2, Alk3, and Alk6 partner with Smads 1, 5, and 8 while activin, TGF-β, and nodal receptors partner with Smads 2 and 3. Implicit in this hypothesis is that each ligand is able to activate only one R-Smad pathway, and information obtained in signaling experiments to date support this idea.[19] Mesenchymal stem cells (MSC), putative BMP target cells within the adult skeleton have a full complement of BMP receptors and *in vitro*, BMPs are potent inducers of osteoblast differentiation from MSC.

BMP activity is tightly regulated prior to receptor recognition. While control points exist for regulation of BMP activity, very little is known about what regulates cell-specific production of BMPs or BMP receptors. What has become clear in recent years is that the amount of BMP available for signaling is tightly regulated by the presence of several structurally distinct BMP antagonists that alter the ability of BMPs to bind to their receptors, effectively blocking BMP action. These antagonists can be loosely divided into three classes: BMP pro-regions that are able to complex back with mature BMP, creating an inactive molecule; ligand antagonists that bind to BMPs and sequester them as inactive; and, receptor antagonists that prevent BMPs from occupying receptors.[20] If we examine the utility of each type of BMP inhibitor in skeletal tissues, we can make some interesting observations that would not be predicted based on what happens in other biological contexts. For example, while the pro-region may be used to potently inhibit GDF8 and thus increase muscle mass *in vivo*, from studies with BMP-2 and BMP-9, it appears that the pro-region is an ineffective blocker of osteogenesis, perhaps in part because the pro-region is easily cleaved by proteases at skeletal sites.[21–23] In contrast, BMP antagonists, synthesized by MSC as they differentiate to osteoblasts, are capable of blocking osteogenesis. To date, noggin, gremlin, follistatin, and sclerostin have all been identified as osteoblast products that bind and sequester BMPs.[24–27] We know that in the embryonic skeleton these antagonists serve an important physiological role and are responsible for tightly regulating BMP activity so that normal skeletal development occurs. In postnatal animals, pertubation of where, when, and the extent of BMP signaling through the misexpression of these same antagonists results in the alteration of bone formation, allowing us to hypothesize that BMP activity is required for the maintenance of bone mass.[28–31] Interestingly, the highly increased osteogenic activity observed with BMP-9 may be due to the fact that it is not bound by BMP antagonists such as noggin, essentially removing the negative feedback normally placed on BMP activity by osteoblasts (FIG. 1). In fact, an area of ongoing interest is the BMP regulatory capacity of

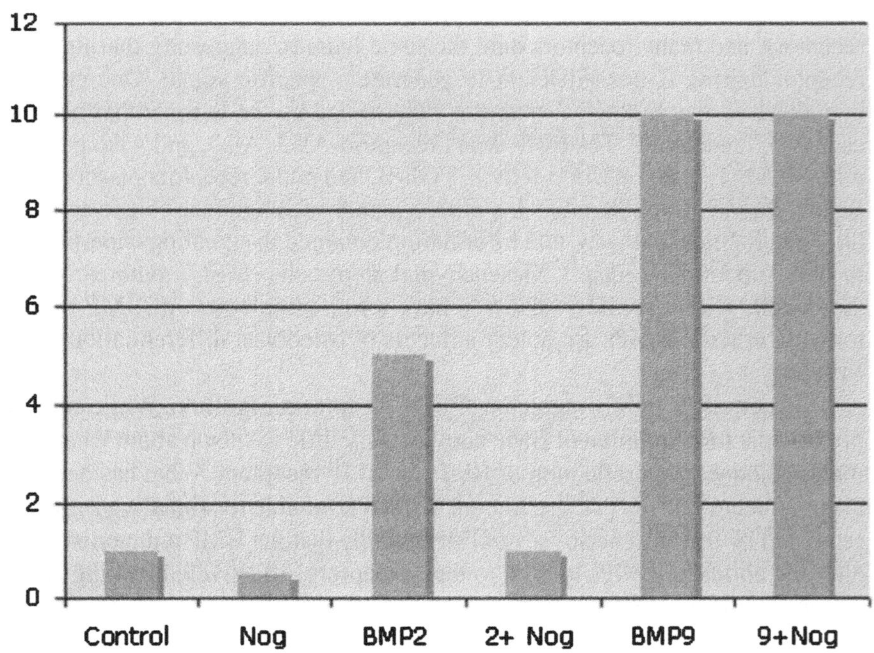

FIGURE 1. BMP-9 is not antagonized by noggin. In W20 cells, a mouse BMS cell line, the ability of BMP-9 to induce alkaline phosphatase is not inhibited by addition of the BMP2 antagonist noggin. *Y* axis = relative amount of alkaline phosphastase after 24 h of treatment; *X* axis = treatment.

bone matrix, which is a depot for both BMPs and BMP antagonists secreted by osteoblasts. As the regulatory capacity of bone matrix changes with aging and is also affected by illness, it is quite possible that the interplay between BMPs and their antagonists within the bone environment is constantly changing.

Another new area that remains to be explored is the physiological role of BMP receptor antagonists in the regulation of bone mass. Two such molecules have been identified to date; BMP-3, an osteoblast product that is stored in bone matrix and acts locally to block BMP signaling through Act RII, and inhibin, a circulating molecule that makes its way to the bone through the vascular system and antagonizes BMP activity in the presence of betaglycan by blocking BMP binding to type II BMP receptors.[32,33] Since mice that are null for BMP-3 show increased bone mass by 5 weeks of age, it is quite likely that these receptor antagonists have the ability to control bone mass in postnatal animals.[34] Our continued understanding of the feedback mechanisms that influence BMP activity should provide us with appropriate information to design novel anabolic agents that can control age-related bone diseases.[35,36]

REFERENCES

1. URIST, M.R. 1965. Bone: formation by autoinduction. Science **150:** 893–899.
2. SAMPATH, T.K. & A.H. REDDI. 1983. Homology of bone-inductive proteins from human, monkey, bovine, and rat extracellular matrix. Proc. Natl. Acad. Sci. USA **80:** 6591–6595.
3. SAMPATH, T.K. & A.H. REDDI. 1984. Importance of geometry of the extracellular matrix in endochondral bone differentiation. J. Cell Biol. **98:** 2192–2197.
4. REDDI, A.H. 1981. Cell biology and biochemistry of endochondral bone development. Coll. Rel. Res. **1:** 209–226.
5. WOZNEY, J.M., V. ROSEN, A.J. CELESTE, *et al.* 1988. Novel regulators of bone formation: molecular clones and activities. Science **242:** 1528–1534.
6. HOGAN, B.L.M. 1996. Bone morphogenetic proteins: multifunctional regulators of vertebrate development. Genes Dev. **10:** 1580–1594.
7. KINGSLEY, D.M. 1994. The TGF-beta superfamily: new members, new receptors, and new genetic tests of function in different organisms. Genes Dev. **8:** 133–146.
8. MACIAS, D., Y. GANAN, T.K. SAMPATH, *et al.* 1997. Role of BMP-2 and OP-1 (BMP-7) in programmed cell death and skeletogenesis during chick limb development. Development **124:** 1109–1117.
9. WOZNEY, J.R. & V. ROSEN. 1998. Bone morphogenetic proteins and the bone morphogenetic protein gene family in bone formation and repair. Clin. Orthop. Rel. Res. **346:** 26–37.
10. ARTEAGA-SOLIS, E., B. GAYRAUD, S.Y. LEE, *et al.* 2001. Regulation of limb patterning by extracellular microfibrils. J. Cell Biol. **154:** 275–281.
11. OHIKAWARA, B., S. LEMURA, P. TEN DIJKE, & N. UENO. 2002. Action range of BMP is defined by its N-terminal basic amino acid core. Current Biol. **12:** 205–209.

12. LANAIN, J., M. OELGESCHLAGER, N.I. KETPURA, *et al.* 2001. Proteolytic cleavage of chordin as a switch for the dual activities of twisted gastrulation in BMP signaling. Development **128:** 4439–4447.
13. MIYAZONO, K. 2002. Bone morphogenetic protein receptors and actions. *In* Principles of Bone Biology, 2nd ed. J.P. Bilezikian, L.G. Raisz & G.A. Rodan, Eds.: 929–942. Academic Press. San Diego, CA.
14. GREENWALD, J., J. GROPPE, P. GRAY, *et al.* 2003. The BMP7/ActRII extracellular domain complex provides new insights into the cooperative nature of receptor assembly. Molec. Cell **11:** 605–617.
15. SUN, P.D. 2003. Conserved in structure but diverse in recognition. Structure **11:** 362–363.
16. ATTISANO, L. & J.L. WRANA. 2002. Signal transduction by the TGF-β superfamily. Science **296:** 1646–1647.
17. MOUSTAKAS, A. & C.-H. HELDIN. 2002. From mono- to oligo-Smads: the heart of the matter in TGF-β signal transduction. Genes Dev. **16:** 1867–1871.
18. MASSAGUE, J. 2000. How cells read TGF-β signals. Nat. Rev. Mol. Cell Biol. **1:** 169–178.
19. KORCHYNSKYI, O., K.J. DECHERING, A.M. SIJBERS, *et al.* 2003. Gene array analysis of bone morphogenetic protein type I receptor-induced osteoblast differentiation. J. Bone Miner. Res. **18:** 1177–1185.
20. GREGORY, K.E., R.N. ONO, N.L. CHARBONNEAU, *et al.* 2005. The prodomain of BMP-7 targets the BMP-7 complex to the ECM. J. Biol. Chem. **280:** 27970–27980.
21. BOGDANOVICH, S., K.J. PERKINS, T.O. KRAG, *et al.* 2005. Myostatin propeptide-mediated amelioration of dystrophic pathophysiology. FASEB J. **19:** 543–549.
22. BROWN, M.A., Q. ZHAO, K.A. BAKER, *et al.* 2005. Crystal structure of BMP-9 and functional interactions with pro-region and receptors. J. Biol. Chem. **280:** 25111–25118.
23. HILLGER, F., G. HERR, R. RUDOLPH & E. SCHWARZ. 2005. Biophysical comparison of BMP-2, pro BMP-2, and the free pro-peptide reveals stabilization of the pro-peptide by the mature growth factor. J. Biol. Chem. **280:** 14974–14980.
24. BRUNET, L.J., J.A. MCMAHON, A.P. MCMAHON, & R.M. HARLAND. 1998. Noggin, cartilage morphogenesis, and joint formation in the mammalian skeleton. Science **280:** 1455–1457.
25. GAZZERRO, E., V. GANGJI & E. CANALIS. 1998. Bone morphogenetic proteins induce the expression of noggin, which limits their activity in cultured rat osteoblasts. J. Clin. Invest. **102:** 2106–2114.
26. KHOKHA, M.K., D. HSU, L.J. BRUNET, *et al.* 2003. Gremlin is the BMP antagonist required for maintenance of Shh and Fgf signals during limb patterning. Nat. Genet. **34:** 303–307.
27. PEREIRA, R.C., A.N. ECONOMIDES & E. CANALIS. 2000. Bone morphogenetic proteins induce gremlin, a protein that limits their activity in osteoblasts. Endocrinology **141:** 4558–4563.
28. GAZZERRO, E., R.C. PEREIRA, V. JORGETTI, *et al.* 2005. Skeletal overexpression of gremlin impairs bone formation and causes osteopenia. Endocrinology **146:** 655–665.
29. SUTHERLAND, M.K., J.C. GEOGHEGAN, C. YU, *et al.* 2004. Sclerostin promotes the apoptosis of human osteoblastic cells: a novel regulation of bone formation. Bone **35:** 828–835.

30. WINKLER, D.G., C. YU, J.C. GEOGHEGAN, *et al.* 2004. Noggin and sclerostin bone morphogenetic protein antagonists form a mutually inhibitory complex. J. Biol. Chem. **279:** 36293–36298.
31. WU, X.B., Y. LI, A. SCHNEIDER, *et al.* 2003. Impaired osteoblastic differentiation, reduced bone formation, and severe osteoporosis in noggin-overexpressing mice. J. Clin. Invest. **112:** 924–934.
32. GAMER, L.W., J. NOVE, M. LEVIN, & V. ROSEN. 2005. BMP-3 is a novel inhibitor of both activin and BMP-4 signaling in Xenopus embryos. Dev. Biol. **285:** 156–168.
33. WINTER, E. & W. VALE. 2003. Inhibin is an antagonist of bone morphogenetic protein signaling. J. Biol. Chem. **278:** 7934–7941.
34. DALUISKI, A., T. ENGSTRAND, M.E. BAHAMONDE, *et al.* 2001. Bone morphogenetic protein-3 is a negative regulator of bone density. Nat. Genet. **27:** 84–88.
35. ROSEN, V. & J.M. WOZNEY. 2002. Bone morphogenetic proteins. *In* Principles of Bone Biology, 2nd ed. J.P. Bilezikian, L.G. Raisz & G.A. Rodan, Eds.: 919–929. Academic Press. San Diego, CA.
36. VALENTIN-OPRAN, A., J. WOZNEY, C. CSIMMA, *et al.* 2002. Clinical evaluation of recombinant human bone morphogenetic protein-2. Clin. Orthop. **395:** 110–120.

BMP Signaling and Skeletogenesis

XUELIN LI AND XU CAO

Department of Pathology, University of Alabama at Birmingham, Birmingham, Alabama 35294, USA

Abstract: Bone morphogenetic proteins (BMPs) are members of the transforming growth factor (TGF)-β superfamily of signal molecules that mediate many diverse biological processes ranging from early embryonic tissue patterning to postnatal tissue homeostasis. BMPs trigger cell responses mainly through the canonical signaling pathway where intracellular Smads play central roles in delivering the extracellular signals to the nucleus. While the same Smads are used by BMPs in all types of cells, different transcription factors account in part for the functional diversity of BMPs. These transcription factors are recruited by Smads to regulate the expression of specific subsets of target genes depending on the cell types. Among the transcription factors are Hox proteins. Experimental gain and loss-of-function studies as well as naturally occurring mutations in *Hox* genes demonstrate their central roles in embryonic skeletal patterning. In addition to the interactions with Smads observed for several Hox proteins, there is also evidence that the expression of a number of *Hox* genes is regulated by BMPs. It is suggested that Hox proteins play an important role in the BMP pathway.

Keywords: BMP; Hox; development; transcription factor; skeletogenesis

INTRODUCTION

The term, *bone morphogenetic protein* (BMP), was first introduced to describe the active component(s) in demineralized bone matrix that can induce ectopic bone formation when implanted intramuscularly or subcutaneously into rodents.[1,2] The identity of BMP was not discovered until the late 1980s when three polypeptides with BMP activity were purified—BMP-1, BMP-2A (later known as BMP-2), and BMP-3.[3] With the exception of BMP-1, the sequences of these molecules revealed that they belong to the transforming growth factor β (TGF-β) superfamily of signal molecules.[3] To date, more than 20 BMP members have been characterized. Evidence from extensive studies indicate

Address for correspondence: Dr. Xu Cao, Department of Pathology, University of Alabama at Birmingham, 1670 University Blvd., VHG003, Birmingham, AL 35294. Voice: 205-934-0162; fax: 205-934-1775.
e-mail: cao@path.uab.edu

**Ann. N.Y. Acad. Sci. 1068: 26–40 (2006). © 2006 New York Academy of Sciences.
doi: 10.1196/annals.1346.006**

that BMPs are multifunctional growth factors whose activities are essential for embryonic development and tissue homeostasis in adults.[4,5]

BMP molecules are first synthesized as large precursors and then cleaved at a dibasic site so that the C-terminal active domain is released. Seven conserved cysteines within the active domain are involved in intramolecular and intermolecular disulfide bonding. The mature BMP molecules are secreted as either heterodimers or homodimers. Although heterodimers have been generated *in vitro* and, in some cases, had more potent effects than the homodimers, their *in vivo* functions have not been well characterized.[4]

SIGNAL TRANSDUCTION BY BMPS

As members of the TGF-β superfamily, BMPs trigger cellular responses mainly through the Smad pathway,[6] although the signal molecules can also activate the mitogen-activated protein kinase (MAPK) pathway.[7] In the Smad pathway, type II and type I transmembrane serine/threonine receptor kinases and intracellular Smad proteins relay the signal from the cell surface to the nucleus. Three type II receptors have been shown to bind BMP ligands: type II BMP receptor (BMPR-II), and type II and IIB activin receptors (ActR-II and ActR-IIB).[8–11] Three type I receptors for BMPs have also been characterized: type IA and IB BMP receptors (BMPIA or ALK3 and BMPIB or ALK6), and type IA activin receptor (ActRIA or ALK2).[12–14] The expression of both type II and type I receptors is required to achieve high-affinity binding with ligands.[6] Upon ligand binding, the activated type I receptors phosphorylate a subgroup of receptor-regulated Smads (R-Smads, Smad1, 5, and 8) at their C-terminal SSXS motif.[15–17] The phosphorylated R-Smads then disassociate from their respective receptors and form complexes with the common partner Smad, Smad4. The Smad complexes then move into the nucleus where they associate with transcription factors to regulate gene transcription in a cell type-specific manner.[18,19] Smad6 and Smad7 represent another subclass of Smads that negatively regulate cell signaling by blocking phosphorylation of R-Smads and preventing complex formation between Smad1 and Smad4.[20–22]

Several transcription factors have been characterized for BMP signaling, among which are Runx2,[23–25] Menin,[26] OAZ,[27] YY1,[28,29] and Hoxc8.[30–33] All of these factors interact with Smad proteins and mediate BMP responses under different biological situations, thereby accounting for the functional diversity of BMPs.

Runx2, also known as polyomavirus enhancer binding protein 2αA/core binding factor 1 (PEBP2αA/CBFA1), is indispensable for bone formation.[34] The interactions between Runx2 and BMP-specific Smads (Smad1 and 5) suggest their intrinsic cooperation in mediating BMP functions.[23,25] While Runx2 alone does not induce osteoblast differentiation, it synergizes with Smad1 and Smad5 in this biological event.[24,25] Mutant Runx2 with a truncated transcription activation domain fails to interact with Smad1 and consistently blocks BMP/Smad-induced osteoblast differentiation.[25] Runx2 has also been shown to interact with TGF-β-specific Smads[23,25] and mimics the common effects of TGF-β1 and BMP-2 to block myogenic differentiation.[24] However, it only synergizes with BMP-2, but not TGF-β1, to induce osteoblast differentiation,[24] suggesting that BMP-specific Smads are required for Runx2 activities.[24] In addition to Runx2, menin, the product of the multiple endocrine neoplasia type 1 (MEN1) gene, is required for BMP-induced osteoblast differentiation.[26] Menin interacts with both Runx2 and Smad1/5 in multipotential mesenchymal cells. When menin is knocked down, the cells fail to differentiate into osteoblast lineage. Conversely, menin interacts with TGF-β specific Smad3 in well-differentiated osteoblasts and inhibits BMP-induced late-stage differentiation.[26] Therefore, the function of menin in BMP signaling is stage-dependent.

Nuclear factor Yin Yang 1 (YY1) interacts with Smad1/4 and, depending on the cell types and target genes can act as both a repressor and an activator in the BMP-signaling pathway.[28,29] It inhibits Smad binding to its consensus DNA target *in vitro* and represses certain BMP target gene transcription. Furthermore, YY1 inhibits BMP-induced cell differentiation but has no effect on BMP-mediated growth inhibition.[28] In a different situation, YY1, in response to BMP, serves as an activator to drive the transcription of a key cardiac regulator (*Nkx2.5*) during mouse embryonic development. YY1 and Smad1/4 form a complex and bind to adjacent promotor elements. Disruption of the YY1/Smad1/4 complex eliminates transactivation by BMP.[29] Again, YY1 seems not to be specific for BMP since it also interacts with Smad2 and Smad3, although the biological consequences are different.[28]

Compared with the factors discussed above, Olf-1/EBF-associated zinc finger protein (OAZ) specifically interacts with activated Smad1, but not activated Smad2.[27] Although OAZ lacks transactivation activity alone, its interaction with Smad1/4 is required for BMP-induced homeobox gene *Xvent-2* transcription, an event that is essential for mesoderm ventralization and neural inhibition. The *Xvent-2* gene promotor element contains separate binding sites for OAZ and Smads. The presence of both OAZ and Smad complexes on the element is necessary and sufficient for *Xvent-2* gene transactivation. OAZ does not activate another BMP target gene *Tlx-2* suggesting that it helps Smad proteins distinguish a specific subset of target genes from others.[27]

Hoxc8, a member of the Hox family, has been demonstrated to act as a repressor in the BMP pathway. It binds to the consensus site of the osteopontin (OPN) gene and silences its transcription. The Smad1/4 complex, in response to BMP, interacts with its DNA-binding domain (known as homeodomain) and dislodges it from the OPN promotor element, thereby initiating gene transcription and inducing osteoblast differentiation.[30,31] Importantly, transgenic mice that overexpress the interaction domain of Smad1 with Hoxc8 in bone tissues have enhanced bone density.[33] In contrast, Smad6 Complexed with Hoxc8 on the promotor element, serves as a transcriptional co-repressor.[32] Similar interactions have also been observed for Hoxa9.[32,35] Recently, Hoxa13 and Hoxd13 have been shown to antagonize Smad-induced transactivation by interacting with both Smad1/5 and Smad2.[36] It seems that Smad/Hox interaction is universal for most Hox proteins. However, due to the molecular diversity of Hox proteins, other interactions are likely to exist. For example, some Hox proteins interact with BMP-specific Smads or TGF-β-specific Smads, or both.

BIOLOGICAL FUNCTIONS OF BMPS

BMPs as Ventralizing Factors during Early Development

As the fertilized egg divides, three germ layers (endoderm, ectoderm, and mesoderm) are established during early development that will ultimately give rise to all types of somatic cells. The ectoderm will become neural tissues and epidermis. The mesoderm will develop into blood, mesenchyme, muscle, and notochord. The endoderm will form the respiratory and digestive tracts.[37,38] Dorsal–ventral patterning of the mesoderm and ectoderm during blastula and early gastrula stages, guided by inductive tissue interactions, is the crucial step for cell-fate determination. Accumulating data indicate that these tissue–tissue interactions are largely mediated by BMP-related factors and their antagonists. Results from *Xenopus laevis* studies will be reviewed according to BMP functions during early development. Tissue patterning of other vertebrates such as zebrafish and mice share similar mechanisms, albeit with certain variations.

Before the blastula stage, the animal and vegetal regions of the embryo are occupied by ectoderm and endoderm, respectively. During blastula and gastrula stages, mesoderm is formed and patterned in the marginal zone between ectoderm and endoderm. Mesoderm patterning is guided by ventralizing and dorsalizing signals. Ventralizing signals induce ventral mesoderm (blood, mesothelium, and mesenchyme), and dorsalizing signals induce dorsal mesoderm (notochord, muscle, and kidney). BMP-4 has been shown to be the most potent ventralizing factor. *Bmp-4* mRNA is present on the ventral side of the developing embryo.[39,40] Overexpression of *Bmp-4* by mRNA injection induces ventral mesoderm structures and marker expression.[39–43] Its effects are strong enough to override dorsalizing factors such as activin and lithium.[39,41,42,44] Conversely, disrupting BMP signaling either by a dominant-negative BMP receptor[40,45–47] or by antisense *Bmp-4* RNA[44] results in dorsalized mesoderm. In addition to BMP-4, BMP-2[48] and BMP-7[49] have also been shown to have ventralizing effects. Interestingly, BMP-3 has dorsalizing activities.[50,51] BMP-3 antagonizes BMP signaling and partially blocks the ventralizing effects of BMP-4, most likely by competing for the receptors since it fails to inhibit the activities of constitutively activated BMP type I receptor (CA-ALK3).[51] The roles of BMPs in ventral mesoderm formation are also strongly supported by studies on the Spemann organizer (the dorsal mesoderm), which secretes BMP antagonists to counteract BMP function and induce dorsal mesoderm. The Spemann organizer was first characterized by Spemann and Mangold in 1924, who successfully demonstrated that the small dorsal lip of the embryo induces dorsal mesoderm as well as dorsal ectoderm formation when grafted to the ventral side.[52] Seventy years later it was found that the dorsalizing activities of the Spemann organizer are mediated by several secreted signals—noggin, chordin, and follistatin.[53–57] Soon thereafter, noggin[58] and chordin[59] were demonstrated to bind to extracellular BMPs with high affinity, blocking signal transduction by preventing ligand–receptor interactions. Noggin and chordin specifically bind to BMP molecules, since they bind to neither TGF-β1 nor activin.[58,59] Follistatin, on the other hand, primarily serves as an activin antagonist.[57] However, it also binds to BMPs and has similar inhibitory effects.[57,60] Therefore, it is believed that BMPs ventralize mesoderm and their antagonists released from the Spemman organizer dorsalize mesoderm. The balance between these two sets of signals in specific mesoderm regions determines their terminal fates.

During relatively later gastrulation stages, BMPs and their antagonists mediate ectoderm patterning in a manner similar to that of mesoderm patterning. At these stages, the ectodermal cells differentiate into either epidermis (ventral fate) or neural tissue (dorsal fate). BMP ligands serve as epidermal inducers and neural inhibitors.[40,49,56,61–64] In contrast, noggin, chordin, and follistatin favor neural fate.[56,57,59,64,65] Moreover, another family of BMP antagonists (DAN family), including Gremlin, Cerberus, and the tumor suppressor, DAN, has been demonstrated to have similar neural-inducing effects.[66,67] Overall, BMPs and their antagonists serve as the central control that guides dorsal–ventral patterning of mesoderm and ectoderm during early development.

BMPs and Skeletogenesis

As their name indicates, BMP molecules are capable of inducing ectopic cartilage and bone formation, a process that mimics embryonic endochondral bone formation.[68,69] Extensive studies demonstrate that BMPs are important factors regulating chondrogenesis and skeletogenesis during normal embryonic development. Individual BMPs exhibit distinct expression patterns in skeletal elements. *Bmp-2* is expressed in areas surrounding the initial cartilage condensations,[70,71] while *Bmp-4* is expressed in perichondrium.[71] *Bmp-2* is also expressed in periosteal and osteogenic zones.[70] *Bmp-5* is expressed in initial cartilage condensations as well as in perichondrium and periosteum at later stages of bone development.[72] *Bmp-6* is expressed in hypertrophic chondrocytes.[70] High levels of *Bmp-7* mRNA have been observed in the perichondrium, but its expression is absent in the zones of joint formation.[73] In contrast to *Bmp-7*, *Gdf-5* transcripts were detected in regions where joints will form.[74] Distinct expression patterns of individual BMPs indicate that different

BMP molecules mediate specific events during skeletogenesis. For example, BMP-2 recruits mesenchymal cells surrounding the initial cartilage condensations into chondrogenic fate, while BMP-4 recruits perichondrial cells.[71] BMP-6 may be essential for terminal chondrocyte differentiation.[70] BMP-7 is a potential inhibitory factor for joint formation[73] while GDF-5 is required for joint formation.[74]

BMPs are potent inducers of cartilage and bone formation *in vitro*. BMP-2 induces cartilage nodule formation in chick limb bud mesenchyme cultures.[75] BMP-6 accelerates hypertrophic chondrocyte differentiation and mineral accretion.[76] In multipotential mesenchymal cells (MMCs) isolated from human bone marrow, BMP-2 and BMP-9 promote chondrogenic differentiation, possibly through activation of *Sox-9*, a chondrogenic-related transcription factor. BMP-9 appears to have stronger effects than BMP-2.[77] BMP-2 and BMP-7 have been shown to induce both chondrocyte and osteoblast differentiation in C3H10T1/2 cells.[78,79] Although the mechanisms are not fully understood, it has been proposed that BMPs can induce both undifferentiated stem cells and more differentiated multipotent cells along chondrogenic or osteogenic pathways.[80–82]

Genetic studies have provided direct evidence of the specific functions of individual BMPs in cartilage and bone formation. First, strong evidence comes from the naturally occurring mutations of *Bmp-5* and *Gdf-5* genes.[72,83,84] Mutations of *Bmp-5* are the direct cause of skeletal defects observed in *short ear* mice, including a reduction in size of the external ear, malformations of the sternum, rib cage and the sixth cervical vertebra, and defective repair of bone fractures.[72,83] Mutations of *Gdf-5* are responsible for the defects in limb skeletons observed in *brachypodism* mice.[84] Experimental mutation studies provide further evidence. *Bmp-7*-deficient mice exhibit skeletal alterations in the rib cage, hind limbs, and skull.[85] *Bmp-6* mutant mice have delayed sternum ossification.[86] The cumulative data indicate that individual BMPs mediate particular stages during skeletogenesis of specific bone elements. BMP-3 represents an exception. Consistent with its opposite roles to BMP-4 during mesoderm patterning,[50,51] BMP-3 inhibits BMP-2-induced osteogenic differentiation.[50] *Bmp-3* knockout mice exhibit increased bone density.[50] Again, the data indicate that BMP-3 antagonizes BMP signaling *in vivo*. BMP-2 and BMP-4 are the most well-characterized ligands within the family. Homozygous *Bmp-2* mutant embryos die between embryonic day 7.5 (E7.5) and 10.5 (E10.5) with cardiac defects.[87] *Bmp-4* null mice die between E6.5 and E9.5 with mesoderm formation defects.[88]

BMPs and Apoptosis

During embryonic development, BMPs mediate programmed cell death, or apoptosis, the process that removes unnecessary tissues, thereby ensuring proper morphogenesis. BMP-4 was first demonstrated to induce apoptosis in rhombomeric neural crest cells.[89] Other evidence for BMP-mediated apoptosis mainly comes from studies on limb bud mesenchymal cell death, a notable feature of limb development. In chick limb bud, apoptosis occurs in the anterior necrotic zone (ANZ), posterior necrotic zone (PNZ), and interdigital necrotic zone (INZ) to ensure proper limb structure. *Bmp-4* is expressed in these regions prior to the onset of apoptosis. Its expression is maintained during the course of apoptosis. *Bmp-2* has a similar but weaker expression pattern.[90] Overexpression of dominant-negative BMP type I receptors[90,91] suppresses apoptosis in chick limb bud, leading to the webbing phenotype. Conversely, overexpression of constitutively active BMP type I receptor promotes apoptosis.[92] When BMP-2- or BMP-7-soaked beads were implanted into chick limb buds, they induced apoptosis in the undifferentiated mesenchymal cells.[73] In transgenic mice overexpressing the BMP antagonist, noggin, the interdigital tissue had incomplete regression.[93] BMP-2 and BMP-4 also induce apoptosis in mesenchymal cells isolated from the interdigital regions of chick limb buds. This effect was not observed for either TGF-β1 or activin.[90] Similar results from separate studies support the notion that the apoptotic effect is specific for BMP molecules.[94]

Several lines of evidence indicate that the homeobox-containing gene *Msx2* acts downstream of BMPs to mediate apoptosis. Overexpression of dominant-negative BMPR-IB leads to downregulation of *Msx2* and suppression of apoptosis in the chick embryos.[91] High level of *Msx2* expression, either by stable transfection or induction by BMP-4, results in increased apoptosis in P19 cells. Furthermore, BMP-4 cannot increase apoptosis in cells with forced expression of *Msx2*.[95] On the other hand, *Msx2* was found to be not essential for BMP-mediated apoptosis in mouse limb.[93] It is possible that different mechanisms exist for different species, or that *Msx2* is only one of the factors responsible for the apoptotic effects of BMPs.[93] Dickkopf-1 (Dkk-1), a potent inhibitor of the Wnt/β-catenin pathway, has also been shown to be involved in BMP-mediated apoptosis.[96] It appears that BMP-mediated Dkk-1 activation is mainly via the MAPK pathway.[96]

HOX PROTEINS AND SKELETOGENESIS

Hox proteins are a family of transcription factors that control anterior–posterior body axis patterning during embryonic development.[97] The fact that Hoxc8,[30–32] Hoxa9,[35] and Hoxa13/d13[36] interact with Smads suggests their potential roles as BMP/TGF-β downstream transcription factors. The *Hox* gene family members share well-conserved homeobox sequences that encode the homeodomains and have a characteristic organization along the genome. In mice and humans, 39 *Hox* genes that are arranged into four genomic clusters (*Hoxa-d*) have been identified. According to the sequence similarities and positions along the clusters, the 39 genes are divided into 13 paralogs.[98] During development, the expression of *Hox* genes follows the rule of temporal and spatial collinearity. The more 3′ *Hox* genes are expressed earlier in anterior regions and the more 5′ *Hox* genes are expressed later in posterior regions. Consistently, unique subsets of *Hox* genes specify the identities of individual regions where they are expressed.[99,100]

The crucial roles of *Hox* genes in embryonic pattern formation have been substantiated by extensive genetic studies. Mutations of *Hox* genes usually result in multiple skeletal defects. *Hoxa2* mutant mice exhibit homeotic transformation of the cranial skeleton. As a result, a set of the first branchial arch derivatives, including malleus and incus, are duplicated at the expense of the second branchial arch elements such as stapes and the lesser horns of hyoid bone.[101,102] *Hoxa2* was further demonstrated to restrict chondrogenic centers as well as inhibit dermal bone formation, likely via excluding *Sox9* expression and downregulating *Cbfa1* expression, thereby providing mechanisms for *Hoxa2*-mediated second branchial arch patterning.[103] In *Hoxc8* null mice, transformations occur in vertebrae (1st lumbar vertebra to 14th thoracic vertebra) and ribs. Additional sternebra is formed. Some ribs attach to the sternum abnormally.[104] Transgenic mice overexpressing *Hoxc8* under its own promotor do not exhibit homeotic transformation, suggesting that the temporal stages of *Hoxc8* expression are critical for its function.[105] Instead of patterning defects, transgenic mice show immature chondrocyte differentiation, with accumulation of proliferating chondrocytes and reduced hypertrophic cartilage. *Hoxd4* expression results in a phenotype similar to that produced by *Hoxc8* expression, indicating comparable functions in chondrocyte differentiation.[105] In contrast to the anteriorized transformations observed in *Hoxa2* and *Hoxc8* mutant mice, *Hoxa11* mutant mice develop a posteriorized phenotype.[106] The 13th thoracic vertebra is transformed into the 1st lumbar vertebra. The possible underlying mechanism is that disruption of the *Hoxa11* gene may change the expression of other *Hox* genes in the regions where *Hoxa11* is normally expressed, thereby changing the Hox code in those regions.[106] In addition to the axial skeletal malformations, limb skeletal defects have also been observed in *Hoxa11* mutant mice.[106]

While most of the evidence for *Hox* gene functions during development comes from experiments with targeted gene disruption, several naturally occurring *Hox* gene mutations have also been identified with distinct developmental defects in mice and humans. The phenotype of Hypodactyly (*Hd*) mice is caused by a 50 base-pair deletion in the first exon of the *Hoxa13* gene; therefore, no functional Hoxa13 protein is made from the mutated allele. As a result, heterozygous *Hd* mice have a shortened digit I in all limbs and homozygous *Hd* mice have more severe defects with only one digit in each limb.[107] The molecular basis of human hand-foot-genital (HFG) syndrome has been found to be a nonsense mutation in the *HOXA13* gene, leading to the production of truncated proteins which, in all likelihood, cannot serve as functional transcription factors. The defects include autopod and urinary tract malformations.[108] Mutation of *HOXD13* is involved in synpolydactyly (SPD) which is characterized by an increase in the number of digits and digit fusions.[109,110] The mutant HOXD13 proteins contain duplicated polyalanines upstream of the homeodomain.

Several studies reveal an intrinsic linkage between Hox factors and BMP signaling. Hox factors participate in BMP functions by two means. First, Hoxc8[30–33] and Hox13 proteins[36] interact with Smads and serve as BMP/TGF-β downstream transcription factors. Second, *Hox* and *Bmp* gene expression is mutually regulated by each other. BMP-2 has been shown to induce ectopic expression of *Hoxd11* and *Hoxd13* in anterior limb mesenchyme.[111] BMP-2 also induces ectopic *Hoxa13* expression in the anterior region of limb muscle masses, while the BMP antagonists, noggin/chordin, downregulate its expression.[112] There is also evidence that BMP signals regulate the initial expression of a number of *Hox* genes during early *Xenopus* development.[113] On the other hand, Hoxa13 and Hoxd13 can activate *Bmp-4* promotor activity.[114] The cumulative results indicate that Hox proteins and BMP signals cooperate with each other to fulfill their biological missions.

REFERENCES

1. URIST, M.R. 1965. Bone: formation by autoinduction. Science **150:** 893–899.
2. URIST, M.R. & B.S. STRATES. 1971. Bone morphogenetic protein. J. Dent. Res. **50:** 1392–1406.
3. WOZNEY, J.M., V. ROSEN, A.J. CELESTE, *et al.* 1988. Novel regulators of bone formation: molecular clones and activities. Science **242:** 1528–1534.
4. HOGAN, B.L. 1996. Bone morphogenetic proteins: multifunctional regulators of vertebrate development. Genes Dev. **10:** 1580–1594.
5. WOZNEY, J.M. & V. ROSEN. 1998. Bone morphogenetic protein and bone morphogenetic protein gene family in bone formation and repair. Clin. Orthop. Relat. Res. **346:** 26–37.
6. MASSAGUE, J. 1998. TGF-beta signal transduction. Annu. Rev. Biochem. **67:** 753–791.
7. VON BUBNOFF, A. & K.W. CHO. 2001. Intracellular BMP signaling regulation in vertebrates: pathway or network? Dev. Biol. **239:** 1–14.
8. YAMASHITA, H., P. TEN DIJKE, D. HUYLEBROECK, *et al.* 1995. Osteogenic protein-1 binds to activin type II receptors and induces certain activin-like effects. J. Cell Biol. **130:** 217–226.
9. NOHNO, T., T. ISHIKAWA, T. SAITO, *et al.* 1995. Identification of a human type II receptor for bone morphogenetic protein-4 that forms differential heteromeric complexes with bone morphogenetic protein type I receptors. J. Biol. Chem. **270:** 22522–22526.
10. ROSENZWEIG, B.L., T. IMAMURA, T. OKADOME, *et al.* 1995. Cloning and characterization of a human type II receptor for bone morphogenetic proteins. Proc. Natl. Acad. Sci. USA **92:** 7632–7636.
11. KAWABATA, M., A. CHYTIL & H.L. MOSES. 1995. Cloning of a novel type II serine/threonine kinase receptor through interaction with the type I transforming growth factor-beta receptor. J. Biol. Chem. **270:** 5625–5630.

12. KOENIG, B.B., J.S. COOK, D.H. WOLSING, *et al.* 1994. Characterization and cloning of a receptor for BMP-2 and BMP-4 from NIH 3T3 cells. Mol. Cell Biol. **14:** 5961–5974.
13. TEN DIJKE, P., H. YAMASHITA, T.K. SAMPATH, *et al.* 1994. Identification of type I receptors for osteogenic protein-1 and bone morphogenetic protein-4. J. Biol. Chem. **269:** 16985–16988.
14. MACIAS-SILVA, M., P.A. HOODLESS, S.J. TANG, *et al.* 1998. Specific activation of Smad1 signaling pathways by the BMP7 type I receptor, ALK2. J. Biol. Chem. **273:** 25628–25636.
15. HOODLESS, P.A., T. HAERRY, S. ABDOLLAH, *et al.* 1996. MADR1, a MAD-related protein that functions in BMP2 signaling pathways. Cell **85:** 489–500.
16. NISHIMURA, R., Y. KATO, D. CHEN, *et al.* 1998. Smad5 and DPC4 are key molecules in mediating BMP-2-induced osteoblastic differentiation of the pluripotent mesenchymal precursor cell line C2C12. J. Biol. Chem. **273:** 1872–1879.
17. CHEN, Y., A. BHUSHAN & W. VALE. 1997. Smad8 mediates the signaling of the ALK-2 [corrected] receptor serine kinase. Proc. Natl. Acad. Sci. USA **94:** 12938–12943.
18. MASSAGUE, J. & Y.G. CHEN. 2000. Controlling TGF-beta signaling. Genes Dev. **14:** 627–644.
19. MASSAGUE, J. & D. WOTTON. 2000. Transcriptional control by the TGF-beta/Smad signaling system. EMBO J. **19:** 1745–1754.
20. IMAMURA, T., M. TAKASE, A. NISHIHARA, *et al.* 1997. Smad6 inhibits signalling by the TGF-beta superfamily. Nature **389:** 622–626.
21. NAKAO, A., M. AFRAKHTE, A. MOREN, *et al.* 1997. Identification of Smad7, a TGFbeta-inducible antagonist of TGF-beta signalling. Nature **389:** 631–635.
22. HATA, A., G. LAGNA, J. MASSAGUE, *et al.* 1998. Smad6 inhibits BMP/Smad1 signaling by specifically competing with the Smad4 tumor suppressor. Genes Dev. **12:** 186–197.
23. HANAI, J., L.F. CHEN, T. KANNO, *et al.* 1999. Interaction and functional cooperation of PEBP2/CBF with Smads. Synergistic induction of the immunoglobulin germline Calpha promoter. J. Biol. Chem. **274:** 31577–31582.
24. LEE, K.S., H.J. KIM, Q.L. LI, *et al.* 2000. Runx2 is a common target of transforming growth factor beta1 and bone morphogenetic protein 2, and cooperation between Runx2 and Smad5 induces osteoblast-specific gene expression in the pluripotent mesenchymal precursor cell line C2C12. Mol. Cell Biol. **20:** 8783–8792.
25. ZHANG, Y.W., N. YASUI, K. ITO, *et al.* 2000. A RUNX2/PEBP2alpha A/CBFA1 mutation displaying impaired transactivation and Smad interaction in cleidocranial dysplasia. Proc. Natl. Acad. Sci. USA **97:** 10549–10554.
26. SOWA, H., H. KAJI, G.N. HENDY, *et al.* 2004. Menin is required for bone morphogenetic protein 2- and transforming growth factor beta-regulated osteoblastic differentiation through interaction with Smads and Runx2. J. Biol. Chem. **279:** 40267–40275.
27. HATA, A., J. SEOANE, G. LAGNA, *et al.* 2000. OAZ uses distinct DNA- and protein-binding zinc fingers in separate BMP-Smad and Olf signaling pathways. Cell **100:** 229–240.
28. KURISAKI, K., A. KURISAKI, U. VALCOURT, *et al.* 2003. Nuclear factor YY1 inhibits transforming growth factor beta- and bone morphogenetic protein-induced cell differentiation. Mol. Cell Biol. **23:** 4494–4510.

29. LEE, K.H., S. EVANS, T.Y. RUAN, *et al.* 2004. SMAD-mediated modulation of YY1 activity regulates the BMP response and cardiac-specific expression of a GATA4/5/6-dependent chick Nkx2.5 enhancer. Development **131:** 4709–4723.
30. SHI, X., X. YANG, D. CHEN, *et al.* 1999. Smad1 interacts with homeobox DNA-binding proteins in bone morphogenetic protein signaling. J. Biol. Chem. **274:** 13711–13717.
31. YANG, X., X. JI, X. SHI, *et al.* 2000. Smad1 domains interacting with Hoxc-8 induce osteoblast differentiation. J. Biol. Chem. **275:** 1065–1072.
32. BAI, S., X. SHI, X. YANG, *et al.* 2000. Smad6 as a transcriptional corepressor. J. Biol. Chem. **275:** 8267–8270.
33. LIU, Z., W. SHI, X. JI, *et al.* 2004. Molecules mimicking Smad1 interacting with Hox stimulate bone formation. J. Biol. Chem. **279:** 11313–11319.
34. KOMORI, T., H. YAGI, S. NOMURA, *et al.* 1997. Targeted disruption of Cbfa1 results in a complete lack of bone formation owing to maturational arrest of osteoblasts. Cell **89:** 755–764.
35. SHI, X., S. BAI, L. LI, *et al.* 2001. Hoxa-9 represses transforming growth factor-beta-induced osteopontin gene transcription. J. Biol. Chem. **276:** 850–855.
36. WILLIAMS, T.M., M.E. WILLIAMS, J.H. HEATON, *et al.* 2005. Group 13 HOX proteins interact with the MH2 domain of R-Smads and modulate Smad transcriptional activation functions independent of HOX DNA-binding capability. Nucleic Acids Res. **33:** 4475–4484.
37. GRAFF, J.M. 1997. Embryonic patterning: to BMP or not to BMP, that is the question. Cell **89:** 171–174.
38. WELLS, J.M. & D.A. MELTON. 1999. Vertebrate endoderm development. Annu. Rev. Cell Dev. Biol. **15:** 393–410.
39. FAINSOD, A., H. STEINBEISSER & E.M. DE ROBERTIS. 1994. On the function of BMP-4 in patterning the marginal zone of the Xenopus embryo. EMBO J. **13:** 5015–5025.
40. SCHMIDT, J.E., A. SUZUKI, N. UENO, *et al.* 1995. Localized BMP-4 mediates dorsal/ventral patterning in the early Xenopus embryo. Dev. Biol. **169:** 37–50.
41. DALE, L., G. HOWES, B.M. PRICE, *et al.* 1992. Bone morphogenetic protein 4: a ventralizing factor in early Xenopus development. Development **115:** 573–585.
42. JONES, C.M., K.M. LYONS, P.M. LAPAN, *et al.* 1992. DVR-4 (bone morphogenetic protein-4) as a posterior-ventralizing factor in Xenopus mesoderm induction. Development **115:** 639–647.
43. RE'EE-KALMA, Y., T. LAMB & D. FRANK. 1995. Competition between noggin and bone morphogenetic protein 4 activities may regulate dorsalization during Xenopus development. Proc. Natl. Acad. Sci. USA **92:** 12141–12145.
44. STEINBEISSER, H., A. FAINSOD, C. NIEHRS, *et al.* 1995. The role of gsc and BMP-4 in dorsal-ventral patterning of the marginal zone in Xenopus: a loss-of-function study using antisense RNA. EMBO J. **14:** 5230–5243.
45. GRAFF, J.M., R.S. THIES, J.J. SONG, *et al.* 1994. Studies with a Xenopus BMP receptor suggest that ventral mesoderm-inducing signals override dorsal signals *in vivo*. Cell **79:** 169–179.
46. SUZUKI, A., R.S. THIES, N. YAMAJI, *et al.* 1994. A truncated bone morphogenetic protein receptor affects dorsal-ventral patterning in the early Xenopus embryo. Proc. Natl. Acad. Sci. USA **91:** 10255–10259.
47. MAENO, M., R.C. ONG, A. SUZUKI, *et al.* 1994. A truncated bone morphogenetic protein 4 receptor alters the fate of ventral mesoderm to dorsal mesoderm: roles

of animal pole tissue in the development of ventral mesoderm. Proc. Natl. Acad. Sci. USA **91:** 10260–10264.
48. HEMMATI-BRIVANLOU, A. & G.H. THOMSEN. 1995. Ventral mesodermal patterning in Xenopus embryos: expression patterns and activities of BMP-2 and BMP-4. Dev. Genet. **17:** 78–89.
49. HAWLEY, S.H., K. WUNNENBERG-STAPLETON, C. HASHIMOTO, *et al.* 1995. Disruption of BMP signals in embryonic Xenopus ectoderm leads to direct neural induction. Genes Dev. **9:** 2923–2935.
50. DALUISKI, A., T. ENGSTRAND, M.E. BAHAMONDE, *et al.* 2001. Bone morphogenetic protein-3 is a negative regulator of bone density. Nat. Genet. **27:** 84–88.
51. GAMER, L.W., J. NOVE, M. LEVIN, *et al.* 2005. BMP-3 is a novel inhibitor of both activin and BMP-4 signaling in Xenopus embryos. Dev. Biol. **285:** 156–168.
52. SPEMANN, H. & H. MANGOLD. 1924. Uber Induktion von embryonanlagen durch implantation artfremder organisatoren. Arch. Mikr. Anat. EntwMech **100:** 599–638.
53. SMITH, W.C. & R.M. HARLAND. 1992. Expression cloning of noggin, a new dorsalizing factor localized to the Spemann organizer in Xenopus embryos. Cell **70:** 829–840.
54. SMITH, W.C., A.K. KNECHT, M. WU, *et al.* 1993. Secreted noggin protein mimics the Spemann organizer in dorsalizing Xenopus mesoderm. Nature **361:** 547–549.
55. SASAI, Y., B. LU, H. STEINBEISSER, *et al.* 1994. *Xenopus chordin*: a novel dorsalizing factor activated by organizer-specific homeobox genes. Cell **79:** 779–790.
56. SASAI, Y., B. LU, H. STEINBEISSER, *et al.* 1995. Regulation of neural induction by the Chd and Bmp-4 antagonistic patterning signals in Xenopus. Nature **376:** 333–336.
57. HEMMATI-BRIVANLOU, A., O.G. KELLY & D.A. MELTON. 1994. Follistatin, an antagonist of activin, is expressed in the Spemann organizer and displays direct neuralizing activity. Cell **77:** 283–295.
58. ZIMMERMAN, L.B., J.M. JESUS-ESCOBAR & R.M. HARLAND. 1996. The Spemann organizer signal noggin binds and inactivates bone morphogenetic protein 4. Cell **86:** 599–606.
59. PICCOLO, S., Y. SASAI, B. LU, *et al.* 1996. Dorsoventral patterning in Xenopus: inhibition of ventral signals by direct binding of chordin to BMP-4. Cell **86:** 589–598.
60. IEMURA, S., T.S. YAMAMOTO, C. TAKAGI, *et al.* 1998. Direct binding of follistatin to a complex of bone-morphogenetic protein and its receptor inhibits ventral and epidermal cell fates in early Xenopus embryo. Proc. Natl. Acad. Sci. USA **95:** 9337–9342.
61. WILSON, P.A. & A. HEMMATI-BRIVANLOU. 1995. Induction of epidermis and inhibition of neural fate by Bmp-4. Nature **376:** 331–333.
62. SUZUKI, A., E. KANEKO, N. UENO, *et al.* 1997. Regulation of epidermal induction by BMP2 and BMP7 signaling. Dev. Biol. **189:** 112–122.
63. XU, R.H., J. KIM, M. TAIRA, *et al.* 1995. A dominant negative bone morphogenetic protein 4 receptor causes neuralization in Xenopus ectoderm. Biochem. Biophys. Res. Commun. **212:** 212–219.
64. CHANG, C. & A. HEMMATI-BRIVANLOU. 1999. Xenopus GDF6, a new antagonist of noggin and a partner of BMPs. Development **126:** 3347–3357.
65. LAMB, T.M., A.K. KNECHT, W.C. SMITH, *et al.* 1993. Neural induction by the secreted polypeptide noggin. Science **262:** 713–718.

66. HSU, D.R., A.N. ECONOMIDES, X. WANG, *et al.* 1998. The Xenopus dorsalizing factor Gremlin identifies a novel family of secreted proteins that antagonize BMP activities. Mol. Cell **1:** 673–683.
67. BOUWMEESTER, T., S. KIM, Y. SASAI, *et al.* 1996. Cerberus is a head-inducing secreted factor expressed in the anterior endoderm of Spemann's organizer. Nature **382:** 595–601.
68. REDDI, A.H. & C. HUGGINS. 1972. Biochemical sequences in the transformation of normal fibroblasts in adolescent rats. Proc. Natl. Acad. Sci. USA **69:** 1601–1605.
69. REDDI, A.H. 1981. Cell biology and biochemistry of endochondral bone development. Coll. Relat. Res. **1:** 209–226.
70. LYONS, K.M., R.W. PELTON & B.L. HOGAN. 1989. Patterns of expression of murine Vgr-1 and BMP-2a RNA suggest that transforming growth factor-beta-like genes coordinately regulate aspects of embryonic development. Genes Dev. **3:** 1657–1668.
71. DUPREZ, D., E.J. BELL, M.K. RICHARDSON, *et al.* 1996. Overexpression of BMP-2 and BMP-4 alters the size and shape of developing skeletal elements in the chick limb. Mech. Dev. **57:** 145–157.
72. KING, J.A., P.C. MARKER, K.J. SEUNG, *et al.* 1994. BMP5 and the molecular, skeletal, and soft-tissue alterations in short ear mice. Dev. Biol. **166:** 112–122.
73. MACIAS, D., Y. GANAN, T.K. SAMPATH, *et al.* 1997. Role of BMP-2 and OP-1 (BMP-7) in programmed cell death and skeletogenesis during chick limb development. Development **124:** 1109–1117.
74. STORM, E.E. & D.M. KINGSLEY. 1996. Joint patterning defects caused by single and double mutations in members of the bone morphogenetic protein (BMP) family. Development **122:** 3969–3979.
75. DUPREZ, D.M., M. COLTEY, H. AMTHOR, *et al.* 1996. Bone morphogenetic protein-2 (BMP-2) inhibits muscle development and promotes cartilage formation in chick limb bud cultures. Dev. Biol. **174:** 448–452.
76. BOSKEY, A.L., E.P. PASCHALIS, I. BINDERMAN, *et al.* 2002. BMP-6 accelerates both chondrogenesis and mineral maturation in differentiating chick limb-bud mesenchymal cell cultures. J. Cell Biochem. **84:** 509–519.
77. MAJUMDAR, M.K., E. WANG & E.A. MORRIS. 2001. BMP-2 and BMP-9 promotes chondrogenic differentiation of human multipotential mesenchymal cells and overcomes the inhibitory effect of IL-1. J. Cell Physiol. **189:** 275–284.
78. WANG, E.A., D.I. ISRAEL, S. KELLY, *et al.* 1993. Bone morphogenetic protein-2 causes commitment and differentiation in C3H10T1/2 and 3T3 cells. Growth Factors **9:** 57–71.
79. ASAHINA, I., T.K. SAMPATH & P.V. HAUSCHKA. 1996. Human osteogenic protein-1 induces chondroblastic, osteoblastic, and/or adipocytic differentiation of clonal murine target cells. Exp. Cell Res. **222:** 38–47.
80. YAMAGUCHI, A., T. KATAGIRI, T. IKEDA, *et al.* 1991. Recombinant human bone morphogenetic protein-2 stimulates osteoblastic maturation and inhibits myogenic differentiation *in vitro*. J. Cell Biol. **113:** 681–687.
81. GIMBLE, J.M., C. MORGAN, K. KELLY, *et al.* 1995. Bone morphogenetic proteins inhibit adipocyte differentiation by bone marrow stromal cells. J. Cell Biochem. **58:** 393–402.
82. KATAGIRI, T., A. YAMAGUCHI, M. KOMAKI, *et al.* 1994. Bone morphogenetic protein-2 converts the differentiation pathway of C2C12 myoblasts into the osteoblast lineage. J. Cell Biol. **127:** 1755–1766.

83. KINGSLEY, D.M., A.E. BLAND, J.M. GRUBBER, *et al.* 1992. The mouse short ear skeletal morphogenesis locus is associated with defects in a bone morphogenetic member of the TGF beta superfamily. Cell **71:** 399–410.
84. STORM, E.E., T.V. HUYNH, N.G. COPELAND, *et al.* 1994. Limb alterations in brachypodism mice due to mutations in a new member of the TGF beta-superfamily. Nature **368:** 639–643.
85. LUO, G., C. HOFMANN, A.L. BRONCKERS, *et al.* 1995. BMP-7 is an inducer of nephrogenesis, and is also required for eye development and skeletal patterning. Genes Dev. **9:** 2808–2820.
86. SOLLOWAY, M.J., A.T. DUDLEY, E.K. BIKOFF, *et al.* 1998. Mice lacking Bmp6 function. Dev. Genet. **22:** 321–339.
87. ZHANG, H. & A. BRADLEY. 1996. Mice deficient for BMP2 are nonviable and have defects in amnion/chorion and cardiac development. Development **122:** 2977–2986.
88. WINNIER, G., M. BLESSING, P.A. LABOSKY, *et al.* 1995. Bone morphogenetic protein-4 is required for mesoderm formation and patterning in the mouse. Genes Dev. **9:** 2105–2116.
89. GRAHAM, A., P. FRANCIS-WEST, P. BRICKELL, *et al.* 1994. The signalling molecule BMP4 mediates apoptosis in the rhombencephalic neural crest. Nature **372:** 684–686.
90. YOKOUCHI, Y., J. SAKIYAMA, T. KAMEDA, *et al.* 1996. BMP-2/-4 mediate programmed cell death in chicken limb buds. Development **122:** 3725–3734.
91. ZOU, H. & L. NISWANDER. 1996. Requirement for BMP signaling in interdigital apoptosis and scale formation. Science **272:** 738–741.
92. ZOU, H., R. WIESER, J. MASSAGUE, *et al.* 1997. Distinct roles of type I bone morphogenetic protein receptors in the formation and differentiation of cartilage. Genes Dev. **11:** 2191–2203.
93. GUHA, U., W.A. GOMES, T. KOBAYASHI, *et al.* 2002. *In vivo* evidence that BMP signaling is necessary for apoptosis in the mouse limb. Dev. Biol. **249:** 108–120.
94. GANAN, Y., D. MACIAS, M. DUTERQUE-COQUILLAUD, *et al.* 1996. Role of TGF-βs and BMPs as signals controlling the position of the digits and the areas of interdigital cell death in the developing chick limb autopod. Development **122:** 2349–2357.
95. MARAZZI, G., Y. WANG & D. SASSOON. 1997. Msx2 is a transcriptional regulator in the BMP4-mediated programmed cell death pathway. Dev. Biol. **186:** 127–138.
96. GROTEWOLD, L. & U. RUTHER. 2002. The Wnt antagonist Dickkopf-1 is regulated by Bmp signaling and c-Jun and modulates programmed cell death. EMBO J. **21:** 966–975.
97. KRUMLAUF, R. 1994. Hox genes in vertebrate development. Cell **78:** 191–201.
98. SCOTT, M.P. 1992. Vertebrate homeobox gene nomenclature. Cell **71:** 551–553.
99. CAPECCHI, M.R. 1997. Hox genes and mammalian development. Cold Spring Harb. Symp. Quant. Biol. **62:** 273–281.
100. CILLO, C., M. CANTILE, A. FAIELLA, *et al.* 2001. Homeobox genes in normal and malignant cells. J. Cell Physiol **188:** 161–169.
101. Gendron-MAGUIRE, M., M. MALLO, M. ZHANG, *et al.* 1993. Hoxa-2 mutant mice exhibit homeotic transformation of skeletal elements derived from cranial neural crest. Cell **75:** 1317–1331.
102. RIJLI, F.M., M. MARK, S. LAKKARAJU, *et al.* 1993. A homeotic transformation is generated in the rostral branchial region of the head by disruption of Hoxa-2, which acts as a selector gene. Cell **75:** 1333–1349.

103. KANZLER, B., S.J. KUSCHERT, Y.H. LIU, *et al.* 1998. Hoxa-2 restricts the chondrogenic domain and inhibits bone formation during development of the branchial area. Development **125:** 2587–2597.
104. Le MOUELLIC, H., Y. LALLEMAND & P. BRULET. 1992. Homeosis in the mouse induced by a null mutation in the Hox-3.1 gene. Cell **69:** 251–264.
105. YUEH, Y.G., D.P. GARDNER & C. KAPPEN. 1998. Evidence for regulation of cartilage differentiation by the homeobox gene Hoxc-8. Proc. Natl. Acad. Sci. USA **95:** 9956–9961.
106. SMALL, K.M. & S.S. POTTER. 1993. Homeotic transformations and limb defects in Hox A11 mutant mice. Genes Dev. **7:** 2318–2328.
107. MORTLOCK, D.P., L.C. POST & J.W. INNIS. 1996. The molecular basis of hypodactyly (Hd): a deletion in Hoxa 13 leads to arrest of digital arch formation. Nat. Genet. **13:** 284–289.
108. MORTLOCK, D.P. & J.W. INNIS. 1997. Mutation of HOXA13 in hand-foot-genital syndrome. Nat. Genet. **15:** 179–180.
109. AKARSU, A.N., I. STOILOV, E. YILMAZ, *et al.* 1996. Genomic structure of HOXD13 gene: a nine polyalanine duplication causes synpolydactyly in two unrelated families. Hum. Mol. Genet. **5:** 945–952.
110. MURAGAKI, Y., S. MUNDLOS, J. UPTON, *et al.* 1996. Altered growth and branching patterns in synpolydactyly caused by mutations in HOXD13. Science **272:** 548–551.
111. DUPREZ, D.M., K. KOSTAKOPOULOU, P.H. FRANCIS-WEST, *et al.* 1996. Activation of Fgf-4 and HoxD gene expression by BMP-2 expressing cells in the developing chick limb. Development **122:** 1821–1828.
112. HASHIMOTO, K., Y. YOKOUCHI, M. YAMAMOTO, *et al.* 1999. Distinct signaling molecules control Hoxa-11 and Hoxa-13 expression in the muscle precursor and mesenchyme of the chick limb bud. Development **126:** 2771–2783.
113. WACKER, S.A., C.L. MCNULTY & A.J. DURSTON. 2004. The initiation of Hox gene expression in *Xenopus laevis* is controlled by Brachyury and BMP-4. Dev. Biol. **266:** 123–137.
114. SUZUKI, M., N. UENO & A. KUROIWA. 2003. Hox proteins functionally cooperate with the GC box-binding protein system through distinct domains. J. Biol. Chem. **278:** 30148–30156.

Function of BMPs and BMP Antagonists in Adult Bone

ETSUKO ABE

Mount Sinai Bone Program, Department of Medicine, Mount Sinai School of Medicine, New York, NY 10029 and Bronx VA Medical Center, Bronx, NY 10468, USA

ABSTRACT: The expression and function of BMPs and BMPs in bone tissues have been studied for a long time because of their remarkable activities. However, their biological functions in normal bone remodeling in adults were not fully understood until recently. Advanced technologies using gene manipulation were used to study their roles in adulthood. In addition, findings of new BMP antagonists and the effect of Wnt-canonical pathways on bone features also provided new insights in bone studies.

KEYWORDS: BMP; BMP antagonist; Wnt; catenin; osteoblast; bone remodeling

INTRODUCTION

Bone morphogenetic proteins (BMPs) belong to the transforming growth factor (TGF)-β superfamily, which includes TGF-β, activin, and inhibin.[1] BMP receptors are present in a variety of tissues and cells, and thus BMPs display numerous functions that are not limited in mesodermal development. Recent studies demonstrated new roles of BMPs in brain, heart, kidney, skin, and intestine.[2,3] In bone tissues comprising two distinct cell types, chondrocytes and osteoblasts, BMPs and their antagonists are temporally and spatially expressed, and both factors are necessary for normal development.[4–6] Thus, if the functions are lost or gained, severe abnormalities are observed and thereby cause death in embryonal stage or in postnatal stage. In bone cells using *ex vivo* system, BMPs and BMP antagonists endogenously express and regulate cell differentiation.[7,8] The addition of those factors accelerates or inhibits cell differentiation in cartilage and bone. Based on the results we have hypothesized that BMPs and BMP antagonists are required for normal bone remodeling in adulthood as well as embryonic development; however, the information

Address for correspondence: Etsuko Abe, Division of Endocrinology, Department of Medicine, Mount Sinai School of Medicine, One Gustave L Levy Place, Box 1055, New York, NY 10029. Voice: 212-241-8735; fax: 212-241-4218.
e-mail: etsuko.abe@mssm.edu

**Ann. N.Y. Acad. Sci. 1068: 41–53 (2006). © 2006 New York Academy of Sciences.
doi: 10.1196/annals.1346.007**

to prove this hypothesis was limited. Current technologies using conditional knockout mice or transgenic mice with inducible system gave new insight into BMPs and BMP antagonists on bone remodeling. Furthermore, the new signals of Wingless type (Wnt)-canonical pathway[9] in addition to BMP signaling[10–12] provide a new strategy to understand bone biology under normal and pathological conditions.

BMP AND BMP SIGNALING

BMPs bind to BMP receptors types I and II, both of which are required for signal transduction.[10–12] There are seven distinct BMP type I receptors, including ALK-2, ALK-3 (BMPRIA), and ALK-6 (BMPRIB), and BMP type I receptor mainly determines the specificity of the intracellular signals. BMP type II also has distinct receptors, such as BMP type II receptor (BMPRII), activin type II receptor (ActR-II), and activin type IIB receptor (ActR-IIB). ALK-3 ubiquitously expressed in most types of cells including osteoblasts. Chondrocytes and osteoclasts are known to express only ALK-6. A common feature of the TGF-β superfamily has seven conserved cysteins. BMPs have two extra-domains in addition to the structure identified by another TGF family, TGF-β, activin/inhibin, or nodal. BMPs are translated as a large molecule with a signal peptide and cleaved to be a dimeric form of mature protein with 25 kD after secretion. Currently, over 20 distinct BMPs have been identified and each of them has slightly different affinities to their receptors. BMP-2, -4, -6, and -7 (OP-1) are widely accepted to have osteoinductive activities *in vivo*.[1] Other BMPs,[12] namely GDF (growth differentiation factor)-5, -6, and -7 (BMP-12) and activin/inhibin family, exhibit low osteoinductive activity by themselves, but modulate BMP action. Both BMP-2 and -4 preferentially bind to ALK-3 and -6 and transduce signals through smad-dependent and/or -independent mechanisms.

After binding to their receptors, BMPs activate intracellular transcription factors, smad-1, -5 and -8 (smad-1/5/8, R-smads) and proceed toward dimerization with smad-4 (Co-smad, common smad) before translocation into nucleus.[10–12] The generated complexes are then translocated in the nucleus to regulate specific gene transcription (FIG. 1). Osteopontin, osteoprotegerin, BMP-7, and smad-1 are known to be activated through The BMP-induced smad activation mechanism.[13–15] Similar to the BMPs, TGF-β and activin stimulate smad-2 and -3 (smad-2/3, R-smads) activation by generating complex with smad-4. Importantly, inhibitory smads, smad-6 and -7 (I-smads) exist to negatively regulate TGF-β superfamily signaling; they do so by binding to the BMP/TGF-β receptors or binding to the smad-1 and thereby intefering with activation.

Smad-independent signals are also required for the maximum activation of BMP signals in osteoblasts and chondrocytes. BMPs stimulate MAP

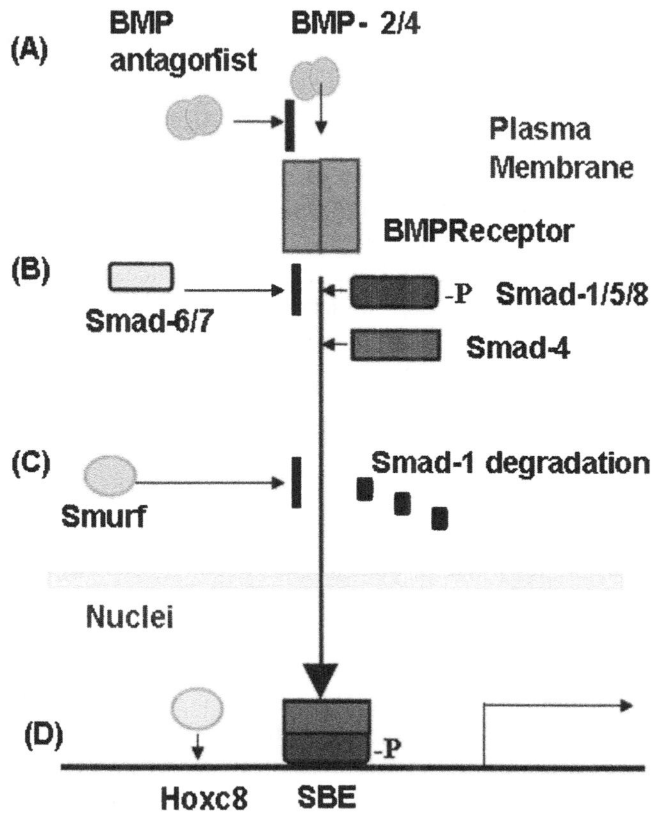

FIGURE 1. BMP signaling and its inhibition. There are four distinct inhibitions for BMP signaling. (**A**) BMP antagonists inhibit binding of BMP-2/4 to their receptors. (**B**) Smad-6/7 completes binding to the BMP receptors with smad-1/5/8 for the inhibition of BMP signaling. (**C**) Smurf 2 promotes smad-1 degradation after ubiquitination. (**D**) Transcription factor (Hoxc8) binds to target genes to inhibit BMP-induced transcription.

kinase, pI3 kinase, and/or JNK (c-Jun N-terminal kinase),[16,17] and application of the specific inhibitors for these signals diminish expression of osteoblast and chondrocyte differentiation markers, supporting the important role of smad-independent mechanism in BMP signaling as well.

BMP ANTAGONISTS

BMP antagonists were found in the Spemann organizer of *Xenopus* embryos as a molecule to inhibit BMP binding to their receptors. They are designated as noggin and chordin.[18,19] After the first discovery of the BMP antagonists, numerous BMP antagonists, which are secretory proteins with cysteine arrangement structure, were found.[20,21] They are now classified into three subfamilies based on the size of the cysteine knot: the DAN family, twisted gastrulation, and noggin/chordin. Dan family is further subdivided to four groups: Gremlin/PRDC, Cerberus/Cer 1, Dan, and sclerostin/USAG-1. Tsukushi (TSK), which recently was found as a new family of the BMP antagonist, cooperates with other BMP antagonists.[22]

Noggin binds to BMP-2 and -4 with high affinity and to BMP-6 and -7 with low affinity to prevent further action of BMP action. Null mice of noggin-displayed impaired joint formation and thinning calvarial formation with small numbers of osteogenic progenitors thereby die just after birth.[23] Heterozygous show normal growth and normal bone formation in mice; however, in human apical joint fusion has appeared,[24] indicating that functional incompletion occurred in the haplotype of the gene. Noggin expression appears in the early stage of embryonal development: node, notochord, dorsal somite, condensing cartilage, and immature chondrocyte are positive for noggin.[23] In adult mice, high expression of noggin expression was observed in the limited tissues and cells, such as osteoblasts, chondrocytes, and cranial suture in vertebrae and calvaria.[25,26] Interestingly, the expression of noggin as well as BMPs and the BMP receptor (ALK-3) in macrophages serve to regulate osteoclast formation.[26] In pathological conditions, such as bone fracture healing, BMPs as well as noggin are intensively expressed at the fracture site[27,28] and serve to stimulate bone formation.

Noggin transgenic mice are generated by two groups using a rat or mouse osteocalcin promotor.[26,29] The use of rat osteocalcin promotor (1.7 kb) displays severely impaired bone formation with bone fracture at young age (4–5 weeks).[29] In contrast, the use of mouse osteocalcin promotor (1.3 kb) displays effectiveness in bone remodeling in adult animals.[26] Bone mineral density was reduced at 6 months old and bone formation rate was significantly suppressed at 8 months old with diminished osteoblast differentiation markers. Notably, osteoclast progenitor numbers are also decreased by noggin overexpression in osteoblasts, indicating that impaired osteoblast function affects osteoclast formation through diminished RANKL (receptor activator of NF-κB) expression in osteoblasts.

Chordin as well as noggin bind BMPs and modulate BMP action.[30,31] The biological activity of chordin resides in the cysteine-rich domains (CRs), specifically in CR1 and CR3 regions, and individual CRs reveal lower activity than that of full-length chordin. As a result, the cleavage of the chordin by an extracellular zinc metalloproteinase, Tolloid/BMP-1, would release a molecule with lower anti-BMP activity than an intact one. Both noggin and chordin display similar BMP antagonistic activity; however, both factors are required for normal head and facial development,[32] since their expression patterns are different from each other in terms of location and timing. The chordin appears in an earlier stage of embryo than noggin. In the mid-gastrula stage, the expression of noggin and chordin patterns overlap, indicating that they might compensate for each other if one of the two is lost. Conversely, if both chordin and noggin are mutated, mice display severe defectiveness in development.

Twisted gastrulation (TSG) is expressed in lung, thymus, and kidney,[20] and binds to BMPs to inhibit BMP action in osteoblasts.[33] This process cooperates with the other BMP antagonist, chordin. TSG null mice show growth retardation after birth, and display dwarfism with delayed endochondral ossification with lymphopenia,[34] followed by death within a month. The bone phenotypes with decreased bone mineral density were similar to those of the transgenic mice expressing a truncated dominant-negative BMP type 1B receptor (ALK-6). Is TSG BMP agonist or antagonist? Interestingly, TSG null mice displayed lymphoid cell depletion with small sizes of spleen and thymus,[34] suggesting that TSG might act as an agonist. However, smad-1 signals in the TSG null thymocyte were intensively activated without stimulation,[34] indicating that TSG does act as an antagonist. Thus, the action of TSG may be determined in cell-specific manner. Recent studies using TSG overexpression system in stromal/preosteoblasts supported antagonistic activity of TSG on BMP signaling[33]; the cells inhibited smad activation and thereby suppressed osteoblast differentiation. How does the same molecule act either as an antagonist or as an agonist? The mechanism for agonist activity of TSG is explained by indirect action: TSG promotes cleavage of co-associator chordin to enhance to BMP activity.

Sclerostin is a protein newly discovered BMP antagonist in the patients showing sclerosteosis, an autosomal recessive disorder.[35] Sclerostin expresses in osteogenic cells including hypertrophic chondrocytes, osteoblasts, and osteocytes, and inhibits action of BMP-6 and -7,[36–38] and in some cases the action of BMP-2 and -4. Thus, loss of sclerostin increases bone mass seen in sclerosteosis; conversely transgenic mice overexpressing sclerostin exhibit low bone mass with reduced osteoblast activity and bone formation activity.[37] Importantly, the intense expression of sclerostin in osteocyte emerges as a negative regulator,[38] which controls the fate of osteoblasts on bone surface through canalicurae. The high expression of sclerostin similar to other BMP antagonists, is also observed in the long-term culture or after treatment with BMPs. A mechanism for how sclerostin inhibits BMP action is still controversial. Both smad-dependent and -independent mechanisms[38,39] are proposed. In the case of a smad-independent mechanism, sclerostin binds to LRP-5 and -6 and inhibits Wnt-canonical signaling pathway and thereby suppresses osteoblast differentiation.[39] Noggin and sclerostin are capable of forming an inhibitory complex as observed for chordin and TSG.

TSK, which was cloned from a chick lens library, is a new BMP antagonist.[22] TSK displays a unique protein structure with small leucine-rich proteoglycan (LRP), which is a different type of molecule from other the BMP antagonists described above. The expression of TSK in chick development is overlapped with chordin, and TSK cooperates with chordin, not noggin or follistatin, to inhibit BMP-4 action *in vitro* and *in vivo*. Thus, TSK similar to chordin, blocks BMP function. The finding of the new protein, which can cooperate with other BMP antagonists, emphasized the important role of BMP antagonists *in vitro* and *in vivo* and some redundancy among BMP antagonists.

A BALANCE OF BMP AND BMP ANTAGONIST IN BONE REMODELING

Endogenous BMP expression appears during cell differentiation from mesenchymal cells to osteoblasts. After differentiation, cells express alkaline phosphatase and osteoblast markers, such as collagen type I, bone sialoprotein (BSP), osterix, and osteocalcin. In the late stage cells deposit minerals stained by Von Kossa or alizarin red. Similar to BMP expression, BMP antagonists also appear and increase their expression levels in the late stage.[8] The expression levels of the BMP antagonists are transcriptionally regulated by BMPs[7] and the balance of BMPs and BMP antagonists determines the rate of cell differentiation. If BMPs or BMP antagonists are exogenously added, cells accelerate or stop further differentiation. Interestingly, mesenchymal cell differentiation is inhibited in aged mice or in aged animal models, which show osteopenia at a young age, namely senescence-accelerated mice (SAM).[40,41] We thought the decreased bone formation might be explained by the imbalance of BMPs and BMP antagonists. We observed that noggin expression is elevated in bone marrow cell cultures and bone tissues of aged animals.[26] Notably, no changes are seen in BMP expression.

A deficiency of BMPs, BMPR, or smads severely disrupts normal development and thereby animals die in an early stage of development. Conversely, but more importantly, if BMP action is more enhanced like in noggin deficiency,[23] cartilage area in long bone is enlarged and endochondral bone formation is not developed. In contrast to the deficient animals of BMP antagonists, overexpression of BMPs and their related proteins sometimes show induction of ectopic ossification. Fibrodysplasia ossificans progressiva (FOP) is a patient exhibiting congenital malformations by progressive heterotopic endochondral ossifications.[42] Transgenic mice using neuron-specific enolase (NSE) promotor displayed severe postnatal heterotopic ossification,[43] which closely recapitulated major FOP patients. Generated double transgenic mice with NSE-noggin rescued ectopic ossification, suggesting that only BMP may be required for ectopic bone formation in FOP patients.

Transgenic mice of dominant negative (truncated) form of BMPIA receptor (ALK-3) using colla promotor (2.3 kb) displayed postnatal osteopenia due to reduced bone formation rate and inhibition of osteoblast differentiation.[44] Similar bone phenotype has been observed in transgenic mice of smurf (smad ubiquitin regulatory factor) 1 using colla1 promotor.[45] Smurf 1/2 is able to inhibit BMP signaling through the enhancement of ubiquitination of smad-1 and also degradation of BMP type I receptor by cooperating with I-smads. These results suggested the critical role of BMP signaling in postnatal stage in addition to embryonic stage. Conditional knockout mice of BMPR1A (ALK-3) in osteoblasts using a cre-loxP system and ostecalcin promotor (OG2) have also demonstrated the crucial role of BMPs specifically in adulthood.[46] The mutant mice at 6 months of age showed decreased bone mineral density, trabecular bone area, and bone formation rate with irregular calcification. Although the bone volume was increased in the mutant mice, this was explained by reduced bone resorption by osteoclasts, rather than increased bone formation. The mutant mice lost more bone after ovariectomy likely resulting from the decreased osteoblast function, which could not overcome ovariectomy-induced bone resorption. Thus, the bone phenotype obtained BMPR1A is very similar to that in the noggin transgenic mice we generated. Mice transgenic for the transcription factors required for BMP signaling may increase osteoblast differentiation and thereby enhance bone formation. Interestingly, mice overexpressing smad-1C using a doxycycline inducible system show enhanced bone mineral density and bone formation parameters, such as increased osteoid surface, thicker trabecular bone with close proximity, and appearance of cuboidal osteoblasts.[47] Smad-1C, which was discovered as a factor that interacts with Hoxc-8 in osteopontin regulation, mimics smad-1 activity in osteoblast differentiation. These results reinforced that the transcription factors required for BMP signaling have a significant role in adult bone remodeling as well.

ROLE OF WNT-CANONICAL PATHWAY IN BONE AND A CROSS-TALK WITH BMP SIGNALS

Recent studies of Wnt-canonical pathway evidenced novel role of new proteins in skeletal development and bone remodeling. Wnts are secreted cysteine-rich glycoproteins that influence cell proliferation and survival, therefore Wnt signaling and their roles are implicated not only in development but also in some diseases.[9] In bone tissues Wnt signaling is designated as a required factor and sometimes a mediator of BMP action during normal bone remodeling. Extracellular Wnt ligands interact with a host of secreted antagonists, secreted frizzled-related protein (sFRP), and Dickkopf (Dkk) family members, preventing the activation process (FIG. 2). Under this condition, β-catenin is degraded in the cytosol and therefore signals are not transduced. Once Wnt concentration is exceeded, Wnts interact with members of the Frizzled (FZ) family, and LDL (low-density lipoprotein receptor-related proteins 5 and 6) leads to further activation through intracellular proteins. This process protects degradation of β-catenin and stimulates translocation of β-catenin to the nucleus where specific gene carrying TCF/LEF (T cell factor) should be transactivated. In pathological conditions, such as myloma and osteosarcoma, Wnt signaling is enhanced.[48,49] Myeloma cell lines and primary myloma cells specifically produce sFRP-2, not sFRP-1, or -3 and thereby impair bone formation *in vivo*.

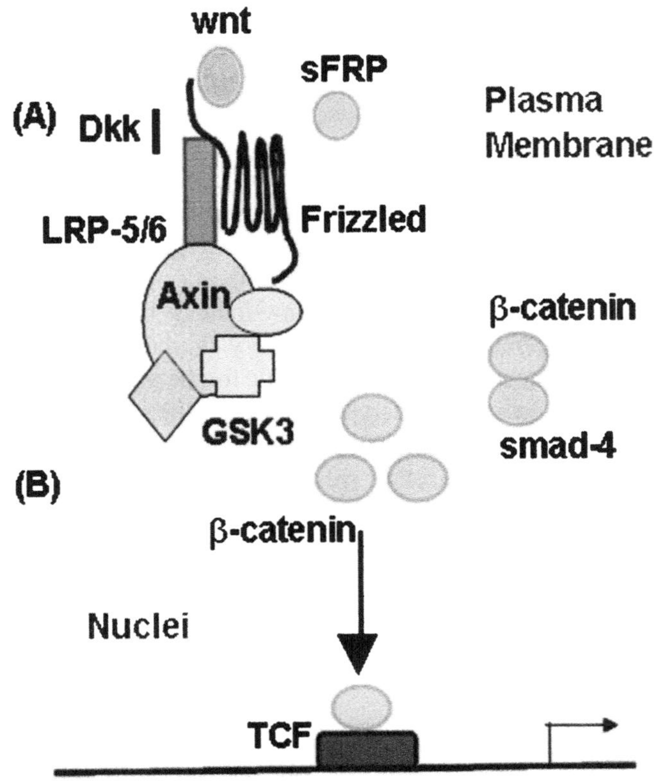

FIGURE 2. Wnt-canonical pathway and cooperation with BMP signaling. Wnt binds to the receptor (Frizzled) with co-receptor (LRP-5/6) and activates downstream signals with GSK3 (glycogen synthase kinase) complex. Under an activation condition, β-catenin is in stable form, translocated into the nucleus to bind to the TCF sequence in target gene. DKK is a protein that can bind to LRP-5/6 to block Wnt signaling.

Osteosarcoma patient revealed LRP-5 expression at a very high rate (50%) in tumor samples, and the presence of LRP-5 correlated with tumor metastasis and the chondroblastic subtype of osteosarcoma.

Mesenchymal/stromal cells (MSC) have the potential to differentiate into osteoblasts in certain culture conditions and are candidates for therapeutic applications. MSC are consistently expressed Wnt signal components, such as LRP-5, Dkk-1, sFRPs, GSK-3b, and TCFs.[50] After treatment with Wnt3a or Litium, cells promote canonical Wnt signaling by inhibiting GSK-3b, reduce phosphorylated β-catenin in cytosol, and increase β-catenin nuclear translocation. Thus, autocrine Wnt signaling operates in primitive MSC populations and regulates self-renewal and lineage-specific differentiation. A similar line of experiments using osteo-progenitors calvarial cells,[51] which expresses collagen 2.3 kb-driven GFP protein, exhibited higher expression of Wnt3a, LRP-5, and TCF1 in addition to BMPs, BMP antagonists, and IGF-1. After a long-term culture under osteogenic factors, the GFP positive cells showed a higher expression in some BMP antagonist, BAMBI (BMP and activin membrane bound inhibitors), as compared to that in GFP negative cells. The expression levels of BMPs and other BMP antagonists, such as Cerberus, follistatin, and TSG protein were not changed.

β-Catenin-targeted null mice show embryonic lethality with a lack of skeletal structures derived from cranial neural crest.[52] Some mutation in Wnt signaling can produce a bone phenotype without growth retardation. Loss-of-function mutation in LRP-5 results in low bone mass in humans and mice.[53] Conversely, a null mutation of the Wnt antagonist sFRPs results in a high bone mass.[54] These data suggest that wnt signaling stimulates osteoblast differentiation and thereby increases bone formation. However, increased bone mass is not only the result of increased bone formation but also of decreased bone resorption. A recent study by the Karsenty group using a stable form of β-catenin and colla promotor-driven β-catenin deletion demonstrated that Wnt signals do not affect bone formation but affect bone resorption by elevating osteoprotegerin expression in osteoblasts.[55] They could not observe any changes in the expression of osteoblast differentiation markers. This result raised the question whether Wnt signals positively affect osteoblast differentiation as proposed from the results obtained from transgenic mice and null mice of LRP. β-catenin deletion in limb and head mesenchyme showed that Wnt signals determine an early stage of mesenchymal cell differentiation to osteoblasts.[56] If Wnt signals are not sufficient, cells prefer to differentiate into chondrocytes rather than osteoblasts. Taken together, these results suggest Wnt signals might intensively affect mesenchymal lineage cell differentiation.

Both BMP and Wnt signaling promote osteogenesis. How do they interact in the signaling process? Wnt signaling promotes osteogenesis by directly stimulating RUNX2 (cbfa-1) gene expression[57] and this process is blocked by sFRP-1. In fact, sFRP-1 null mice displayed high bone mass with higher expression of RUNX2, TCF1, and osteocalcin in bone samples.[54] Interestingly, Wnt signaling similar to BMP signaling directly regulates RUNX2 transcriptional activity. In fact, an RUNX2 promotor retains a TCF consensus sequence with proximity to a RUNX2 regulatory site. Cooperative signaling pathways of BMP and Wnt are proved by the use of Wnt signal inhibitors.[58] The ability of BMP2 to induce alkaline phophatase was inhibited by blocking Wnt/LRP-5 signals in mesenchymal cells, and conversely, sclerostin and noggin inhibit Wnt-induced osteoblast differentiation. An independent mechanism between BMPs and Wnt signaling has been shown by others.[59] A β-catenin mutant with constitutive transcriptional activity (deltaN151) did not stimulate osteogenic differentiation in multipotent embryonic cells C3H10T1/2 by itself; however, in the presence of BMP-2 osteoblast marker expression along with matrix mineralization was markedly enhanced. Similarly, the active mutants of the β-catenin and BMP-2 synergistically stimulated new bone formation in mouse calvaria, and also the synergism was explained by a RUNX2-independent mechanism.

Thus, it may be suggested that BMP and Wnt signals use a common pathway at some points and depend on cells and systems used.

Close interplay between the Wnt signal and BMP signals has been reported.[60] Noggin interacts with Dkk1 in head induction, and BMPs regulate the expression level of Dkk1 in limb development. These interactions may explain that the binding sites of BMP antagonists to Wnt protein are different from the site bound to BMPs.

SUMMARY

We believe that current studies have provided some evidence that BMPs and BMP antagonists play an important role in adult bone remodeling. To study the interaction mechanism of BMP and Wnt signaling may give rise to new therapeutic means for osteoporosis.

ACKNOWLEDGMENT

We thank Dr. Zaidi for reading this manuscript and making the discussion focused for writing.

REFERENCES

1. Wozney, J.M. 1993. Bone morphogenetic proteins and their gene expression. *In* Cellular and Molecular Biology of Bone. M. Noda, Ed.: 131–167. Academic Press. San Diego, CA.
2. Botchkarev, V. & A.A. Sharov. 2004. BMP signaling in the control of skin development and hair follicle growth. Differentiation **72:** 512–526.
3. Needham, S.J., M. Brittan, S.A. McDonald, *et al.* 2005. Intestinal stem cells. J. Cell Mol. Med. **9:** 11–24.
4. Kishigami, S. & Y. Mishina. 2005. BMP signaling and early embryonic patterning. Cytokine Growth Factor Rev. **16:** 265–278.
5. Tsumaki, N. & H. Yoshikawa. 2005. The role of bone morphogenetic proteins in endochondral bone formation. Cytokine Growth Factor Rev. **16:** 279–285.
6. Kugimiya, F., H. Kawaguchi, S. Kamekura, *et al.* 2005. Involvement of endogenous bone morphogenetic protein (BMP) 2 and BMP6 in bone formation. J. Biol. Chem. **280:** 35704–35712.
7. Gazzerro, E., V. Gangji & E. Canalis. 1998. Bone morphogenetic proteins induce the expression of noggin, which limits their activity in cultured rat osteoblasts. J. Clin. Invest. **10:** 2106–2114.
8. Abe E, M. Yamamoto, Y. Taguchi, *et al.* 2000. Essential requirement of BMPs-2/4 for both osteoblast and osteoclast formation in murine bone marrow cultures from adult mice: antagonism by noggin. J. Bone Miner. Res. **15:** 663–673.
9. Moon, R.T., A.D., Kohn, G.V. DeFerrari, *et al.* 2004. Wnt and β-catenin signaling: diseases and therapies. Nature Rev. Genet. **5:** 689–699.

10. Miyazono, K., S. Maeda & T. Imamura. 2005. BMP receptor signaling: transcriptional targets, regulation of signals and signaling cross-talk. Cytokine Growth Factor Rev. **16:** 251–263.
11. Nohe, A., E. Keating, P. Knaus, *et al.* 2004. Signal transduction of bone morphogenetic protein receptors. Cell. Signal. **16:** 291–299.
12. Chen, D.I., M. Zhao & G.R. Mundy. 2004. Bone morphogenetic proteins. Growth Factors **22:** 233–241.
13. Hullinger, T.G., Q. Pan, H. Viswanathan, *et al.* 2001. TGFbeta and BMP2 activation of the OPN promoter: roles of smad-and hox-binding elements. Exp. Cell Res. **262:** 69–74.
14. Wan, M., X. Shi, X. Feng, *et al.* 2001. Stimulation of osteoprotegerin (OPG) gene expression by transforming growth factor-beta (TGF-beta). Mapping of the OPG promoter region that mediates TGF-beta effects. J. Biol. Chem. **276:** 36241–36250.
15. Stopa, M., D. Anhuf, L. Trestegen, *et al.* 2000. Participation of smad2, smad3, and smad4 in transforming growth factor beta (TGF-beta) induced activation of smad7. J. Biol. Chem. **275:** 29308–29317.
16. Guicheux, J., J. Lemonnier, C. Ghayor, *et al.* 2003. Activation of p38 mitogen-activated protein kinase and c-Jun-NH2-terminal kinase by BMP-2 and their implication in the stimulation of osteoblastic cell differentiation. J. Bone Miner. Res. **18:** 2060–2068.
17. Osyczka, A.M. & P.S. Leboy. 2005. Bone morphogenetic protein regulation of early osteoblast genes in human marrow stromal cells is mediated by extracellular signal-regulated kinase and phosphatidylinositol 3-kinase signaling. Endocrinology **146:** 3428–3437.
18. Zimmerman, L.B., J.M. De Jesus-Escobar & R.M. Harland. 1996. The Spemann organizer signal, noggin, binds and inactivates bone morphogenetic protein 4. Cell **86:** 599–606.
19. Larrain, J., D. Bachiller, B. Lu, *et al.* 2000. BMP-binding modules in chordin: a model for signalling regulation in the extracellular space. Development **127:** 821–830.
20. Canalis, E., A. Economides & E. Gazzerro. 2003. Bone morphogenetic proteins, their antagonists, and the skeleton. Endocr. Rev. **24:** 218–235.
21. Yanagita, M. 2005. BMP antagonists: their roles in development and involvement in pathophysiology. Cytokine Growth Factor Rev. **16:** 309–317.
22. Kuriyama, S., G. Lupo, K. Ohta, *et al.* 2006. Tsukushi controls ectodermal patterning and neural crest specification in Xenopus by direct regulation of BMP4 and X-delta-1 activity. Development **133:** 75–88.
23. Brunet, L.J., J.A. McMahon, A.P. McMahon, *et al.* 1998. Noggin, cartilage morphogenesis, and joint formation in the mammalian skeleton. Science **280:** 1455–1457.
24. Gong, Y., D. Krakow, J. Marcelino, *et al.* 1999. Heterozygous mutations in the gene encoding noggin affect human joint morphogenesis. Nat. Genet. **21:** 302–304.
25. Gazzerro, E., V. Gangji & E. Canalis. 1998. Bone morphogenetic proteins induce the expression of noggin, which limits their activity in cultured rat osteoblasts. J. Clin. Invest. **10:** 2106–2114.
26. Wu, X.B., Y. Li, A. Schneider, *et al.* 2003. Impaired osteoblastic differentiation, reduced bone formation, and severe osteoporosis in noggin-overexpressing mice. J. Clin. Invest. **112:** 924–934.

27. NAKASE, T., S. NOMURA, H. YOSHIKAWA, *et al.* 1994. Transient and localized expression of bone morphogenetic protein 4 messenger RNA during fracture healing. J. Bone Miner. Res. **9:** 651–659.
28. KLOEN, P., M. DI PAOLA, O. BORENS, *et al.* 2003. BMP signaling components are expressed in human fracture callus. Bone **33:** 362–371.
29. DEVLIN, R.D., Z. DU, R.C. PEREIRA, *et al.* 2003. Skeletal overexpression of noggin results in osteopenia and reduced bone formation. Endocrinology **144:** 1972–1978.
30. BACHILLER, D., J. KLINGENSMITH, N. SHNEYDER, *et al.* 2003. The role of chordin/Bmp signals in mammalian pharyngeal development and DiGeorge syndrome. Development **130:** 3567–3578.
31. ZHANG, D., C.M. FERGUSON, R.J. O'KEEFE, *et al.* 2002. A role for the BMP antagonist chordin in endochondral ossification. J. Bone Miner. Res. **17:** 293–300.
32. ANDERSON, R.M., A.R. LAWRENCE, R.W. STOTTMANN, *et al.* 2002. Chordin and noggin promote organizing centers of forebrain development in the mouse. Development **129:** 4975–4987.
33. GAZZERRO, E., V. DEREGOWSKI, S. VAIRA, *et al.* 2005. Overexpression of twisted gastrulation inhibits bone morphogenetic protein action and prevents osteoblast cell differentiation in vitro. Endocrinology **146:** 3875–3882.
34. NOSAKA. T., S. MORITA, H. KITAMURA, *et al.* 2003. Mammalian twisted gastrulation is essential for skeleto-lymphogenesis. Mol. Cell Biol. **23:** 2969–2980.
35. STAEHLING-HAMPTON, K., S. PROLL, B.W. PAEPER, *et al.* 2002. A 52-kb deletion in the SOST-MEOX1 intergenic region on 17q12-q21 is associated with van Buchem disease in the Dutch population. Am. J. Med. Genet. **110:** 144–152.
36. KASU, N., J. LAURIKKALA, M. IMANISHI, *et al.* 2003. Sclerostin is a novel secreted osteoclast-derived bone morphogenetic protein with unique ligand specificity. J. Biol. Chem. **278:** 24113–24117.
37. WINKLER, D., M.K. SUTHERL, J. GEOGHEGAN, *et al.* 2003. Osteocyte control of bone formation via sclerostin, a novel BMP antagonist. EMBO J. **22:** 6267–6276.
38. VAN BEZOOIJEN, R.L., B.A.J. ROELEN, A. VISSER, *et al.* 2004. Sclerostin is an osteocyte negative regulator of bone formation, but not a classical BMP antagonist. J. Exp. Med. **199:** 805–814.
39. LI, X., Y. ZHANG, H. KANG, *et al.* 2005. Sclerostin binds to LRP5/6 and antagonizes canonical Wnt signaling. J. Biol. Chem. **280:** 19883–19887.
40. JILKA, R.L., R.S. WEINSTEIN, K. TAKAHASHI, *et al.* 1996. Linkage of decreased bone mass with impaired osteoblastogenesis in a murine model of accelerated senescence. J. Clin. Invest. **97:** 1732–1740.
41. SILVA, M.J., M.D. BRODT & B.A. UTHGENANNT. 2004. Morphological and mechanical properties of caudal vertebrae in the SAMP6 mouse model of senile osteoporosis. Bone **35:** 425–431.
42. KAPLAN, F.S. & E.M. SHORE. 2000. Progressive osseous heteroplasia. J Bone Miner Res. **15:** 2084–2094.
43. KAN L., M. HU, WA. GOMES, *et al.* 2004. Transgenic mice overexpressing BMP4 develop a fibrodysplasia ossificans progressiva (FOP)-like phenotype. Am. J. Pathol. **165:** 1107–1115.
44. ZHAO, M., S.E. HARRIS, D. HORN, *et al.* 2002. Bone morphogenetic protein receptor signaling is necessary for normal murine postnatal bone formation. J. Cell Biol. **157:** 1049–1060.

45. YAMASHITA, M., Y. SAI-XIA, Z. GEN-MU, *et al.* 2005. Ubiquitin ligase smurf1 control osteoblast activity and bone homeostases by targeting MEKK2 for degradation. Cell **121:** 101–113.
46. MISHINA, Y., M.W. STARBUCK, M.A. GENTILE, *et al.* 2004. Bone morphogenetic protein 1A receptor signaling regulates postnatal osteoblast function and bone remodeling. J. Biol. Chem. **279:** 27560–27566.
47. LIU, Z., W. SHI, X. JI, *et al.* 2004. Molecules mimicking smad1 interacting with Hox stimulate bone formation. J. Biol. Chem. **279:** 11313–11319.
48. OSHIMA, T., M. ABE, J. ASANO, *et al.* 2005. Myeloma cells suppress bone formation by secreting a soluble Wnt inhibitor, sFRP-2. Blood. **106:** 3160–3165.
49. HOANG, B.H., T. KUBO, J.H. HEALEY, *et al.* 2004. Expression of LDL receptor-related protein 5 (LRP5) as a novel marker for disease progression in high-grade osteosarcoma. Int. J. Cancer **109:** 106–111.
50. ETHERIDGE, S.L., G.J. SPENCER, D.J. HEATH, *et al.* 2004. Expression profiling and functional analysis of wnt signaling mechanisms in mesenchymal stem cells. Stem Cells **22:** 849–860.
51. BILIC-CURCIC, I., M. KRONENBERG, X. JIANG, *et al.* 2005. Visualizing levels of osteoblast differentiation by a two-color promoter-GFP strategy: type I collagen-GFPcyan and osteocalcin-GFPtpz. Genesis **43:** 87–98.
52. HAEGEL, H., L. LARUE, M. OHSUGI, *et al.* 1995. Lack of beta-catenin affects mouse development at gastrulation. Development **121:** 3529–3537.
53. KATO, M., M.S. PATEL, R. LEVASSEUR, *et al.* 2002. Cbfa1-independent decrease in osteoblast proliferation, osteopenia, and persistent embryonic eye vascularization in mice deficient in Lrp5, a Wnt coreceptor. J. Cell Biol. **157:** 303–314.
54. BODINE, P.V., W. ZHAO, Y.P. KHARODE, *et al.* 2004. The Wnt antagonist secreted frizzled-related protein-1 is a negative regulator of trabecular bone formation in adult mice. Mol. Endocrinol. **18:** 1222–1237.
55. GLASS, D.A. II, P. BIALEK, J.D. AHN, *et al.* 2005. Canonical Wnt signaling in differentiated osteoblasts controls osteoclast differentiation. Dev. Cell **8:** 751–764.
56. HILL, T.P., D. SPATER, M.M. TAKETO, *et al.* 2005. Canonical Wnt/beta-catenin signaling prevents osteoblasts from differentiating into chondrocytes. Dev. Cell **8:** 727–738.
57. GAUR, T., C.J. LENGNER, H. HOVHANNISYAN, *et al.* 2005. Canonical WNT signaling promotes osteogenesis by directly stimulating Runx2 gene expression. J. Biol. Chem. **280:** 33132–33140.
58. RAWADI, G., B. VAYSSIERE, F. DUNN, *et al.* 2003. BMP-2 controls alkaline phosphatase expression and osteoblast mineralization by a Wnt autocrine loop. J. Bone Miner. Res. **18:** 1842–1853.
59. MBALAVIELE, G., S. SHEIKH, J.P. STAINS, *et al.* 2005. Beta-catenin and BMP-2 synergize to promote osteoblast differentiation and new bone formation. J. Cell. Biochem. **94:** 403–418.
60. MUKHOPADHYAY, M., S. SHTROM, C. RODRIGUEZ-ESTEBAN, *et al.* 2001. Dickkopf1 is required for embryonic head induction and limb morphogenesis in the mouse. Dev. Cell **1:** 423–434.

Dysregulation of the BMP-4 Signaling Pathway in Fibrodysplasia Ossificans Progressiva

FREDERICK S. KAPLAN,[a,b,d] JENNIFER FIORI,[a,d]
LOURDES SERRANO DE LA PEÑA,[a,d] JAIMO AHN,[a,d]
PAUL C. BILLINGS,[a,d] AND EILEEN M. SHORE[a,c,d]

[a]*Department of Orthopaedic Surgery, The University of Pennsylvania School of Medicine, Philadelphia, Pennsylvania 19104, USA*

[b]*Department of Medicine, The University of Pennsylvania School of Medicine, Philadelphia, Pennsylvania 19104, USA*

[c]*Department of Genetics, The University of Pennsylvania School of Medicine, Philadelphia, Pennsylvania 19104, USA*

[d]*The Center for Research in FOP and Related Disorders, The University of Pennsylvania School of Medicine, Philadelphia, Pennsylvania 19104, USA*

ABSTRACT: **Identification of gene mutations in Mendelian disorders is often determined by linkage analysis and positional cloning, an approach that is difficult for fibrodysplasia ossificans progressiva (FOP) due to a low reproductive fitness that results in a small number of multigenerational families showing inheritance of the disease. Altered signaling pathways can be investigated as a complementary method to identify the consequences of the mutated gene responsible for FOP and to identify potential therapeutic targets. Candidate signaling pathways for FOP are those that malfunctioning could account for the malformation of the great toes during embryonic development and could explain the postnatal progressive heterotopic endochondral ossification. Signaling pathways that fit these criteria are the BMP signaling pathway and its interacting pathways. A large body of data suggest that the BMP-4 signaling pathway is dysregulated in FOP.**

KEYWORDS: **fibrodysplasia ossificans progressiva; BMP-4**

INTRODUCTION

Effective treatments for inherited disorders will result from identifying the gene mutations and dysregulated signaling pathways that cause these disorders.

Address for correspondence: Frederick S. Kaplan, M.D., Isaac & Rose Nassau Professor of Orthopaedic Molecular Medicine, University of Pennsylvania School of Medicine, Department of Orthopaedic Surgery, Silverstein Two, 3400 Spruce Street, Philadelphia, PA 19104. Voice: 215-349-8727; fax: 215-349-5928.
e-mail: frederick.kaplan@uphs.upenn.edu

Ann. N.Y. Acad. Sci. 1068: 54–65 (2006).
doi: 10.1196/annals.1346.008

For extremely rare developmental disorders, family pedigrees are scarce and genetic linkage and positional cloning are difficult, inefficient, and/or impossible. An alternate strategy to identify the primary pathology in these conditions is to study the dysregulated signaling pathway(s).

FIBRODYSPLASIA OSSIFICANS PROGRESSIVA AND CANDIDATE SIGNALING PATHWAYS

Fibrodysplasia ossificans progressiva (FOP), a rare genetic disorder, is the most severe disorder of heterotopic osteogenesis in humans and results in the postnatal formation of an ectopic skeleton.[1–6] In FOP, congenital malformation of the great toes is common. Progressive, disabling heterotopic endochondral osteogenesis occurs in characteristic anatomic patterns.[1–6]

In searching for candidate signaling pathways, we postulated that bone morphogenetic protein 4 (BMP-4) signaling was dysregulated in FOP cells and that a dysregulated BMP-4 signaling pathway could account for a congenital loss-of-function phenotype causing malformed great toes as well as a postnatal gain-of-function phenotype resulting in progressive heterotopic ossification in an endochondral pattern.[7–17] Bone morphogenetic proteins (BMPs), members of the transforming growth factor (TGF)-β superfamily of proteins, act as gradient morphogens in a concentration-dependent manner to specify cell fate in embryogenesis and in postnatal bone regeneration.[18–20] BMPs are potent osteogenic morphogens capable of inducing ectopic bone formation in animal models.[21–23] Recombinant BMPs induce mesenchymal cells to differentiate bone through cellular events that parallel the stages of bone formation in FOP lesions.[14] BMPs are both necessary and sufficient to induce the complete cellular program of endochondral osteogenesis at heterotopic sites *in vivo.*[18–22] BMP-4 is essential for the morphogenesis of the great toe[17] and transgenic mice overexpressing BMP-4 manifest an FOP-like phenotype of progressive heterotopic ossification.[23]

BMP-4 SIGNALING PATHWAYS

BMPs interact with specific BMP receptors (BMPRs) that phosphorylate downstream effector molecules.[18–20] Several combinations of type I (BMPRIA, BMPRIB, or ActRI) and type II (BMPRII or ActR-II) BMPRs mediate the BMP-4 signaling pathway.[18–20] After ligand binding, constitutively activated type II BMPRs attract and phosphorylate type I BMPRs. The activated type I BMPRs subsequently phosphorylate intracellular proteins of at least two interacting signaling cascades: the canonical Smad pathway and the p38 mitogen-activated protein kinase (p38 MAPK) pathway. Both pathways regulate nuclear transcriptional activity through cofactor association in a cell and developmental-specific context.[18–20,24,25]

BMP-4 OVEREXPRESSION IN FOP

A lymphocyte-derived cell system was developed for studying the molecular pathogenesis and candidate signaling pathways in FOP.[8,11,12,26] Although lymphocytes are present in the earliest stages of FOP lesion formation, their role as causative factors in the disease process remains uncertain. Nevertheless, the lymphoblastoid cell (LCL) is a useful model system for investigating the BMP pathway in FOP patients, and is one of the few cell types that is readily and safely available for study in this patient population.[26] Early studies on the BMP-4 pathway in FOP LCLs showed that BMP-4 messenger RNA (mRNA) and protein are overexpressed in lymphocytes and lesional cells from patients who have FOP.[4,8,9] The BMP-4 gene, however, is not mutated in FOP and the BMP-4 locus is exonerated by linkage exclusion studies in the few available multigenerational families with FOP.[16,27–30]

AN AUTOREGULATORY FEEDBACK LOOP BETWEEN BMP-4 AND BMP-4 ANTAGONISTS

Autoregulatory feedback loops in gene circuits provide stability, thereby limiting the range over which the concentration of vital network components fluctuate. Negative feedback provides a mechanism to ensure optimal concentrations of highly regulated morphogens such as BMP-4.[31,32] BMP-4 upregulates the expression of multiple secreted antagonists (noggin, gremlin, chordin, and follistatin) in various cell types and in various developmental contexts.[12,33] These antagonists bind BMP ligands, prevent them from engaging their cognate cell membrane receptors, and thus establish an autoregulatory negative feedback loop for BMP-4 to exert spatio-temporal control over its own activity.[12,33–35]

A DEFECT IN THE NEGATIVE AUTOREGULATORY FEEDBACK LOOP BETWEEN BMP-4 AND BMP-4 ANTAGONISTS IN FOP CELLS

We postulated that a defect in a negative feedback loop between BMP-4 and one or more of the genes encoding its antagonists could contribute to elevated BMP-4 activity in FOP cells.[12] We investigated basal- and BMP-4-stimulated mRNA expression of BMP-4-induced antagonists in LCLs derived from patients with FOP and from normal controls. Upon stimulation with exogenous BMP-4, control LCLs exhibited marked increases in noggin and gremlin mRNA levels whereas LCLs derived from patients with FOP exhibited a dramatically attenuated response.[12] These data support that FOP cells have dysregulation of negative feedback by which cells normally regulate the magnitude and boundaries of ambient BMP-4 morphogenetic signals.[12] The genes encoding the BMP antagonists noggin and gremlin are not mutated in FOP, and the corresponding loci are exonerated by linkage exclusion studies.[16]

DYSREGULATED BMP-4 HOMEOSTASIS IN FOP CELLS

In order to determine the fidelity of BMP-4 homeostasis in FOP cells, we examined the ability of FOP cells to maintain constant endogenous BMP-4 mRNA levels during growth in cell culture. In addition to overexpressing BMP-4, most FOP LCLs appear unable to stably regulate endogenous BMP-4 levels. Thus, FOP LCLs are unable to autoregulate BMP-4 levels during growth in cell culture, a finding that suggested that the defect in BMP-4 regulation in FOP LCLs might arise at or near the BMPRs.[26]

BMPR PROTEIN PROFILE IN FOP AND CONTROL CELLS

To examine the expression of BMP-4 receptor protein on the cell surface, we performed fluorescent-activated cell sorting (FACS) analysis using monoclonal antibodies to an array of possible BMPRs expressed in the LCLs. We found that control and FOP LCLs expressed BMPRIA and activin receptor type II (ActR-II), while neither BMPRIB nor BMPRII protein was detected.[26] Immunoprecipitation experiments revealed that BMPR complexes formed between BMPRIA and ActR-II (a recognized receptor complex for both BMP-4 and growth differentiation factor 5 [GDF5]). FACS analysis revealed similar levels of ActR-II on all cells but a sixfold overabundance of BMPRIA on the cell surface of FOP LCLs compared to control LCLs.[26]

Steady-state mRNA levels for BMPRIA and ActR-II receptors were evaluated by reverse transcription polymerase chain reaction (RT-PCR). The similar ActR-II mRNA levels in FOP and control LCLs are consistent with similar levels of ActR-II protein detected in those cell lines. By comparison, no differences in BMPRIA mRNA levels were observed between affected and unaffected individuals, despite clear differences in BMPRIA protein levels at the cell surface.[26]

MUTATIONAL ANALYSIS OF BMPRIA IN FOP CELLS

In order to determine if the overabundance of BMPRIA on the surface of FOP LCLs was due to genetic mutations, the BMPRIA gene was sequenced from genomic DNA of the FOP LCLs. No mutations were detected.[26] Regulatory mutations in BMPRIA were exonerated by linkage exclusion analysis in four small multigenerational families with FOP.[16,30]

BMPR BIOSYNTHESIS IN FOP CELLS

Dysregulation of receptor protein synthesis or degradation could give rise to increased BMPRIA protein on the surface of FOP cells despite similar steady-state levels of BMPRIA mRNA in affected and unaffected cells. To examine receptor synthesis, cellular proteins were metabolically labeled with [^{35}S]-methionine, and newly synthesized BMPRIA was detected by immunoprecipitation from total cell extracts at defined time points. BMPRIA was synthesized at similar rates in control and FOP lymphocytes.[26]

BMPR DEGRADATION IN FOP CELLS

To investigate BMPRIA degradation, cell proteins were steady-state labeled with [^{35}S]-methionine, then incubated in chase medium (without label) in the absence or presence of ligand. At defined time points during the chase period, BMPRIA was immunoprecipitated from total cell extracts and quantified. No difference in the rate of receptor turnover was observed in control or FOP LCLs in the absence of ligand.[26] Following treatment of control cells with BMP-4, BMPRIA was rapidly degraded. By contrast, BMP treatment of FOP LCLs did not stimulate BMPRIA degradation. No significant differences in BMPRIA levels were observed in the absence or presence of BMP-4 in FOP cells.[26]

BMPR INTERNALIZATION IN FOP CELLS

In order to determine whether the observed failure of receptor turnover was due to a defect in receptor degradation or due to a failure of receptor delivery to the intracellular degradation machinery, we examined ligand-mediated receptor internalization in FOP and control LCLs. Quantification of surface receptor was performed prior to and after BMP-4 treatment using FACS analysis with BMPRIA antibodies. For these experiments, cycloheximide blocked new receptor synthesis during BMP-4 treatment. In control cells, BMPRIA was significantly reduced after 1 h in the presence of the ligand. However, in FOP cells, BMP-4 treatment did not affect the abundance of surface receptor. These data suggest a defect in ligand-mediated internalization of BMPRIA in FOP lymphocytes.[26]

BMP-4 DOWNSTREAM SIGNAL TRANSDUCTION PATHWAYS IN FOP CELLS

To evaluate the functional activity of BMPRIA in FOP and control LCLs, BMP-4 downstream signal transduction through the Smad pathway was analyzed by immunoblotting total cell protein extracts with antibodies for phosphorylated Smad1. Control C2C12 cells demonstrated enhanced Smad1 phosphorylation in response to BMP-4 as expected. In control and FOP LCLs, however, no increase in Smad1 phosphorylation was observed after BMP-4 stimulation in any cell line.[26]

To determine whether BMP-4 treatment induced translocation of phosphorylated Smad1 to the nucleus, we isolated cytoplasmic and nuclear protein fractions from FOP and control LCLs. No differences were observed in the levels of phosphorylated Smad1 in either protein fraction, suggesting that little or no signaling through the Smad1 pathway occurs in control or FOP lymphocytes.[26]

Despite this apparent inactivity of Smad signaling in response to BMP-4, we observed, by quantitative real-time RT-PCR, that BMP-4 increases ID-1 mRNA levels in control and FOP LCLs. Furthermore, BMP-4-induced ID-1 mRNA levels in FOP cells were significantly greater than those in control cells. BMP-4-induced ID-1 synthesis was effectively blocked by SB203580, a specific inhibitor of p38 MAPK. These data show that the p38 MAPK signaling pathway is the predominant BMP signaling pathway in both control and FOP LCLs, and suggest that activation of the p38 MAPK pathway is downstream of BMPRIA in FOP lymphocytes.[26]

DYSREGULATION OF BMPRIA TRAFFICKING AT THE CELL MEMBRANE: A MODEL FOR EMBRYONIC SKELETAL MALFORMATION AND POSTNATAL HETEROTOPIC OSSIFICATION IN FOP

These and other studies suggested the likelihood of a fundamental disruption of BMP-4 signaling in FOP LCLs and pointed to a possible defect in BMPR physiology. These data suggested that the proximate cause of overabundant BMPRIA at the cell surface in FOP LCLs may reside in a defect in ligand-mediated receptor internalization.[26]

Flexibility in the pattern of receptor oligomerization and receptor-ligand interactions could be an important functional determinant of BMP action. Recent studies have shown that the specific pattern of receptor oligomerization preceding ligand binding determines the selection of downstream BMP signal transduction pathways.[36,37] The Smad signaling pathway is activated following BMP binding to preformed BMP type I/II heterodimeric receptor complexes, whereas the p38 MAPK signaling pathway is activated following BMP-induced hetero-oligomerization of homodimeric complexes.[37] Our data suggest that FOP cells may contain a set of ligand-independent phosphorylated BMPRIA/ActR-II heterodimers as well as an excessive pool of BMPRIA homodimers capable of being recruited into ligand-stimulated heterotetrameric complexes that signal through p38 MAPK.[26]

The BMPR trafficking defect in FOP could affect both p38 MAPK and Smad signaling pathways. We do not suggest that Smad signaling plays an unimportant role in FOP, but rather that it does not appear to play an important role in the LCL experimental system that we have used to study BMP signaling. Although not active in our experimental LCL test system, the Smad pathway could be important in other cell types, such as connective tissue cells, muscle cells, peripheral nerve cells, or other physiologic targets of BMP-4 pathway overactivity in FOP.[38] However, such cell types are not currently available for analysis due to the restrictions imposed by the natural history of FOP and by the consequent dangers to the patients.[6]

Constitutive phosphorylation of BMPRIA in the absence of exogenous ligand in FOP cells suggests that at least some BMPRIA receptors are constitutively activated in these cells, predicting that transcriptional targets of BMPRIA activation would be overexpressed in FOP LCLs and that little or no additional increase in transcription of these targets would occur upon BMP treatment. Unexpectedly, our data show that ID-1 mRNA is not expressed at high basal levels (without ligand) in FOP cells. However, ID-1 transcription is stimulated by treatment with BMP to higher levels in FOP as compared to control LCLs. We hypothesize that a subset of BMPRIA receptors is constitutively phosphorylated in FOP cells in the absence of ligand, but that these receptors fail to signal due to the lack of an intact BMP-specific Smad signaling pathway in the LCLs. Upon ligand stimulation of the receptors, robust signaling occurs through the p38 MAPK pathway. We propose that it is this ligand-sensitive pool of excessive BMPRIA that fails to properly internalize in the presence of BMP-4 and that it is responsible for excessive downstream signaling through the p38 MAPK pathway in FOP LCLs.[26]

Failure to internalize the type I BMPRs in FOP could have important developmental implications that are consistent with many of the clinical features of FOP. While increased BMP signaling can explain the heterotopic ossification in FOP, the great toe malformation is best explained by decreased activity of the BMP signaling pathway, specifically by decreased activity of BMPRIB, a target gene reciprocally regulated by BMPRIA in various developmental systems.[38–40] BMPRIB is not detected in the LCL test system, but is expressed in cartilage cells during early embryonic limb development.[38] Interestingly, homozygous loss of BMPRIB results in skeletal malformations that are largely restricted to the distal limbs.[39] Skeletal anomalies associated with knockout of BMPRIB includes short metacarpals and metatarsals, malformation of the proximal phalanges, absence of the interphalangeal joints, and fusions of the metacarpophalangeal and metatarsophalangeal joints with the most severe malformations noted in the anterior digits of the hind limb.[39] This highly specific pattern of skeletal malformations due to homozygous loss of BMPRIB is strikingly similar to the malformations of the great toe in patients with FOP.

So, how could failure to internalize BMPRIA following ligand stimulation lead to both a loss-of-function phenotype developmentally and gain-of-function phenotype postnatally? While the definitive answers are not yet known, an important clue may reside in how the two different BMP signaling pathways are activated following BMPRIA internalization.

BMPs can signal through either the Smad or the p38 MAPK pathways. Although there is abundant "cross-talk" between these two pathways, the canonical Smad signaling pathway is predominant developmentally, while the p38 MAPK signaling pathway may be predominant in stress-related processes such as response to tissue injury.[18–20] While little is presently known about the specific effects of BMPR internalization,[41] recent studies on TGF-β receptor internalization through the Smad pathway demonstrate that receptor internalization is necessary for downstream signaling to occur.[42–44] The TGF-β type I receptor is internalized through clathrin-mediated endocytosis, and binds SARA and Smads 2 or 3 in the early endosomes where downstream signaling is initiated.[42] Blocking internalization of the TGF-β type I receptors by blocking clathrin-mediated endocytosis diverts the receptors to the caveolin-mediated endocytosis and degradation pathway, which leads to decreased downstream TGF-β signaling.[42] While similar studies have not yet been published for the BMP signaling pathway, preliminary data suggest that similar trafficking and signaling paradigms prevail. While the canonical BMP-Smad signaling pathway does not appear to be functional in our LCL test system, the p38 MAPK system is a predominant signaling pathway in the LCLs.[26] Preliminary data in the control LCLs have demonstrated that endocytosis of ligand-stimulated BMPRIA through the p38 MAPK pathway is correlated with downregulation of downstream BMP signaling.

Thus, one could hypothesize that normal ligand-mediated internalization of BMPRIA leads to increased signaling through the developmentally mediated Smad signaling pathway and to signaling with rapid attenuation of the p38 MAPK inflammatory-related pathway. Conversely, if a primary physiologic abnormality in FOP cells is the failure to properly internalize BMPRIA following ligand stimulation, then the converse might prevail; that is, decreased signaling through the Smad pathway and failure to downregulate signaling through the p38 MAPK pathway. Such a scenario would be predicted to lead to congenital malformations of the great toes due to decreased signaling through the Smad pathway (as in the BMPRIB knockout and in the BMPRIB-type brachydactyly in humans), and to heterotopic ossification due to failure to downregulate BMP signaling through the p38 MAPK pathway. Preliminary data in our model system support this hypothesis (FIG. 1).

CANDIDATE GENES IN THE DYSREGULATED BMP-4 SIGNALING PATHWAY

The proximate genetic cause of the dysregulated BMP-4 signaling pathway in FOP cells remains unknown, but studies to date suggest the possibility of

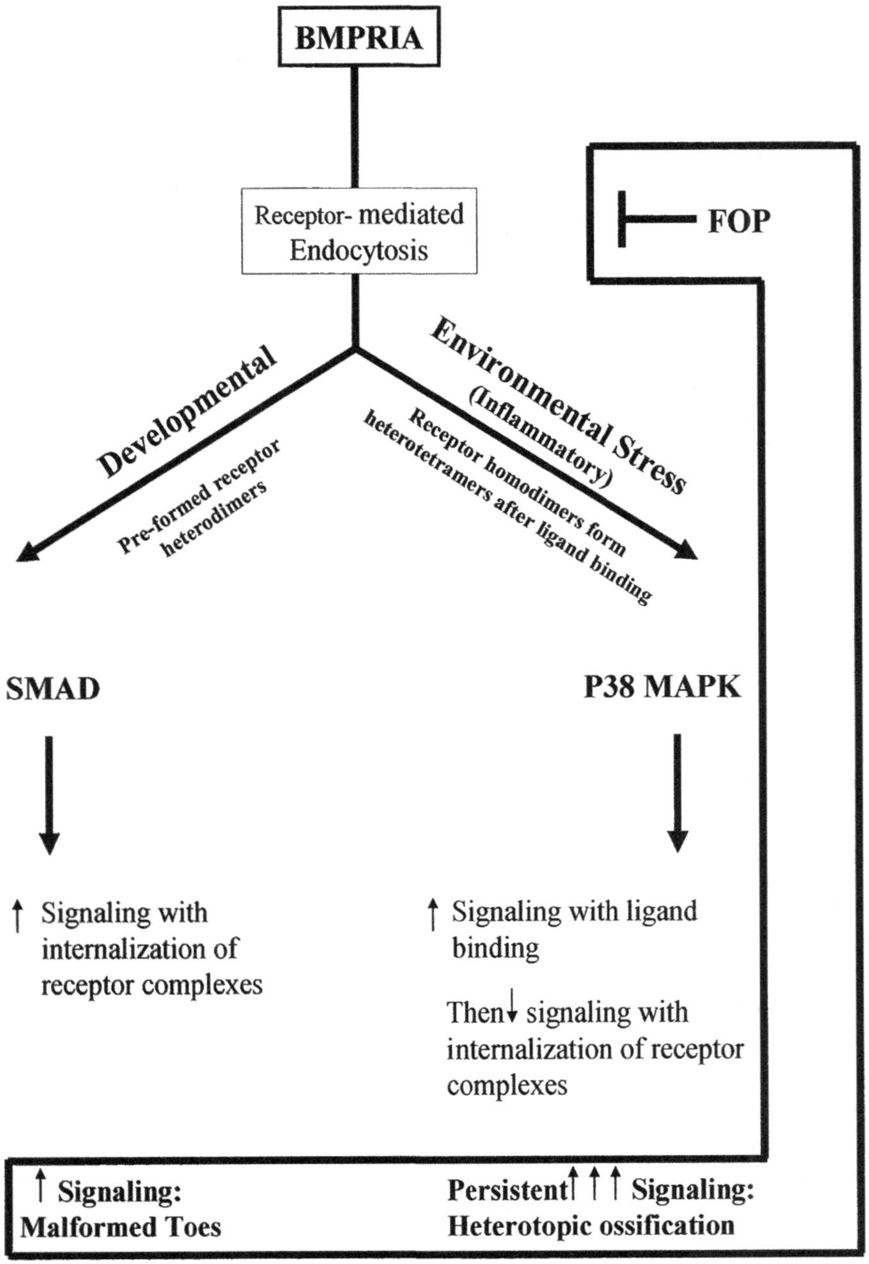

FIGURE 1. Hypothetical schema of BMP signaling in FOP cells. Long arrows indicate signaling pathways. Short arrows: up arrows indicate increased activity; down arrows indicate decreased activity.

a primary disorder in the inability to internalize BMPRs in the presence of ligand. Two reasonable postulates are that a BMPR-associated protein is deficient in quantity or activity in FOP cells, or that mutation of a BMPR or receptor-associated protein impairs ligand-mediated endocytosis of the receptor. Intensive work is presently focused in this area.

CONCLUSION

In conclusion, the BMP-4 signal transduction pathway is dysregulated in FOP cells and predicts a BMP-4 loss-of-function phenotype developmentally and a BMP-4 gain-of-function phenotype postnatally. Detailed analysis of the molecular pathophysiology of the BMP-4 signaling pathway in FOP cells is beginning to provide critical insight not only into the candidate genes for FOP, but also into the molecular mechanisms of BMPR trafficking and signaling that subserve the earliest events in the pathogenesis of FOP. Such an approach will reveal the most therapeutically appropriate locations in the pathway to target for therapy and will hasten the development of treatment strategies for FOP and other disorders of heterotopic ossification in humans.

REFERENCES

1. COHEN, R.B., G.V. HAHN, J. TABAS, *et al.* 1993. The natural history of heterotopic ossification in patients who have fibrodysplasia ossificans progressiva. J. Bone Joint Surg. **75-A:** 215–219.
2. KAPLAN, F.S., J. TABAS, F.H. GANNON, *et al.* 1993. The histopathology of fibrodysplasia ossificans progressiva: an endochondral process. J. Bone Joint Surg. **75-A:** 220–230.
3. KAPLAN, F.S., W. MCCLUSKEY, G. HAHN, *et al.* 1993. Genetic transmission of fibrodysplasia ossificans progressiva. J. Bone Joint Surg. **75-A:** 1214–1220.
4. GANNON, F.H., B.A. VALENTINE, E.M. SHORE, *et al.* 1998. Acute lymphocytic infiltration in an extremely early lesion of fibrodysplasia ossificans progressiva. Clin. Orthop. Relat. Res. **346:** 19–25.
5. KAPLAN, F.S., D.L. GLASER, N. HEBELA & E.M. SHORE. 2004. Heterotopic ossification. J. Am. Acad. Orthop. Surg. **12:** 116–125.
6. KAPLAN, F.S., E.M. SHORE & J.M. CONNOR. 2002. Fibrodysplasia ossificans progressiva (FOP). *In* Connective Tissue and Its Heritable Disorders: molecular, Genetic, and Medical Aspects, 2nd ed. P.M. Royce & B. Steinmann, Eds.: 827–840. Wiley-Liss; John Wiley & Sons, Inc., New York.
7. KAPLAN, F.S., J.A. TABAS & M.A. ZASLOFF. 1990. Fibrodysplasia ossificans progressiva: a clue from the fly? Calcif. Tissue. Int. **47:** 117–125.
8. SHAFRITZ, A.B., E.M. SHORE, F.H. GANNON, *et al.* 1996. Over-expression of an osteogenic morphogen in fibrodysplasia ossificans progressiva. N. Engl. J. Med. **335:** 555–561.

9. GANNON, F.H., F.S. KAPLAN, E. OLMSTED, *et al.* 1997. Bone morphogenetic protein 2/4 in early fibromatous lesions of fibrodysplasia ossificans progressiva. Hum. Pathol. **28:** 339–343.
10. KAPLAN, F.S. & E.M. SHORE. 1998. Encrypted morphogens of skeletogenesis: biological errors and pharmacological patients. Biochem. Pharmacol. **55:** 373–382.
11. OLMSTED, E.A., F.S. KAPLAN & E.M. SHORE. 2003. Bone morphogenetic protein 4 regulation in fibrodysplasia ossificans progressiva. Clin. Orthop. Relat. Res. **408:** 331–343.
12. AHN, J., L. SERRANO DE LA PEÑA, E.M. SHORE & F.S. KAPLAN. 2003. Paresis of a bone morphogenetic protein-antagonist response in a genetic disorder of heterotopic skeletogenesis. J. Bone Joint Surg. **85-A:** 667–674.
13. HEGYI, L., F.H. GANNON, D.L. GLASER, *et al.* 2003. Stromal cells of fibrodysplasia ossificans progressiva lesions express smooth muscle lineage markers and the osteogenic transcription factor Runx2/Cbfa-1: clues to a vascular origin of heterotopic ossification. J. Pathol. **201:** 141–148.
14. GLASER, D.L., A.N. ECONOMIDES, L. WANG, *et al.* 2003. In vivo somatic cell gene transfer of an engineered noggin mutein prevents BMP-4-induced heterotopic ossification. J. Bone Joint Surg. **85-A:** 2332–2342.
15. KAPLAN, F.S. 1998. The metamorphosis. Clin. Orthop. Relat. Res. **346:** 2–3.
16. XU, M.Q., G. FELDMAN, M. LE MERRER, *et al.* 2000. Linkage exclusion and mutational analysis of the noggin gene in patients with fibrodysplasia ossificans progressiva. Clin. Genet. **58:** 291–298.
17. DUNN, N.R., G.E. WINNIER, L.K. HARGETT, *et al.* 1997. Haploinsufficient phenotypes in BMP-4 heterozygous null mice and modification by mutations in Gli3 and Alx4. Dev. Biol. **188:** 235–247.
18. MASSAGUÉ, J., S.W. BLAIN & R.S. LO. 2000. TGF beta signaling in growth control, cancer, and heritable disorders. Cell **103:** 295–309.
19. SHI, Y. & J. MASSAGUE. 2003. Mechanisms of TGF-beta signaling from cell membrane to the nucleus. Cell **113:** 685–700.
20. CANALIS, E., A.N. ECONOMIDES & E. GAZZERRO. 2003. Bone morphogenetic proteins, their antagonists, and the skeleton. Endocr. Rev. **24:** 218–235.
21. URIST, M.R. 1965. Bone formation by autoinduction. Science **150:** 893–899.
22. WOZNEY, J.M., V. ROSEN, A.J. CELESTE, *et al.* 1988. Novel regulators of bone formation: molecular clones and activities. Science **242:** 1528–1534.
23. KAN, L., M. HU, W.A. GOMES & J.A. KESSLER. 2004. Transgenic mice overexpressing BMP-4 develop a fibrodysplasia ossificans progressiva (FOP)-like phenotype. Am. J. Pathol. **165:** 1107–1115.
24. ATTISANO, L. & J.L. WRANA. 2000. Smads as transcriptional co-modulators. Curr. Opin. Cell Biol. **12:** 235–243.
25. DERYNCK, R. & Y.E. ZHANG. 2003. Smad-dependent and Smad-independent pathways in TGF-beta family signalling. Nature **425:** 577–584.
26. DE LA PEÑA, L.S., P.C. BILLINGS, J.L. FIORI, *et al.* 2005. Fibrodysplasia ossificans progressiva (FOP), a disorder of ectopic osteogenesis, misregulates cell surface expression and trafficking of BMPRIA. J. Bone Miner. Res. **20:** 1168–1176.
27. TABAS, J.A., G.V. HAHN, R.B. COHEN, *et al.* 1993. Chromosomal assignment of the human gene for bone morphogenetic protein 4. Clin. Orthop. Relat. Res. **293:** 310–316.

28. SHORE, E.M., M.Q. XU, P.B. SHAH, *et al.* 1998. The human bone morphogenetic protein (BMP-4) gene: molecular structure and transcriptional regulation. Calcif. Tissue. Int. **63:** 221–229.
29. XU, M.Q. & E.M. SHORE. 1998. Mutational screening of the bone morphogenetic protein 4 gene in a family with fibrodysplasia ossificans progressiva. Clin. Orthop. **346:** 53–58.
30. FELDMAN, G., M. LI, S. MARTIN, *et al.* 2000. Fibrodysplasia ossificans progressiva, a heritable disorder of severe heterotopic ossification, maps to human chromosome 4q27-31. Am. J. Hum. Genet. **66:** 128–135.
31. ZIMMERMAN, L.B., J.M. DE JESUS-ESCOBAR & R.M. HARLAND. 1996. The Spemann organizer signal noggin binds and inactivates bone morphogenetic protein 4. Cell **86:** 599–606.
32. BRUNET, L.J., J.A. MCMAHON, A.P. MCMAHON & R.M. HARLAND. 1998. Noggin, cartilage morphogenesis, and joint formation in the mammalian skeleton. Nature **280:** 1455–1457.
33. GAZZERRO, E., V. GANGJI & E. CANALIS. 1998. Bone morphogenetic proteins induce the expression of noggin, which limits their activity in cultured rat osteoblasts. J. Clin. Invest. **102:** 2106–2114.
34. GROPPE, J., J. GREENWALD, E. WIATER, *et al.* 2002. Structural basis of BMP signaling inhibition by the cystine knot protein Noggin. Nature **420:** 636–642.
35. GROPPE, J., J. GREENWALD, E. WIATER, *et al.* 2003. Structural basis of BMP signaling inhibition by Noggin, a novel twelve-membrane cystine knot protein. J. Bone Joint Surg. Am. **85-A**(Suppl 3): 52–58.
36. GILBOA, L., A. NOHE, T. GEISSENDORFER, *et al.* 2000. Bone morphogenetic protein receptor complexes on the surface of live cells: a new oligomerization mode for serine/threonine kinase receptors. Mol. Biol. Cell **11:** 1023–1035.
37. NOHE, A., S. HASSEL, M. EHRLICH, *et al.* 2002. The mode of bone morphogenetic protein (BMP) receptor oligomerization determines different BMP-2 signaling pathways. J. Biol. Chem. **277:** 5330–5338.
38. HOFFMANN, A. & G. GROSS. 2001. BMP signaling pathways in cartilage and bone formation. Crit. Rev. Eukaryot. Gene Expr. **11:** 23–45.
39. YI, S.E., A. DALUISKI, R. PEDERSON, *et al.* 2000. The type I BMP receptor BMPRIB is required for chondrogenesis in the mouse limb. Development **127:** 621–630.
40. PANCHISON, D.M., J.M. PICKEL, L. STUDER, *et al.* 2001. Sequential actions of BMP receptors control neural precursor cell production and fate. Genes Dev. **15:** 2094–2110.
41. RAUCH, C., A.C. BRUNET, J. DELEULE & E. FARGE. 2002. C2C12 myoblast/osteoblast transdifferentiation steps enhanced by epigenetic inhibition of BMP2 endocytosis. Am. J. Physiol. Cell Physiol. **283:** C235–C243.
42. DI GUGLIELMO, G.M., C. LE ROY, A.F. GOODFELLOW & J.L. WRANA. 2003. Distinct endocytic pathways regulate TGF-beta receptor signalling and turnover. Nat. Cell Biol. **5:** 410–421.
43. GILBOA, L., R.G. WELLS, H.F. LODISH & Y.I. HENIS. 1998. Oligomeric structure of type I and type II transforming growth factor beta receptors: homodimers form in the ER and persist at the plasma membrane. J. Cell Biol. **140:** 767–777.
44. DORE, J.J. JR., M. EDENS, N. GARAMSZEGI & E.B. LEOF. 1998. Heteromeric and homomeric transforming growth factor-beta receptors show distinct signaling and endocytic responses in epithelial cells. J. Biol. Chem. **273:** 31770–31777.

Hypoxia and HIF-1α in Chondrogenesis

ERNESTINA SCHIPANI

Endocrine Unit, Massachusetts General Hospital and Harvard Medical School, Boston, Massachusetts 02114, USA

ABSTRACT: In endochondral bone development chondrocytes undergo proliferation, hypertrophic differentiation, mineralization of the surrounding matrix, death, blood vessel invasion, and finally replacement of cartilage with bone. The chondrocytic growth plate is a unique mesenchymal tissue, as it is avascular but it requires blood vessel invasion in order to be replaced by bone. We have recently provided evidence that the growth plate is hypoxic during fetal development. Adaptation to hypoxia is a critical event in numerous pathological settings, such as tumor progression and survival of tissues in which blood flow has been suddenly interrupted. One of the hallmarks of the response to hypoxia is activation of the transcription factor HIF-1α. The von Hippel–Lindau (VHL) tumor suppressor protein is a component of a ubiquitin ligase promoting proteolysis of HIF-1α. By using a genetic approach, we have demonstrated that VHL and HIF-1α are critical regulators of endochondral bone development.

KEYWORDS: hypoxia; HIF-1α; VHL; VEGF; chondrocytes

INTRODUCTION

Low oxygen tension has an important role not only in tumor progression or in the biology of tissues in which blood flow has been suddenly interrupted, but also in development and differentiation.[1–2] We have recently provided evidence that the fetal growth plate is hypoxic. In this article, we will briefly summarize our current knowledge about hypoxia, HIF-1α, and chondrocytes.

Skeletal development depends on two mechanisms, intramembranous and endochondral.[3–5] Intramembranous bone formation, in which mesenchymal cells after condensation differentiate directly into osteoblasts, is involved in the formation of the flat bones of the skull. Endochondral bone formation, which accounts for the development of most of the other bones, is a two-stage mechanism: first chondrocytes shape a template, on which osteoblasts then differentiate to form bone.

Address for correspondence: Ernestina Schipani, M.D., Ph.D., Massachusetts General Hospital, Endocrine Unit, 50 Blossom Street, Wellman 501, Boston, MA 02114-2696. Voice: 617-726-3966; fax: 617-726-7543.

e-mail: schipani@helix.mgh.harvard.edu

Ann. N.Y. Acad. Sci. 1068: 66–73 (2006). © 2006 New York Academy of Sciences.

doi: 10.1196/annals.1346.009

Endochondral bone development is a fascinating story of proliferation, differentiation, and death.[6] The process recapitulates basic and fundamental mechanisms of cell biology with a highly specific spatial and temporal pattern, and it thus constitutes a useful model for the analysis of such mechanisms. In recent years, the tools provided by modern genetic further strengthened the validity and usefulness of the model. Round proliferative chondrocytes synthesize type II collagen and form a columnar layer; they, then, differentiate into post-mitotic hypertrophic cells, which express predominantly type X collagen. This unique differentiation process is followed by mineralization of the surrounding matrix, death of hypertrophic chondrocytes, blood vessel invasion, and finally replacement of cartilage with primary spongiosa. The fetal growth plate is a unique mesenchymal tissue, since it is virtually avascular, but it requires blood vessel invasion for being substituted by bone.[7] Differently from the cartilage anlage, the perichondrium, a mesenchymal structure that surrounds the growth plate, is vascularized.[8]

By using EF-5, which is a marker for bioreductive activity,[9] we have demonstrated that the fetal growth plate has a hypoxic central region, the most highly hypoxic chondrocytes being located in the round proliferative layer near the joint space, in the center of the columnar proliferative layer and in the upper portion of the hypertrophic zone.[10] Notably, the late hypertrophic chondrocytes, which are adjacent to the primary spongiosa, do not appear to be hypoxic; this finding is consistent with the extensive vascularization of the primary spongiosa.

The EF5 data prompted us to ask two basic questions: (1) How do chondrocytes survive in a hypoxic milieu? and; (2) Does hypoxia play any role in chondrogenesis *per se*? The transcription factor HIF-1α plays a critical role in adaptive responses to hypoxia.[11–15] HIF-1 belongs to the PAS (PER-ARNT-SIM) subfamily of basic helix-loop-helix (bHLH) transcription factors. The basic domain is responsible for binding to DNA, while the helix-loop-helix and PAS domains are critical for heterodimerization.[12–15] HIF-1 is a heterodimer that consists of the subunit HIF-1α, and the subunit HIF-1β. Both HIF-1α, and HIF-1β (also known as ARNT, aryl hydrogen receptor nuclear translocator) are constitutively expressed in cells. However, whereas HIF-1β is stable in normoxic conditions, HIF-1α is extremely unstable. Under normoxic conditions, HIF-1α is hydroxylated on specific proline residues by a recently discovered family of prolyl-hydroxylases.[16–18] The von Hippel–Landau (VHL) tumor suppressor protein, which is a recently identified novel E3 ubiquitin ligase,[19] recognizes the proline-hydroxylated form of HIF-1α and targets it for polyubiquitination and degradation by the proteasome. Conversely, under hypoxic conditions, oxygen-sensitive prolyl-hydroxylase activity is reduced, and HIF-1α translocates into the nucleus. Within the nucleus, the complex binds to specific hypoxic responsive elements (HRE), thereby initiating the transcription of specific genes. HIF-1α regulates the hypoxic response by controlling genes with metabolic functions, such as glucose transport and metabolism, angiogenic factors like vascular endothelial growth factor (VEGF), proteins of the cell cycle and of the apoptotic machinery.[18,20–22]

Mice that lack HIF-1α or VHL as result of homologous recombination die around embryonic days 8.5/9.5.[23–26] Therefore, in order to study the role of HIF-1α and VHL in endochondral bone development, we performed conditional knockout of HIF-1α and VHL, respectively, in chondrocytes, using a Cre-LoxP system in which Cre recombinase expression is driven by a fragment of the collagen type II promotor. Mice lacking HIF-1α die at birth; the causes of death are cleft palate (E. Zelzer and B. Olsen, unpublished data), and respiratory distress, which is a consequence of partial trachea collapse and malformation of the rib cage.[10] Conversely, absence of VHL in cartilage is not lethal, but leads to severe dwarfism.[27] Interestingly, the conditional deletion of both genes generates a phenotype that is apparently indistinguishable from the phenotype caused by lack of HIF-1α alone, suggesting that HIF-1α is epistatic to VHL in growth plate chondrocytes.[27]

HIF-1α IS A SURVIVAL FACTOR FOR HYPOXIC CHONDROCYTES

Conditional knockout of HIF-1α has demonstrated that this transcription factor is critical for the survival of hypoxic chondrocytes and negatively regulates chondrocyte proliferation.[10,28] Conditional knockout of VHL in chondrocytes has confirmed an important role of both HIF-1α and VHL in chondrocyte biology.[27]

Embryos lacking HIF-1α in chondrocytes have shorter and deformed bones; furthermore, the growth plates are wider and misshapen. In particular, the center of the null growth plate is hypocellular, and occupied by abnormal cellular elements with picnotic nuclei. This spatially localized loss of cell viability, which was confirmed by Tunel and *in situ* hybridization analysis, was particularly striking in the "core" of the cartilaginous elements, whereas within few microns from the surface, the specimens appear to be histologically normal.[10] Interestingly, the central cell death phenotype of the HIF-1α null growth plates was not preceded by ectopic or premature hypertrophic differentiation, as shown by the analysis of collagen type X expression, which is a marker of hypertrophic chondrocytes, at different stages of fetal development.[10] This finding indicates that chondrocyte death secondary to lack of HIF-1α does not require hypertrophic differentiation, and, therefore, it is likely to be different at the molecular level from the chondrocyte apoptosis that precedes blood vessel invasion and replacement of cartilage with bone.

A potential downstream effector of HIF-1α action as "survival" factor is VEGF, one of the most potent pro-angiogenic factors characterized by data.[29,30,37] We have clearly demonstrated by *in situ* hybridization analysis that the fetal growth plate has two distinct domains of VEGF expression.[27,31] The domain with "low" level of VEGF expression comprises the center of the round, columnar and upper hypertrophic layers, respectively, that is the hypoxic zones of the growth plate. Conversely, the domain with "high" level of VEGF expression overlaps with the nonhypoxic layer of late hypertrophic chondrocytes that is adjacent to bone. Numerous studies have elegantly shown that expression of VEGF by late hypertrophic chondrocytes is critical for blood vessel invasion and replacement of cartilage by bone.[7,8,32–34] Notably, conditional knockout of HIF-1α in chondrocytes still allows detectable expression of VEGF in this domain.[10,31] This finding indicates that VEGF expression by late hypertrophic chondrocytes is largely HIF-1α independent. The transcription factor Runx2 has been reported to be an important regulator of VEGF in this zone.[35] Conversely, VEGF expression in the hypoxic areas of the fetal growth plate is controlled by HIF-1α.[27,31] Numerous lines of evidence support this conclusion. In particular, in fetal growth plates in which HIF-1α transcriptional activity is upregulated as a result of VHL conditional knockout, the outside-inside pattern of expression of VEGF mRNA is perturbed, VEGF mRNA is now detectable throughout the whole growth plate, and is upregulated in round proliferative chondrocytes located in close proximity to the articular surface.[27] Furthermore, consistent with these data, hypoxia strikingly increases VEGF accumulation in chondrocytes *in vitro*, and this increase is HIF-1α dependent.[28] Lastly, conditional knockout of VEGF in chondrocytes, using the same Cre-LoxP strategy adopted for both HIF-1α and VHL conditional knockouts, leads to a central cell death phenotype in the center of the proliferative and upper hypertrophic zones that very closely mimics the one observed in HIF-1α conditional knockout.[31] However, this phenotype is definitively "milder" than that observed in HIF-1α null growth plates.[10,31] Similar findings were also reported in mice that lack the soluble forms of VEGF, i.e., VEGF120 and VEGF164, as result of a universal knockout.[36] Taken together, these data show that VEGF is important, even if very likely not the only mediator of the central cell death observed in HIF-1α-deficient growth plates.

Consistent with the hypoxic status of the growth plate, mRNA encoding phosphoglycerokinase (PGK), a key enzyme of the anaerobic glycolysis pathway, is strikingly upregulated in the fetal chondrocytic growth plate in comparison with the surrounding tissues, and, like VEGF mRNA, it displays and outside-inside pattern of expression that resembles the pattern of oxygenation of the growth plate.[10] Notably, lack of HIF-1α decreases PGK mRNA expression to virtually background levels;[10] conversely, increased HIF-1α transcriptional activity in VHL null growth plates leads to a generalized expression of PGK mRNA throughout the whole growth plate, with loss of the outside–inside pattern and upregulation in the round proliferative chondrocytes adjacent to the articular surface.[27] Impaired regulation of metabolic pathways is likely thus to be an additional pathogenetic event causing the central cell death phenotype observed in HIF-1α-deficient growth plates. Glycolysis and apoptosis have been considered for long time to be major but independent pathways that are critical for cell survival. A recent report provides evidence that the proapoptotic protein BAD and the glycolytic enzyme glucokinase integrate both pathways as part of a common network in mitochondria.[38] HIF-1α may contribute to integrating glucose metabolism and apoptosis by regulating expression of enzymes of the glycolytic pathway.

HIF-1α directly regulates both pro- and anti-apoptotic genes.[21,29–42] It will thus be interesting to now investigate which other molecular mechanisms mediate the role of HIF-1α as a survival factor in chondrocytes, in addition to its regulation of anaerobic glycolysis and VEGF.

HIF-1α AND VHL SHAPE UP THE GROWTH PLATE BY MODULATING CHONDROCYTE SIZE AND PROLIFERATION

It has been reported that hypoxia leads to the arrest of cell proliferation, at least in part through upregulation of HIF-1α transcriptional activity.[43,44] VHL-deficient growth plates display both hypocellularity and a striking decrease in chondrocyte proliferation rate, which are concomitant to an increase of HIF-1α accumulation, and are likely to be the cause of the dwarfism of mice lacking VHL in chondrocytes.[27] It is paradoxical that in fetal growth plates the lack of this tumor-suppressor gene inhibits chondrocyte proliferation.[27,45] However, the finding is consistent with the inhibition of cell proliferation by the upregulation of HIF-1α transcriptional activity.[44] Interestingly, to further support the model that lack of VHL inhibits cell proliferation through upregulation of HIF-1α transcriptional activity, viable chondrocytes at the periphery of HIF-1α-deficient fetal growth plates show a striking increase in the proliferation rate, i.e., chondrocytes lacking HIF-1α die if located in the center of the fetal growth plate, whereas they proliferate at a higher rate at its periphery.[10]

Proliferating chondrocytes of VHL-deficient growth plates are enlarged with a high cytoplasm to nucleus ratio.[27] Our current working hypothesis is that decreased proliferation is the pathogenetic event causing both the hypocellularity and the appearance of these somehow atypical cells in the VHL null growth plates.[27] Interestingly, this unique cartilage phenotype of VHL null growth plates, which is characterized by reduced proliferation rate and increased cell size, has not been described for other knockout models in which chondrocyte proliferation has been reported to be severely impaired.[46] Therefore, it is possible that the increased cell size observed in the VHL-deficient growth plates is not the necessary or direct consequence of the decreased cell proliferation *per se*. Lack of VHL in chondrocytes may uncouple cell size from cell proliferation by regulating a yet unknown molecular mechanism, which may or may not involve HIF-1α.

CONCLUSIONS

Taken together, our findings lead to a working model in which hypoxia-dependent upregulation of HIF-1α transcriptional activity is critical for the survival of hypoxic chondrocyte, and also significantly contributes to shaping up the growth plate by inhibiting chondrocyte proliferation and probably regulating cell size. It will be now interesting to investigate the potential role of hypoxia and HIF-1α in early chondrogenesis.

In any event, HIF-1α may not be the only downstream effector of hypoxia in cartilage, and *vice versa* hypoxia may not be the only regulator of HIF-1α transcriptional activity in chondrocytes, as also shown in other cellular systems.[31] In addition, it is still an open and important question whether indeed VHL actions in cartilage are exclusively mediated by HIF-1α.

One of the best-documented consequences of hypoxia *in vivo* is increased angiogenesis. It is thus quite intriguing and almost paradoxical that hypoxia and HIF-1α are of critical importance in the biology of cartilage, which is a classical avascular tissue. The apparent paradox suggests that the hypoxia/VHL/HIF pathway may turn on a genetic program that could be specific for chondrocytes, as also shown for other cell types.[43] Unveiling this program could expand our understanding of the molecular mechanisms underlying the hypoxic response in development.

ACKNOWLEDGMENT

This work was supported by NIH grant AR048191.

REFERENCES

1. Giaccia, A., M. Simon & R. Johnson. 2004. The biology of hypoxia: the role of oxygen sensing in development, normal function, and disease. Genes Dev. **18:** 2183–2194.
2. Chen, E., M. Fujinaga & A. Giaccia. 1999. Hypoxic microenvironment within an embryo induces apoptosis and is essential for proper morphological development. Teratology **60:** 215–225.
3. Karsenty, G. 2003. The complexities of skeletal biology. Nature **423:** 316–318.
4. Erlebacher, A. *et al.* 1995. Toward a molecular understanding of skeletal development. Cell **80:** 371–378.
5. Kronenberg, H. 2003. Developmental regulation of the growth plate. Nature **423:** 332–336.
6. Schipani, E. & S. Provot. 2003. PTHrP, PTH, and the PTH/PTHrP receptor in endochondral bone development. Birth Defects Res. Part C Embryo Today **69:** 352–362.
7. Vu, T. *et al.* 1998. MMP-9/gelatinase B is a key regulator of growth plate angiogenesis and apoptosis of hypertrophic chondrocytes. Cell. **93:** 411–422.

8. ZELZER, E. *et al.* 2002. Skeletal defects in VEGF120/120 mice reveal multiple roles for VEGF in skeletogenesis. Development **129:** 1893–1904.
9. LEE, J. *et al.* 1996. Direct relationship between radiobiological hypoxia in tumors and monoclonal antibody detection of EF5 cellular adducts. Int. J. Cancer **67:** 372–378.
10. SCHIPANI, E. *et al.* 2001. Hypoxia in cartilage: HIF-1alpha is essential for chondrocyte growth arrest and survival. Genes Dev. **15:** 2865–2876.
11. BUNN, H. & R. POYTON. 1996. Oxygen sensing and molecular adaptation to hypoxia. Physiol. Rev. **76:** 839–885.
12. LIU, L. & M. SIMON. 2004. Regulation of transcription and translation by hypoxia. Cancer Biol. Ther. **3:** 492–497.
13. SEMENZA, G. 2003. Targeting HIF-1 for cancer therapy. Nat. Rev. Cancer **3:** 721–732.
14. GIACCIA, A., B. SIIM & R. JOHNSON. 2003. HIF-1 as a target for drug development. Nat. Rev. Drug. Discov **2:** 803–811.
15. KAELIN, W.G.J. 2002. How oxygen makes its presence felt. Genes Dev. **16:** 1441–1445.
16. JAAKKOLA, P. *et al.* 2001. Targeting of HIF-alpha to the von Hippel-Lindau ubiquitylation complex by O2-regulated prolyl hydroxylation. Science. **292:** 468–472.
17. MIN, J. *et al.* 2002. Structure of an HIF-1alpha-pVHL complex: hydroxyproline recognition in signaling. Science **296:** 1886–1889.
18. CHAN, D. *et al.* 2002. Role of prolyl hydroxylation in oncogenically stabilized hypoxia-inducible factor-1alpha. J. Biol. Chem. **277:** 40112–40117.
19. OHH, M. & W.G. KAELIN. 1999. The von Hippel-Lindau tumor suppressor protein: new perspectives. Mol. Med. Today **5:** 257–263.
20. KALLIO, P.J. *et al.* 1998. Signal transduction in hypoxic cells: inducible nuclear translocation and recruitment of the CBP/p300 coactivator by the hypoxia-inducible factor-1alpha. EMBO J. **17:** 6573–6586.
21. LEO, C., A. GIACCIA & N. DENKO. 2004. The hypoxic tumor microenvironment and gene expression. Semin. Radiat. Oncol. **14:** 207–214.
22. BISHOP, T. *et al.* 2004. Genetic analysis of pathways regulated by the von Hippel-Lindau tumor suppressor in *Caenorhabditis elegans*. PLoS. Biol. **2:** e289.
23. RYAN, H.E., J. LO & R.S. JOHNSON. 1998. HIF 1 alpha is required for solid tumor formation and embryonic vascularization. EMBO J. **17:** 3005–3015.
24. IYER, N. *et al.* 1998a. Cellular and developmental control of O2 homeostasis by hypoxia-inducible factor 1alpha. Genes Dev. **12:** 149–162.
25. CARMELIET, P. *et al.* 1998. Role of HIF-1alpha in hypoxia mediated apoptosis, cell proliferation and tumour angiogenesis. Nature **394:** 485–490.
26. GNARRA, J. *et al.* 1997. Defective placental vasculogenesis causes embryonic lethality in VHL-deficient mice. Proc. Natl. Acad. Sci. USA **94:** 9102–9107.
27. PFANDER, D. *et al.* 2004. Deletion of Vhlh in chondrocytes reduces cell proliferation and increases matrix deposition during growth plate development. Development **131:** 2497–2508.
28. PFANDER, D. *et al.* 2003. HIF-1alpha controls extracellular matrix synthesis by epiphyseal chondrocytes. J. Cell. Sci. **116:** 1819–1826.
29. FERRARA, N., H. GERBER & J. LECOUTER. 2003. The biology of VEGF and its receptors. Nat. Med. **9:** 669–676.
30. CRAMER, T. *et al.* 2004. Expression of VEGF isoforms by epiphyseal chondrocytes during low-oxygen tension is HIF-1 alpha dependent. Osteoarthritis Cartilage **12:** 433–439.

31. ZELZER, E. *et al.* 2004. VEGFA is necessary for chondrocyte survival during bone development. Development **131:** 2161–2171.
32. GERBER, H. *et al.* 1999. VEGF couples hypertrophic cartilage remodeling, ossification and angiogenesis during endochondral bone formation. Nat. Med. **5:** 623–628.
33. MAES, C. *et al.* 2002. Impaired angiogenesis and endochondral bone formation in mice lacking the vascular endothelial growth factor isoforms VEGF164 and VEGF188. Mech. Dev. **111:** 61–73.
34. IVKOVIC, S., B. YOON, *et al.* 2003. Connective tissue growth factor coordinates chondrogenesis and angiogenesis during skeletal development. Development **130:** 2779–2791.
35. ZELZER, E. *et al.* 2001. Tissue specific regulation of VEGF expression during bone development requires Cbfa1/Runx2. Mech. Dev. **106:** 97–106.
36. MAES, C. *et al.* 2004. Soluble VEGF isoforms are essential for establishing epiphyseal vascularization and regulating chondrocyte development and survival. J. Clin. Invest. **113:** 188–199.
37. TANG, N. *et al.* 2004. Loss of HIF-1alpha in endothelial cells disrupts a hypoxia-driven VEGF autocrine loop necessary for tumorigenesis. Cancer Cell **6:** 485–495.
38. DANIAL, N. *et al.* 2003. BAD and glucokinase reside in a mitochondrial complex that integrates glycolysis and apoptosis. Nature **424:** 952–956.
39. BIJU, M. *et al.* 2004. Vhlh gene deletion induces Hif-1-mediated cell death in thymocytes. Mol. Cell. Biol. **24:** 9038–9047.
40. LEE, M. *et al.* 2004. Identification of the hypoxia-inducible factor 1 alpha-responsive HGTD-P gene as a mediator in the mitochondrial apoptotic pathway. Mol. Cell. Biol. **24:** 3918–3927.
41. BRUICK, R. 2000. Expression of the gene encoding the proapoptotic Nip3 protein is induced by hypoxia. Proc. Natl. Acad. Sci. USA **97:** 9082–9087.
42. ALARCON, R., N. DENKO, A. GIACCIA. 2001. Genetic determinants that influence hypoxia-induced apoptosis. Novartis Found. Symp. **240:** 115–128.
43. DENKO, N. *et al.* 2003. Investigating hypoxic tumor physiology through gene expression patterns. Oncogene **22:** 5907–5914.
44. GODA, N. *et al.* 2003. Hypoxia-inducible factor 1alpha is essential for cell cycle arrest during hypoxia. Mol. Cell. Biol. **23:** 359–369.
45. KAELIN, W. & E. MAHER. 1998. The VHL tumor-suppressor gene paradigm. Trends Genet. **14:** 423–426.
46. ST-JACQUES, B., M. HAMMERSCHMIDT & A. MCMAHON. 1999. Indian hedgehog signaling regulates proliferation and differentiation of chondrocytes and is essential for bone formation. Genes Dev. **15:** 2072–2086.

Cellular and Molecular Mechanisms of Synovial Joint and Articular Cartilage Formation

MAURIZIO PACIFICI, EIKI KOYAMA, YOSHIHIRO SHIBUKAWA, CHANGSHAN WU, YOSHIHIRO TAMAMURA, MOTOMI ENOMOTO-IWAMOTO, AND MASAHIRO IWAMOTO

Department of Orthopaedic Surgery, Thomas Jefferson University College of Medicine, Philadelphia, Pennsylvania 19107, USA

ABSTRACT: Synovial joints and articular cartilage play crucial roles in the skeletal function, but relatively little is actually known about their embryonic development. Here we first focused on the interzone, a thin mesenchymal cell layer forming at future joint sites that is widely thought to be critical for joint and articular cartilage development. To determine interzone cell origin and fate, we microinjected the vital fluorescent dye DiI at several peri-joint sites in chick limbs and monitored the behavior and fate of labeled cells over time. Peri-joint mesenchymal cells located immediately adjacent to incipient joints migrated, became part of the interzone, and were eventually found in epiphyseal articular layer and joint capsule. Interzone cells isolated and reared *in vitro* expressed typical phenotypic markers, including GDF-5, Wnt-14, and CD-44, and differentiated into chondrocytes over time. To determine the molecular mechanisms of articular chondrocyte formation, we carried out additional studies on the *ets* transcription factor family member ERG and its alternatively spliced variant C-1-1 that we previously found to be expressed in developing avian articular chondrocytes. We cloned the human counterpart of avian C-1-1 (ERGp55Δ81) and conditionally expressed it in transgenic mice under cartilage-specific Col2 gene promotor-enhancer control. The entire transgenic mouse limb chondrocyte population exhibited an immature articular-like phenotype and a virtual lack of growth plate formation and chondrocyte maturation compared to wild-type littermate. Together, our studies reveal that peri-joint mesenchymal cells take part in interzone and articular layer formation, interzone cells can differentiate into chondrocytes, and acquisition of a permanent articular chondrocyte phenotype is aided and perhaps dictated by *ets* transcription factor ERG.

Address for correspondence: Maurizio Pacifici, Department of Orthopaedic Surgery, Thomas Jefferson University College of Medicine, 501 Curtis Bldg. 1015, Walnut Street, Philadelphia, PA 19107. Voice: 215-955-7352; fax: 215-955-9159.
e-mail: maurizio.pacifici@jefferson.edu

**Ann. N.Y. Acad. Sci. 1068: 74–86 (2006). © 2006 New York Academy of Sciences.
doi: 10.1196/annals.1346.010**

KEYWORDS: synovial joint formation; articular cartilage; progenitor stem cells; transcription factor ERG

INTRODUCTION

Synovial joints are crucial for skeletal function and quality of life. As a result, joint biology remains the subject of extensive research activity and much is known about structure and composition of articular cartilage, ligaments, synovium, and joint capsule and the specific roles each of these tissues plays in joint function in growing and adult organisms.[1] Much is also known about the susceptibility of joint tissues to damage and malfunction during natural aging and congenital or acquired skeletal conditions. Natural aging is characterized by a progressive loss of extracellular matrix and resilience in articular cartilage, resulting in stiffness and reduced biomechanical properties. Osteoarthritis and rheumatoid arthritis are common conditions that affect billions of patients worldwide. These pathologies target and disrupt structure and function in articular cartilage, synovium, or meniscus and result in painful, debilitating, and costly conditions for patients and society.

In comparison, there continues to be only limited information on the processes and mechanisms by which synovial joints actually form during embryogenesis and become established in fetal and early postnatal life.[2–4] Classic studies showed several years ago that the blueprint of the developing skeleton in early embryonic limbs consists at first of an interrupted condensed mesenchymal structure (FIG. 1A).[5–7] The first oversign of joint formation is the appearance of an interzone at each future joint site, be it a phalangeal, carpal, tarsal, or elbow site (FIG. 1B). The interzone is made of mesenchymal cells that are in close proximity to each other and connected by gap junctions.[3] The adjacent condensed mesenchymal cells differentiate into chondrocytes and produce the cartilaginous framework of the skeleton (FIG. 1B). With time, the interzone thickens (FIG. 1C), and the interzone cells are thought to give rise to chondrocytes that are apposed at each epiphyseal end and participate in initial longitudinal lengthening of the long bone anlagen. The cells are also thought to take part in morphogenetic processes that begin to mold the opposing interlocking sides of each joint (FIG. 1D) and to eventually give rise to articular chondrocytes and other components including ligaments (FIG. 1E).

Clearly, joint formation is a complex multistep process, but details about the process remain rather scanty. It would be highly desirable to know precisely how each step is regulated, how the joint formation sites are initially determined and committed, how interzone cells form and whether they indeed give rise to articular chondrocytes and/or other joint tissues, and how each joint acquires its characteristic three-dimensional configuration and structure, be it an elbow or a hip. This information would be of obvious biological interest and would

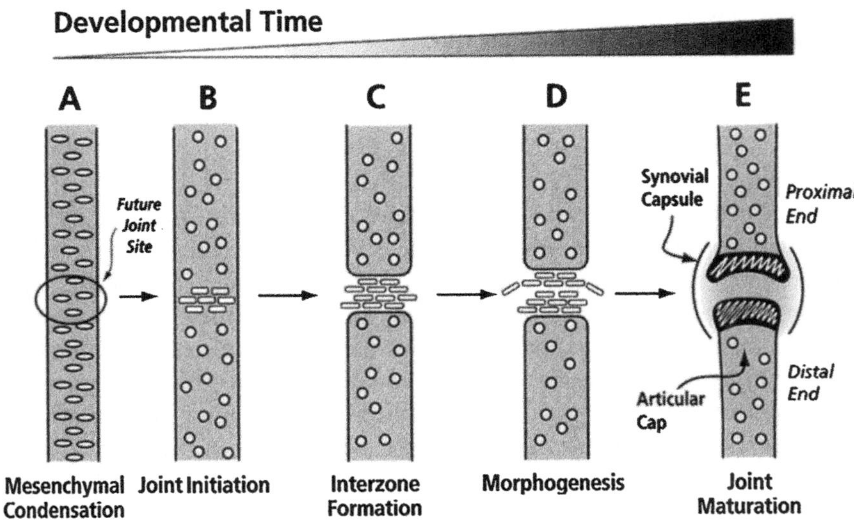

FIGURE 1. Schematic representation of the major steps of synovial joint formation. Mesenchymal pre-chondrogenic condensations appearing in the early limb bud exhibit no oversigns of joint formation and are thus uninterrupted (**A**). Unknown upstream mechanisms possibly involving *Hox* genes determine the exact location for joint initiation, and the first overt morphological sign of joint formation is a further gathering and condensation of cells (**B**). Soon after, the interzone becomes readily recognizable as a thicker and quite compact structure oriented perpendicularly to long axis of the long bone anlagen and composed of mesenchymal cells connected by gap junctions (**C**). The interzone and its adjacent epiphyseal cartilaginous tissue initiate a morphogenetic process that will eventually mold the joint into its required and distinct three-dimensional configuration (**D**). Eventually, all the distinct components of a mature joint form and include articular cartilage and capsule (**E**).

also have major biomedical relevance, value, and implications. The present study was conducted to address some of these questions and issues.

INTERZONE CELLS

Origin and Differentiation Potential

Given that an interzone forms at each future joint site, one could assume that its cells are merely part of the initial uninterrupted mesenchymal condensations; that is, interzone cells would simply be those cells that chance to occupy each future joint site. An alternative possibility is that interzone cells are a distinct population of cells that originate in peri-joint sites and migrate into the nascent joint. To test these possibilities, we made use of a popular approach consisting of microinjection of the fluorescent vital dye DiI that allows monitoring of behavior and fate of labeled cells over developmental time. For

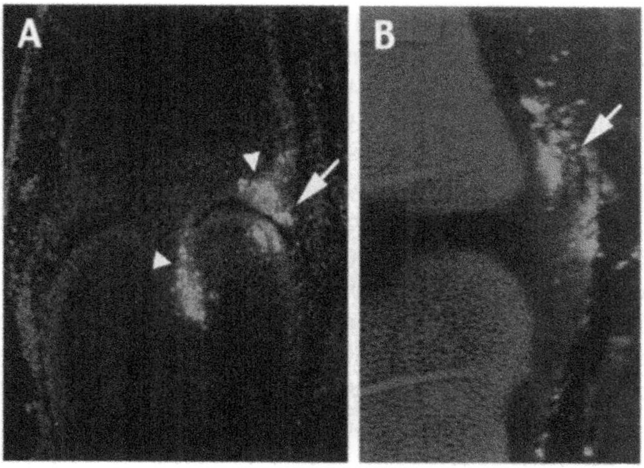

FIGURE 2. Fluorescent images of chick embryo digit joints. DiI was microinjected in the immediate vicinity of incipient metatarsal-phalangeal joint in day 6.5 chick embryos *in ovo* (**A**, arrow) or at a distance of about 0.3 mm (**B**, arrow). Eggs were re-incubated and examined 48 h later. Limbs were sectioned and examined under fluorescence microscopy. Note in (**A**) that labeled peri-joint cells appear to have migrated into the developing joint from the point of injection (arrow) to locations along the joint (arrowheads). Note in (**B**) that labeled cells have migrated from the injection point (arrow) as well, but not into the joint.

instance, this approach has been used to identify and characterize the multiple sites from which progenitor cells originate and give rise to the heart and its distinct portions in the early chick embryo.[8] Accordingly, we microinjected DiI at sites close to incipient joints in the leg buds of day 6.5 chick embryos *in ovo* (one site per leg bud), and eggs were re-incubated and analyzed over time. At each time point parasagittal sections of the limbs (i.e., along the dorsal–ventral axis) were prepared and processed for fluorescence microscopy. DiI-labeled cells had apparently migrated from the site of DiI microinjection (FIG. 2A, arrow) and were present all along the developing epiphyseal joint and interzone by 2 days after injection (FIG. 2A, arrowheads). When we examined companion specimens on day 3 from injection, we found that labeled cells occupied the articulating layer of the long bone anlagen (not shown). To assess specificity, we microinjected DiI at sites what were 0.2–0.7 mm further away from incipient joints in companion embryos (FIG. 2B, arrow). Interestingly, the labeled cells appeared to have migrated in these embryos as well, but had done so up and down the limb mesenchyme and not into the nascent joint (FIG. 2B).

Understanding of interzone cell biology would be greatly aided if the cells could be isolated and studied in tissue culture directly, a rather difficult task given the interzone's small size and cell number. Using our expertise with embryonic tissue microdissection and micromanipulation, we set out to develop methods for interzone isolation and cell culturing. To monitor the dissection process, we first microinjected DiI into the interzones of day 7.5 chick embryo or E14.5 mouse embryo autopods. At these stages, the metatarsal-phalangeal and inter-phalangeal interzones are visible under a dissecting scope by virtue of their differential optical characteristics and are readily distinguishable from adjacent cartilaginous tissue. DiI-labeled interzones were then dissected off the adjacent cartilage. Interzones were trypsinized into single cell suspensions, and cells were plated in monolayer in medium 199 containing fetal bovine serum. During the first 2 days of culture, the cells exhibited a fibroblastic morphology and proliferated actively (FIG. 3A) and expressed a number of characteristic marker genes, including Wnt-14, GDF-5, Gli-3, CD-44, and collagen IIA as revealed by RT-PCR (FIG. 3C, lanes 1–5, respectively). By day 7 of culture,

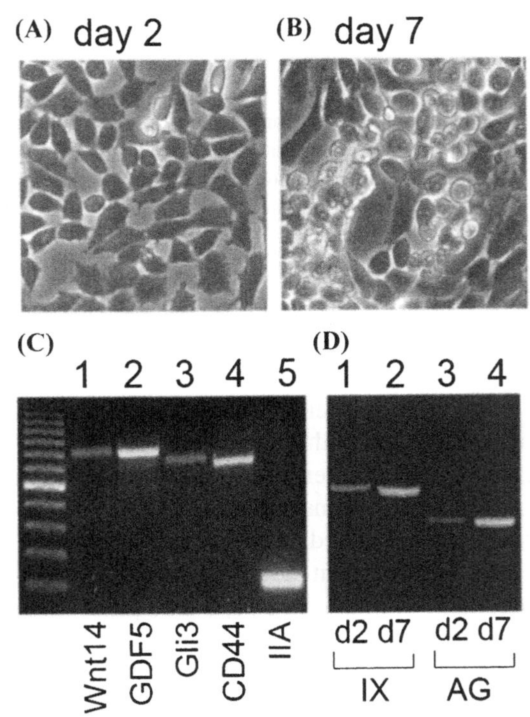

FIGURE 3. Microscopic and RT-PCR analyses of interzone cells in culture. Interzones were microsurgically isolated from day 6.5 chick embryo digits, dissociated into single cell suspensions, and grown in monolayer: (**A**) phase microscopic image of day 2 cells exhibiting a fibroblastic and flat morphology; (**B**) phase microscopic image of day 7 cells exhibiting a round and refractile morphology; (**C**) RT-PCR analysis of genes expressed by freshly-isolated interzone cells, namely Wnt-14 (lane 1), GDF-5 (lane 2), Gli-3 (lane 3), CD-44 (lane 4), and collagen IIA (lane 5); and (**D**) RT-PCR analysis showing that the day 7 cultures express high levels of collagen IX and aggrecan genes (lanes 2 and 4) compared to lower levels in day 2 cultures (lanes 1 and 3).

many of the cells had acquired a round architecture and a highly refractile contour typical of chondrocytes in culture (FIG. 3B) and now expressed high levels of markers of the chondrocyte phenotype including collagen IX and aggrecan (FIG. 3D, lanes 2 and 4) compared to the much lower levels expressed on day 2 (FIG. 3D, lanes 1 and 3).

ARTICULAR CARTILAGE DEVELOPMENT

Ets Transcription Factors

The cell fate maps shown above suggest that interzone cells give rise to the articular layer of long bone anlagen. The data imply that articular chondrocytes have a distinct embryonic origin as compared to the chondrocytes that derive from the mesenchymal condensations and form the bulk of the anlagen (heretofore called transient chondrocytes) (see FIG. 1). There are other substantial differences between articular chondrocytes and transient chondrocytes. Articular chondrocytes maintain joint function throughout life and do so by retaining a stable and permanent phenotype, producing all the macromolecular components of articular cartilage including aggrecan and collagen II, and providing the tissue with functional resilience. Instead, transient chondrocytes display a very dynamic phenotype, undergo proliferation, maturation, hypertrophy, and apoptosis in the growth plates, and are eventually replaced by bone cells during endochondral ossification. Interestingly, even articular chondrocytes can potentially undergo maturation and hypertrophy and display traits of transient growth plate chondrocytes. Such a shift from a permanent to a transient chondrocyte phenotype is often seen in the joints of osteoarthritic patients and can even be provoked in healthy adult articular chondrocytes once isolated from the joint and reared in monolayer culture.[9,10] Traits expressed by such abnormal articular chondrocytes include an enlarged hypertrophic cell size, metalloproteases, and alkaline phosphatase, traits that are incompatible with joint function and lead/cause joint disease. What the above studies and findings imply is that articular chondrocytes must actively safeguard their permanent stable phenotype with strong mechanisms to exert function throughout life and to prevent maturation and hypertrophy and hence loss of cell and tissue function. In other words, as soon as articular chondrocytes form at the epiphyseal ends of long bone anlagen (presumably as the descendants of interzone cells), the cells must acquire mechanisms to establish and retain a permanent phenotype.

In searching for such mechanisms, we previously focused on the transcription factor ERG. This factor belongs to the *ets* family of transcription factors.[11] The current 26 family members are involved in a variety of cellular and developmental processes and when mutated or misexpressed, can cause pathologies.[11–13] *Ets* proteins share a highly conserved 85-amino acid domain (the ETS domain) that binds to the consensus DNA core sequence 5′-GGA(A/T)-3′.[14] Individual *ets* proteins can select specific nucleotides over an 11-bp sequence centered around the core sequence. Additional specificity is provided by interactions with other transcription factors, including members of the bHLH, Pax, bZIP, Id, and nuclear hormone receptor families.[15,16] The biological function of *ets* proteins is also modulated by phosphorylation and by conformational changes consequent to interactions with other factors.[11,17,18] Structural studies have revealed that the ETS domain is a variant of the winged helix-turn-helix motif and displays three α-helices and four β-sheets, with the main protein-DNA contacts provided by residues located along the third α-helix. Though the ETS domain is highly conserved with the family, *ets* proteins differ from each other in other domains (such as absence or presence of the pointed/SAM domain) and are thus distinguished in subfamilies.[11,16]

ERG and Articular Cartilage Development

The *ets* member ERG has been widely studied for its involvement in human cancers following gene translocation to form fusion proteins.[19] ERG is also linked to normal processes, such as mesoderm formation and has been found to form functional complexes with Jun/Fos, with the resulting ternary complexes regulating expression of genes, such as metalloprotease-1 (MMP-1) and MMP-3.[20,21] The seminal studies of Dhordain *et al.* provided the initial evidence that ERG is expressed at sites of future synovial joint formation in chick embryo limbs.[22] Following that lead, we previously carried out additional studies in chick embryos. We found that ERG was not only expressed at the onset of joint formation, but also persisted once the articular layer had developed further. Aware of the fact that *ets* proteins can be alternatively spliced,[23] we carried out additional analyses and cloned a variant missing an 81-bp segment (exon 5) in the central portion of ERG. We named this variant C-1-1 and found that it was preferentially, albeit not exclusively, expressed in most epiphyseal chondrocytes in developing long bones. When we misexpressed C-1-1 within the developing chick limb, we observed that C-1-1 was able to impose a stable, immature, and articular-like phenotype onto the entire limb chondrocyte population, effectively blocking maturation and endochondral ossification.[24,25]

To extend our previous studies, we asked whether mammalian cartilages express ERG variants similar to those in chick. Briefly, 5′ and 3′ RACE procedures were applied using cDNA templates prepared with RNA from human cartilage and brain. Clones were sequenced and were used to design new primers for the amplification and cloning of entire coding sequences. Clones characterized so far are schematically represented in FIGURE 4 and are named according to Duterque–Coquillaud *et al.*[23] : (*a*) a prototypical ERGp38 containing the 81-bp exon and a short N terminus; (*b*) ERG-1 containing the 81-bp exon and a longer N terminus; (*c*) ERGp55 containing both the 81- and 72-bp exons and

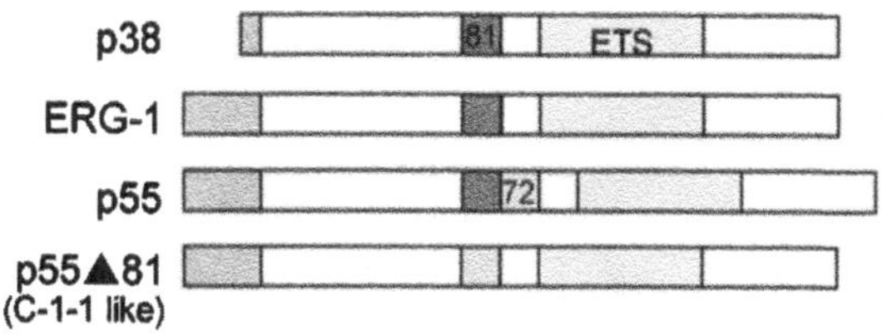

FIGURE 4. Schematic representation of the major human ERG variants. Indicated are the major domains, including the N-terminal region, the alternatively spliced 81- and 72-bp exons in the middle region, the *est* DNA-binding domain (ETS), and the C-terminal region.

corresponding to chick ERG; and (*d*) ERGp55Δ81 that contains the 72-bp exon but lacks the 81-bp exon and corresponds to chick C-1-1. In related experiments, we screened a mouse embryo cDNA library by GeneTrapper (Invitrogen, Carlsbad, CA) and cloned several ERG variants; however, we have not found a mouse C-1-1-like variant so far.[26]

Having identified a human C-1-1 homologue, it became important to ask whether this variant has biological properties similar to those of chick C-1-1. Thus, we prepared a construct encoding human ERGp55Δ81 under the control of mouse Col2a1 promotor/enhancer sequences and used it to produce transgenic mice. The Col2 promotor/enhancer sequences were those that we previously used to misexpress RUNX2 (or a dominant-negative form of it) in transgenic mice.[27] Constructs were microinjected into the pronuclei of fertilized eggs by a centralized facility. Transgenic mice were identified by polymerase chain reaction (PCR) and southern hybridization. The expression of the transgene product was monitored by immunostaining of tissue sections with ERG antibodies that recognize the C-terminal 20 amino acid-long portion of the protein (common to all ERG variants). We obtained 6 mice that overexpressed ERGp55Δ81 in cartilage out of a total of 45 transgene-positive mice. Since ERGp55Δ81-overexpressing mice die at birth due to respiratory failure, we carried out most of our analyses on E18.5 embryos.

The skeletal phenotype of the transgenic (TG) embryos was severe. Anatomical analysis showed that TG embryos were much smaller than wild-type (WT) littermates (FIGS. 5A,B) and their trunk and limb skeletal elements were hypomineralized and stained poorly with alizarin red (not shown). Limbs were sectioned longitudinally and serial sections were processed for histology and *in situ* hybridization. Wild-type E18.5 tibias displayed a typical histology with a prominent epiphyseal cartilaginous cap (where a secondary ossification center will eventually form) and a well-organized growth plate in the metaphysis followed by abundant bone and marrow (FIG. 5C). Instead, the transgenic tibias were completely disorganized (FIG. 5D). Typical growth plates were absent and most of the chondrocytes, regardless of location, had a small and fairly uniform diameter; there was bone-like tissue in the diaphysis, but it appeared fibrotic and poor in marrow (FIG. 5D). Gene expression of aggrecan

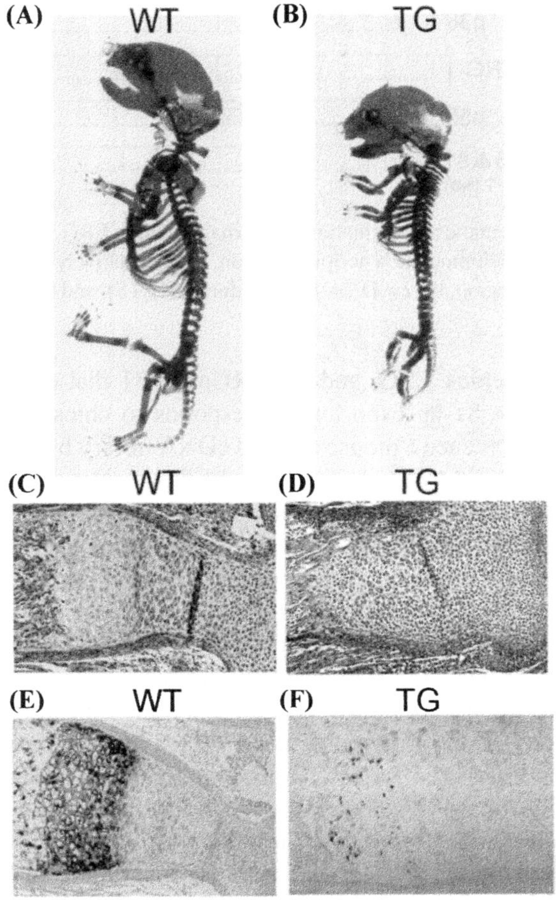

FIGURE 5. Anatomical, histological, and *in situ* hybridization analyses of developing skeleton. Day E18.5 wild-type (WT) mouse embryos and transgenic embryos expressing human ERGp55Δ81 in cartilage (TG) were processed for skeletal anatomical analysis (**A** and **B**), histology (**C** and **D**), and *in situ* hybridization analysis of collagen X gene expression (**E** and **F**). Note in WT that the skeleton and long bone growth plate have normal appearances (**A** and **C**) and normal strong expression of maturation marker collagen X (**E**). In contrast, the transgenic skeleton is much smaller (**B**), the growth plate is completely disorganized and composed of uniform chondrocytes displaying an immature small-sized appearance (**D**), and there is poor expression of collagen X (**F**).

and other cartilage-characteristic matrix genes was equally strong in wild-type and transgenic cartilages, indicating that transgene expression was not causing de-differentiation or having other unwanted side effects (not shown). However, expression of gene markers of maturation, such as collagen X was markedly reduced in transgenic (FIG. 5F) versus wild-type cartilage (FIG. 5E). On the other hand, the epiphyseal articular cartilage marker tenascin-C was widely expressed throughout the transgenic cartilage, but was confined to the epiphyseal cartilaginous ends in wild type as to be expected (not shown).

CONCLUSIONS AND IMPLICATIONS

Synovial joint formation has attracted the interest of developmental biologists for decades, but its mechanisms of regulation have remained poorly understood, a reflection of the sheer complexity and intricacies of the process. The interzone has long been regarded as a critical player in joint formation. Studies by Holder originally revealed that microsurgical removal of incipient interzones in developing limb skeletal elements leads to joint ablation and fusion of contiguous cartilaginous elements.[5] Work by Archer and collaborators showed that interzone cells have special cytological characteristics, particularly gap junctions, which distinguish them from the condensed mesenchyme of neighboring skeletal anlagen.[3] Interest in interzone cells has been reinvigorated by recent studies showing that the cells are characterized by expression of specific genes, including GDF-5, GDF-6, Wnt-14, chordin, and Cux1.[28–32] These genes are thought to be crucial for joint formation, but their exact functions are not fully clear. The DiI data from chick embryos we present here now indicate that mesenchymal cells present at peri-joint sites migrate into the incipient joint and interzone and become part of the epiphyseal articulating layer of long bone anlagen. However, cells located a few hundred microns away appear to be equally mobile and are found scattered up and down the limb mesenchyme, but do not become part of the developing joint. In addition, our studies with cultured interzone cells provide the most direct evidence to date that interzone cells are indeed chondrogenic and can differentiate into chondrocytes. Together, our data suggest that: the cell origin of the interzone is more complex than previously realized; the interzone is constituted in part or entirely of cells that originate from peri-joint sites; and interzone cells may thus be endowed with special and unique characteristics, including the ability to give rise to articular chondrocytes. The fact that only peri-joint cells immediately adjacent to a nascent joint appear to be able to migrate and take part in interzone formation suggests that those cells and the microenvironment in which they reside may both be special. The microenvironment could represent a niche. Thus, a detailed understanding of those cells and their niche could be of great significance and could be used therapeutically to create new cell-based means for joint and articular cartilage reconstruction and repair.

A critical issue in developmental biology is the identification of genes, and master genes in particular, that are critical for establishing the differentiated phenotype of specific cell types. However, it has long remained unclear what genes regulate the genesis of articular chondrocytes and allow the cells to retain a permanent differentiated phenotype throughout life. As pointed out above, such longevity and phenotypic stability are remarkably different from the characteristics of transient growth plate chondrocytes that instead have a limited life span, undergo major changes in phenotype, and eventually die and are replaced by bone cells. The data presented here provide further support for our earlier studies that the transcription factor ERG is part of the molecular mechanisms by which articular chondrocytes form and function.[24,25,33] Both avian and human C-1-1 are able to block the maturation of chondrocytes *in vitro* and *in vivo*, and the ERG overexpressing chondrocytes all display an immature articular-like phenotype characterized by a small cell size, expression of aggrecan and other typical cartilage matrix components, and expression of articular cartilage-specific traits, such as tenascin-C.[34] It is possible then that when chondrocytes first emerge in a developing joint (likely as the progeny of interzone cells), local expression of ERG may direct the cells toward a permanent path and immediately set them apart from neighboring transient chondrocytes in the cartilaginous anlagen that instead become organized in growth plates and undergo endochondral ossification.[26,33] It will be of great interest to establish whether ERG expression is altered in disease states, such as osteoarthritis and whether this may be associated to, or even cause, loss of a permanent phenotype and initiation of a pathological process of maturation and hypertrophy in articular chondrocytes. If so, therapies directed to revamping ERG expression or boosting its function could represent novel means of intervention on joint disease. Ongoing studies should provide answers to these intriguing possibilities.

ACKNOWLEDGMENTS

This study was supported by NIH grants AR46000 and AG25868. We would like to express our gratitude to Dr. William Abrams (University of Pennsylvania) for help with informatics and other colleagues who participated in related work upon which the present study is based.

REFERENCES

1. Archer, C.W. *et al.* 1999. The Biology of the Synovial Joint. Harwood Academics. London.
2. Pacifici, M., M. Liu & E. Koyama. 2002. Joint formation: new findings shed more light on this critical process in skeletogenesis. Curr. Opin. Orthop. **13:** 339–344.
3. Archer, C.W., G.P. Dowthwaite & P. Francis-West. 2003. Development of synovial joints. Birth Defects Res. **69:** 144–155.
4. Lamb, K.J. *et al.* 2003. Defining boundaries during joint cavity formation: going out on a limb. Int. J. Exp. Path. **84:** 55–67.
5. Holder, N. 1977. An experimental investigation into the early development of the chick elbow joint. J. Embryol. Exp. Morphol. **39:** 115–127.

6. MITROVIC, D. 1978. Development of the diathrodial joints in the rat embryo. Am. J. Anat. **151:** 475–485.
7. HINCHLIFFE, J.R. & D.R. JOHNSON. 1980. The Development of the Vertebrate Limb: 72–83. Oxford University Press. New York.
8. HOCHGREB, T. *et al.* 2003. A caudorostral wave of RALDH2 conveys anteroposterior information to the cardiac field. Development **130:** 5363–5374.
9. PFANDER, D., B. SWOBODA & T. KIRSCH. 2001. Expression of early and late differentiation markers (proliferating cell nuclear antigen, syndecan-3, annexin VI and alkaline phosphatase) by human ostearthritic chondrocytes. Am. J. Path. **159:** 1777–1783.
10. PACIFICI, M. *et al.* 1993. Tenascin is associated with articular cartilage development. Dev. Dynam. **198:** 123–134.
11. SHARROCKS, A.D. 2001. The ets-domain transcription factor family. Nature Reviews Mol. Cell Biol. **2:** 827–837.
12. TONDRAVI, M.M. *et al.* 1997. Osteopetrosis in mice lacking haematopoietic transcription factor PU.1. Nature **386:** 81–84.
13. MUTHUSAMY, N., K. BARTON & J.M. LEIDEN. 1995. Defective activation and survival of T cells lacking Ets-1 transcription factor. Nature **377:** 639–642.
14. KARIM, F.D. *et al.* 1990. The ETS-domain: a new DNA-binding motif that recognizes a purine-rich core DNA sequence. Genes Dev. **4:** 1451–1453.
15. VERGER, A. *et al.* 2001. Identification of amino acid residues in the ETS transcription factor Erg that mediate Erg-Jun/Fos-DNA ternary complex formation. J. Biol. Chem. **276:** 17181–17189.
16. VERGER, A. & M. DUTERQUE-COQUILLAUD. 2002. When Ets transcription factors meet their partners. BioEssays **24:** 362–370.
17. YANG, S.H. *et al.* 2001. Temporal recruitment of the mSin3A-histone deacetylase corpressor complex to the ETS domain of transcription factor Elk-1. Mol. Cell. Biol. **21:** 2802–2814.
18. PAPOUTSOPOLOU, S. & R. JANKNECHT. 2000. Phosphorylation of ETS transcription factor ER81 in complex with its coactivators CREB-binding protein and p300. Mol. Cell. Biol. **20:** 7300–7310.
19. SHIMIZU, K. *et al.* 1993. An Ets-related gene, Erg, is rearranged in human myeloid leukemia with t(16;21) chromosomal translocation. Proc. Natl. Acad. Sci. USA **90:** 10280–10284.
20. VLAEMINCK-GUILLEM, V. *et al.* 2000. The Ets member Erg gene is expressed in mesodermal tissues and neural crests at fundamental steps during mouse embryogenesis. Mech. Dev. **91:** 331–335.
21. BUTTICE, G. *et al.* 1996. Erg, an Ets-family member, differentially regulates human collagenase 1 (MMP1) and stromelysin (MMP3) gene expression by physically interacting with the Fos/Jun complex. Oncogene **13:** 2297–2306.
22. DHORDAIN, P. *et al.* 1995. Mesodermal expression of the chicken *erg* gene associated with precartilaginous condensation and cartilage differentiation. Mech. Dev. **50:** 17–28.
23. DUTERQUE-COQUILLAUD, M. *et al.* 1993. New human erg isoforms generated by alternative splicing are transcriptional activators. Oncogene **8:** 1865–1873.
24. IWAMOTO, M. *et al.* 2000. Transcription factor ERG variants and functional diversification of chondrocytes during long bone development. J. Cell Biol. **150:** 27–39.
25. IWAMOTO, M. *et al.* 2001. The role of ERG (ets related gene) in cartilage development. Osteoarthr. Cart. **9:** S41–S47.

26. IWAMOTO, M. *et al.* 2004. Maintenance of limb articular cartilage and negative regulation of endochondral bone formation by ETS-related gene [abstract]. Trans. Orthop. Res. Soc. Meeting **29:** 67.
27. UETA, C. *et al.* 2001. Skeletal malformations caused by overexpression of Cbfa1 or its dominant negative form in chondrocytes. J. Cell Biol. **153:** 87–99.
28. STORM, E.E. & D.M. KINGSLEY. 1996. Joint patterning defects caused by single and double mutations in members of the bone morphogenetic protein (BMP) family. Development **122:** 3969–3979.
29. SETTLE, S.H. *et al.* 2003. Multiple joint and skeletal patterning defects caused by single and double mutations in the mouse *Gdf5* and *Gdf6* genes. Dev. Biol. **254:** 116–130.
30. HARTMANN, C. & C.J. TABIN. 2001. Wnt-14 plays a pivotal role in inducing synovial joint formation in the developing appendicular skeleton. Cell **104:** 341–351.
31. GUO, X. *et al.* 2004. Wnt/β-catenin signaling is sufficient and necessary for synovial joint formation. Genes Dev. **18:** 2404–2417.
32. LINZARRAGA, G. *et al.* 2002. Studies on the role of Cux1 in regulation of the onset of joint formation in the developing limb. Dev. Biol. **243:** 44–54.
33. IWAMOTO, M. *et al.* 2005. The balancing act of transcription factors C-1-1 and Runx2 in articular cartilage development. Biochem. Biophys. Res. Commun. **328:** 777–782.
34. PACIFICI, M. 1995. Tenascin-C and the development of articular cartilage. Matrix Biol. **14:** 689–698.

Multiple Roles of *Hoxc8* in Skeletal Development

ASTER H. JUAN,[a] HAIYAN LEI,[a] PRERNA BHARGAVA,[b] MARIELLE LEBRUN,[c] AND FRANK H. RUDDLE[a]

[a]*Department of Molecular, Cellular and Developmental Biology, Yale University New Haven, Connecticut 06511, USA*

[b]*Department of Biology, Brandeis University, Waltham, Massachusetts 02454, USA*

[c]*Laboratory of Virology and Immunology, Institute of Pathology, University of Liège, B-4000 Liège, Belgium*

ABSTRACT: We are interested in investigating the function of *Hoxc8* in skeletogenesis during mouse development. Previous studies have shown that deregulation of *Hoxc8* expression in the mouse leads to several skeletal defects, such as homeotic transformation in the thoracic vertebrae, abnormal development of the rib cage, and overproliferation of chondrocytes in the hypertrophic area. By deleting a crucial enhancer of *Hoxc8 in vivo*, we found that precise temporal expression of *Hoxc8* is important for determining the correct identity of the vertebral column in early embryos. We also identified downstream targets of *Hoxc8* relevant to osteoblast differentiation at later developmental stages.

KEYWORDS: *Hoxc8*; axial skeleton; osteoblast; Wnt; BMP

INTRODUCTION

In early embryos, *Hoxc8* is expressed in the presomitic mesoderm and paraxial mesoderm, which give rise to axial skeleton. The precise spatial and temporal expression of *Hoxc8* is important for determining the correct identity of the vertebral column.[1–4] In postnatal skeletal development, it has been shown that *Hoxc8* is involved in regulating osteogenesis via bone morphogenic protein (BMP) pathways.[5] To define the role of *Hoxc8* in skeletal morphogenesis more precisely, we focused on the transcriptional regulation of *Hoxc8* in mouse embryos by deleting *in vivo* a crucial enhancer that regulates the transcription of *Hoxc8* in early development. We also identified downstream targets of *Hoxc8* relevant to osteoblast differentiation.

Address for correspondence: Aster Juan, 50 South Drive, Room 1148, NIH, Bethesda, MD 20892. Voice: 301-594-9617; fax: 301-402-0009.
e-mail: juana2@mail.nih.gov

Ann. N.Y. Acad. Sci. 1068: 87–94 (2006). © 2006 New York Academy of Sciences.
doi: 10.1196/annals.1346.046

MATERIAL AND METHODS

We have deleted *Hoxc8* early enhancer (EE) *in vivo* using transgenic and knockout technologies. A 30-kb genomic region of the Hoxc cluster including the intact *Hoxc8* gene and flanked by truncated coding segments of *Hoxc9* and *Hoxc6* genes was used for transgenic study. A LacZ cassette was inserted, in frame, into the second exon of *Hoxc8* to generate a reporter gene. EE was then deleted from this transgene by homologous recombination in yeast. Mouse lines carrying EE-deleted transgene were generated and the expression pattern of LacZ at different embryonic stages was analyzed. For the knockout study, the endogenous EE was deleted using embryonic stem cell technology. The expression pattern of endogenous *Hoxc8* and skeletal morphology were compared to knockout mice. To identify the downstream targets of *Hoxc8*, we overexpressed *Hoxc8* in the mouse embryonic fibroblast (MEF) cells and applied microarray and chromatin immunoprecipitation (ChIP) assays to determine the potential target genes of *Hoxc8*.

RESULTS

Previous studies showed that EE is necessary and sufficient to reconstitute the endogenous pattern of *Hoxc8* expression in mesodermal derivatives at early

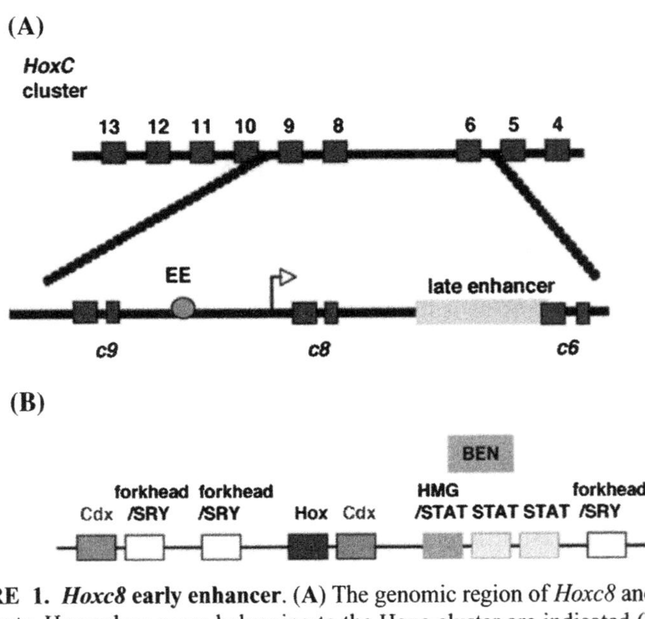

FIGURE 1. ***Hoxc8* early enhancer.** (**A**) The genomic region of *Hoxc8* and its regulatory elements. Homeobox genes belonging to the Hoxc cluster are indicated (boxes). The position of the EE (oval) and *Hoxc8* late enhancer (rectangular) are shown. Arrow represents transcription start site of *Hoxc8*. (**B**) Putative transcription factor-binding sites within *Hoxc8* early enhancer.

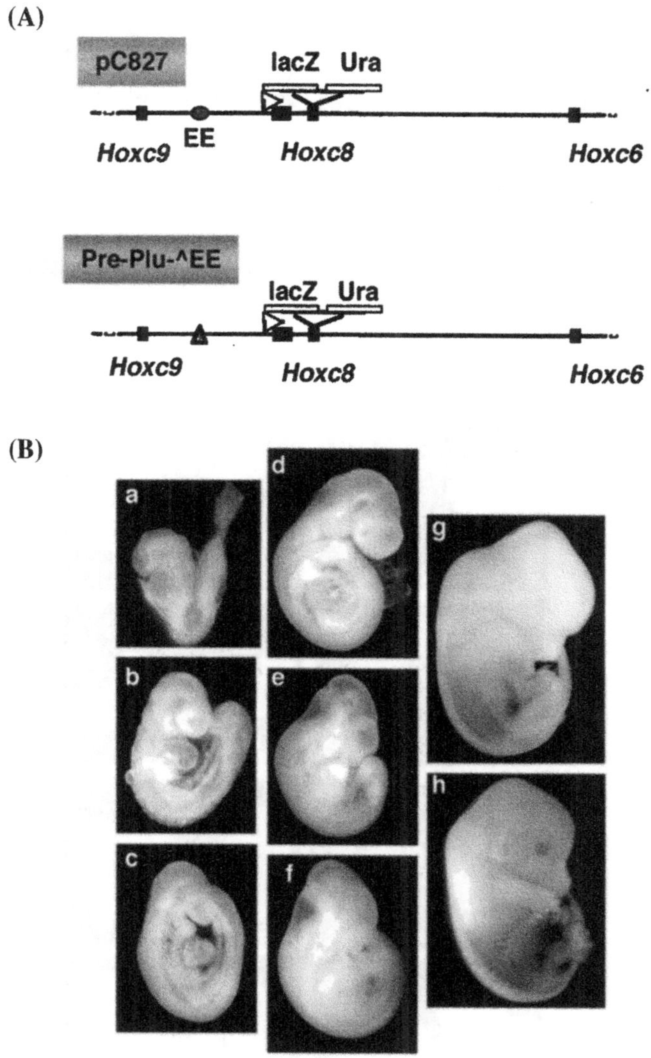

FIGURE 2. Transgenic analysis of EE using large genomic constructs. (**A**) Schematic representation of pC827, the 30-kb transgene including whole *Hoxc8* intergenic region. LacZ-Ura cassette is inserted in the second exon of *Hoxc8*. The oval represents the location of the EE. In the second transgene, Pre-Plu-^EE, the EE is deleted from pC827. (**B**) Reporter gene analysis of Pre-Plu-^EE in transgenic embryos. No LacZ expression is detected in 8.5 dpc (a) and 9 dpc (b) embryos. LacZ expression is only detected in the neural tube but not in somites at 9.5 dpc (c), 10 dpc (d), 10.5 dpc (e), and 11.5 dpc (f) embryos. In 12.5 dpc (g) and 13.5 dpc (h) embryos, LacZ expression is resumed in the neural tube, somites, and developing ribs.

embryonic stages.[3] It is located 3 kb 5' to the *Hoxc8* promotor and is highly conserved between species (FIG. 1A). To investigate the regulatory mechanism of *Hoxc8* in skeletal development, we first deleted EE from a 30-kb transgene that contains full transcriptional regulatory units of *Hoxc8* (FIG. 2A). When

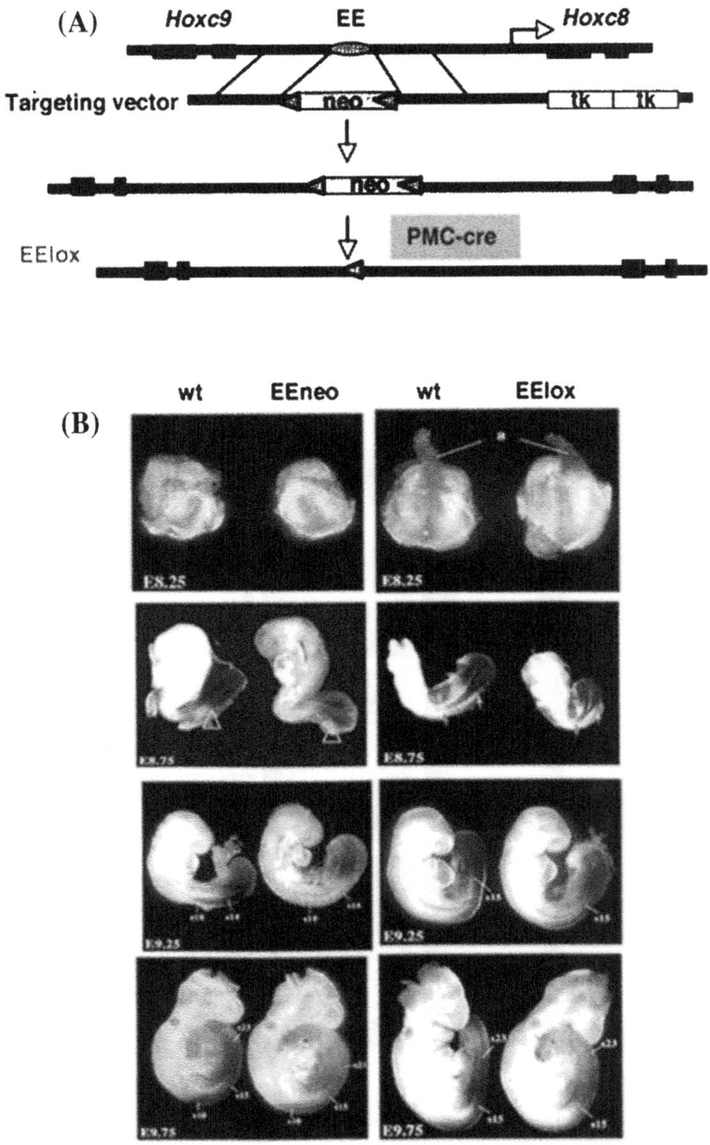

FIGURE 3. ***In vivo* inactivation of *Hoxc8* early enhancer**. (**A**) Schematic representation of the EE (oval), targeted vector, disrupted EE locus (EEneo), and deleted EE locus (EElox). (**B**) *Hoxc8* expression in the wild-type and EE knockout embryos from 8.25 to 9.75 dpc by whole mount *in situ* hybridization. At 8.25 dpc, *Hoxc8* expression is not detected or delayed in the allantois (a) in EEneo or EElox embryo, respectively. At 8.75 and 9.25 dpc, EEneo embryo shows delay of *Hoxc8* expression in both neural tube and presomitic mesoderm (arrows), while EElox embryo shows comparable expression pattern of the wild-type embryo. From 9.75 dpc and beyond, EEneo and EElox embryos establish normal anterior boundaries of *Hoxc8* expression domains in the somites and neural tube.

(A)

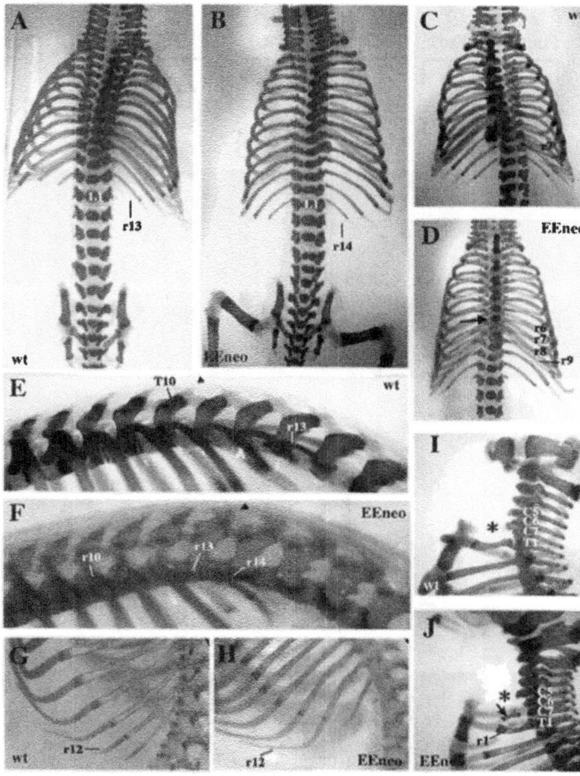

(B)

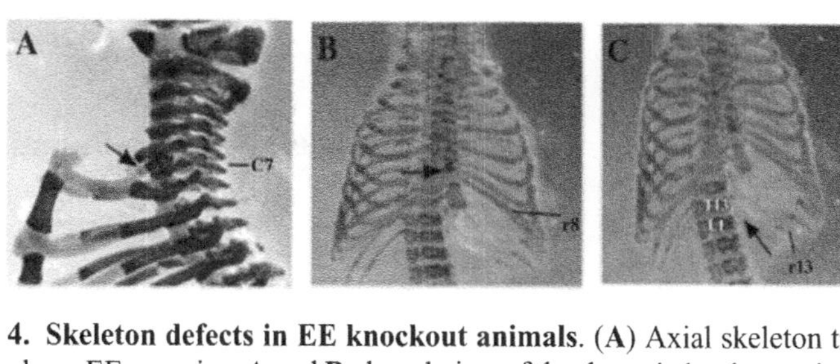

FIGURE 4. Skeleton defects in EE knockout animals. (**A**) Axial skeleton transformations in newborn EEneo mice. **A** and **B**, dorsal view of the thoracic lumbar region shows the first lumbar vertebra transformed to the 13th thoracic vertebra. **C** and **D**, ventral view of the thorax shows anterior transformations in the 7th, 8th, and 9th thoracic vertebrae. Arrow indicates an extra sternebra developed between the 6th and 7th ribs. **E** and **F** show the 11th and 12th vertebrae are transformed anteriorly to the identity of the 10th vertebra. Arrowheads indicate the position of the transitional vertebra. **G** and **H** show anterior transformation in the 12th rib. **I** and **J**, lateral view of the cervico-thoracic transition shows posterior transformations in the 5th, 6th , and 7th cervical vertebrae. Asterisk indicates the position of the anterior tuberculum. Arrow indicates an ectopic rib developed in the 7th cervical vertebra. (**B**) Skeletal phenotypes in newborn EElox mice. **A** shows posterior transformation in the 7th cervical vertebra. **B** shows anterior transformations in the 7th and 8th vertebrae. **C** shows anterior transformation in the 1st lumbar vertebra.

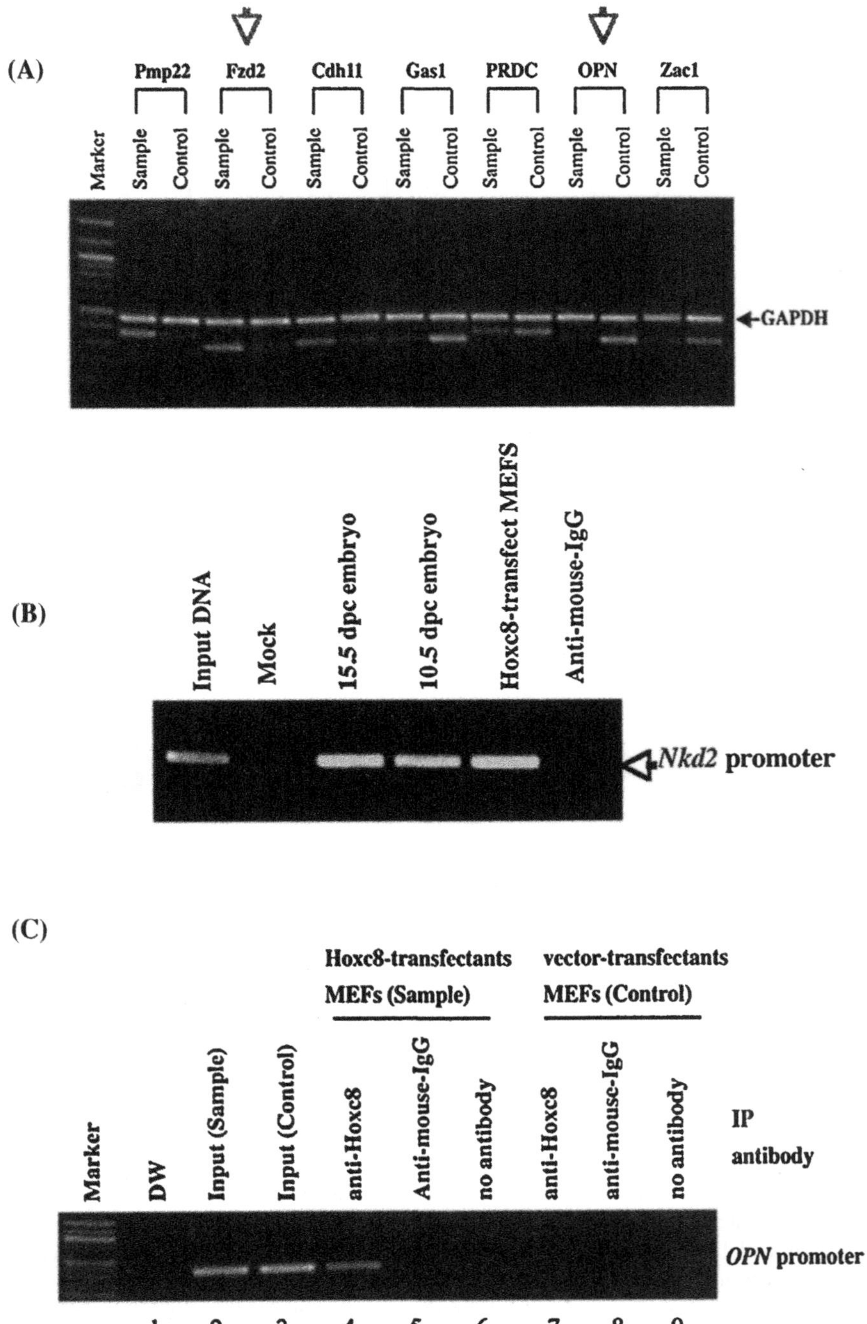

FIGURE 5. Downstream targets of *Hoxc8*. (**A**) Semiquantitative RT-PCR analysis of gene expression of *Pmp22, Fzd2, Cdh11, Gas1, PRDC, and Zac1* in *Hoxc8* transfectant cells (sample) and empty-vector transfectant cells (control). Arrows indicate *Fzd2* is upregulated

EE was deleted from this transgene, the expression of LacZ was abolished in the presomitic and paraxial mesoderm (FIG. 2B; A.J., unpublished data). This suggests EE is important for regulating *Hoxc8* expression in mesodermal derivatives. Indeed, endogenously deleted EE results in significant delay in the temporal expression of *Hoxc8* in presomitic mesoderm in early mouse embryos (FIG. 3).[4] This temporal delay is sufficient in reproducing many of the axial skeletal defects associated with the *Hoxc8*-null mutation (FIG. 4).[4] Sequence and transgenic analysis of EE revealed a minimum of seven putative transcription factor-binding motifs including CDX- and Hox-binding sites (FIG. 1B).[3] Furthermore, microarray assay identified 34 candidate genes as *Hoxc8* downstream targets.[6] ChIP assay confirmed that genes involved in osteogenesis, such as OPN, frizzed homolog 2 (Fzd2), and Nkd2 are direct transcriptional targets of *Hoxc8 in vivo* (FIG. 5; H.L., unpublished data).[6]

CONCLUSION

Our studies have discovered that strict control of *Hoxc8* expression timing in embryonic mesodermal lineages is critical for the establishment of correct skeleton identities in early mouse development. EE is important for precise regulation of *Hoxc8* temporal expression, possibly via several transacting factors, such as CDX and Hox family members. In later developmental stages, *Hoxc8* regulates osteogenesis via BMP and Wingless type (Wnt)-mediated signaling pathways. Overall, our data suggest that *Hoxc8* plays a pivotal but divergent role in skeletogenesis in different developmental stages. In early embryonic stages, *Hoxc8* is important for laying down correct patterning of the axial skeleton along the rostrocaudal axis. In late embryonic and postnatal stages, *Hoxc8* is involved in osteoblast differentiation by regulating genes that are crucial players in BMP and Wnt-signaling pathways.

REFERENCES

1. VAN DEN AKKER, E. *et al.* 2001. Axial skeletal patterning in mice lacking all paralogous group 8 Hox genes. Development **128:** 1911–1921.

and *OPN* is downregulated in *Hoxc8* overexpressing cells. (**B**) ChIP assay shows *Hoxc8* binds to *Nkd2* regulatory region in *Hoxc8* overexpressing cells and 10.5 and 15.5 dpc embryos. Input chromatin represents a portion of the sonicated chromatin before immunoprecipitation. Mock sample contains only dialysis buffer instead of chromatin and immunoprecipitate with anti-mouse IgG are used for controls. (**C**) ChIP assay shows *Hoxc8* binds to the promotor region of *OPN* in *Hoxc8* overexpressing cells. Immunoprecipitate with anti-mouse IgG or in the absence of antibody are used for controls. DW is a no-template control.

2. YUEH, Y.G. *et al.* 1998. Evidence for regulation of cartilage differentiation by the homeobox gene Hoxc-8. Proc. Natl. Acad. Sci. USA **95:** 9956–9961.
3. BELTING, H.G. *et al.* 1998. Multiple phases of expression and regulation of mouse *Hoxc8* during early embryogenesis. J. Exp. Zool. **282:** 196–222.
4. JUAN, A.H. & F.H. RUDDLE. 2003. Enhancer timing of Hox gene expression: deletion of the endogenous *Hoxc8* early enhancer. Development **130:** 4823–4834.
5. WAN, M. *et al.* 2001. Transcriptional mechanisms of bone morphogenetic protein-induced osteoprotegrin gene expression. J. Biol. Chem. **276:** 10119–10125.
6. LEI, H. *et al.* 2005. The identification of *Hoxc8* target genes. Proc. Natl. Acad. Sci. USA **102:** 2420–2424.

Osteoclasts and Integrins

STEVEN L. TEITELBAUM

Department of Pathology, Washington University School of Medicine, St. Louis, Missouri 63110, USA

ABSTRACT: The osteoclast is the unique bone resorptive cell that accomplishes its mission by forming an isolated acidified microenvironment between itself and the bone surface. Creation of this compartment is the first step in bone degradation and establishes that an intimate physical relationship must exist between the osteoclast and bone. Thus, identification of the mechanisms by which the osteoclast attaches to bone is essential to understanding how the cell degrades skeletal tissue. Our studies have investigated whether absence of the αvβ3 integrin modifies the ability of c-Fms to induce Rho GTPases, and the implications for formation of the osteoclast cytoskeleton.

KEYWORDS: osteoclast; integrin; bone mass; bone resorption, Rho GTPase; Rac; Vav

The osteoclast is the unique bone resorptive cell that accomplishes its mission by forming an isolated acidified microenvironment between itself and the bone surface. Creation of this compartment is the first step in bone degradation and establishes that an intimate physical relationship must exist between the osteoclast and bone.[1] Thus, identification of the mechanisms by which the osteoclast attaches to bone is essential to understanding how the cell degrades skeletal tissue.

Integrins are transmembrane αβ heterodimers that mediate cell/cell and cell/matrix recognition.[2] In the case of integrins mediating matrix attachment, their extracellular domains are liganded by extracellular proteins, while their intracellular tails interact with signaling molecules and the cytoskeleton.

A series of *in vitro* experiments established that αvβ3 is the principal osteoclast integrin. αvβ3, like all members of the αv family of integrins, recognizes the arginine-glycine-aspartic acid (RGD) motif. This ligand is present in a number of bone-residing proteins, such as osteopontin and bone sialoprotein. To determine if the αvβ3 integrin, identified *in vitro*, has physiological significance, we deleted the β3 integrin gene in mice.[3] The net result of this

Address for correspondence: Steven L. Teitelbaum, Department of Pathology, Washington University School of Medicine, St. Louis, MO 63110. Voice: 314-454-8463; fax: 314-454-5505.
e-mail: teitelbs@wustl.edu

Ann. N.Y. Acad. Sci. 1068: 95–99 (2006).
doi: 10.1196/annals.1346.017

exercise is an age-dependent increase in bone mass, consistent with osteoclast dysfunction.

Because increased skeletal mass may reflect either enhanced bone formation or arrested osteoclast function, we turned our attention to the osteoclast in these mutant animals. Thus, we generated osteoclasts from wild-type, heterozygous, and β3 integrin-deficient bone marrow macrophages. Osteoclasts derived from wild-type and heterozygous mice appear identical as they are well spread, multi-nucleated, and express the osteoclast characteristic enzyme, TRAP.[3] On the other hand, osteoclasts generated from bone marrow macrophages of β3 integrin-deficient mice, have a crenated appearance indicative of a cytoskeletal abnormality. The appearance of the actin cytoskeleton of bona fide osteoclasts, isolated directly from bones of both β3 knockout and heterozygous animals, confirms this hypothesis. Specifically, osteoclastic bone resorption requires the formation of an actin ring-like structure, known as the "sealing zone," which isolates the resorptive microenvironment from the general extracellular space.[4] While such ring-like structures are apparent in osteoclasts isolated from heterozygous mice, β3 integrin-deficient osteoclasts fail to form actin rings.[3] These cytoskeletal abnormalities are functionally manifest by the inability of β3 integrin–/– osteoclasts to excavate normal resorptive lacunae when placed on dentin slices.

Cytoskeletal organization, in general, is mediated by the Rho family of GTPases, which include RhoA, Rac, and CDC42.[5] We, therefore, asked if the cytoskeletal abnormalities present in β3-deficient osteoclasts reflect insufficient Rho GTPase activation. Because Rho GTPases, in osteoclasts, are activated by the occupancy of the macrophage colony-stimulating factor (M-CSF) receptor, c-Fms, we addressed this issue in the context of c-Fms activation.[6] Specifically, we determined if the absence of the αvβ3 integrin modifies the ability of c-Fms to induce Rho GTPases. We find that in wild-type pre-osteoclasts, 15 minutes exposure to M-CSF leads to substantial activation of RhoA by transiting it from its GDP to its GTP bound form. In contrast, there is no evident activation of RhoA in the absence of the β3 integrin. Similarly, Rac is activated within 5 minutes of exposure to M-CSF in authentic pre-osteoclasts but not in those lacking αvβ3.

Activation of Rho GTPases such as Rac, is under the influence of guanine nucleotide exchange factors (GEFs). Specifically, GEFs transit small GTPase from their GDP to the GTP bound form eventuating in cytoskeletal organization. Because the Vav family of GEFs activates Rac, we asked if Vavs play a role in αvβ3 mediated stimulation of this Rho GTPase.[7]

Vav is known to exist in three isoforms, namely Vav1, Vav2, and Vav3. In fact, Vav1 is the dominant isoform in T-lymphocytes. On the other hand, Vav3, which is virtually undetectable in T cells, predominates in osteoclasts and is expressed, constitutively, in marrow macrophages as they differentiate into these bone-resorbing polykaryons under the influence of RANK ligand and

M-CSF. In contrast, Vav1 is present in moderate amounts and Vav2, in only small quantities, in osteoclasts.

The abundance of Vav3 in the bone-resorbing cells raises the possibility that this GEF is activated by αvβ3 occupancy. To address this issue, we placed wild-type pre-osteoclasts in suspension or on osteopontin. Tyrosine phosphorylation of Vav3 is markedly enhanced in cells on the αvβ3 ligand indicating that activation of the integrin leads to activation of Vav3. In keeping with this observation, Vav3 knockout osteoclasts fail to form normal sealing zones and similar to those lacking the β3 integrin, fibrillar actin is distributed diffusely throughout the cell. This is also manifest by the inability of the cells to spread effectively when derived from mice lacking Vav3 or both Vav1 and Vav3. Furthermore, whereas resorption lacunae generated by Vav1-deficient osteoclasts are relatively normal, those cells derived from Vav3 and Vav1,3 null mice are incapable of degrading mineralized matrix, *in vitro*. Extending these observations to the *in vivo* state, radiographic analysis documents a substantial increase in bone density in the Vav3–/– and Vav1,3–/– animals. This observation is in keeping with histomorphometric analysis, which reveals that the absence of Vav3 and/or Vav1 and Vav3 eventuates in a 2.5- to 3-fold increase in trabecular bone mass. Furthermore, whereas parathyroid hormone (PTH) administration to wild-type mice leads to an approximate doubling of serum pyridinoline cross-link content, a global indicator of bone resorption, the absence of Vav3 and Vav1 and Vav 3, completely blunts PTH-stimulated skeletal degradation.

The data presented, thus far, document that osteoclastic bone resorption is deficient in the absence of Vav3. On the other hand, they do not establish whether this defect in osteoclast function reflects a primary osteoclast abnormality or is a manifestation of dysfunctional osteoblasts producing insufficient quantities of bone resorptive cytokines such as RANK ligand (RANKL). To address this issue, we asked if Vav3 osteoclasts are capable of responding to the osteoclastogenic cytokine, with the rationale that a primary defect in these cells would result in their failure to respond to osteoblast-produced RANKL. The significance of this question is buttressed by our finding that osteoblasts, like osteoclasts, express abundant Vav3 and Vav1. The appearance of the RANKL-induced bone loss in Vav3-deficient mice, despite stimulated osteoclastogenesis, establishes that their subnormal bone resorption reflects a primary defect in the osteoclast, and not osteoblast, function. In keeping with this posture, bone formation rates remain unchanged in the absence of Vav3.

Having established that the αvβ3 integrin activates Vav3 and Vav3-deficient osteoclasts are defective, prompted us to determine if Vav3 modulates the osteoclast cytoskeleton by impacting integrin-induced Rho GTPases. Whereas M-CSF activates Rac in a Vav3-dependent manner, such is not the case regarding RhoA. Therefore, both the αvβ3 integrin and M-CSF occupancy of c-Fms partner to phosphorylate Vav3, which in turn activates Rac, leading to organization in the osteoclast cytoskeleton.

The c-Src-deficient mouse demonstrates the surprising phenotype in that osteopetrosis, due to dysfunctional osteoclasts, is its principal abnormality.[8] Indeed, c-Src-deficient osteoclasts appear reminiscent of those lacking the β3 integrin as cytoskeletal abnormalities are extant in both.[3,9] This observation suggests that c-Src and αvβ3 enjoy a common signaling pathway. In fact, c-Src associates with αvβ3 in the osteoclast and does so in the context of the tyrosine kinase, Syk. In this regard, Syk–/– osteoclasts also demonstrate the phenotype of cytoskeletal disorganization common to both c-Src and αvβ3 null cells.[10] Thus, Syk-deficient osteoclasts also fail to form actin rings.

Because Syk associates with a β3 integrin in the context of the platelet,[11] we asked if the same obtains in the osteoclast. Syk does recognize the β3 integrin in osteoclasts and does so independently of c-Src. Once Syk and Src bind to the β3 integrin cytoplasmic domain, c-Src phosphorylates, and thus activates Syk, which, in turn, induces Vav3 and ultimately reorganizes the actin cytoskeleton. Finally, the common signaling pathway of Syk and Vav3 is documented by the crossing of haplo-insufficient mice.[7] Thus, Syk/Vav3 compound heterozygous animals have a trabecular bone mass, which is greater than that of the Vav3 knockout mouse, and their osteoclasts fail to normally resorb dentin *in vitro*.

Because the αvβ3 integrin is an essential component of optimal osteoclast function, it presents as a candidate anti-resorptive therapeutic target for treatment of osteopenic diseases such as postmenopausal osteoporosis.[12] In fact, small molecule inhibitors of the integrin are bone-sparing in oophorectomized rats. Most importantly, the same holds in osteoporotic women as such compounds are presently in clinical trials.[13]

REFERENCES

1. Teitelbaum, S.L. 2000. Bone resorption by osteoclasts. Science **289:** 1504–1508.
2. Hynes, R.O. 1992. Integrins: versatility, modulation, and signaling in cell adhesion. Cell **69:** 11–25.
3. McHugh, K.P. *et al.* 2001. Cloning and characterization of the murine β3 integrin gene promoter: identification of an interleukin-4 responsive element and regulation by Stat-6. J. Cell. Biochem. **81:** 320–325.
4. Vaananen, H.K. & M. Horton. 1995. The osteoclast clear zone is a specialized cell-extracellular matrix adhesion structure. J. Cell Sci. **108:** 2729–2732.
5. Machesky, L.M. & A. Hall. 1997. Role of actin polymerization and adhesion to extracellular matrix in Rac- and Rho-induced cytoskeletal reorganization. J. Cell Biol. **138:** 913–926.
6. Faccio, R. *et al.* 2003. Dynamic changes in the osteoclast cytoskeleton in response to growth factors and cell attachment are controlled by β3 integrin. J. Cell Biol. **162:** 499–509.
7. Faccio, R. *et al.* 2005. Vav3 regulates osteoclast function and bone mass. Nat. Med. **11:** 284–290.
8. Soriano, P. *et al.* 1991. Targeted disruption of the c-src proto-oncogene leads to osteopetrosis in mice. Cell **64:** 693–702.

9. BOYCE, B.F. *et al.* 1992. Requirement of pp60c-src expression for osteoclasts to form ruffled borders and resorb bone in mice. J. Clin. Invest. **90:** 1622–1627.
10. FACCIO, R. *et al.* 2003. High dose M-CSF partially rescues the Dap12−/− osteoclast phenotype. J. Cell Biochem. **90:** 871–883.
11. WOODSIDE, D.G. *et al.* 2002. The N-terminal SH2 domains of Syk and ZAP-70 mediate phosphotyrosine-independent binding to integrin beta cytoplasmic domains. J. Biol. Chem. **277:** 39401–39408.
12. ENGLEMAN, V.W. *et al.* 1997. A peptidomimetic antagonist of the αvβ3 integrin inhibits bone resorption *in vitro* and prevents osteoporosis *in vivo*. J. Clin. Invest. **99:** 2284–2292.
13. MURPHY, M.G. *et al.* 2005. Effect of L-000845704, an αVβ3 integrin antagonist, on markers of bone turnover and bone mineral density in postmenopausal osteoporotic women. J. Clin. Endocrinol. Metab. **90:** 2022–2028.

Regulation of Osteoclast Differentiation

G. DAVID ROODMAN

University of Pittsburgh, School of Medicine/Hematology-Oncology, and VA Pittsburgh Healthcare System, Medicine/Hematology-Oncology, Pittsburgh, Pennsylvania 15240, USA

ABSTRACT: The osteoclast (OCL) is derived from the cells in monocyte-macrophage lineage. The earliest identifiable OCL precursor is the granulocyte-macrophage colony-forming unit (CFU-GM), which gives rise to granulocytes, monocytes, and OCL. CFU-GM–derived cells then differentiate to committed OCL precursors, which are post-mitotic cells, and fuse to form multinucleated OCL. A variety of factors both positively and negatively regulate OCL formation and activity. These include growth factors, such as macrophage colony-simulating factor, which simulates the proliferation and prevents apoptosis of early OCL precursors, and RANK ligand (RANKL), which is the primary mediator of OCL formation. Most factors that induce OCL differentiation, such as PTHrP, IL-11, and prostaglandins, do so by inducing expression of RANKL on the surface of immature osteoblasts. Osteoprotegerin is a decoy receptor that blocks RANKL activity. In addition, OCL produce autocrine-paracrine factors that regulate OCL formation, such as IL-6, which is produced at high levels by OCL in Paget's disease and increases OCL formation. We screened human and murine OCL cDNA libraries to identify autocrine-paracrine factors that regulate OCL activity. We identified annexin-II, MIP-1α, ADAM8, eosinophil chemotactic factor, and OCL inhibitor factors 1 and 2 as factors involved in OCL formation. Most recently, we have identified the receptor for ADAM8, $\alpha_9\beta_1$ integrin, which appears to be critical for normal OCL activity. OCL differentiation is controlled by exogenous hormones and cytokines as well as autocrine-paracrine factors that positively or negatively regulate OCL proliferation and differentiation.

KEYWORDS: **osteoclasts; differentiation; autocrine; paracrine; cytokines**

INTRODUCTION

The osteoclast (OCL) is the primary bone-resorbing cell. It is formed from mononuclear precursors, which fuse to form multinucleated OCL.[1] Increases

Address for correspondence: G. David Roodman, M.D., Ph.D., University of Pittsburgh, School of Medicine/Hematology-Oncology, VA Pittsburgh Healthcare System, R&D (151-U), Room 2E-113, University Drive C, Pittsburgh, PA 15240. Voice: 412-688-6571; fax: 412-688-6960.
e-mail: roodmangd@upmc.edu

Ann. N.Y. Acad. Sci. 1068: 100–109 (2006).
doi: 10.1196/annals.1346.013

in the degree of multinucleation enhance the capacity of OCL to resorb bone[2] so that OCL, which contains greater numbers of nuclei, has an increased bone-resorbing capacity compared to OCL with fewer nuclei. OCL differentiation and activity are regulated by both systemic hormones and cytokines produced locally in the bone microenvironment. In addition, other cells in the marrow microenvironment can influence OCL formation and activity. These cells include T and B lymphocytes, marrow stromal cells, osteoblasts, and osteocytes. All of these cell types produce cytokines and chemokines that stimulate or inhibit OCL formation and activity. Studies by Walker and colleagues[3,4] clearly showed that the OCL was hematopoietic in origin. Using parabiotic experiments with mice, Walker and co-workers demonstrated that the OCL precursor was blood borne and could cure osteopetrosis in the parabiotic littermate. Further, transplantation of normal marrow, peripheral blood, or spleen cells into osteopetrotic animals have clearly shown that OCL precursors reside in hematopoietic organs. Previously, the tissue origin of the OCL was in dispute and was thought to be mesenchymal[5] but now it is clear that the OCL is hematopoietic in origin.

LINEAGE OF THE OCL

The OCL is derived from the pluripotent hematopoietic stem cell. Kurihara and co-workers[6] have shown that the multipotent hematopoietic precursor, CFU-blast, can form OCL when cultured in the appropriate cytokine milieu *in vitro*. More recently, Miyamoto and co-workers have clearly identified the different stages of OCL differentiation and the surface phenotype of these cells.[7] The pluripotent hematopoietic stem cell gives rise to a myeloid stem cell, which can further differentiate to megakaryocytes, granulocytes, monocyte-macrophages, and OCL. The earliest identifiable hematopoietic precursor that can form OCL is the granulocyte-macrophage colony-forming cell (CFU-GM).[8] A variety of studies have confirmed that CFU-GM can form OCL and that CFU-M, the more differentiated monocyte precursor, forms OCL at a much lower efficiency. Kerby and co-workers have shown, using single-cell manipulation studies with the progeny of hematopoietic stem cells, that CFU-GM rather than CFU-M form greater numbers of OCL.[9] Early OCL precursors are proliferative cells, which can increase in numbers in response to hematopoietic growth factors such as IL-3, GM-colony-stimulating factor (CSF), and M-CSF.[10] M-CSF prevents apoptosis of these precursors as well.[11] The critical role of M-CSF in OCL differentiation has been demonstrated in rodent models in which mutations in the M-CSF gene result in severe osteopetrosis.[12] The op/op mouse has a mutation in the M-CSF gene that results in a stop codon and a truncated M-CSF protein.[12] These animals develop osteopetrosis at an early age, but if maintained on an appropriate diet, survive to adult age and

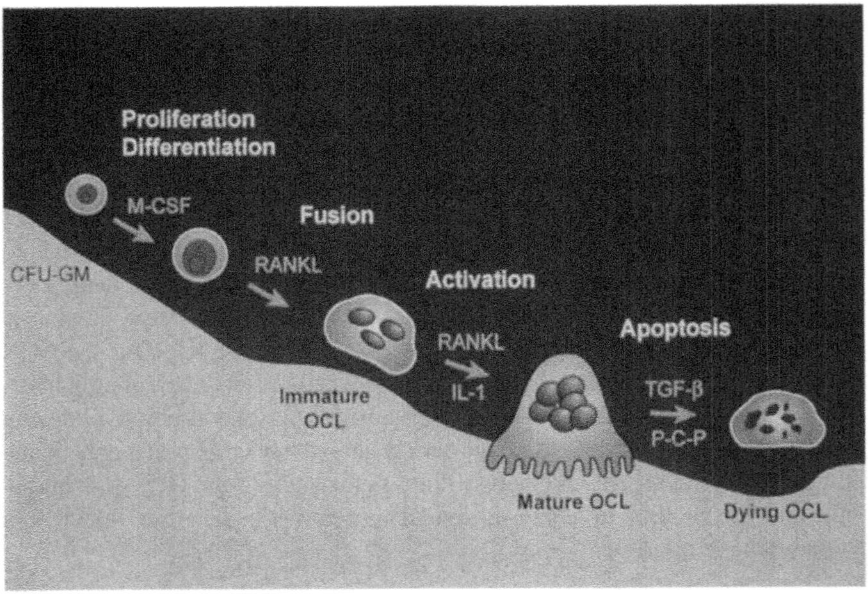

FIGURE 1. Osteoclast life cycle from proliferation and differentiation to apoptosis.

partially cure their osteopetrosis. The factors that appear to supplant M-CSF in the op/op mouse include IL-3, GM-CSF, or vascular endothelial growth factor (VEGF).[12]

The early OCL precursors proliferate and differentiate to form a post-mitotic committed OCL precursor (FIG. 1). These committed OCL precursors under the influence of RANK ligand (RANKL) or 1,25-$(OH)_2D_3$ then differentiate and fuse to form immature multinucleated OCL. These immature OCL then must be activated to form bone-resorbing OCL. Factors that can activate OCL include RANKL and IL-1.[13,14] The activated OCL survive for approximately 2 weeks in the marrow[15] and then undergo apoptosis. Factors that can enhance OCL apoptosis include bisphosphonates and TGF-β.[16,17]

KEY REGULATORS OF OCL DIFFERENTIATION

There are a number of key regulators of OCL differentiation and function. The earliest transcription factor that has been linked to OCL differentiation is PU.1. Loss of PU.1 by homologous recombination results in severe osteopetrosis in animals that lack both granulocytes and cells in the monocyte-macrophage lineage.[18] Another critical transcription factor for OCL differentiation is *c-fos*. Animals lacking *c-fos* can still develop monocyte macrophages but cannot form OCL and are osteopetrotic.[19] *c-fos* appears to regulate OCL differentiation through induction of a second transcription factor NFAT-c1, which is critical for OCL differentiation.[20] Other genes, which have been shown to be critical for OCL function, include *c-Src* that regulates OCL attachment and the OCL cytoskeleton. Loss of *c-Src* results in formation of OCL that cannot resorb bone.[21]

RANKL AND OCL FORMATION

Both systemic and local factors can regulate OCL formation including parathyroid hormone, IL-11, prostaglandins, and 1,25-$(OH)_2D_3$. All of these factors induce OCL formation through increasing expression of RANKL on the surface of immature osteoblasts and marrow stromal cells.[22,23] RANKL is a member of the tumor necrosis factor (TNF) gene family and is predominantly expressed as a membrane-bound protein. RANKL also can be cleaved from the cell surface and expressed as a soluble protein by the action of TNF-α converting enzyme-like proteins. RANKL then binds its cognate receptor RANK on OCL precursors and induces OCL formation. RANKL has a natural occurring decoy receptor, osteoprotegerin (OPG) that can bind RANKL and block its interaction with RANK. Proof that RANKL is a critical factor in osteoclastogenesis was shown by several laboratories using the techniques of homologous recombination to delete the RANKL gene or its receptor RANK in mice.[24,25] These mice develop severe osteopetrosis. Furthermore, overexpression of RANKL or deletion of OPG resulted in severe osteoporosis, with the animals sustaining spontaneous fractures because of their markedly decreased bone mass.[26] RANKL induces OCL formation by signaling through several pathways (FIG. 2). Boyce and co-workers have clearly shown that the NF-κB pathway is critical for osteoclastogenesis.[27] Animals lacking both the p50 and p52 subunit of NF-κB develop severe osteopetrosis.[27]

The pathways that are involved in induction of RANKL in marrow stromal cells are just beginning to be identified. Both the PKA and PKC pathways and the MAP kinase pathway have all been implicated in the upregulation of RANKL by hormones and cytokines.[28,29] However, it is clear that the NF-κB signaling pathway is not required for production of RANKL, because marrow stromal cells from animals lacking both the p50 and p52 subunits of NF-κB can still produce RANKL.[28] The relative ratio of RANKL to OPG determines whether there is increased or decreased OCL formation. Normally, the levels of OPG are much higher than RANKL so there are few OCL present in normal bone. However, with inflammatory conditions, such as rheumatoid arthritis, bone metastasis, myeloma, and Paget's disease, the ratio of RANKL to OPG is increased, favoring increased osteoclastogenesis.[30–32]

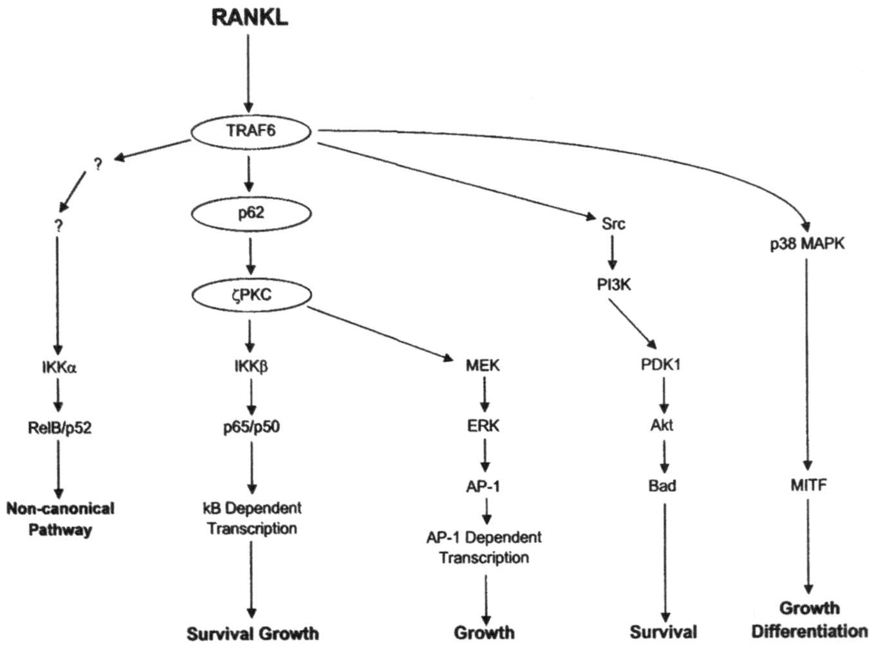

FIGURE 2. RANK signaling pathways. When RANK ligand binds RANK on OCL precursors, multiple signaling pathways are activated, which affect OCL formation and survival. These pathways include NF-κB, p38 MAP kinase, JNK, and *c-Src*.

AUTOCRINE FACTOR REGULATION OF OCL FORMATION

As noted above, both systemic and local factors can regulate OCL formation and activity. In addition, the OCL itself can produce regulatory cytokines, which stimulate and inhibit OCL formation. Initial studies in our laboratory demonstrated that the OCL was a secretory cell, which produced cytokines that simulated its own activity. In studies with giant cell tumors from bone, which are osteoclastomas, we demonstrated that these cells expressed IL-6 mRNA and protein.[33] Further, when these cells are treated with anti-IL-6, bone resorption was inhibited. The addition of IL-6 to OCL treated with anti-IL-6 restored their bone-resorbing capacity. Similarly, antisense constructs to IL-6 blocked bone resorption by OCL from giant cell tumors of bone.[34] Thus, IL-6 was the first OCL autocrine factor identified, which simulated OCL formation. We have shown that OCL from patients with Paget's disease also produced high levels of IL-6.[35] Kurihara and co-workers had previously demonstrated that IL-6 was a potent inducer of OCL formation.[36] IL-6 induces human OCL formation by acting directly on OCL precursors and does not induce RANKL expression.[37,38]

Based on these initial results, we initiated studies to identify other autocrine-paracrine factors produced by OCL that regulated OCL formation and bone resorption. To undertake these studies, we produced large numbers of human marrow-derived OCL *in vitro*, purified the OCL, and then isolated mRNA from the highly purified OCL. An OCL cDNA expression library was established in 293 cells, and the conditioned media from these 293 cells was screened for the factors that stimulated or inhibited human or murine OCL formation in bone marrow cultures.[39] We identified nine pools that contained an OCL stimulatory or inhibitory activity. Three of these nine pools contained known factors that regulated OCL activity, including IL-6 and IL-1. The first novel autocrine-paracrine stimulator of OCL formation that we identified was Annexin-II (AX-II).[39] AX-II was previously thought to be an intracellular protein, which had no known physiologic function except as a possible calcium channel. AX-II purified from human placenta or bovine lung simulated human and murine OCL formation in a dose-dependent fashion at concentrations of 1–100 ng/mL. Furthermore, AX-II was expressed by multinucleated OCL in murine marrow cultures and in OCL from giant cell tumors of bone. Studies on the mechanism of action of AX-II demonstrated that AX-II induced OCL formation indirectly by increasing production of both GM-CSF and RANKL by marrow stromal cells.[40,41] Most recently, we have identified a putative AX-II receptor on marrow stromal cells, which binds AX-II and does not bind AX-III or AX-V. The kDa for this receptor is about 10^{-9} M, and there are about 300,000 receptors per cell.[42] The receptor specifically binds the p11 subunit of AX-II, which is heterotetramer containing two p11 and two p36 subunits.

In addition to autocrine-paracrine stimulators of OCL formation, we identified two novel inhibitors of OCL formation. One was osteoclast inhibitory peptide (OIP)-1, which is identical to the human SCA protein.[43] OIP-1 is a GPI-linked protein that can be cleaved from the cell surface to inhibit OCL formation. The mechanism of action of OIP-1 is under intensive investigation and appears to involve interferon induction.[44] The second inhibitor of osteoclastogenesis produced by OCL we identified was OIP-2, which is identical to the enzyme legumain.[44] Structure–function studies of OIP-2 showed that the C-terminal peptide, which is cleaved by autocatalysis when the protein is secreted, mediates the inhibition of osteoclastogenesis.[45] Choi and co-workers have shown that OIP-2 can block increased OCL formation and hypercalcemia *in vivo* in mice treated with parathyroid hormone-related peptide.[46]

To identify genes that were overexpressed in OCL compared to OCL precursors, which might be involved in OCL differentiation, we developed a murine OCL precursor cell line by targeting Bcl-xL and large T-antigen to the OCL lineage in transgenic mice using the tartrate acid phosphatase (TRAP) promoter.[47] Bone marrow from these mice was then cultured and an OCL precursor cell line was developed. All the cells were OCL precursors and OCL formation by these cells was 500-times greater than normal marrow. These cells responded appropriately to most osteoclastogenic factors[48] and were used to obtain large numbers of OCL precursors and OCL. Using the techniques of subtractive hybridization, we generated an OCL cDNA library enriched for genes that were upregulated in mature OCL compared to the precursors, and detected several genes that regulated OCL formation. One of these genes was ADAM8, which is a member of the ADAM (A Disintegrin And Metalloproteinase) family of genes.[49] There are over 30 members of the ADAM gene family, and only ADAM8 mRNA is overexpressed in OCL compared to OCL precursors. Soluble ADAM8 induced formation of bone-resorbing OCL and acted at the later stages of OCL differentiation and precursor fusion. Structure-function studies showed that the disintegrin domain of ADAM8 mediated its effects on OCL formation. Recently, we have identified the receptor for ADAM8, which is $\alpha_9\beta_1$ integrin.[50] $\alpha_9\beta_1$ integrin is expressed in OCL precursors at higher levels than in mature OCL but it is predominantly expressed at late stages of OCL precursor differentiation. Neutralizing antibodies to α_9 integrin inhibits human OCL formation, and OCL from mice lacking the α_9 integrin subunit gene are abnormal. These α_9 –/– OCL are small and contracted, resorb bone poorly and are similar to OCL that lack β_3 integrin. In addition, OCL from α_9 knockout mice do not form actin rings, consistent with their impaired capacity to resorb bone. Other genes which are upregulated during OCL differentiation that effect OCL formation include eosinophil chemotactic factor,[51] C3 component of complement, which was reported by Sato *et al.* to be critical for OCL formation,[52] and macrophage inflammatory peptide (MIP)-1α.[37]

CONCLUSION

OCL formation and activity are controlled by both systemic hormones, local factors, and factors produced by the OCL themselves. These factors both positively and negatively regulate OCL formation. Local factors in the bone microenvironment, such as RANKL and OPG are critical regulators of OCL formation. In addition, factors produced by OCL include IL-6, TNF-α, MIP-1α, ADAM8, AX-II, OIP-1, and OIP-2, which appear to be important regulators of OCL formation in normal and pathologic states.

REFERENCES

1. Roodman, G.D. *et al.* 1985. 1,25-Dihydroxyvitamin D3 causes formation of multinucleated cells with several osteoclast characteristics in cultures of primate marrow. Proc. Natl. Acad. Sci. USA **82:** 8213–8217.
2. Fallon, M.D., S.L. Teitelbaum & A.J. Kahn. 1983. Multinucleation enhances macrophage-mediated bone resorption. Lab. Invest. **49:** 159–164.
3. Marks, S.C., Jr. & D.G. Walker. 1981. The hematogenous origin of osteoclasts: experimental evidence from osteopetrotic (microphthalmic) mice treated with spleen cells from beige mouse donors. Am. J. Anat. **161:** 1–10.
4. Walker, D.G. 1975. Spleen cells transmit osteopetrosis in mice. Science **190:** 785–787.

5. HANAOKA, H., H. YABE & H. BUN. 1989. The origin of the osteoclast. Clin. Orthop. Relat. Res. **239:** 286–298.
6. KURIHARA, N. *et al.* 1989. Generation of osteoclasts from isolated hematopoietic progenitor cells. Blood **74:** 1295–1302.
7. MIYAMOTO, T. *et al.* 2001. Bifurcation of osteoclasts and dendritic cells from common progenitors. Blood **98:** 2544–2554.
8. MENAA, C., N. KURIHARA & G.D. ROODMAN. 2000. CFU-GM-derived cells form osteoclasts at a very high efficiency. Biochem. Biophys. Res. Commun. **267:** 943–946.
9. KERBY, J.A. *et al.* 1992. Derivation of osteoclasts from hematopoietic colony-forming cells in culture. J. Bone Miner. Res. **7:** 353–362.
10. LORENZO, J.A. *et al.* 1987. Colony-stimulating factors regulate the development of multinucleated osteoclasts from recently replicated cells in vitro. J. Clin. Invest. **80:** 160–164.
11. GLANTSCHNIG, H. *et al.* 2003. M-CSF, TNF-alpha and RANK ligand promote osteoclast survival by signaling through mTOR/S6 kinase. Cell Death Differ. **10:** 1165–1177.
12. TOLAR, J., S.L. TEITELBAUM & P.J. ORCHARD. 2004. Osteopetrosis [review]. N. Engl. J. Med. **351:** 2839–2849.
13. UDAGAWA, N. 2002. Mechanisms involved in bone resorption [review]. Biogerontology **3:** 79–83.
14. ARMSTRONG, A.P. *et al.* 2002. A RANK/TRAF6-dependent signal transduction pathway is essential for osteoclast cytoskeletal organization and resorptive function. J. Biol. Chem. **277:** 44347–44356.
15. MARKS, S.C., JR. & M.F. SEIFERT. 1985. The lifespan of osteoclasts: experimental studies using the giant granule cytoplasmic marker characteristic of beige mice. Bone **6:** 451–455.
16. WEITZMANN, M.N. *et al.* 2000. B lymphocytes inhibit human osteoclastogenesis by secretion of TGF beta. J. Cell Biochem. **78:** 318–324.
17. MUNDY, G.R., T. YONEDA & T. HIRAGA. 2001. Preclinical studies with zoledronic acid and other bisphosphonates: impact on the bone microenvironment [review]. Semin. Oncol. **282:** 35–44.
18. TONDRAVI, M.M. *et al.* 1997. Osteopetrosis in mice lacking haematopoietic transcription factor PU.1. Nature **386:** 81–84.
19. MATSUO, K. *et al.* 2000. Fosl1 is a transcriptional target of c-Fos during osteoclast differentiation. Nat. Genet. **24:** 184–187.
20. TAKATSUNA, H. *et al.* 2005. Inhibition of RANKL-induced osteoclastogenesis by (-)-DHMEQ, a novel NF-kappaB inhibitor, through downregulation of NFATc1. J. Bone Miner. Res. **20:** 653–662.
21. MIYAZAKI, T. *et al.* 2004. Src kinase activity is essential for osteoclast function. J. Biol. Chem. **279:** 17660–17666.
22. LEE, S.K. & J.A. LORENZO. 1999. Parathyroid hormone stimulates TRANCE and inhibits osteoprotegerin messenger ribonucleic acid expression in murine bone marrow cultures: correlation with osteoclast-like cell formation. Endocrinology **140:** 3552–3561.
23. HORWOOD, J.J. *et al.* 1998. Osteotropic agents regulate the expression of osteoclast differentiation factor and osteoprotegerin in osteoblastic stromal cells. Endocrinology **139:** 4743–4746.
24. ODGREN, P.R. *et al.* 2003. The role of RANKL (TRANCE/TNFSF11), a tumor necrosis factor family member, in skeletal development: effects of gene knockout and transgenic rescue. Connect. Tissue Res. **44:** 264–271.

25. Dougall, W.C. *et al.* 1999. RANK is essential for osteoclast and lymph node development. Genes Dev. **13:** 2412–2424.
26. Bucay, N. *et al.* 1998. Osteoprotegerin-deficient mice develop early onset osteoporosis and arterial calcification. Genes Dev. **12:** 1260–1268.
27. Li, X. *et al.* 2002. p38 MAPK-mediated signals are required for inducing osteoclast differentiation but not for osteoclast function. Endocrinology **143:** 3105–3113.
28. Xing, L. *et al.* 2002. NF-kappaB p50 and p52 expression is not required for RANK-expressing osteoclast progenitor formation but is essential for RANK- and cytokine-mediated osteoclastogenesis. J. Bone Miner. Res. **17:** 1200–1210.
29. Lee, S.K. & J.A. Lorenzo. 2002. Regulation of receptor activator of nuclear factor-kappa B ligand and osteoprotegerin mRNA expression by parathyroid hormone is predominantly mediated by the protein kinase a pathway in murine bone marrow cultures. Bone **31:** 252–259.
30. Vanderborght, A. *et al.* 2004. Osteoprotegerin and receptor activator of nuclear factor-kappaB ligand mRNA expression in patients with rheumatoid arthritis and healthy controls. J. Rheumatol. **31:** 1483–1490.
31. Terpos, E. *et al.* 2003. Soluble receptor activator of nuclear factor kappaB ligand-osteoprotegerin ratio predicts survival in multiple myeloma: proposal for a novel prognostic index. Blood **102:** 1064–1069.
32. Hofbauer, L.C., C.A. Kuhne & V. Viereck. 2004. The OPG/RANKL/RANK system in metabolic bone diseases [review]. J. Musculoskelet. Neuronal. Interact. **4:** 268–275.
33. Ohsaki, Y. *et al.* 1992. Evidence for an autocrine/paracrine role for interleukin-6 in bone resorption by giant cells from giant cell tumors of bone. Endocrinology **131:** 2229–2234.
34. Reddy, S.V. *et al.* 1994. Interleukin-6 antisense deoxyoligonucleotides inhibit bone resorption by giant cells from human giant cell tumors of bone. J. Bone Miner. Res. **9:** 753–757.
35. Roodman, G.D. *et al.* 1992. Interleukin 6. A potential autocrine/paracrine factor in Paget's disease of bone. J. Clin. Invest. **89:** 46–52.
36. Kurihara, N. *et al.* 1990. IL-6 stimulates osteoclast-like multinucleated cell formation in long term human marrow cultures by inducing IL-1 release. J. Immunol. **144:** 4226–4230.
37. Han, J.H. *et al.* 2001. Macrophage inflammatory protein-1alpha is an osteoclastogenic factor in myeloma that is independent of receptor activator of nuclear factor kappaB ligand. Blood **97:** 3349–3353.
38. Hofbauer, L.C. *et al.* 1999. Interleukin-1beta and tumor necrosis factor-alpha, but not interleukin-6, stimulate osteoprotegerin ligand gene expression in human osteoblastic cells. Bone **25:** 255–259.
39. Takahashi, S. *et al.* 1994. Cloning and identification of annexin II as an autocrine/paracrine factor that increases osteoclast formation and bone resorption. J. Biol. Chem. **269:** 28696–28701.
40. Menaa, C. *et al.* 1999. Annexin II increases osteoclast formation by stimulating the proliferation of osteoclast precursors in human marrow cultures. J. Clin. Invest. **103:** 1605–1613.
41. Li, F. *et al.* 2005. Annexin II stimulates RANKL expression through a MAPK-dependent pathway in primary human bone marrow cells. J. Bone Miner. Res. **20**(7): 1161–7.
42. Lu, G. *et al.* 2005. Cloning and Characterization of the annexin II receptor [abstract]. ASBMR 27th Annual Meeting, Nashville, TN.

43. CHOI, S.J. *et al.* 1998. Cloning and identification of human Sca as a novel inhibitor of osteoclast formation and bone resorption. J. Clin. Invest. **102:** 1360–1368.
44. KOIDE, M. *et al.* 2003. Cytokine regulation and the signaling mechanism of osteoclast inhibitory peptide-1 (OIP-1/hSca) to inhibit osteoclast formation. J. Bone Miner. Res. **18:** 458–465.
45. CHOI, S.J. *et al.* 2001. Osteoclast inhibitory peptide 2 inhibits osteoclast formation via its C-terminal fragment. J. Bone Miner. Res. **16:** 1804–1811.
46. CHOI, S.J. *et al.* 1999. Identification of human asparaginyl endopeptidase (legumain) as an inhibitor of osteoclast formation and bone resorption. J. Biol. Chem. **274:** 27747–27753.
47. HENTUNEN, T.A. *et al.* 1998. Immortalization of osteoclast precursors by targeting Bcl -XL and Simian virus 40 large T antigen to the osteoclast lineage in transgenic mice. J. Clin. Invest. **102:** 88–97.
48. HENTUNEN, T.A. *et al.* 1999. Characterization of immortalized osteoclast precursors developed from mice transgenic for both bcl-X(L) and simian virus 40 large T antigen. Endocrinology **140:** 2954–2961.
49. CHOI, S.J., J.H. HAN & D.G. ROODMAN. 2001. ADAM8: a novel osteoclast stimulating factor. J. Bone Miner. Res. **16:** 814–822.
50. RAO, H. *et al.* 2005. $\alpha_9\beta_1$ integrin signaling is required for normal osteoclast activity [abstract]. ASBMR 27th Annual Meeting, Nashville, TN.
51. OBA, Y. *et al.* 2003. Eosinophil chemotactic factor-L (ECF-L): a novel osteoclast stimulating factor. J. Bone Miner. Res. **18:** 1332–1341.
52. SATO, T. *et al.* 1993. The biological roles of the third component of complement in osteoclast formation. Endocrinology **133:** 397–404.

M-CSF, c-Fms, and Signaling in Osteoclasts and their Precursors

F. PATRICK ROSS

Department of Pathology and Immunology, Washington University School of Medicine, St. Louis, Missouri 63110, USA

ABSTRACT: Prevention of conditions, such as osteoporosis, requires an understanding of the molecular mechanisms of bone resorption. The understanding that cells of the myeloid lineage are osteoclast precursors suggests that macrophage colony-stimulating factor (M-CSF) plays an important role in osteoclast biology. Signals generated by the binding of M-CSF to the cell-surface receptor c-Fms appear to trigger events leading to osteoclast differentiation. We have created a chimeric variant of the c-Fms receptor, which has allowed study of downstream events activated by M-CSF in a model more relevant to normal physiology than prior studies, which have relied on myeloid tissues. Our studies suggest novel regulatory signaling pathways initiated via the c-Fms receptor.

KEYWORDS: osteoporosis; osteoclast; RANKL; M-CSF; c-Fms; macrophages

INTRODUCTION

Bone mass is determined by the net activity of osteoblasts, which form bone, and osteoclasts, which degrade the matrix. An increase in the rate of skeletal degradation relative to formation results in pathological bone loss. Thus, prevention of conditions such as osteoporosis requires an understanding of the molecular mechanisms of bone resorption.

The osteoclast, the exclusive bone resorptive cell, is a member of the monocyte/macrophage family and a polykaryon that can be generated *in vitro* from myeloid precursors, with bone marrow macrophages representing the largest reservoir. Two cytokines are essential and sufficient for osteoclastogenesis, receptor activator of NF-kB ligand (RANKL)[1,2] and macrophage colony-stimulating factor (M-CSF), also called colony-stimulating factor-1 (CSF-1), with the former being the key osteoclastogenic cytokine and the latter contributing to proliferation, survival, and differentiation of early precursors.

Address for correspondence: F. Patrick Ross, Ph.D., Department of Pathology and Immunology, Washington University School of Medicine, 660 South Euclid Avenue, Campus Box 8118, St. Louis, MO 63110. Voice: 314-454-8079; fax: 314-454-5505.
e-mail: rossf@wustl.edu

Ann. N.Y. Acad. Sci. 1068: 110–116 (2006). © 2006 New York Academy of Sciences.
doi: 10.1196/annals.1346.014

Each is produced by mesenchymal cells and their derivatives in the marrow environment.[3]

M-CSF was identified as a molecule that mediates the survival and proliferation of precursors of the monocyte/macrophage lineage and their differentiation into mature phagocytes.[4] The understanding that cells of the myeloid lineage are osteoclast precursors[5] suggested that M-CSF plays an important role in osteoclast biology. This hypothesis was confirmed by the observation that *op/op* mice, which fail to express functional M-CSF as a result of a point mutation in the *Csf1* gene, are osteopetrotic.[6,7] Furthermore, administration of soluble M-CSF to either *op/op* mice or to rats bearing an analogous mutation (*tl*, *toothless*) rescues their osteopetrosis.[6–8] While M-CSF is produced constitutively by a range of cells, regulated secretion of the cytokine has pathological consequences in the context of the osteoclast. Thus, absence of estrogen, the cause of postmenopausal osteoporosis, is due to enhanced bone resorption caused at least in part by increased secretion of M-CSF from marrow stromal cells in response to higher circulating levels of the inflammatory molecules IL-1 and TNF-α.[9,10]Finally, raised serum levels of parathyroid hormone stimulate osteoblasts to secrete the same cytokine.[11]

c-Fms, the cellular homologue of the feline transforming virus v-Fms, is the sole receptor for M-CSF.[12,13] The functional linkage between M-CSF and its receptor is established by the fact that mice lacking *csf1r,* the gene coding for c-Fms, exhibit the same major phenotype as the op/op mouse, namely a marked decrease in tissue macrophages and severe osteopetrosis due to a lack of osteoclasts.[14]

Binding of M-CSF to c-Fms results in dimerization and hence activation of the receptor tyrosine kinase, leading to auto-phosphorylation of the dimer in trans on selected tyrosine residues. Each phosphorylated species then acts as a binding site for SH2 or PTB domain-containing proteins, which amplify and transduce the original signal.[15] The identity of the key tyrosine residues in c-Fms, which are targets of its autokinase activity, has been the subject of considerable study. Almost all experiments exploring this issue involved the use of fibroblastic or myeloid cell lines. The fact that the myeloid lines cannot differentiate into osteoclasts raises further questions about their physiological significance. While these studies contain conflicting data regarding the importance of individual c-Fms tyrosine residues in M-CSF-dependent proliferation and/or differentiation,[4] it is generally agreed that seven tyrosine residues in the cytoplasmic tail of murine c-Fms undergo phosphorylation following activation; Y559, Y697, Y706, Y721, Y807, Y921, and Y974.

c-Fms autophosphorylation prompts downstream signals by mechanisms largely common to the RTK superfamily as a whole. In brief, the initial adaptors binding to the phosphorylated tyrosine sites associate with other kinases or adaptors.[15] Using a proteomic approach coupled to extensive purification under nondenaturing conditions, Stanley and colleagues established that in excess of 150 proteins bind to c-Fms following its activation.[16] Members of this complex are involved in cell proliferation, survival, differentiation, and cytoskeletal reorganization, thus encompassing all known functions of the M-CSF/c-Fms pathway. While a massive macromolecular complex assembled on c-Fms in response to M-CSF suggests that the downstream signals may be complicated. Nevertheless, the available data, including our experiments using primary macrophages, indicate that PI3K and p42/44 ERK and c-Cbl, are key transducers of M-CSF signaling. Additionally, phospholipase C gamma (PLCγ) may also mediate c-Fms signals. This lipase generates diacylglycerol and releases stores of intracellular Ca^{2+}, thus activating a host of pathways.[17] Knowledge concerning the significance of the PLC pathway in mediating the response of macrophages and/or osteoclasts to M-CSF is limited[18] and further studies are required to clarify this issue.

MACROPHAGE PROLIFERATION

ERK1/2 (p42/44) and PI3K/Akt are established mediators of macrophage proliferation.[4] Binding of M-CSF to c-Fms recruits the adaptor protein complex Grb2/Sos toY697 in the cytoplasmic tail of c-Fms and Sos, acting as a guanosine exchange factor for Ras, stimulates the Ras/Raf/MEK/ERK pathway. In contrast, there is almost no activation of the ERK family members such as JNK or p38. M-CSF also robustly stimulates the PI3K/Akt pathway in macrophages. In general, PI3K/Akt regulates cell proliferation via GSK3β and the FOXO family of transcription factors.[19] In brief, GSK3β phosphorylates cyclin D1, leading to its rapid proteosomal degradation, while FOXO inhibits transcription of the same cyclin and increases the cell cycle inhibitors, p27 and p130. By phosphorylating GSK3β and FOXO, Akt suppresses their capacity to inhibit entry into the cell cycle.

The SHIP1-deficient mouse provides evidence that the PI3K/Akt stimulates preosteoclast to divide. SHIP1, as a 5″ lipid phosphatase, diminishes phosphatidylinositol 3, 4, 5 trisphosphate (PIP_3), and hence deactivates Akt. Animals lacking SHIP1 have large hyper-resorptive osteoclasts and in consequence are severely osteoporotic. Moreover, these terminally differentiated polykaryons are protected from programmed cell death, reflecting prolonged and more robust activation of the PI3K/Akt axis.[20] Given our observations on decreased apoptosis in SHIP1 null osteoclasts, we hypothesized that macrophages from the same animals would be similarly longer lived, thus contributing to the massive increase in macrophages in these mice.[21] Surprisingly, we find that the major difference from wild-type (WT) macrophages is not decreased apoptosis, but rather increased proliferation and accelerated entry into the cell cycle. Moreover, in accord with the current model,[19] PI3K mediates M-CSF-induced macrophage proliferation via a process involving suppression of p27 (Zhou *et al.*, unpublished data). Overexpression studies suggest that c-Fms Y721 binds the p85 subunit of the PI3K complex, leading to activation of the catalytic moiety and hence enhanced function of Akt.[4]

Following cytokine treatment, rapid induction (via ERKs) and stabilization (via PI3K/Akt/GSK3β) of D-type cyclins, typically forms a cyclin D/cdk4 complex, which hyper-phosphorylates the pocket protein Rb.[22] Once again, consistent with present dogma, our data indicate that exposure of primary macrophages to M-CSF enhances expression of all three D cyclins and simulates Rb phosphorylation.

Receptor internalization, followed by lysosomal degradation provides an alternative mechanism by which c-Fms signaling may be suppressed. The proteosome may also participate in c-Fms degradation as M-CSF induces rapid c-Cbl phosphorylation in primary macrophages.[23] Consistent with the fact that Cbl family members are ubiquitin ligases for RTKs,[24] M-CSF treatment is also followed by multi-ubiquitination and endocytosis of c-Fms, c-Src, and c-Cbl itself, events that regulate macrophage proliferation.[25]

c-Fms SIGNALING IN AUTHENTIC OSTEOCLAST PRECURSORS

As discussed, the use of cells other than authentic members of the monocyte/macrophage family to study the M-CSF/c-Fms axis has often yields conflicting information. Because primary macrophages express native c-Fms, which is activated by M-CSF, it was not possible to express specific mutants of the receptor as a means of assessing their impact on pre-osteoclast function. To address this issue we retrovirally transduced primary macrophages, which contain endogenous c-Fms, with a chimeric receptor comprising the external domain of the erythropoietin (Epo) receptor coupled to either the WT or mutated forms of the cytosolic domain of c-Fms (FIG. 1). In this circumstance,

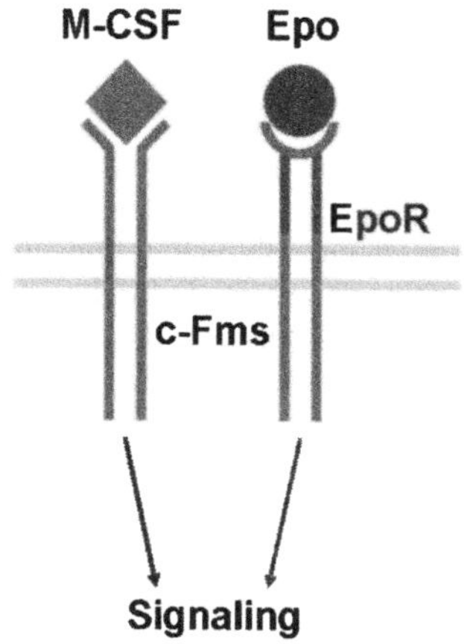

FIGURE 1. Chimeric receptor strategy for studying c-Fms signaling in primary macrophages or osteoclasts. Cells containing endogenous murine c-Fms (red line) were transduced retrovirally with a plasmid coding for the external domain of the Epo receptor (blue) linked to the transmembrane and WT cytoplasmic domains of murine c-Fms.

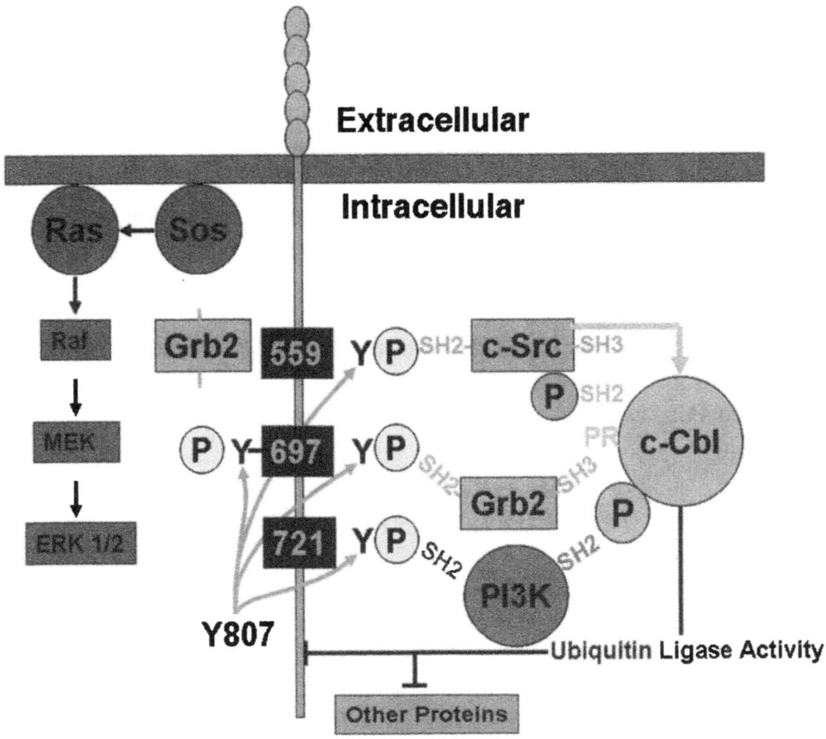

FIGURE 2. Model for major pathways leading to activation of the ERK1/2 and PI3K/Akt pathways following treatment of macrophages or osteoclasts with M-CSF. Binding of M-CSF to the extracellular domain of c-Fms stimulates phosphorylation of Y807 in the activation loop, leading to a change in its conformation and increased kinase activity of the receptor. Consequently, tyrosines 559, 697, and 721 also are phosphorylated (yellow), and become docking sites for proteins containing SH2 domains. A c-Src family kinase (in the case of the osteoclast c-Src itself) ligates position 559, resulting in autophosphorylation (position 416 in c-Src, green). c-Cbl, binding through its proline rich and SH2 domains to the SH3 and phospho-tyrosine domains of c-Src, becomes a substrate for the nonreceptor tyrosine kinase (light blue, purple). c-Cbl contains docking sites for PI3K and Grb2, each of which also interacts simultaneously with c-Fms by SH2 and/or SH3 interactions (red and orange, respectively). Shown separately for the sake of clarity, Grb2 in the same complex is constitutively bound to Sos (via a second SH3 domain in Grb2) which recruits Ras to the plasma membrane. This multimeric complex would be predicted to be highly stable and hence generate prolonged signaling, with the subsequent Raf/MEK/ERK and PI3K/PDK1/Akt cascades acting as downstream effectors. Finally, c-Cbl also acts as an E3 ubiquitin ligase for a number of proteins, including c-Fms, thereby attenuating the original signal (light blue).

M-CSF activates endogenous c-Fms and Epo the chimeric receptor. The fact that Epo, in the presence of RANKL, effectively promotes osteoclastogenesis via the chimeric receptor bearing the WT c-Fms cytoplasmic domain validates the model.[26] Furthermore, the same chimera provides proliferative stimuli equivalent to those derived from endogenous c-Fms (data not shown). Thus, transducing specific mutants of the chimera permits structure-function analysis of the c-Fms cytoplasmic domain in the context of authentic macrophages and osteoclasts.

Using a range of single Y to F point mutants and combinations thereof, we identified the tyrosine residues in c-Fms that mediate its function. In contrast to the classical model,[4,27] we find that activation of the downstream effectors p42/44 or PI3K by chimeric receptors containing the single-point mutants Y697F or Y721F is virtually unaltered. In contrast, c-FmsY559F attenuates both Erk and Akt activity. These observations and additional studies involving M-CSF-stimulated association of adaptor proteins with its receptor suggest a novel model of c-Fms-mediated ERK and PI3K activation. Thus, we propose that binding of a src family kinase to the phosphorylated form of c-FmsY559 recruits and phosphorylates c-Cbl, which attracts and stabilizes a multi-protein complex that augments signal transduction. The complex contains Grb2/Sos and PI3K, with the former activating p42/44 and the latter the PDK1/Akt axis. As discussed previously c-Cbl is also a ubiquitin ligase,[24] in which mode it enhances receptor degradation (for further details see FIG. 2 and its legend).

REFERENCES

1. LACEY, D.L. *et al.* 1998. Osteoprotegerin ligand is a cytokine that regulates osteoclast differentiation and activation. Cell. **93:** 165–176.
2. YASUDA, H. *et al.* 1998. Osteoclast differentiation factor is a ligand for osteoprotegerin/osteoclastogenesis-inhibitory factor and is identical to TRANCE/RANKL. Proc. Natl. Acad. Sci. USA **95:** 3597–3602.
3. UDAGAWA, N. *et al.* 1990. Origin of osteoclasts: mature monocytes and macrophages are capable of differentiating into osteoclasts under a suitable microenvironment prepared by bone marrow-derived stromal cells. Proc. Natl. Acad. Sci. USA **87:** 7260–7264.
4. LOTZE, M.T. & J.A. HAMILTON. 2003. Macrophage colony stimulating factor [CSF-1]. *In* The Cytokine Handbook. A.W. Thomson & M.T. Lotze, Eds.: Elsevier Science. London.
5. TEITELBAUM, S.L. & F.P. ROSS. 2003. Genetic regulation of osteoclast development and function. Nat. Rev. Genet. **4:** 638–649.
6. KODAMA, H. *et al.* 1991. Essential role of macrophage colony-stimulating factor in the osteoclast differentiation supported by stromal cells. J. Exp. Med. **173:** 1291–1294.
7. WIKTOR-JEDRZEJCZAK, W. *et al.* 1990. Total absence of colony-stimulating factor 1 in the macrophage-deficient osteopetrotic (op/op) mouse. Proc. Natl. Acad. Sci. USA **87:** 4828–4832.
8. MARKS, S.C., JR., *et al.* 1992. Administration of colony stimulating factor-1 corrects some macrophage, dental, and skeletal defects in an osteopetrotic mutation (toothless, tl) in the rat. Bone **13:** 89–93.
9. KIMBLE, R.B. *et al.* 1996. Estrogen deficiency increases the ability of stromal cells to support murine osteoclastogenesis via an interleukin-1 and tumor necrosis

factor-mediated stimulation of macrophage colony-stimulating factor production. J. Biol. Chem. **271:** 28890–28897.

10. Isaacs, S.D. *et al.* 1999. Role of NFkappaB in the regulation of macrophage colony stimulating factor by tumor necrosis factor-alpha in ST2 bone stromal cells. J. Cell Physiol. **179:** 193–200.
11. Weir, E.C. *et al.* 1996. Colony stimulating factor-1 plays a role in osteoclast formation and function in bone resorption induced by parathyroid hormone and parathyroid hormone-related protein. J. Bone Miner. Res. **11:** 1474–1481.
12. Sherr, C.J., H. Matsushime & M.F. Roussel. 1992. Regulation of CYL/cyclin D genes by colony-stimulating factor 1. CIBA Found. Symp. **170:** 209–219.
13. Stanley, E.R. *et al.* 1997. Biology and action of colony-stimulating factor-1. Mol. Reprod. Dev. **46:** 4–10.
14. Dai, X.-M. *et al.* 2002. Targeted disruption of the mouse colony-stimulating factor 1 receptor gene results in osteopetrosis, mononuclear phagocyte deficiency, increased primitive progenitor cell frequencies, and reproductive defects. Blood **99:** 111–120.
15. Schlessinger, J. 2000. Cell signaling by receptor tyrosine kinases. Cell **103:** 211–225.
16. Yeung, Y.G. *et al.* 1998. Colony-stimulating factor-1 stimulates the formation of multimeric cytosolic complexes of signaling proteins and cytoskeletal components in macrophages. J. Biol. Chem. **273:** 17128–17137.
17. Newton, A.C. 2001. Protein kinase C: structural and spatial regulation by phosphorylation, cofactors, and macromolecular interactions. Chem. Rev. **101:** 2352–2364.
18. Nakamura, I., G.A. Rodan & L.T. Duong. 2003. Distinct roles of p130Cas and c-Cbl in adhesion-induced or macrophage colony-stimulating factor-mediated signaling pathways in prefusion osteoclasts. Endocrinology **144:** 4739–4741.
19. Luo, J., B.D. Manning & L.C. Cantley. 2003. Targeting the PI3K-Akt pathway in human cancer: rationale and promise. Cancer Cell **4:** 257–262.
20. Takeshita, S. *et al.* 2002. SHIP-deficient mice are severely osteoporotic due to increased numbers of hyper-resorptive osteoclasts. Nat. Med. **8:** 943–949.
21. Helgason, C.D. *et al.* 1998. Targeted disruption of SHIP leads to hemopoietic perturbations, lung pathology, and a shortened life span. Genes Dev. **12:** 1610–1620.
22. Cobrinik, D. 2005. Pocket proteins and cell cycle control. Oncogene **24:** 2796–2809.
23. Wang, Y., Y.-G. Yeung & E.R. Stanley. 1999. CSF-1 stimulated multiubiquitination of the CSF-1 receptor and of Cbl follows their tyrosine phosphorylation and association with other signaling proteins. J. Cell. Biochem. **72:** 119–134.
24. Duan, L. *et al.* 2004. The Cbl family and other ubiquitin ligases: destructive forces in control of antigen receptor signaling. Immunity **21:** 7–17.
25. Lee, P.S.W. *et al.* 1999. The Cbl protooncoprotein stimulates CSF-1 receptor multiubiquitination and endocytosis, and attenuates macrophage proliferation. EMBO J. **18:** 3616–3628.
26. Feng, X. *et al.* 2002. Tyrosines 559 and 807 in the cytoplasmic tail of the M-CSF receptor play distinct roles in osteoclast differentiation and function. Endocrinology **143:** 4868–4874.
27. Pixley, F.J. & E.R. Stanley. 2004. CSF-1 regulation of the wandering macrophage: complexity in action. Trends Cell Biol. **14:** 628–638.

Canonical Wnt Signaling in Osteoblasts Is Required for Osteoclast Differentiation

DONALD A. GLASS, II AND GERARD KARSENTY

Department of Molecular and Human Genetics, Bone Disease Program of Texas, Baylor College of Medicine, Houston, Texas 77030, USA

ABSTRACT: Inactivation of *Lrp5*, a gene encoding a likely Wnt co-receptor, results in low bone mass (osteopenia) by decreasing bone formation, suggesting that Wnt signaling in osteoblasts regulates bone formation. Here we show that *Tcf1* and *Tcf4* are expressed in osteoblasts during development and after birth; stabilization of β-catenin, an essential component of canonical Wnt signaling, in differentiated osteoblasts results in high bone mass while its deletion from differentiated osteoblasts leads to osteopenia. Histological analysis showed that these mutations affect bone resorption. Cellular and molecular studies showed that β-catenin together with TCF proteins regulates in osteoblasts the expression of *Osteoprotegerin*, a major inhibitor of osteoclast differentiation. These findings demonstrate that, in differentiated osteoblasts, β-catenin and presumably Wnt signaling are negative regulators of osteoclast differentiation; thus they broaden our knowledge about functions that Wnt proteins may have at various stages of skeletogenesis.

KEYWORDS: osteoblast; osteoclast; Wnt; β-catenin; osteoprotegerin; skeletogenesis

INTRODUCTION

The osteoblast is defined by three major functions. First, it is responsible for the secretion of the bone extracellular matrix (ECM); second, it also expresses genes that are necessary and sufficient for the mineralization of this ECM; and third, the osteoblast is required for osteoclast differentiation and thereby for bone resorption. The two main genes required for osteoclast differentiation, *M-Csf* and *Rankl*, are expressed in osteoblasts. Osteoblasts also express a third gene that negatively regulates osteoclast differentiation. *Osteoprotegerin* (*Opg*) is a soluble tumor necrosis factor (TNF-α) receptor that functions as a decoy receptor for RANKL. However, the molecular mechanisms regulating the expression of *Opg* remain largely unidentified.

Address for correspondence: Gerard Karsenty, Department of Molecular and Human Genetics, Bone Disease Program of Texas, Baylor College of Medicine, One Baylor Plaza, Houston, TX 77030, Voice: 713-798-5489; fax:713-798-1530.

e-mail: karsenty@bcm.tmc.edu

Ann. N.Y. Acad. Sci. 1068: 117–130 (2006). © 2006 New York Academy of Sciences.
doi: 10.1196/annals.1346.015

The *LDL receptor-related protein 5* (*Lrp5*) gene encodes the vertebrate homologue of arrow, which is a co-receptor for wingless, the *Drosophila* homologue of Wnt proteins. Loss-of-function mutations in *Lrp5* in humans and in mice result in a low bone mass, associated with a decrease in bone formation, while gain-of-function mutations lead to high bone mass in humans. Forced expression of *Lrp5* in NIH3T3 fibroblasts enhances canonical Wnt signaling; however, Wnt signaling can still occur in the absence of LRP5, suggesting that *Lrp5* loss-of-function mutations may not represent a complete loss of canonical Wnt signaling. Collectively these data suggested, but did not establish, that canonical Wnt signaling regulates bone formation.

The role of canonical Wnt signaling in a given cell type can be achieved by studying the function of intracellular proteins required for signaling. Unlike the multiple Wnts, Frizzled receptors, and LRP co-receptors, there is only one β-catenin molecule, which is the molecular node of the canonical Wnt signaling pathway. Thus studying its function is a powerful approach to discover the role of canonical Wnt signaling in a cell type of interest.

To this end, we engineered mice harboring either gain-of-function or loss-of-function mutations in *β-catenin* in differentiated osteoblasts. These mice had a high bone mass and low bone mass phenotype, respectively. These phenotypes, however, were primarily caused by altering bone resorption, and molecular analyses revealed that the canonical Wnt signaling pathway controls this process by regulating the expression of *Opg* in differentiated osteoblasts. These results establish the canonical Wnt signaling pathway in osteoblasts as a major inhibitor of bone resorption.

OSTEOPETROSIS IN MICE EXPRESSING A STABILIZED FORM OF β-CATENIN IN OSTEOBLASTS

Mutations of serine and threonine residues within β-catenin prevent its phosphorylation by glycogen synthase kinase 3β (GSK-3β) and its subsequent degradation via ubiquitination, which in effect creates a constitutively active molecule.[6] Mice expressing a mutant allele of β-catenin in which exon 3, the exon encoding the aforementioned serine and threonine residues, is flanked by loxP sequences had been generated previously ($Catnb^{+/lox(ex3)}$)[3] and were crossed with $\alpha_1(I)$ *Collagen-Cre* transgenic mice that express *Cre* in osteoblasts starting at E14.5[2] to generate $\alpha_1(I)$ *Collagen-Cre;* $Catnb^{+/lox(ex3)}$ mice [called hereafter $\beta cat(ex3)_{ob}$ mice].

$\beta cat(ex3)_{ob}$ mice were born at the expected Mendelian ratio, and appeared normal at birth and during the first 2 weeks of life; however, these mutant mice failed to thrive and died a few days after weaning. Physical examination of the $\beta cat(ex3)_{ob}$ mice revealed that they lacked their lower incisors (FIG. 1A); this failure of tooth eruption was suggestive of a defect in osteoclast differentiation.[10] Histological analysis of 3-week-old $\beta cat(ex3)_{ob}$ mice revealed in their vertebrae a massive increase in bone mass (FIG. 1B). The

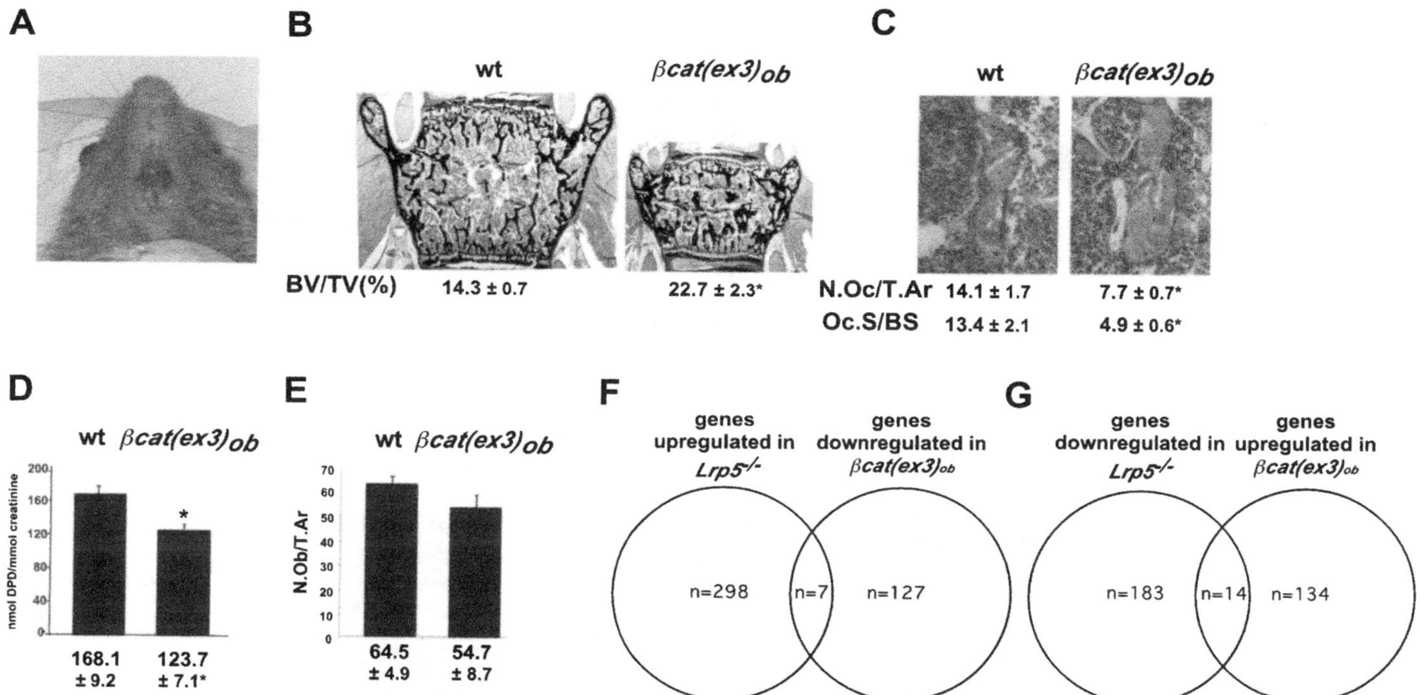

FIGURE 1. Activation of β-catenin in osteoblasts leads to osteopetrosis. **(A)** Lower incisors are absent in a 3-week-old β$cat(ex3)_{ob}$ mouse. **(B)** Histomorphometric analysis of 3-week-old wild type (wt) and β$cat(ex3)_{ob}$ vertebrae shows increased bone volume (BV/TV) in β$cat(ex3)_{ob}$ mice (25×). **(C)** TRAP staining in 3-week-old wt and β$cat(ex3)_{ob}$ vertebrae. β$cat(ex3)_{ob}$ osteoclasts are smaller and fewer in number (400×). **(D)** Urinary deoxypyridinoline crosslinks (Dpd) elimination is decreased in 3-week-old β$cat(ex3)_{ob}$ mice compared to wt. **(E)** Number of osteoblasts in 3-week-old wt and β$cat(ex3)_{ob}$ vertebrae is similar. **(F and G)** Calvarial RNA microarray results identify many genes dysregulated greater than twofold in β$cat(ex3)_{ob}$ mice and in $Lrp5^{-/-}$ mice compared to wt. The asterisk (∗) indicates statistical significance. t-test value <0.05.

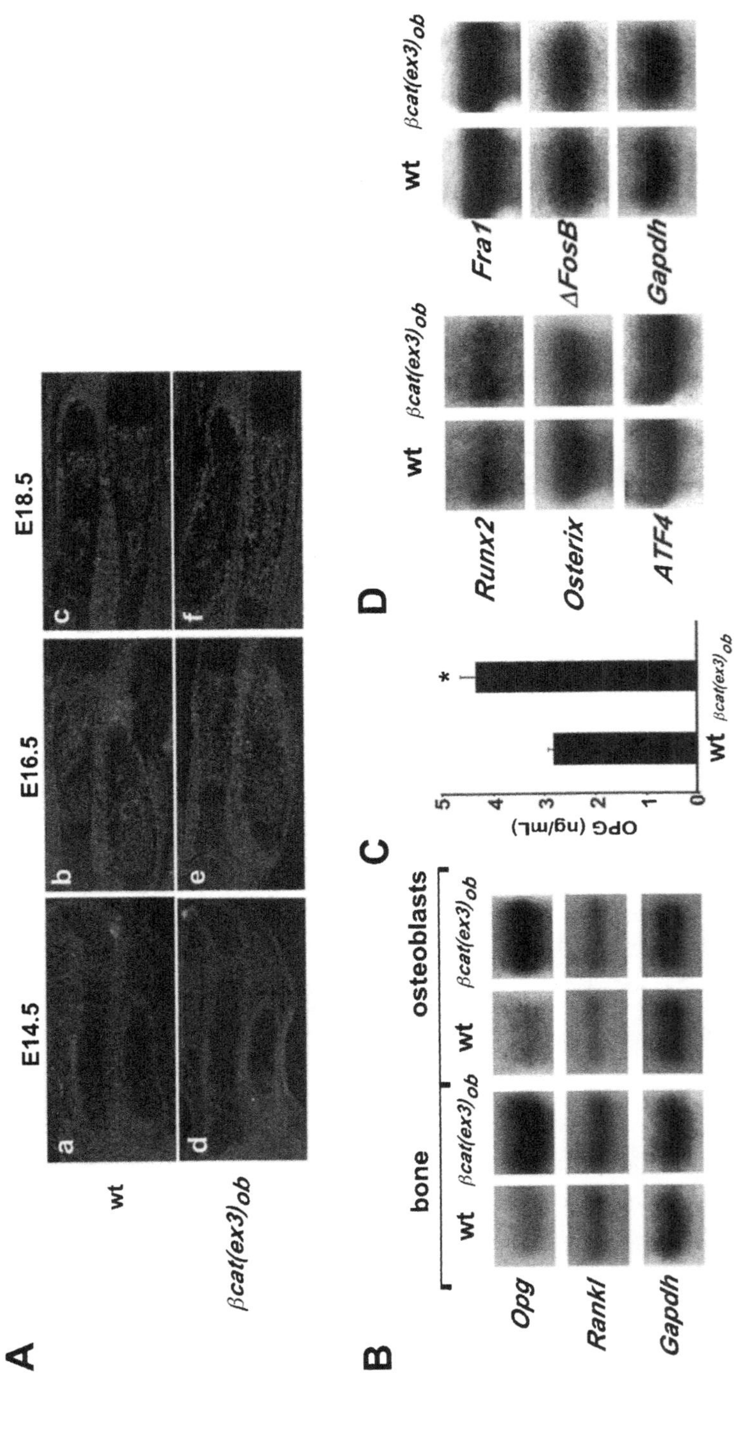
A
E14.5
E16.5
E18.5
wt
βcat(ex3)ob
B
bone
osteoblasts
Opg
Rankl
Gapdh
C
OPG (ng/mL)
D
Runx2
Osterix
ATF4
Fra1
ΔFosB
Gapdh

number of multinucleated osteoclasts (identified by positive Tartrate Resistant Acid Phosphatase [TRAP] staining) was decreased in β*cat*(*ex3*)$_{ob}$ mice (FIG. 1C), as was the urinary excretion of the collagen degradation byproduct deoxypyridinoline (Dpd) (FIG. 1D). In contrast, the number of osteoblasts was normal (FIG. 1E). Thus, the analysis of the β*cat*(*ex3*)$_{ob}$ mice revealed, unexpectedly, a defect in osteoclast differentiation leading to osteopetrosis.

Opg AS A WNT SIGNALING TARGET GENE

The unexpected differences between the bone phenotype of *Lrp5*$^{-/-}$ mice (a low bone mass due to a defect in bone formation) and that observed in β*cat*(*ex3*)$_{ob}$ mice (a high bone mass due to a failure of bone resorption) raised the hypothesis that the genes regulated by β-catenin-dependent signaling may differ from those regulated by LRP5-dependent signaling. Microarray analysis of wt, *Lrp5*$^{-/-}$, and β*cat*(*ex3*)$_{ob}$ osteoblast RNA was used to test this hypothesis.

A total of 12,000 genes were analyzed between the mutant and wt osteoblasts. A number of genes were upregulated (more than twofold) in *Lrp5*$^{-/-}$ compared to wt osteoblasts, and a number of genes were downregulated in β*cat*(*ex3*)$_{ob}$ compared to wt osteoblasts (FIG. 1F). However, very few of these genes were common to both groups, that is, upregulated in *Lrp5*$^{-/-}$ and downregulated in β*cat*(*ex3*)$_{ob}$ osteoblasts. Similar results were obtained when looking for genes downregulated in *Lrp5*$^{-/-}$ osteoblasts and/or upregulated in β*cat*(*ex3*)$_{ob}$ osteoblasts (FIG. 1G). Furthermore, none of the genes common to both groups had an obvious function in osteoblasts. These data suggest that the β-catenin mutation has a more drastic effect on bone mass regulation; for this reason we focused our analysis on the molecular bases of the osteopetrosis of the β*cat*(*ex3*)$_{ob}$ mice.

A review of the microarray data identified only one gene regulating osteoclast differentiation whose expression was noticeably increased (3.2 fold) in β*cat*(*ex3*)$_{ob}$ mice. This gene, *Opg*, encodes a soluble TNF-α receptor that inhibits osteoclast differentiation by functioning as a decoy receptor for RANKL, an osteoclast differentiation factor.[8,13] *Opg* expression was markedly

←

FIGURE 2. *Opg* expression is increased in β*cat*(*ex3*)$_{ob}$ osteoblasts. **(A)** *In situ* hybridization for *Opg* (a–f) at E14.5 (a, b), E16.5 (c, d), and E18.5 (e, f) on wt (a, c, e) and β*cat*(*ex3*)$_{ob}$ (b, d, f) forelimbs. *Opg* expression is increased in β*cat*(*ex3*)$_{ob}$ forelimbs (50×). **(B)** Northern blot analysis of wt and β*cat*(*ex3*)$_{ob}$ long bone and osteoblast RNA. Blots were sequentially probed for *Opg*, *Rankl*, and *Gapdh* as a control. Note the increased *Opg* expression in β*cat*(*ex3*)$_{ob}$ cells. **(C)** ELISA analysis for OPG in serum from wt and β*cat*(*ex3*)$_{ob}$ mice. Note the increased levels of OPG in β*cat*(*ex3*)$_{ob}$ mice. **(D)** Northern blot analysis of wt and β*cat*(*ex3*)$_{ob}$ long bone RNA. Blots were sequentially probed for *Runx2*, *Osterix*, *ATF4*, *Fra1*, and Δ*FosB*. No detectable change in expression levels of these genes is observed. The asterisk (∗) indicates statistical significance. *t*-test value <0.05.

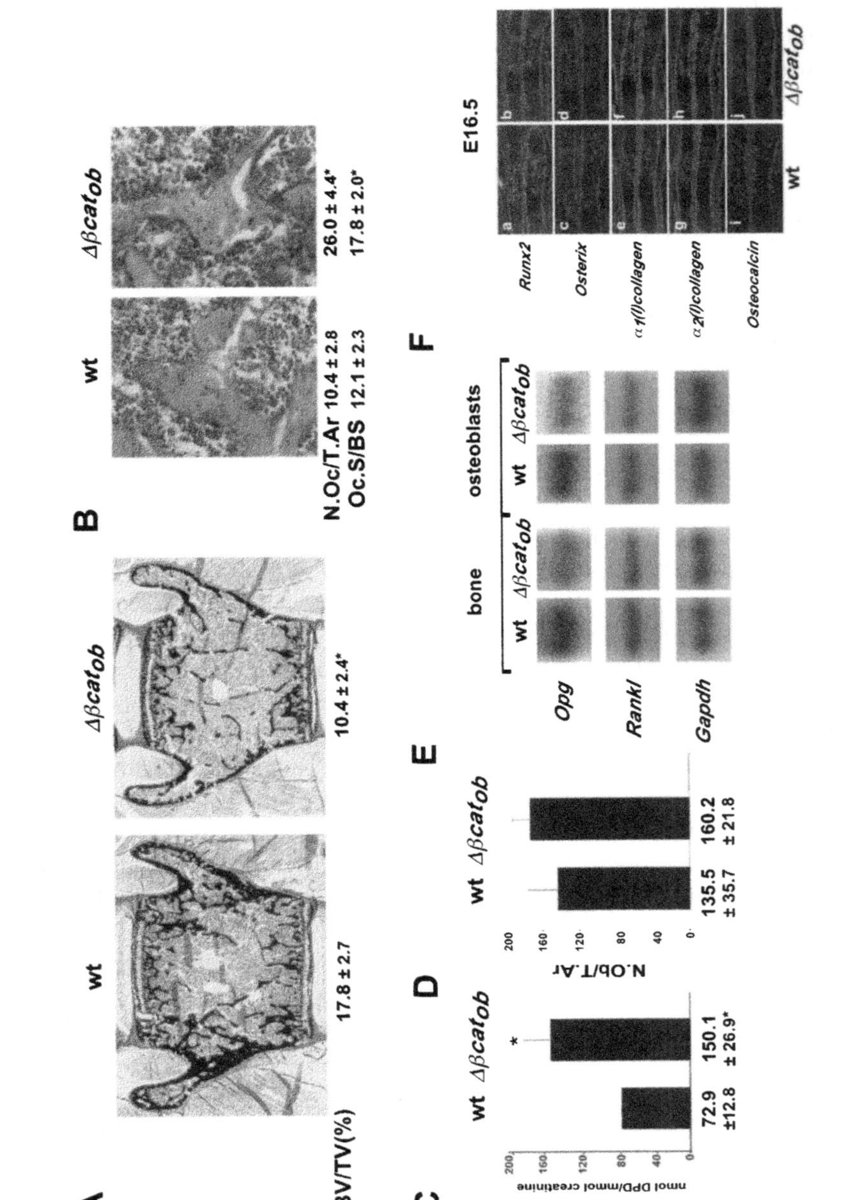
A
wt
Δβcat$_{ob}$
BV/TV(%)
17.8 ± 2.7
10.4 ± 2.4*
B
wt
Δβcat$_{ob}$
N.Oc/T.Ar 10.4 ± 2.8
26.0 ± 4.4*
Oc.S/BS 12.1 ± 2.3
17.8 ± 2.0*
C
nmol DPD/mmol creatinine
wt Δβcat$_{ob}$
72.9 ±12.8
150.1 ± 26.9*
D
N.Ob/T.Ar
wt Δβcat$_{ob}$
135.5 ± 35.7
160.2 ± 21.8
E
bone
osteoblasts
wt Δβcat$_{ob}$ wt Δβcat$_{ob}$
Opg
Rankl
Gapdh
F
E16.5
Runx2
Osterix
α1(I)collagen
α2(I)collagen
Osteocalcin
wt
Δβcat$_{ob}$

increased in β*cat(ex3)*$_{ob}$ developing long bones (FIG. 2Aa–Af) and Northern blot analysis using 1-week-old bones or primary osteoblast RNA showed that *Opg* overexpression in β*cat(ex3)*$_{ob}$ bones persisted beyond birth (FIG. 2B). Consequently, higher levels of OPG were detected in β*cat(ex3)*$_{ob}$ serum than in wt serum (FIG. 2C). In contrast to the expression of *Opg*, the expression of transcription factors implicated in osteoblast differentiation was unchanged between wt and β*cat(ex3)*$_{ob}$ osteoblasts (FIG. 2D).

LOW BONE MASS AND HIGH BONE RESORPTION IN MICE LACKING β-CATENIN IN OSTEOBLASTS

To more accurately determine the biological function(s) of β-catenin in osteoblasts we utilized a loss-of-function approach. Mice in which loxP sites had been introduced 5′ of exon 2 and 3′ of exon 6 of β-*catenin* had been generated previously to achieve cell-specific inactivation of β-catenin.[1] These mutant mice were crossed with $\alpha_1(I)$ *Collagen-Cre* mice to generate $\alpha_1(I)$ *Collagen-Cre*; β-*catenin*$^{floxed/floxed}$ mice (hereafter called Δβ*cat*$_{ob}$ mice).

Δβ*cat*$_{ob}$ mice were born at the expected Mendelian ratio, had a normal lifespan, and showed no overt phenotypic abnormalities. Histological analysis of 1-month-old mutant mice revealed a low bone mass phenotype (FIG. 3A). A marked increase in the number of TRAP-positive multinucleated osteoclasts along with an increase in urinary Dpd elimination demonstrated an increase in bone resorption (FIG. 3B, C), while the number of osteoblasts was similar between wt and Δβ*cat*$_{ob}$ mice (FIG. 3D). *Opg* expression was decreased in Δβ*cat*$_{ob}$ bones and osteoblasts while *Rankl* expression was not (FIG. 3E). *In situ* hybridization analysis showed that the expression of molecular markers of osteoblast differentiation and function was similar in wt and Δβ*cat*$_{ob}$ developing long bones (FIG. 3Fa–Fj). Thus, the molecular and histological analyses concur to establish that an important function of β-catenin occurring in osteoblasts once they are differentiated is to control osteoclast differentiation by regulating *Opg* expression.

FIGURE 3. Osteoporosis and decreased *Opg* expression in mice lacking β-catenin in osteoblasts. **(A)** Histomorphometric analysis of 1-month-old wt and Δβ*cat*$_{ob}$ vertebrae shows decreased bone volume (BV/TV) in Δβ*cat*$_{ob}$ mice (25×). **(B)** TRAP staining of 1-mo-old wt and Δβ*cat*$_{ob}$ vertebrae. Δβ*cat*$_{ob}$ osteoclasts are larger and greater in number (400×). **(C)** Dpd elimination is increased in Δβ*cat*$_{ob}$ mice. **(D)** Number of osteoblasts in 1-mo-old wt and Δβ*cat*$_{ob}$ vertebrae is similar. **(E)** Northern blot analysis of wt and Δβ*cat*$_{ob}$ long bone and osteoblast RNA. Blots were sequentially probed for *Opg*, *Rankl,* and *Gapdh* as a control. *Opg* expression is decreased in Δβ*cat*$_{ob}$ cells. **(F)** *In situ* hybridization for *Runx2* (a, b), *Osterix* (c, d), $\alpha_1(I)$ *Collagen* (e, f), $\alpha_2(I)$ *Collagen* (g, h), and *Osteocalcin* (i, j) at E16.5 (a–j) on forelimbs of wt (a, c, e, g, i) and Δβ*cat*$_{ob}$ (b, d, f, h, j) embryos. The asterisk (∗) indicates statistical significance. *t*-test value <0.05.

OSTEOBLAST-AUTONOMOUS NATURE OF THE OSTEOCLAST DIFFERENTIATION DEFECT IN β-CATENIN MUTANT MICE

To determine if the bone resorption phenotype in these mutant mice was osteoblast autonomous we performed co-culture experiments using wt, $\Delta\beta cat_{ob}$, or $\beta cat(ex3)_{ob}$ osteoblasts to trigger osteoclast differentiation of bone marrow macrophages (BMMs) of various genotypes.[9] Co-culture of wt osteoblasts with wt or mutant BMMs generated a similar number of TRAP-positive multinucleated osteoclasts (FIG.4A,B), indicating that the defect in osteoclast differentiation in the $\beta cat(ex3)_{ob}$ and $\Delta\beta cat_{ob}$ mice did not originate in cells of the osteoclast lineage. Co-culture of $\Delta\beta cat_{ob}$ osteoblasts with wt BMMs generated a greater number of TRAP-positive multinucleated osteoclasts (FIG. 4A) while co-culture of $\beta cat(ex3)_{ob}$ osteoblasts with wt BMMs failed to generate any osteoclasts (FIG. 4B); this latter defect could be rescued by adding a neutralizing antibody against OPG to the co-culture (FIG. 4C). These data establish that the origin of the osteoclast differentiation defect observed in both β-catenin mutant mice resides in osteoblasts.

EXPRESSION OF CANONICAL WNT SIGNALING PATHWAY COMPONENTS IN OSTEOBLASTS

If β-catenin controls *Opg* expression by signaling through the canonical Wnt signaling pathway, then one or more of the *Lef/Tcf* transcription factor family members must be present in osteoblasts to interact with β-catenin. To identify which *Lef/Tcf* transcription factors were expressed in cells of the osteoblast lineage we performed *in situ* hybridization on wt embryos at various stages of development and compared their expression to that of *Runx2* and $\alpha_1(I)$ *Collagen*, two osteoblast-specific markers.[4] Of the four family members, only *Tcf1* and *Tcf4* were detected in *Runx2*-expressing and $\alpha_1(I)$ *Collagen*-expressing cells: *Tcf1* expression was detected as early as E14.5 while *Tcf4* expression became detectable at E16.5.

Tcf1 has various isoforms; some contain a β-catenin interacting domain (BID) and act as transcriptional activators while others lack this domain and serve as repressors.[11] To assess whether the *Tcf1* transcripts present in osteoblasts could belong to the former category we performed Real-Time PCR (polymerase chain reaction) using osteoblast and thymus RNA and appropriate primers and probes. The percentage of BID-containing *Tcf1* transcripts is 40-fold higher in osteoblasts than in thymus (FIG. 5A), suggesting that TCF-1 proteins in osteoblasts can activate transcription. Thus, at least two members of the *Lef/Tcf* family of transcription factors are expressed in osteoblasts during development and after birth, and can serve as transcriptional activators of the canonical Wnt signaling pathway.

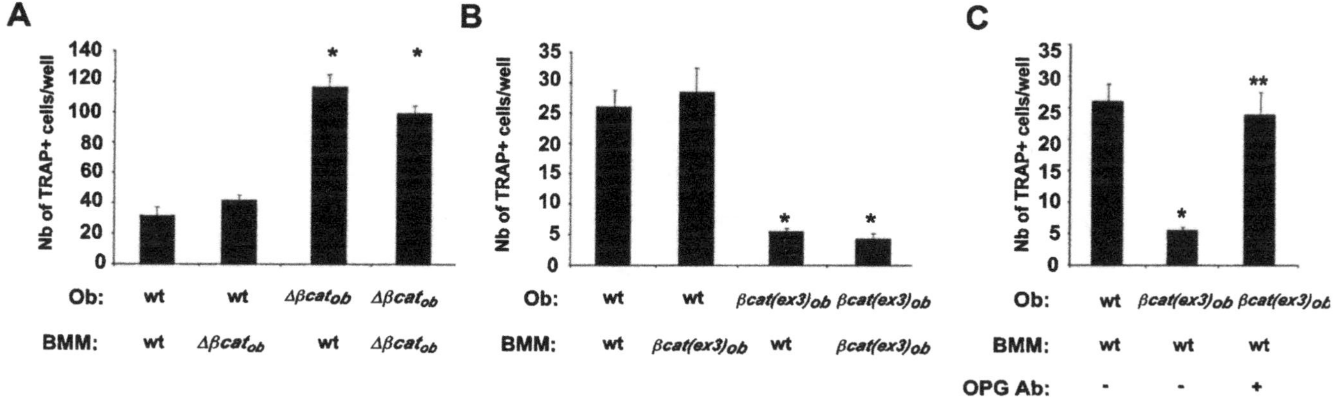

FIGURE 4. The bone resorption defect in β-catenin mutant mice is osteoblast autonomous. **(A)** TRAP (+) cells were counted following co-culture of wt or $\Delta\beta cat_{ob}$ BMMs with either wt or $\Delta\beta cat_{ob}$ osteoblasts. $\Delta\beta cat_{ob}$ osteoblasts increased osteoclast differentiation regardless of the BMM genotype. **(B)** TRAP (+) cell number following co-culture of wt or $\beta cat(ex3)_{ob}$ BMMs with either wt or $\beta cat(ex3)_{ob}$ osteoblasts. $\beta cat(ex3)_{ob}$ osteoblasts decreased osteoclast differentiation regardless of the BMM genotype. **(C)** TRAP (+) cell number following co-culture of wt BMMs with either wt or $\beta cat(ex3)_{ob}$ osteoblasts, with or without an anti-OPG antibody. $\beta cat(ex3)_{ob}$ osteoblasts decreased osteoclast differentiation of wt BMMs while the anti-OPG antibody prevented this effect. The (*) asterisk indicates statistical significance. *t*-test value <0.05.

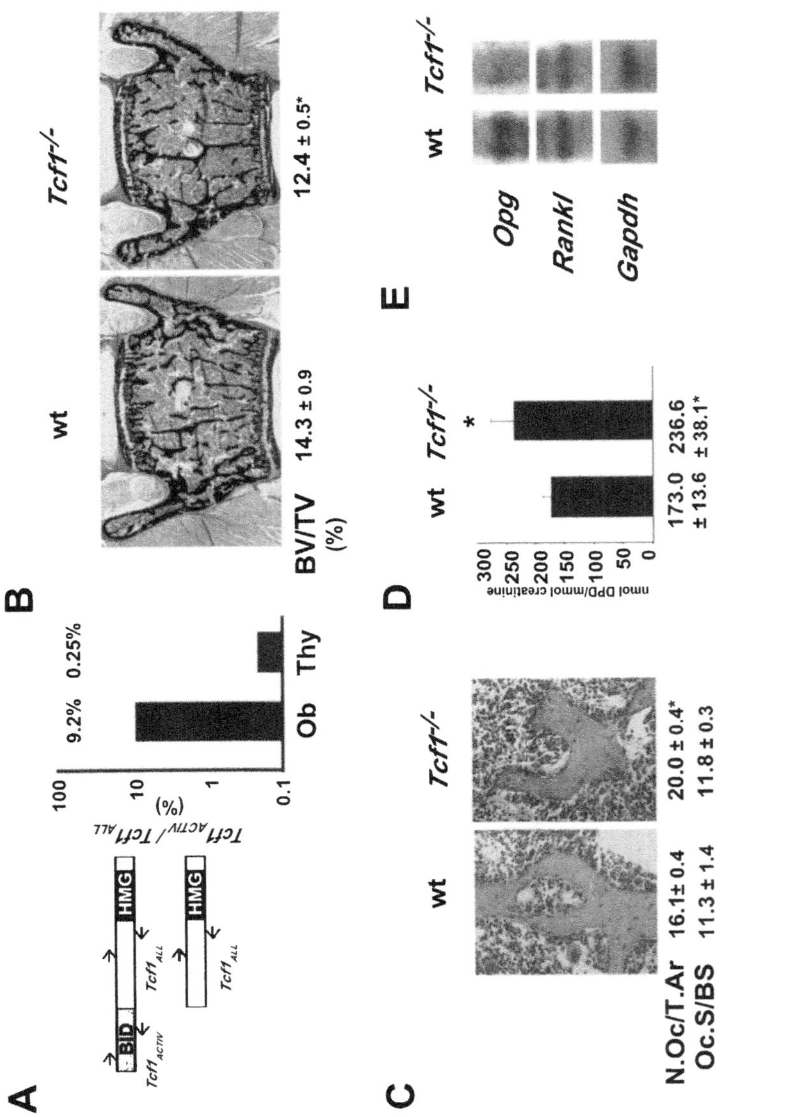
A
B
C
D
E
wt
Tcf1-/-
BV/TV (%)
14.3 ± 0.9
12.4 ± 0.5*
9.2%
0.25%
Ob
Thy
N.Oc/T.Ar 16.1 ± 0.4
Oc.S/BS 11.3 ± 1.4
20.0 ± 0.4*
11.8 ± 0.3
nmol DPD/mmol creatinine
173.0 ± 13.6
236.6 ± 38.1*
Opg
Rankl
Gapdh
HMG
BID

To confirm that the bone phenotype observed in the β-catenin mutant mice was due to alterations in the canonical Wnt signaling pathway we analyzed mice lacking *Tcf1*, one of the two *Tcf* genes expressed in osteoblasts. $Tcf1^{-/-}$ mice have a normal life span and have no apparent abnormalities,[12,7] yet bone histological analysis revealed a low bone mass in 1-month-old $Tcf1^{-/-}$ mice (FIG. 5B). As with the $\Delta\beta cat_{ob}$ mice there was an increase in the number of TRAP-positive multinucleated osteoclasts and in urinary elimination of Dpd (FIG. 5C, D) while the number of osteoblasts was normal. Northern blot analysis confirmed that, similar to $\Delta\beta cat_{ob}$ osteoblasts, the expression of *Opg* was decreased in $Tcf1^{-/-}$ osteoblasts (FIG. 5E).

REGULATION OF *Opg* EXPRESSION BY β-CATENIN AND TCF PROTEINS

The phenotypic abnormalities present in $\beta cat(ex3)_{ob}$, $\Delta\beta cat_{ob}$, and $Tcf1^{-/-}$ mice and the changes in *Opg* expression in these mice suggest that *Opg* might be a direct target of TCF-1 and/or TCF-4 in osteoblasts. Consistent with this hypothesis, there are three potential TCF-binding sites within the *Opg* promotor (FIG. 6A).

The respective importance of these sites was first tested by electric mobility shift assay (EMSA) using osteoblast nuclear extracts and labeled double stranded oligonucleotides, containing the sequences of each of these three sites, as probes. Of the three possible sites, a protein-DNA complex only formed upon incubation of osteoblast nuclear extracts with site 1 oligonucleotides (FIG. 6B). To verify this finding *in vivo*, chromatin immunoprecipitation analysis using wt primary osteoblasts was performed. Antibodies against TCF-1, TCF-4, or β-catenin could immunoprecipitate site 1 of the *Opg* promotor but could not immunoprecipitate the first exon of *Opg* or the *Osteocalcin* promotor (FIG. 6C).

To assess the relative importance of sites 1, 2 and 3 in determining *Opg* promotor activity DNA co-transfection experiments in COS cells, which do

←

FIGURE 5. Regulation of osteoclast differentiation by β-catenin requires TCF proteins. **(A)** (*left*) Schematic representation of activating (*top*) and repressive (*bottom*) isoforms of *Tcf1*. The location of the BID and Real-Time PCR primers for detection of *Tcf1* isoforms ($Tcf1_{ACTIV}$ and $Tcf1_{ALL}$, respectively) are indicated. (*right*) Real-Time PCR analysis for $Tcf1_{ACTIV}$ and $Tcf1_{ALL}$ expression in osteoblasts and thymus, normalized to *18S*. Activating isoforms of *Tcf1* are more abundant in osteoblasts than in thymus. **(B)** Histomorphometric analysis of 1-month-old wt and $Tcf1^{-/-}$ vertebrae shows decreased bone volume (BV/TV) in $Tcf1^{-/-}$ mice (25×). **(C)** TRAP staining of 1-month-old wt and $Tcf1^{-/-}$ vertebrae. Osteoclast number is increased in $Tcf1^{-/-}$ mice (400×). **(D)** Dpd elimination is increased in $Tcf1^{-/-}$ mice. **(E)** Northern blot analysis of wt and $Tcf1^{-/-}$ osteoblast RNA. Blots were sequentially probed for *Opg, Rankl,* and *Gapdh*. Note the decreased *Opg* expression in $Tcf1^{-/-}$ cells. The asterisk (∗) indicates statistical significance. *t*-test value <0.05.

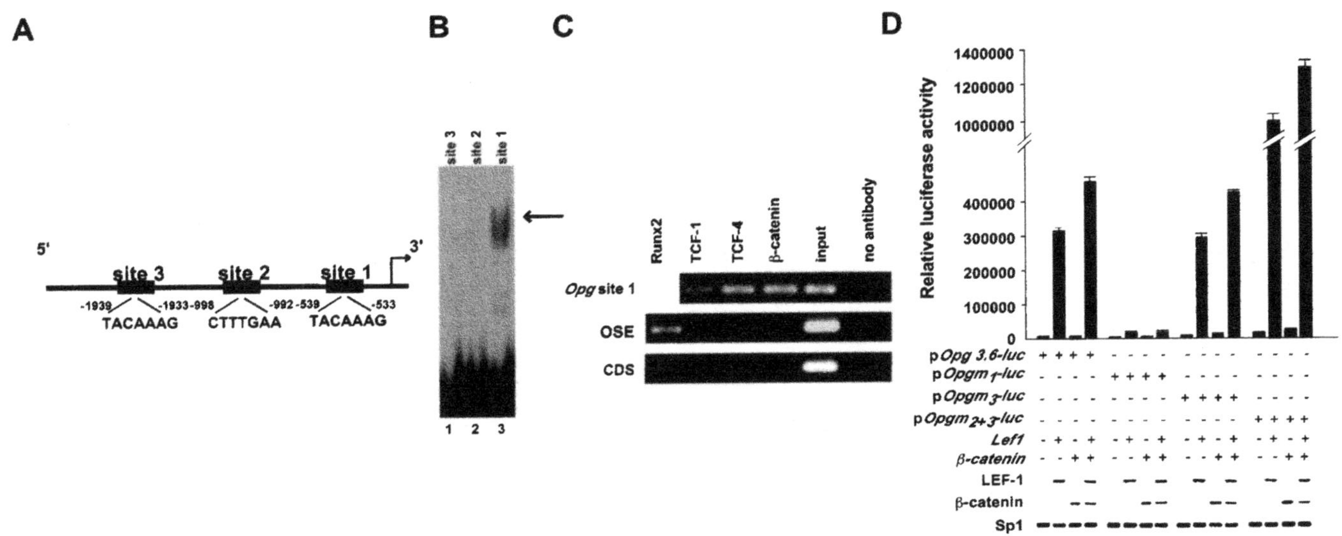

FIGURE 6. *Opg* expression is regulated by β-catenin and TCF proteins. **(A)** Schematic representation of the proximal 3.6 kb of the *Opg* promotor with the three putative canonical TCF-binding sites indicated by black boxes. **(B)** EMSA. Osteoblast nuclear extracts (NE) were incubated with labeled *Opg* site 3 (lane 1), site 2 (lane 2), or site 1 oligonucleotides (lane 3). A protein-DNA complex formed only upon incubation of NE with site 1. **(C)** Chromatin immunoprecipitation for Runx2, TCF-1, TCF-4, and β-catenin from primary osteoblast cultures. Antibodies against the TCF proteins and β-catenin can immunoprecipitate site 1 of the *Opg* promotor, but not the *Osteocalcin* promotor (OSE) or the *Opg* coding sequence (CDS). **(D)** COS7 cells were transiently transfected with p*Opg 3.6-luc*, p*Opgm*$_1$*-luc*, p*Opgm*$_3$*-luc*, or p*Opgm*$_{2+3}$*-luc* reporter constructs with or without HA-*Lef1* and/or β-*catenin*^{activ}. LEF-1 increased transactivational activity of p*Opg 3.6-luc*, p*Opgm*$_3$*-luc*, and p*Opgm*$_{2+3}$*-luc*, which was further increased with β-catenin. In contrast, LEF-1 failed to activate p*Opgm*$_1$*-luc*.

not express *Tcf* genes, were performed. We used as a reporter construct a vector containing 3.6 kb of the *Opg* promotor fused to the *luciferase* gene (p*Opg 3.6-luc*). An expression vector for LEF-1, used here as a prototype of LEF/TCF proteins, strongly transactivated p*Opg 3.6-luc* and this transactivation by *Lef1* was further enhanced by co-transfection with a β-*catenin* expression vector. In control experiments, *Lef1* alone, or in combination with β-*catenin,* could not transactivate p$Opgm_1$-*luc*, which has a mutation at site 1 of the *Opg* promotor. Moreover, the transactivational function of *Lef1* was conserved when mutations were introduced in sites 2 and 3, further establishing the importance of site 1 in the regulation of *Opg* expression by TCF proteins (FIG. 6D).

DISCUSSION

Here we present evidence demonstrating that β-catenin and presumably canonical Wnt signaling have a novel function in differentiated osteoblasts: the control of osteoclast differentiation. The evidence supporting this role includes: (1) the expression of two TCF proteins in osteoblasts that arise from intramembranous or endochondral ossification; (2) the decrease or increase in bone resorption caused by stabilizing or deleting β-catenin in differentiated osteoblasts, respectively; (3) the increase in bone resorption detected in $Tcf1^{-/-}$ mice; and (4) the identification of *Opg*, an inhibitor of osteoclast differentiation, as a target gene for β-catenin and TCF proteins. This function of β-*catenin* was revealed by stabilizing or deleting it from osteoblasts once differentiation has occurred through the use of $\alpha_1(I)$ *Collagen-Cre* mice, which do not express *Cre* in osteoblasts before E14.5.[2]

Why does the loss of β-*catenin* in differentiated osteoblasts not result in the same phenotypic and molecular abnormalities in bone cells as the loss of *Lrp5*? One possibility is that the lack of β-catenin has a more dramatic effect on osteoblast function than the lack of LRP5, which is only one of two likely co-receptors for Wnt proteins present in osteoblasts. LRP6, another protein with homology to Arrow, is also expressed in osteoblasts, and hypomorphic mutations in *Lrp6* lead to a low bone mass in mice.[5] Thus it is possible that mice lacking both *Lrp5* and *Lrp6* in osteoblasts would display the bone resorption abnormalities characterizing the $\Delta\beta cat_{ob}$ mice. Another possibility is that the ligand for LRP5 controlling bone formation may not be a Wnt protein. To address this possibility the identification of one or more Wnts whose deletion in differentiated osteoblasts reproduces the phenotype caused by the deletion of β-catenin could be used to determine if they also act through LRP5.

REFERENCES

1. BRAULT, V., R. MOORE, S. KUTSCH, *et al.* 2001. Inactivation of the beta-catenin gene by Wnt1-Cre-mediated deletion results in dramatic brain malformation and failure of craniofacial development. Development **128**(8): 1253–1264.

2. DACQUIN, R., M. STARBUCK, T. SCHINKE, *et al.* 2002. Mouse alpha1(I)-collagen promoter is the best known promoter to drive efficient Cre recombinase expression in osteoblast. Dev. Dyn. **224**(2): 245–251.
3. HARADA, N., Y. TAMAI, T. ISHIKAWA, *et al.* 1999. Intestinal polyposis in mice with a dominant stable mutation of the beta-catenin gene. EMBO J. **18**(21): 5931–5942.
4. KARSENTY, G. & E.F. WAGNER. 2002. Reaching a genetic and molecular understanding of skeletal development. Dev. Cell **2**(4): 389–406.
5. KOKUBU, C., U. HEINZMANN, T. KOKUBU, *et al.* 2004. Skeletal defects in ringelschwanz mutant mice reveal that Lrp6 is required for proper somitogenesis and osteogenesis. Development **131**(21): 5469–5480.
6. LOGAN, C.Y. & R. NUSSE. 2004. The Wnt signaling pathway in development and disease. Annu. Rev. Cell. Dev. Biol. **20:**781–810.
7. ROOSE, J., G. HULS, M. VAN BEEST, *et al.* 1999. Synergy between tumor suppressor APC and the beta-catenin-Tcf4 target Tcf1. Science **285**(5435): 1923–1926.
8. SIMONET, S., D.L. LACEY, C.R. DUNSTAN, *et al.* 1997. Osteoprotegerin: a novel secreted protein involved in the regulation of bone density. Cell **89:** 309–319.
9. TAKAHASHI, N., T. AKATSU, N. UDAGAWA, *et al.* 1988. Osteoblastic cells are involved in osteoclast formation. Endocrinology **123**(5): 2600–2602.
10. TEITELBAUM, S.L. & F.P. ROSS. 2003. Genetic regulation of osteoclast development and function. Nat. Rev. Genet. **4**(8): 638–649.
11. VAN DE WETERING, M., J. CASTROP, V. KORINEK, *et al.* 1996. Extensive alternative splicing and dual promoter usage generate Tcf-1 protein isoforms with differential transcription control properties. Mol. Cell Biol. **16**(3): 745–752.
12. VERBEEK, S., D. IZON, F. HOFHUIS, *et al.* 1995. An HMG-box-containing T-cell factor required for thymocyte differentiation. Nature **374**(6517): 70–74.
13. YASUDA, H., N. SHIMA, N. NAKAGAWA, *et al.* 1998. Identity of osteoclastogenesis inhibitory factor (OCIF) and osteoprotegerin (OPG): a mechanism by which OPG/OCIF inhibits osteoclastogenesis in vitro. Endocrinology **139:** 442–444.

Inhibition of ATF4 Transcriptional Activity by FIAT/γ-Taxilin Modulates Bone Mass Accrual

VIONNIE W.C. YU,[a] CLAUDE GAUTHIER,[a] AND RENÉ ST-ARNAUD[a,b]

[a]*Genetics Unit, Shriners Hospital for Children, Montreal (Quebec), Canada H3G 1A6*

[b]*Departments of Medicine, Surgery, and Human Genetics, McGill University, Montreal (Quebec), Canada H3A 2T5*

ABSTRACT: The basic domain-leucine zipper protein, activating transcription factor 4 (ATF4), was recently shown to control key aspects of osteoblast biology. ATF4 regulates the timely onset of osteoblast differentiation, the synthesis of type I collagen, and the transcription of the osteocalcin and RANKL (receptor activator of NFκ-B ligand) genes. Accordingly, the levels and activity of ATF4 are under tight control through mechanisms that include protein stability and phosphorylation. We have uncovered yet another mode of inhibition of ATF4 through its interaction with the leucine zipper protein FIAT (Factor Inhibiting ATF4-mediated Transcription, also described as γ-taxilin). FIAT/γ-taxilin localizes to the nucleus in osteoblasts and dimerizes with ATF4 to form inactive dimers, because it does not contain a DNA-binding basic domain moiety. The interaction of FIAT/γ-taxilin with ATF4 thus inhibits ATF4-mediated transcription. Transgenic mice overexpressing FIAT/γ-taxilin show osteopenia and reduced expression of the ATF4 target gene, osteocalcin. Interestingly, FIAT/γ-taxilin also interacts with the transcriptional coactivator αNAC (Nascent polypeptide associated complex And Coactivator alpha), suggesting alternative, non-mutually exclusive mechanisms contributing to the inhibition of ATF4-dependent osteocalcin gene transcription by FIAT/γ-taxilin.

KEYWORDS: ATF4; FIAT/γ-taxilin; αNAC; osteocalcin; osteoblast gene transcription

The basic domain-leucine zipper (bZIP) transcription factor ATF4 (Activating Transcription Factor 4) was recently characterized as a key regulator of osteoblast biology. ATF4 is a member of the activating transcription factor (ATF)/cyclic adenosine monophosphate responsive element-binding (CREB)

Address for correspondence: René St-Arnaud, Genetics Unit, Shriners Hospital for Children, 1529 Cedar Avenue, Montreal (Quebec), Canada H3G 1A6. Voice: 514-282-7155; fax: 514-842-5581. e-mail: rst-arnaud@shriners.mcgill.ca

Ann. N.Y. Acad. Sci. 1068: 131–142 (2006). © 2006 New York Academy of Sciences.
doi: 10.1196/annals.1346.027

family. This gene family consists of transcription factors that bind the cyclic adenosine monophosphate (cAMP) response element (CRE) through highly related, carboxy-terminal bZIP dimerization domains.[1] ATF4 is a 381 amino acid protein containing three acidic regions, hallmark of transactivation domains.[2–7] ATF4 can form homodimers[7,8] but can also heterodimerize with a variety of partners.[2,7,8] The dimerization partner appears to influence specificity of DNA binding[7] as well as transcriptional activity.[9,10]

One line of investigation that led to the identification of the role of ATF4 in osteoblasts was the study of the molecular etiology of Coffin-Lowry Syndrome (CLS), an X-linked mental retardation condition accompanied by delayed bone age, short stature, and delayed fontanelle closure.[11,12] The gene responsible for CLS is the growth factor-regulated Ribosomal S6 Kinase 2 (RSK2),[13] a serine/threonine kinase sporting dual kinase domains.[14] Activation of the C-terminal kinase domain of RSK2 by extracellular signal regulated kinase (ERK) leads to phosphorylation and activation of its N-terminal domain,[15,16] which can then phosphorylate its downstream targets, which include the bZIP family members CREB and c-Fos.[17,18] It was surmised, however, that RSK2 has additional substrates because CREB phosphorylation and c-Fos regulation are not perturbed in RSK2-deficient mice.[19] The search for additional substrates focused on bZIP family members (because the two known substrates belong to that large gene family) and *in vitro* kinase assays with recombinant bZIP molecules revealed that ATF4 was efficiently phosphorylated by the RSK2 enzyme.[20] This was confirmed through structure-function analyses in which the phosphoacceptor site was mapped, and the demonstration that ATF4 was hypophosphorylated in RSK2-deficient cells.[20] Thus, ATF4 rapidly appeared as a significant, physiologically relevant RSK2 target.

In a parallel line of experiments, the same team of investigators had previously purified a protein binding to the osteoblast specific element 1 (OSE1) motif within the proximal promotor region of the osteocalcin gene.[21,22] The physico-chemical properties of this protein suggested that it belonged to the bZIP family. The two seemingly unrelated projects merged when ATF4 was shown to bind the OSE1 element and to control transcription from the osteocalcin promotor in an RSK2-dependent manner.[20] Following this elegant biochemical analysis, the study of the bone phenotype of ATF4-deficient mice revealed that lack of ATF4 delays osteoblast differentiation and impacts on the expression of markers of terminal osteoblast differentiation such as bone sialoprotein and osteocalcin. Interestingly, while ATF4 gene inactivation does not affect the transcription of the alpha 1 chain of type I collagen, it severely inhibits the synthesis of the type I collagen protein. This is most likely due to the role of ATF4 in the transcription of genes involved in amino acid transport and metabolism, as the abnormality in type I collagen synthesis of ATF4-deficient osteoblasts was corrected when the cells were supplemented with non-essential amino acids. Thus, ATF4 was shown to control the timely onset of osteoblast differentiation, to be essential for terminal osteoblast differentiation, and to be required for proper synthesis of the major protein product of osteoblasts, i.e., type I collagen. The genetic cascade linking RSK2 and ATF4 was confirmed through the analysis of compound heterozygotes for both genes, which displayed similar phenotypic and cellular abnormalities as those described in RSK2- and ATF4-deficient mice.[20]

One additional transcriptional target of ATF4 in osteoblasts was subsequently identified as the osteoclast differentiation factor RANKL.[23] Thus ATF4 is involved in both aspects of bone remodeling, formation, and resorption.

With such a plethora of critical functions, it is not surprising that the transcriptional activity of ATF4 is tightly regulated. A first level of control involves protein stability. While ATF4 mRNA can be detected in all tissues and cell lines, the ATF4 protein is degraded via ubiquitinylation, except in osteoblasts where it can accumulate.[24,25] Post-translational modifications such as phosphorylation by protein kinase A or RSK2 can activate its transcriptional function.[20,23] An additional level of control is provided through interaction with specific partners. For example, heterodimerization of ATF4 with C/EBP changes its binding site specificity from a CRE to a CAAT element.[7] Interaction of ATF4 with Runx2 has also been suggested as necessary for osteoblast-specific activity of ATF4.[26]

Partnering can also inhibit ATF4 activity, as we have recently demonstrated through the cloning and characterization of the leucine zipper containing protein FIAT (Factor Inhibiting ATF4-mediated Transcription).[27] We identified FIAT in a yeast two-hybrid screen for proteins interacting with the αNAC transcriptional co-activator (FIG. 1).[28,29] A protein identical to FIAT (named γ-taxilin) was characterized on the basis of its extended C-terminal coiled-coil, and shown to interact with syntaxin family members.[30] In subsequent work, γ-taxilin was also shown to interact with αNAC.[31] Combined with our results that FIAT/γ-taxilin heterodimerizes with ATF4,[27] it appears that FIAT/γ-taxilin can interact with a variety of partners. Sorting out the physiological relevance of all interactions represents an active area of future investigation.

Primary sequence analysis revealed the presence of putative leucine zipper motifs within FIAT/γ-taxilin, and we sought to determine whether the protein could homodimerize or interact with other zipper-containing factors. Yeast two-hybrid assays with reciprocal FIAT/γ-taxilin "bait" and "target" vectors demonstrated that it does not homodimerize.[27] Using the FIAT/γ-taxilin bait to screen cDNA expression libraries from primary osteoblasts or osteoblastic cell lines, however, identified several independent clones for ATF4. FIAT/γ-taxilin and ATF4 are co-expressed in osteoblasts, and we successfully performed reciprocal co-immunoprecipitation of both endogenous proteins from osteoblastic cells, confirming that they interact in mammalian osteoblasts.[27]

We have shown that FIAT/γ-taxilin localizes to the nucleus in calvarial osteoblasts[27] and in ROS 17/2.8 (FIG. 2) and MC3T3-E1 osteoblastic cells (data not shown). Using overexpression of epitope-tagged FIAT, Yoshida *et al.*[31]

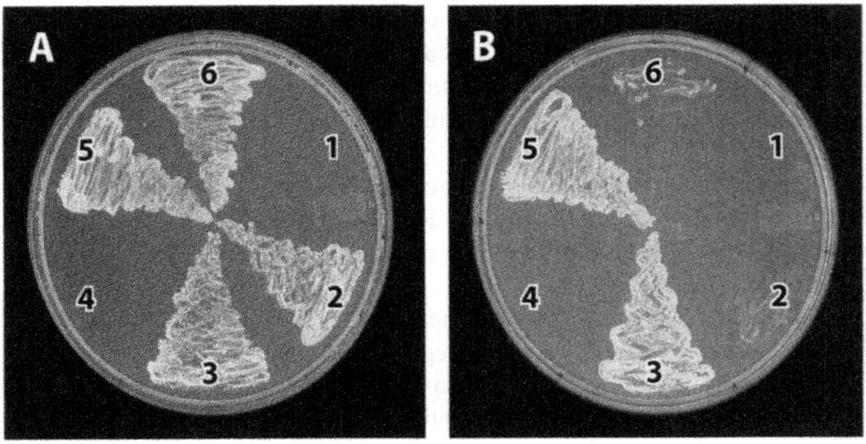

FIGURE 1. FIAT/γ-taxilin interacts with the αNAC transcriptional co-activator. Yeast two-hybrid protein interaction assays. (**A**) Growth on two-minus selection media. (**B**) Growth on three-minus selection media. Position 1: empty; position 2: αNAC "bait" + deletion mutant of the FIAT/γ-taxilin "target"; position 3: αNAC bait + FIAT/γ-taxilin target; position 4: empty; position 5: positive control; position 6: negative control. Note that on three-minus selection media, only yeasts at position 3 (αNAC + FIAT/γ-taxilin) and position 5 (positive control) grow, demonstrating a positive interaction between αNAC and FIAT/γ-taxilin. The failure of yeasts to grow at position 2 on three-minus selection media demonstrate that the C-terminal domain of FIAT/γ-taxilin is needed for its interaction with αNAC.

have described a cytoplasmic subcellular localization for the protein. Whether this observation is an artifact due to overexpression, epitope tagging, or the non-osteoblastic nature of the cell line used remains to be determined. Our data, which include immunofluorescence detection of the protein in the nucleus of primary osteoblasts and osteoblastic cells, Western blot detection in MC3T3-E1 nuclear extracts, and immunoprecipitation from ROS 17/2.8 nuclear extracts, clearly establish that FIAT/γ-taxilin can reside in the nucleus of osteoblastic cells.[27]

We examined whether the interaction of FIAT/γ-taxilin with ATF4 would impact the transcriptional activity of ATF4 and regulate osteocalcin gene transcription in mammalian cells. Osteoblastic MC3T3-E1 cells were transfected with the OSE1-luc reporter construct in which six copies of the OSE1-regulatory element from the osteocalcin gene promoter are subcloned upstream of the adenovirus type II major late promoter fused to the luciferase reporter gene.[21] The cells also received expression vectors for ATF4 and FIAT/γ-taxilin. As previously reported,[20] ATF4 induced the transcription of the OSE1-luc reporter (FIG. 3). FIAT/γ-taxilin suppressed the transcriptional activity of ATF4 (FIG. 3). We have also confirmed that FIAT/γ-taxilin by itself is without effect, and that the inhibition of ATF4-mediated transcription by FIAT/γ-taxilin can

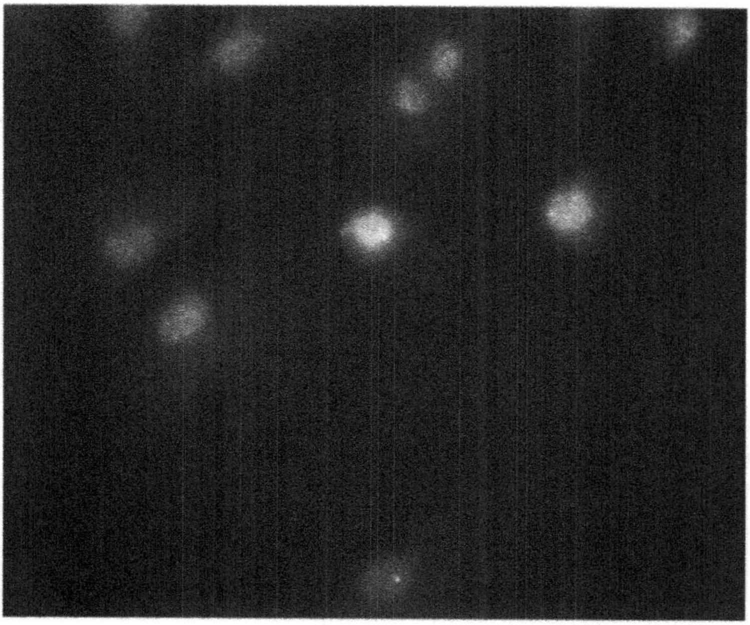

FIGURE 2. FIAT/γ-taxilin localizes to the nucleus of osteoblastic cells. ROS 17/2.8 osteoblastic cells were stained with the anti-FIAT antibody. The endogenous FIAT protein is detected in the nucleus of the cells.

occur in the presence or absence of an expression vector for constitutively active RSK2. These experiments show that FIAT/γ-taxilin is able to inhibit ATF4-mediated transcription.[27]

Considering the critical role of ATF4 in osteoblast biology,[20] we sought to determine whether FIAT/γ-taxilin could modulate the activity of ATF4 in bone and what impact this putative interaction would have on bone physiology. We engineered strains of transgenic mice expressing a human FIAT/γ-taxilin transgene under the control of the 2.3-kb osteoblast-specific fragment of the α1(I) collagen promoter.[32,33] The transgene was specifically expressed in osteoblasts and showed moderate levels of overexpression.[27] Interestingly, perturbed expression of FIAT/γ-taxilin in osteoblasts of transgenic mice inhibited osteocalcin gene transcription and decreased bone mass (FIG. 4).[27] The osteopenic phenotype manifested itself as a decreased number and size of trabeculae (FIG. 4). The same phenotype was observed in two independent strains of transgenic mice, confirming that it was not integration site dependent (data not shown).

The osteopenic phenotype of the FIAT/γ-taxilin transgenic mice was further characterized by histomorphometry. While the number of osteoblasts was not statistically different between transgenic animals and littermate controls, measurements of double-labeled tetracycline tibial sections revealed a

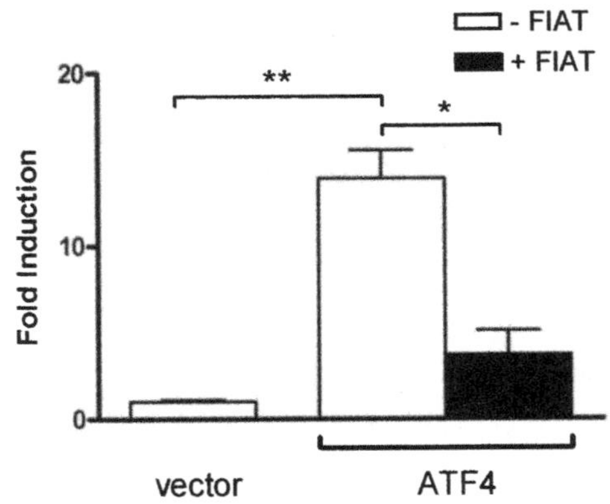

FIGURE 3. FIAT/γ-taxilin inhibits ATF4-mediated transcription. MC3T3-E1 osteoblastic cells were transfected with a reporter construct in which the luciferase gene was regulated by six copies of the wild-type OSE1 sequence from the mouse osteocalcin gene promoter. Co-transfected plasmids included expression vectors for ATF4 or FIAT/γ-taxilin (FIAT). Reporter gene activity was measured with a luminometer 24 hours post-transfection. The activity measured in cells transfected with the reporter construct and empty expression vectors was arbitrarily ascribed a value of 1. FIAT/γ-taxilin significantly inhibited ATF4-induced transcription. $^{*}P < 0.05$; $^{**}P < 0.01$.

lowered mineral apposition rate (46%) and bone formation rate (52%) in the transgenics.[27]

We also assessed differentiation using a battery of markers and RNA isolated from bones, primary calvarial cultures from newborn animals, and marrow

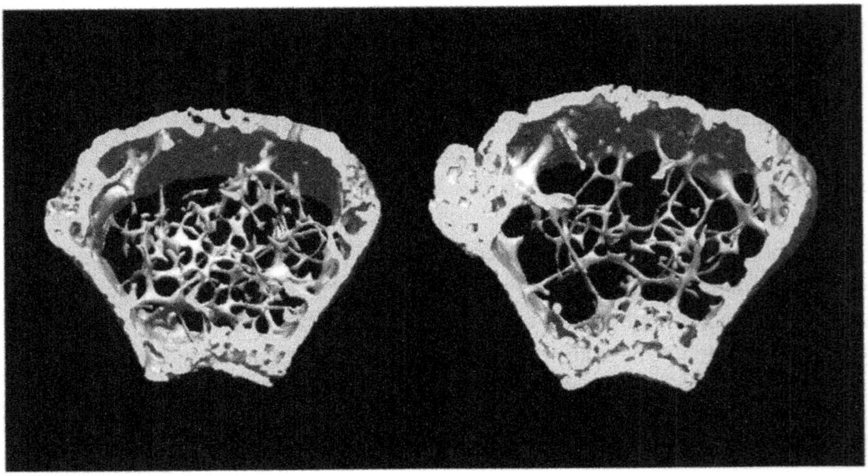

FIGURE 4. Reduced bone mass in FIAT/γ-taxilin transgenic mice. Femurs from a FIAT transgenic mouse (right) and a wild-type littermate control (left) were analyzed by micro-Computed Tomography. The reduced bone mass and lower number of trabeculae are evident in the FIAT transgenic sample.

stromal cell cultures. Relative expression of all markers assessed was similar irrespective of the source of the RNA. Our results show no differences between the expressions of most of the markers tested, except for the decrease in osteocalcin gene transcription. We interpret these results to mean that FIAT/γ-taxilin transgene expression has no major effect on osteoblast differentiation.

We ascertained that osteoblast proliferation or apoptosis was not affected by FIAT/γ-taxilin transgene expression.[27] However, when primary osteoblast cultures were analyzed for the production of alkaline phosphatase, a marker of osteoblastic differentiation and function, a statistically significant 34% decrease was observed in transgenic mice compared to wild type. This demonstrated that the bone phenotype of FIAT/γ-taxilin transgenic mice was primarily due to a cell-autonomous decrease in osteoblast activity.[27]

As previously mentioned, ATF4-deficient osteoblasts exhibit a marked reduction in the synthesis of type I collagen,[20] the most abundant protein produced by osteoblasts, which constitutes the majority of the mineralization scaffold in bone. We did not observe any change in type I collagen synthesis in FIAT/γ-taxilin transgenic osteoblasts.[27] It remains possible that FIAT/γ-taxilin does not modulate this activity of the ATF4 protein and thus would not impact on type I collagen synthesis. Alternatively, our results could be due to the relatively low level of FIAT/γ-taxilin overexpression achieved in the transgenics.[27] Additional studies (including structure-function analysis of the FIAT/γ-taxilin-ATF4 interaction and its impact on the transcription of other ATF4 target genes besides osteocalcin, FIAT/γ-taxilin RNA knock-down, or FIAT/γ-taxilin ablation by gene targeting) will be required to elucidate this point.

Several mechanisms could be involved in mediating the observed FIAT/γ-taxilin-dependent inhibition of ATF4-mediated transcription (FIG. 5). First and foremost, we have shown that the interaction of FIAT/γ-taxilin with ATF4 leads to the creation of inactive dimers that cannot bind the OSE1 response element (FIG. 5B).[27] This is not surprising because FIAT/γ-taxilin contains leucine zippers but no basic domain that would contribute to DNA binding. Other non-mutually exclusive possible mechanisms include FIAT/γ-taxilin interference with the interaction between ATF4 and other transcription factors, or FIAT/γ-taxilin recruitment of co-repressors to the chromatin (FIG. 5D). Interestingly, the c-Jun transcriptional coactivator αNAC was recently shown to potentiate c-Jun-dependent osteocalcin gene transcription by binding to the osteocalcin proximal promoter element near the OSE1 site.[34] The observed interaction of FIAT/γ-taxilin with αNAC (FIG. 1)[31] could lead to steric hindrance at the proximal promotor and contribute to the FIAT/γ-taxilin-mediated inhibition of osteocalcin gene transcription (FIG. 5C). Additional experiments will be required to determine whether these mechanisms contribute to the FIAT/γ-taxilin repressor activity.

It also remains a formal possibility that FIAT/γ-taxilin may disturb bone formation by other mechanisms in addition to suppressing ATF4 transcriptional

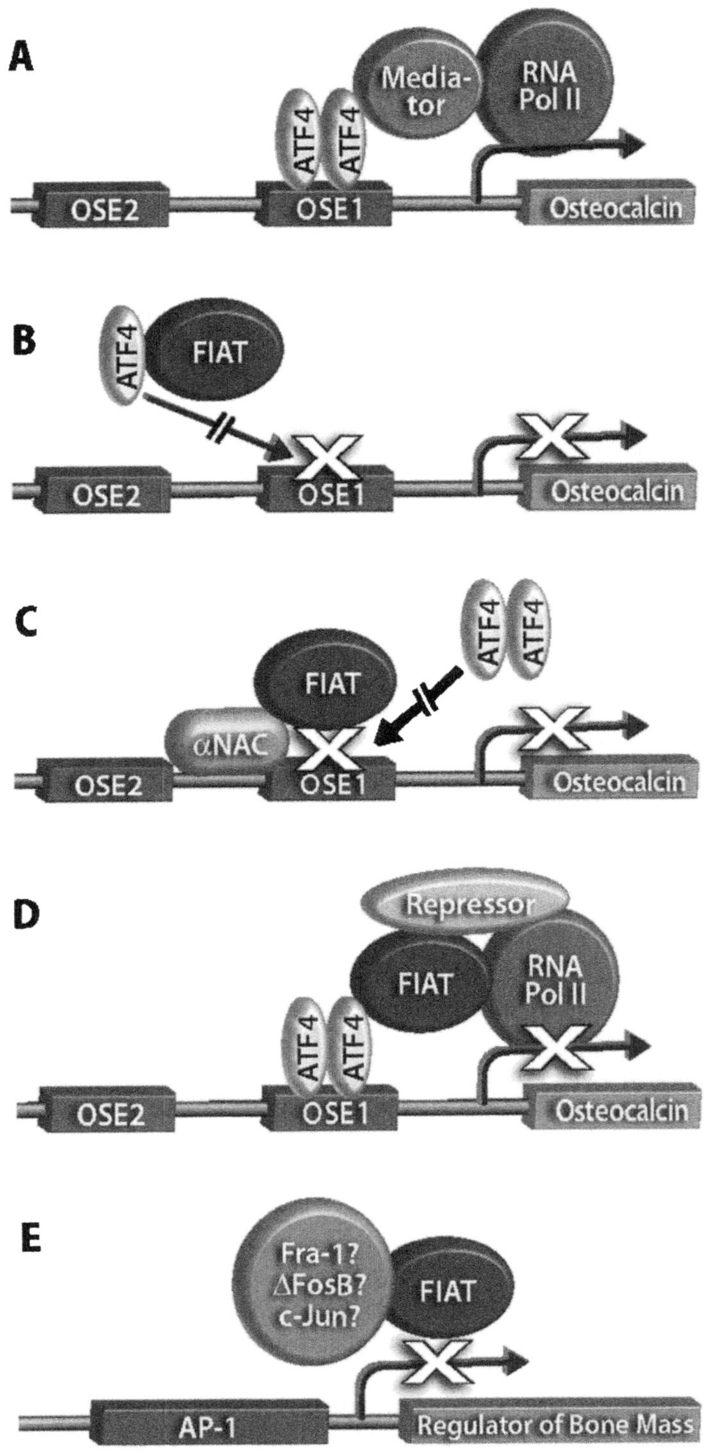
A
Media-
tor
RNA
Pol II
ATF4
ATF4
OSE2
OSE1
Osteocalcin
B
ATF4
FIAT
OSE2
OSE1
Osteocalcin
C
ATF4
ATF4
FIAT
αNAC
OSE2
OSE1
Osteocalcin
D
Repressor
FIAT
RNA
Pol II
ATF4
ATF4
OSE2
OSE1
Osteocalcin
E
Fra-1?
ΔFosB?
c-Jun?
FIAT
AP-1
Regulator of Bone Mass

activity. FIAT/γ-taxilin could heterodimerize with additional bZip transcription factors, such as Fra-1 or ΔFosB, that have been shown to regulate bone mass accrual.[35,36] Alternatively, FIAT/γ-taxilin could sequester dimerization partners away from b2Ip factors that control bone mass. A likely candidate remains c-Jun, a well-established dimerization partner of Fos family members (FIG. 5E).[37]

Several interesting questions remain unanswered. Will the targeted inactivation of FIAT/γ-taxilin generate the mirror image phenotype of the overexpression study and lead to a gain in bone mass, or even to an osteopetrosis? Which structural motifs within FIAT/γ-taxilin are responsible for its interaction with different partners? Does FIAT/γ-taxilin interact with additional bZIP transcription factors, such as c-Jun, Fra-1, or ΔfosB? What is the physiological relevance of the interaction of FIAT/γ-taxilin with syntaxin family members or αNAC? Is the subcellular localization of FIAT/γ-taxilin regulated post-translationally? Answers to these questions will further our knowledge of the physiological role of FIAT/γ-taxilin and determine whether it could represent an interesting target for pharmacological intervention in the treatment of bone diseases such as osteoporosis.

ACKNOWLEDGMENTS

R.St-A. thanks Dr. Mone Zaidi for participation in the Symposium and invitation to contribute to this volume of the Annals. Work from the laboratory of the authors was supported by the Shriners of North America. V.W.C.Y. was a Canada Bone Scholar from the Canadian Institutes for Health Research. The figures were prepared by Mark Lepik and Guylaine Bedard.

FIGURE 5. Putative mechanisms of FIAT/γ-taxilin-mediated transcriptional repression. (**A**) Schematized osteocalcin gene transcription. The binding of the ATF4 homodimer at the OSE1 (Osteoblast Specific Element 1) site recruits accessory factors (here globally represented as mediator) and leads to increased RNA polymerase II (RNA Pol II)-mediated transcription. (**B**) Interaction of ATF4 with FIAT/γ-taxilin (FIAT) forms inactive dimers that cannot bind the OSE1 site. (**C**) FIAT/γ-taxilin also interacts with the αNAC transcriptional co-activator, which has been shown to bind the osteocalcin promotor near the OSE1 site. This could lead to steric hindrance at the proximal promotor and prevent ATF4 dimers from binding to the OSE1 element. (**D**) Alternatively, FIAT/γ-taxilin could recruit repressor molecules to the promotor. (**E**) An additional, non-mutually exclusive mechanism would involve the interaction of FIAT/γ-taxilin with AP-1 bZIP family members that control the transcription of target genes regulating bone mass. We have accumulated experimental evidence demonstrating the validity of the mechanism depicted in panel B; the other mechanisms remain to be proven experimentally. OSE2, Osteoblast Specific Element 2, which is the binding site for Runx2.

REFERENCES

1. HAI, T. & M.G. HARTMAN. 2001. The molecular biology and nomenclature of the activating transcription factor/cAMP responsive element binding family of transcription factors: activating transcription factor proteins and homeostasis. Gene **273:** 1–11.
2. CHEVRAY, P.M. & D. NATHANS. 1992. Protein interaction cloning in yeast: identification of mammalian proteins that react with the leucine zipper of Jun. Proc. Natl. Acad. Sci. USA **89:** 5789–5793.
3. HAI, T.W. *et al.* 1989. Transcription factor ATF cDNA clones: an extensive family of leucine zipper proteins able to selectively form DNA-binding heterodimers. Genes Dev. **3:** 2083–2090.
4. KARPINSKI, B.A. *et al.* 1992. Molecular cloning of human CREB-2: an ATF/CREB transcription factor that can negatively regulate transcription from the cAMP response element. Proc. Natl. Acad. Sci. USA **89:** 4820–4824.
5. MIELNICKI, L.M. & S.C. PRUITT. 1991. Isolation and nucleotide sequence of a murine cDNA homologous to human activating transcription factor 4. Nucleic Acids Res. **19:** 6332.
6. TSUJIMOTO, A. *et al.* 1991. Isolation of cDNAs for DNA-binding proteins which specifically bind to a tax-responsive enhancer element in the long terminal repeat of human T-cell leukemia virus type I. J. Virol. **65:** 1420–1426.
7. VALLEJO, M. *et al.* 1993. C/ATF, a member of the activating transcription factor family of DNA-binding proteins, dimerizes with CAAT/enhancer-binding proteins and directs their binding to cAMP response elements. Proc. Natl. Acad. Sci. USA **90:** 4679–4683.
8. HAI, T. & T. CURRAN. 1991. Cross-family dimerization of transcription factors Fos/Jun and ATF/CREB alters DNA binding specificity. Proc. Natl. Acad. Sci. USA **88:** 3720–3724.
9. FAWCETT, T.W. *et al.* 1999. Complexes containing activating transcription factor (ATF)/cAMP-responsive-element-binding protein (CREB) interact with the CCAAT/enhancer-binding protein (C/EBP)-ATF composite site to regulate Gadd153 expression during the stress response. Biochem. J. **339:** 135–141.
10. LIM, C. *et al.* 2000. Latency-associated nuclear antigen of Kaposi's sarcoma-associated herpesvirus (human herpesvirus-8) binds ATF4/CREB2 and inhibits its transcriptional activation activity. J. Gen. Virol. **81:** 2645–2652.
11. COFFIN, R. *et al.* 1966. Treatment of lead encephalopathy in children. J. Pediatr. **69:** 198–206.
12. LOWRY, B., J.R. MILLER & F.C. FRASER. 1971. A new dominant gene mental retardation syndrome. Association with small stature, tapering fingers, characteristic facies, and possible hydrocephalus. Am. J. Dis. Child. **121:** 496–500.
13. TRIVIER, E. *et al.* 1996. Mutations in the kinase Rsk-2 associated with Coffin-Lowry syndrome. Nature **384:** 567–570.
14. FRODIN, M. *et al.* 2000. A phosphoserine-regulated docking site in the protein kinase RSK2 that recruits and activates PDK1. EMBO. J. **19:** 2924–2934.
15. DALBY, K.N. *et al.* 1998. Identification of regulatory phosphorylation sites in mitogen-activated protein kinase (MAPK)-activated protein kinase-1a/p90rsk that are inducible by MAPK. J. Biol. Chem. **273:** 1496–1505.
16. FISHER, T.L. & J. BLENIS. 1996. Evidence for two catalytically active kinase domains in pp90rsk. Mol. Cell. Biol. **16:** 1212–1219.

17. BRUNING, J.C. *et al.* 2000. Ribosomal subunit kinase-2 is required for growth factor-stimulated transcription of the c-Fos gene. Proc. Natl. Acad. Sci. USA **97:** 2462–2467.
18. XING, J., D.D. GINTY & M.E. GREENBERG. 1996. Coupling of the RAS-MAPK pathway to gene activation by RSK2, a growth factor-regulated CREB kinase. Science **273:** 959–963.
19. DUFRESNE, S.D. *et al.* 2001. Altered extracellular signal-regulated kinase signaling and glycogen metabolism in skeletal muscle from p90 ribosomal S6 kinase 2 knockout mice. Mol. Cell. Biol. **21:** 81–87.
20. YANG, X. *et al.* 2004. ATF4 is a substrate of RSK2 and an essential regulator of osteoblast biology; implication for Coffin-Lowry syndrome. Cell **117:** 387–398.
21. DUCY, P. & G. KARSENTY. 1995. Two distinct osteoblast-specific cis-acting elements control expression of a mouse osteocalcin gene. Mol. Cell. Biol. **15:** 1858–1869.
22. SCHINKE, T. & G. KARSENTY. 1999. Characterization of Osf1, an osteoblast-specific transcription factor binding to a critical cis-acting element in the mouse Osteocalcin promoters. J. Biol. Chem. **274:** 30182–30189.
23. ELEFTERIOU, F. *et al.* 2005. Leptin regulation of bone resorption by the sympathetic nervous system and CART. Nature **434:** 514–520.
24. LASSOT, I. *et al.* 2001. ATF4 degradation relies on a phosphorylation-dependent interaction with the SCF(betaTrCP) ubiquitin ligase. Mol. Cell. Biol. **21:** 2192–2202.
25. YANG, X. & G. KARSENTY. 2004. ATF4, the osteoblast accumulation of which is determined post-translationally, can induce osteoblast-specific gene expression in non-osteoblastic cells. J. Biol. Chem. **279:** 47109–47114.
26. XIAO, G. *et al.* 2005. Cooperative interactions between activating transcription factor 4 and Runx2/Cbfa1 stimulate osteoblast-specific osteocalcin gene expression. J. Biol. Chem. **280:** 30689–30696.
27. YU, V.W. *et al.* 2005. FIAT represses ATF4-mediated transcription to regulate bone mass in transgenic mice. J. Cell. Biol. **169:** 591–601.
28. MOREAU, A. *et al.* 1998. Bone-specific expression of the alpha chain of the nascent polypeptide-associated complex, a coactivator potentiating c-Jun-mediated transcription. Mol. Cell. Biol. **18:** 1312–1321.
29. YOTOV, W.V., A. MOREAU & R. ST-ARNAUD. 1998. The alpha chain of the nascent polypeptide-associated complex functions as a transcriptional coactivator. Mol. Cell. Biol. **18:** 1303–1311.
30. NOGAMI, S. *et al.* 2004. Identification and characterization of taxilin isoforms. Biochem. Biophys. Res. Commun. **319:** 936–943.
31. YOSHIDA, K. *et al.* 2005. Interaction of the taxilin family with the nascent polypeptide-associated complex that is involved in the transcriptional and translational processes. Genes Cells **10:** 465–476.
32. DACQUIN, R. *et al.* 2002. Mouse alpha1(I)-collagen promoter is the best known promoter to drive efficient Cre recombinase expression in osteoblast. Dev. Dyn. **224:** 245–251.
33. ROSSERT, J., H. EBERSPAECHER & B. DE CROMBRUGGHE. 1995. Separate cis-acting DNA elements of the mouse pro-alpha 1(I) collagen promoter direct expression of reporter genes to different type I collagen-producing cells in transgenic mice. J. Cell. Biol. **129:** 1421–1432.

34. AKHOUAYRI, O., I. QUELO & R. ST-ARNAUD. 2005. Sequence-specific DNA binding by the {alpha}NAC coactivator is required for potentiation of c-Jun-dependent transcription of the osteocalcin gene. Mol. Cell. Biol. **25:** 3452–3460.
35. JOCHUM, W. *et al.* 2000. Increased bone formation and osteosclerosis in mice overexpressing the transcription factor Fra-1. Nat. Med. **6:** 980–984.
36. SABATAKOS, G. *et al.* 2000. Overexpression of DeltaFosB transcription factor(s) increases bone formation and inhibits adipogenesis. Nat. Med. **6:** 985–990.
37. HARTL, M., A.G. BADER & K. BISTER. 2003. Molecular targets of the oncogenic transcription factor jun. Curr. Cancer Drug Targets **3:** 41–55.

Paget's Disease of Bone and Genetic Disorders of RANKL/OPG/RANK/ NF-κB Signaling

MICHAEL P. WHYTE

Center for Metabolic Bone Disease and Molecular Research, Shriners Hospitals for Children and Division of Bone and Mineral Diseases, Washington University School of Medicine, St. Louis, Missouri 63131 USA

ABSTRACT: Identification of the RANKL/OPG/RANK/NF-kB (receptor activator of nuclear factor κ-B ligand / osteoprotegerin) signaling pathway as the major regulatory system for osteoclastogenesis began with discovery of these ligands and receptors in the tumor necrosis factor (TNF) superfamily. Subsequently, genetically altered mice revealed physiologic roles for these proteins in bone biology. However, full appreciation of their significance for the human skeleton came from clinical characterization of several extremely rare, heritable disorders followed by discovery of their genetic bases. Familial expansile osteolysis (FEO) is an autosomal dominant disorder featuring constitutive activation of RANK due to an 18-bp tandem duplication in its gene (*TNFRSF11A*). A similar, 27-bp duplication causes what has been called a familial form of early-onset Paget's disease of bone (PDB2). Expansile skeletal hyperphosphatasia (ESH) is allelic to FEO and PDB2 and involves a 15-bp tandem duplication in *TNFRSF11A*. Autosomal recessive inheritance of deactivating mutations of the gene encoding OPG (*TNFRSF11B*) causes most cases of juvenile Paget disease. These disorders feature high bone turnover, deafness during early childhood, "idiopathic external lysis" of adult teeth, and sometimes focal lesions in appendicular bones that mimic active PDB. Biochemical markers indicate rapid skeletal remodeling. In FEO, osteolysis progresses to fat-filled bone rather than to osteosclerosis. Antiresorptive therapy with bisphosphonates can be effective for each disorder.

KEYWORDS: alkaline phosphatase; bisphosphonate; bone resorption; familial expansile osteolysis; fracture; hyperostosis; hyperphosphatasia; juvenile Paget's disease; NF-κB; osteoclast; osteolysis; osteoprotegerin; osteosarcoma; skeletal remodeling; tumor necrosis factor

Address for correspondence: Michael P. Whyte, M.D., Shriners Hospitals for Children, 2001 South Lindbergh Boulevard, St. Louis, MO 63131. Voice: 314-872-8305; fax: 314-872-7844.
e-mail: mwhyte@shrinenet.org

Ann. N.Y. Acad. Sci. 1068: 143–164 (2006). © 2006 New York Academy of Sciences.
doi: 10.1196/annals.1346.016

INTRODUCTION

Discovery of the RANKL/OPG/RANK/NF-κB signaling pathway as the major regulatory system for osteoclast formation and action began with the identification of new ligands and receptors in the tumor necrosis factor (TNF) superfamily and culminated with studies of genetically altered mice.[1] Nevertheless, full appreciation of its importance for humans came from revelation of the genetic bases of several extremely rare, mendelian disorders shown to be caused by either constitutive enhancement of receptor activator of nuclear factor-kappa B (RANK) or deficiency of osteoprotegerin (OPG).[2] In 2000, Hughes and colleagues found a heterozygous, gain-of-function, 18-bp tandem duplication in the gene encoding RANK (*TNFRSF11A*) in three kindreds with *familial expansile osteolysis* (FEO), and a similar insertional 27-bp duplication in *early-onset Paget's disease of bone* in Japan (PDB2).[3] In 2002, Whyte and Hughes reported that *expansile skeletal hyperphosphatasia* (ESH) was caused by a 15-bp tandem duplication in *TNFRSF11A*.[4] Also that year, Whyte and co-workers uncovered homozygous deletion of the gene encoding OPG (*TNFRSF11B*) in Navajo patients with *juvenile Paget's disease* (JPD).[5] Mutation of *TNFSF11* encoding RANKL has not been encountered in humans.

In this article, FEO, ESH, PDB2, and JPD are reviewed to illustrate the significance RANKL/OPG/RANK/NF-κB signaling has for skeletal homeostasis in man. However, a brief synopsis of Paget's disease of bone (PDB) is presented beforehand to appreciate how these unusual disorders resemble this far more common condition that is increasingly understood to involve heritable predisposition and other genes.[7]

PAGET'S DISEASE OF BONE

PDB (MIM 167250)[6] is common in the United States (i.e., overall prevalence at least 1%, and perhaps as much as 2%).[8,9] The disorder features focally increased skeletal remodeling within axial or appendicular bones in the adult skeleton.[10,11] Any site can be affected. Initially, a "wave" of osteoclast-mediated osteolysis moves slowly but relentlessly through an individual bone, and is then followed by disorganized skeletal repair.[10,11] This leads to bony expansion as well as to osteosclerosis and hyperostosis (excessive woven and lamellar bone in trabeculae and cortices, respectively).[10,11] Pagetic bone is unsound and predisposes to pain, fracture, and deformity. Eventually, "mosaic bone" from repeated episodes of resorption and formation characterizes lesional trabeculae.[10,11] Deafness of multifactorial pathogenesis is common,[12] and dental problems include loosening and migration of teeth.[13] Rarely, there is malignant transformation to, for example, osteosarcoma and chondrosarcoma.[10,11].

Despite its obvious focal attributes, PDB has traditionally been called the second most common metabolic bone disease (after osteoporosis), where generalized acceleration in skeletal remodeling has been attributed to increments in circulating parathyroid hormone levels.[14]

Although the precise etiology and pathogenesis of PDB remains unknown,[7,15,16] research advances are considerable. Accumulating evidence supports the name for the disorder suggested by Paget in 1877, that is, *osteitis deformans.*[17] Mindful that pathogens have not been cultured from pagetic bone, paramyxovirus infection of osteoclasts and their precursor cells at some remote time seems associated with the foci of excessive osteoclast action.[15,16,18–20] The marrow microenvironment is especially osteoclastogenic with viral proteins, transcripts, and inclusion bodies in osteoclasts from lesional tissue[15,16] and is reproduced in a measles virus-based mouse model for PDB.[21]

Additionally, increasing evidence indicates that PDB is often heritable and, therefore, develops on a systemic background. McKusick, in 1960,[22] recognized the familial predisposition and sometimes multigenerational occurrence of PDB and assigned to it autosomal dominant inheritance (MIM 167250).[6] Nevertheless, reports of large kindreds with PDB have, until recently, been rare.[10,23] Now, the prevalence of PDB in first-degree relatives of affected individuals is appreciated to be 12–40%, representing a sevenfold enhanced risk.[9,24]

In 1977, investigation of a few, small, PDB families suggested involvement of chromosome 6p21.3, near the HLA loci (PDB1) (MIM 167250).[6,25] Subsequent clinical characterization of FEO with its similarity to PDB (see below) and obvious autosomal dominant inheritance strengthened interest in a possible genetic basis for PDB.[26,27] In 1994, FEO was mapped to chromosome 18q21–22[28] and some studies of familial PDB indicated that this was also a chromosomal region predisposing to PDB (PDB2) (MIM 602080).[29,30] In fact, the as yet unpublished Japanese family with the 27-bp tandem duplication (75dup27) of *TNFRSF11A* in this region[3] (see below) was said to have an early-onset form of PDB (PDB2).[31] Linkage studies of PDB families were then consistent[29,32] or inconsistent[33,34] with a gene defect at 18q21.1–q22, but no further gene mutations were identified within this chromosomal region.[34] Additionally, PDB was shown not to involve *TNFRSF11A* mutations.[35,36]

Now, PDB is understood to be genetically heterogeneous,[6] and considerable effort is under way to identify all associated genes.[27] In 2001, PDB was mapped to chromosome 5q35-qter and 5q31 in French Canadian families (i.e., PDB3 and PDB4, respectively)[37,38] leading to discovery in 2002 of a mutation at 5q35-qter in the gene encoding sequestosome (*SQSTM1*).[39] *SQSTM1* defects were then identified worldwide in a large number of familial and some sporadic PDB cases.[39–43] Furthermore, the rare, autosomal dominant syndrome called inclusion body myopathy with early-onset Paget disease and frontotemporal dementia[44] (MIM 167320) was shown in 2004 to involve mutations in the gene encoding valosin containing protein (VCP).[45] SQSTM1 and VCP both seem to participate intracellularly in "ubiquination."[46] VCP may regulate nuclear factor-kappa B (NF-κB) signaling by binding to Inhibitor-κBα (IκBα) affecting ubiquitin-dependent proteosome degradation of IκBα[46]; however, this syndrome too seems to be genetically heterogeneous.[45] Additional genetic loci for PDB include 2q36, 10p13, and 18q23 (PDB 5,6, and 7, respectively).[7,38,47] Perhaps a factor from 18q21 acts as a modifier for PDB in some families, reflecting digenic inheritance.[34]

RANK EXCESS

Familial Expansile Osteolysis

Constitutive activation of RANK from tandem duplications in exon 1 of *TNFRSF11A* has been found in three seemingly distinctive disorders: FEO, ESH, and PDB2.[2]

FEO (hereditary expansile polyostotic osteolytic dysplasia) is an extremely rare yet remarkably instructive, autosomal dominant disorder (MIM 174810).[6] Patients manifest deafness early in life that is followed by destruction of adult dentition and focal, lytic expansion of major appendicular bones causing pain, fracture, and deformity.[26,48,49] The skeletal lesions initially resemble the clinical, radiographic, and histopathologic findings of PDB in its osteolytic phase,[10,11] but without treatment affected bones eventually become expanded, shell-like, and fat-filled rather than coarse and dense.[50,51] Histology of FEO osteolysis reveals an extreme local excess of bone turnover. Furthermore, generalized osteopenia and a coarse trabecular pattern suggest that FEO is a systemic bone disease. Elevation in serum alkaline phosphatase activity and other biochemical markers of bone remodeling surely derives from focal lesions, but is also contributed to, in some patients, by diffuse acceleration in skeletal remodeling.[2]

History

FEO was briefly reported in 1976 by Osterberg from Northern Ireland.[52] However, detailed characterization commenced with two 1979 publications by Enderle and colleagues from Germany concerning two brothers with an unusual,[10,11,53] severe, "osteolytic-expansive" PDB.[51,54] Subsequently, nine papers beginning in 1987 delineated the Northern Ireland kindred.[49,50,55–61] In 2002, Whyte and co-workers recounted 30 years' experience with an American FEO family demonstrating relatively mild disease.[62] That same year, a Spanish kindred was described.[63]

After FEO was mapped by Hughes and colleagues to chromosome 18q21.1–q22[28] in 1994, they discovered the molecular defect in 2000 using a candidate gene approach.[3] RANKL, a paracrine factor from mesenchyme and activated T cells in the bone marrow, binds to RANK promoting osteoclastogenesis.[1,64–73]

In the Northern Ireland, German, and American kindreds, Hughes and coworkers uncovered an identical, in-frame, 18-bp tandem duplication of bases 84–101 (84dup18) in exon 1 of the *TNFRSF11A* gene (MIM 603499) encoding RANK.[3] Transfection studies showed increased NF-κB activity presumably due to extension of the RANK signal peptide trapping the receptor intracellularly.[3]

Kindred Reports

Multigenerational FEO comprises the two-generation family in Germany [$FEO_{(Ger)}$], the five-generation kindred in Northern Ireland [$FEO_{(NI)}$], the five-generation kindred in the United States [$FEO_{(Am)}$], and the four-generation kindred in Spain [$FEO_{(Sp)}$] with 3, 46, 8, and 20 affected individuals, respectively.[26,50–52,54–62] Two additional, unrelated, American cases were reported in 2003.[74] Description separately of $FEO_{(Ger)}$, $FEO_{(NI)}$, $FEO_{(Am)}$, and $FEO_{(Sp)}$ is worthwhile because their *TNFRSF11A* defects are identical,[3] yet the families are unrelated and their disease expressivities differ.

$FEO_{(Ger)}$: For $FEO_{(Ger)}$, it is not know if all of the principal clinical features of FEO were manifest. The proband had expansile disease in his fibulae, radii, and patellae at ~age 25 years.[51,54] His brother had similar clinical and radiographic findings.[51] Their father was probably affected because he died at 47 years of age from metastatic osteosarcoma[51]—a cancer with increased prevalence in PDB and in $FEO_{(NI)}$[26] (see below). However, hearing loss or dental problems were not mentioned.[54] Detailed light microscopy of the bone lesions showed filigree-like trabeculae of woven bone (yet no mosaic pattern), abundant osteoclasts and osteoblasts lining trabeculae, giant osteoclasts with bizarre shapes and numerous nuclei, fibrous marrow, and subperiosteal bone formation.[51] Electron microscopy was not performed. The authors believed $FEO_{(Ger)}$ to be an "osteolytic-expansive" form of familial PDB in the "initial" "active" phase that failed to progress to the "inactive" phase.[51]

$FEO_{(NI)}$: Beginning in 1987,[49] eight reports by several groups of investigators[50,55–61] documented nearly complete penetrance of $FEO_{(NI)}$.[26,58] Severity differed among family members yet, with few exceptions, each suffered three major manifestations: osteolytic lesions, deafness, and tooth loss.[58] Osteolytic defects (90% prevalence) were especially common in the lower extremities and typically multifocal,[26,59] and a tibia was always involved.[59,61] Most of these lesions started at or near the end of a long bone[61] and then advanced[55] 6.5–22 mm yearly[56,59,61] like in PDB.[10,11] Pain in $FEO_{(NI)}$ began between the ages of 18 and 44 years, sometimes leading to limb amputation.[55–57] The maximum number of such defects in one patient was 12.[61] Conductive deafness (ultimately mixed-type) troubled at least 95% of affected individuals.[59,61] Deafness presented as young as the age of 4 years,[59] but more commonly in the

second decade.[56] The long process of the incus was usually absent or fibrous tissue.[26,59,61] Loosening and pain and/or fracture of adult teeth distressed most patients.[59,61] Primary teeth, however, were not involved.[57,58]

Radiographic findings in $FEO_{(NI)}$ included altered long bone modeling, particularly involving the humerus, radius, ulna, and tibia.[55,59] Furthermore, a tightly meshed ("fish net") trabecular pattern was variably present throughout the skeleton,[55] especially in metaphyses,[59] and sometimes in the mandible.[56] Bone scanning showed greater isotope uptake in tibias compared to femurs.[59] Osteopenia, however, was not mentioned as a significant feature of $FEO_{(NI)}$ [or $FEO_{(Ger)}$].[26]

Biochemical markers indicated accelerated skeletal turnover.[59]

Histopathology of the osteolytic lesions in $FEO_{(NI)}$ resembled PDB.[50,59,61] Osteoclasts were particularly large with increased numbers of nuclei.[61] On electron microscopy, all had microcylindrical nuclear inclusion bodies with ultrastructure resembling measles, canine distemper, or respiratory syncytial virus[50,59,61] similar to PDB.[15,50,56] Intermediate-stage disease, however, showed scanty skeletal matrix with abundant fibrous tissue and vascularity. Advanced disease featured almost total loss of cortical and trabecular bone, and fat occupied medullary spaces.[26,59,61] Nevertheless, light microscopy of the iliac crest was unremarkable.[61]

In $FEO_{(NI)}$ dentition, major histological disturbances included extensive root resorption, reduction in the size of pulp chambers and root canals, and patchy narrowing of the periodontal ligament.[57–59,61] However, the most remarkable finding was "idiopathic external resorption" that destroyed teeth (see below).[57,58]

$FEO_{(Am)}$: Experience with $FEO_{(Am)}$ supported heritable, excessive RANK/NF-κB activity[3] as a cause of early-onset deafness, loss of adult dentition, and expansile osteolysis of major appendicular bones (FIG. 1),[62] but it also underscored its systemic consequences on the skeleton. Osteopenia in adults was a common and sometimes clinically important feature of $FEO_{(Am)}$. Pathologic fractures occurred at sites other than those with osteolysis. Iliac crest histology did not show PDB. Notably, trauma seemed to incite osteolytic lesions in several patients (see below).[62] Additionally, $FEO_{(Am)}$ showed that the severity of FEO differed significantly among affected kindreds despite the identical 84dup18 *TNFRSF11A* duplication.[62]

$FEO_{(Am)}$ was not always fully penetrant. The four major features of FEO manifested in only some individuals. Deafness from degeneration of middle ear bones[75] was the first symptom—generally presenting from early childhood to young adult life, but perhaps not until the seventh decade. Premature tooth loss was exceptional, although idiopathic external resorption of secondary dentition could develop suddenly.[58]

In $FEO_{(Am)}$, it was not known if elevated serum alkaline phosphatase activity could occur at birth. The skeletal disease seemed to first disturb middle ear ossicles causing deafness. Accelerated bone remodeling then accompanied generalized osteopenia and probably accounted also for the characteristic, coarse trabecular pattern.[62] Osteolysis had the radiographic features of active PDB[10,11] and could begin any time during young adult life or middle age, and might present without hyperphosphatasemia. Osteopenia or a coarse trabecular pattern was not always evident when lytic lesions were discovered.[62]

Light microscopy of osteolytic lesions was always typical of active PDB.[10,23,50,51] Dr. Barbara Mills concluded that nothing distinguished the findings from "garden variety" PDB.[62] There was osteoclast-mediated bone resorption, membranous bone, and marrow fibrosis. However, "mosaic bone," the hallmark of advanced PDB,[10] was rare—perhaps reflecting the extreme rates of bone remodeling. Instead, woven bone seemed to be rapidly deposited and did not mature or remodel into cortical bone. This helped explain the

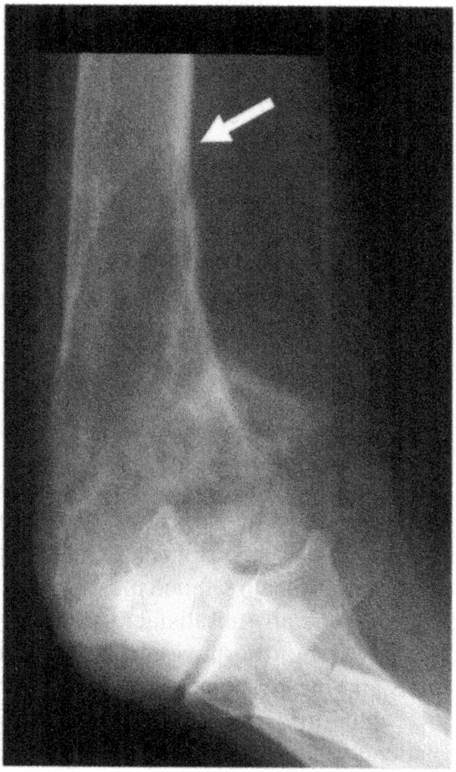

FIGURE 1. Lateral radiograph of the left elbow of the proband with $FEO_{(Am)}$ at the age of 71 years, a few years before elective amputation of the limb because of severe pain, shows an expansile ("soap bubble") osteolytic lesion in his distal left humerus, which is moving proximally (arrow) and predisposing to fracture. The radius and ulna are osteopenic with coarse trabeculae, probably reflecting disuse atrophy of muscle and bone distally as well as generalized acceleration of bone turnover.

absence of osteosclerosis.[62] Generally, little radiographic osteosclerosis occurred unless antiresorptive pharmaceuticals were administered.[62] Osteolytic lesions contained fibrous structures often involving most of the bone. Eroded areas of fibrous trabeculae had foci of numerous, enlarged, giant cells with many nuclei, some nestled in Howship's lacunae and appearing like PDB osteoclasts. Unlike in PDB, the lesions in untreated $FEO_{(Am)}$ had no "cutting cones" containing extensive concentrations of osteoclasts. On electron microscopy, paramyxovirus inclusions in osteoclasts were rare.[62] Iliac crest specimens showed no features of PDB. Advanced, appendicular, lesional disease has not been examined histologically in $FEO_{(Am)}$.

$FEO_{(Sp)}$. The fourth kindred with FEO, reported in Spain in 2002, had 20 affected individuals in four generations troubled by deafness, tooth loss, and osteoporosis with variable penetrance, but osteolytic lesions were a distinctly uncommon finding.[63] $FEO_{(Sp)}$ is caused by the same 18-bp *TNFRSF11A* insertional duplication (84dup18) found in the other three FEO kindreds.[3]

Additional Cases

The two unrelated American patients with FEO[74] manifested early-onset deafness and dental disease and especially tibial expansion. They had an 18-bp duplication in exon 1 of *TNFRSF11A* that is shifted proximally by one base-pair compared to the FEO kindreds yet resulted in substitution of the same six amino acids in the signal peptide of RANK.[74]

Treatment

The earliest report concerning FEO described, in 1976, symptomatic relief for Northern Ireland patients after injections of synthetic salmon calcitonin.[52] Later, dichloromethylene diphosphonate treatment was ineffective.[61] Intravenous infusions of pamidronate seemed most helpful.[26,61] Nevertheless, antiresorptive therapy was generally unsatisfactory.[26]

$FEO_{(Am)}$ patients received antiresorptive pharmaceuticals as they became available.[62] Calcitonins and bisphosphonates diminished biochemical markers of skeletal turnover with different efficacies. Radiographic changes were observed for some patients, including partial healing of osteolytic fronts with remineralization and cortical thickening of lesions,[62] as occurs in PDB.[11,76] Additionally, some $FEO_{(Am)}$ patients showed improvements in their generalized osteoporosis. Nevertheless, outcomes still seemed fair at best until recent excellent successes using oral alendronate treatment. Alendronate could mineralize early osteolytic lesions when they most closely resembled active PDB,[50,51] halt external tooth resorption (unpublished), increase bone density from the associated osteoporosis (unpublished), and fully correct biochemical markers of bone turnover using at first a 6-month course of 40 mg orally each day as for PDB.[62] Sustained clinical, biochemical, and radiographic improvement with increased bone mineral density could persist 2 years after stopping this therapy. However, refractoriness to antiresorptive treatment seemed to be present when bones were fat filled and lytic disease had "burnt-out."[50,51] No $FEO_{(Am)}$ patient, however, reported any change in deafness.

Pamidronate therapy administered intravenously was very helpful for another patient.[74]

In case overt or microscopic fracture incites osteolysis in FEO, avoiding trauma and orthodonture seemed prudent (see below).[62]

Phenotypic Variation

Delineation, thus far, of only four kindreds worldwide with FEO limits conclusions about the range of interfamilial variation despite sharing the identical 84dup18 in *TNFRSF11A*. In retrospect, $FEO_{(Ger)}$ seems to be the most severe FEO—featuring unusually symmetrical, polyostotic disease in the extremities; particularly the radii and fibulae.[51] $FEO_{(Am)}$ and $FEO_{(Sp)}$ appear more mild than $FEO_{(NI)}$[26] and especially $FEO_{(Ger)}$.[51,54] In $FEO_{(Am)}$, only one long bone per patient (humerus, ulna, femur, tibia, or fibula) developed lytic disease, and only one patient suffered tooth loss.[62] Additionally, despite elective amputation for one patient, no one with $FEO_{(Am)}$ developed an osteosarcoma[10,11] which affected one patient with $FEO_{(Ger)}$ and several with $FEO_{(NI)}$. Perhaps genetic background or greater dietary calcium and higher vitamin D levels suppressing circulating parathyroid hormone concentrations diminished the severity of $FEO_{(Am)}$ and $FEO_{(Sp)}$ compared to $FEO_{(Ger)}$ and $FEO_{(NI)}$.[62] Furthermore, we identified nuclear inclusions in only 1 of the 12 osteoclasts examined within the lytic lesion of one individual with $FEO_{(Am)}$.[62] Two other $FEO_{(Am)}$ patients studied had no inclusion bodies. The difference in clinical severity between relatively mild $FEO_{(Am)}$ versus $FEO_{(NI)}$ suggests that paramyxovirus infection could be an inciting or exacerbating factor.[62] Finally, different dental care (see below) and earlier therapeutic intervention or compliance with medical therapy could also influence the FEO phenotype among affected families. The two additional American patients, with 83dup18, also seemed to have FEO. Remarkably, the consequences of RANK activation appeared to differ according to the subtle mutation heterogeneity in exon 1 in *TNFRSF11A* demonstrated by the Japanese family with PDB2 caused by a 27-bp *TNFRSF11A* tandem duplication (75dup27)[3,31] and ESH caused by a 15-bp tandem duplication (84dup15)[4] (see below).

For one FEO patient, pregnancy (when skeletal turnover accelerates) seemed to exacerbate her osteolytic disease.[55] Additionally, the mother with ESH (see below) experienced worsening of her disorder during lactation.[4,77]

Nevertheless, any effect of birth control pills or hormone replacement therapy on FEO remains unknown.

In FEO, there seems to be no significant gender influence or physical effect from left-right dominance conditioning where osteolysis appears. However, it may be that skeletal injury partly explains the focal nature of FEO. Indeed, trauma has been considered as an initiating factor for some PDB lesions.[10,11] Several $FEO_{(Am)}$ patients gave histories consistent with this hypothesis.[62] Microscopic or macroscopic fracture in FEO might initiate skeletal repair that becomes deranged because excessive RANK effect enhances osteoclast numbers and activity. Osteolysis could then progress until an entire bone is involved.[10,11] In fact, the best evidence for this hypothesis could be the remarkable tooth loss of one young man with $FEO_{(Am)}$ who was fitted with orthodontic braces at about 11 years of age.[62] Only he among his kindred lost adult teeth by idiopathic external osteolysis and had undergone orthodonture, which moves teeth by activating osteoclasts to cause osteolysis within alveolar bone.[58] However, destruction of the dentition has been common over generations in $FEO_{(NI)}$ and other American patients[74] and, therefore, orthodontic trauma is not always the trigger.[57,58] Against the notion of pathogenetic trauma in FEO, some $FEO_{(Am)}$ patients suffered fractures or had iliac crest biopsies, yet did not develop osteolytic lesions at those sites. Furthermore, osteolytic disease was not a major disease feature in ESH or the Japanese family with PDB2[31] (see below). Pathogenetic trauma seems more plausible in FEO than PDB, because PDB often affects the skull, pelvis, scapula, and other axial bones that are not often fractured.[10,11]

FEO illustrates how focal bone lesions can be an important manifestation of an autosomal dominant skeletal disease.[62] Similar changes can also appear in ESH, PDB2, and JPD (see below) showing that such seemingly isolated skeletal defects are characteristic of disorders of RANKL/OPG/RANK/NF-κB signaling. The paucity of skeletal repair in untreated FEO (osteolytic expansion of the entire long bone without osteosclerosis or hyperostosis) compared to PDB is not understood.[10,11] The RANK abnormality seems to disproportionately compromise osteoblasts compared to osteoclasts. Once osteolysis runs its course, expanded bone becomes fat-filled, perhaps because the mesenchymal stem cell pool has differentiated excessively to adipocytes.[50,51] Understandably, advanced expansile lesions are then unresponsive to antiresorptive therapy.[62]

Expansile Skeletal Hyperphosphatasia

ESH was characterized clinically in 2000 in a mother and daughter at 36 and 11 years of age, respectively, and, like FEO and PDB2, is inherited as an autosomal dominant trait.[77] ESH features early-onset deafness, accelerated bone remodeling, premature loss of teeth, progressive hyperostotic widening of long bones causing painful phalanges in the hands, and episodic hypercalcemia. At first, hypercalcemia together with absence of large osteolytic lesions or cortical thinning—rather hyperostosis in major long bones—suggested that ESH was not a variant of FEO.[77] However, in 2002, Whyte and Hughes identified a 15-bp tandem duplication (84dup15) in *TNFRSF11A* in both ESH patients that is remarkably similar to the mutations causing FEO and PDB2 (see before).[4] Hence FEO, ESH, and PDB2 are allelic conditions involving RANK, and ESH likely reflects systemic activation of NF-κB, etc., in the skeleton.[4]

As in FEO,[62] deafness was the earliest problem in ESH.[77,78] Both affected individuals developed hearing loss in infancy that was complete by early childhood. The mother had absence of middle ear bones[77] perhaps resembling the necrotic degeneration with other abnormalities causing deafness in FEO.[26,75]

Premature tooth loss during childhood or early adult life also seemed to characterize ESH.[79] In the daughter, adult teeth were lost early, and the dental pathology was that of FEO.[79]

When the generalized skeletal disturbance of ESH first manifests is not known. Skeletal symptoms began at about 10 years of age.[77] The hands suffered the greatest discomfort and deformity.[77] Bony expansion at proximal interphalangeal joints of the daughter was understandably worrisome for "arthritis," and the joints could be tender, but they did not feel hot. Possibly, pain was due to periosteal stretching or joint distortion.[77]

Radiographs of the mother revealed that ESH eventually disturbs the entire skeleton.[77] Her calvarium, thoracic vertebral bodies, and long bones were affected most. The major abnormalities were expansion (undertubulation) as well as hyperostosis (cortical thickening) of large and small long bones.[77] However, the most remarkable changes involved her fingers where phalangeal widening could be extraordinary. The daughter's radiographic changes were similar, but less severe.[77] Of interest, the daughter's bone pain and finger swelling started soon after a forearm fracture.[77]

Reportedly,[78] the mother's skeletal disease was influenced by intercurrent illnesses and pregnancies.[4] Episodic hypercalcemia occurred spontaneously during childhood and later during illnesses and lactation. Hypercalcemia eventually manifested in her daughter.[4] Chosich and colleagues[78] postulated that the mother's hypercalcemia could reflect hormonal changes, including lowered estrogen levels during lactation, acting on rapid skeletal turnover in JPD.[78]

Serum alkaline phosphatase activity and additional markers of bone remodeling in ESH were considerably elevated before antiresorptive therapy.[77] Coarse trabeculae on X ray were possibly explained by thin cortices providing a "window" on thicker trabecular struts.[77]

Histological studies of the iliac crest showed increased numbers of osteoblasts and osteoclasts.[77] In the daughter, accelerated bone turnover was suggested by the appearance of enhanced cellular resorptive activity.[77] Total bone volume was substantially increased in the mother.[77] There were reversal lines, but not in the mosaic pattern of PDB.[77] Peritrabecular fibrosis was not seen. Electron microscopy of the mother's bone showed disorganized collagen bundles as well as necrotic and apoptotic bone cells, but no osteocytic osteolysis.[77] Measles virus gene transcripts were not detected in her peripheral blood monocytes.[77]

Because ESH and FEO had sufficient likeness for a shared etiology and pathogenesis,[20,38] Whyte and Hughes tested *TNFRSF11A* as a candidate gene and reported in 2002 that ESH was caused by a 15-bp, tandem duplication (84dup15) in exon 1.[4] Despite the phenotypic differences, the mutations causing FEO, PDB2, and ESH proved remarkably similar, i.e., overlapping, in-frame, duplications in exon 1 of *TNFRSF11A* (84dup18, 75dup27, and 84dup15, respectively)[4] predicted to lengthen the signal peptide of RANK by six, nine, and five amino acids, respectively.[4] Hence, FEO,[3] PDB2, and ESH probably reflect excessive RANK activity, although transfection studies of the 84dup15 mutation have not been performed.[4] Additional genetic, epigenetic, or environmental factors could explain differences between FEO and ESH. Finally, after we studied these ESH patients, our suggestion for a trial of alendronate therapy,[4,77] based on the favorable observations in $FEO_{(Am)}$,[62] led to symptomatic relief and normalization of the biochemical disturbances.[77,79]

It is of interest that in 2005 we briefly described an American teenager who seems to be a phenocopy for ESH.[80] She manifested sporadic, early-onset, conductive deafness, tooth loss after orthodonture, expansion of a metacarpal, lytic lesion in a humerus, and hyperphosphatasemia, but had no defect in exon 1 of *TNFRSF11A*.

Early-Onset PDB2

"PDB2" was found to reflect a 27-bp tandem duplication (75dup27) in *TNFRSF11A* in 2000,[3] but the disease features of the six affected individuals spanning three generations in this Japanese family were first reported in 2003.[31] Subjects with PDB2 suffered hearing impairment and tooth loss beginning in the second or third decade of life, and later pain and deformity in the lower limbs.[31] One individual developed transient hypercalcemia during immobilization. Radiographs of affected bones revealed lytic and sclerotic lesions as well as bony enlargement and deformity. The majority of PDB2 patients also had pelvic and skull involvement, and all had particularly striking disease of the maxilla and mandible.[31] Most had enlargement of the small joints of the hands. Serum alkaline phosphatase activity was elevated 2–17 times above the normal range. An iliac crest specimen from an affected 51-year-old man showed evidence of increased bone turnover and woven bone considered typical of PDB.[31] The 75dup27 was believed to cause a PDB-like phenotype that is distinct from, but overlaps, FEO and ESH.[3,4]

Phenotype/Genotype Correlations

Unlike FEO, both ESH and PDB2 share episodic hypercalcemia, involvement of the skull and pelvis, bony enlargement in the hands, and absence of osteolytic lesions of major long bones.[31,77] The RANK transfection studies in 2000 had demonstrated that 75dup27 causing PDB2 increased NF-κB signaling, but perhaps not as much as the 84dup18 causing FEO.[3] How this difference in NF-κB stimulation could account for these differences is not clear. Studies (*in silico*) of the secondary structure of exon 1 of *TNFSF11A* revealed a possible explanation for its inclination to duplicate.[74]

Until additional kindreds or individuals with various activating *TNFRSF11A* defects are reported, ESH and PDB2 might best be considered allelic to FEO[4]—broadening the phenotype for *TNFRSF11A* enhancement due to lengthening of RANK's signal peptide[3] and adding similarity to PDB.[10,11] However, with so few patients described worldwide with RANK activation, future reports may show greater phenotypic overlap leading to their classification as a single disorder of "RANK excess."[2]

OPG DEFICIENCY

JPD [MIM 23900], also called idiopathic or hereditary hyperphosphatasia,[6] is typically diagnosed in infants or young children.[81,82] Exceptionally, the bone disease does not manifest until later in childhood.[5,82] There also seems to be an especially unusual and relatively mild type of JPD associated with mental retardation.[82] Approximately 40 case reports have been published.[6,82] Unlike autosomal dominant FEO, ESH, and PDB2, all forms of JPD are considered autosomal recessive disorders.[6]

JPD affects the entire skeleton.[83] This, in part, has prompted objection in the past to the disorder being called a type of PDB. However, JPD as well as FEO, ESH, and PDB2 seem increasingly to share a number of features with PDB (see below).

Severe JPD causes bone pain, fracture, and deformity.[6,81,84–86] Premature loss of teeth and deafness[87] are also typical manifestations. Radiographs show marked undertubulation of long bones with osteopenic cortices.[87,88] Rapid rates of skeletal remodeling are inferred from substantially elevated biochemical markers of bone turnover,[81] including extreme hyperphosphatasemia,[5,81,82] but confirmed by histopathologic findings.[82] However, a mosaic pattern in bone characteristic of PDB is not found in JPD (or FEO, ESH, or PDB2). Measles virus transcripts, commonly identified in PDB,[20] were not detected in peripheral blood in mild[89] or severe cases of JPD (M.P. Whyte, R.V. Reddy, G.D. Roodman; unpublished).

Mild JPD is characterized by fewer fractures and less bony deformity.[5,82] Radiographs show diffuse, acquired hyperostosis and osteosclerosis associated with biochemical and histological evidence of accelerated skeletal turnover.[82] Understandably, Chosich and colleagues suggested that the mother with ESH had a mild form of JPD (see before).[78]

We have encountered a further similarity between JPD, PDB, FEO, and ESH—a focal, expansile, osteolytic ("soap bubble") defect. This occurred in the distal humerus of our Navajo proband with JPD.[2] Furthermore, even severe JPD will respond to antiresorptive therapy, including human calcitonin[5] and bisphosphonates.[84,90]

In 2002, we speculated that OPG deficiency could explain JPD, because OPG suppresses bone turnover by functioning as a decoy receptor for RANKL,[1] and reported complete, selective deletion of the gene encoding OPG (*TNFRSF11B*) in two apparently unrelated Navajos with this disorder.[5] Both affected individuals had identical break points on chromosome 8q24.2.[5] The deletion spanned ~ 100 kb, but neighboring genes seemed to be intact. Serum levels of OPG and RANKL were undetectable and markedly increased, respectively. However, we detected no splice-site or exon mutations in the OPG gene in two unrelated women with relatively mild JPD.[5] Soon after, Cundy and colleagues reported an Iraqi sibship with JPD and a homozygous 3-bp *TNFRSF11B* deletion.[91] In 2003, they described a variety of deactivating *TNFRSF11B* mutations in additional JPD patients. The nature of the gene defect seemed to these investigators to predict the severity of the skeletal disease.[92] In 2003, in a preliminary communication,[93] we reported that homozygous defects in *TNFRSF11B* are found in nearly all JPD patients having OPG deficiency—illustrating the importance of a founder as a cause for this inborn error of metabolism.[93] Homozygous deletion of *TNFRSF11B* in our two Navajo patients likely reflected a unique founder defect emerging in this "bottleneck" population that had decreased to approximately 6,000 people in 1868[94] and then re-expanded to approximately 225,000 by 1990.[95] Because three JPD patients from apparently separate Navajo families had been identified since the 1960s, we calculated that ~1 in 100 Navajo is a carrier of this *TNFRSF11B* deletion.[96]

Discovery of OPG deficiency in JPD provided both a cause and a mechanism for this osteopathy.[5] OPG is normally secreted into the marrow space by cells derived from mesenchyme, such as preosteoblasts and osteoblasts.[1] Absence of OPG biosynthesis would in turn cause high circulating (and presumably marrow space) unbound RANKL, which would bind to RANK and markedly accelerate osteoclastogenesis and bone turnover. In fact, mice lacking OPG from homozygous *TNFRSF11B* knockout had been reported to have osteoporosis, yet numerous osteoclasts and rapidly remodeling woven bone were noted.[97] Accordingly, these animals can be said to manifest JPD. Mice heterozygous for OPG deficiency could be osteopenic,[97] but the carrier parents of our JPD proband had unremarkable skeletal radiographs and bone densitometry (DXA) (unpublished), although their serum OPG levels were ~ 50% of control values.[5]

Furthermore, observations in JPD complement *TNFRSF11B* knockout mouse model studies that suggest a potential role for OPG in vascular calcification.[97–99] In these mice, aorta and renal artery mineralization was detected by histopathologic methods.[97] Nevertheless, no calcification was observed in these same arteries of our second Navajo JPD patient at age 23 years using computed tomography and "bone windows" sensitive for mineral.[5] Initially, our proband with marked hypercalciuria had nephrocalcinosis and perhaps several small calculi[5] on renal sonography; however, computed tomography of his kidneys, including renal arteries, was recently negative for calcifications (unpublished). Hence, we found no evidence for permanent, macroscopic, ectopic mineralization in severe JPD, at least during childhood or early adulthood. Nevertheless, the literature concerning JPD includes "calcifying arteriopathy," on histopathologic analysis of the internal elastic membrane of a temporal artery, in a 6-year-old boy.[100] Striking changes consistent with *pseudoxanthoma elasticum* (MIM 177850, 264800),[6] including granular and coarse deposits of calcium in the membranes and intima of the muscular arteries and arterioles, were reported in all autopsy tissues from a 26-year-old man with JPD.[101] It is of interest that the prevalence of vascular calcification also seems increased in PDB.[11,102]

JPD belongs among the disorders characterized by excessive signaling in the RANKL/OPG/RANK/NF-κB pathway leading to increased osteoclast action and accelerated rates of osseous tissue turnover. Nevertheless, JPD has some clinical and radiographic features (see above) that seem different compared to autosomal dominant FEO, ESH, and PDB2, suggesting that additional factors modify the effects of this pathway when activated by mutation in the genes encoding OPG verses RANK. Chromosome 8q24.2 contains *TNFRSF11B*, but has not been reported as one of the susceptibility loci for PDB. The coding sequence of *TNFRSF11B* is not mutated in PDB.[35]

Antiresorptive treatment using calcitonin or bisphosphonates has been beneficial for JPD.[5,84,89] Furthermore, if recombinant OPG will not be rejected as a foreign protein when there is deletion or mutated *TNFRSF11B*, OPG replacement might be helpful.[5] Recently, recombinant OPG has been effective for adult siblings with JPD.[103] Anti-RANKL antibody could be another therapeutic approach.

Finally, although similarities continue to emerge for PDB and disorders of the RANKL/OPG/RANK/NF-κB signaling pathway, henceforth it might be best to call JPD, caused by *TNFRSF11B* deactivation, "OPG deficiency."

CONCLUSIONS

FEO, ESH, and PDB2 (caused by constitutive activation of RANK) and JPD (due to OPG deficiency leading to increased access of RANKL to RANK) are the heritable disorders identified to date, which directly involve the RANKL/OPG/RANK/NF-κB signaling pathway.[2] Activating mutations in *TNFRSF11A* encoding RANK and deactivating mutations in *TNFRSF11B* encoding OPG cause systemic bone disease featuring accelerated bone turnover, low bone mass, deafness early in life, and loss of dentition by enhancing signaling by the RANK pathway. Osteopetrosis, featuring defective osteoclast action and dense, "marble bones," would be expected for deactivation of RANK or activation of OPG, but has not yet been reported in humans. Similarly, defects in the *TNFSF11* gene encoding RANKL have not been identified in humans.[6]

Despite the heritable (systemic) basis for these three autosomal dominant (FEO, ESH, PDB2) and one autosomal recessive (JPD) conditions, focal expansile osteolytic lesions may variably manifest in the phenotype. Perhaps such defects can begin from further local activation of osteoclast-mediated bone resorption due to skeletal trauma and are influenced by past exposure of bone to paramyxovirus. In fact, these four disorders resemble PDB, which can be inherited as an autosomal dominant trait featuring focal osteolytic disease, sometimes with deafness and tooth loss, and increasingly associated with mutations in other genes involved in intracellular processes. Although rare, further investigation of similar entities will likely reveal additional heritable disorders involving the RANKL/OPG/RANK/ NF-κB signaling pathway.

ACKNOWLEDGMENTS

Supported by Shriners Hospitals for Children and The Clark and Mildred Cox Inherited Metabolic Bone Disease Research Fund.

REFERENCES

1. MARTIN, T.J. 2004. Paracrine regulation of osteoclast formation and activity: milestones in discovery. J. Musculoskel. Neuron. Interact. **4:** 243–253.
2. WHYTE, M.P. & S. MUMM. 2004. Heritable disorders of the RANKL/OPG/RANK signaling pathway. J. Musculoskel. Neuron. Interact. **4:** 254–267.
3. HUGHES, A.E., S.H. RALSTON, J. MARKEN, *et al.* 2000. Mutations in TNFRSF11A, affecting the signal peptide of RANK, cause familial expansile osteolysis. Nat. Genet. **24:** 45–48.
4. WHYTE, M.P. & A.E. HUGHES. 2002. Expansile skeletal hyperphosphatasia is caused by a 15-base pair tandem duplication in TNFRSF11A encoding RANK and is allelic to familial expansile osteolysis. J. Bone Miner. Res. **17:** 26–29.
5. WHYTE, M.P., S.E. OBRECHT, P.M. FINNEGAN, *et al.* 2002. Osteoprotegerin deficiency and juvenile Paget's disease. N. Engl. J. Med. **347:** 174–184.
6. MCKUSICK, V.A. 1998 Mendelian Inheritance in Man: catalogs of Human Genes and Genetic Disorders, 12th ed. Johns Hopkins University Press, Baltimore and Online Mendelian Inheritance in Man 2000, OMIM (TM) McKusick-Nathans Institute for Genetic Medicine, Johns Hopkins University, Baltimore and National Center for Biotechnology Information, National Library of Medicine. Bethesda.
7. LUCAS, G.J.A., L.J. HOCKING, A. DAROSZEWSKA, *et al.* 2005. Ubiquitin-associated domain mutations of *SQSTM1* in Paget's disease of bone: evidence for a founder effect in patients of British descent. J. Bone Miner. Res. **20:** 227–231.
8. ALTMAN, R.D., D.A. BLOCH, M.C. HOCHBERG, *et al.* 2000. Prevalence of pelvic Paget's disease of bone in the United States. J. Bone Miner. Res. **15:** 461–465.

9. SIRIS, E.S. 1994. Epidemiological aspects of Paget's disease: family history and relationship to other medical conditions. Semin. Arthritis Rheum. **23:** 222–225.
10. HAMDY, R.C. 1981. Paget's Disease of Bone: assessment and Management. Praeger, New York.
11. KANIS, J.A. 1998. Pathophysiology and Treatment of Paget's Disease of Bone, 2nd ed. Blackwell Science. Malden, MA.
12. NAGER, G.T. 1975. Paget's disease of the temporal bone. Ann. Otol. Rhinol. Laryngol. **84:** 1–32.
13. SMITH, B.J. & J.W. EVESON. 1981. Paget's disease of bone with particular reference to dentistry. J. Oral Pathol. **10:** 233–247.
14. MEUNIER, P.J., J. COINDRE, C.M. EDOUARD, *et al.* 1980. Bone histomorphometry in Paget's disease. Quantitative and dynamic analysis of Paget's disease and non-pagetic bone tissue. Arthritis Rheum. **23:** 1095–1103.
15. REDDY, S.V., C. MENAA, F.R. SINGER, *et al.* 1999. Cell biology of Paget's disease. J. Bone Miner. Res. **14:** 3–8.
16. SINGER, F.R. 1999. Update on viral etiology of Paget's disease of bone. J. Bone Miner. Res. **14:** 29–33.
17. PAGET, J. 1877. On a form of chronic inflammation of bones (osteitis deformans). Med. Chir. Trans. **60:** 37–63.
18. REDDY, S.V. N. KURIHARA, C. MENAA, *et al.* 2001. Osteoclasts formed by measles virus-infected osteoclast precursors from hCD46 transgenic mice express characteristics of pagetic osteoclasts. Endocrinology **142:** 2898–2905.
19. ROSS, F.P. 2000. RANKing the importance of measles virus in Paget's disease. J. Clin. Invest. **105:** 555–558.
20. REDDY, S.V., F.R. SINGER, L. MALLETTE, *et al.* 1996. Detection of measles virus nucleocapsid transcripts in circulating blood cells from patients with Paget disease. J. Bone Miner. Res. **11:** 1602–1607.
21. ROODMAN, G.D. & J.J. WINDLE. 2005. Paget disease of bone. J Clin Invest. **115:** 200–208.
22. MCKUSICK, V.A. 1960. Heritable Disorders Of Connective Tissue, 2nd ed. Mosby, St. Louis, MO.
23. KIM, G.S., S.H. KIM, J.K. CHO, *et al.* 1997. Paget bone disease involving young adults in 3 generations of a Korean family. Medicine (Baltimore) **76:** 157–169.
24. MORALES-PIGA, A.A., J.S. REY-REY, J. CORRES-GONZALEZ, *et al.* 1995. Frequency and characteristics of familial aggregation of Paget's disease of bone. J. Bone Miner. Res. **10:** 663–670.
25. LEACH, R.J., F.R. SINGER & G.D. ROODMAN. 2001. Genetics of endocrine disease – The genetics of Paget's disease of the bone. J. Clin. Endocrinol. Metab. **86:** 24–28.
26. HUGHES, A.E. & J. BARR. 1996. Familial expansile osteolysis: a genetic model of Paget's disease. *In* The Molecular Biology of Paget's Disease. P.T. Sharpe, Ed.: 179–199. RG Landes. Heidelberg.
27. SINGER, F.R. & R.J. LEACH. 2000. Genetics of Paget's disease of bone. *In* The Genetics of Osteoporosis and Metabolic Bone Disease. M.J. Econs, Ed.: 309–318. Humana Press. Totowa, NJ.
28. HUGHES, A.E., A.M. SHEARMAN, J.L. WEBER, *et al.* 1994. Genetic linkage of familial expansile osteolysis to chromosome 18q. Hum. Mol. Genet. **3:** 359–361.
29. CODY, J.D., F.R. SINGER, G.D. ROODMAN, *et al.* 1997. Genetic linkage of Paget disease of the bone to chromosome 18q. Am. J. Hum. Genet. **61:** 1117–1122.

30. Haslam, S.I., W. Van Hul, A. Morales-Piga, *et al.* 1998. Paget's disease of bone: evidence for a susceptibility locus on chromosome 18q and for genetic heterogeneity. J. Bone Miner. Res. **13:** 911–917.
31. Nakatsuka, K., Y. Nishizawa & S.H. Ralston. 2003. Phenotypic characterization of early onset Paget's disease of bone caused by a 27-bp duplication in the TNFRSF11A gene. J. Bone Miner. Res. **18:** 1381–1385.
32. Leach, R.J., F.R. Singer, J.D. Cody, *et al.* 1999. Variable disease severity associated with a Paget's disease predisposition gene. J. Bone Miner. Res. **14:** 17–20.
33. Good, D., F. Busfield, D. Duffy, *et al.* 2001. Familial Paget's disease of bone: nonlinkage to the PDB1 and PDB2 loci on chromosomes 6p and 18q in a large pedigree. J. Bone Miner. Res. **16:** 33–38.
34. Johnson-Pais, T.L., J.H. Wisdom, K.S. Weldon, *et al.* 2003. Three novel mutations in SQSTM1 identified in familial Paget's disease of bone. J. Bone Miner. Res. **18:** 1748–1753.
35. Wuyts, W., L. Van Wesenbeeck, A. Morales-Piga, *et al.* 2001. Evaluation of the role of RANK and OPG genes in Paget's disease of bone. Bone **28:** 104–107.
36. Sparks, A.B., S.N. Peterson, C. Bell, *et al.* 2001. Mutation screening of the TNFRSF11A gene encoding receptor activator of NFκB (RANK) in familial and sporadic Paget's disease of bone and osteosarcoma. Calcif. Tissue Int. **68:** 151–155.
37. Laurin, N., J.P. Brown, A. Lemainque, *et al.* 2001. Paget disease of bone: mapping of two loci at 5q35–qter and 5q3. Am. J. Hum. Genet. **69:** 528–543.
38. Hocking, L.H., C.A. Herbert, R.K. Nicholls, *et al.* 2001. Genomewide search in familial Paget disease of bone shows evidence of genetic heterogeneity with candidate loci on chromosomes 2q36, 10p13, and 5q35. Am. J. Hum. Genet. **69:** 1055–1061.
39. Laurin, N., J.P. Brown, J. Morissette, *et al.* 2002. Recurrent mutation of the gene encoding sequestosome 1 (SQSTM1/p62) in Paget disease of bone. Am. J. Hum. Genet. **70:** 1582–1588.
40. Hocking, L.J., G.J.A. Gucas, A. Daroszewksa, *et al.* 2002. Domain-specific mutations in sequestosome 1 (SQSTM1) cause familial and sporadic Paget's disease. Hum. Mol. Genet. **11:** 2735–2739.
41. Falchetti, A., M. Di Stefano, F. Marini, *et al.* 2004. Two novel mutations at exon 8 of the sequestosome 1 (SQSTM1) gene in an Italian series of patients affected by Paget's disease of bone (PDB). J. Bone Miner. Res. **19:** 1013–1017.
42. Hocking, L.J., G.J. Lucas, A. Daroszewska, *et al.* 2004. Novel UBA domain mutations of SQSTM1 in Paget's disease of bone: genotype–phenotype correlation, functional analysis, and structural consequences. J. Bone Miner. Res. **19:** 1122–1127.
43. Good, D.A., F. Busfield, B.H. Fletcher, *et al.* 2004. Identification of SQSTM1 mutations in familial Paget's disease in Australian pedigrees. Bone **35:** 277–282.
44. Kimonis, V.E., M.J. Kovach, B. Waggoner, *et al.* 2000. Clinical and molecular studies in a unique family with autosomal dominant limb-girdle muscular dystrophy and Paget disease of bone. Genet. Med. **2:** 232–241.
45. Watts, G.D.J., J. Wymer, M.J. Kovach, *et al.* 2004. Inclusion body myopathy associated with Paget disease of bone and frontotemporal dementia is caused by mutant valosin-containing protein. Nat. Genet. **10:** 1–5.

46. Asai, T., Y. Tomita, S. Nakatsuka, *et al.* 2002. VCP (p97) regulates NFkappaB signaling pathway, which is important for metastasis of osteosarcoma cell line. Jpn. J. Cancer Res. **93:** 296–304.
47. Good, D.A., F. Busfield, B.H. Fletcher, *et al.* 2002. Linkage of Paget disease of bone to a novel region on human chromosome 18q23. Am. J. Hum. Genet. **70:** 517–525.
48. Nance, M.A., F.Q. Nuttall, M.J. Econs, *et al.* 2000. Heterogeneity in Paget disease of the bone. Am. J. Med. Genet. **92:** 303–307.
49. Wallace, R.G.H. 1987. A study of a familial bone dysplasia. MD thesis. The Queen's University of Belfast, Northern Ireland.
50. Dickson, G.R., P.V. Shirodria, J.A. Kanis, *et al.* 1991. Familial expansile osteolysis: a morphological, histomorphometric and serological study. Bone **12:** 331–338.
51. Enderle, A. & H.G. Willert. 1979. Osteolytic-expansive type of familial Paget's disease (osteitis deformans). Pathol. Res. Pract. **166:** 131–139.
52. Osterberg, P.H. 1976. An unusual familial disorder of bone: response to calcitonin. *In* Bone Disease and Calcitonin. J.A. Kanis, Ed.: 181–185. Armour Pharmaceutical Company. London, UK.
53. Altman, R.D. 1992. Paget's disease of bone. *In* Disorders of Bone and Mineral Metabolism. F.L. Coe & M.J. Favus, Eds.: 1027–1064. Raven Press. New York.
54. Enderle, A. & S. von Gumppenberg. 1979. Osteitis deformans (Paget) — or a tarda-type of a hereditary hyperphosphatasia. [German]. Arch. Orthop. Trauma Surg. **94:** 127–134.
55. Barr, R.J., A.E. Hughes, R.A. Mollan, *et al.* 1989. Idiopathic multicentric osteolysis: report of two new cases and a review of the literature. [letter]. Am. J. Med. Genet. **32:** 556.
56. Crone, M.D. & R.G. Wallace. 1990. The radiographic features of familial expansile osteolysis. Skeletal Radiol. **19:** 245–250.
57. Mitchell, C.A., J.G. Kennedy & R.G. Wallace. 1990. Dental abnormalities associated with familial expansile osteolysis: a clinical and radiographic study. Oral Surg. Oral Med. Oral Pathol. Oral Radiol. Endod. **70:** 301–307.
58. Mitchell, C.A., J.G. Kennedy & P.D. Owens. 1990. Dental histology in familial expansile osteolysis. J. Oral Pathol. Med. **19:** 65–70.
59. Osterberg, P.H., R.G. Wallace, D.A. Adams, *et al.* 1988. Familial expansile osteolysis: a new dysplasia. J. Bone Joint. Surg. Br. **70:** 255–260.
60. Wallace, R. 1988. McCabe's disease: hereditary expansile polyostotic osteolytic dysplasia [abstract]. J. Med. Genet. **25:** 276.
61. Wallace, R.G., R.J. Barr, P.H. Osterberg, *et al.* 1989. Familial expansile osteolysis. Clin. Orthop. **248:** 265–277.
62. Whyte, M.P., W.R. Reinus, M.N. Podgornik, *et al.* 2002. Familial expansile osteolysis (excessive RANK effect) in a 5-generation American kindred. Medicine (Baltimore) **81:** 101–121.
63. Palenzuela, L., C. Vives-Bauza, I. Fernández-Cadenas, *et al.* 2002. Familial expansile osteolysis in a large Spanish kindred resulting from an insertion mutation in the *TNFRSF11A* gene. J. Med. Genet. **39:** e67.
64. Dunstan, C.R. 2000. Osteoprotegerin and osteoprotegerin ligand mediate the local regulation of bone resorption. Endocrinologist **10:** 18–26.
65. Hofbauer, L.C., S. Khosla, C.R. Dunstan, *et al.* 2000. The roles of osteoprotegerin and osteoprotegerin ligand in the paracrine regulation of bone resorption. J. Bone Miner. Res. **15:** 2–12.

66. HOFBAUER, L.C. & A.E. HEUFELDER. 1998. Osteoprotegerin and its cognate ligand: a new paradigm of osteoclastogenesis. Eur. J. Endocrinol. **139:** 152–154.
67. HOFBAUER, L.C. & A.E. HEUFELDER. 2000. The role of receptor activator of nuclear factor-κB ligand and osteoprotegerin in the pathogenesis and treatment of metabolic bone diseases. J. Clin. Endocrinol. Metab. **85:** 2355–2363.
68. HSU, H., D.L. LACEY, C.R. DUNSTAN, *et al.* 1999. Tumor necrosis factor receptor family member RANK mediates osteoclast differentiation and activation induced by osteoprotegerin ligand. Proc. Natl. Acad. Sci. USA **96:** 3540–3545.
69. LACEY, D.L., E. TIMMS, *et al.* 1998. Osteoprotegerin ligand is a cytokine that regulates osteoclast differentiation and activation. Cell **93:** 165–176.
70. LORIES, R.J.U. & F.P. LUYTEN. 2001. Osteoprotegerin and osteoprotegerin-ligand balance: a new paradigm in bone metabolism providing new therapeutic targets. Clin. Rheumatol. **20:** 3–9.
71. YASUDA, H., N. SHIMA, N. NAKAGAWA, *et al.* 1998. Osteoclast differentiation factor is a ligand for osteoprotegerin/osteoclastogenesis-inhibitory factor and is identical to TRANCE/RANKL. Proc. Natl. Acad. Sci. USA **95:** 3597–3602.
72. LI, J., I. SAROSI, X.Q. YAN, *et al.* RANK is the intrinsic hematopoietic cell surface receptor that controls osteoclastogenesis and regulation of bone mass and calcium metabolism. Proc. Natl. Acad. Sci. USA **97:** 1566–1571.
73. NAKAGAWA, N., M. KINOSAKI, K. YAMAGUCHI, *et al.* 1998. RANK is the essential signaling receptor for osteoclast differentiation factor in osteoclastogenesis. Biochem. Biophys. Res. Commun. **253:** 395–400.
74. JOHNSON-PAIS, T.L., F.R. SINGER, H.G. BONE, *et al.* 2003. Identification of a novel tandem duplication in exon 1 of the TNFRSF11A gene in two unrelated patients with familial expansile osteolysis. J. Bone Miner. Res. **18:** 376–380.
75. ESSELMAN, G.H., J.A. GOEBEL & F.J. WIPPOLD 2nd. 1996. Conductive hearing loss caused by hereditary incus necrosis: a study of familial expansile osteolysis. Otolaryngol. Head Neck Surg. **114:** 639–641.
76. WHYTE, M.P. WIPPOLD, W.A. MURPHY, J.G., JR., *et al.* 1981. Arrest and healing of osteolytic Paget's bone disease with synthetic human calcitonin therapy: radiographic documentation. *In* Calcitonin: Chemistry, Physiology, Pharmacology, and Clinical Aspects. A. Pecile, Ed.: 217–224. Excerpta Medica, Princeton, NJ.
77. WHYTE, M.P., B.G. MILLS, G.D. ROODMAN, *et al.* 2000. Expansile skeletal hyperphosphatasia: a new familial metabolic bone disease. J. Bone Miner. Res. **15:** 2330–2344.
78. CHOSICH, N., F. LONG, R. WONG, *et al.* 1991. Post-partum hypercalcemia in hereditary hyperphosphatasia (juvenile Paget's disease). J. Endocrinol. Invest. **14:** 591–597.
79. OLSEN, C.B., K. TANGHAITRONG, I. CHIPPENDALE, *et al.* 1999. Tooth root resorption associated with a familial bone dysplasia affecting mother and daughter. Pediatr. Dent. **21:** 363–367.
80. WENKERT, D., W.H. MCALISTER, X. ZHANG, *et al.* Phenocopy for familial expansile osteolysis–expansile skeletal hyperphosphatasia [abstract]. J. Bone Miner. Res. In press.
81. COLE, D.E.C. & M.P. WHYTE. 1996. Hyperphosphatasia syndromes. *In* Studies in Stomatology and Craniofacial Biology. M.M. Cohen, JR., B.J. Baum, Eds.: 245–272. IOS Press. Amsterdam.
82. GOLOB, D.S., W.H. MCALISTER, B.G. MILLS, *et al.* 1996. Juvenile Paget disease: life-long features of a mildly affected young woman. J. Bone Miner. Res. **11:** 132–142.

83. CAFFEY, J. 1972. Familial hyperphosphatasemia with ateliosis and hypermetabolism of growing membranous bone; review of the clinical, radiographic and chemical features. Bull. Hosp. Joint Dis. **33:** 81–110.

84. CASSINELLI, H.R., C.A. MAUTALEN, J.J. HEINRICH, *et al.* 1992. Familial idiopathic hyperphosphatasia (FIH): response to long-term treatment with pamidronate (APD). Bone. Miner. **19:** 175–184.

85. EYRING, E.J. & E. EISENBERG. 1968. Congenital hyperphosphatasia: a clinical, pathological, and biochemical study of two cases. J. Bone Joint Surg. Am. **50:** 1099–1117.

86. BLANCO, O., M. STIVEL, C. MAUTALEN, *et al.* 1977. Familial idiopathic hyperphosphatasia: a study of two young siblings treated with porcine calcitonin. J. Bone Joint Surg. Br. **59:** 421–427.

87. RESNICK, D. 1995. Diagnosis of Bone and Joint Disorders, 3rd ed. WB Saunders, Philadelphia.

88. IANCU, T.C., G. ALMAGOR, E. FRIEDMAN, *et al.* 1978. Chronic familial hyperphosphatasia. Radiology **129:** 669–676.

89. WHYTE, M.P., R. LEELAWATTANA, S.V. REDDY, *et al.* 1996. Absence of paramyxovirus transcripts in juvenile Paget bone disease [letter]. J. Bone Miner. Res. **11:** 1041.

90. PETTIFOR, J.M., E.D. DANIELS, C.M. SCHNITZLER, *et al.* 1996. Hereditary hyperphosphatasia: response to oral alendronate in a 5-year-old child [abstract]. J. Bone Miner. Res. **11**(Suppl 1): S254.

91. CUNDY, T., M. HEGDE, D. NAOT, *et al.* 2002. A mutation in the gene TNFRSF11B encoding osteoprotegerin causes an idiopathic hyperphosphatasia phenotype. Hum. Mol. Genet. **11:** 2119–2127.

92. CHONG, B., M. HEGDE, M. FAWKNER, *et al.* 2003. International Hyperphosphatasia Collaborative Group. Idiopathic hyperphosphatasia and TNFRSF11B mutations: relationships between phenotype and genotype. J. Bone Miner. Res. **18:** 2095–2104.

93. MUMM, S., S. BANZE, J. PETTIFOR, *et al.* 2003. Juvenile Paget's disease: molecular analysis of TNFRSF11B encoding osteoprotegerin indicates homozygous deactivating mutations from consanguinity as the predominant etiology. J. Bone Miner. Res. **18**(Suppl 2): S388.

94. GIPIN, L. 1968. The Enduring Navaho. University of Texas Press. Austin.

95. 1994. 1990 Census of population and housing: characteristics of American Indians by tribe and language. Bureau of the Census. Washington, DC.

96. WHYTE, M.P., M.N. PODGORNIK & S. MUMM. 2002. Osteoprotegerin deficiency and juvenile Paget's disease [reply letter]. N. Engl. J. Med. **347:** 1622–1623.

97. BUCAY, N., I. SAROSI, C.R. DUNSTAN, *et al.* 1998. Osteoprotegerin-deficient mice develop early onset osteoporosis and arterial calcification. Genes Dev. **12:** 260–268.

98. WALLIN, R., N. WAJIH, G.T. GREENWOOD, *et al.* 2001. Arterial calcification: a review of mechanism, animal models, and the prospects for therapy. Med. Res. Rev. **21:** 274–301.

99. TINTUT, Y. & L.L. DEMER. 2001. Recent advances in multifactorial regulation of vascular calcification. Curr. Opin. Lipidol. **12:** 555–560.

100. SILVE, C., B. GROSSE, C. TAU, *et al.* 1986. Response to parathyroid hormone and 1,25-dihydroxyvitamin D3 of bone-derived cells isolated from normal children and children with abnormalities in skeletal development. J. Clin. Endocrinol. Metab. **62:** 583–590.

101. MITSUDO, S.M. 1971. Chronic idiopathic hyperphosphatasia associated with pseudoxanthoma elasticum. J. Bone Joint Surg. Am. **53:** 303–314.
102. STRICKBERGER, S., S. SCHULMAN & G. HUTCHINS. 1987. Association of Paget's disease of bone with calcific aortic valve disease. Am. J. Med. **82:** 953–956.
103. CUNDY, T., J. DAVIDSON, M.D. RUTLAND, *et al.* 2005. Recombinant osteoprotegerin for juvenile Paget's disease. N. Engl J. Med. **353:** 918–923.

The Regulation of Cathepsin K Gene Expression

BRUCE R. TROEN

Geriatrics Research, Education, and Clinical Center & Research Service, Miami Veterans Affairs Medical Center, and Geriatrics Institute, Department of Medicine, Miller School of Medicine, University of Miami, Miami, Florida 33125, USA

ABSTRACT: Cathepsin K is essential for normal bone resorption. Osteoclasts synthesize and secrete cathepsin K into the extracellular compartment at the attachment site between osteoclasts and the bone surface, wherein the organic matrix is subsequently degraded by cathepsin K. RANKL, NFAT, Mitf, and various components of AP-1 enhance osteoclast formation and bone resorption, whereas IFN-γ, calcitonin, estradiol, and calcium inhibit it. These agents appear to act correspondingly to alter cathepsin K mRNA and protein expression in order to stimulate and suppress the osteoclast's resorbing potential. RANKL signaling via the calcineurin-calcium-NFAT signaling cascade plays a significant role in the regulation of cathepsin K expression. Activation via p38 and the micropthalmia transcription factor also enhances cathepsin K expression. Future studies will be needed to elucidate the relative roles of various signaling pathways at different stages of osteoclast formation and activation and to determine whether genetically disrupting these pathways can modulate bone resorption with or without impeding other osteoclast functions.

KEYWORDS: cathepsin K; osteoclast; RANK ligand; NFAT; bone resorption

INTRODUCTION

Osteoporosis is one of the leading causes of morbidity in the elderly, resulting in over 1 million fractures each year and increased mortality for those with hip fractures. Osteoporosis is characterized by a progressive loss of total bone mass and bone density due to an imbalance between osteoclastic bone resorption and osteoblastic bone formation. Normal bone resorption and remodeling critically depend upon the synthesis and secretion of cathepsin K by osteoclasts (for reviews, please see Refs. 1–3). Humans with mutations in the cathepsin

Address for correspondence: Bruce R. Troen, M.D., Miami VA Medical Center, 11 GRC, 1201 NW 16th Street, Miami, FL 33125. Voice: 305-575-3388; fax: 305-575-3365.
e-mail: troen@miami.edu

**Ann. N.Y. Acad. Sci. 1068: 165–172 (2006). © 2006 New York Academy of Sciences.
doi: 10.1196/annals.1346.018**

K (CTSK) gene exhibit skeletal abnormalities (pycnodysostosis), and mice that are completely CTSK-deficient develop osteopetrosis. Therefore, CTSK plays an important role in the development of osteoporosis. Indeed, a number of pharmaceutical firms are actively developing inhibitors of cathepsin K enzymatic activity and are conducting clinical trials to use these agents as an antiresorptive to treat osteoporosis. Despite these clinically translational studies, the regulation of CTSK expression during osteoclast differentiation and in active bone-resorbing osteoclasts is incompletely understood. However, in the past several years, an increasing number of studies have begun to delineate the mechanisms underlying the transcription of the CTSK gene. Here we briefly review the progress made in understanding the regulation of CTSK gene expression.

REGULATION OF CTSK EXPRESSION

The regulation of bone resorption involves a complicated set of hormonal and/or cytokine interactions that initially stimulate osteoblasts and stromal cells, which then elaborate factors that signal osteoclasts to degrade bone (for reviews see Refs. 1,4–6). Key to understanding the regulation of cathepsin K expression is that many of the agents that have been shown to induce osteoclast formation and activation or to inhibit osteoclast activity enhance and suppress, respectively, cathepsin K gene expression. The receptor activator of NF-κB ligand (RANKL) plays a critical role in osteoclast differentiation and activation.[7–9] RANKL is a membrane-bound factor that is produced by osteoblasts and stromal cells in response to a variety of signals. RANKL binds to the cytoplasmic membrane receptor RANK (receptor activator of NF-κB), which is a member of the tumor necrosis factor (TNF) receptor superfamily, and subsequently induces both osteoclast differentiation and activation. RANKL stimulates CTSK mRNA and protein expression in human osteoclasts.[10] We have demonstrated that RANKL acutely stimulates cathepsin K mRNA expression in isolated mature rat osteoclasts.[11] In addition, as bone marrow-derived osteoclasts differentiate and fuse, they produce increasing amounts of cathepsin K.[11] RANKL also enhances CTSK mRNA expression in murine myeloid RAW 264.7 cells concomitant with induction toward an osteoclastic phenotype.[12] This phenomenon is both concentration- and time-dependent over a 7-day period [13] and also occurs at earlier times.[14] TNF-α, which is a member of the same superfamily of ligands as RANKL, also stimulates CTSK mRNA expression in pre-osteoclasts.[15] A large number of agents regulate the production of RANKL by osteoblasts and stromal cells.[1] Stimulators include vitamin D, parathyroid hormone, TNF-α, glucocorticoids, interleukins 1 (IL-1) and 11 (IL-11), thyroid hormone, prostaglandin E2, lipopolysaccharide, fibroblast growth factor-2, histamine, insulin-like growth factor-1, histamine, and low gravity. Inhibitors of RANKL expression include estrogen and transforming growth factor-β. Consequently, these agents indirectly exert corresponding stimulatory and inhibitory effects upon cathepsin K expression.

RANKL appears to stimulate the transcription of the cathepsin K gene via a number of mechanisms. An early and proximal event in RANKL-mediated signaling involves activation of TRAF6 (TNF receptor-associated factor 6), which is a critical adaptor molecule for the cognate receptor of RANKL. Overexpression of TRAF6 stimulates cathepsin K promotor activity, and RANKL stimulation of cathepsin K promotor activity is inhibited by the overexpression of dominant negative TRAF6.[14] More distally in the signaling pathway, Matsumoto *et al.* demonstrated that RANKL leads to the phosphorylation of NFAT2 by p38, thereby inducing translocation of NFAT2 into the nucleus and subsequent transactivation of the human CTSK promotor.[16] Overexpression of MKK6, which enhanced p38 activity also stimulated CTSK gene expression and promotor activity. This phosphorylation of NFAT2 contrasts with the classical paradigm whereby calcineurin dephosphorylates both NFAT1 and NFAT2, leading to nuclear translocation and subsequent promotor activation of a spectrum of genes.[17,18] However, it is possible that both dephosphorylation and phosphorylation of different moieties of NFAT2 may induce translocation and subsequent transactivation of transcription. We have shown that RANKL treatment induces NFAT2 mRNA expression in pre-osteoclasts.[19] Cyclosporine, which inhibits the phosphatase activity of calcineurin, thereby inhibiting NFAT activation, suppresses the stimulation of cathepsin K mRNA expression by RANKL.[14] Furthermore, inhibitors of calcium signaling, such as BAPTA-AM and the calcium ionophore, A23187, reduce RANKL-induced cathepsin K mRNA expression, NFAT2 overexpression markedly stimulates the rat cathepsin K promotor, and NFAT2 binds to a region of the rat cathepsin K promotor within 100 bp of the transcription initiation site.[14] Transfection of pre-osteoclasts with siRNA specific to NFAT2 markedly reduces the induction of cathepsin K mRNA by RANKL (Troen *et al.*, unpublished results).

RANKL treatment of cells also induces phosphorylation of the micropthalmia transcription factor (Mitf) via p38.[20] Mutations in the micropthalmia (*mi*) gene selectively affect osteoclast development and/or function and lead to osteopetrosis.[21,22] Mitf is the *mi* gene product and is a member of the helix-loop-helix (HLH) leucine zipper family and directly regulates cathepsin K gene transcription. Dominant negative mutations of Mitf exhibit osteopetrosis and lack cathepsin K mRNA and protein.[22] Overexpression of wild-type Mitf or TFE3, a member of the same transcription factor family, in cultured osteoclasts significantly enhanced cathepsin K expression. Mitf binds to three E-box motifs in the human CTSK promotor, of which mutation of any one significantly impairs stimulation of promotor activity by Mitf or TFE3.[22] In addition, both Mitf and PU.1 synergistically potentiated NFAT2 stimulation of human CTSK promotor activity.[16] Interestingly, osteoclasts that are null for both p21 and p27 fail to exhibit the marked stimulation of cathepsin K mRNA by RANKL.[23] Since p21 and p27 are cyclin-dependent kinase inhibitors that appear to modulate the activity of HLH transcription factors, the authors hypothesize that p21 and p27 regulate cathepsin K expression via Mitf.

Additional agents active in bone physiology stimulate cathepsin K expression. Retinoic acid can increase transcription of cathepsin K in isolated mature rabbit osteoclasts.[24] Extracellular matrix proteins that bind to osteoclast integrins, such as collagen type I, fibronectin, vitronectin, and osteopontin increased cathepsin K mRNA expression.[25] Carbaprostacyclin, which is a ligand for peroxisome proliferator-activated receptor delta/beta (PPARδ/β), induced cathepsin K mRNA expression in mature rabbit osteoclasts concomitantly with bone resorption.[26] Antisense oligonucleotide to the PPARδ/β blocked the response to carbaprostacyclin. Intermittent mechanical stretching of mature osteoclasts stimulated cathepsin K mRNA expression along with bone resorption.[27]

Physiological inhibitors of osteoclast differentiation and activation can also directly suppress cathepsin K expression. Osteoprotegerin (OPG) inhibited cathepsin K expression in purified rabbit osteoclasts.[28] Estrogen inhibited bone resorption by isolated rabbit osteoclasts and also downregulated cathepsin K mRNA.[29] In calvarial culture, estradiol suppressed cathepsin K mRNA expression and cathepsin K activity in response to PTH, IL-1α, IL-6, TNF-α.[30] Calvarial cultures from ovariectomized mice exhibited enhanced synthesis of cathepsin K, likely due to an indirect effect mediated by increased expression of RANKL by osteoblasts. Interferon-γ (INF-γ) inhibits cathepsin K mRNA and protein expression in murine osteoclasts, while IL-1α stimulated cathepsin K expression.[31] These same investigators were unable to observe a significant effect of either TNF-α or IL-6 on cathepsin K expression. We have also demonstrated that INF-γ also inhibits the stimulation of cathepsin K mRNA and protein expression by RANKL in the preosteoclast RAW 264.7 cell line.[13] Sodium butyrate and trichostain, which are both histone deacetylase inhibitors, reduced the stimulation of cathepsin K by RANKL.[32]

Members of the AP-1 family of transcription factors also regulate cathepsin K transcription. A mouse homologue of Jun dimerization protein 2 activates the cathepsin K promotor in transient transfections of mouse preosteoclast RAW cells.[33] Overexpression of junB stimulates cathepsin K promotor activity and also potentiates the stimulation by NFAT2.[14]

CONCLUSION

Therefore, activators and inhibitors of osteoclast bone resorption act, at least in part, through the stimulation and/or inhibition of cathepsin K gene expression. Some of these known and potential regulators of cathepsin expression are depicted in FIGURE 1. RANKL, NFAT, Mitf, and various components of AP-1 enhance osteoclast formation and bone resorption, whereas IFN-γ, calcitonin,

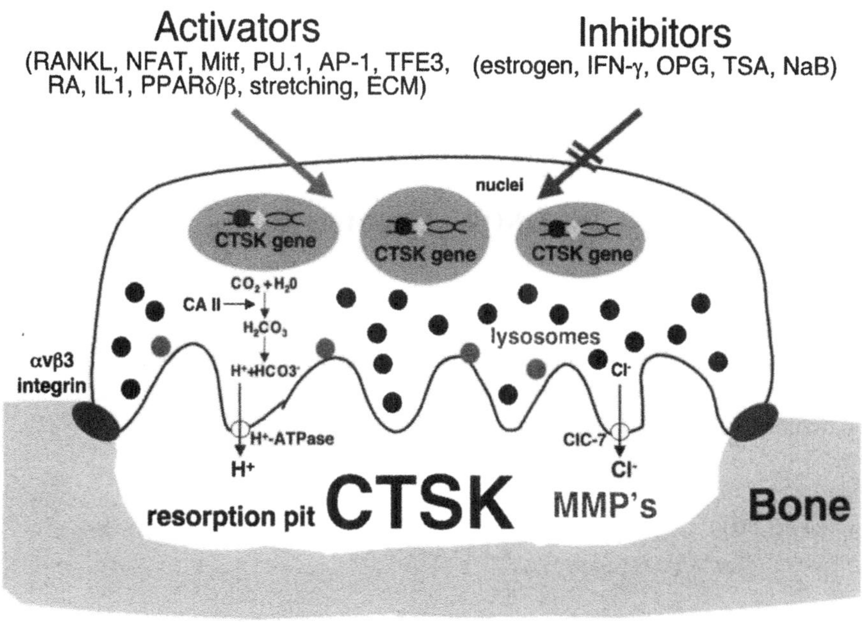

FIGURE 1. Regulation of cathepsin K expression. The osteoclast adheres tightly to bone, in part via binding of the $\alpha_V\beta_3$ integrin to the extracellular matrix proteins. The extracellular compartment is acidified by the H^+-ATPase proton pump that actively transports H^+ ions produced by the action of carbonic anhydrase (CAII) in the cell. Chloride channels (ClC-7) maintain electrical balance within the cell by extruding Cl^- ions into the resorption lacuna. Cathepsin K and other matrix metalloproteineases (MMPs) are transported from the endoplasmic reticulum within lysosomes and other secretory vesicles and then released into the resorption lacuna. Agents that stimulate the osteoclast to produce increased amounts of cathepsin K include RANKL, NFAT, Mitf, TNF, PU.1, AP-1, TFE3 (a member of the microphthalmia transcription factor family), retinoic acid (RA), IL-1, peroxisome proliferator-activator receptor δ/β (PPARδ/β), stretching, and extracellular matrix proteins (ECM). Inhibitors of cathepsin K expression include estrogen, interferon-γ (IFN-γ), osteoprotegerin, trichostatin A (TSA), sodium butyrate (NaB). (Adapted from reference 1, with permission.)

estradiol, and calcium inhibit it. These agents appear to act correspondingly to alter cathepsin K gene and protein expression in order to stimulate and suppress the osteoclast's resorbing potential. However, it is not known whether the regulation of cathepsin K gene expression is inextricably linked with osteoclast formation and activation. It would be intriguing to investigate the possibility of genetically enhancing or reducing cathepsin K gene transcription without facilitating or impeding other osteoclast functions. Furthermore, while multiple RANKL-associated signaling cascades and nuclear-binding factors play a role in cathepsin K gene expression, additional studies are needed to determine whether specific pathways and transcriptional stimulators preferentially act at different stages during osteoclastogenesis and bone resorption by mature osteoclasts. Such investigations will both enhance our understanding of osteoclast function and bone resorption and potentially provide new targets for developing therapies for osteoporosis treatment.

ACKNOWLEDGMENTS

Dr. Troen is supported by grants from the Department of Veterans Affairs (Merit Review) and the Indian Trail Foundation and also receives support from the Miami VA Geriatrics, Research, Education, and Clinical Center and the Gerontology and Geriatrics Division at the Miller School of Medicine at University of Miami.

REFERENCES

1. TROEN, B.R. 2004. The role of cathepsin K in normal bone resorption. Drug News Perspect. **17:** 19–28.
2. YASUDA, Y., J. KALETA & D. BROMME. 2005. The role of cathepsins in osteoporosis and arthritis: rationale for the design of new therapeutics. Adv. Drug Deliv. Rev. **57:** 973–993.
3. MOTYCKOVA, G. & D.E. FISHER. 2002. Pycnodysostosis: role and regulation of cathepsin K in osteoclast function and human disease. Curr. Mol. Med. **2:** 407–421.
4. TROEN, B.R. 2003. Molecular mechanisms underlying osteoclast formation and activation. Exp. Gerontol. **38:** 605–614.
5. KATAGIRI, T. & N. TAKAHASHI. 2002. Regulatory mechanisms of osteoblast and osteoclast differentiation. Oral Dis. **8:** 147–159.
6. TAKAYANAGI, H. 2005. Mechanistic insight into osteoclast differentiation in osteoimmunology. J. Mol. Med. **83:** 170–179.
7. LACEY, D.L. *et al.* 1998. Osteoprotegerin ligand is a cytokine that regulates osteoclast differentiation and activation. Cell **93:** 165–176.
8. SIMONET, W.S. *et al.* 1997. Osteoprotegerin: a novel secreted protein involved in the regulation of bone density [see comments]. Cell **89:** 309–319.
9. YASUDA, H. *et al.* 1998. Osteoclast differentiation factor is a ligand for osteoprotegerin/osteoclastogenesis-inhibitory factor and is identical to TRANCE/RANKL. Proc. Natl. Acad. Sci. USA **95:** 3597–3602.
10. SHALHOUB, V. *et al.* 1999. Osteoprotegerin and osteoprotegerin ligand effects on osteoclast formation from human peripheral blood mononuclear cell precursors. J. Cell Biochem. **72:** 251–261.
11. CORISDEO, S. *et al.* 2001. New insights into regulation of cathepsin K gene expression by osteoprotegerin ligand. Biochem. Biophys. Res. Commun. **285:** 335–339.
12. HSU, H. *et al.* 1999. Tumor necrosis factor receptor family member RANK mediates osteoclast differentiation and activation induced by osteoprotegerin ligand. Proc. Natl. Acad. Sci. USA **96:** 3540–3545.

13. PANG, M. *et al.* 2005. RANK ligand and interferon gamma differentially regulate cathepsin gene expression in pre-osteoclastic cells. Biochem. Biophys. Res. Commun. **328:** 756–763.
14. TROEN, B.R. *et al.* 2005. c-Jun Suppresses and JunB Enhances the stimulation of Cathepsin K promoter activity by NFAT2 in RAW 264.7 Cells. J. Bone Miner. Res. **20:** S372.
15. KOMINE, M. *et al.* 2001. Tumor necrosis factor-alpha cooperates with receptor activator of nuclear factor kappaB ligand in generation of osteoclasts in stromal cell-depleted rat bone marrow cell culture. Bone **28:** 474–483.
16. MATSUMOTO, M. *et al.* 2004. Essential role of p38 mitogen-activated protein kinase in cathepsin K gene expression during osteoclastogenesis through association of NFATc1 and PU.1. J. Biol. Chem. **279:** 45969–45979.
17. MACIAN, F., C. LOPEZ-RODRIGUEZ & A. RAO. 2001. Partners in transcription: NFAT and AP-1. Oncogene **20:** 2476–2489.
18. IM, S.H. & A. RAO. 2004. Activation and deactivation of gene expression by Ca2+/calcineurin-NFAT-mediated signaling. Mol. Cells **18:** 1–9.
19. ZHU, L.L. *et al.* 2004. RANK-L induces the expression of NFATc1, but not of NFkappaB subunits during osteoclast formation. Biochem. Biophys. Res. Commun. **326:** 131–135.
20. MANSKY, K.C. *et al.* 2002. Microphthalmia transcription factor is a target of the p38 MAPK pathway in response to receptor activator of NF-kappa B ligand signaling. J. Biol. Chem. **277:** 11077–11083.
21. WEILBAECHER, K.N. *et al.* 1998. Age-resolving osteopetrosis: a rat model implicating microphthalmia and the related transcription factor TFE3. J. Exp. Med. **187:** 775–785.
22. MOTYCKOVA, G. *et al.* 2001. Linking osteopetrosis and pycnodysostosis: regulation of cathepsin K expression by the microphthalmia transcription factor family. Proc. Natl. Acad. Sci. USA **98:** 5798–5803.
23. SANKAR, U. *et al.* 2004. RANKL coordinates cell cycle withdrawal and differentiation in osteoclasts through the cyclin-dependent kinase inhibitors p27KIP1 and p21CIP1. J. Bone Miner. Res. **19:** 1339–1348.
24. SANESHIGE, S. *et al.* 1995. Retinoic acid directly stimulates osteoclastic bone resorption and gene expression of cathepsin K/OC-2. Biochem. J. **309:** 721–724.
25. UEMURA, T., Y.K. LIU & Y. KUBOKI. 2000. Preliminary communication. MRNA expression of MT1-MMP, MMP-9, cathepsin K, and TRAP in highly enriched osteoclasts cultured on several matrix proteins and ivory surfaces. Biosci. Biotechnol. Biochem. **64:** 1771–1773.
26. MANO, H. *et al.* 2000. Cloning and function of rabbit peroxisome proliferator-activated receptor delta/beta in mature osteoclasts. J. Biol. Chem. **275:** 8126–8132.
27. KURATA, K. *et al.* 2001. Mechanical strain effect on bone-resorbing activity and messenger RNA expressions of marker enzymes in isolated osteoclast culture. J. Bone Miner. Res. **16:** 722–730.
28. WITTRANT, Y. *et al.* 2002. Osteoprotegerin differentially regulates protease expression in osteoclast cultures. Biochem. Biophys. Res. Commun. **293:** 38–44.
29. MANO, H. *et al.* 1996. Mammalian mature osteoclasts as estrogen target cells. Biochem. Biophys. Res. Commun. **223:** 637–642.
30. FURUYAMA, N. & Y. FUJISAWA. 2000. Regulation of collagenolytic cysteine protease synthesis by estrogen in osteoclasts. Steroids **65:** 371–378.

31. KAMOLMATYAKUL, S., W. CHEN & Y.P. LI. 2001. Interferon-gamma down-regulates gene expression of cathepsin K in osteoclasts and inhibits osteoclast formation. J. Dent. Res. **80:** 351–355.
32. RAHMAN, M.M. *et al.* 2003. Two histone deacetylase inhibitors, trichostatin A and sodium butyrate, suppress differentiation into osteoclasts but not into macrophages. Blood **101:** 3451–3459.
33. KAWAIDA, R. *et al.* 2003. Jun dimerization protein 2 (JDP2), a member of the AP-1 family of transcription factor, mediates osteoclast differentiation induced by RANKL. J. Exp. Med. **197:** 1029–1035.

Direct and Indirect Estrogen Actions on Osteoblasts and Osteoclasts

ALBERTA ZALLONE

Department of Human Anatomy and Histology, University of Bari 70124, Italy

ABSTRACT: Cells of osteoblast and osteoclast lineage are provided with the receptor for sex steroids, but discrepancies concerning mechanism of action still exist. Skeletal estrogen (ER) agonists induce osteoblastic osteoprotegerin (OPG) production through ER receptor-α activation *in vitro*, while immune cells appear to overexpress RANKL in ER deficiency *in vivo*, not reproduced in *in vitro* study. It has also been evident that the effects of ER on bone to a large extent are mediated via its action on immune cells. We know now that ER regulates the expression of cytokines that target cell types involved in modulating bone turnover, as IL-1 and IL-6, and the latest findings confirm and expand the concept that T cells are key mediators of bone loss following gonadal failure. Although early work demonstrated that tumor necrosis factor-α plays an important role in regulating bone mass, recent studies also implicate the lymphopoietic molecule IL-7: it suppresses the bone-forming osteoblasts, while stimulating formation and function of osteoclasts. More recent *in vitro* studies, however, indicate a stimulating effect of ER on osteoclastogenesis, which could have a positive effect on maintaining a high level of bone cell activity.

KEYWORDS: osteoclasts; osteoblast; estrogen; bone remodeling

The deleterious effect on the skeleton of loss of ovarian function at menopause has been known for long time, being followed by massive bone loss. The male skeleton also appears sensitive to the decline of androgen (AR) levels, albeit not as abrupt as menopause, and has been demonstrated that locally androgens are transformed by aromatase in estrogens active on bone cells.[1] Sex steroid deficiency causes bone loss unbalancing the normal process of bone regeneration, where bone resorption is followed by a proportional amount of bone formation. It has been suggested that this final effect is not dependent on the different activities of each single cell, but from the total number of cells working on a given surface.[2] The mechanism directly responsible for bone

Address for correspondence: Alberta Zallone, Dipartimento di Anatomia Umana e Istologia, Facoltà Medicina e Chirurgia, Policlinico P.za Giulio Cesare, Università di Bari 70124, Bari, Italy. Voice: 080-5478307; fax: 080-5478308.

e-mail: a.zallone@anatomia.uniba.it

Ann. N.Y. Acad. Sci. 1068: 173–179 (2006).
doi: 10.1196/annals.1346.019

loss observed after the reduction of sex steroids, is still a matter of passionate debate that has been recently intensified by the discovery of the alternative nongenomic pathway mediated by ER receptors.[3]

Classical receptors for estrogens (ERα and ERβ) or AR are present in bone marrow stromal cells, osteoblasts, and osteoclasts and their progenitors[4,5] indicating that the effects of sex steroids on bone are mediated, at least in part, directly. The level of receptor expression in bone cells is low (at least 10-fold) compared to reproductive organs and does not vary by gender, as similar levels of ER and AR have been found in both sexes.[6,7]

Osteoclasts derive from hematopoietic progenitor of the myeloid lineage, while osteoblasts are derived from mesenchymal stem cells, as well as stromal cells and adipocytes. Osteoclast formation is dependent on a delicate equilibrium between several cytokines, acting in an autocrine and paracrine manner and produced by osteoblasts, stromal cells, and as more recently evidenced, by lymphocytes. It has been proposed by Manolagas and Jilka[8] that ER regulates the production of these osteoclastogenic cytokines and that ER deficiency is responsible for the unbalance of this delicate equilibrium.

An increase in the circulating levels of interleukin 1 (IL-1) and interleukin 6 (IL-6) has been observed after menopause and these cytokines have been the first indicated in order of time as being responsible for bone loss in ER deficiency. High levels of IL-6 have been indeed associated with increased osteoclastogenesis.[9,10] They found that ER and AR suppress the production of IL-6 as well as the expression of the two subunits of IL-6 receptor in the bone marrow stromal/osteoblastic lineage. In *ex vivo* bone marrow cultures of ovariectomized mice, the expression of the ligand-binding subunit of the IL-6 receptor (gp80) and the signal-transducing subunit (gp130), was found to be increased.[11] Furthermore, estrogens regulate not only the receptor expression, but also the release of the cytokine from peripheral blood monocytes (PBMCs)[12] and its gene expression through NF-κB and C/EBP beta in human osteoblasts.The effect of estrogen/AR on IL-6 production was mediated by their classical receptors. The IL-6 promotor is also inhibited by ER in the absence of a functional ER receptor (ER)-binding site. This inhibition is mediated by the transcription factors NF-κB and C/EBP beta with a direct interaction between these two factors and ER.[13] Knockout of the IL-6 gene in mice prevents the loss of bone following the loss of sex steroids.[4]

A similar inhibitory effect of ER was observed on the production of the cytokines macrophage colony-stimulating factor (M-CSF) and tumor necrosis factor (TNF).[14,15] Early studies found enhanced biochemical indices of *in vivo* bone resorption associated with elevated levels of IL-1 and TNF-α, produced by mononuclear cells derived from ovariectomized women, but not from ovariectomized women after hormone replacement therapy.[16,17] The ovariectomy-induced bone loss in an animal model of postmenopausal osteoporosis is driven by TNF-α derived from T cells. Interestingly, ER deficiency does not lead to an upregulation of TNF-α production per cell, but stimulates the expansion of the population of TNF-α-producing T cells.[18] The capacity of TNF-α to stimulate osteoblastic RANKL production is, however, according to a very recent study, mediated by IL-1 and IL-1 may directly stimulate the differentiation of osteoclast precursors.[19]

After ovariectomy, T cell expansion and prolonged life span are driven through the interferon-γ (IFN-γ)-mediated induction of the multitarget immune modulator class II transactivator (CIITA), enhancing antigen presentation by macrophages that, in turn, secrete more IL-12 and IL-18. This T cell expansion and activation further drives up additional IFN-γ production.[20] IFN-γ inhibitory action on osteoclastogenesis has also been suggested in normal physiological conditions, by interfering with the RANKL-RANK signaling pathway. IFN-γ induces rapid degradation of the RANK adaptor protein, TRAF-6 (TNF receptor-associated factor 6), which results in strong inhibition of the RANKL-induced activation of the transcription factor NF-κB and c-Jun N-terminal kinase (JNK).[21] In the presence of ER deficiency, however, the negative effect prevails. Moreover, recent studies demonstrate that preexposure of osteoclast precursors to RANKL renders them resistant to the suppressive effect of IFN-γ.[22]

More recently also IL-7 has been associated to postmenopausal bone loss. IL-7 is significantly upregulated after ovariectomy: *in vivo* neutralization of IL-7 completely prevents ovariectomy-induced bone loss in mice and, moreover, suppressing the rise in IL-7 levels after ovariectomy also leads to upregulated bone formation.[23] IL-7 stimulates bone destruction when injected into intact wild-type mice, but fails to induce bone loss in T cell-deficient nude mice. Reconstitution of nude mice with T cells rescues IL-7-induced bone destruction, demonstrating that IL-7-dependent bone resorption is a T cell-mediated phenomenon.[24]

All together there seems to be good evidence of a regulatory link between the immune system and bone homeostasis, while there are severe discrepancies concerning the suggested direct effects of ER on bone cells.

Solid evidence arises from a group of researches indicating that the positive effect of the sexual hormone in maintaining bone homeostasis is exerted keeping a balance between cell activity and apoptosis. It has been demonstrated that, following the loss of sex steroids, osteoclasts erode deeper than normal cavities, leading to the removal of entire cancellous elements and loss of connection between the remaining ones.[25,26] This phenomenon is most likely due to the removal of the pro-apoptotic effect of estrogens on osteoclasts.[27] Ovariectomy or orchidectomy causes a dramatic increase in the prevalence of osteoblast and osteocyte apoptosis in mice. Addition of estrogens or androgens suppressed *in vitro* the apoptotic effect of several stimuli.[3] The authors elegantly demonstrated that the signals acted through a novel nongenotropic pathway leading to mitogen-activated protein kinase (MAPK) activation. The receptors are localized in the membrane in caveolae and appear to be derived from the same transcripts as nuclear transcripts.[3]

It has been hypothesized by several authors that the protective action of sex steroid on bone can be due to a change of the ratio among RANKL, cytokine-inducing osteoclastogenesis, and osteoprotegerin (OPG), its decoy-soluble receptor. This statement, however, has not yet been clearly proven, due to the discrepancy between the data in the literature. Different experimental models, different kinds or lengths of treatment, and culture systems not sufficiently estrogen-free gave rise to very different results obtained from different groups. Doubtless, OPG prevents bone loss when administered to ovariectomized rats, but RANKL can be produced by many sources. 17β-Estradiol dose dependently increased OPG mRNA and protein levels in human osteoblasts transfected with ER receptors.[28] The authors could not assess RANKL-mRNA expression, because the human osteoblastic-transfected cell lines used, produced very low levels of RANKL mRNA and were unable to give information about the estrogen-dependent RANKL/OPG ratio.

Using a primary cell line from young human donors, other authors prompted the first report of estrogen-induced changes in OPG, RANKL, and ER expression measured in the same human osteoblast cultures. They quantified cellular synthesis of OPG in response to ER and showed that the protein expression was significantly increased in estrogen-treated cells at both 24 and 48 h. At the same time either low or high doses of ER produced an increase in RANKL protein synthesis at 24 h that, however, returned to basal values at 48 h and an increase of OPG at 24 h returned back to control level at 48 h. These data were obtained by immunohistochemistry while more precise real-time polymerase chain reaction data indicate an increase in both OPG and RANKL at 24 h, with no data at a later time.[29] Other authors [30] in a more complex research with *in vivo* and *in vitro* data, analyzed the antiresorptive effect of ER or raloxifene *in vitro* and the concomitant gene expression profile of RANKL/RANK/OPG culturing not fractionated PBMCs. Bone resorption was decreased in all the treatment, while they found a decrease of mRNA for OPG and an increased or decreased RANKL mRNA depending on the concentration used.

Finally, two very recent reports indicate that ER can enhance osteoclastogenesis. In the first, Kousteni *et al.*[31] found that osteoclastogenesis, assayed in *ex vivo* bone marrow cultures, was upregulated after 1 h of treatment of ovx mice with estradiol. At the same time, E2 increased apoptosis of mature osteoclasts and protected calvaria-derived osteoblastic cells from etoposide-induced apoptosis. Another report shows, similarly, an increased osteoclastogenesis after a short *in vitro* or *ex vivo* treatment with estradiol, coupled with a decreased bone resorption.[31,32] The authors demonstrated that these effects were mediated by the autocrine/paracrine secretion of oxytocin, induced by ER treatment as in the presence of oxytocin inhibitors there were no changes in comparison with controls. In parallel, osteoblasts treated with a pulse of ER sufficient to induce oxytocin secretion, decreased OPG level and increased RANKL mRNA. These effects were for a short duration and followed a return to the basal condition at 48 h. Partially, in contradiction with these two reports are the data from

Garcia-Palacios *et al.*[33] The authors found an increase in the precursor's proliferation, not followed by an increased differentiation. They used, however, raw 264.7 cells and not primary cultures and the discrepancy could depend upon the experimental system used and the experimental planning: Data from Colaianni *et al.*[32] were obtained after 12 days of culture, while raw cells were observed after 5 days.

From our viewpoint, while the results regarding the role of the immune system in the unbalance of the cytokine and the direct nongenotropic effects on bone cells are clear, more researches, performed with more accurate and comparable protocols are necessary to uncover completely the role of sex hormones on bone cells. It is also clear that the mechanisms by which the presence of ER maintains a good bone quality imply several actors and mediators that are not necessarily related to the complex cascade of events that start in the absence of sex hormones.

REFERENCES

1. Bilezikian, J.P. 2002. Sex steroids, mice and men: when androgens and estrogens get very close to each other. J. Bone Miner. Res. **17:** 563–566.
2. Manolagas, S.C. 1999. Cell number versus cell vigor—what really matters to a regenerating skeleton? Endocrinology **140:** 4377–4381.
3. Kousteni, S., T. Bellido, L.I. Plotkin, *et al.* 2001. Nongenotropic, sex-nonspecific signaling through the ER or AR receptors: dissociation from transcriptional activity. Cell **104:** 719–730.
4. Bellido, T., R.L. Jilka, B.F. Boyce, *et al.* 1995. Regulation of interleukin-6, osteoclastogenesis, and bone mass by androgens. The role of the AR receptor. J. Clin. Invest. **95:** 2886–2895.
5. Couse, J.F. & K.S. Korach. 1999. ER receptor null mice: what have we learned and where will they lead us? Endocr. Rev. **20:** 358–417.
6. Benz, D.J., M.R. Haussler, M.A. Thomas, *et al.* 1991. High-affinity AR binding and androgenic regulation of alpha 1(I)-procollagen and transforming growth factor-beta steady state messenger ribonucleic acid levels in human osteoblast-like osteosarcoma cells. Endocrinology **128:** 2723–2730.
7. Braidman, I.P., C. Baris, P.L. Selby, *et al.* 2000. Preliminary report of impaired oER receptor-alpha expression in bone, but no involvement of AR receptor, in male idiopathic osteoporosis. J. Pathol. **192:** 90–96.
8. Manolagas, S.C. & R.L. Jilka. 1992. Cytokines, hematopoiesis, osteoclastogenesis, and estrogens. Calcif. Tissue Int. **50:** 199–202.
9. Manolagas, S.C.1995. Role of cytokines in bone resorption. Bone **17:** 63S–67S.
10. Manolagas, S.C., S. Kousteni & R.L. Jilka. 2002. Sex steroids and bone. Recent Prog. Horm. Res. **57:** 385–409.
11. Lin, S.C., T. Yamate, Y. Taguchi, *et al.* 1997. Regulation of the gp80 and gp130 subunits of the IL-6 receptor by sex steroids in the murine bone marrow. J. Clin. Invest. **100:** 1980–1990.
12. Bernard-Poenaru, O., C. Roux, R. Blanque, *et al.* 2001. Bone-resorbing cytokines from peripheral blood mononuclear cells after hormone replacement therapy: a longitudinal study. Osteoporos. Int. **12:** 769–776.

13. STEIN, B. & M.X. YANG. 1995. Repression of the interleukin-6 promoter by ER receptor is mediated by NF-kappa B and C/EBP beta. Mol. Cell Biol. **15:** 4971–4979.
14. SRIVASTAVA, S., M.N. WEITZMANN, R.B. KIMBLE, *et al.* 1998. Pacifici RER blocks M-CSF gene expression and osteoclast formation by regulating phosphorylation of Egr-1 and its interaction with Sp-1. J. Clin. Invest. **15:** 1850–1859.
15. SRIVASTAVA, S., M.N. WEITZMANN, S. CENCI, *et al.* 1999. ER decreases TNF gene expression by blocking JNK activity and the resulting production of c-Jun and JunD. J. Clin. Invest. **104:** 503–513.
16. PACIFICI, R., A. CARANO, S.A. SANTORO, *et al.* 1991. Bone matrix constituents stimulate interleukin-1 release from human blood mononuclear cells. J. Clin. Invest. **87:** 221–228.
17. KITAZAWA, R., R.B. KIMBLE, J.L. VANNICE, *et al.* 1994. Interleukin-1 receptor antagonist and tumor necrosis factor binding protein decrease osteoclast formation and bone resorption in ovariectomized mice. J. Clin. Invest. **94:** 2397–2406.
18. CENCI, S., M.N. WEITZMANN, C. ROGGIA, *et al.* 2000. ER deficiency induces bone loss by enhancing T-cell production of TNF-alpha. J. Clin. Invest. **106:** 1229–1237.
19. WEI, S., H. KITAURA, P. ZHOU, *et al.* 2005. IL-1 mediates TNF-induced osteoclastogenesis. J. Clin. Invest. **115:** 282–290.
20. CENCI S., G. TORALDO, M.N. WEITZMANN, *et al.* 2003. ER deficiency induces bone loss by increasing T cell proliferation and lifespan through IFN-gamma-induced class II transactivator. Proc. Natl. Acad. Sci. USA **100:** 10405–10410.
21. TAKAYANAGI, H., S. KIM & T. TANIGUCHI. 2002. Signaling crosstalk between RANKL and interferons in osteoclast differentiation. Arthritis Res. **4:** S227–S232.
22. HUANG, W., R.J. O'Keefe, M.F. SCHWARZ. 2003. Exposure to receptor activator of NF-Kb ligand renders pre-osteoclasts resistant to IFN-gamma by inducing terminal differentiation. Arthritis Res. Ther. **5:** 49–59.
23. WEITZMANN, M.N., C. ROGGIA, G. TORALDO, *et al.* 2002. Increased production of IL-7 uncouples bone formation from bone resorption during ER deficiency. J. Clin. Invest. **110:** 1643–1650.
24. TORALDO, G., C. ROGGIA, W.P. QIAN, *et al.* 2003. IL-7 induces bone loss in vivo by induction of receptor activator of nuclear factor kappa B ligand and tumor necrosis factor alpha from T cells. Proc. Natl. Acad. Sci. USA **100:** 125–130.
25. PARFITT, A.M., G.R. MUNDY, G.D. ROODMAN, *et al.* 1996. Boyce BFA new model for the regulation of bone resorption, with particular reference to the effects of bisphosphonates. J. Bone Miner. Res. **11:** 150–159.
26. ERIKSEN, E.F., B. LANGDAHL, A. VESTERBY, *et al.* 1999. Hormone replacement therapy prevents osteoclastic hyperactivity: a histomorphometric study in early postmenopausal women. J. Bone Miner. Res. **14:** 1217–1221.
27. HUGHES, D.E., A. DAI, J.C. TIFFEE, *et al.* 1996. ER promotes apoptosis of murine osteoclasts mediated by TGF-beta. Nat. Med. **2:** 1132–1136.
28. HOFBAUER, L.C., S. KHOSLA, C.R. DUNSTAN, *et al.* 1999. ER stimulates gene expression and protein production of osteoprotegerin in human osteoblastic cells. Endocrinology **140:** 4367–4370.
29. BORD, S., D.C. IRELAND, S.R. BEAVAN & J.E. COMPSTON. 2003. The effects of ER on osteoprotegerin, RANKL, and ER receptor expression in human osteoblasts. Bone **32:** 136–141.

30. Bashir, A., Y.T. Mak, S. Sankaralingam, *et al.* 2005. Changes in RANKL/OPG/RANK gene expression in peripheral mononuclear cells following treatment with ER or raloxifene. Steroids **70:** 847–855.
31. Kousteni, S., A. Almeida, L. Han, *et al.* 2005. Estrogens control the birth and apoptosis of bone cells in mice in which ERα cannot interact with DNA ($ER\alpha^{NERKI/-}$). J. Bone Miner. Res. **20:** 1098, S26.
32. Colaianni, G., R. Tamma, N. Patano, *et al.* 2005. Lack of oxytocin in KO mice results in a denser appendicular skeleton. Role of the hormone as autocrine-paracrine enhancer of osteoblast and osteoclast activity. J. Bone Miner. Res. **20:**(Suppl) 1102, S27.
33. Garcia-Palacios V., L.J. Robinson, C.W. Borysenko, *et al.* 2005. Negative regulation of RANKL-induced osteoclastic differentiation in RAW264.7 Cells by ER and phytoestrogens. J. Biol. Chem. **280:** 13720–13727.

Molecular Mechanism of the Life and Death of the Osteoclast

SAKAE TANAKA, TSUYOSHI MIYAZAKI, AKIRA FUKUDA, TORU AKIYAMA, YUHO KADONO, HIDETOSHI WAKEYAMA, SHINJIRO KONO, SHINYA HOSHIKAWA, MASAKI NAKAMURA, YASUSHI OHSHIMA, ATSUHIKO HIKITA, ICHIRO NAKAMURA, AND KOZO NAKAMURA

Department of Orthopaedic Surgery, Faculty of Medicine, The University of Tokyo, Tokyo 113-0033, Japan

ABSTRACT: The life span of osteoclasts is critically regulated by various cytokines, and therapeutics such as bisphosphonates act directly on osteoclasts and induce apoptosis of the cells. This article will focus on the molecular mechanism of osteoclast apoptosis and summarize the recent advances in this field with an emphasis on the role of intracellular signaling pathways.

KEYWORDS: osteoclast; cytokines; apoptosis; Akt; TNF; Rac1; Bim

THE ROLE OF APOPTOSIS IN BONE HOMEOSTASIS

The homeostasis of the skeletal tissues is maintained by a well-organized regulation of bone formation and bone resorption. This process is called remodeling, in which osteoblasts are mainly involved in the bone formation process and osteoclasts are essential for bone resorption. Osteoclasts are terminally differentiated cells primarily involved in physiological and pathological bone resorption. The life span of osteoclasts is relatively short both *in vitro* and *in vivo*, and once differentiated, they rapidly die in the absence of supporting cells such as osteoblasts or bone marrow stromal cells, or growth factors, such as interleukin (IL)-1, receptor activator of NF-κB ligand (RANKL), and macrophage colony-stimulating factor (M-CSF).[1] Antiresorptive drugs such as estrogen, raloxifene, and bisphosphonates are known to reduce the life span of osteoclasts.[2]

Recent studies have revealed that the rapid cell death of osteoclasts is caused by apoptosis. Apoptosis is a form of programmed cell death that is characterized

Address for correspondence: Sakae Tanaka, Department of Orthopaedic Surgery, Faculty of Medicine, The University of Tokyo, 7-3-1 Hongo, Bunkyo-ku, Tokyo 113-0033, Japan. Voice: 81-3-3815-5411 ext 33376; fax: 81-3-3818-4082.
e-mail: TANAKAS-ORT@h.u-tokyo.ac.jp

**Ann. N.Y. Acad. Sci. 1068: 180–186 (2006). © 2006 New York Academy of Sciences.
doi: 10.1196/annals.1346.020**

by specific morphological and biochemical properties.[3,4] Morphologically, apoptosis is characterized by a series of structural changes in dying cells: blebbing of the plasma membrane, condensation of the cytoplasm and the nucleus, and cellular fragmentation into membrane apoptotic bodies. Biochemically, apoptosis is characterized by the degradation of chromatin, initially into large fragments of 50–300 kilobases and subsequently into smaller fragments that are monomers and multimers of 200 bases.[4,5] Not only does apoptosis regulate various aspects of the biological activity but also can trigger cancer, autoimmune diseases, and degenerative disorders.

INTRACELLULAR SIGNALING PATHWAYS REGULATING OSTEOCLAST SURVIVAL

Using adenovirus vector-mediated gene transduction system, we found that the activation of extracellular-regulating kinase (Erk) by transducing a constitutively active mutant of Mek1 markedly promoted the survival of osteoclasts.[6] Conversely, inhibiting Erk activation by overexpressing a dominant negative *ras* gene mutant rapidly induced apoptotic cell death of the cells.[6] These results, combined with the fact that anti-apoptotic factors such as RANKL, IL-1, and M-CSF also induce Erk activation in osteoclasts, suggest that the Ras-Erk pathway plays an essential role in osteoclast survival.

Recent studies have revealed the critical role of phosphatidylinositol 3-kinase (PI3K)/Akt pathways in the survival of osteoclasts. Xing *et al.* reported that the transgenic overexpression of Src251, which contains SH2 and SH3 domains of chicken c-Src but lacks the kinase domain, under the control of the tartrate-resistant acid phosphatase (TRAP) promotor-induced osteopetrosis in mice.[7] The mice had a reduced number of osteoclasts, and osteoclasts of Src251 transgenic mice showed apoptotic phenotypes. The activity of Akt in the Src251 transgenic mouse osteoclasts was significantly reduced, and the RANKL-induced Akt activation was impaired in the cells, suggesting that Akt plays a critical role in osteoclast survival, and Akt activation is at least partly mediated by c-Src.[7] Using a mouse osteoclast formation system, Lee *et al.* showed that tumor necrosis factor-α (TNF-α) prolonged the survival of osteoclasts, which was abrogated by either PI3K inhibitors or Mek/Erk inhibitors.[8] They also revealed the involvement of Grb2 and ceramide in TNF-α-induced Erk activation in osteoclasts.[8] A similar effect was observed in IL-1α.[9] Glantschnig *et al.* exhibited the central role of mTOR/S6K pathways in M-CSF- and RANKL-induced osteoclast survival, which was suppressed by rapamycin.[10] We also found that the adenovirus vector-mediated overexpression of constitutively active Akt stimulates osteoclast survival (Fukuda *et al.*, unpublished observation). These results clearly suggest the crucial role of Akt pathways on the osteoclast survival. However, contrary to these previous observations, Sugatani and Ross recently reported that silencing of Akt1 and/or Akt2 by small interfering RNA (siRNA) suppressed osteoclast differentiation but did not affect osteoclast survival.[11] The reason of this discrepancy remains unknown, and further investigation is required to clarify the exact role of PI3K/Akt pathways in the osteoclast survival.

Rac1 is a member of the Rho family small G-proteins and recent studies have revealed that it mediates anti-apoptotic signals in some types of cells. Rac1 is reported to be required for the cytoskeletal organization and bone-resorbing activity of osteoclasts, but its role in the osteoclast survival and function is not fully elucidated yet. To examine the role of Rac1 in the osteoclast survival and function, we examined the effect of dominant negative Rac1 and constitutively active Rac1 expression on osteoclasts.[12] Adenovirus vector-mediated dominant negative Rac1 ($Rac1^{DN}$) expression significantly reduced the pit formation of osteoclasts, and promoted their apoptosis. M-CSF rapidly activated Rac1, consistent with the results reported by Faccio *et al.*,[13] and the pro-survival effect of M-CSF for osteoclasts was abrogated by $Rac1^{DN}$ overexpression. Constitutively active Rac1 enhanced osteoclast survival, which was completely suppressed by PI3K inhibitors, while Mek inhibitors had only partial effects. Using time-lapse video-microscopy, we found that $Rac1^{DN}$ expression reduced the membrane ruffling and spreading of osteoclasts in response to M-CSF. These results strongly suggest that small GTPase Rac1 is critically involved in M-CSF receptor signaling, and mediates survival signaling of osteoclasts primarily by modulating PI3K/Akt pathways. Faccio *et al.* recently demonstrated that a guanine nucleotide exchange factor Vav3 lies upstream of Rac1 activation in osteoclasts, and is involved in the signaling of M-CSF receptor and $\alpha v\beta 3$ integrin.[14]

ROLE OF BCL-2 FAMILY MEMBERS IN OSTEOCLAST APOPTOSIS

The apoptotic process can be divided into two different pathways, the extrinsic (death receptor) pathway and the intrinsic (mitochondrial) pathway.[15] In the death receptor pathway, the activation of a receptor belonging to the TNF receptor family leads to the induction of apoptosis through the activation of aspartate-specific cystein proteases (caspases) 8, and the anti- and pro-apoptotic Bcl-2 family members are critically involved in the intrinsic pathway by regulating cytochrome *c* release from the mitochondrial intermembrane into the cytosol, which leads to the activation of caspase 9.[15] The anti-apoptotic members include mammalian Bcl-2, Bcl-x_L, Bcl-w, Mcl-1, A1, and *C. elegans* CED-9. So far, more than 20 members of pro-apoptotic Bcl-2 family proteins have been identified in mammals, and they can be further divided into two groups; multi-domain members possessing homology in BH1-3 domains and BH3 domain-only members. Multi-domain members include Bax, Bak, Bok and *D. melanogaster* DEBCL/DROB, and BH3-only members include mammalian Bad, Bik/Nbk, Bid, Hrk/DP5, Blk, Bim, Noxa and *C. elegans* EGL-1.[16] It was recently reported that BH3-only proteins failed to induce cytochrome *c* release and apoptosis in *bax-/-bak-/-* cells, suggesting that Bax-like proteins mediate death signals from various BH3-only proteins.[17] Bax-like proteins are ubiquitously expressed, while BH3-only members have a more tissue-specific distribution, indicating that the latter may play a role in the tissue/cell-specific regulation of apoptosis.

The activity of BH3-only proteins is strictly regulated to prevent unorganized cell death. The expression of Noxa is induced by p53 or IRF-1, and DP5/Hrk is transcriptionally upregulated in the absence of growth factors or when the cells are exposed to β-amyloid protein. The other mechanism regulating their activity is the posttranslational modification. It is now widely recognized that Bad is phosphorylated at various sites and sequestrated away from Bcl-2 family members by binding to 14-3-3.[18,19] Many kinases including Akt, mitochondrial-anchored protein kinase A, Pak1, Rsk and Raf1, and phosphatase calcineurin have been reported to modify Bad, although the physiological importance of these regulations still remains unclear.

Depolarization of mitochondrial transmembrane potential, chromatin condensation, and cytochrome *c* release from mitochondria into cytoplasm were observed in the apoptotic osteoclasts, implying that cytokine deprivation triggers osteoclast apoptosis through the mitochondrial pathway.

A possible role of anti-apoptotic Bcl-2 family members for osteoclast survival has been reported. Okahashi *et al.*[20] showed that M-CSF treatment increased Bcl-2 mRNA expression in osteoclasts, and we found that adenovirus vector-mediated overexpression of Bcl-xL markedly prolonged survival of the cells *in vitro*.[21] Roodman and co-workers developed a transgenic mouse, in which Bcl-xL together with Simian Virus 40 large T antigen were specifically overexpressed in osteoclasts under the control of the TRAP promotor, and successfully established an immortalized osteoclast precursor cell line from the mice.[22] These results clearly showed that anti-apoptotic Bcl-2 family proteins play important roles in osteoclast survival both *in vitro* and *in vivo*. The mechanism regulating Bcl-2 and Bcl-xL expression in osteoclasts has not been clarified yet, but McGill *et al.* showed that Bcl-2 expression is regulated by Mitf, a transcription factor essential for osteoclast and melanocytes development.[23]

PRO-APOPTOTIC BH3-ONLY PROTEIN BIM IS INVOLVED IN OSTEOCLAST APOPTOSIS

Bim, one of the BH3-only proteins, was first identified as a Bcl-2 interacting protein by screening a λ-phage expression library constructed from a mouse thymic lymphoma.[24] Bim is expressed in hematopoietic, epithelial, neuronal, and germ cells,[25] and alternative splicing generates various Bim isoforms, including Bim_S, Bim_L, and Bim_{EL}. Bim has been shown to play an essential role in the induction of apoptosis in T lymphocytes, B lymphocytes, and neurons, mainly based on the finding in Bim-deficient animals,[26,27] and its activity is also regulated both transcriptionally and posttranscriptionally.[28] The expression of Bim is downregulated at the transcriptional level by IL-3 signaling through Raf/Erk pathways and/or PI3K/mammalian target of rapamycin (mTOR) pathways in IL-3-dependent Ba/F3 cells.[29] The transcription of Bim is also upregulated in neonatal sympathetic neurons in response to the nerve growth factor deprivation. In healthy cells, Bim_L is bound to LC8, a component of the microtubule-associated dynein motor complex, and is sequestered to this complex in the cytoplasm. Certain apoptosis stimuli induce dissociation of this complex, and allow the translocation of Bim_L-LC8 complex to Bcl-xL on the mitochondrial membrane.[28]

We demonstrated a novel and unique regulation of apoptosis by ubiquitylation-dependent degradation of Bim in osteoclasts.[21] In the presence of M-CSF, Bim is constitutively ubiquitylated and degraded, and cytokine deprivation induced rapid upregulation of Bim due to the reduced level of its ubiquitylation.[21] C-Cbl is a possible candidate of E3 ubiquitin ligase of Bim in osteoclasts because Bim ubiquitylation was reduced in c-Cbl-deficient osteoclasts. Bim-deficient osteoclasts exhibited prolonged survival both *in vitro* and *in vivo*, and however, their bone-resorbing activity was significantly reduced, consistent with the *in vivo* observation that *bim–/–* animals showed mild osteosclerosis.[21] Consistent with our observation, Sugatani and Hruska reported that silencing the *bim* gene by siRNA prolonged the survival of osteoclasts.[11] These results clearly demonstrate that pro-apoptotic protein Bim plays a crucial role in osteoclast apoptosis and activation. Further studies are required to elucidate the mechanism of action and the regulation of Bim in osteoclasts, and the role of Bim in skeletal disorders.

CONCLUSIONS AND PERSPECTIVES

The success of bisphosphonates as the therapeutics of osteoporosis, together with the finding that bisphosphonates act directly on osteoclasts to induce their apoptosis, has attracted a great deal of attention to the molecular mechanism of osteoclast apoptosis, and remarkable progress has been made during the last several years to reveal the signaling pathways involved in the process. Therapeutics targeting these signaling pathways (molecules) will provide a novel treatment for bone diseases.

ACKNOWLEDGMENTS

This work was in part supported by Grants-in-Aid from the Ministry of Education, Culture, Sports, Science and Technology of Japan and Health Science research grants from the Ministry of Health, Labour and Welfare of Japan to ST.

REFERENCES

1. TANAKA, S. *et al.* 2003. Signal transduction pathways regulating osteoclast differentiation and function. J. Bone Miner. Metab. **21:** 123–133.
2. ROGERS, M.J. 2003. New insights into the molecular mechanisms of action of bisphosphonates. Curr. Pharm. Des. **9:** 2643–2658.
3. KERR, J.F., A.H. WYLLIE & A.R. CURRIE. 1972. Apoptosis: a basic biological phenomenon with wide-ranging implications in tissue kinetics. Br. J. Cancer **26:** 239–257.
4. WYLLIE, A.H., J.F. KERR & A.R. CURRIE. 1980. Cell death: the significance of apoptosis. Int. Rev. Cytol. **68:** 251–306.
5. OBERHAMMER, F. *et al.* 1993. Condensation of the chromatin at the membrane of an apoptotic nucleus is not associated with activation of an endonuclease. J. Cell Sci. **104**(Pt 2): 317–326.
6. MIYAZAKI, T. *et al.* 2000. Reciprocal role of ERK and NF-kappaB pathways in survival and activation of osteoclasts. J. Cell Biol. **148:** 333–342.
7. XING, L. *et al.* 2001. Genetic evidence for a role for Src family kinases in TNF family receptor signaling and cell survival. Genes Dev. **15:** 241–253.
8. LEE, S.E. *et al.* 2001. Tumor necrosis factor-alpha supports the survival of osteoclasts through the activation of Akt and ERK. J. Biol. Chem. **276:** 49343–49349.
9. LEE, Z.H. *et al.* 2002. IL-1alpha stimulation of osteoclast survival through the PI 3-kinase/Akt and ERK pathways. J. Biochem. (Tokyo) **131:** 161–166.
10. GLANTSCHNIG, H. *et al.* 2003. M-CSF, TNFalpha and RANK ligand promote osteoclast survival by signaling through mTOR/S6 kinase. Cell Death Differ. **10:** 1165–1177.
11. SUGATANI, T. & K.A. HRUSKA. 2005. Akt1/Akt2 and mTOR/Bim play critical roles in osteoclast differentiation and survival, respectively, while Akt is dispensable for cell survival in isolated osteoclast precursors. J. Biol. Chem. **280:** 3583–3589.
12. FUKUDA, A. *et al.* 2005. Regulation of osteoclast apoptosis and motility by small GTPase binding protein Rac1. J. Bone Miner. Res. **20:** 2245–2253.
13. FACCIO, R. *et al.* 2003. Dynamic changes in the osteoclast cytoskeleton in response to growth factors and cell attachment are controlled by beta3 integrin. J. Cell. Biol. **162:** 499–509.
14. FACCIO, R. *et al.* 2005. Vav3 regulates osteoclast function and bone mass. Nat. Med. **11:** 284–290.
15. HENGARTNER, M.O. 2000. The biochemistry of apoptosis. Nature **407:** 770–776.
16. HUANG, D.C. & A. STRASSER. 2000. BH3-Only proteins-essential initiators of apoptotic cell death. Cell **103:** 839–842.
17. CHENG, E.H. *et al.* 2001. BCL-2, BCL-X(L) sequester BH3 domain-only molecules preventing BAX- and BAK-mediated mitochondrial apoptosis. Mol. Cell **8:** 705–711.
18. FRANKE, T.F. & L.C. CANTLEY. 1997. Apoptosis. A bad kinase makes good. Nature **390:** 116–117.
19. MASTERS, S.C. *et al.* 2002. Survival-promoting functions of 14-3-3 proteins. Biochem. Soc. Trans. **30:** 360–365.

20. OKAHASHI, N. *et al.* 1998. Caspases (interleukin-1beta-converting enzyme family proteases) are involved in the regulation of the survival of osteoclasts. Bone **23:** 33–41.
21. AKIYAMA, T. *et al.* 2003. Regulation of osteoclast apoptosis by ubiquitylation of proapoptotic BH3-only Bcl-2 family member Bim. EMBO J. **22:** 6653–6664.
22. HENTUNEN, T.A. *et al.* 1998. Immortalization of osteoclast precursors by targeting Bcl -XL and Simian virus 40 large T antigen to the osteoclast lineage in transgenic mice. J. Clin. Invest. **102:** 88–97.
23. MCGILL, G.G. *et al.* 2002. Bcl2 regulation by the melanocyte master regulator Mitf modulates lineage survival and melanoma cell viability. Cell **109:** 707–718.
24. O'CONNOR, L. *et al.* 1998. Bim: a novel member of the Bcl-2 family that promotes apoptosis. Embo J. **17:** 384–395.
25. O' REILLY, L.A. *et al.* 2000. The proapoptotic BH3-only protein bim is expressed in hematopoietic, epithelial, neuronal, and germ cells. Am. J. Pathol. **157:** 449–461.
26. BOUILLET, P. *et al.* 1999. Proapoptotic Bcl-2 relative Bim required for certain apoptotic responses, leukocyte homeostasis, and to preclude autoimmunity. Science **286:** 1735–1738.
27. BOUILLET, P. *et al.* 2002. BH3-only Bcl-2 family member Bim is required for apoptosis of autoreactive thymocytes. Nature **415:** 922–926.
28. PUTHALAKATH, H. *et al.* 1999. The proapoptotic activity of the Bcl-2 family member Bim is regulated by interaction with the dynein motor complex. Mol. Cell **3:** 287–296.
29. SHINJYO, T. *et al.* 2001. Downregulation of Bim, a proapoptotic relative of Bcl-2, is a pivotal step in cytokine-initiated survival signaling in murine hematopoietic progenitors. Mol. Cell. Biol. **21:** 854–864.

Understanding Parathyroid Hormone Action

NICOLA C. PARTRIDGE, XIN LI, AND LING QIN

Department of Physiology and Biophysics, UMDNJ-Robert Wood Johnson Medical School, Piscataway, New Jersey 08854, USA

ABSTRACT: In addition to regulating serum calcium and stimulating bone resorption, parathyroid hormone (PTH) is known to stimulate bone formation under certain conditions. The mechanisms behind this counterintuitive anabolic action are largely unknown. We have set out to identify PTH-regulated genes that might be responsible for each of these contrasting effects of the hormone. This article describes our work on a select number of those genes.

KEYWORDS: parathyroid hormone; PTH; EGF; amphiregulin; bone resorption; bone formation; osteoblast; MCP-1

The classical regulation of serum calcium by parathyroid hormone (PTH) is well known.[1] In this situation low serum calcium levels cause the release of PTH from the four parathyroid glands behind the thyroid gland. The hormone has two sites of action; it acts on kidney to stimulate calcium re-absorption and also enhance the hydroxylation of 25 hydroxy vitamin D_3 to 1,25-dihydroxy vitamin D_3, which then has its subsequent actions on increasing calcium uptake by the small intestine. The other site of PTH action is bone, where it stimulates bone resorption. The actions on the kidney and bone then cause the restoration of serum calcium. This is our classical concept of PTH action, to have minute-to-minute responsibility for the regulation of serum calcium levels.

However, for some time now, we have known that PTH can also stimulate bone formation.[2] In other words, it has an anabolic action in contrast to its catabolic action to cause bone resorption. In fact, clinical trials where PTH (1-34) or (1-84) were given to postmenopausal women have shown significant increases in bone mineral density (BMD).[3] This then led to the approval by the FDA for the use of PTH 1-34 (teraparatide, Forteo™, Eli Lilly and Company) to treat osteoporosis. However, we still do not know why PTH has this anabolic effect in contrast to its catabolic effect.

Address for correspondence: Nicola C. Partridge, Department of Physiology and Biophysics, UMDNJ-Robert Wood Johnson Medical School, 675 Hoes Lane, Piscataway, NJ 008854. Voice: 732-235-4552; fax:732-235-3977.

e-mail: partrinc@umdnj.edu

Ann. N.Y. Acad. Sci. 1068: 187–193 (2006). © 2006 New York Academy of Sciences.
doi: 10.1196/annals.1346.024

In our laboratory, we set out to identify PTH-regulated genes that might be responsible for each of these two effects of the hormone. Initially, we performed microarray analysis on UMR 106-01 cells, which is a rat osteoblastic osteosarcoma cell line. These cells had been treated with PTH (1-34) for 4, 12, or 24 hours. The RNA was isolated, converted to cDNA, then to biotinylated cRNA, and this was hybridized to Affymetrix rat chips containing 8,799 probes, which consisted of 7% known genes and 23% expressed sequence tags (ESTs). Our results showed that there are 125 known genes and 30 ESTs that were regulated by PTH.[4]

This article describes our work on a select number of those genes. The first genes investigated were members of the epidermal growth factor (EGF) family, which contains 10 ligands and four receptors. The ligands that we found stimulated were amphiregulin and transforming growth factor α (TGF-α). These together with EGF, heparin-binding EGF, β-cellulin, and epiregulin bind to the first-known EGF receptor, also known as ErbB1. This can heterodimerize with ErbB2 to form the most potent heterodimer. The four other ligands of this family, the neuregulins, bind to ErbB3 and ErbB4. We found that stimulation of amphiregulin was very rapid in the UMR cells with peak stimulation of the mRNA at 1 hour, which then declined by 12 hours. The stimulation was dose-dependent at 1 hour with a significant increase at 10^{-10} M rat PTH (1-34) and maximal effect at 10^{-8} M. The PTH regulation of amphiregulin was also observed in differentiating primary rat osteoblastic cultures. These cells are cultured so that they proceed through three stages, proliferation, differentiation, and mineralization. We found that rat PTH (1-34, 10^{-8} M) stimulated amphiregulin expression in all three stages; however, the greatest effect was seen in the mineralizing cells. In every case, there was maximal stimulation at 1 hour with a decline back to basal levels by 24 hours. We have also examined another osteoblastic cell line, the MC3T3 cells, and we also found rat PTH (1-34, 10^{-8} M) stimulated amphiregulin expression in these cells with a maximal regulation at 1 hour, which also declined by 6 hours; however, the stimulation was about fourfold in contrast to the 20- to 25-fold stimulations that we had seen in both the UMR and primary osteoblastic cells. Finally, we injected human PTH (1-38) into 4-week-old rats and then analyzed the metaphyseal region of the distal femur at various times after injection. We observed a very rapid and pronounced 13- to 14-fold increase in amphiregulin expression 1 hour after injection of PTH (1-38). This also declines very rapidly, reaching basal levels again 8 hours after the injection. Therefore, PTH stimulates the growth factor very rapidly, both in osteoblastic cultures *in vitro* as well as *in vivo*.[5]

The conclusion that we have reached from the work that we have performed on amphiregulin is that PTH acting on the mature osteoblast stimulates the expression of amphiregulin and this then acts on mesenchymal stem cells as well as committed preosteoblasts to stimulate their replication and also block their differentiation. This effect is likely to be transient since the injection of PTH is transient and, as we have seen, results in transient expression of amphiregulin.

Therefore, we suggest that there should be a transient stimulation of replication and an increased pool of preosteoblasts. When the amphiregulin signal dissipates, the increased number of preosteoblasts will be able to progress into differentiated, mature osteoblasts.

The next set of genes we examined is related to growth of osteoblasts. The mitogen-activated protein kinase (MAPK) family is well known for being involved with the proliferation of cells. The major families are the ERKs, JNKs, and p38 families. All of these are activated by phosphorylation on both threonine and tyrosine. They must also be inactivated by dephosphorylation of those phosphates, and this is achieved by specific phosphatases. The phosphatases in question are the MAPK phosphatases. These are dual-specificity phosphatases in the overall protein tyrosine phosphatase superfamily. We found that one of these MAPK phosphatases, MKP1, was also another gene that was highly regulated by PTH (1-34) and was one of the genes shown to be increased in the microarray of the UMR cells. We examined its regulation in greater detail and found that this too was increased very rapidly, with the mRNA reaching maximal levels at 1 hour after PTH treatment in the UMR cells, dissipating back to lower levels by 7 hours. We found similar regulation in the normal differentiating osteoblasts, again with very rapid regulation. Maximum stimulation was observed at 1 hour and this again was reduced to basal within 4 hours. This regulation of MKP1 is protein kinase A dependent. We found that PTH (1-31), which does not activate the protein kinase C (PKC) pathway, was able to mimic the effect of PTH (1-34), whereas PTH (13-34), which activates the PKC pathway, did not stimulate MKP1 expression.

Eight bromo-cAMP could also stimulate MKP1 mRNA levels, as could the PKC stimulator, PMA. However, when we treated cells with PTH (1-34) in the presence of the protein kinase A inhibitor, H-89, we completely blocked the stimulation of MKP1 in contrast to when we used the protein kinase C inhibitor, GF109203X, there was no inhibition of the PTH stimulation of MKP1. These data would indicate that protein kinase A is necessary for PTH (1-34) stimulation of MKP1 whereas protein kinase C is not. This work was done on UMR cells but, as well, we injected PTH (1-34), (1-31), or (3-34) into 4-week-old rats, removed the metaphyseal region of the distal femur 1 hour after the injection, and found that PTH (1-34) and (1-31) had identical robust stimulation of MKP1 *in vivo* whereas PTH (3-34) had no effect at all. This would indicate that *in vivo* we have similar results to those seen *in vitro*; in other words, it is likely that the protein kinase A (PKA) pathway is responsible for the stimulation of MKP1 expression *in vivo*.

The activation of ERKs increases cyclin D1 mRNA expression while inhibition of ERKs (by MKP1) would block cyclin D1's induction. The cell cycle requires the transition from G1 to S and one of the first genes that is switched on is cyclin D1. Some of the inhibitors that regulate both cyclin D1 and cyclin E are p16, p15, p21, p27, and p57. We investigated the PTH regulation of some of these genes *in vivo*. Injection of PTH (1-38) into 4-week-old male rats

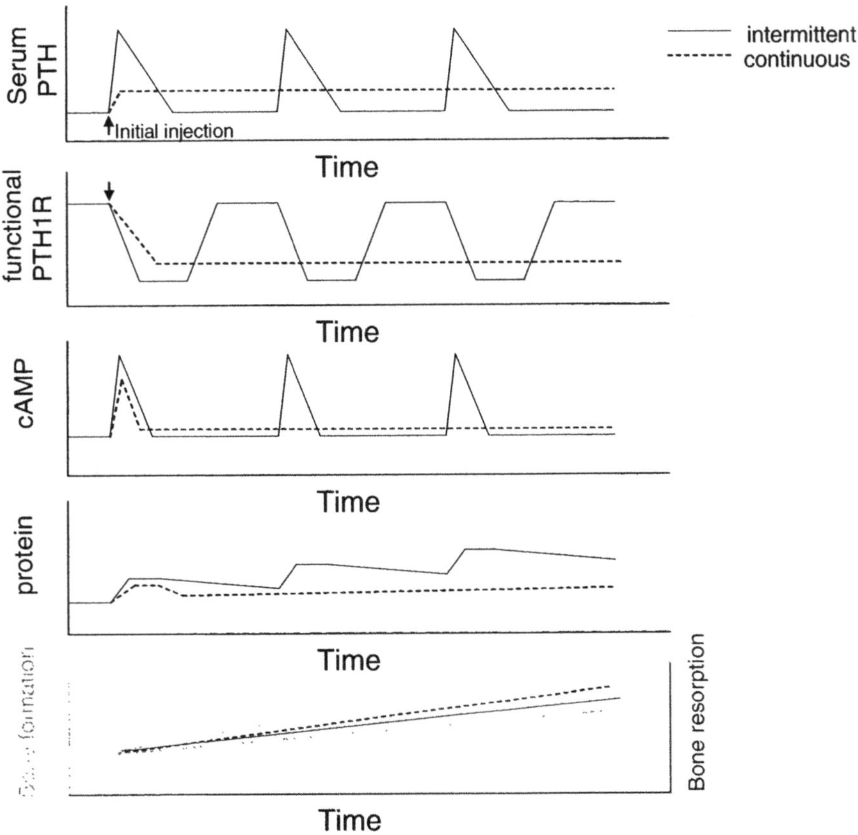

FIGURE 1. A theoretical time course of serum PTH (1-34), functional PTH receptor (PTH1R) level, cAMP amount, a putative PTH-regulated protein amount, and bone formation versus bone resorption after intermittent or continuous PTH (1-34) treatment. The arrow indicates the initial PTH administration. (Modified from Qin *et al.*[7] Reprinted with permission from Elsevier.)

showed a very rapid stimulation of MKP1; in fact, the maximal stimulation is at 30 minutes, which then declines back to basal by 4 hours. At the same time, we observe a decrease in cyclin D1 mRNA expression, but this actually rebounds to basal levels by 4 hours. PTH also regulates the inhibitor of the cyclins, p21. This is also very rapid with maximal effects at 30 minutes, which again declines to basal by 4 hours. It is noteworthy that we found that if 3-month-old female rats are given either one injection versus daily PTH (1-34) injections for 14 days we found that there were substantially greater increases in p21 mRNA with 14 days of injections compared with one injection.[6] This suggests that there is a ramping up in the sensitivity of cells to PTH. It should be noted this is not the case for all genes; for instance, MKP1 showed the same fold stimulation for one injection versus 14 daily injections.

Comparing the intermittent versus continuous effects of PTH we have proposed FIGURE 1. With daily intermittent injection the hormone levels are transient and return to baseline whereas with continuous infusion they rise, but not to the same levels as with the intermittent injection, but then stay constant throughout the whole period. We think that the PTH receptor will be transiently downregulated with the intermittent injections but that this then will rebound to normal within 24 hours prior to the next injection, whereas we think that with continuous treatment that the PTH receptor may decrease in abundance, both mRNA and protein, and may stay at lower levels throughout the time of infusion. With respect to cAMP, the intermittent injection would cause a transient increase after each injection whereas we think with the continuous injection that there may be a lesser increase in cAMP and that it may never return to basal but will stay slightly above basal throughout the infusion period. What we then think might happen in terms of particular proteins is that, with intermittent injections, there is a ramping up with each injection adding to the previous increase until there are much greater levels with each repeated injection. This is in contrast, we think, to the continuous infusion where you may observe an initial increase and this might slowly decline over the infusion period. We think these are the reasons that the result is greater bone formation versus resorption in the intermittent injections compared with the continuous infusion where there is more bone resorption than formation.[7]

To address this, we have undertaken microarrays in animals receiving PTH peptides by intermittent injections. We have also used different PTH peptides to try to determine which signal transduction pathway is responsible for the hormone's anabolic effect. We found that PTH (1-34) and (1-31) had the same stimulation of bone formation, in contrast to PTH (3-34), which does not. The histomorphometry shows a significant increase in the number of osteoblasts for both PTH (1-34) and (1-31) as well as an increase in the number of osteoclasts. This does not happen with PTH (3-34) although there is a trend toward an increase in both these parameters compared with the controls. When we examine the microarrays for the animals that have been injected for 14 days with these peptides we find that the set of genes that were regulated with PTH (1-34) and (1-31) overlap extensively, while there are much fewer numbers of genes regulated by PTH (3-34). Most of these are common to PTH (1-31) and (1-34). Thus, it appears that the gene expression profile is regulated by the PKA pathway rather than the intracellular calcium or PKC pathway. We have now done some preliminary work where we have given continuous injections of PTH (1-34) to 3-month-old female rats and undertaken microarrays. This is work in progress at present. We find that there are large numbers of genes regulated by the intermittent injections as well as to the continuous infusion. This is in significant contrast to the work published recently by Onyia *et al.*[8] where they found very few genes regulated by the intermittent injection. However, it should be noted that in that latter work they removed the bones of the rats 24 hours after the last PTH (1-34) injection and, it seems, immediately after

TABLE 1. Unique Genes Regulated by Intermittent PTH(1-34)

Amphiregulin	Jagged1	Syndecan 4
MKP1	CREM	Cox-2
p21	CEB/Pβ	MCP-1

the cessation of the PTH (1-34) infusion. In contrast, we removed the bones immediately after the end of the infusion and 1 hour after the last intermittent injection of PTH peptides. This is because of the observations we already had from the microarray of the osteoblastic cells as well as the data studying specific genes *in vivo*. We found significant stimulation of many genes 1 hour after the last PTH injection.

There are many genes also commonly regulated by both intermittent and continuous PTH treatment. Generally, we find that intermittent injection stimulates the expression of these genes to a greater level than that seen by the continuous injection. It is notable that RANK ligand is stimulated about sevenfold 1 hour after the last injection with PTH whereas with continuous infusion we see about a twofold stimulation.

Intermittent injection of PTH (1-34) also regulates many unique genes, and some of these are shown in TABLE 1. It should be noted that there are a number of genes that we have observed in the microarrays of the UMR cells, for instance, amphiregulin, MKP1, p21, Jagged1, and others. A notable gene, monocyte chemoattractant protein-1 (MCP-1), is regulated in the microarrays from the intermittent injections.

MCP-1 is a member of the CC chemokine family and has a characteristic –C-C- structure. We examined its expression in rats injected with PTH daily for 14 days. We found, again, that this is a gene that is regulated extremely rapidly with maximal expression 1 hour after the injection. As the injections increase we found a ramping-up of MCP-1 expression. After 14 days we observe almost a 250-fold stimulation of expression of MCP-1 mRNA in the distal femur. We think PTH acts on the mature osteoblast and stimulates the production of MCP-1 but this then has a number of roles acting on monocytes, preosteoclasts, and mature osteoclasts.

In summary, we think that PTH acts on the differentiating or mature osteoblast through the PKA pathway, causing expression of genes such as MKP1 or p21 that cause the differentiating or mature osteoblast to become terminally differentiated and to cease further proliferation. The hormone causes this post-proliferative osteoblast to express a number of growth factors and chemokines, amphiregulin, for instance, which stimulates the proliferation of bone marrow stem cells and preosteoblasts to then increase their numbers and at a later stage to become larger numbers of osteoblasts. We also think that PTH stimulates the expression of Jagged1, which has already been shown to stimulate the proliferation of hematopoietic stem cells and we think this is also involved with stimulating the proliferation of mesenchymal stem cells and preosteoblasts.

Finally, we think PTH stimulates the production of the chemokine MCP-1, which we think is involved with recruiting preosteoclasts, fusion of mononuclear osteoclasts, and possibly activating resident osteoclasts. This latter effect may be part of the requirement for osteoclast action in PTH's anabolic effect since it has been implicated in the clinical trials where bisphosphonates attenuated the anabolic effect of PTH.[9,10]

REFERENCES

1. STREWLER, G.J. & M. ROSENBLATT. 1995. Mineral metabolism. *In* Endocrinology and Metabolism. P. Felig, *et al.* Eds.: 1407–1516. McGraw-Hill. New York.
2. DEMPSTER, D.W., F. COSMAN, M. PARISIEN, *et al.* 1993. Anabolic actions of parathyroid hormone. Endo. Rev. **14:** 690–709.
3. NEER, R.M., C.D. ARNAUD, J.R. ZANCHETTA, *et al.* 2001. Effect of parathyroid hormone (1-34) on fractures and bone mineral density in postmenopausal women with osteoporosis. N. Engl. J. Med. **344:** 1434–1441.
4. QIN, L., P. QIU, L. WANG, *et al.* 2003. Gene expression profiles and transcription factors involved in parathyroid hormone signaling in osteoblasts revealed by microarray and bioinformatics. J. Biol. Chem. **278:** 19723–19731.
5. QIN, L., J. TAMASI, L.J. RAGGATT, *et al.* 2005a. Amphiregulin is a novel growth factor involved in normal bone development and in the cellular response to parathyroid hormone stimulation. J. Biol. Chem. **280:** 3104–3111.
6. QIN, L., L. XIN, J.-K KO, *et al.* 2005b. Parathyroid hormone uses multiple mechanisms to arrest the cell cycle progression of osteoblastic cells from G1 to S phase. J. Biol. Chem. **280:** 3974–3981.
7. QIN, L., L.J. RAGGATT & N.C. PARTRIDGE. 2004. Parathyroid hormone: a double-edged sword for bone metabolism. Trends Endocrinol. Metab. **15:** 60–65.
8. ONYIA, J.E., L.M. HELVERING, L. GELBERT, *et al.* 2005. Molecular profile of catabolic versus anabolic treatment regimes of parathyroid hormone in rat bone: an analysis by DNA microarray. J. Cell. Biochem. **95:** 403–418.
9. BLACK, D.M., S.L. GREENSPAN, K.E. ENSRUD, *et al.*, PaTH Study Investigators. 2003. The effects of parathyroid hormone and alendronate alone or in combination in postmenopausal osteoporosis. N. Engl. J. Med. **349:** 1207–1215.
10. FINKELSTEIN, J.S., A. HAYES, J.L. HUNZELMAN, *et al.* 2003. The effects of parathyroid hormone, alendronate, or both in men with osteoporosis. N. Engl. J. Med. **349:** 1216–1226.

New Insights into the Mechanisms Involved in the Pleiotropic Actions of 1,25Dihydroxyvitamin D_3

SYLVIA CHRISTAKOS, PUNEET DHAWAN, QI SHEN, XIAORONG PENG, BRYAN BENN, AND YAN ZHONG

Department of Biochemistry and Molecular Biology, UMDNJ-New Jersey Medical School, Newark, New Jersey 07103, USA

ABSTRACT: Vitamin D functions to regulate calcium homeostasis in intestine, kidney, and bone. Vitamin D deficiency during bone development causes rickets and in adults vitamin D deficiency, which has been shown to be common in the elderly population, can cause secondary hyperparathyroidism that can result in osteomalacia and increased risk of fracture. Recent evidence has suggested that vitamin D can have numerous other physiological functions including protection against certain autoimmune diseases, such as diabetes and multiple sclerosis and inhibition of proliferation of a number of malignant cells including breast and prostate cancer cells. Exactly how vitamin D affects numerous different systems is a subject of continuing investigation. This article will review new developments related to the function and regulation of vitamin D target proteins in classic vitamin D target tissues that have provided novel insight into the mechanism of vitamin D action.

KEYWORDS: vitamin D; 1,25-dihydroxyvitaminD_3; osteopontin; TRPV6 and TRPV5 epithelial calcium channels; 25-hydroxyvitaminD_3 24-hydroxylase; calbindin; CCAAT enhancer-binding protein

INTRODUCTION

Vitamin D Metabolism

The hormonally active form of vitamin D, 1,25-dihydroxyvitaminD_3 [1,25$(OH)_2D_3$] is produced by two sequential hydroxylations of vitamin D by 25-hydroxylase in the liver and by 25-hydroxyvitaminD_3 1α hydroxylase [1α(OH)ase] in the proximal tubules of the kidney.[1] It is known that 1α(OH)ase activity is stimulated by the parathyroid hormone (PTH) and by low serum

Address for correspondence: Sylvia Christakos, Department of Biochemistry and Molecular Biology, UMDNJ-New Jersey Medical School, 185 South Orange Ave., Newark, NJ 07103. Voice: 973-972-4033; fax: 973-972-5594.
e-mail: christak@umdnj.edu

**Ann. N.Y. Acad. Sci. 1068: 194–203 (2006). © 2006 New York Academy of Sciences.
doi: 10.1196/annals.1346.025**

calcium and phosphorus.[1] PTH has been reported to stimulate 1α(OH)ase by acting on the promotor of the 1α(OH)ase gene.[2] 1,25$(OH)_2D_3$ can negatively regulate renal 1α(OH)ase.[1] It is of interest that calcitonin has been reported to stimulate the expression of 1α(OH)ase in the proximal tubule of the kidney under normocalcemic conditions.[3] Since serum calcitonin is known to be elevated in the fetus, it has been suggested that calcitonin may have a role (in addition to its known role in shrinking osteoclasts under high calcium conditions) in regulating 1α(OH)ase during development. FGF23 has recently been identified as an essential physiological regulator of vitamin D metabolism, which reduces 1α(OH)ase mRNA.[4] Further studies are needed to determine the mechanism by which FGF23 acts to affect vitamin D metabolism. An additional novel finding is that megalin, a member of low-density lipoprotein (LDL) receptor superfamily present in the renal proximal tubular cells, is essential for the uptake of 25$(OH)D_3$ for the generation of 1,25$(OH)_2D_3$. The role of megalin in the uptake of 25$(OH)D_3$ was initially found using megalin knockout mice in which there was abnormal urinary calcium excretion of 25$(OH)D_3$, vitamin D deficiency, and bone disease.[5] Although the proximal tubules of the kidney are one site of synthesis of 1α(OH)ase, it has been suggested that 1α(OH)ase is present in other cell types and that local production of 1,25$(OH)_2D_3$ could play a role in the function of extra renal tissues. It is of interest that besides rachitic abnormalities, immune and reproductive dysfunction has been noted in the 1α(OH)ase null mutant mouse.[6] Further studies are needed to test the hypothesis of a physiological role of 1α(OH)ase in extra renal tissues that has been a matter of debate.

Inactivation of 1,25$(OH)_2D_3$ is initiated by the enzyme 25-hydroxyvitaminD_3-24-hydroxylase [24(OH)ase]. Studies using 24(OH)ase null mutant mice provided the first *in vivo* evidence for a role for 24(OH)ase in the catabolism of 1,25$(OH)_2D_3$.[7] 24(OH)ase null mutant mice were unable to clear 1,25$(OH)_2D_3$ from the blood in response to 1,25$(OH)_2D_3$ treatment, confirming the physiological importance of the 24(OH)ase enzyme for maintaining vitamin D homeostasis.

Mechanism of 1,25$(OH)_2D_3$ Action

The actions of 1,25$(OH)_2D_3$ are mediated by the vitamin D receptor (VDR), which heterodimerizes with the retinoid X receptor and interacts with DNA sequences in the promotor of target genes. The mechanisms involved in VDR-mediated transcription are now being defined. TFIIB, several TATA-binding protein-associated factors (TAFs) as well as the p160 co-activators SRC-1, SRC-2, and SRC-3 that have histone acetylase (HAT) activity, have been reported to be involved in VDR-mediated transcription.[8] The p160 coactivators can recruit CBP (CREB-binding protein) that also has HAT activity, resulting in a multisubunit complex.[8] VDR-mediated transcription is also mediated by the coactivator complex DRIP (vitamin D receptor-interacting protein, also called TRAP and ARC) that is known to act through the recruitment of RNA polymerase II holoenzyme.[8] In addition, a number of promotor-specific transcription factors have been reported by our lab and others to modulate VDR-mediated transcription including YY1- and CCAAT enhancer-binding protein (C/EBP) β and δ.[9–12] It has been suggested that cell- and promotor-specific functions of VDR may be mediated through differential recruitment of coactivators and that increased interaction between VDR and coactivators may be a major mechanism that couples extracellular signals to vitamin D action.[13–16]

NEW INSIGHTS INTO THE ROLE OF $1,25(OH)_2D_3$ IN CLASSIC TARGET TISSUES AND THE MOLECULAR MECHANISM OF $1,25(OH)_2D_3$ ACTION

Can $1,25(OH)_2D_3$ act directly on bone or is the anti-rachitic effect of $1,25(OH)_2D_3$ indirectly resulting from an increase in calcium and phosphorus absorption from the intestine permitting normal skeletal mineralization? Studies in VDR null mutant mice indicate that the rickets and osteomalacia observed in the VDR null mutant mice can be prevented when these mice are fed a high calcium, high phosphorus, lactose-supplemented diet by the third week of life, suggesting that the major effect of $1,25(OH)_2D_3$ is on intestinal calcium and phosphorus absorption.[17] However, it should be noted that direct effects of $1,25(OH)_2D_3$ on bone have been noted in *in vitro* studies. $1,25(OH)_2D_3$ can stimulate the formation of osteoclasts by a mechanism that involves induction by $1,25(OH)_2D_3$ in osteoblastic cells of osteoprotegerin ligand (RANKL).[18] $1,25(OH)_2D_3$ also induces the transcription of osteocalcin and osteopontin (OPN), two calcium-binding proteins in bone. The synthesis of osteocalcin is associated with new bone formation.[19] OPN is a glycosylated phosophoprotein that comprises about 2% of the noncollagenous protein in bone.[20] Studies in OPN null mice indicate that these mice are resistant to bone resorption and have impaired activation of osteoclasts, suggesting the requirement of OPN in bone resorption.[21,22] Recent studies in our lab provided novel insight regarding the regulation of OPN by $1,25(OH)_2D_3$. We found cooperative effects between RUNX2, VDR, and Hes-1 (a downstream factor in the notch signaling pathway) in the transcriptional regulation of OPN (FIGS. 1,2).[23] $1,25(OH)_2D_3$ was found to enhance both VDR and RUNX2 recruitment to the OPN promotor *in vivo* as determined by chromatin immunoprecipitation assays (FIG. 1B). Hes-1 was found to act as an enhancer of basal and $1,25(OH)_2D_3$-induced OPN transcription (FIG. 2A). Hes-1 and RUNX2 can interact and $1,25(OH)_2D_3$ enhances that interaction (FIG. 2B). These findings define novel mechanisms involving the interaction of VDR, RUNX2, and the notch signaling pathway that cooperate in the transcriptional regulation of OPN in osteoblastic cells and therefore in the process of bone remodeling. In addition, novel analogs of

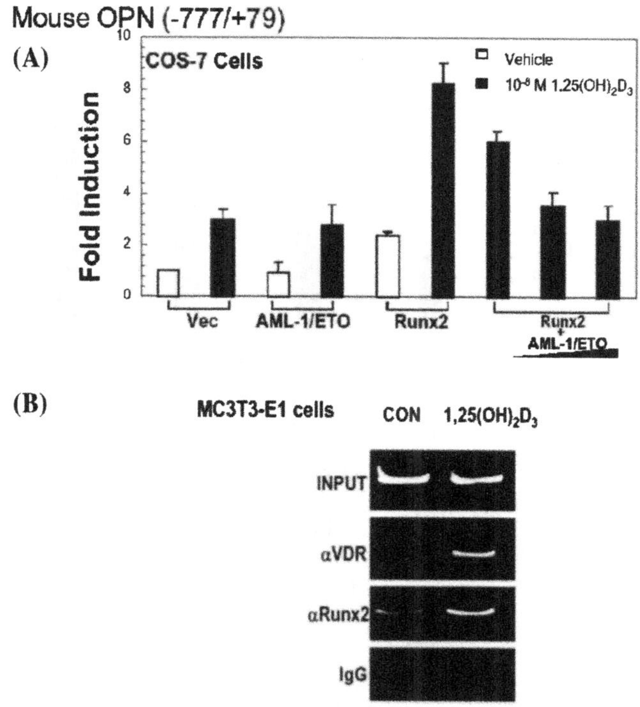

FIGURE 1. Cooperative effects between VDR and RUNX2 in the regulation of OPN transcription. (**A**) In $1,25(OH)_2D_3$-treated VDR-transfected COS-7 cells, OPN transcription (mouse OPN promotor-777/+79) was induced 3.0 ± 0.4-fold. OPN transcription was induced 2.4 ± 0.1-fold by co-transfection of RUNX2 in the absence of $1,25(OH)_2D_3$. Co-expression of RUNX2 and VDR and treatment with $1,25(OH)_2D_3$ resulted in an 8.3 ± 0.8-fold induction of OPN transcription, suggesting functional cooperation between RUNX2 and VDR. The chimeric protein AML-1/ETO, which can block RUNX2-mediated transcriptional activation, inhibited the inductive action of $1,25(OH)_2D_3$ and RUNX2 in a dose dependent manner. Inhibition by AML-1/ETO of $1,25(OH)_2D_3$-induced OPN transcription was observed in MC3T3 osteoblastic cells and in ROS17/2.8 cells (not shown). (**B**) ChIP analysis showed that $1,25(OH)_2D_3$ can enhance both VDR and RUNX2 recruitment to the OPN promotor in MC3T3-E1 cells.

$1,25(OH)_2D_3$ have been developed that have selective actions in osteoblasts and induce bone formation.[24] Thus these studies indicate that $1,25(OH)_2D_3$ does have direct effects on bone. $1,25(OH)_2D_3$ can affect resorption or formation. It is possible that direct effects of $1,25(OH)_2D_3$ on bone may not be observed in VDR null mutant mice due to similar actions by other factors.

$1,25(OH)_2D_3$ is the principal factor controlling intestinal calcium transport. $1,25(OH)_2D_3$ affects calcium entry across the brush border membrane, intracellular calcium diffusion, and extrusion of calcium across the basolateral membrane via the calcium ATPase. It is thought that $1,25(OH)_2D_3$ affects intracellular calcium diffusion by inducing the calcium-binding protein,

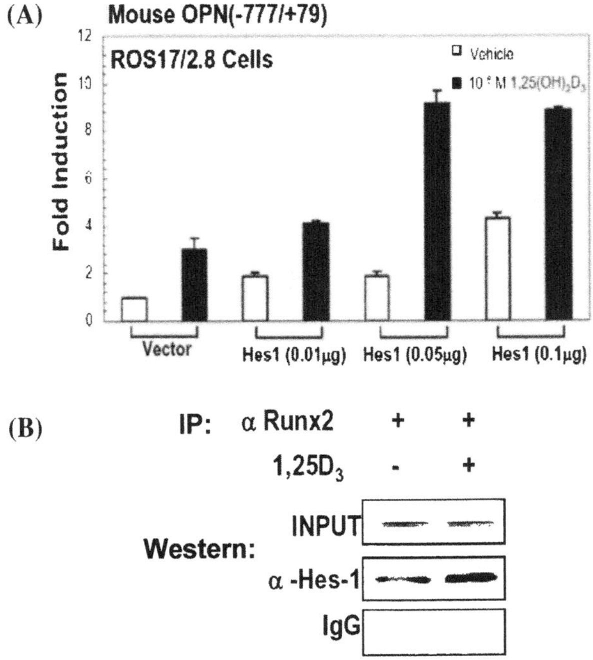

FIGURE 2. Hes-1 enhancement of $1,25(OH)_2D_3$-induced OPN transcription in osteoblastic cells. (**A**) In ROS17/2.8 cells transfection of Hes-1 resulted in an enhancement of both basal and $1,25(OH)_2D_3$-induced OPN transcription. (**B**) Immunoprecipitation assay using ROS17/2.8 cells indicated that Hes-1 and RUNX2 interact and that $1,25(OH)_2D_3$ can increase this interaction.

calbindin, that acts to facilitate the diffusion of calcium through the cell interior toward the basolateral membrane. Findings in VDR null mutant mice showing a 50% reduction in intestinal calbindin-D_{9k} mRNA, which accompanies the defect in intestinal calcium absorption, support this hypothesis.[25] However, it has been reported by our lab and others that the induction of calbindin does not always correlate to an increase in intestinal calcium absorption.[26,27] The generation of calbindin-D_{9k} null mutant mice will be important to address the role of calbindin in $1,25(OH)_2D_3$-mediated intestinal calcium transport.

Although it has been known that the rate of calcium entry into the intestinal cell can be increased by $1,25(OH)_2D_3$, the existence of a calcium channel at the brush border membrane responsible for this process had been controversial. Only recently have apical calcium channels in $1,25(OH)_2D_3$ responsive epithelia [in duodenum and jejunum co-localized with calbindin (TRPV6) and in the distal nephron also co-localized with calbindin (TRPV5)] been identified, suggesting a mechanism for calcium entry.[28] Studies by us and others have reported $1,25(OH)_2D_3$ sensitivity of TRPV5 and TRPV6.[29,30] In our studies (FIG. 3) we found a similar regulation of TRPV6 and calbindin-D_{9k} [both are

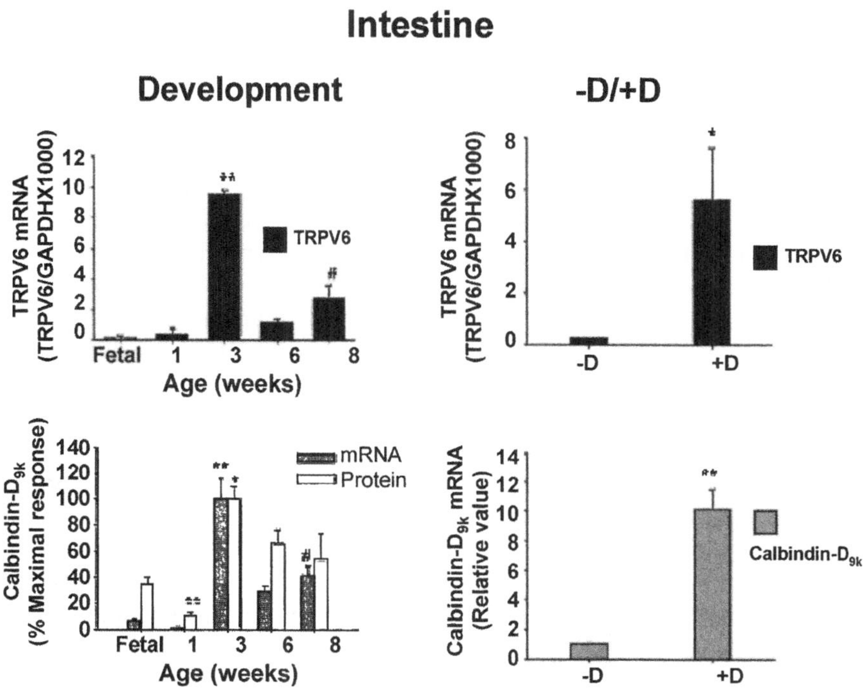

FIGURE 3. Note similar regulation of TRPV6 mRNA and calbindin-D_{9k} mRNA in mouse intestine.

induced at weaning, the time of onset of active intestinal calcium transport and intestinal responsiveness to $1,25(OH)_2D_3$, and both are induced under low dietary calcium conditions and after $1,25(OH)_2D_3$ injection], suggesting an important role for TRPV6 in $1,25(OH)_2D_3$-mediated intestinal calcium absorption. TRPV6 null mutant mice have recently been generated and will allow for the first time *in vivo* studies of the mechanisms underlying $1,25(OH)_2D_3$-mediated intestinal calcium absorption.

Besides calbindin, TRPV5, and TRPV6 the other major target of $1,25(OH)_2D_3$ in intestine and kidney is 24(OH)ase. 24(OH)ase is the most transcriptionally responsive $1,25(OH)_2D_3$-inducible gene identified to date. We have identified YY1 as a transcription factor involved in the repression of $1,25(OH)_2D_3$-induced transcription.[9] We also reported that C/EBPβ is a major transcriptional activator and we found cooperation between activation of protein kinase C and VDR in the regulation of 24(OH)ase.[11,31] We found that the enhanced activation of $1,25(OH)_2D_3$-induced 24(OH)ase transcription by activation of protein kinase C is due, in part, to an increase in VDR concentration and may also be mediated through the phosphorylation of VDR co-regulators.[31] Future studies examining phosphorylation of co-activators by different cellular signaling pathways and the physiological effects of phosphorylation of co-activators will be important for understanding how coactivators may mediate different physiological functions of VDR. Recently, we found, in addition to p160 coactivators, that SWI/SNF (complexes that facilitate transcription by remodeling chromatin using the energy of ATP hydrolysis) are also involved in VDR-mediated 24(OH)ase transcription. The SWI/SNF complex contains one of two homologous ATPases [Brahma (Brm) and Brahma-related gene (Brg-1)]. We found, using Brm- and Brg-1-deficient cell lines (SW13 and C33A) that VDR-mediated activation of 24(OH)ase transcription is markedly reduced but can be restored preferentially by Brm (FIG. 4). We also found that Brm can bind to C/EBPβ and that mutant Brm, that acts as a dominant negative inhibitor, inhibits C/EBP enhancement of 24(OH)ase transcription, suggesting that Brm may functionally cooperate with C/EBPβ in regulating 24(OH)ase (Shen, Dhawan, and Christakos, preliminary studies). These findings reveal a role for SWI/SNF in VDR-mediated transcription and suggest that the interaction between C/EBPβ and SWI/SNF may also be an important determinant in the C/EBPβ enhancement of 24(OH)ase transcription. Future studies examining the role of SWI/SNF in addition to p160 coactivators as well as the kinetics of the molecular events in the functional cross-talk between VDR, CBP, C/EBPβ, and SWI/SNF will be of fundamental importance in understanding the mechanisms involved in VDR-mediated transcription.

Thus an understanding of the functional significance of target proteins as well as studies related to the molecular mechanism of $1,25(OH)_2D_3$ action are needed in order to obtain new insight into mechanisms by which $1,25(OH)_2D_3$ mediates its multiple biological effects. An understanding of the mechanisms of $1,25(OH)_2D_3$ action may lead in the future to the development of specific analogs of $1,25(OH)_2D_3$ that could selectively modulate the $1,25(OH)_2D_3$ response in specific target tissues.

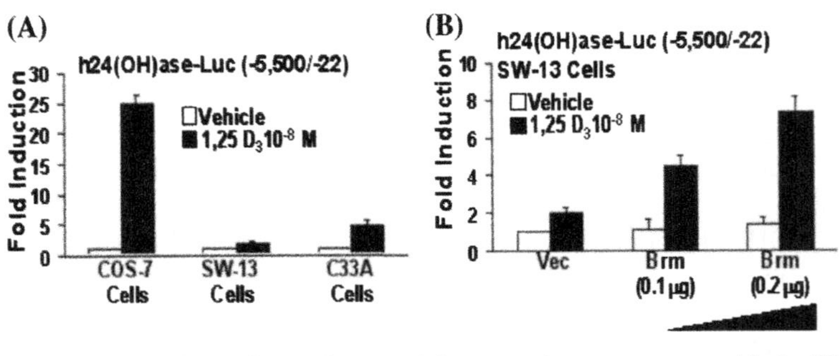

FIGURE 4. SWI/SNF chromatin-remodeling complexes cooperate with the VDR in the regulation of 24(OH)ase transcription. (**A**) COS-7 cells were transfected with the same concentration of VDR. Note the responsiveness of the 24(OH)ase promotor to $1,25(OH)_2D_3$ is markedly reduced in the Brm- and Brg-1-deficient cell lines SW13 and C33A. (**B**) $1,25(OH)_2D_3$-induced 24(OH)ase transcription can be restored by Brm.

REFERENCES

1. HORST, R.L. & T.A. REINHARDT. 1997. Vitamin D metabolism. *In* Vitamin D. D. Feldman, F.H. Glorieux & J.W. Pike, Eds.: 13–32. Academic Press. San Diego, CA.
2. BRENZA, H.L. & H.F. DELUCA. 2000. Regulation of 25-hydroxyvitamin D3 1alpha-hydroxylase gene expression by parathyroid hormone and 1,25-dihydroxyvitamin D3. Arch. Biochem. Biophys. **381:** 143–152.
3. SHINKI, T., Y. UENO, H.F. DELUCA, & T. SUDA. 1999. Calcitonin is a major regulator for the expression of renal 25-hydroxyvitamin D3-1alpha-hydroxylase gene in normocalcemic rats. Proc. Natl. Acad. Sci. USA **96:** 8253–8258.
4. SHIMADA, T., M. KAKITANI, Y. YAMAZAKI, *et al.* 2004. Targeted ablation of FGF23 demonstrates an essential physiological role of FGF23 in phosphate and vitamin D metabolism. J. Clin. Invest. **113:** 561–568.
5. NYKJAER, A., D. DRAGUN, D. WALTHER, *et al.* 1999. An endocytic pathway essential for renal uptake and activation of the steroid 25-(OH) vitamin D3. Cell **96:** 507–515.
6. PANDA, D.K., D. MIAO, M.L. TREMBLAY, *et al.* 2001. Targeted ablation of the 25-hydroxyvitamin D 1alpha -hydroxylase enzyme: evidence for skeletal, reproductive, and immune dysfunction. Proc. Natl. Acad. Sci. USA **98:** 7498–7503.
7. ST-ARNAUD, R., A. ARABIAN, R. TRAVERS, *et al.* 2000. Deficient mineralization of intramembranous bone in vitamin D-24-hydroxylase-ablated mice is due to elevated 1,25-dihydroxyvitamin D and not to the absence of 24,25-dihydroxyvitamin D. Endocrinology **141:** 2658–2666.
8. RACHEZ, C. & L.P. FREEDMAN. 2000. Mechanisms of gene regulation by vitamin D(3) receptor: a network of coactivator interactions. Gene **246:** 9–21.
9. RAVAL-PANDYA, M., P. DHAWAN, F. BARLETTA, *et al.* 2001. YY1 represses vitamin D receptor-mediated 25-hydroxyvitamin D(3)24-hydroxylase transcription: relief of repression by CREB-binding protein. Mol. Endocrinol. **15:** 1035–1046.
10. GUO, B., F. ASLAM, A.J. VAN WIJNEN, *et al.* 1997. YY1 regulates vitamin D receptor/retinoid X receptor mediated transactivation of the vitamin D responsive osteocalcin gene. Proc. Natl. Acad. Sci. USA **94:** 121–126.
11. DHAWAN, P., X. PENG, A.L. SUTTON, *et al.* 2005. Functional cooperation between CCAAT/enhancer-binding proteins and the vitamin D receptor in regulation of 25-hydroxyvitamin D3 24-hydroxylase. Mol. Cell. Biol. **25:** 472–487.
12. GUTIERREZ, S., A. JAVED, D.K. TENNANT, *et al.* 2002. CCAAT/enhancer-binding proteins (C/EBP) beta and delta activate osteocalcin gene transcription and synergize with Runx2 at the C/EBP element to regulate bone-specific expression. J. Biol. Chem. **277:** 1316–1323.
13. BARLETTA, F., L.P. FREEDMAN & S. CHRISTAKOS. 2002. Enhancement of VDR-mediated transcription by phosphorylation: correlation with increased interaction between the VDR and DRIP205, a subunit of the VDR-interacting protein coactivator complex. Mol. Endocrinol. **16:** 301–314.
14. LIU, Y., Q. SHEN, P.J. MALLOY, *et al.* 2005. Enhanced coactivator binding and transcriptional activation of mutant vitamin D receptors from patients with hereditary 1,25-dihydroxyvitamin D-resistant rickets by phosphorylation and vitamin D analogs. J. Bone Miner. Res. **20:** 1680–1691.
15. ELLISON, T.I., D.R. DOWD, P.N. MACDONALD. 2005. Calmodulin-dependent kinase IV stimulates vitamin D receptor-mediated transcription. Mol. Endocrinol. **19:** 2309–2319.

16. ISSA, L.L., G.M. LEONG, J.B. BARRY, *et al.* 2001. Glucocorticoid receptor-interacting protein-1 and receptor-associated coactivator-3 differentially interact with the vitamin D receptor (VDR) and regulate VDR-retinoid X receptor transcriptional cross-talk. Endocrinology **142:** 1606–1615.
17. AMLING, M., M. PRIEMEL, T. HOLZMANN, *et al.* 1999. Rescue of the skeletal phenotype of vitamin D receptor-ablated mice in the setting of normal mineral ion homeostasis: formal histomorphometric and biomechanical analyses. Endocrinology **140:** 4982–4987.
18. YASUDA, H., N. SHIMA, N. NAKAGAWA, *et al.* 1998. Osteoclast differentiation factor is a ligand for osteoprotegerin/osteoclastogenesis-inhibitory factor and is identical to TRANCE/RANKL. Proc. Natl. Acad. Sci. USA **95:** 3597–3602.
19. PRICE, P.A. & S.A. BAUKOL. 1980. 1,25-Dihydroxyvitamin D3 increases synthesis of the vitamin K-dependent bone protein by osteosarcoma cells. J. Biol. Chem. **255:** 11660–11663.
20. DENHARDT, D.T., C.M. GIACHELLI, S.R. RITTLING. 2001. Role of osteopontin in cellular signaling and toxicant injury. Annu. Rev. Pharmacol. Toxicol. **41:** 723–749.
21. IHARA, H., D.T. DENHARDT, K. FURUYA, *et al.* 2001. Parathyroid hormone-induced bone resorption does not occur in the absence of osteopontin. J. Biol. Chem. **276:** 13065–13071.
22. SHAPSES, S.A., M. CIFUENTES, L. SPEVAK, *et al.* 2003. Osteopontin facilitates bone resorption, decreasing bone mineral crystallinity and content during calcium deficiency. Calcif. Tissue Int. **73:** 86–92.
23. SHEN, Q. & S. CHRISTAKOS. 2005. The vitamin D receptor, RUNX2 and the notch signaling pathway cooperate in the transcriptional regulation of osteopontin. J. Biol. Chem. **280:** 40589–40598.
24. SHEVDE, N.K., L.A. PLUM, M. CLAGETT-DAME, *et al.* 2002. A potent analog of 1 alpha 25-dihydroxyvitamin D3 selectively induces bone formation. Proc. Natl. Acad. Sci. USA **99:** 13487–13491.
25. LI, Y.C., M. AMLING, A.E. PIRRO, *et al.* 1998. Normalization of mineral ion homeostasis by dietary means prevents hyperparathyroidism, rickets, and osteomalacia, but not alopecia in vitamin D receptor-ablated mice. Endocrinology **139:** 4391–4396.
26. WANG, Y.Z., H. LI, M.E. BRUNS, *et al.* 1993. Effect of 1,25,28-trihydroxyvitamin D_2 and 1,24,25-trihydroxyvitamin D_3 on intestinal calbindin-D_{9k} mRNA and protein: is there a correlation with intestinal calcium transport? J. Bone Miner. Res. **8:** 1483–1490.
27. KRISINGER, J., M. STROM, H.D. DARWISH, *et al.* 1991. Induction of calbindin-D_{9k} mRNA but not calcium transport in rat intestine by 1,25dihydroxyvitamin D_3 24-homologs. J. Biol. Chem. **266:** 1910–1913.
28. HOENDEROP, J.G., B. NILIUS, R.J. BINDELS. 2005. Calcium absorption across epithelia. Physiol. Rev. **85:** 373–422.
29. SONG, Y., X. PENG, A. PORTA, *et al.* 2003. Calcium transporter 1 and epithelial calcium channel messenger ribonucleic acid are differentially regulated by 1,25 dihydroxyvitamin D3 in the intestine and kidney of mice. Endocrinology **144:** 3885–3894.
30. VAN CROMPHAUT, S.J., M. DEWERCHIN, J.G. HOENDEROP, *et al.* 2001. Duodenal calcium absorption in vitamin D receptor-knockout mice: functional and molecular aspects. Proc. Natl. Acad. Sci. USA **98:** 13324–13329.

31. BARLETTA, F., P. DHAWAN, S. CHRISTAKOS. 2004. Integration of hormone signaling in the regulation of human 25(OH)D3 24-hydroxylase transcription. Am. J. Physiol. Endocrinol. Metab. **286:** E598-E608.

Mechanism of Vitamin D Receptor Action

MARIE B. DEMAY

Endocrine Unit, Massachusetts General Hospital and Harvard Medical School, Boston, Massachusetts 02114, USA

ABSTRACT: Studies in humans and in animal models have demonstrated that the receptor-dependent actions of 1,25-dihydroxyvitamin D are required for normal skeletal growth and maturation. Investigations were undertaken to address which consequences of vitamin D receptor deficiency are a direct result of impaired receptor-dependent hormone actions versus being due to metabolic changes. Vitamin D receptor (VDR) knockout mice were therefore generated. Investigations were performed in mice with abnormal mineral ion homeostasis, as well as in mice in which the development of abnormal mineral ion homeostasis was prevented by dietary means. VDR null mice had hypocalcemia, hyperparathyroidism, and hypophosphatemia in the first month of life. Rickets and osteomalacia are observed as well. Institution of a high-calcium, high-phosphorus, lactose-supplemented diet by the third week of life prevents abnormalities in mineral ion homeostasis. The bones of the VDR null mice with normal mineral ion homeostasis are indistinguishable from those of their wild-type littermates. The rachitic changes in the growth plates are also prevented by maintenance of normal mineral ion homeostasis. Investigations into the pathophysiological basis for the growth plate abnormalities in the VDR null mice with abnormal mineral ion homeostasis demonstrated that impaired apoptosis of hypertrophic chondrocytes due to hypophosphatemia was the cause of rachitic changes. Studies investigating the cause of the alopecia demonstrate novel ligand-independent VDR actions in the keratinocyte. The skeletal effects of VDR ablation are therefore indirect and reflect absence of ligand-dependent receptor actions in the intestine. In contrast, the cutaneous phenotype of VDR ablation is a direct consequence of absence of ligand-independent VDR actions in epidermal keratinocytes.

KEYWORDS: vitamin D receptor; keratinocyte; knockout; alopecia; rickets; osteomalacia

Address for correspondence: Marie B. Demay, Endocrine Unit, Massachusetts General Hospital and Harvard Medical School, 50 Blossom St, Boston, MA 02114. Voice: 617-726-3966; fax: 617-726-7543.

e-mail: demay@helix.mgh.harvard.edu

Ann. N.Y. Acad. Sci. 1068: 204–213 (2006).
doi: 10.1196/annals.1346.026

INTRODUCTION

The biologically active metabolite of vitamin D, 1,25-dihydroxyvitamin D, is thought to exert its principal actions by binding to the vitamin D receptor (VDR), which is a ligand-dependent transcription factor.[1] Several investigations have been performed in animals and humans to examine the biological effects of vitamin D and its metabolites. These studies have demonstrated that deficiency of vitamin D leads to hypocalcemia, presumably secondary to a defect in intestinal calcium absorption (reviewed in Refs. 2,3). This hypocalcemia is associated with secondary hyperparathyroidism and hypophosphatemia due to a parathyroid hormone (PTH)-dependent increase in urinary phosphate excretion. The skeletal manifestations of vitamin D deficiency include a widened, expanded, disorganized growth plate known as rickets (in growing animals) and osteomalacia, characterized by an increase in unmineralized bone matrix. However, considerable uncertainty remained as to whether the manifestations of vitamin D deficiency reflect impaired hormone-dependent VDR actions, or are a consequence of the resultant metabolic changes.

VDR ABLATED MICE

To address this question, we generated mice with targeted ablation of the VDR.[4] Although these mice are born phenotypically normal, by the third week of life they develop abnormalities in mineral ion homeostasis. Serum calcium, PTH, and phosphate levels are normal at 16 days of age; however, by 21 days of age, secondary hyperparathyroidism is observed. This secondary hyperparathyroidism is due to an increase in parathyroid cellular proliferation, which results in a marked increase in parathyroid gland volume by 70 days of age, as well as an increase in PTH mRNA levels, assessed by *in situ* hybridization analyses. The development of hyperparathyroidism leads to hypophosphatemia by 21 days of age. Rickets is evident by 24 days of age and osteomalacia by 35 days of age.[4] To determine whether these abnormalities were directly related to the absence of VDR in the parathyroid cell, the osteoblast, and the chondrocyte, or were a result of the metabolic consequences of VDR ablation, a diet high in calcium, phosphorus, and lactose was instituted before abnormalities in mineral ions were observed in the VDR null mice (day 19). This diet, which was originally formulated to prevent the development of hyperparathyroidism in vitamin D-deficient rats,[5] prevented the development of hypocalcemia, hyperparathyroidism, and hypophosphatemia in the VDR null mice, thus permitting studies aimed at clarifying those aspects of the VDR null phenotype that were a consequence of the absence of VDR in a specific cell/tissue versus a consequence of the absence of VDR actions in the intestine, leading to impaired intestinal calcium absorption and the resultant metabolic abnormalities observed with VDR ablation.[6]

The parathyroid glands of the VDR null mice with normal mineral ion levels were indistinguishable from those of their wild-type littermates at 70 days of age, suggesting that in the presence of normocalcemia, the receptor-dependent actions of 1,25-dihydroxyvitamin D are not essential for the prevention of hyperparathyroidism.[6] In addition, normalization of mineral ions prevented the development of the skeletal abnormalities in the VDR null mice. Histomorphometric and biomechanical analyses of the long bones of the VDR null mice with normal mineral ion levels were indistinguishable from those of their wild-type littermates.[6,7] These *in vivo* studies suggest that the VDR is not required for skeletal homeostasis. However, they do not exclude the possibility that the role of the VDR in the skeleton is redundant and that, in its absence a second receptor or alternative homeostatic pathway performs its normal functions. Notable in this respect is the effect of 1,25-dihydroxyvitamin D on the regulation of RANK ligand production by the osteoblast. Although 1,25-dihydroxyvitamin D has been shown to play an important role in inducing the synthesis of this key regulator of osteoclast differentiation,[8] histomorphometric analyses in the normocalcemic VDR null mice have demonstrated normal osteoclast numbers and resorption surfaces.[7] However, *in vitro* studies have demonstrated that osteoblasts from VDR null mice cannot support osteoclastogenesis when co-cultured with normal spleen cells and 1,25-dihydroxyvitamin D, although when these co-cultures were performed with PTH and interleukin 1α, osteoclasts with resorbing activity were formed.[9] These data clearly support the hypothesis that the VDR plays a key role in the skeleton; however, in its absence, other regulatory molecules (hormones, cytokines) may be called upon to maintain skeletal homeostasis.

GROWTH PLATES IN VDR-DEFICIENT MICE

Studies were then undertaken to identify the basis for the growth plate phenotype in the VDR null mice. Vitamin D metabolites have been shown to promote chondrocyte differentiation, suggesting that the absence of the VDR in the chondrocyte may underlie the rachitic phenotype.[10] However, calcium has been shown to promote chondrocyte differentiation and mineralized matrix formation, suggesting that hypocalcemia may play a role in the etiology of the growth plate abnormality.[11] Alternatively, since inorganic phosphate has been shown to promote chondrocyte apoptosis in an avian culture system, hypophosphatemia may lead to the expansion of the growth plate by impairing hypertrophic chondrocyte apoptosis.[12] The observation that the growth plate of the VDR null mice is indistinguishable from that of their wild-type littermates prior to the onset of abnormal mineral ion levels suggests that the receptor-dependent actions of 1,25-dihydroxyvitamin D in the chondrocyte are not required for normal growth plate maturation. In support of this hypothesis, prevention of abnormal mineral ion homeostasis prevents the development of rickets.[13] Investigations were, therefore, undertaken to address the hypothesis that hypophosphatemia, hypocalcemia, and/or hyperparathyroidism were the cause of the growth plate abnormalities observed in the VDR null mice fed a regular diet and to clarify the cellular and pathophysiological basis for the growth plate abnormality in the VDR null mice.

Initial studies focused on further characterization of the growth plate phenotype of the VDR null mice. Based on the work of other investigators demonstrating that extracellular calcium promotes expression of markers of terminal chondrocyte differentiation and increases the production of mineralized matrix[11] the hypothesis we initially addressed was that abnormal mineral ion homeostasis led to impaired chondrocyte differentiation. These investigations demonstrated that within 2 days of the development of hyperparathyroidism, there was an expansion in the hypertrophic chondrocyte layer of the VDR null mice. However, BrdU incorporation into proliferative chondrocytes was unaltered, as was the transition of the BrdU-labeled cells to the hypertrophic layer.[13] These studies demonstrated that there was no detectable abnormality in chondrocyte proliferation nor was there accelerated maturation of the proliferative chondrocytes into hypertrophic cells. Markers of chondrocyte differentiation were unaltered as was production of vascular endothelial growth factor (VEGF), a key signal for vascular invasion, a process that is required for the replacement of the terminally differentiated hypertrophic chondrocytes by bone. Of particular interest, however, was the marked increase in the expression of matrix gla protein (MGP). MGP is a potent inhibitor of matrix mineralization, as shown by knockout and transgenic overexpression models. We therefore proposed the hypothesis that overexpression of MGP may play a pathophysiological role, leading to impaired matrix mineralization and expansion of the late hypertrophic chondrocytes. Unfortunately, making the VDR knockout mice null for MGP as well, did not normalize the growth plate phenotype. Since the cellular basis for the rachitic changes involved expansion of the terminally differentiated, osteopontin-expressing late hypertrophic chondrocytes, we then addressed the hypothesis that there was impaired apoptosis of these cells. Histological analyses demonstrated a marked decrease in apoptosis of the late hypertrophic chondrocytes in the rachitic VDR null mice, thus clarifying the cellular basis for this abnormality. Notably, prevention of rickets by maintaining normal mineral ion homeostasis was associated with the normalization of hypertrophic chondrocyte apoptosis.[13] Subsequent analyses pointed to hypophosphatemia as being the critical pathophysiological abnormality responsible for impaired hypertrophic chondrocyte apoptosis in the VDR null mice.[14]

VITAMIN D AND CUTANEOUS HOMEOSTASIS

The VDR null mice develop total body alopecia by 100 days of age.[4,15,16] Alopecia totalis is also seen in some human kindreds with VDR mutations and has been thought to correlate with a more severe metabolic phenotype. Interestingly, normalization of mineral ion homeostasis fails to prevent the development of alopecia in the VDR null mice. These studies suggest that, unlike the skeletal phenotype of the VDR null mice, which reflects impaired VDR action in an unrelated organ (the intestine), the cutaneous phenotype of VDR ablation is a consequence of lack of VDR expression in the skin. We, therefore, undertook studies to address whether the hair loss was a consequence of an abnormality in keratinocytes, which give rise to the epidermal component of the hair follicle or in the dermal papilla cells, which make up the mesodermal component of the hair follicle. The VDR null mice develop a normal first coat of hair, reflecting normal hair follicle morphogenesis, which in mice extends to the second week of life. This suggested that the defect may be secondary to an abnormality in postnatal hair cycling. The hair follicle goes through cyclical periods of growth (anagen), involution (catagen), and quiescence (telogen) during postnatal life. Therefore, to address the hypothesis that the alopecia in the VDR null mice was due to a defect in the hair cycle, the response of the VDR knockout (KO) mice to anagen initiation was examined at 18 days of age, a time when there is no histological difference in the skin of the wild-type and VDR null mice. Because these animals are in the C57BL/6J background, this procedure should result in progressive skin pigmentation and thickening within 5 to 6 days, correlating with induction of anagen. As expected, marked BrdU incorporation, correlating with the marked proliferation that characterizes early anagen, was observed in the wild-type animals 6 days post depilation. By 10 days post depilation, the wild-type mice had visible hair shafts. In contrast, rare BrdU positive cells were seen among the hair follicle keratinocytes of the VDR null mice 6 days post depilation.[17] Also notable was the absence of skin thickening at 10 days post depilation and the absence of hair shafts even 13 weeks post depilation. These studies demonstrated that the alopecia in the VDR null mice was secondary to a defect in anagen initiation and set the stage for subsequent experiments aimed at determining whether this was due to a cellular or a metabolic defect.

Because, 1,25-dihydroxyvitamin D downregulates its own biosynthesis by repressing the 25-hydroxyvitamin D-1α-hydroxylase gene[18,19] and increases its metabolism by upregulating the 24-hydroxylase gene through VDR-dependent actions,[20,21] VDR ablated mice have very high levels of 1,25-dihydroxyvitamin D even when mineral ion levels are normalized. Since 1,25-dihydroxyvitamin D has been shown to inhibit keratinocyte proliferation and to promote keratinocyte differentiation,[22,23] very high levels of this hormone or its metabolites may lead to alopecia by toxic interactions with an alternative receptor. To address the hypothesis that 1,25-dihydroxyvitamin D toxicity was the cause of the alopecia, VDR null mice were raised in an ultraviolet (UV)-free environment, for seven generations, on a diet lacking vitamin D metabolites. The absence of detectable circulating levels of 25-hydroxyvitamin D and 1,25-dihydroxyvitamin D failed to prevent the development of alopecia in the VDR null mice, demonstrating that this phenotype was not a consequence of toxic levels of circulating hormone.[24]

In parallel, the alternative hypothesis, that the alopecia in the VDR knockout mice was secondary to the absence of hormone-dependent VDR actions, was addressed by examining the skin of wild-type mice with undetectable circulating levels of 25-hydroxyvitamin D and 1,25-dihydroxyvitamin D. Although local hormone production in the skin could not be definitively excluded, the absence of UV radiation (which is required for local synthesis of pre-vitamin D from 7-dehydrocholesterol) and of detectable circulating levels of 25-hydroxyvitamin D, would leave the keratinocyte 1α hydroxylase without a substrate on which to act. Therefore, based on our current understanding of vitamin D metabolism, it is extremely unlikely that these animals would have local production of vitamin D metabolites. Alopecia was not observed in these mice, demonstrating that absence of receptor per se rather than absence of ligand-dependent receptor effects was responsible for the alopecia in the VDR null mice. This suggests that the VDR may have ligand-independent effects on the skin and that absence of these ligand-independent effects is the cause of alopecia in humans and mice lacking functional VDRs.[24]

To identify whether absence of the VDR in the epidermal keratinocyte or the mesodermal dermal papilla cell was responsible for the hair cycle abnormality, hair reconstitution assays were performed in nude mice. Implantation of a mixture of keratinocytes and activated dermal papilla cells into a nude mouse host recapitulates the process of hair follicle morphogenesis,[25] leading to a functional hair follicle. The use of early anagen dermal papilla cells is critical since these cells do not retain their ability to induce the formation of functional hair follicles during all stages of the hair cycle. By implanting wild-type or VDR null neonatal dermal papilla cells with wild-type or VDR null keratinocytes, hair follicles with epidermal and mesodermal components differing in VDR status (+/+ versus –/–) were formed and permitted analyses directed at determining which cell population was responsible for the defect in anagen initiation. Epithelial cells are easily purified from the epidermis of neonatal mice; however, dissection of a pure population of dermal papilla cells is impracticable, since the dermal papilla is embedded in the ensheathing epidermal component. To overcome this problem the VDR null mice were cross-bred with transgenic mice expressing green fluorescent protein (GFP) under the control of the versican promotor. In these transgenic mice, GFP is specifically expressed in the dermal papilla cells[25]; therefore, these cells are easily isolated from the dermis of neonatal mice by fluorescence-activated cell sorting. These studies demonstrated that VDR null keratinocytes and dermal papilla cells were capable of reconstituting functional hair follicles, recapitulating morphogenesis. However, when the reconstituted follicles were subjected to anagen initiation, follicles containing VDR null keratinocytes were unable to initiate a new hair cycle. In contrast, when dermal papilla cells of either genotype were mixed with wild-type keratinocytes, a normal response to anagen induction, including generation of new hair shafts, was observed. These studies identified the VDR null keratinocyte as the cell responsible for the hair cycle defect in the VDR null mice.[24]

Although these investigations demonstrated that expression of the VDR in the epidermal keratinocyte was necessary for the prevention of alopecia, they did not address whether VDR expression solely in keratinocytes was sufficient for the maintenance of the normal hair cycle. Transgenic mice, expressing the VDR in keratinocytes under the keratin 14 promotor were generated. Several transgenes have been successfully targeted to epidermal keratinocytes using this promotor.[26–30] Mice expressing this transgene were mated into the VDR null background to obtain mice that expressed the VDR only in keratinoctyes. These studies demonstrated that the VDR null mice expressing the transgene do not develop alopecia by 100 days and respond to anagen induction with the re-growth of hair, analogous to what is observed in wild-type mice.[31]

Studies were then performed to address whether maintenance of the hair cycle required ligand-dependent actions of the VDR. Initial investigations in VDR null mice and their wild-type littermates, rendered deficient for both 25-hydroxyvitamin D and 1,25-dihydroxyvitamin D, demonstrated that absence of detectable circulating vitamin D metabolites did not result in abnormal hair cycling, nor did it prevent the development of alopecia in VDR null mice. These studies suggested that maintenance of the hair cycle was a novel ligand-independent action of the VDR, since the absence of detectable circulating ligand and the absence of functional receptor had different phenotypes. However, these investigations could not exclude the possibility that novel locally produced ligands could activate the VDR, analogous to the effects of lithocholic acid on VDR-dependent transactivation in the gut.[32]

Based on the hypothesis that the effects of the VDR on hair follicle homeostasis were ligand-independent and thus, perhaps did not involve transactivation but rather transcriptional repression, transgenic mice expressing VDRs with mutations that either abolished ligand binding (L233S-VDR) or abolished recruitment of nuclear receptor co-activators (L417S-VDR) were engineered. Studies by Erben *et al.* in their line of VDR knockout mice in which the first zinc finger of the VDR DNA-binding domain was targeted for homologous recombination, definitively demonstrated that receptor–DNA interactions were essential for the prevention of alopecia.[16] These mice express a truncated VDR that is unable to bind to vitamin D response elements, but contains intact hormone-binding and nuclear receptor co-activator (AF2) domains. Mice homozygous for this truncated transcript are a phenocopy of the VDR knockout mice that lack the entire VDR protein, thus demonstrating that the DNA-binding domain of the receptor is essential for the prevention of alopecia.

Investigations in VDR null mice with keratinocyte-specific expression of the mutant transgenes demonstrated that mice expressing the transgene with the ligand-binding domain mutation had normal postmorphogenic hair cycles and thus revealed critical ligand-independent effects of the VDR *in vivo.*

Intriguingly, the transgene with the AF2 domain mutation affected a partial rescue: initially postmorphogenic hair cycles were normal, but with age, progressive hair loss, dermal cysts, and dilated piliary canals were seen, analogous to the skin phenotype of the transgene negative VDR null mice.[33]

These investigations raised the question as to whether transcriptional repression by the VDR is required for cutaneous homeostasis. Unlike other closely related nuclear receptors, the VDR does not associate strongly with the co-repressors NcoR and SMRT.[34,35] However, more recent investigations revealed that the nuclear receptor co-repressor, Hairless, interacts with the VDR and that these interactions are unaffected by the presence of ligand.[36] Mutations in the hairless gene in mice and humans result in a cutaneous phenotype similar to that observed with VDR ablation,[37,38] suggesting that Hairless–VDR actions may be required for normal hair cycling. Thus studies were undertaken to address whether the two mutant VDRs were able to interact with Hairless. These investigations demonstrated that neither the L233S nor the L417S mutation interfered with Hairless–VDR interactions. Thus, the effects of the VDR on the hair cycle are ligand-independent and may involve transcriptional repression. The partial rescue obtained with the AF2 domain mutant transgene suggests that VDR interactions with nuclear receptor co-modulators are required for long-term hair follicle homeostasis and that, although Hairless–VDR interactions may be required, they are not sufficient. Thus additional factors are likely recruited to the VDR complex, modulating the effects of Hairless–VDR on the hair follicle keratinocyte.

CONCLUSIONS

These investigations illustrate that the two major target organs in which the actions of VDR are critical are the intestine and the skin. The skeletal effects of VDR ablation are indirect and reflect the absence of ligand-dependent receptor actions in the intestine. In contrast, the cutaneous phenotype of VDR ablation is a direct consequence of absence of ligand-independent VDR actions in epidermal keratinocytes.

ACKNOWLEDGMENTS

These investigations were supported by grants from the National Institutes of Health (DK46974) and from the National Alopecia Areata Foundation.

REFERENCES

1. HAUSSLER, M.R. & G.K. WHITFIELD, *et al.* 1998. The nuclear vitamin D receptor: biological and molecular regulatory properties revealed. J. Bone Miner. Res. **13:** 325–349.

2. THOMAS, M. & M. DEMAY. 2000. Vitamin D Deficiency and disorders of Vitamin D metabolism. Endocrinol. Metab. Clinics N. Am. **29:** 611–627.
3. WALTERS, M. & U. KOLLENKIRCHEN, *et al.* 1992. What is Vitamin D deficiency? Proc. Soc. Exp. Biol. Med. **199:** 385–393.
4. LI, Y.C. & A.E. PIRRO, *et al.* 1997. Targeted ablation of the vitamin D receptor: an animal model of vitamin D-dependent rickets type II with alopecia. Proc. Natl. Acad. Sci. USA **94:** 9831–9835.
5. KOLLENKIRCHEN, U. & J. FOX, *et al.* 1991. Normocalcemia without hyperparathyroidism in vitamin D-deficient rats. J. Bone Miner. Res. **6:** 273–278.
6. LI, Y.C. & M. AMLING, *et al.* 1998. Normalization of mineral ion homeostasis by dietary means prevents hyperparathyroidism, rickets, and osteomalacia, but not alopecia in vitamin D receptor-ablated mice. Endocrinology **139:** 4391–4396.
7. AMLING, M. & M. PRIEMEL, *et al.* 1999. Rescue of the skeletal phenotype of vitamin D receptor ablated mice in the setting of normal mineral ion homeostasis: formal histomorphometric and biomechanical analyses. Endocrinology **140:** 4982–4987.
8. YASUDA, H. & N. SHIMA, *et al.* 1998. Osteoclast differentiation factor is a ligand for osteoprotegerin/osteoclastogenesis-inhibitory factor and is identical to TRANCE/RANKL. Proc. Natl. Acad. Sci. USA **95:** 3597–3602.
9. TAKEDA, S. & T. YOSHIZAWA, *et al.* 1999. Stimulation of osteoclast formation by 1,25-dihydroxyvitamin D requires its binding to vitamin D receptor (VDR) in osteoblastic cells: studies using VDR Knockout mice. Endocrinology **140:** 1005–1008.
10. BOYAN, B. & V. SYLVIA, *et al.* 2001. 25,25-(OH)2D3 regulates cartilage and bone via autocrine and endocrine mechanisms. Steroids **66:** 363–374.
11. CHANG, W. & C. TU, *et al.* 2002. Extracellular Ca(2+)-sensing receptors modulate matrix production and mineralization in chondrogenic RCJ3.1C5.18 cells. Endocrinology **143:** 1467–1474.
12. MANSFIELD, K. & R. RAJPUROHIT, *et al.* 1999. Extracellular phosphate ions cause apoptosis of terminally differentiated epiphyseal chondrocytes. J. Cell. Physiol. **179:** 276–286.
13. DONOHUE, M. & M. DEMAY. 2002. Rickets in VDR null mice is secondary to decreased apoptosis of hypertrophic chondrocytes. Endocrinology **143:** 3691–3694.
14. SABBAGH, Y. & T. CARPENTER, *et al.* 2005. Hypophosphatemia leads to rickets by impairing caspase-mediated apoptosis of hypertrophic chondrocytes. PNAS. **102:** 9637–9642.
15. YOSHIZAWA, T. & Y. HANDA, *et al.* 1997. Mice lacking the vitamin D receptor exhibit impaired bone formation, uterine hypoplasia and growth retardation after weaning. Nat. Genet. **16:** 391–396.
16. ERBEN, R.G. & D.W. SOEGIARTO, *et al.* 2002. Deletion of deoxyribonucleic acid binding domain of the vitamin D receptor abrogates genomic and nongenomic functions of vitamin D. Mol. Endocrinol. **16:** 1524–1537.
17. SAKAI, Y. & M. DEMAY. 2000. Evaluation of keratinocyte proliferation and differentiation in vitamin D receptor knockout mice. Endocrinology **141:** 2043–2049.
18. ST-ARNAUD, R. & S. MESSERLIANN, *et al.* 1997. The 25-hydroxyvitamin D 1-alpha hydroxylase gene maps to the pseudovitamin D-deficiency rickets (PDDR) disease locus. J. Bone Miner. Res. **12:** 1552–1559.
19. TAKEYAMA, K. & S. KITANAKA, *et al.* 1997. 25-Hydroxyvitamin D3 1alpha-hydroxylase and vitamin D synthesis. Science **277:** 1827–1830.

20. KERRY, D.M. & P.P. DWIVEDI, *et al.* 1996. Transcriptional synergism between vitamin D-responsive elements in the rat 25-hydroxyvitamin D3 24-hydroxylase (CYP24) promoter. J. Biol. Chem. **271:** 29715–29721.
21. OHYAMA, Y. & M. NOSHIRO, *et al.* 1991. Cloning and expression of cDNA encoding 25-hydroxyvitamin D_3 24-hydroxylase. FEBS. **278:** 195–198.
22. BIKLE, D.D. & S. PILLAI. 1993. Vitamin D, calcium, and epidermal differentiation. Endocr. Rev. **14:** 3–19.
23. ECKERT, R.L. & J.F. CRISH, *et al.* 1997. The epidermal keratinocyte as a model for the study of gene regulation and cell differentiation. Physiol. Rev. **77:** 397–424.
24. SAKAI, Y. & J. KISHIMOTO, *et al.* 2001. Metabolic and cellular analysis of alopecia in vitamin D receptor knockout mice. J. Clin. Invest. **107:** 961–966.
25. KISHIMOTO, J. & R. EHAMA, *et al.* 1999. Selective activation of the versican promoter by epithelial-mesenchymal interactions during hair follicle development. Proc. Natl. Acad. Sci. USA **96:** 7336–7341.
26. CHENG, J. & K. TURKSEN, *et al.* 1992. Cachexia and graft- vs. host-disease-type skin changes in keratino promoter-driven TNF alpha transgenic mice. Genes. Dev. **6:** 1444–1456.
27. ROSSITER, H. & S. BEISSERT, *et al.* 2001. Targeted expression of bcl-2 to murine basal epidermal keratinocytes results in paradoxical retardation of ultraviolet and chemical induced tumorigenesis. Cancer Res. **61:** 3619–3626.
28. VASSAR, R. & M. ROSENBERG, *et al.* 1989. Tissue-specific and differentiation-specific expression of a human K14 keratin gene in transgenic mice. PNAS. **86:** 1563–1567.
29. WILLIAMS, I. & R. ORT, *et al.* 1994. Keratinocyte expression of B7-1 in transgenic mice amplifies the primary immune response to cutaneous antigens. PNAS. **91:** 12780–12784.
30. WYSOLMERSKI, J.J. & A.E. BROADUS, *et al.* 1994. Overexpression of parathyroid hormone-related protein in the skin of transgenic mice interferes with hair follicle development. Proc. Natl. Acad. Sci. USA **91:** 1133–1137.
31. CHEN, C. & Y. SAKAI, *et al.* 2001. Targeting expression of the human vitamin D receptor to the keratinocytes of vitamin D receptor null mice prevents alopecia. Endocrinology **142:** 5386–5389.
32. MAKISHIMA, M. & T. LU, *et al.* 2002. Vitamin D receptor as an intestinal bile acid sensor. Science **296:** 1313–1316.
33. SKORIJA, K. & M. COX, *et al.* 2005. Ligand-independent actions of the vitamin D receptor maintain hair follicle homeostasis. Mol. Endocrinol. **19:** 855–862.
34. CHEN, J.D. & K. UMESONO, *et al.* 1996. SMRT isoforms mediate repression and anti-repression of nuclear receptor heterodimers. Proc. Natl. Acad. Sci. USA **93:** 7567–7571.
35. TAGAMI, T. & W.H. LUTZ, *et al.* 1998. The interaction of the vitamin D receptor with nuclear receptor corepressors and coactivators. Biochem. Biophys. Res. Commun. **253:** 358–363.
36. HSIEH, J.C. & J.M. SISK, *et al.* 2003. Physical and functional interaction between the vitamin D receptor and hairless corepressor, two proteins required for hair cycling. J. Biol. Chem. **278:** 38665–38674.
37. MANN, S.J. 1971. Hair loss and cyst formation in hairless and rhino mutant mice. Anat. Rec. **170:** 485–499.
38. AHMAD, W. & M. FAIYAZ UL HAQUE, *et al.* 1998. Alopecia universalis associated with a mutation in the human hairless gene. Science **279:** 720–724.

Role of Connexin43 in Osteoblast Response to Physical Load

SUSAN K. GRIMSTON,[a] JOANNE SCREEN,[a] JENNIFER H. HASKELL,[a] DONG JIN CHUNG,[a,c] MICHAEL D. BRODT,[b] MATTHEW J. SILVA,[b] AND ROBERTO CIVITELLI[a]

[a]*Division of Bone and Mineral Diseases, Department of Internal Medicine, Washington University School of Medicine, St. Louis, Missouri 63110, USA*

[b]*Department of Orthopaedic Surgery, Washington University School of Medicine, St. Louis, Missouri 63110, USA*

[c]*Department of Internal Medicine, Chonnam University Research Institute for Medical Sciences, Chonnam National University Medical School, Gwangju, South Korea*

ABSTRACT: Gap junctions are hexameric transmembrane channels formed by connexins, and are responsible for direct cell-to-cell communication. The most abundant gap junction protein in bone is connexin43 (Cx43), although connexin45 (Cx45) is also expressed. In the present study, we tested the hypothesis that bone cell responses to mechanical stimulation are dependent on the type of gap junction communication provided by Cx43 *in vitro* and in an *in vivo* model of physical load. Application of cyclic stretch to calvaria osteoblasts results in a modest but detectable increase in PGE_2 levels, and the amount of PGE_2 produced was lower in cells isolated from Cx43 null mice. Mice with an osteoblast-specific deletion of the *Cx43* gene were subjected to an *in vivo* four-point bending protocol on the tibia. This resulted in fast and exuberant formation of woven bone at the region directly below the loading fulcrum in both osteoblast *Cx43*-deleted and wild-type mice. However, indirect measurement of endosteal bone apposition suggested a less pronounced effect of physical load in *Cx43*-deficient than in wild-type mice. Taken together, these results indicate that deficiency of *Cx43* in osteoblasts attenuates but does not abolish anabolic responses to mechanical strain.

KEYWORDS: gap junctions; connexin43; mechanical strain; bone formation

Address for correspondence: Roberto Civitelli, M.D., Division of Bone and Mineral Diseases, Washington University in St. Louis, 660 S. Euclid Avenue, St. Louis, MO 63110. Voice: 314 -454-7765; fax: 314-454-5047.
e-mail: rcivitel@im.wustl.edu

Ann. N.Y. Acad. Sci. 1068: 214–224 (2006). © 2006 New York Academy of Sciences.
doi: 10.1196/annals.1346.023

INTRODUCTION

The mechanisms underlying bone's response to mechanical stimulation are only partially understood. It is generally believed that the osteocyte is the mechanosensory cell and that it transmits signals to bone-forming cells in the surface via gap junctions, to initiate an anabolic response. Gap junctions consist of intercellular channels that serve to connect adjacent cells and allow for the diffusion of small molecules and ions from cell to cell.[1,2] Each cell contributes half of the intercellular channel (hemichannel), which is composed of six subunits, connexins, arranged in a hexagonal array to form the hemichannel, also called connexon. More than two dozen connexins exist, but the most abundant in bone is connexin43 (Cx43), although osteoblasts also express connexin45 (Cx45).

A variety of approaches have been used to understand how mechanical signals are received and transduced into biochemical responses at the cell surface, and in particular how mechanically generated signals are elaborated and diffused among bone cells. Both cell stretching by substrate deformation and shear stress induced by fluid flow have been applied to cell models of osteoblasts and osteocytes.[3–10] *In vivo* loading of bone itself has also been used,[11–13] and although none of these models exactly replicate the stress and strain that bone receives *in vivo* during daily living activities, they do result in osteoblastic responses that can be considered to reflect an anabolic response. In particular, release of prostaglandin E_2 (PGE_2) has been frequently reported following application of fluid flow shear stress.[14–16] PGE_2 has many functions and it can stimulate bone mass in mammals, when used at the appropriate doses. It has also been shown to be involved in mediating new bone formation in response to mechanical load *in vivo*.[17,18]

New bone formation requires the coordinated activity of osteoblasts at a bone formation site. This coordination can be achieved through direct communication among cells provided by gap junctions, which are also believed to mediate direct osteoblast-osteocyte signaling.[19,20] Because of the unique microstructural features of osteocytes and their intimate relationship with their microenvironment, it is very difficult to model osteoblast-osteocyte relationships *in vitro*. A more powerful approach is to genetically modify animals to delete one connexin gene specifically in bone cells, and then test whether response to skeletal load is altered in these conditionally connexin-deleted animals. We have recently developed such a mouse model of osteoblast-specific ablation of the Cx43 gene using the Cre/loxP approach.[21] Because in this model, Cre-induced recombination and gene deletion is driven by a fragment of the $\alpha_1(I)$ *collagen* promotor, gene ablation occurs in differentiated osteoblasts and thus, osteocytes will be deleted as well.

The goal of the present work was to assess the applicability of one methodology to mechanically load the skeleton in the conditionally *Cx43* ablated mice, and begin testing whether this stimulus is dependent on the presence of Cx43 in osteoblasts. We also used Cx43-deficient cells to assess their sensitivity to mechanical stretch *in vitro* in terms of PGE_2 production.

METHODS

Animals

A mouse line with a germline null mutation of Cx43 was used for calvaria cell isolation, as described.[22] For the *in vivo* studies, we used a tissue-specific Cx43 gene deletion using Cre-mediated replacement of a "floxed" Cx43 allele ($Cx43^{fl}$) with a *LacZ* reporter cassette.[21] Homozygous $Cx43^{fl/fl}$ mice were crossed with mice expressing Cre driven by a 2.3 kb-fragment of the $\alpha_1(I)$ *collagen* promotor (abbreviated as *ColCre*) in a heterozygous Cx43 null background, yielding $ColCre;Cx43^{-/fl}$ mice. Effectiveness of *in vivo* recombination event was demonstrated in osteoblasts by X-gal staining of tibia sections and calvaria cell cultures, and by detection of the deleted allele in osteoblastic cell extracts.[21]

Cell Stretching Studies

Cells from newborn mice calvaria were harvested as described,[22] and used for the *in vitro* studies. The Flexercell Strain unit (Flexcell Corporation, McKeesport, PA) was used for application of mechanical stretch to osteoblasts, as detailed in a previous study.[6] Cells were grown on specially designed 6-well tissue culture dishes with collagen-coated flexible silicone bottoms (Flex I, Flexcell Corporation). Through an air pump, a negative pressure of 22 kPa was applied to the flexible bottomed wells at three cycles/min. During each cycle, a 10-s stretch period was followed by a 10-s relaxation. As the flexible bottoms are pulled downward by the negative pressure, the cells attached to their upper surface are stretched by the deformation of the rubber. Control cultures were grown under the same conditions on dishes of the same size containing a collagen-coated silicone disc identical to the one used for stretch, sitting on a rigid plastic bottom (Flex II, Flexcell Corporation). We optimized this system using two cell lines—ROS 17/2.8 (a rat osteosarcoma cell line) and ROS/Cx45, which stably express the gap junction protein Cx45.[23] Experiments with newborn calvaria cells were performed by applying stretch at a rate of three 10-s cycles/min for different lengths of time. An ELISA assay was used to detect PGE_2 levels in the medium (Cayman Chemical Company Ann Arbor, MI).

In Vivo *Studies*

To apply mechanical stimulation *in vivo*, a method based on a four-point bending loading paradigm established for mouse tibia was used.[24] For these experiments, 18 mice were used, 6 of each of three genotype, $Cx43^{+/fl}$ (wild-type equivalent), $Cx43^{-/fl}$ (heterozygous equivalent), and $ColCre;Cx43^{-/fl}$ (conditional knockout). The right tibia of each mouse was subjected to a bending force of 17N (50% of the breaking point of the bones, calculated experimentally for each genotypes in a pilot study) for 200 cycles, with each cycle comprising 1 s of force application followed by a 10-s rest.[25,26] This regimen was carried out 5 days per week for 2 weeks. Calcein (5-mg calcein/mL solution at a dose of body weight x 4 μL) was injected on days 7 and 2 before sacrifice, and the right (loaded) and left (unloaded) tibia used for microstructural and histologic analysis.

Micro computed tomography (μCT 40, Scanco Medical AG, Basserdorf, Switzerland) was used for three-dimensional reconstruction of the tibial diaphysis and to calculate the internal marrow area, with the assumption that the area is inversely proportional to new endosteal bone apposition. Histomorphometry was used to calculate labeled surface and mineral apposition rate, as described in detail elsewhere.[27]

RESULTS

In Vitro *Studies*

Release of PGE_2 has been found to occur rapidly after cell exposure to mechanical stimuli.[14–16] In pilot studies, we tested the medium from cultures of rat osteosarcoma cells (ROS 17/2.8) and cells stably transfected with Cx45 during and after cyclic stretch (10-s cycles of 1–2% deformation applied for 1–4 h). We could not detect significant PGE_2 release by these cells, either in resting conditions or during stretch (data not shown). Further investigation revealed that these cells did not express inducible *COX-2* (the enzyme isoform that produces PGE_2 upon stimulation), and were thus deemed an inappropriate model to test our hypothesis.

Therefore, we moved to calvaria cells isolated from wild-type and *Cx43*-deficient mice for these studies. We first verified that cell isolates from *Cx43*-deficient mice did not express Cx43-specific immunostain, whereas those obtained from wild-type animals did express abundant Cx43-specific stain at appositional membranes (a location suggestive of functional channels), as well as in perinuclear areas (FIG. 1). More to the point, there was a tendency toward a minimal increase in PGE_2 production over time with mechanical stretch. Nonetheless, lower amounts of PGE_2 were produced in the *Cx43* knockout cells compared to the wild type (FIG. 2).

In Vivo *Studies*

The most striking result from the μCT analysis was an exuberant subperiosteal formation of woven bone in the areas directly underneath the

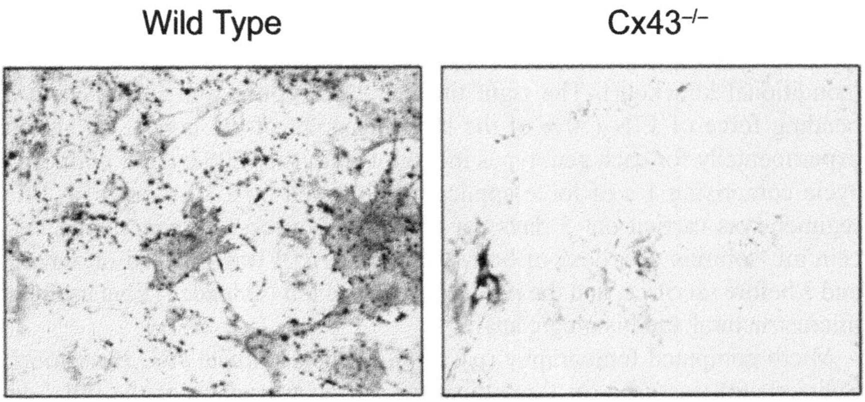

FIGURE 1. Immunohistochemistry of calvaria cells isolated from either wild type of Cx43 null mice. Cells were grown to confluence, fixed with formalin, and immunostained for Cx43. Punctate peroxidase stain was visible only in the wild-type cells, primarily at appositional membranes and in cytoplasmic areas.

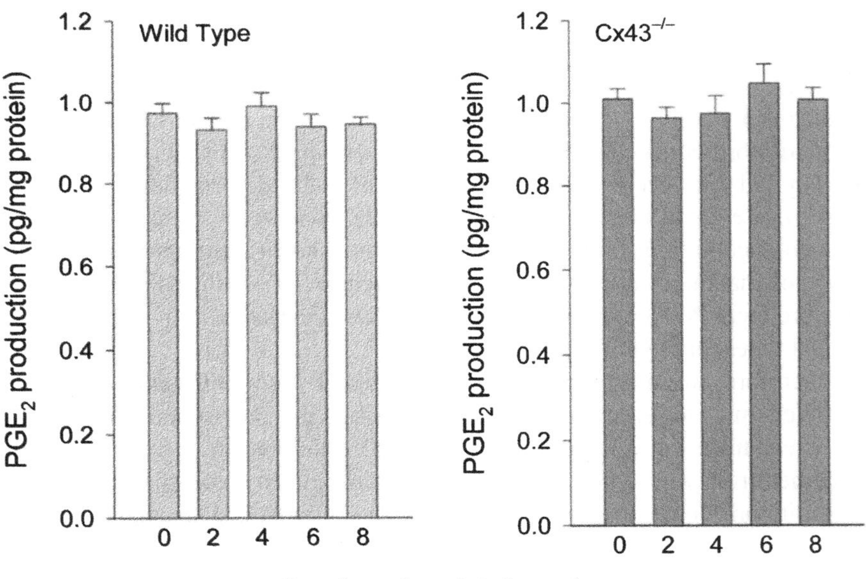

FIGURE 2. Mechanical stretch was applied to primary calvaria cells obtained from wild-type (WT) and Cx43 knockout (KO) mice over a total time course of 8 h with mechanical stretch being applied for durations of 0, 2, 4, 6, and 8 h. Conditioned medium was analyzed for PGE_2 concentration and total protein content, to control for cell density. In this experiment, there was a minimal change of PGE_2 over time.

FIGURE 3. Three-dimensional reconstruction of μCT images of mice tibiae. Tibiae of wild-type mice were subjected to a four-point bending regimen for 16 days under anesthesia, before euthanasia. Note the exuberant subperiosteal bone formation in loaded as compared to unloaded bones.

bending fulcrum in the loaded tibia that was evident in animals of all genotypes (FIG. 3). The degree and extension of this rapid new bone formation may reflect a direct effect of mechanical pressure rather than a reaction to loading, which would be expected to occur primarily on the endosteal surface. Therefore, two types of analyses were performed to determine whether endosteal bone formation was also affected. The first measurement was based on histomorphometry. Qualitative analysis of fluorescent calcein labeling showed higher single and double labeling in the $Cx43^{+/fl}$ relative to the other genotypes (FIG. 4). However, quantitative histomorphometry did not reveal significant differences in labeled surfaces or mineral apposition rates, probably because of the high variability of these results (not shown).

We then measured the area of the medullary space in a zone not directly below the pressure points of the loading device. Assuming that the bone would not expand within the short testing period, this area should be inversely proportional to endosteal new bone apposition. While no differences were detected between loaded and unloaded $Cx43^{-/fl}$ and $ColCre;Cx43^{-/fl}$ tibiae, cross-sectional medullary area was about 5% lower in $Cx43^{+/fl}$ bones relative to the other genotypes, although the difference was not statistically significant (FIG. 5).

DISCUSSION

The underlying hypothesis of this exploratory study was to assess two different approaches to determine whether the anabolic response of the skeleton to mechanical loading is dependent on Cx43. The development of an osteoblast-specific *Cx43* conditional knockout mouse now allows testing this hypothesis *in vivo*. This is a much more powerful model for assessing the sensitivity of responses to stimulators, such as mechanical load compared to chemical inhibition of gap junctional communication. A recent, very pertinent example of this premise has been provided by the finding of a normal skeletal response

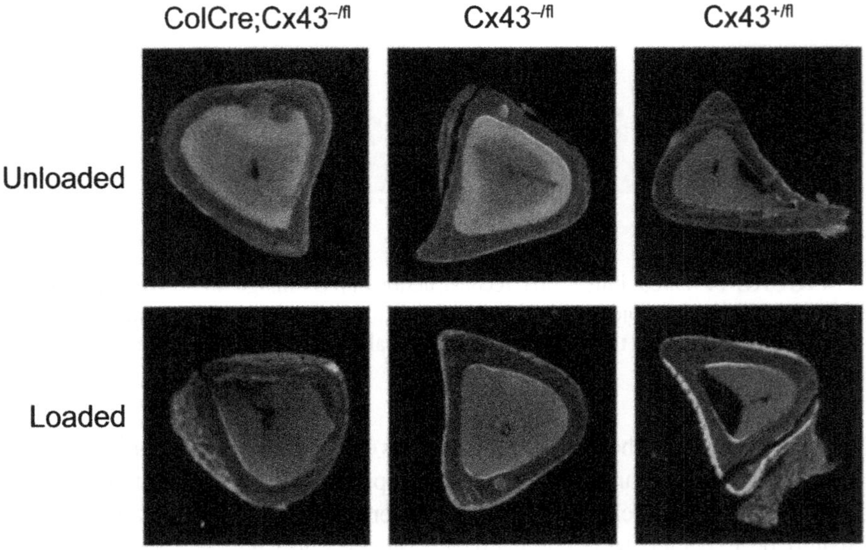

FIGURE 4. Tibiae were subjected to a four-point bending regimen for 16 days under anesthesia. Mice were given double tetracycline labeling injections. Bones were plastic embedded and observed under fluorescent microscopy. Note the exuberant subperiosteal labeling and the more intense fluorescence signal in loaded $Cx43^{+/fl}$ bones as compared to the other genotypes and unloaded bones.

to mechanical load in mice genetically deficient of the *COX-2* gene,[28] despite previous *in vitro* evidence of the contrary.[14] One caveat here is that animals with germline inactivating mutations may develop compensatory mechanisms if they tolerate the gene defect and survive, whereas short-term inhibition of a gene function may not allow the time for such a compensatory mechanism to develop. We designed this study based on the assumption that long-term responsiveness of bone cells to mechanical stimulation, *in vitro* and *in vivo*, was more biologically relevant than short-term responsiveness.

In the *in vitro* experiments, we used cells harvested from *Cx43* null mice to assess the role of this connexin in biochemical responsiveness of osteoblasts to the application of cyclic stretch. Although *Cx43*-deficient cells did release PGE_2, the amount released was lower than in wild-type cells, and the increased PGE_2 secretion in response to mechanical stretch was lower than that observed for the wild-type cells. The conclusion of these cell biology experiments is that application of cyclic stretch *in vitro* results in minimal changes in PGE_2 production and this is only marginally affected by the presence of Cx43 in these cells. The finding of minimal changes in PGE_2 as a result of cell stretching would appear to be in contrast to the findings of previous studies looking at the response of osteoblasts to fluid flow stress, where significant increases in PGE_2 were demonstrated.[14–16] However, cell stretching causes a different type

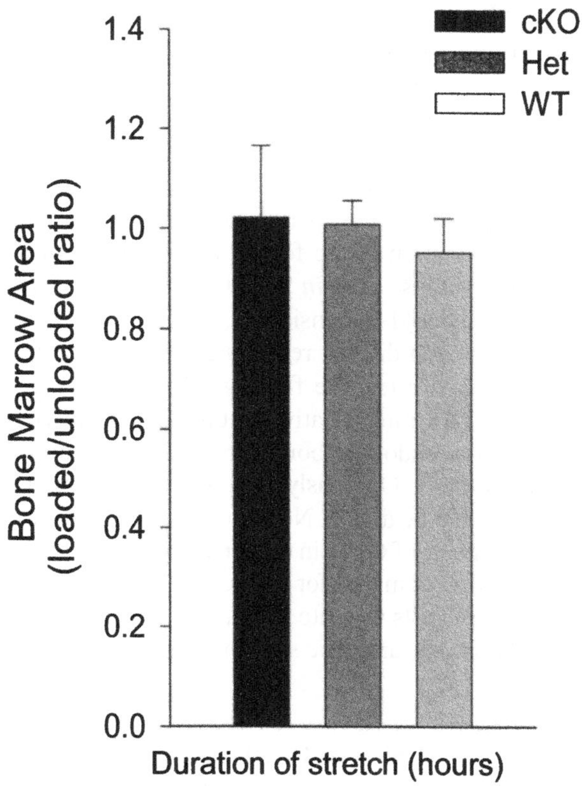

FIGURE 5. Effect of *in vivo* loading on the cross-sectional area of medullary space. μCT sections of mouse tibiae were obtained after 16 days of *in vivo* loading, and expressed as the ratio of loaded versus unloaded limb in each animal. The bars represent the average ±95% confidence interval of three genotypes, *ColCre;Cx43*$^{-/fl}$ (cKO); *Cx43*$^{-/fl}$ (Het); and *Cx43*$^{+/fl}$ (WT). Note the slight decline of the ratio in "wild-type" equivalent mice, indicative of increased new endosteal bone apposition.

of mechanical stimulation than does fluid flow shear stress that may not involve prostaglandin release. For example, cell stretching by substrate deformation induces a reorganization of the cytoskeleton and opening of stretch-activated ion channels.[29] Thus, different physical stimuli may differentially affect cellular response.

The most important aspect of our study is the application of a four-point bending model to conditionally *Cx43*-deleted mice. The remarkable and extensive subperiosteal formation of woven bone was certainly not expected to the extent that it was observed. Because this occurred in the area directly below the loading fulcrum of the tibia, it is very likely that this may represent a response to direct pressure from the loading fulcrum on the bone itself. Previous studies using four-point bending methods have used much lower loads than the 17N used in this study.[24] Earlier studies in Sprague–Dawley rats used loads corresponding to 37% of the load necessary to break the tibia.[30] In our study, we used a load that was 50% of that required to break the bone, as an initial effort to stimulate a clear response. It is possible that the force applied was in excess of what was required to generate an adaptive response on the periosteal surface, generating instead a reactive response to the pressure of the bending fulcrum. This appears to be a limitation of four-point tibial bending in mice, as others have observed the robust periosteal response.[24] In any case, these results emphasize a limitation of the four-point bending model for applying mechanical load to the mouse tibia *in vivo*. Consequently, either lower force or a different approach should be considered.

Although calcein labeling did not reveal significant differences in labeled surfaces or mineralization rates, the finding of slightly larger marrow area in conditionally *Cx43* null mice relative to the other genotypes may reflect a slightly attenuated new endosteal bone apposition after mechanical load in the *Cx43* deficient animals. Obviously, this was a small pilot study and no definitive conclusions can be drawn. Nonetheless, the results are encouraging for a potential involvement of Cx43 in response to load and a full-scale study is ongoing, using a different model for application of mechanical load.

Taken together these results indicate that lack of osteoblast *Cx43* may attenuate but does not abolish the anabolic skeletal response to mechanical load *in vitro* or *in vivo*.

ACKNOWLEDGMENT

This study was supported by NIH grant R01 AR41255 (to RC).

REFERENCES

1. YEAGER, M. & B.J. NICHOLSON. 1996. Structure of gap junction intercellular channels. Curr. Opin. Struct. Biol. **6:** 183–192.
2. GOODENOUGH, D.A., J.A. GOLIGER & D.L. PAUL. 1996. Connexins, connexons, and intercellular communication. Annu. Rev. Biochem. **65:** 475–502.
3. DUNCAN, R.L. & S. MISLER. 1989. Voltage-activated and stretch activated Ba^{2+}-conducting channels in an osteoblast-like cell line (UMR 106-01). FEBS Lett. **251:** 17–21.
4. MIKUNI-TAKAGAKI, Y., Y. SUZUKI, T. KAWASE & S. SAITO. 1996. Distinct responses of different populations of bone cells to mechanical stress. Endocrinology **137:** 2028–2035.
5. SAKAI, K., M. MOHTAI & Y. IWAMOTO. 1998. Fluid shear stress increases transforming growth factor beta 1 expression in human osteoblast-like cells: modulation by cation channel blockades. Calcif. Tissue Int. **63:** 515–520.
6. ZIAMBARAS, K., F. LECANDA, T.H. STEINBERG & R. CIVITELLI. 1998. Cyclic stretch enhances gap junctional communication between osteoblastic cells. J. Bone Miner. Res. **13:** 218–228.

7. CHENG, B., S. ZHAO, J. LUO, *et al.* 2001. Expression of functional gap junctions and regulation by fluid flow in osteocyte-like MLO-Y4 cells. J. Bone Miner. Res. **16:** 249–259.
8. YOU, J., G.C. REILLY, X. ZHEN, *et al.* 2001. Osteopontin gene regulation by oscillatory fluid flow via intracellular calcium mobilization and activation of mitogen-activated protein kinase in MC3T3-E1 osteoblasts. J. Biol. Chem. **276:** 13365–13371.
9. ALFORD, A.I., C.R. JACOBS & H.J. DONAHUE. 2003. Oscillating fluid flow regulates gap junction communication in osteocytic MLO-Y4 cells by an ERK1/2 MAP kinase-dependent mechanism. Bone **33:** 64–70.
10. CHERIAN, P.P., A.J. SILLER-JACKSON, S. GU, *et al.* 2005. Mechanical strain opens connexin 43 hemichannels in osteocytes: a novel mechanism for the release of prostaglandin. Mol. Biol. Cell **16:** 3100–3106.
11. RUBIN, C.T., T.S. GROSS, K.J. MCLEOD & S.D. BAIN. 1995. Morphologic stages in lamellar bone formation stimulated by a potent mechanical stimulus. J. Bone Miner. Res. **10:** 488–495.
12. GROSS, T.S., S. SRINIVASAN, C.C. LIU, *et al.* 2002. Noninvasive loading of the murine tibia: an in vivo model for the study of mechanotransduction. J. Bone Miner. Res. **17:** 493–501.
13. WARDEN, S.J. & C.H. TURNER. 2004. Mechanotransduction in the cortical bone is most efficient at loading frequencies of 5-10 Hz. Bone **34:** 261–270.
14. KLEIN-NULEND, J., E.H. BURGER, C.M. SEMEINS, *et al.* 1997. Pulsating fluid flow stimulates prostaglandin release and inducible prostaglandin G/H synthase mRNA expression in primary mouse bone cells. J. Bone Miner. Res. **12:** 45–51.
15. REICH, K.M. & J.A. FRANGOS. 1991. Effect of flow on prostaglandin E2 and inositol trisphosphate levels in osteoblasts. Am. J. Physiol. **261:** C428–C432.
16. GENETOS, D.C., D.J. GEIST, D. LIU, *et al.* 2005. Fluid shear-induced ATP secretion mediates prostaglandin release in MC3T3-E1 osteoblasts. J. Bone Miner. Res. **20:** 41–49.
17. JEE, W.S. & Y.F. MA. 1997. The in vivo anabolic actions of prostaglandins in bone. Bone **21:** 297–304.
18. FORWOOD, M.R. 1996. Inducible cyclo-oxygenase (COX-2) mediates the induction of bone formation by mechanical loading in vivo. J. Bone Miner. Res. **11:** 1688–1693.
19. CHERIAN, P.P., B. CHENG, S. GU, *et al.* 2003. Effects of mechanical strain on the function of gap junctions in osteocytes are mediated through the prostaglandin EP2 receptor. J. Biol. Chem. **278:** 43146–43156.
20. DONAHUE, H.J. 2000. Gap junctions and biophysical regulation of bone cell differentiation. Bone **26:** 417–422.
21. CASTRO, C.H., J.P. STAINS, S. SHEIKH, *et al.* 2003. Development of mice with osteoblast-specific connexin43 gene deletion. Cell Commun. Adhes. **10:** 445–450.
22. LECANDA, F., P.M. WARLOW, S. SHEIKH, *et al.* 2000. Connexin43 deficiency causes delayed ossification, craniofacial abnormalities, and osteoblast dysfunction. J. Cell Biol. **151:** 931–944.
23. STEINBERG, T.H., R. CIVITELLI, S.T. GEIST, *et al.* 1994. Connexin43 and connexin45 form gap junctions with different molecular permeabilities in osteoblastic cells. EMBO J. **13:** 744–750.
24. AKHTER, M.P., D.M. CULLEN, E.A. PEDERSEN, *et al.* 1998. Bone response to in vivo mechanical loading in two breeds of mice. Calcif. Tissue Int. **63:** 442–449.

25. HSIEH, Y.F., A.G. ROBLING, W.T. AMBROSIUS, *et al.* 2001. Mechanical loading of diaphyseal bone in vivo: the strain threshold for an osteogenic response varies with location. J. Bone Miner. Res. **16:** 2291–2297.
26. SRINIVASAN, S., S.C. AGANS, K.A. KING, *et al.* 2003. Enabling bone formation in the aged skeleton via rest-inserted mechanical loading. Bone **33:** 946–955.
27. CASTRO, C.H., C.S. SHIN, J.P. STAINS, *et al.* 2004. Targeted expression of a dominant-negative N-cadherin in vivo delays peak bone mass and increases adipogenesis. J. Cell Sci. **117:** 2853–2864.
28. ALAM, I., S.J. WARDEN, A.G. ROBLING & C.H. TURNER. 2005. Mechanotransduction in bone does not require a functional cyclooxygenase-2 (COX-2) gene. J. Bone Miner. Res. **20:** 438–446.
29. DUNCAN, R.L. & K.A. HRUSKA. 1992. Parathyroid hormone (PTH) exhibits dual actions on the stretch-activated cation channel (SA-cat) of the osteoblast. J. Bone Miner. Res. **7:** S242.
30. TURNER, C.H., I. OWAN & Y. TAKANO. 1995. Mechanotransduction in bone: role of strain rate. Am. J. Physiol. **269:** E438–E442.

Interactive Effect of Interleukin-6 and Prostaglandin E_2 on Osteoclastogenesis via the OPG/RANKL/RANK System

XIN-HUA LIU,[a] ALEXANDER KIRSCHENBAUM,[b] SHEN YAO,[a] AND ALICE C. LEVINE[a]

[a]*Department of Medicine, Mount Sinai School of Medicine, New York, New York 10029, USA*

[b]*Department of Urology, Mount Sinai School of Medicine, New York, New York 10029, USA*

ABSTRACT: The OPG/RANKL/RANK system regulates osteoclastogenesis. Both cyclooxygenase-2 (COX-2)/prostaglandin E2 (PGE2) and interleukin-6 (IL-6) are reported to induce osteoclast differentiation. The mechanisms underlying these signaling pathways on the OPG/RANKL/RANK system are not fully understood. We herein demonstrate that COX-2 and PGE2 stimulated osteoclastogenesis through inhibition of OPG secretion, stimulation of RANKL production by osteoblasts, and upregulation of RANK expression in osteoclasts. PGE2 also stimulated IL-6 production, and IL-6, in turn, increased PGE2 secretion, COX-2, and EP_4/EP_2 expression in bone cells. These findings provide evidence of interactive effect of PGE2 and IL-6 signaling pathways in osteoclastogenesis via effect on the OPG/RANKL/RANK system.

KEYWORDS: COX-2; PGE2; IL-6; OPG/RANKL/RANK system; osteoclastogenesis

INTRODUCTION

The cellular interactions of osteoblasts and osteoclasts determine the type and extent of bone remodeling. These interactions are mediated by receptor activator of NFκB (RANK) on the osteoclast surface and a balance in RANK ligand (RANKL) and osteoprotegerin (OPG) production by osteoblasts, i.e., the RANKL/OPG/RANK system.[1] COX-2, the highly inducible enzyme that catalyzes the conversion of arachidonic acid to prostaglandins, and PGE2, a major eicosanoid product of the COX-2-catalyzed reaction, play important roles in osteoclast formation *in vitro* and *in vivo.*[2] COX-2 and PGE2 have also

Address for correspondence: Xin-Hua Liu, Department of Medicine, Box 1055, Mount Sinai School of Medicine, New York, NY 10029. Voice: 212-241-4130; fax: 212-241-4218.
e-mail: Liu.Xinhua@mssm.edu

Ann. N.Y. Acad. Sci. 1068: 225–233 (2006). © 2006 New York Academy of Sciences.
doi: 10.1196/annals.1346.047

been demonstrated to be critical regulators of osteoblast differentiation and essential elements in bone repair.[3] These data indicate that COX-2 and PGE2 are involved in both osteoclast activation and osteoblastic bone formation. IL-6 is a major proinflammatory cytokine that regulates osteoclastic bone resorption and has been implicated in the pathogenesis of several bone diseases.[4] IL-6 has also been reported to stimulate bone formation.[5] Although interactions between the COX-2/PGE2 and IL-6 systems have been described, the mechanisms underlying these cooperative signaling pathways and the possible involvement of the OPG/RANKL/RANK system in bone remodeling have not been fully elucidated. In the present study, we demonstrate interactive roles of COX-2/PGE2 and IL-6 signaling pathways in the regulation of osteoclastogenesis via the OPG/RANKL/RANK system.

MATERIALS AND METHODS

The osteoblastic MC3T3 cell line was co-cultured with osteoclastic precursor cells (RAW264.7) in DMEM containing 10% FBS, and treated with or without various factors. Protein expression of OPG, COX-2, RANK, and subtypes of EP receptor was determined by using Western blotting. The secretion of soluble factors, i.e., IL-6, PGE2, OPG, and RANKL was measured by enzyme-linked immuno-sorbent assay (ELISA). The expression of subtypes of EP receptor mRNA was evaluated by using polymerase chain reaction (PCR).

RESULTS AND DISCUSSION

Effects of PGE2 and IL-6 on Osteoblast Growth and Osteoclast Formation

We initially determined the effects of PGE2 and IL-6 on osteoblast growth and osteoclast differentiation. As shown in FIGURE 1A, PGE2 and IL-6 stimulated MC3T3 cell growth. In contrast, two COX-2 inhibitors, NS398 and meloxicam, decreased osteoblast cell growth, which could be reversed by the addition of exogenous PGE2, suggesting a COX-2-dependent mechanism. The growth of osteoblasts was also stimulated by PGE_1-alcohol, a selective EP_4 receptor agonist, but not Butaprost, a selective EP_2 receptor agonist, indicating that PGE2-stimulated osteoblast cell proliferation is mediated by the EP_4 receptor subtype. FIGURE 1B demonstrates a significant increase in RAW264.7 cell differentiation induced by PGE2 and IL-6 when the cells were grown in co-culture with MC3T3 cells, compared to RAW264.7 cells grown alone.

Effects of PGE2 and IL-6 on RANKL and OPG Production and RANK Expression

Our initial studies suggested that the effects of PGE2 and IL-6 on osteoclasts are mediated by the RANK/RANKL/OPG system. Next we determined

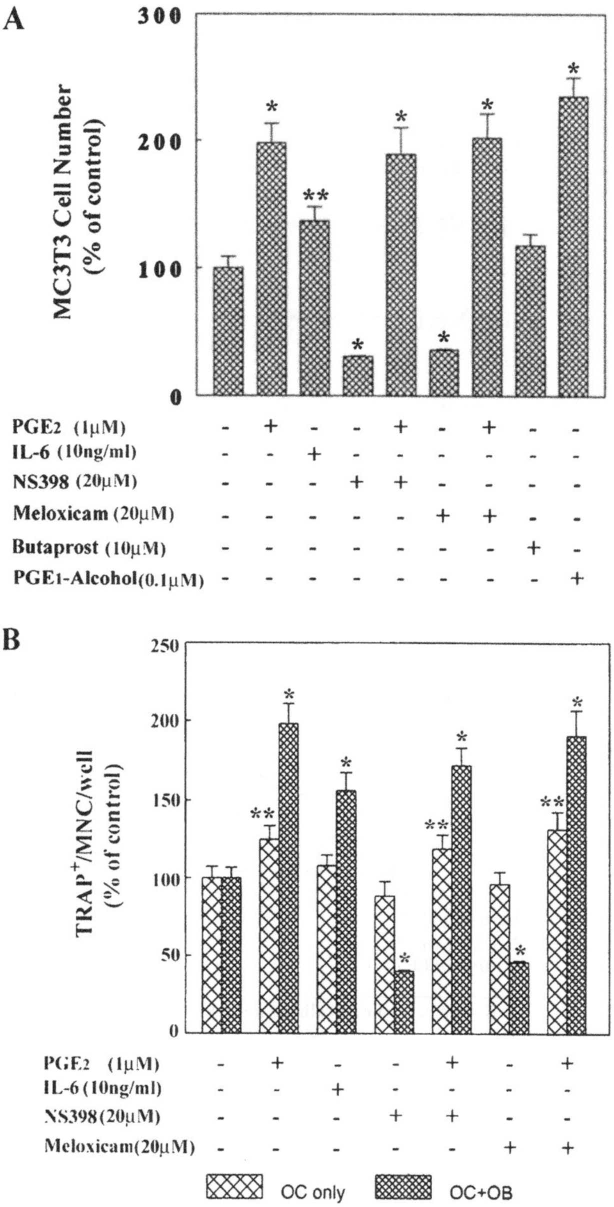

FIGURE 1. Effects of PGE2 and IL-6 on osteoblast growth and osteoclast formation. (**A**) Effects of PGE2 and IL-6 on osteoblast growth. MC3T3 cells were cultured in serum-free medium (SFM) and subjected to various treatments, as indicated, for 3 days. The number of living cells was counted. (**B**) Effects of PGE2, IL-6, and COX-2 inhibitors on osteoclast formation in the presence or absence of osteoblasts. RAW264.7 cells were cultured, either alone or co-cultured with MC3T3 cells, in SFM and treated with various of treatments for 3 days. The number of osteoclast-like cells was determined by counting TRAP-positive multinucleated cells (MNCs) with three or more nuclei. The data represent means ± SEM from three separate experiments. $^{**}P < 0.05$, $^{*}P < 0.01$.

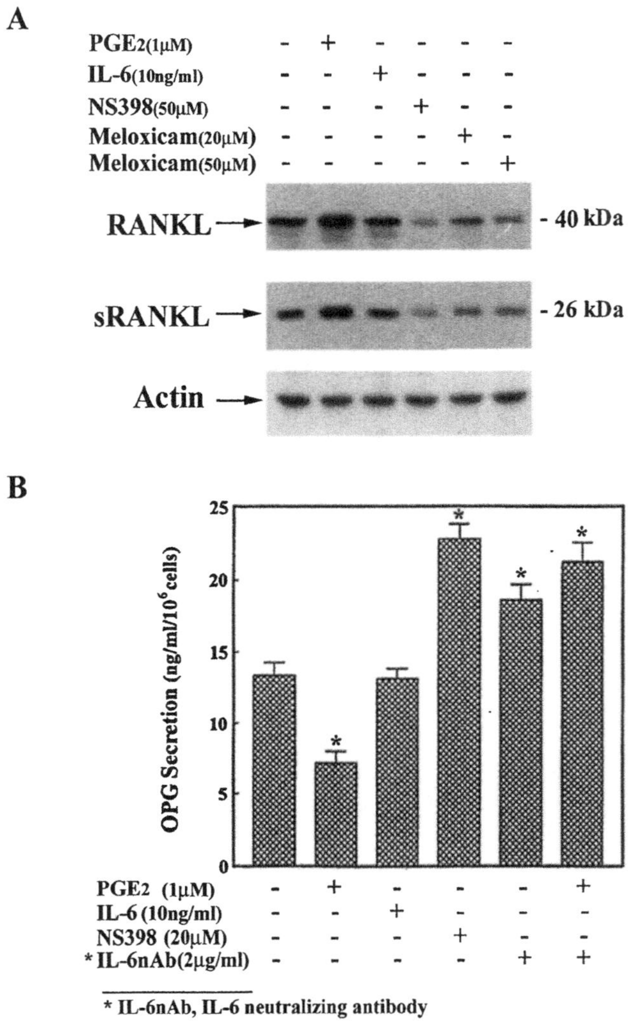

FIGURE 2. Effects of PGE2 and IL-6 on RANKL/OPG secretion and RANK expression. (**A**) *Effects of PGE2 and IL-6 on RANKL/OPG production by osteoblasts.* Cultured cells were treated with vehicle or various factor or compound as indicated for 2 days. Cell lysates were prepared for the analysis of membrane-bound RANKL, and culture medium was concentrated 100-fold for the analysis of sRANKL by Western blotting (WB). (**B**) *Effects of PGE2 and IL-6 on RANK expression by RAW264.7 cells.* Cells were treated with vehicle or various compounds, as indicated, for 2 days. Protein was extracted and subjected to WB. Data shown are representative of three separate experiments. (**C**) *Effects of PGE2 and IL-6 on OPG protein expression.* MC3T3 cells were cultured in serum free medium and treated with or without various factors or compounds, as indicated, for 2 days (*Upper panel*). Dose-dependent effects of PGE2 and IL-6 on OPG expression (*low panel*). In these experiments, cells were treated with various doses of PGE2 and IL-6, as indicated, for 2 days. After the various treatments, protein was extracted and subjected to Western blot analysis. Equal amounts of total protein (30μg) were loaded in each lane. Data shown are representative of three separate experiments.

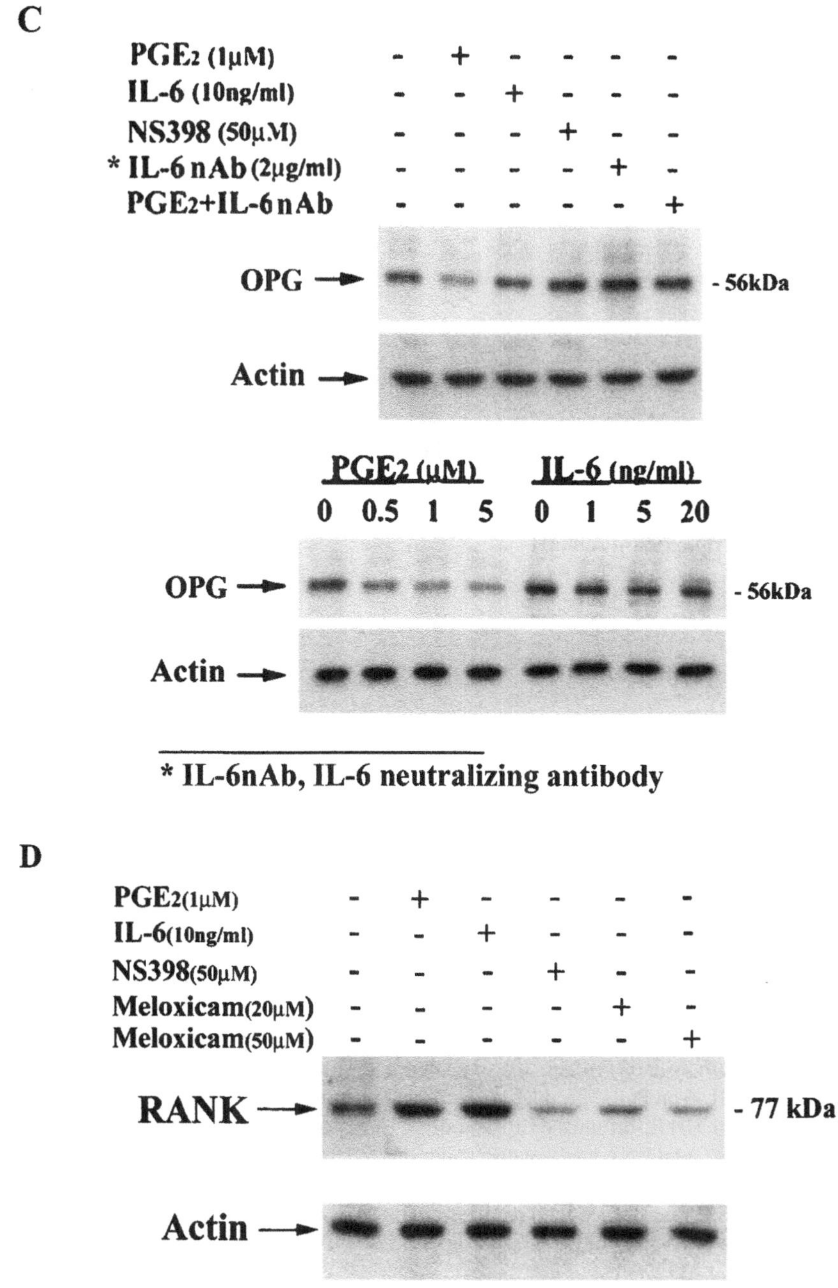

FIGURE 2. (**D**) *Effects of PGE2 and IL-6 on RANK expression in osteoclast.* RAW264.7 cells were cultured in serum free medium and treated with vehicle or various factors or compounds, as indicated, for 2 days. After treatment, protein was extracted and subjected to Western blotting. Equal amounts of total protein (30μg) were loaded in each lane. Data shown are representative of three separate experiments.

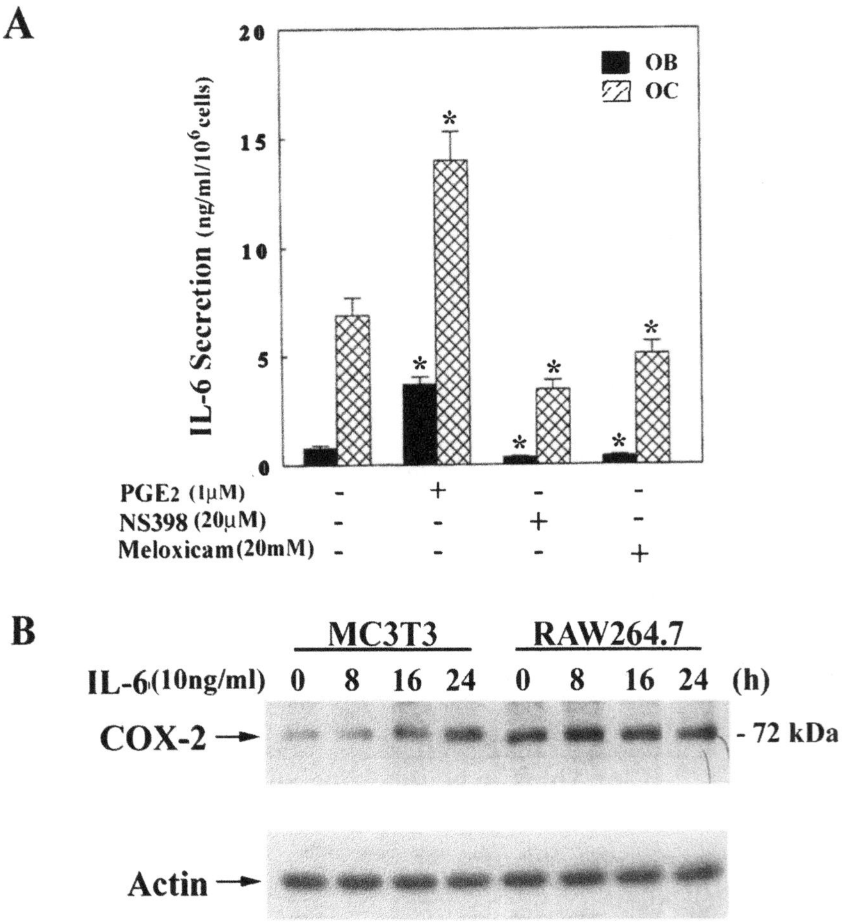

FIGURE 3. Reciprocal interactions between the IL-6 and PGE2 systems on bone cells. (**A**) Effects of PGE2 and COX-2 inhibitors on IL-6 secretion. Cultured cells were treated with PGE2, NS398, or meloxicam for 3 days. Medium was collected and subjected to ELISA. (**B**) Effect of IL-6 on COX-2 expression. Cells were treated with 10 ng/ml IL-6 for the times indicated. Protein was extracted and subjected to Western blotting (WB). (**C**) Effect of IL-6 on PGE2 production. Cells were treated with IL-6 for 3 days and ELISA was performed. (**D**) IL-6-induced EP_4 and EP_2 expression. Cells were treated with 10 ng/ml IL-6 for 2 days. Total RNA was isolated, and subjected to Rt-PCR. (**E**) Protein was extracted and subjected to WB. The results from ELISA were normalized to cell number and represent as mean ± SEM from three separate experiments. Data shown derived from WB and PCR are representative of three separate experiments. $^*P < 0.01$.

the effects of COX-2, PGE2, and IL-6 on the production of RANKL and OPG by MC3T3 cells, as well as RANK expression in RAW264.7 cells. As shown in FIGURE 2A, IL-6 had no detectable effect on the expression of RANKL production, whereas PGE2 increased the expression of both membrane-bound

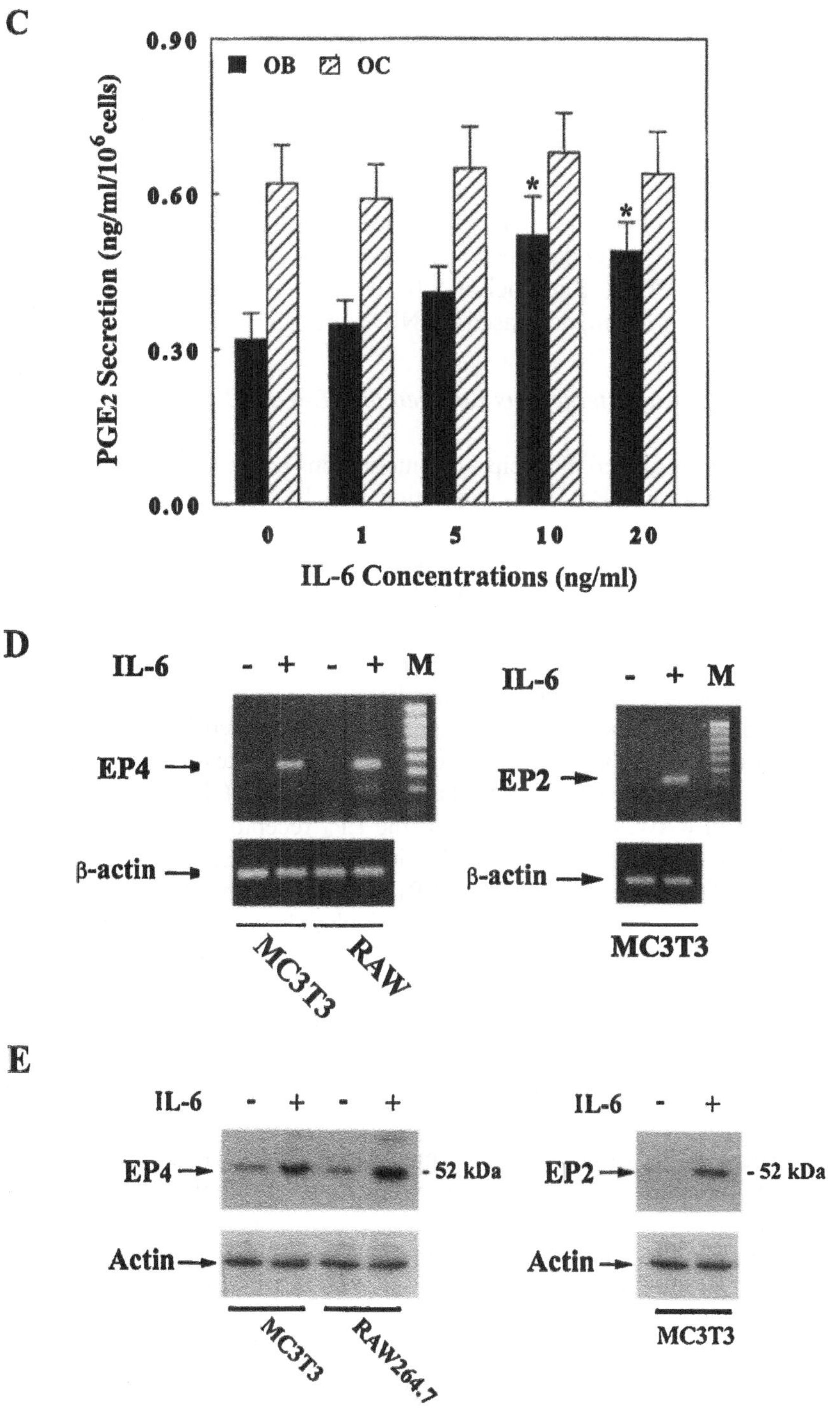

FIGURE 3. *Continued*

RANKL and soluble RANKL (sRANKL) levels in MC3T3 cells, while COX-2 inhibitors decreased RANKL expression in these same cells. FIGURE 2B–C demonstrates that PGE2 addition decreases OPG secretion, while COX-2 inhibition significantly stimulates OPG secretion by MC3T3 cells. Although exogenous IL-6 alone had no detectable effect, IL-6-neutralizing antibodies increased OPG secretion and reversed the observed suppression of OPG secretion by PGE2, suggesting that IL-6 mediates the inhibitory effects of PGE2 on OPG production. FIGURE 2D demonstrates that both IL-6 and PGE2 increased the expression of RANK in osteoclast cells. The two distinct COX-2 inhibitors, NS398 and meloxicam, decreased RANK expression in these same cells.

Reciprocal Interactions Between the IL-6 and PGE2 Systems

We next investigated the reciprocal interactions between the IL-6 and PGE2 signaling pathways. Basal IL-6 secretion was 10-fold higher in RAW264.7 cells than in MC3T3 cells. PGE2 significantly stimulated IL-6 production in both cell lines and COX-2 inhibitors inhibited IL-6 secretion (FIG. 3A). RAW264.7 cells expressed a high basal level, but a modest inducible level, of COX-2. Although basal expression was barely detectable, a significant, time-dependent increase in COX-2 expression was observed in MC3T3 cells after IL-6 addition (FIG. 3B). On the other hand, IL-6 had little effect on PGE2 secretion in RAW264.7 cells but significantly increased PGE2 secretion in MC3T3 cells (FIG. 3C). The effect of IL-6 on the expression of EP receptor subtypes in the bone cell lines was also assessed. Rt-PCR revealed that both MC3T3 and RAW264.7 cells express the EP_4 receptor subtype mRNA and protein, but do not express detectable levels of EP_1, EP_2, and EP_3 subtype receptors. The expression levels of EP_4 in both cell lines were increased by IL-6 (FIG. 3D–E). Although MC3T3 cells had undetectable basal levels of the EP_2, the levels were also induced by IL-6.

In the present study, we demonstrate that COX-2 and PGE2 increase both osteoblast proliferation and osteoclast differentiation. These results confirm the observation that disruption of COX-2 gene expression results in defective osteoblast secretion of RANKL and OPG. We also observed that IL-6 stimulation of osteoclast differentiation appears to be due to its effects on PGE2-induced OPG secretion by osteoblasts and upregulation of RANK expression by osteoclasts. In addition, IL-6 increases the expression of both the EP_4 and EP_2 receptor subtypes in bone cells. All of these events induced by an interaction between the COX-2/PGE2 and IL-6 systems, coordinately, tip the balance of the RANKL/OPG/RANK system in favor of increased osteoclastogenesis and enhanced bone resorption.

REFERENCES

1. KHOSLA, S. 2001. The OPG/RANKL/RANK system. Endocrinology **142:** 5050–5055.

2. ZHANG, X., S.G. MORHAM, R. LANGENBACH, *et al.* 2001. Evidence for a direct role of cyclooxygenase-2 in implant wear debris induced osteolysis. J. Bone Mineral Res. **16:** 660–669.
3. ZHANG, X., E.M. SCHWARZ, D.A. YOUNG, *et al.* 2002. COX-2 regulates mesenchymal cell differentiation into the osteoblast lineage and is critically involved in bone repair. J. Clin. Invest. **109:** 1405–1415.
4. OHSHIMA, S., Y. SAEKI, T. MIMA, *et al.* 1998. IL-6 plays a key role in the development of antigen-induced arthritis. Proc. Natl. Acad. Sci. USA **95:** 8222–8226.
5. SIMS, N.A., B.J. JENKINS, J.M.W. OUINN, *et al.* 2004. Glycoprotein 130 regulates bone turnover and bone size by distinct downstream signaling pathways. J. Clin. Invest. **113:** 379–389.

Does TNF Have Anti-Osteoclastogenic Actions?

JAMEEL IQBAL

Department of Endocrinology and Mount Sinai Bone Program, Mount Sinai School of Medicine, New York, New York 10029, USA

ABSTRACT: Tumor necrosis factor (TNF) is a pro-resorption agent that leads to bone degradation. Several mechanisms of action have been proposed to account for these effects: TNF directly inhibits osteoblast differentiation; TNF augments osteoclast formation by inducing stromal cells to increase expression of RANKL and macrophage colony-stimulating factor (M-CSF) and decrease that of osteoprotegerin (OPG); and TNF serves to synergize with pathways downstream of RANK to directly increase osteoclast differentiation. All of these actions in sum suggest that TNF dramatically induces osteopenia upon overexpression or injection. However, that the osteopenia seen with TNF is significantly milder than that seen in OPG$^{-/-}$ animals prompted a reevaluation of existing paradigms on TNF action. The hypothesis that TNF directly enhances osteoclast differentiation was tested by examining the effects of TNF on RANKL-induced osteoclast formation and marker expression. The data show that TNF decreased RANKL-induced expression of the osteoclast markers, TRAP and cathepsin K. Furthermore, the addition of 10–60 ng/mL TNF failed to significantly increase RANKL-induced osteoclast differentiation. Instead, data are presented to suggest that the pro-osteoclastogenic actions of TNF are mediated through increases in the number of available osteoclast precursors (macrophages).

KEYWORDS: TNF; osteoclast; M-CSF; inhibition; RANKL

INTRODUCTION

Tumor necrosis factor (TNF) was co-discovered as a lipepolysaccharide (LPS)-induced protein that caused tumor necrosis and as a soluble factor that caused the cachexia of systemic infections with *T. brucei.*[1,2] TNF is now recognized as the key cytokine-mediating inflammation. It is produced primarily by macrophages and acts in an autocrine manner to stimulate macrophage activation and differentiation.[3] Osteoprotegerin (OPG) was discovered as an agent

Address for correspondence: Jameel Iqbal, Department of Endocrinology (1055), Mount Sinai School of Medicine, One Gustave L. Levy Place New York, NY 10029. Voice: 212-241-8797; fax: 212-426-8312.
e-mail: jameel.iqbal@mssm.edu

Ann. N.Y. Acad. Sci. 1068: 234–239 (2006).
doi: 10.1196/annals.1346.056

whose overexpression leads to increases in bone mineral density.[4] The OPG receptor was expression-cloned and identified as Receptor Activator of NF-κB (RANK).[5] RANK and its ligand, RANKL, are members of the TNF receptor-ligand superfamily, although they are relatively far apart evolutionarily in the TNF superfamily tree.

The similarities in the signaling pathways between TNF and RANKL have lead to the notion that the two can cooperatively increase osteoclast formation. Indeed, TNF is generally appreciated to be strongly osteoclastogenic and its actions lead to bone degradation.[6,7] In addition to its pro-osteoclastogenic actions, TNF serves to inhibit osteoblast differentiation, while inducing stromal cells to increase their expression of RANKL and decrease their expression of OPG.[8,9] All of these actions in sum suggest that TNF would induce dramatic osteopenia upon overexpression or injection. However, the major osteologic manifestation seen in TNF-transgenic mice is joint inflammation and erosion. While these mice do develop osteopenia, their osteopenia is mild in comparison to OPG$^{-/-}$ animals.[10] This suggests that the existing paradigm on TNF action is not correct, and promoted a reexamination of the hypothesis that TNF directly affects osteoclast differentiation. Data presented here show that TNF does not augment differentiation, but that TNF does increase the proliferation of osteoclast precursors and therefore their availability for differentiation.

RESULTS AND DISCUSSION

This study aimed to examine the effects of TNF on osteoclast differentiation. To carry out this analysis, the effects of TNF on inducing the expression of the osteoclast markers, TRAP and cathepsin K, were examined by real-time polymerase chain reaction (PCR) of primary Ficoll-purified bone marrow cultures. These cultures were not contaminated with stromal cells or osteoblasts, and lymphocytes were removed prior to the addition of RANKL or TNF by media exchange. FIGURE 1A, B shows that RANKL treatment for 2 days expectedly increased the expression of TRAP and cathepsin K. TNF treatment alone led to a mild elevation of these markers; however, when TNF was added together with RANKL, osteoclast marker expression was significantly decreased compared to RANKL treatment Alone (FIG. 1A, B). This action of TNF is in contrast with other established pro-osteoclastogenic stimuli, such as follicle-stimulating hormone (FSH), which dramatically increases osteoclast marker expression at 2 days of culture (Li *et al.*, In press).

The fact that TNF decreased osteoclastogenic marker expression contradicts the existing paradigm that TNF serves to augment osteoclast differentiation induced by RANKL.[6,7,11] To reexamine the hypothesis that TNF augments RANKL-induced differentiation, Ficoll-purified bone marrow osteoclast progenitors were treated with 50 ng/mL and 0, 10, 30, or 60 ng/mL soluble murine TNF. As shown in FIGURE 1C, TNF failed to significantly augment RANKL-induced osteoclast formation when TNF and RANKL are added together.

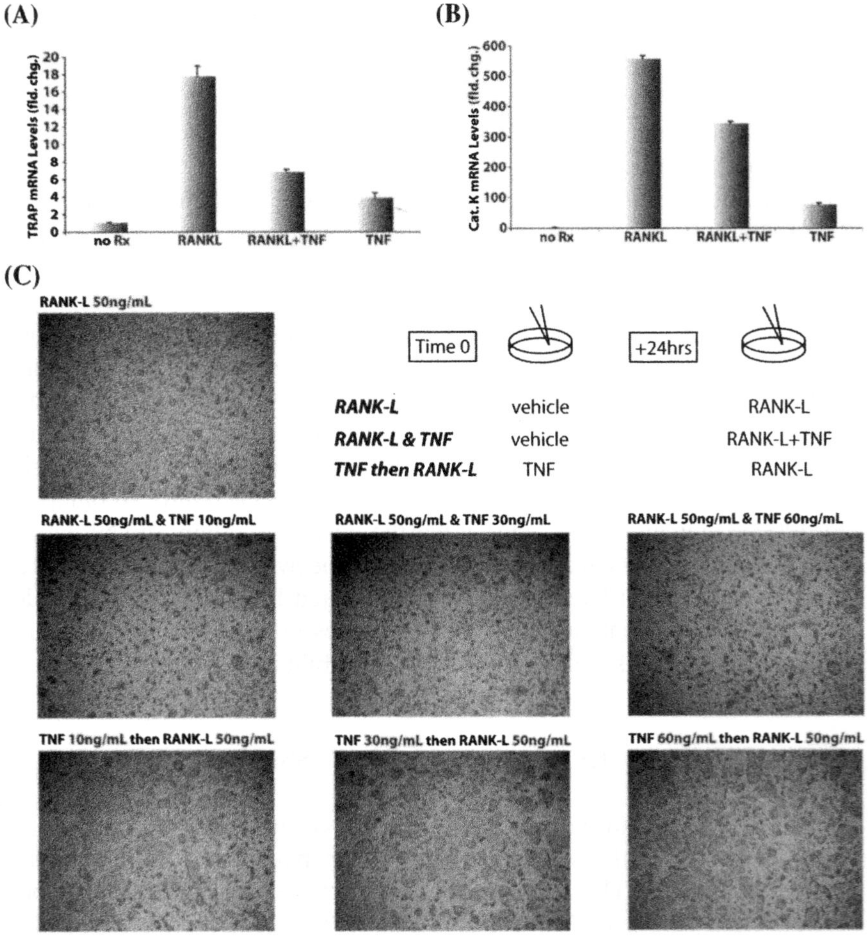

FIGURE 1. Anti-osteoclastogenic actions of TNF. (**A** and **B**) TRAP and cathepsin K mRNA expression, respectively, in primary bone marrow osteoclast cultures treated with vehicle, 50 ng/mL soluble RANKL (sRANKL), 40 ng/mL murine sTNF with 50 ng/mL sRANKL, or with 40 ng/mL sTNF alone. Each sample was run in triplicate and the values quantitatively adjusted to levels of β_2-microglobulin. Each experiment was repeated twice. $^*P$ value < 0.05 as assessed by unpaired Student's t-test. (**C**) Osteoclast formation in primary bone marrow osteoclast cultures treated with 50 ng/mL sRANKL, 10–60 ng/mL murine sTNF added at the same time as 50 ng/mL sRANKL, or in cultures pre-exposed to 10–60 ng/mL sTNF for 24 h prior to the addition of sRANKL. Digital photographs were taken on 4× power using a Nikon TS200 equipped with a Nikon digital camera.

Thus far, the results suggested that TNF does not increase osteoclastogenesis. Work from Abu-Amer, Teitelbaum, and colleagues suggests that TNF signaling through TNFR1 is "pro-osteoclastogenic," while signaling through TNFR2 is not.[12,13] Furthermore, transgenic mice overexpressing soluble TNF (sTNF) or treated with daily injections of TNF develop osteopenia.[8,9] One characteristic of these animals is that there is a marked increase in CD11b^{+} monocyte-macrophages. To test whether TNF induced the proliferation of Ficoll-purified bone marrow cells, MTT metabolism assays were preformed. The data indicated that TNF significantly increased the number of viable cells (data not shown). A possible explanation for the *in vivo* phenotype of transgenic mice overexpressing sTNF or treated with daily injections of TNF is that TNF has no effects on osteoclast differentiation, but merely induces the number of available osteoclast precursors available for differentiation. To test this hypothesis, Ficoll-purified bone marrow cells were incubated with 0, 10, 30, or 60 ng/mL TNF for 24 h, following which 50 ng/mL RANKL was added. FIGURE 1C shows that all three doses of TNF (10, 30, and 60 ng/mL) lead to significant increases in osteoclast formation. All together, the data suggests that the pro-osteoclastogenic actions of TNF are mediated through increases in the number of available osteoclast precursors (macrophages), and not through augmentation of osteoclast differentiation.

EXPERIMENTAL PROCEDURES

Animals and Cell Culture

To obtain cells for the real-time PCR and osteoclastogenesis experiments the following protocol was used: both tibiae and femora of 2–3 C57 Black mice (age 3 months) were extracted. The marrow flushed out into α-MEM (with 10% Invitrogen Select FBS, Carlsbad, CA and 5% penicillin/streptomycin) using a syringe with a 21-gauge needle. The cells were cultured for 1 day with recombinant murine macrophage colony-stimulating factor (M-CSF) (5 ng/mL, Peprotech, Rocky Hill, NJ). Nonadherent cells were collected, filtered on a 70-μM cell strainer (BD Bioscience, San Jose, CA), and purified using Ficoll-Plus (Amersham GE, Piscataway, NJ) by centrifuging for 15 min at 1500 rpm. Cells from the interface layer were counted, diluted, and used for experimentation following the addition of M-CSF (40 ng/mL). Some experiments used murine-soluble TNF (10–60 ng/mL; R&D systems Minneapolis, MN) or soluble RANKL (20–60 ng/mL; R&D systems).

cDNA Production/Real-Time PCR

Total RNA was purified using an RNeasy Mini kit (Qiagen, Valencia, CA). The cells (2×10^6 in count) were grown for 1 day in the presence of 40 ng/mL M-CSF, following which, all media was aspirated and adherent cells were washed with PBS. New media and M-CSF (40 ng/mL) were added; the cells were stimulated with varying combinations of 50 ng/mL RANKL and 40 ng/mL TNF for 2 days. Following this time, the cells were washed twice with ice-cold PBS, disrupted by the addition of Buffer RLT, scraped, and shredded using QIA shredder spin columns. Purification of RNA was then carried out following the manufacturer's directions.

Following manufacturer's directions, 2.0-μg total RNA for each sample was converted into cDNA using BD Sprint PowerScipt preprimed single-shot tubes. The reaction products were diluted in water and then used for 40-cycle two-step (95°C/60°C) PCR in an ABI Prism 7900HT (Applied Biosystems, Foster City, CA) using ABsolute SYBR Green mix with ROX (ABGene AB-1163/A) and 200-nM gene-specific primers (Synthesized by Invitrogen). CT values were determined after setting the baseline reference range using the ABI-suggested guidelines. Amplicon size and reaction specificity was confirmed by agarose gel electrophoresis and melting curve analysis. Each transcript in each sample was assayed three times and the fold-change ratios between experimental and control samples for each gene used in the analysis were calculated using murine β_2-microglobulin levels as reference.

Osteoclast Formation Assays

Cells from the interface layer, prepared as described above, were plated in quadruplicate at 3×10^4 cells/well in a 96-well tissue culture plate (BD Biosciences). Murine-soluble TNF (10–60 ng/mL) was added for 24 h prior to the addition of 50 ng/mL RANKL, or was added at the same time as 50 ng/mL RANKL. The cells were grown for a total of 5 days. Cultures were fixed and stained using the protocol and materials detailed in Technical Bulletin 445 from BD Biosciences. Digital photographs of osteoclast cultures were obtained on an inverted Nikon TS200 microscope outfitted with a digital camera (Morrell Instrument Co., Melville, NY).

ACKNOWLEDGMENTS

JI is supported by an NIH Training Grant and an MSTP Institutional Grant to Mt. Sinai.

REFERENCES

1. Beutler, B. & A. Cerami. 1988. Cachectin (tumor necrosis factor): a macrophage hormone governing cellular metabolism and inflammatory response. Endocr. Rev. **9:** 57–66.
2. Carswell, E.A., L.J. Old, R.L. Kassel, *et al.* 1975. An endotoxin-induced serum factor that causes necrosis of tumors. Proc. Natl. Acad. Sci. USA **72:** 3666–3670.
3. Fong, Y., S.F. Lowry, & A. Cerami. 1988. Cachetin/TNF: a macrophage protein that induces cachexia and shock. JPEN J. Parenter. Enteral Nutr. **12:** 72S–77S.

4. SIMONET, W.S., D.L. LACEY, C.R. DUNSTAN, *et al.* 1997. Osteoprotegerin: a novel secreted protein involved in the regulation of bone density. Cell **89:** 309–319.
5. LACEY, D.L., E. TIMMS, H.L. TAN, *et al.* 1998. Osteoprotegerin ligand is a cytokine that regulates osteoclast differentiation and activation. Cell **93:** 165–176.
6. FULLER, K., C. MURPHY, B. KIRSTEIN, *et al.* 2002. TNFalpha potently activates osteoclasts, through a direct action independent of and strongly synergistic with RANKL. Endocrinology **143:** 1108–1118.
7. LAM, J., S. TAKESHITA, J.E. BARKER, *et al.* 2000. TNF-alpha induces osteoclastogenesis by direct stimulation of macrophages exposed to permissive levels of RANK ligand. J. Clin. Invest. **106:** 1481–1488.
8. ABU-AMER, Y., S. ABBAS & T. HIRAYAMA. 2004. TNF receptor type 1 regulates RANK ligand expression by stromal cells and modulates osteoclastogenesis. J. Cell Biochem. **93:** 980–989.
9. KITAURA, H., M.S. SANDS, K. AYA, *et al.* 2004. Marrow stromal cells and osteoclast precursors differentially contribute to TNF-alpha-induced osteoclastogenesis *in vivo*. J. Immunol. **173:** 4838–4846.
10. BUCAY, N., I. SAROSI, C.R. DUNSTAN, *et al.* 1998. Osteoprotegerin-deficient mice develop early onset osteoporosis and arterial calcification. Genes Dev. **12:** 1260–1268.
11. KOMINE, M., A. KUKITA, T. KUKITA, *et al.* 2001. Tumor necrosis factor-alpha cooperates with receptor activator of nuclear factor kappaB ligand in generation of osteoclasts in stromal cell-depleted rat bone marrow cell culture. Bone **28:** 474–483.
12. ABU-AMER, Y., J. ERDMANN, L. ALEXOPOULOU, *et al.* 2000. Tumor necrosis factor receptors types 1 and 2 differentially regulate osteoclastogenesis. J. Biol. Chem. **275:** 27307–27310.
13. ZHANG, Y.H., A. HEULSMANN, M.M. TONDRAVI, *et al.* 2001. Tumor necrosis factor-alpha (TNF) stimulates RANKL-induced osteoclastogenesis via coupling of TNF type 1 receptor and RANK signaling pathways. J. Biol. Chem. **276:** 563–568.

Eosinophil Chemotactic Factor-L (ECF-L) Enhances Osteoclast Formation by Increasing ICAM-1 Expression

VERONICA GARCIA-PALACIOS,[a] HO YEON CHUNG,[b] SUN JIN CHOI,[a] NORIYOSHI KURIHARA,[a] JUN WON LEE,[a] LORI A. EHRLICH,[a] ROBERT COLLINS,[c] AND G. DAVID ROODMAN,[a,d]

[a]*Department of Medicine/Hematology-Oncology, University of Pittsburgh, Pittsburgh, Pennsylvania, USA*

[b]*Department of Medicine, Sungkyunkwan University, Korea*

[c]*Department of Molecular and Human Genetics, Baylor College of Medicine, Houston, Texas, USA*

[d]*VA Pittsburgh Healthcare System, Pittsburgh, Pennsylvania, USA*

ABSTRACT: Eosinophil chemotactic factor-L (ECF-L) is a novel stimulator of osteoclast (OCL) formation that acts at the differentiation/fusion stage of OCL formation, and is a cofactor for RANK ligand (RANKL). We examined the effects of ECF-L on the intracellular signaling pathways utilized by RANKL, and on the expression of ICAM-1/LFA-1 to determine its mechanism of action. RAW 264.7 and bone marrow cells were treated with RANKL and/or ECF-L Fc protein to determine their effect on NF-κB and AP-1 activity. ECF-L by itself only modestly increased NF-κB binding and JNK activity in RAW 264.7 cells, which were further enhanced by RANKL. In contrast, ECF-L Fc increased LFA-1α and ICAM-1 mRNA levels 1.8-fold in mouse marrow cultures, and anti-ICAM-1 almost completely inhibited OCL formation induced by 10^{-10} M 1,25-$(OH)_2D_3$, and ECF-L Fc. Furthermore, ECF-L Fc did not enhance OCL formation by ICAM-1 knockout (KO) cells. Increased expression of ICAM-1 by ECF-L appears to be critical for its effects on OCL formation.

KEYWORDS: ECF-L; osteoclasts; ICAM-1

We recently identified eosinophil chemotactic factor-L (ECF-L) as an autocrine stimulatory factor produced by osteoclasts (OCLs).[1] ECF-L was originally purified from the culture supernatant of splenocytes of mice and was shown to have eosinophil chemotactic activity.[2] ECF-L is also chemotactic for T

Address for correspondence: G. David Roodman, M.D., Ph.D., Medicine/Hematology-Oncology (111-H), University Drive C, Pittsburgh, PA 15240. Voice: 412-688-6571; fax: 412-688-6960. e-mail: roodmangd@upmc.edu

Ann. N.Y. Acad. Sci. 1068: 240–243 (2006). © 2006 New York Academy of Sciences. doi: 10.1196/annals.1346.048

lymphocytes and bone marrow polymorphonuclear leukocytes *in vitro.* ECF-L is identical to Ym1 but not its isomer Ym2 on GenBank analysis.[3] Ym1 is a macrophage protein that is transiently synthesized and secreted by activated macrophages during inflammation.[4] Furthermore, prominent expression of Ym1 has been observed in early myeloid precursor cells of hematopoietic tissues such as fetal liver, spleen, and bone marrow.[5]

ECF-L increases OCL formation in a dose-dependent manner in mouse bone marrow cultures in the presence of 10^{-10} M 1,25-$(OH)_2D_3$.[1] Chemotactic assays showed that ECF-L increased migration of OCL precursors. Interestingly, a neutralizing antibody to ECF-L blocked RANKL or 10^{-9} M 1,25-$(OH)_2D_3$-induced OCL formation in mouse marrow cultures, suggesting that it is an important cofactor for RANKL even though it does not induce RANKL. ECF-L increased the average nuclei per OCL compared with cultures treated with control Fc protein. Time course studies demonstrated that ECF-L acted at the later differentiation/fusion stages of OCL formation. However, the exact mechanisms responsible for the effects of ECF-L on osteoclastogenesis are unknown. Since ECF-L enhances the effects of RANKL and acts on the fusion stage of OCL formation, we investigated the effects of ECF-L on the RANK signaling pathway, including NF-κB, JNK, and on the expression of cell adhesion molecules such as ICAM-1 and LFA-1.

Binding of RANKL to RANK results in activation of NF-κB and the protein kinase JNK, which increases transcription of AP-1.[6,7] ECF-L significantly enhanced NF-κB, AP-1 binding activity and JNK kinase activity. ECF-L did not increase p44/42 or p38 MAPK activity. However, these effects were modest and could not by themselves account for the effects of ECF-L on OCL formation.

Cell-to-cell contact is necessary for OCL formation since OCL forms by cell fusion rather than endomitosis.[8,9] Cell adhesion molecules such as ICAM-1 and E-cadherin have been reported to be involved in these cell-to-cell interactions.[10,11] ICAM-1 is expressed on osteoblasts,[12] OCL precursors, and OCL.[7] ICAM-1 is increased in response to a number of inflammatory cytokines,[11] and ICAM-1 expression can be regulated by intracellular signaling through NF-κB and JNK.[12] ICAM-1 is one of the ligands for LFA-1,[13] and ICAM-1 and LFA-1 interactions have been reported to be involved in cell-to-cell contact between OCL precursors and osteoblastic/stromal cells or between OCL precursors.[9,14] Our previous time course studies suggested that ECF-L acted at the later differentiation/fusion stages of OCL formation rather than inducing proliferation of OCL precursors.[1] Therefore, we determined if integrins known to be expressed on OCL precursors or OCL were increased by treatment with ECF-L. ECF-L increased ICAM-1 and LFA-1 expression in nonadherent mouse marrow cells but had no effect in ST2 stromal cells. These data suggest that ECF-L increased ICAM-1 expression on OCL precursors. We also tested the effect of ECF-L on β1 and β3 integrin mRNA expression in mouse marrow cells and found no increases in these experiments. Furthermore, anti-ICAM-1 blocked ECF-L-induced OCL formation, suggesting that increasing ICAM-1 levels was important for mediating the effects of ECF-L on OCL formation.

To confirm these findings, we tested the effects of ECF-L on marrow cultures from ICAM-1 KO mice. OCL formation was decreased in ICAM-1 marrow in response to 1,25-$(OH)_2D_3$ (10^{-10} M). This result is consistent with previous findings in LFA-1 KO mice, which also showed decreased OCL formation *in vitro*.[8] Importantly, ECF-L increased OCL in control littermates but had no effect on OCL formation in ICAM-1 KO mice. These data support that ECF-L increases ICAM-1/LFA-1 interactions to enhance OCL formation.

ACKNOWLEDGMENTS

This study was supported by R01-AR41336 from NIAMS. We thank Arthur Beaudet for providing ICAM-1-deficient mice with the support of NIH grant AI-32177. We thank Donna Gaspich for preparing the manuscript.

REFERENCES

1. Oba, Y. *et al.* 2003. Eosinophil chemotactic factor-L (ECF-L): a novel osteoclast stimulating factor. J. Bone Miner. Res. **18:** 1332–1341.
2. Owhashi, M., H. Arita & N. Hayai. 2000. Identification of a novel eosinophil chemotactic cytokine (ECF-L) as a chitinase family protein. J. Biol. Chem. **275:** 1279–1286.
3. Webb, D.C., A.N.J. McKenzie & P.S. Foster. 2001. Expression of the Ym2 lectin-binding protein is dependent on interleukin-4 and IL-13 signal transduction. J. Biol. Chem. **276:** 41969–41976.
4. Chang, N.A. *et al.* 2001. A macrophage protein, Ym1, transiently expressed during inflammation is a novel mammalian lectin. J. Biol. Chem. **276:** 17497–17506.
5. Hung, S.I. *et al.* 2002. Transient expression of Ym1, a heparin-binding lectin, during developmental hematopoiesis and inflammation. J. Leuko. Biol. **72:** 72–82.
6. Hsu, H. *et al.* 1999. Tumor necrosis factor receptor family member RANK mediates osteoclast differentiation and activation induced by osteoprotegerin ligand. Proc. Natl. Acad. Sci. USA **96:** 3540–3545.
7. Wong, B.R. *et al.* 1997. TRANCE is a novel ligand of the tumor necrosis factor receptor family that activates c-Jun N-terminal kinase in T cells. J. Biol. Chem. **272:** 25190–25194.
8. Suda, T., N. Takahashi & T.J. Martin. 1992. Modulation of osteoclast differentiation. Endocr. Rev. **13:** 66–80.
9. Harada, H. *et al.* 1998. Involvement of lymphocyte function-associated antigen-1 and intercellular adhesion molecule-1 in osteoclastogenesis: a possible role in direct interaction between osteoclast precursors. Endocrinology **139:** 3967–3975.
10. Kurachi, K., I. Morita & S. Murota. 1993. Involvement of adhesion molecules LFA-1 and ICAM-1 in osteoclast development. Biochim. Biophys. Acta **1178:** 259–266.

11. MBALAVIELE, G. *et al.* 1995. The role of cadherin in the generation of multinucleated osteoclasts from mononuclear precursors in murine marrow. J. Clin. Invest. **95:** 2757–2765.
12. OKADA, Y. *et al.* 2002. Cell-to-cell adhesion via intercellular adhesion molecule-1 and leukocyte function-associated antigen-1 pathway is involved in 1, $25(OH)_2D_3$, PTH and IL-1α-induced osteoclast differentiation and bone resorption. Endocr. J. **49:** 483–495.
13. SPRINGER, T.A. 1990. Adhesion receptors of the immune system. Nature **346:** 425–434.
14. TANI-ISHII, N. *et al.* 2002. The role of LFA-1 in osteoclast development induced by co-cultures of mouse bone marrow cells and MC3T3-G2/PA6 cells. J. Periodontal Res. **37:** 184–191.

Bone Cell Precursors and the Pathophysiology of Bone Loss

HARRY C. BLAIR,[a] AND JILL L. CARRINGTON[b]

[a]*Departments of Pathology and Cell Biology, University of Pittsburgh, and Pittsburgh VA Medical Center, Pittsburgh, Pennsylvania 15213, USA*

[b]*Biology of Aging Program, National Institute on Aging, National Institutes of Health, Bethesda, Maryland 20892, USA*

ABSTRACT: In health, changes in bone formation and degradation rates are coupled and adequate cellular resources are available in the bone so that a change in bone formation rate occurs with an opposing change in resorption. On the other hand, the regulation of bone volume, particularly in pathological conditions, is dependent not only on the pathways that mediate terminal pathways of bone cell differentiation, but also on the availability of stem cells for allowing the differentiation to occur. Regulation of cell numbers in stem cell compartments and release of stem cells for differentiation of osteoblast or osteoclast precursors are not well understood, although it is clear that changes in stem cell numbers underlie pathological changes in bone mass. This may include effects of aging, fracture, metastatic disease, and autoimmune diseases on the precursor cell pools available for bone formation and degradation. Increases in osteoclast precursors or decreases in osteoblast precursors are common features of bone-losing states; increases in precursors may conversely occur during growth or repair processes. Rational therapy based on modifying stem cell populations may, when the processes are better understood, help prevent chronic bone-losing states and may also be of use in preventing or treating aplastic anemia and related conditions.

KEYWORDS: stem cells; osteoblast; osteoclast; mesenchymal cells; aplastic anemia; autoimmunity; differentiation

MARROW MESENCHYMAL AND HEMATOPOIETIC STEM CELLS

Bone matrix is derived from mesenchymal stem cells that can produce chondrocytes and osteoblasts or which may undergo alternative differentiation to vascular, adipose, connective, and muscular tissue. These cells can be isolated

Address for correspondence: Harry C. Blair, 705 Scaife Hall, University of Pittsburgh, Pittsburgh, PA 15261. Voice: 412-383-9616; fax: 412-647-8567.
e-mail: hcblair@imap.pitt.edu

**Ann. N.Y. Acad. Sci. 1068: 244–249 (2006). © 2006 New York Academy of Sciences.
doi: 10.1196/annals.1346.028**

from marrow[1] and are generally regarded as the durable, self-replicating pool of adherent bone cell precursors for turnover and repair. Bone degradation requires vascular endothelial growth factor (VEGF) and vascular ingrowth,[2] this is generally regarded as necessary for supplying marrow cell precursors to enter into bone, which ultimately will produce circulating monocytic cells that further differentiate to osteoclasts.[3]

The relation between the most primitive mesenchymal cell precursors and the least differentiated hematological stem cells is uncertain, but recent work suggests that parathyroid hormone related protein (PTHrP) or parathyroid hormone (PTH) influence factors including the Notch ligand jagged-1,[4] which supports early hematopoietic precursor proliferation, and helps explain earlier observations pointing to the production of other early myeloid growth factors including stem cell factor,[5] by osteoblasts and osteoblast-related mesenchymal cells. Thus, an important auxiliary function of bone is maintenance of the self-renewing marrow stem cell populations that are required for lifelong bone turnover and also for hematopoiesis.[6] The relationship of marrow stem cell failure in aplastic anemia with mesenchymal cell growth factors is a poorly defined function of great importance, but is not further considered here, where we will concentrate on the more distal osteoclast and osteoblast precursor populations.

The bone marrow contains mononuclear-precursor and mesenchymal stem cells capable, respectively, of producing CD11b osteoclast precursors[7] and committed osteoblasts.[1] There are some indications that there may be earlier primitive monocyte-like stem cells capable of replacing both hematopoietic and mesenchymal stem cells, although these may also represent separate populations of small monocyte-like antigen negative null cells. For example, nonadherent null monocytic cells replace both osteoblasts and marrow elements after lethal irradiation.[8] The issue as to whether this represents the effects of broadly pluripotent cells or multiple committed stem cells of different lineages has been difficult to resolve due to the lack of distinguishing antigens on the cells and to lack of understanding of conditions for the maintenance and differentiation of the earliest stem cells of either the hematopoietic or mesenchymal stem cell components. However, common clonal origin of osteoblastic and marrow cells in mice[8] suggests that a limited number of such pluripotent cells capable of producing mesenchymal stem cells and hematopoietic stem cells may exist in some adult mammals.

CIRCULATING STEM CELLS

Parabiotic experiments with green fluorescent protein mice have established that stem cells, which are abundant in specialized environments such as the marrow, also circulate in appreciable numbers and can regenerate cells in several mesenchymal tissues including muscle.[9] There are also more mature adherent cells that function well as osteoblastic precursors.[10] Kuznetsov *et al.*[11]

reported that these cells are very rare in humans, on the order of one in 10^8 circulating nucleated cells, calling into question their physiological relevance in most cases.

On the other hand, there are no specific data as to earlier mesenchymal/osteoblast precursors that may not be captured by the analysis of adherent cells. Eghbali-Fatourechi *et al.* have addressed this issue of the circulating committed osteoblastic precursor in humans by analyzing osteocalcin expressing, or BAP expressing, nonadherent monocytic circulating cells in adolescent and adult humans and after fracture. This recently published work[12] suggests that osteocalcin-expressing CD15 negative cells circulate in quantity, and may represent nearly 1% of monocytic cells in healthy adults and even higher circulating quantities during rapid growth or after fracture. This work, discussed in detail by Khosla in a subsequent paper in this volume, suggests that a circulating pool of nonadherent cells is an important mechanism in repair and possibly also during normal modeling. Problems still to be resolved include determination of whether some or all of the sorted osteocalcin expressing cells are capable of bone formation, and the specific signals required to cause localization and differentiation of the circulating cells to skeletal sites where osteoblast precursors are required. This work may also help explain the variation in primitive monocytic cells detected on automated screening of leukocytes in blood, which has long been a problem for clinicians, and suggests that in the case of questionable results that sensitive analysis for the early osteoblast product osteocalcin may differentiate between true pluripotent blasts versus dedicated bone cell precursors.

REGULATION OF BONE PRECURSORS BY TUMOR CELLS

Osteolytic and sclerotic lesions occur in varying frequency with different types of cancer. The most common tumors producing bone metastasis are prostate or breast cancer and multiple myeloma, with prostate cancer characteristically producing sclerotic or osteoblastic lesions, while breast cancer and myeloma typically produce lytic lesions. Of the several types of metastatic tumors leading to osteolytic lesions, myeloma is one of the most problematic as well as the most consistent tumor type, with characteristically "punch out" lesions that are almost purely osteolytic. Specific mechanisms center around the production of growth factors by tumor cells; these include PTHrP and several interleukins as well as TNF-family proteins and tyrosine kinase agonists including colony stimulating factor-1 (CSF-1).[13,14] In addition, tumor cells produce immune cell response via multiple mechanisms and result in secondary changes including stromal proliferation, and vascular growth factors that can contribute to osteolytic mechanisms.

Specific tumor cytokines produce additional dramatic specific effects on bone loss, as is most well characterized in myeloma by the work of Abe *et al.*[15]

and independently by Choi *et al.*,[16] who found that the metastatic myeloma cells produce high levels of macrophage inflammatory peptide-1 (MIP-1) alpha or beta. This subject is reviewed in detail by Matsumoto in a subsequent paper in this volume. In brief, antibodies to MIP-1 or its receptor CCR5 greatly diminish the effects of myeloma supernatants on osteoclast formation, validating this as a key mechanism. Mechanisms that are involved include direct RANKL-based mechanisms, tumor cell adhesion effects via VLA4/VCAM, production of VEGF by tumor cells, production by bone cells of growth factors including IL-6 that foster tumor survival, and suppression of bone formation by tumor production of wnt-related inhibitors including sFRP-2. There are reports of other wnt-related proteins in myeloma cells including Dkk1,[17] although the sFRP-2 mechanism appears to be more commonly expressed by this tumor type.

IMMUNITY, AUTOIMMUNITY, AND BONE LOSS REFLECT CHANGES IN BONE CELL PRECURSOR POOLS

Aside from the view that gonadal failure produces a pseudo-autoimmune state, as particularly described[18] and discussed by Pacifici elsewhere in this volume, there are substantive data related both to autoimmune states and bone loss and to secondary bone loss related to therapy-targeting autoimmune states. In some cases, such as rheumatoid arthritis, the predominant mechanism is related to the effects of the key cytokine tumor necrosis factor-α (TNF-α), which has both direct and indirect effects on bone and joint cell survival via effects on macrophages and T lymphocytes with potential downstream effects on bone and joint destruction via macrophages, chondroclasts, and osteoclasts.[19] Understanding of autoimmune and rheumatological states clinically is complicated by the uniformly unfavorable effects of immunosuppressive agents, either in the corticosteroid or more specific calmodulin-inhibitor families, on bone mass. Nonetheless, it is clear that numerous autoimmune or rheumatological states result in a direct deleterious effect on bone mass by more indirect mechanisms.

A key indirect mechanism related to bone loss in autoimmune disorders is modification of the precursor pool for the formation of bone and joint-degrading cells. As is considered in detail by Schwartz later in this volume, studies of autoimmune polyarthritis, which is associated with TNF-α overexpression, relative to systemic lupus erythematosus, in which interferon production is predominant, reveal that in disease with mainly TNF-α overexpression, there is a specific increase in CD11b/CD14 pools of osteoclast precursors that enhance osteoclastic response.[20] In diseases such as systemic lupus erythematosus (SLE), with elevated interferon-α, CD11b pools are depleted in differentiation toward CD11c-expressing dendritic cells,[21] the autoimmune diseases do not have a natural bone- or joint-destructive phenotype but rather are subject to bone loss, largely in response to the anti-inflammatory therapy used to manage the autoimmune disease. The cells involved may be quite early in the macrophage-monocyte precursor series, and may be Fms-positive or -negative and RANK-positive or -negative. These mechanisms, it should be emphasized, being relatively early in the differentiation process of bone cells, affect the precursor cell populations, an entirely separate issue from the late effects of TNF-α on osteoclastic differentiation, which have been described by several workers *in vitro*.[22,23] Effects of TNF family proteins other than RANKL in osteoclast differentiation are complex, and may have both pro- and antiresorptive effects depending on cell maturation, time of exposure, and microenvironmental factors.

ACKNOWLEDGMENTS

This research was supported by the New York Academy of Sciences, by the Department of Veteran's Affairs, and by National Institutes of Health (USA) grants AR47700 and AG12951.

REFERENCES

1. Pittenger, M.F. *et al.* 1999. Multilineage potential of adult human mesenchymal stem cells. Science **284:** 143–147.
2. Engsig, M.T. *et al.* 2000. Matrix metalloproteinase 9 and vascular endothelial growth factor are essential for osteoclast recruitment into developing long bones. J. Cell Biol. **151:** 879–889.
3. Fujikawa, Y. *et al.* 1996. The human osteoclast precursor circulates in the monocyte fraction. Endocrinology **137:** 4058–4060.
4. Calvi, L.M. *et al.* 2003. Osteoblastic cells regulate the haematopoietic stem cell niche. Nature **425:** 841–846.
5. Blair, H.C. *et al.* 1999. Parathyroid hormone-regulated production of stem cell factor in human osteoblasts and osteoblast-like cells. Biochem. Biophys. Res. Commun. **255:** 778–784.
6. Visnjic, D. *et al.* 2004. Hematopoiesis is severely altered in mice with an induced osteoblast deficiency. Blood **103:** 3258–3264.
7. Ferron, M. & J. Vacher. 2005. Targeted expression of Cre recombinase in macrophages and osteoclasts in transgenic mice. Genesis **41:** 138–145.
8. Dominici, M. *et al.* 2004. Hematopoietic cells and osteoblasts are derived from a common marrow progenitor after bone marrow transplantation. Proc. Natl. Acad. Sci. USA **101:** 11761–11766.
9. Sherwood, R.I. *et al.* 2004. Determinants of skeletal muscle contributions from circulating cells, bone marrow cells, and hematopoietic stem cells. Stem Cells **22:** 1292–1304.
10. Long, M.W. *et al.* 1995. Regulation of human bone marrow-derived osteoprogenitor cells by osteogenic growth factors. J. Clin. Invest. **95:** 881–887.
11. Kuznetsov, S.A. *et al.* 2001. Circulating skeletal stem cells. J. Cell Biol. **153:** 1133–1140.

12. EGHBALI-FATOURECHI, G.Z. *et al.* 2005. Circulating osteoblast-lineage cells in humans. N. Engl. J. Med. **352:** 1959–1966.
13. BENDRE, M. *et al.* 2003. Breast cancer metastasis to bone: it is not all about PTHrP. Clin. Orthop. Relat. Res. **415 S:** S39–45.
14. LIN, E.Y. & J.W. POLLARD. 2004. Macrophages: modulators of breast cancer progression. Novartis Found. Symp. **256:** 158–168.
15. ABE, M. *et al.* 2004. Osteoclasts enhance myeloma cell growth and survival via cell-cell contact: a vicious cycle between bone destruction and myeloma expansion. Blood **104:** 2484–2491.
16. CHOI, S.J. *et al.* 2000. Macrophage inflammatory protein 1-alpha is a potential osteoclast stimulatory factor in multiple myeloma. Blood **96:** 671–675.
17. TIAN, E. *et al.* 2003. The role of the Wnt-signaling antagonist DKK1 in the development of osteolytic lesions in multiple myeloma. N. Engl. J. Med. **349:** 2483–2494.
18. CENCI, S. *et al.* 2003. Estrogen deficiency induces bone loss by increasing T cell proliferation and lifespan through IFN-gamma-induced class II transactivator. Proc. Natl. Acad. Sci. USA **100:** 10405–10410.
19. GOLDRING, S.R. 2003. Pathogenesis of bone and cartilage destruction in rheumatoid arthritis. Rheumatology (Oxford). **42** [Suppl 2]: ii11–16.
20. LI, P. *et al.* 2004. RANK signaling is not required for TNFalpha-mediated increase in CD11(hi) osteoclast precursors but is essential for mature osteoclast formation in TNFalpha-mediated inflammatory arthritis. J. Bone Miner. Res. **19:** 207–213.
21. ROBAK, E. *et al.* 2004. Relationship between peripheral blood dendritic cells and cytokines involved in the pathogenesis of systemic lupus erythematosus. Eur. Cytokine Netw. **15:** 222–230.
22. KUDO, O. *et al.* 2002. Proinflammatory cytokine (TNFalpha/IL-1alpha) induction of human osteoclast formation. J. Pathol. **198:** 220–227.
23. LAM, J. *et al.* 2000. TNF-alpha induces osteoclastogenesis by direct stimulation of macrophages exposed to permissive levels of RANK ligand. J. Clin. Invest. **106:** 1481–1488.

Autosomal-Dominant Pseudohypoparathyroidism Type Ib is Caused by Different Microdeletions Within or Upstream of the *GNAS* Locus

HARALD JÜPPNER,[a,b] AGNÈS LINGLART,[a] LEOPOLD F. FRÖHLICH,[a] AND MURAT BASTEPE[a]

[a]*Endocrine Unit, Department of Medicine, Massachusetts General Hospital and Harvard Medical School, Boston, Massachusetts 02114, USA*

[b]*Pediatric Nephrology Unit, MassGeneral Hospital for Children, Massachusetts General Hospital and Harvard Medical School, Boston, Massachusetts 02114, USA*

ABSTRACT: The term *pseudohypoparathyroidism* (PHP) refers to the different disorders that are caused by mutations within *GNAS* or upstream of this complex genetic locus. *GNAS* gives rise to several different transcripts, including Gsα (α-subunit of heterotrimeric stimulatory G protein), XLαs (extra-large variant of Gsα), and several additional sense and antisense transcripts. The complexity of the *GNAS* locus is furthermore reflected by a parent-specific methylation pattern of most of its different promotors. PHP can be divided into two major groups, PHP type Ia (PHP-Ia) and PHP type Ib (PHP-Ib). PHP-Ia is caused by heterozygous mutations affecting one of the 13 *GNAS* exons encoding Gsα or by large intragenic deletions. In contrast, PHP-Ib is caused by heterozygous deletions within *STX16,* the gene-encoding syntaxin 16, which is located more than 220 kb upstream of *GNAS*, or by deletions within *GNAS* involving exon NESP55 and two of the antisense exons. In either form of PHP, hormonal resistance develops only after maternal inheritance of the mutation, while paternal inheritance of the same molecular defect is not associated with endocrine abnormalities. In most familial cases of PHP-Ib, there is a loss of exon A/B methylation combined with active A/B transcription from both parental alleles, which leads to suppression of Gsα transcription in the proximal renal tubules and, therefore, PTH resistance.

KEYWORDS: pseudohypoparathyroidism (PHP); PHP type Ia (PHP-Ia); pseudo-PHP (pPHP); PHP type Ib (PHP-Ib); progressive osseous heteroplasia (POH); α-subunit of stimulatory G protein (Gsα); syntaxin 16 (*STX16*)

Address for correspondence: Harald Jüppner, M.D., Endocrine Unit, Thier 5, Massachusetts General Hospital, Boston, MA 02114. Voice: 617-726-3966; fax: 617-726-7543.
e-mail: hjueppner@partners.org

Ann. N.Y. Acad. Sci. 1068: 250–255 (2006). © 2006 New York Academy of Sciences.
doi: 10.1196/annals.1346.029

INTRODUCTION

The term *pseudohypoparathyroidism* (PHP) was first introduced by Fuller Albright,[1] describing patients who presented with hypocalcemia and hyperphosphatemia that was caused by resistance toward parathyroid hormone (PTH) rather than PTH deficiency. Besides PTH resistance, individuals affected by PHP type Ia (PHP-Ia) show distinctive but variable features such as round face, short stature, obesity, brachydactyly, heterotopic ossifications, and often mental retardation; these clinical findings are now termed *Albright's hereditary osteodystrophy* (AHO). Furthermore, patients with PHP-Ia often present with resistance toward additional hormones, such as thyroid-stimulating hormone (TSH), gonadotropins, growth hormone-releasing hormone, and calcitonin.[2–6] In contrast, patients affected by PHP type Ib (PHP-Ib) typically present only with hypocalcemia and hyperphosphatemia due to PTH resistance, but without AHO features. No hormonal resistance is observed in patients affected by pseudo-PHP (pPHP) or progressive osseous heteroplasia (POH), two disorders often caused by the same molecular defect as observed in patients affected by PHP-Ia, only that the mutations reside on the paternal, rather than the maternal allele.[2,6,7]

Pseudohypoparathyroidism Type Ia (PHP-Ia), Pseudo-Pseudohypoparathyroidism (pPHP), and Progressive Osseous Heteroplasia (POH)

PHP-Ia is caused by maternally inherited heterozygous mutations in one of the 13 exons encoding Gsα. These include nucleotide changes, deletions, and insertions affecting exons, introns, or splice junctions,[4,5,8–10] or deletions comprising the entire *GNAS* region.[11] Although Gsα-mediated signaling is required for a large variety of G protein–coupled receptors, such as the PTH/PTHrP receptor, it remains largely unknown why only few of these receptors are affected by the different *GNAS* mutations.

Over the last few years, it has become apparent that the *GNAS* locus is much more complex than initially thought.[2,6,12–16] For example, there are several different upstream exons that are differentially methylated in a parent-specific manner and transcription occurs only from the nonmethylated promotors (FIG. 1). Because exons A/B (also termed exon 1A or exon 1′), XL (extra-large stimulatory G protein, XLαs), and AS (antisense), as well as their promotors, are methylated on the maternal allele and transcripts are derived only from the paternal allele. In contrast, the NESP55 promotor is methylated on the paternal allele and active transcription occurs only from the maternal allele. With the exception of the AS transcript, all other alternative first exons splice onto exons 2 through 13. The promotor for *GNAS* exon 1 is not methylated and transcripts encoding Gsα are therefore derived from both parental

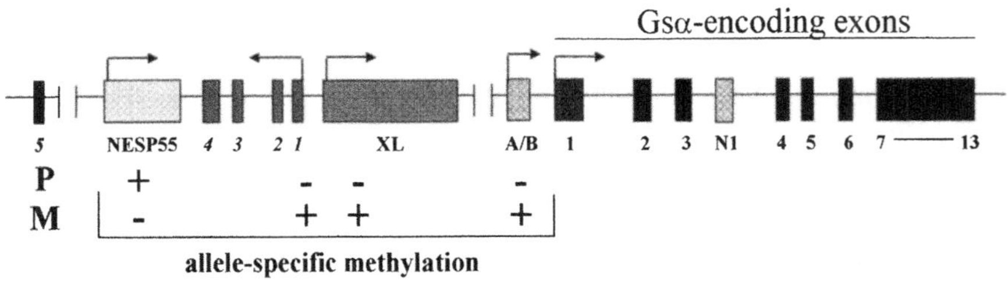

FIGURE 1. Due to allele-specific methylation of the alternative *GNAS* exons upstream of the exons encoding Gsα, several transcripts are generated either from the paternal or the maternal allele. Although exon 1 is not methylated, expression of Gsα occurs only from the maternal allele in proximal renal tubules and possibly few other tissues or cells; the underlying mechanisms are unknown. When mutations affecting the *GNAS* exons encoding Gsα are inherited maternally, individuals develop hormonal resistance and AHO (PHP-Ia). If such mutations are inherited paternally, affected individuals develop AHO without hormone resistance (PHP or POH).

alleles in most tissues. In few tissues, however, such as the proximal renal tubules, Gsα is only expressed from the maternal allele, but the underlying mechanisms leading to this allele-specific expression are still unknown. As a result, heterozygous mutations affecting *GNAS* exons encoding Gsα cause a complete loss of functional protein and thus hormonal resistance in some tissues. This explains why PHP-Ia develops only after maternal inheritance of *GNAS* mutations, while pPHP or POH occurs only after paternal inheritance of such mutations.

Pseudohypoparathyroidism Type Ib (PHP-Ib)

PHP-Ib is characterized by hypocalcemia, hyperphosphatemia, high PTH levels, and in some cases, elevated serum levels of TSH, in the absence of evidence for AHO.[2,3,6] When tested, Gsα activity in readily accessible cells was shown to be normal, and genetic mutations affecting Gsα or the PTH/PTHrP receptor were excluded.[2,3,6] Subsequent linkage studies mapped the genetic locus for an autosomal dominant form of PHP-Ib (AD-PHP-Ib) to a region on chromosome 20q13.3 that comprises, at its telomeric end, the *GNAS* locus (Fig. 2) and, just like PHP-Ia, AD-PHP-Ib was shown to be a paternally imprinted disorder.[17] Furthermore, most patients affected by AD-PHP-Ib show a loss of methylation at *GNAS* exon A/B but not at other promotors of this locus; in contrast, most sporadic cases of PHP-Ib appear to have broader methylation defects.[18,19] Most patients affected by AD-PHP-Ib were recently shown to carry a 3-kb or a 4.4-kb microdeletion within *STX16*, the gene-encoding syntaxin 16.[20–23] This finding suggested that the region bearing the deletions, about 220 kb upstream of the *GNAS* locus, contains an imprinting control element required for establishment and/or maintenance of normal methylation at the A/B locus. When disrupted, the loss

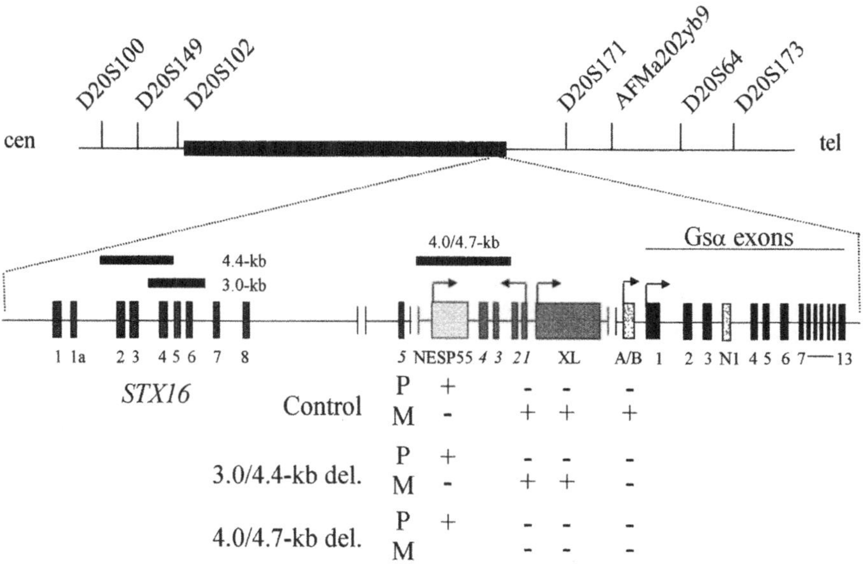

FIGURE 2. After mapping the genetic locus for AD-PHP-Ib to a region on chromosome 20q, a heterozygous 3.0-kb microdeletion was identified in affected individuals and unaffected carriers of numerous families. A 4.4-kb microdeletion that overlaps the 3.0-kb deletion was identified in the affected individuals and unaffected carriers of one single family. Both microdeletions are located about 220 kb upstream of the differentially methylated *GNAS* exon A/B. Additional, almost identical microdeletions within *GNAS* were identified in two other AD-PHP-Ib families; these eliminate exon NESP55 and two antisense exons, and appear to be responsible for the loss of all maternal methylation imprints in affected individuals.

of the methylation is thought to result in enhanced A/B transcription from the maternal allele, thereby silencing Gsα transcription, but only in a small number of tissues including the proximal renal tubules. Both microdeletions were excluded in several sporadic PHP-Ib individuals and in healthy controls. In two other families with AD-PHP-Ib, in which affected members show a broader methylation defect, different microdeletions of about 4 kb have been discovered, each removing exon NESP55 and two of the antisense exons.[24] Another family, which mapped to the *GNAS* locus, has been reported to carry methylation defects that show variable penetrance, but a genetic mutation has not yet been identified.[25]

In summary, PHP-Ia and PPHP/POH are caused by heterozygous mutations in those *GNAS* exons that encode Gsα; these mutations are maternally and paternally inherited, respectively. In contrast, autosomal dominant forms of PHP-Ib are caused by microdeletions located either upstream of or within the *GNAS* locus. As a result, there is a loss of methylation at exon A/B alone or at several additional *GNAS* DMRs, leading to the silencing of Gsα transcription in the proximal renal tubules and possibly few other tissues or cells.

REFERENCES

1. Albright, F., C.H. Burnett, P.H. Smith & W. Parson. 1942. Pseudohypoparathyroidism—an example of "Seabright-Bantam syndrome." Endocrinology **30:** 922–932.
2. Weinstein, L., S. Yu, D. Warner & J. Liu. 2001. Endocrine manifestations of stimulatory G Protein alpha-subunit mutations and the role of genomic imprinting. Endocr. Rev. **22:** 675–705.
3. Jan de Beur, S.M. & M.A. Levine. 2001. Pseudohypoparathyroidism: clinical, biochemical, and molecular features. *In* The Parathyroids: Basic and Clinical Concepts. J.P. Bilezikian, R. Markus & M.A. Levine, Eds.: 807–825. Academic Press. New York.
4. Ahrens, W., O. Hiort, P. Staedt, *et al.* 2001. Analysis of the GNAS1 gene in Albright's hereditary osteodystrophy. J. Clin. Endocrinol. Metab. **86:** 4630–4634.
5. Linglart, A., J. Carel, M. Garabédian, *et al.* 2002. GNAS1 lesions in pseudohypoparathyroidism Ia and Ic: genotype phenotype relationship and evidence of the maternal transmission of the hormonal resistance. J. Clin. Endocrinol. Metab. **87:** 189–197.
6. Bastepe, M. & H. Jüppner. 2005. The GNAS locus and pseudohypoparathyroidism. Hormone Res. **63:** 65–74.
7. Shore, E., J. Ahn, S. Jan de Beur, *et al.* 2002. Paternally-inherited inactivating mutations of the *GNAS1* gene in progressive osseous heteroplasia. N. Engl. J. Med. **346:** 99–106.
8. Patten, J.L., D.R. Johns, D. Valle, *et al.* 1990. Mutation in the gene encoding the stimulatory G protein of adenylate cyclase in Albright's hereditary osteodystrophy. N. Engl. J. Med. **322:** 1412–1419.
9. Weinstein, L.S., P.V. Gejman, E. Friedman, *et al.* 1990. Mutations of the Gs a-subunit gene in Albright hereditary osteodystrophy detected by denaturing gradient gel electrophoresis. Proc. Natl. Acad. Sci. USA **87:** 8287–8290.
10. Dixon, P., S. Ahmed, D. Bonthron, *et al.* 1996. Mutational analysis of the GNAS1 gene in pseudohypoparathyroidism [abstract]. J. Bone Miner. Res. **11:** S494.
11. Aldred, M., S. Aftimos, C. Hall, *et al.* 2002. Constitutional deletion of chromosome 20q in two patients affected with Albright hereditary osteodystrophy. Am. J. Med. Genet. Suppl. **113:** 167–172.
12. Hayward, B.E., V. Moran, L. Strain & D.T. Bonthron. 1998. Bidirectional imprinting of a single gene: GNAS1 encodes maternally, paternally, and biallelically derived proteins. Proc. Natl. Acad. Sci. USA **95:** 15475–15480.
13. Hayward, B., M. Kamiya, L. Strain, *et al.* 1998. The human GNAS1 gene is imprinted and encodes distinct paternally and biallelically expressed G proteins. Proc. Natl. Acad. Sci. USA **95:** 10038–10043.
14. Peters, J., S.F. Wroe, C.A. Wells, *et al.* 1999. A cluster of oppositely imprinted transcripts at the Gnas locus in the distal imprinting region of mouse chromosome 2. Proc. Natl. Acad. Sci. USA **96:** 3830–3835.
15. Hayward, B. & D. Bonthron. 2000. An imprinted antisense transcript at the human GNAS1 locus. Hum. Mol. Genet. **9:** 835–841.
16. Hayward, B., A. Barlier, M. Korbonits, *et al.* 2001. Imprinting of the G(s)alpha gene GNAS1 in the pathogenesis of acromegaly. J. Clin. Invest. **107:** R31–R36.
17. Jüppner, H., E. Schipani, M. Bastepe, *et al.* 1998. The gene responsible for pseudohypoparathyroidism type Ib is paternally imprinted and maps in four unrelated kindreds to chromosome 20q13.3. Proc. Natl. Acad. Sci. USA **95:** 11798–11803.

18. LIU, J., D. LITMAN, M. ROSENBERG, *et al.* 2000. A GNAS1 imprinting defect in pseudohypoparathyroidism type IB. J. Clin. Invest. **106:** 1167–1174.
19. BASTEPE, M., J.E. PINCUS, T. SUGIMOTO, *et al.* 2001. Positional dissociation between the genetic mutation responsible for pseudohypoparathyroidism type Ib and the associated methylation defect at exon A/B: evidence for a long-range regulatory element within the imprinted *GNAS1* locus. Hum. Mol. Genet. **10:** 1231–1241.
20. BASTEPE, M., L. FRÖHLICH, G. HENDY, *et al.* 2003. Autosomal dominant pseudohypoparathyroidism type-Ib is associated with a heterozygous microdeletion that likely disrupts a putative imprinting control element of GNAS. J. Clin. Invest. **112:** 1255–1263.
21. MAHMUD, F., A. LINGLART, M. BASTEPE, *et al.* 2005. Molecular diagnosis of pseudohypoparathyroidism type Ib in a family with presumed paroxysmal dyskinesia. Pediatrics **115:** e242–e244.
22. LINGLART, A., R. GENSURE, R. OLNEY, *et al.* 2005. A novel STX16 deletion in autosomal dominant pseudohypoparathyroidism type Ib delineates a cis-acting imprinting control element of GNAS. Am. J. Hum. Genet. **76:** 804–814.
23. LIU, J., J. NEALON & L. WEINSTEIN. 2005. Distinct patterns of abnormal GNAS imprinting in familial and sporadic pseudohypoparathyroidism type IB. Hum. Mol. Genet. **14:** 95–102.
24. BASTEPE, M., L. FRÖHLICH, A. LINGLART, *et al.* 2005. Deletion of the NESP55 differentially methylated region causes loss of maternal GNAS imprints and pseudohypoparathyroidism type Ib. Nat. Genet. **37:** 25–27.
25. JAN DE BEUR, S., C. DING, E. GERMAIN-LEE, *et al.* 2003. Discordance between genetic and epigenetic defects in pseudohypoparathyroidism type 1b revealed by inconsistent loss of maternal imprinting at *GNAS1*. Am. J. Hum. Genet. **73:** 314–322.

Estrogen Regulation of Immune Cell Bone Interactions

M. NEALE WEITZMANN AND ROBERTO PACIFICI

Division of Endocrinology, Metabolism and Lipids, Emory University School of Medicine, Atlanta, Georgia 30322, USA

ABSTRACT: Estrogen deficiency is one of the most frequent causes of osteoporosis in women and a possible cause of bone loss and insufficient skeletal development in men. Estrogen deficiency results from menopause but also by a number of conditions, such as stress, excessive physical activity, and low body weight. The mechanism by which estrogen deficiency causes bone loss remains largely unknown. Estrogen deficiency leads to an increase in the immune function, which culminates in an increased production of TNF by activated T cells. TNF increases osteoclast formation and bone resorption both directly and by augmenting the sensitivity of maturing osteoclasts to the essential osteoclastogenic factor RANKL. Increased T cell production of TNF is induced by estrogen deficiency via a complex mechanism mediated by antigen-presenting cells and involving the cytokines IFN-γ, IL-7, and TGF-β. Herein we review the experimental evidence that suggests that estrogen prevents bone loss by regulating T cell function and immune cell bone interactions.

KEYWORDS: estrogen; T cells; immune response; bone cells; bone loss

INTRODUCTION

It has long been recognized that infection, inflammation, and autoimmune disorders are associated with systemic and local bone loss. Yet, it is only recently that T lymphocytes and their products have been recognized as key regulators of osteoclast and osteoblast formation, life span, and activity.[1,2] Similarly, sex steroids and aging have been known to regulate the immune system and T cells for decades. In spite of the abundance of clinical and physiological clues, it is only in the last few years that investigators have linked immune cells to the etiology of postmenopausal and senile osteoporosis, as well as to the bone loss caused by a variety of endocrine conditions.

Address for correspondence: Roberto Pacifici, M.D., Division of Endocrinology, Metabolism and Lipids, Emory University School of Medicine, 1639 Pierce Drive, Room 1309, Atlanta, GA 30322. Voice: 404-712-8420; fax: 404-727-1300.

e-mail: roberto.pacifici@emory.edu

Ann. N.Y. Acad. Sci. 1068: 256–274 (2006). © 2006 New York Academy of Sciences.
doi: 10.1196/annals.1346.030

Herein we review the evidence that supports the hypothesis that one of the main mechanisms by which estrogen deficiency causes bone loss is by inducing the production of osteoclastogenic factors by activated T cells.

T CELLS AND OSTEOCLASTOGENIC CYTOKINES

Osteoclasts arise by cytokine-driven proliferation and differentiation of monocyte precursors that circulate within the hematopoietic cell pool.[3] This process is facilitated by bone marrow stromal cells, a population that provides a physical support for nascent osteoclasts and produces soluble and membrane-associated factors essential for the proliferation and/or the differentiation of osteoclast precursors (FIG. 1).

The cytokines required for osteoclast formation are receptor activator of NF-κB ligand (RANKL) and macrophage colony-stimulating factor (M-CSF).

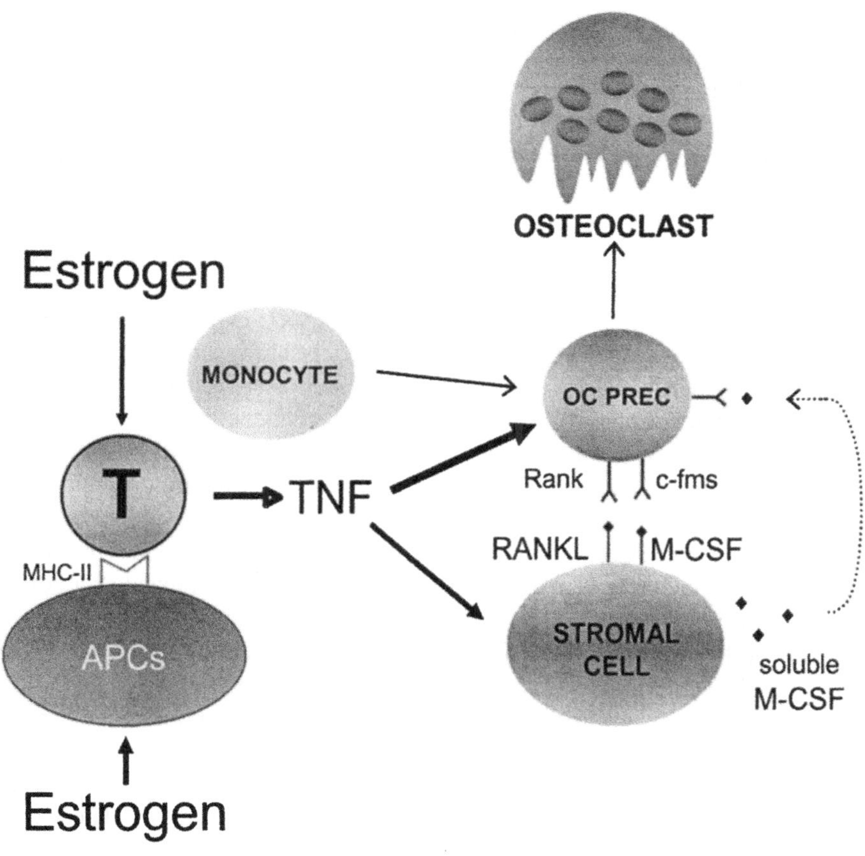

FIGURE 1. Cells and cytokines relevant for osteoclast formation.

These factors are produced primarily by bone marrow stromal cells,[4–6] osteoblasts,[2,5] and activated T cells.[7–10] RANKL is a member of the tumor necrosis factor (TNF) superfamily, which exists in both membrane bound and soluble forms. RANKL binds to the transmembrane receptor RANK, which is expressed on the surface of osteoclasts and osteoclast precursors. RANKL also binds to osteoprotegerin (OPG), a soluble decoy receptor produced by numerous hematopoietic cells. Thus, OPG, by sequestering RANKL and preventing its binding to RANK, functions as a potent anti-osteoclastogenic cytokine.[3,6] M-CSF induces the proliferation of early osteoclast precursors, the differentiation of more mature osteoclasts, the fusion of mononucleated pre-osteoclasts, and increases the survival of mature osteoclasts. RANKL does not induce cell proliferation, but promotes the differentiation of osteoclast precursors from an early stage of maturation into fully mature multinucleated osteoclasts. RANKL is also capable of activating mature osteoclasts, thus stimulating the capacity of these cells to resorb bone.

While RANKL and M-CSF are essential for physiologic osteoclast renewal, additional cytokines either produced by, or regulated by, T cells are responsible for the upregulation of osteoclast formation observed in a variety of conditions, such as inflammation and estrogen deficiency.[11,12] One such factor is TNF, a cytokine that enhances osteoclast formation by upregulating the stromal cell production of RANKL and M-CSF,[13,14] and by augmenting the responsiveness of osteoclast precursors to RANKL.[15,16] The ability of TNF to increase the osteoclastogenic activity of RANKL is due to synergistic interactions at the level of NF-κB and AP-1 signaling.[16,17] Furthermore, TNF stimulates osteoclast activity and inhibits osteoblastogenesis,[18] thus further driving an imbalance between bone formation and bone resorption.

Like TNF, interleukin-1 (IL-1) promotes RANKL expression by marrow stromal cells and osteoblasts and enhances osteoclast life span and activity. IL-1 directly targets osteoclast precursors and promotes osteoclast differentiation in the presence of permissive levels of RANKL. Recent evidence has also shown that IL-1 mediates, in part, the osteoclastogenic effect of TNF by enhancing stromal cell expression of RANKL and by directly stimulating differentiation of osteoclast precursors.[19]

Another cytokine relevant for osteoclast formation is IL-7.[20] IL-7 is a powerful lymphopoietic cytokine that has previously been recognized as a potent inducer of bone destruction *in vivo*.[21] How IL-7 leads to bone loss is controversial, and its mechanisms of action are only now beginning to be elucidated. IL-7 regulates multiple stages of T cell development and homeostasis.[22] IL-7 knockout mice are severely lymphopenic[23] and IL-7 receptor knockout mice have been reported to display increased bone volume and bone mineral density.[21] In contrast, IL-7 transgenic mice have expanded bone marrow cavities with focal osteolysis of cortical bone and eroded bone surfaces.[24] These data suggest that IL-7 may induce bone loss by a T cell-mediated mechanism. Indeed, IL-7 has been reported to induce production of RANKL by human

T cells,[25] and injection of IL-7 into mice *in vivo* induces bone destruction[21,26] by eliciting the secretion by T cells of the key osteoclastogenic cytokines RANKL and TNF.[26] In addition, levels of IL-7 are significantly elevated following ovariectomy. Attesting to the key role of IL-7 in the bone destruction associated with estrogen deficiency, *in vivo* IL-7 blockade, using neutralizing antibodies, is effective in preventing ovariectomy-induced bone destruction.[27] Furthermore, IL-7-induced osteoclastogenesis and bone loss, is compounded by suppression of bone formation leading to uncoupling of bone formation from resorption.

An important, yet controversial, osteoclast regulating factor is interferon-γ (IFN-γ). This factor was initially described as an antiosteoclastogenic cytokine because it is a potent inhibitor of osteoclastogenesis *in vitro*.[28] The notion that IFN-γ is an inhibitor of bone resorption was reinforced by the finding that silencing of IFN-γ signaling leads to a more rapid onset of collagen-induced arthritis and bone resorption[29] as compared to wild-type (wt) controls, and by the report that IFN-γ decreases serum calcium and osteoclastic bone resorption in nude mice.[30,31]

However, observations in humans and in experimental models of disease indicate that IFN-γ promotes bone resorption and causes bone loss in a variety of conditions. For examples, mice lacking IFN-γ production are protected against infection-induced alveolar bone loss,[32] while in erosive tuberculoid leprosy and psoriatic arthritis, IFN-γ production correlates positively with bone destruction.[33,34] In addition, randomized controlled trials have shown that IFN-γ does not prevent bone loss in patients with rheumatoid arthritis (RA),[35,36] nor the bone wasting effect of cyclosporin A.[37] Furthermore, IFN-γ has been reported to be efficacious in the treatment of osteopetrosis through restoration of bone resorption, both in humans[38] and rodents.[39] These latter findings conclusively demonstrate that in some conditions the net effect of IFN-γ *in vivo* is that of stimulating osteoclastic bone resorption.

The complex effects of IFN-γ can be explained by the fact that IFN-γ influences osteoclast formation both via direct and indirect effects. IFN-γ directly blocks osteoclast formation through targeting of maturing osteoclasts. This effect is best observed *in vitro*. However, IFN-γ is also a potent inducer of antigen presentation and thus of T cell activation. Therefore, when IFN-γ levels are increased *in vivo*, activated T cells secrete pro-osteoclastogenic factors and this activity offsets the anti-osteoclastogenic effect of IFN-γ. While in certain conditions the net effect of IFN-γ is that of inhibiting osteoclast formation, in other conditions the balance is in favor of a stimulatory activity on osteoclast differentiation. In general, when T cell activation occurs in response to an innate immune response, such as in response to lipopolysaccharide (LPS) stimulation, IFN-γ functions as an antiresorptive agent. Conversely, when T cell activation occurs as a part of an adaptive immune response, as in conditions of estrogen deficiency, IFN-γ stimulates bone resorption. The inability of IFN-γ to blunt the differentiation of maturing osteoclasts arising in a microenvironment such as bone, where RANKL is abundant,[40] contributes to the explanation of why in some conditions the *in vivo* pro-resorptive effects of IFN-γ are more potent than the suppression of osteoclastogenesis induced by IFN-γ *in vitro*.

A factor that plays a key, complex, and controversial role in osteoclastogenesis is transforming growth factor-β (TGF-β), a pleiotropic growth factor with wide-ranging effects.[41–43] Three isoforms (TGF-β1, -2, and -3) have been described in mammals and all use the same receptor complex for signaling.[42] TGF-β1 is the predominant TGF-β isoform in lymphoid organs and is the major species in serum. Conversely, TGF-β2 and 3 are predominantly expressed in mesenchymal tissues and bone.[44–46] TGF-β is produced by a large number of cells including bone marrow cells, osteoblasts, and stromal cells and is secreted in a latent form and must be activated to mediate its effects.[47] Latent TGF-β is abundantly present in the bone matrix and is noncovalently linked to a precursor molecule termed latency-associated peptide. Mature TGF-β can be released from the latent complex in an active form *in vitro* by a variety of treatments such as heating or acidification.[47] Although several mechanisms of activation *in vivo* have been proposed, the precise mechanism is not known, though a role for extreme pH, free radicals, transglutaminase, or thrombospondin, has been suggested.[48] The mechanism by which TGF-β is activated in the bone marrow and lymphoid organs is presently unknown. Both *in vitro* and *in vivo* studies have shown that TGF-β1, 2, and 3 have complex effects on bone. For example, these growth factors stimulate or repress proliferation of osteoblasts and osteoclasts depending on cell lines and culture conditions *in vitro*.[49–52] Furthermore, mice with osteoblast-specific overexpression of TGF-β2 develop high turnover osteoporosis[53] while administration of TGF-β2 into periosteum increases bone formation locally, but has no effects systemically.

TGF-β has been implicated in the pathogenesis of ovx-induced bone loss because local injection of TGF-β1 and TGF-β2 prevent bone loss at the site of the injection in ovx rats.[54,55] Furthermore, estrogen is known to upregulate the expression of TGF-β in murine osteoblasts, bone extracts, and bone marrow cells,[56,57] and long-term *in vivo* estrogen treatment has been shown to increase serum TGF-β1 and 2 levels in humans.[58]

T CELLS AND OVARIECTOMY-INDUCED BONE LOSS

It is now accepted that the antiresorptive activity of estrogen is a result of multiple genomic and nongenomic effects on bone marrow and bone cells, which leads to decreased osteoclast formation, increased osteoclast apoptosis, and decreased capacity of mature osteoclasts to resorb bone.[59] It is also recognized that stimulation of bone resorption in response to estrogen deficiency is mainly due to cytokine-driven increased osteoclast formation.[60,61] One of the cytokines responsible for the augmented osteoclastogenesis of estrogen deficiency is TNF.[62] The relevance of TNF in the mechanism by which estrogen causes bone loss has been demonstrated using multiple animal models. For example, ovariectomy fails to induce bone loss in TNF knockout mice,[63] transgenic mice insensitive to TNF due to the overexpression of a soluble TNF receptor,[64] and in mice treated with the TNF inhibitor TNF-binding protein.[65]

The presence of increased levels of TNF in the bone marrow of ovariectomized animals and in the conditioned media of peripheral blood cells of postmenopausal women is well documented.[66–71] However, the cells responsible for this phenomenon had not been conclusively identified. Early studies revealed that estrogen does not regulate TNF expression in cultured stromal cells, some osteoblastic cell lines, and bone biopsies.[67,72,73] At the same time evidence began to emerge that estrogen deficiency increases TNF production by monocyte-enriched peripheral blood mononuclear cells, and unfractionated human and murine bone marrow cells.[68,69] Based on these data, the source of upregulated TNF production was ascribed to bone marrow monocytes. Surprisingly, some of these earlier observations showed that phytohemaglutinin stimulation was required to demonstrate increased TNF production from monocytes. Because phytohemaglutinin promotes T cell activation, this requirement remained unexplained. Recent studies on highly purified cells have revealed that ovariectomy increases the production of TNF by T cells, but not by bone marrow monocytes.[15] The earlier identifications of TNF production by monocytes were due, in part, to the fact that adherent bone marrow cells contain T cells ($\sim$10% of total cells) and that ovariectomy increases T cell content by approximately twofold. Thus, the ovariectomy-induced increase in TNF levels is likely to be due to T cell TNF production. These findings in the mouse are concordant with those of others[74] in humans, who demonstrated that adherent mononuclear blood cells contain $CD3^+$ $CD56^+$ lymphocytes, a TNF-producing subset of adherent T cells. In that study, the number of $CD3^+$ $CD56^+$ T cells was decreased by estrogen treatment and inversely correlated with bone density. Thus, earlier findings in bone marrow and adherent cell cultures are consistent with the stimulatory effect of estrogen deficiency on the T cell production of TNF observed in more recent studies.

These results are not surprising as T cells have the capacity to secrete a wide repertoire of cytokines, some pro-osteoclastogenic and some anti-osteoclastogenic. In the absence of strong activation signals T cells appear to repress osteoclast formation,[75] but the relevance of this phenomenon *in vivo* has not been established. In contrast, activated T cells play a key role in the regulation of osteoclast formation through increased production of RANKL, TNF,[8,10,75–77] and OPG.[75] Activated T cells also produce IFN-α and IFN-γ[28] which limits, in part, RANKL-induced bone resorption by repressing NF-κB and C-Jun N-terminal kinase (JNK) signaling pathways. The net effect of T cells on osteoclast formation may consequently represent the prevailing balance of anti- and pro-osteoclastogenic T cell cytokine secretion. However, it appears that during stimulated conditions such as inflammation[76] and during estrogen deficiency,[15,27,63,78] pro-osteoclastogenic cytokines prevail.

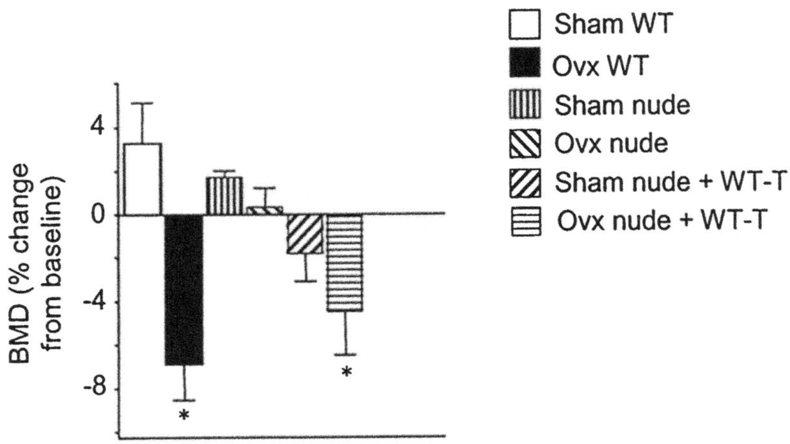

FIGURE 2. T cell-deficient mice are protected against ovariectomy-induced bone loss. BMD (mean ± SEM) as measured by Dexa. *$P < 0.05$ compared to base line. BMD, bone mineral density.

Attesting to the relevance of T cells in estrogen deficiency induced bone loss *in vivo*, athymic T cell-deficient nude mice are completely protected against the bone loss and the increase in bone turnover induced by ovariectomy[15,63] (FIG. 2 and TABLE 1). Reconstitution studies in nude mice have subsequently shown the relevant T cell subpopulation is $CD4^+$ cells. In fact, nude ovariectomized mice lose bone like wt ovariectomized mice after adoptive transfer of $CD4^+$ cells. In contrast, adoptive transfer of $CD8^+$ cells does not reinduce the capacity of ovariectomy to cause bone loss and elevate bone turnover. Together these data establish $CD4^+$ T cells as essential mediators of the bone-wasting effect of estrogen deficiency *in vivo*. T cells play a key role as inducers of bone wasting because ovariectomy increases T cell production of TNF to a level sufficient to augment RANKL-induced osteoclastogenesis.[15] An additional relevant effect of TNF is that of inhibiting bone formation. *In vitro*, TNF

TABLE 1. μCT analysis of femur morphometric indices (mean ± SD) 4 weeks after sham operation or ovx

	BV/TV (%)	Tb.Th (μm)
Sham WT	21.81 ± 1.91	58.12 ± 2.89
Ovx WT	18.19 ± 1.51†	51.38 ± 4.11†
Sham nude	18.63 ± 0.7	45.50 ± 1.98
Ovx nude	17.95 ± 1.37	43.13 ± 1.06
Sham nude + WT T cells	18.83 ± 1.35	46.23 ± 3.49
Ovx nude + WT T cells	14.22 ± 2.32‡	40.50 ± 2.75‡

BV/TV, relative bone volume; Tb.Th, trabecular thickness.
$^\dagger P < 0.05$ from sham-operated WT group.
$^\ddagger P < 0.05$ from sham-operated nude group.

inhibits osteocalcin production and mineralization of osteoblastic cell lines.[79] Consistent with these observations TNF has been reported to repress the differentiation of osteoblast precursors into mature osteoblasts[80] by inhibiting the critical osteoblastic transcription factor Runx2. TNF blocks the expression of Runx2 at multiple levels including destabilization of its mRNA and suppression of its transcription.[81] Thus TNF appears to be a critical cytokine in the etiology of osteoporosis resulting from estrogen deficiency.

Although TNF may be derived from many sources in the bone marrow, the specific relevance of T cell produced TNF *in vivo* was demonstrated by the finding that while reconstitution of nude recipient mice with T cells from wt mice restores the capacity of ovariectomy to induce bone loss, reconstitution with T cells from TNF-deficient mice does not.[63] It should be noted that nude mice reconstituted with TNF –/– T cells are protected against ovariectomy-induced bone loss even though they possess monocytes and other TNF-producing cells, a finding that establishes a specific role for T cell-derived TNF. TNF has been reported to stimulate RANKL in osteoblasts. Thus, the possibility that T cell-produced TNF further augments bone loss by stimulating T cell production of RANKL cannot be formally excluded, although TNF has not been reported to induce RANKL production by murine T cells.

MECHANISMS OF ESTROGEN REGULATION OF T CELL TNF PRODUCTION

Ovariectomy upregulates T cell TNF production by increasing the number of TNF-producing T cells without altering the amount of TNF produced by each T cell.[63] This is the result of a complex pathway summarized in FIGURE 3. Ovariectomy causes an expansion of the T cell pool in the bone marrow by increasing T cell activation, a phenomenon that results in increased T cell proliferation and life span. Ovariectomy increases T cell activation by enhancing antigen presentation by bone marrow macrophages[78] and dendritic cells and this is a result of the ability of estrogen deficiency to upregulate the expression of major histocompatibility complex II (MHCII). Although the mechanism of T cell activation elicited by estrogen deficiency is similar to that triggered by infections, the intensity of the events that follows estrogen withdrawal is significantly less severe and this process should be envisioned as a partial increase in T cell autoreactivity to self-peptides resulting in a modest expansion in the pool of effector $CD4^+$ cells.

The mechanism described above hinges on the ability of antigen-presenting cells to present antigenic peptides, bound to MHCII molecules, to T cells. The question thus arises concerning the source of the antigens involved. Estrogen deficiency is likely to increase the reactivity of T cells to a pool of self and foreign antigens physiologically present in healthy animals and humans. This is consistent with the fact that clones of T cells expressing T cell receptor (TCR)

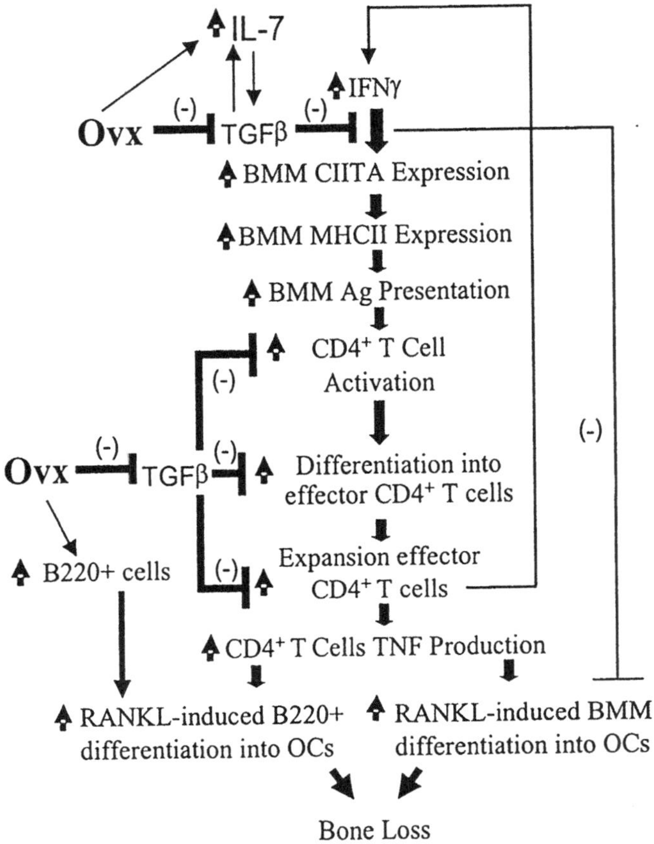

FIGURE 3. Summary of mechanism by which estrogen represses T cell TNF production and prevents bone loss.

directed against self-antigens not expressed in the thymus survive negative selection during T cell maturation.[82–85] Such clones are known as autoreactive or self-reactive T cells and reside in peripheral lymphatic organs of adult individuals.[86] In addition, foreign antigens of bacterial origin are physiologically absorbed in the gut. As these peptides come into contact with immune cells locally and systemically, they induce a low-grade T cell activation.[86] Thus, a moderate immune response is constantly in place in healthy humans and rodents due to presentation by MHCII and MHCI molecules of both self- and foreign peptides to $CD4^+$ and $CD8^+$ T cells.[87] This autoreactive response is thought to be essential for immune cell survival and renewal.[88]

The effects of ovariectomy on antigen presentation and the resulting changes in T cell, activation, proliferation, and life span are explained by a stimulatory effect of ovariectomy on the expression of the gene encoding class II transactivator (CIITA). The product of the CIITA gene is a non-DNA-binding factor that functions as a transcriptional co-activator when recruited to the MHC-II promotor by interaction with promotor-bound factors.[89,90] CIITA expression is, indeed, required and sufficient for the stimulation of antigen presentation in bone marrow monocytes.

The physiologic inducer of CIITA in bone marrow monocytes is IFN-γ. Increased CIITA expression in ovariectomized mice results from the ability of ovariectomy to increase both the T cell production of IFN-γ and the responsiveness of the CIITA gene to IFN-γ in bone marrow monocytes.[78] This second regulatory mechanism is revealed by the greater expression of CIITA and MHCII by bone marrow monocytes from ovariectomized mice as compared to bone marrow monocytes harvested from estrogen replete animals, in response to an equal stimulation with IFN-γ.[78]

That ovariectomy increases T cell production of IFN-γ was demonstrated both by measuring the level of the cytokine in the culture media of purified T cells and by fluorescent activated cell sorting (FACS) analysis of unfractionated bone marrow. The relevance of IFN-γ is shown by the failure of IFN-γR −/− to undergo T cell activation and to sustain bone loss in response to ovariectomy[78] (FIG. 4).

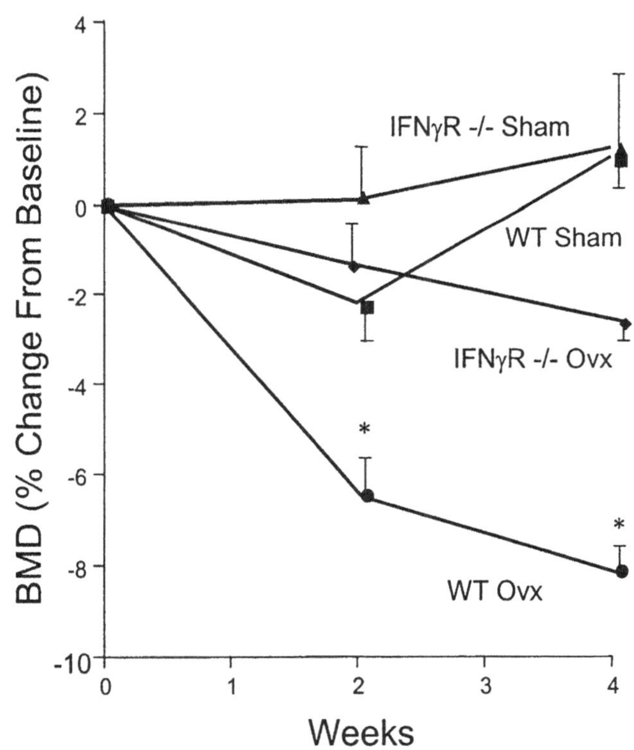

FIGURE 4. IFN-γR −/− mice are protected against ovariectomy-induced bone loss. BMD (mean ± SEM) as measured by Dexa. *$P < 0.05$ compared to base line. BMD, bone mineral density.

A mechanism by which estrogen deficiency upregulates the production of IFN-γ is through TGF-β. Estrogen has a direct stimulatory effect on the production of this factor, which is mediated through direct binding of estrogen/estrogen receptor complex to a estrogen-responsive element region in the TGF-β promotor.[91]

TGF-β is recognized for its potent immunosuppressive activity. TGF-β is a powerful repressor of T cell activation. Indeed, TGF-β exerts strong immunosuppressive effects by inhibiting the proliferation of T cells and inhibiting the expression of proinflammatory cytokines such as IL-1 and TNF.[92] TGF-β blocks the differentiation of naïve T cells into effectors cells.[93] Furthermore, TGF-β blocks T cell proliferation via inhibition of IL-2 production.[94,95] Therefore, TGF-β blunts the differentiation of naïve $CD4^+$ cells into Th1 and Th2 cells and the expansion of these populations, which follow T cell activation. Thus, by blocking Th1 differentiation and proliferation, TGF-β blunts the production of IFN-γ. In agreement with these effects of TGF-β and with earlier studies *in vitro*,[96] TGF-β was found to repress T cell production of IFN-γ *in vivo*.[97] TGF-β is also known to decrease the responsiveness of the CIITA gene to IFN-γ,[97–100] via a mechanism involving Smad3.[101]

As expected, based on the potent blunting effects on IFN-γ production and activity, estrogen was found to repress Ag presentation and CIITA expression via a TGF-β-dependent mechanism.[97]

In addition to its capacity to block antigen presentation, TGF-β is a powerful repressor of T cell activation. Indeed, TGF-β exerts strong immunosuppressive effects by inhibiting the proliferation of T cells and inhibiting the expression of proinflammatory cytokines such as IL-1 and TNF.[92]

Studies with a transgenic mouse, which expresses a dominant negative form of the TGF-β receptor in T cells, have allowed the significance of the repressive effects of the cytokine on T cell function for the bone loss associated with estrogen deficiency to be established.[1,97] This strain, known as CD4dnTGFβRII, is severely osteopenic due to increased bone resorption (FIG. 5). More importantly, mice with T cell-specific blockade of TGF-β signaling are completely insensitive to the bone-sparing effect of estrogen.[97] This phenotype results from a failure of estrogen to repress IFN-γ production which, in turn, leads to increased T cell activation and T cell TNF production.

As a proof of principle, a somatic gene therapy approach was used to induce the overexpression of TGF-β1 in ovx mice. These experiments confirmed that elevation of the systemic levels of TGF-β prevents the bone loss and the increase in bone turnover induced by ovx.[97]

Another mechanism by which estrogen regulates IFN-γ production is through IL-7, a cytokine known to act as a potent stimulator of IFN-γ production.[102] This suggests that the elevated levels of IL-7 during estrogen deficiency may be responsible for driving the T cell proliferation that expands the TNF-producing T cell population, by direct stimulatory and trophic effects on T cells, as well as by indirectly stimulating IFN-γ and antigen presentation.

(A)

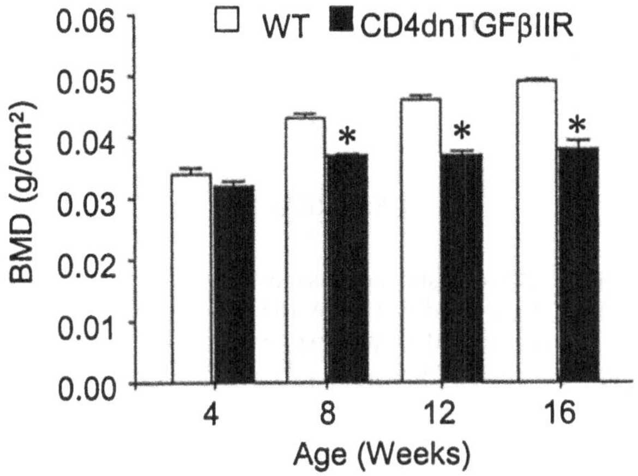

(B)

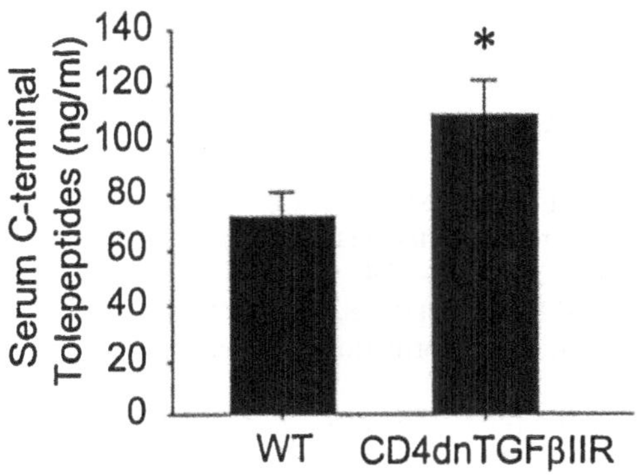

FIGURE 5. Mice with T cells insensitive to TGF-β have lower bone density and increased bone resorption than WT controls. (**A**) BMD (mean ± SEM) as measured by Dexa. (**B**) Serum levels (mean ± SEM) of c-terminal telopeptides. $^*P < 0.05$. BMD, bone mineral density.

The elevated response of bone marrow monocytes and antigen presentation to T cells further drives and sustains T cell activation, proliferation, and osteoclastogenic cytokine secretion. Finally, as IL-7 and TGF-β are reported to inversely regulate each others' production,[103–105] the reduction in TGF-β signaling, characteristic of estrogen deficiency may serve to further stimulate IL-7 production, thus driving the cycle of osteoclastogenic cytokine production and ultimately of bone destruction.

CONCLUSIONS

Remarkable progress has been made in elucidating the cross-talk between the immune system and bone and in uncovering the mechanism by which sex steroids, infection, and inflammation lead to bone loss by dysregulating T lymphocyte function in animal models. If the findings in experimental animals are confirmed in humans, it will, perhaps, be appropriate to classify osteoporosis as an inflammatory, or even an autoimmune condition and certainly new therapeutic "immune" targets will emerge.

REFERENCES

1. TEITELBAUM, S.L. 2004. Postmenopausal osteoporosis, T cells, and immune dysfunction. Proc. Natl. Acad. Sci. USA **101:** 16711–16712.
2. RIFAS, L., S. ARACKAL & M.N. WEITZMANN. 2003. Inflammatory T cells rapidly induce differentiation of human bone marrow stromal cells into mature osteoblasts. J. Cell. Biochem. **88:** 650–659.
3. TEITELBAUM, S.L. 2000. Bone resorption by osteoclasts. Science **289:** 1504–1508.
4. LACEY, D.L., E. TIMMS, H.L. TAN, *et al.* 1998. Osteoprotegerin ligand is a cytokine that regulates osteoclast differentiation and activation. Cell **93:** 165–176.
5. YASUDA, H., N. SHIMA, N. NAKAGAWA, *et al.* 1998. Osteoclast differentiation factor is a ligand for osteoprotegerin/osteoclastogenesis-inhibitory factor and is identical to TRANCE/RANKL. Proc. Natl. Acad. Sci. USA **95:** 3597–3602.
6. KHOSLA, S. 2001. Minireview: the OPG/RANKL/RANK system. Endocrinology **142:** 5050–5055.
7. ANDERSON, D.M., E. MARASKOVSKY, W.L. BILLINGSLEY, *et al.* 1997. A homologue of the TNF receptor and its ligand enhance T-cell growth and dendritic-cell function. Nature **390:** 175–179.
8. HORWOOD, N.J., V. KARTSOGIANNIS, J.M. QUINN, *et al.* 1999. Activated T lymphocytes support osteoclast formation in vitro. Biochem. Biophys. Res. Commun. **265:** 144–150.
9. KONG, Y.Y., H. YOSHIDA, I. SAROSI, *et al.* 1999. OPGL is a key regulator of osteoclastogenesis, lymphocyte development and lymph-node organogenesis. Nature **397:** 315–323.
10. WEITZMANN, M.N., S. CENCI, L. RIFAS, *et al.* 2001. T cell activation induces human osteoclast formation via receptor activator of nuclear factor kappaB ligand-dependent and -independent mechanisms. J. Bone. Miner. Res. **16:** 328–337.
11. GREY, A., M.A. MITNICK, U. MASIUKIEWICZ, *et al.* 1999. A role for interleukin-6 in parathyroid hormone-induced bone resorption in vivo. Endocrinology **140:** 4683–4690.
12. PACIFICI, R. 1998. Cytokines, estrogen and postmenopausal osteoporosis. The second decade. Endocrinology **139:** 2659–2661.
13. HOFBAUER, L.C., D.L. LACEY, C.R. DUNSTAN, *et al.* 1999. Interleukin-1beta and tumor necrosis factor-alpha, but not interleukin-6, stimulate osteoprotegerin ligand gene expression in human osteoblastic cells. Bone **25:** 255–259.
14. SHERMAN, M.L., B.L. WEBER, R. DATTA & D.W. KUFE. 1990. Transcriptional and posttranscriptional regulation of macrophage-specific colony stimulating factor gene expression by tumor necrosis factor. J. Clin. Invest. **85:** 442–447.
15. CENCI, S., M.N. WEITZMANN, C. ROGGIA, *et al.* 2000. Estrogen deficiency induces bone loss by enhancing T-cell production of TNF-alpha. J. Clin. Invest. **106:** 1229–1237.

16. LAM, J., S. TAKESHITA, J.E. BARKER, *et al.* 2000. TNF-alpha induces osteoclastogenesis by direct stimulation of macrophages exposed to permissive levels of RANK ligand. J. Clin. Invest. **106:** 1481–1488.
17. ZHANG, Y.H., A. HEULSMANN, M.M. TONDRAVI, *et al.* 2001. Tumor necrosis factor-alpha (TNF) stimulates RANKL-induced osteoclastogenesis via coupling of TNF type 1 receptor and RANK signaling pathways. J. Biol. Chem. **276:** 563–568.
18. NANES, M.S. 2003. Tumor necrosis factor-alpha: molecular and cellular mechanisms in skeletal pathology. Gene **321:** 1–15.
19. WEI, S., H. KITAURA, P. ZHOU, *et al.* 2005. IL-1 mediates TNF-induced osteoclastogenesis. J. Clin. Invest. **115:** 282–290.
20. ROSS, F.P. 2003. Interleukin 7 and estrogen-induced bone loss. Trends Endocrinol. Metab. **14:** 147–149.
21. MIYAURA, C., Y. ONOE, M. INADA, *et al.* 1997. Increased B-lymphopoiesis by interleukin 7 induces bone loss in mice with intact ovarian function: similarity to estrogen deficiency. Proc. Natl. Acad. Sci. USA **19:** 9360–9365.
22. FRY, T.J. & C.L. MACKALL. 2001. Interleukin-7: master regulator of peripheral T-cell homeostasis? Trends Immunol. **22:** 564–571.
23. VON FREEDEN-JEFFRY, U., P. VIEIRA, L.A. LUCIAN, *et al.* 1995. Lymphopenia in interleukin (IL)-7 gene-deleted mice identifies IL-7 as a nonredundant cytokine. J. Exp. Med. **181:** 1519–1526.
24. VALENZONA, H.O., R. POINTER, R. CEREDIG & D.G. OSMOND. 1996. Prelymphomatous B cell hyperplasia in the bone marrow of interleukin-7 transgenic mice: precursor B cell dynamics, microenvironmental organization and osteolysis. Exp. Hematol. **24:** 1521–1529.
25. WEITZMANN, M.N., S. CENCI, L. RIFAS, *et al.* 2000. Interleukin-7 stimulates osteoclast formation by up-regulating the T- cell production of soluble osteoclastogenic cytokines. Blood **96:** 1873–1878.
26. TORALDO, G., C. ROGGIA, W.P. QIAN, *et al.* 2003. IL-7 induces bone loss in vivo by induction of receptor activator of nuclear factor kappa B ligand and tumor necrosis factor alpha from T cells. Proc. Natl. Acad. Sci. USA **100:** 125–130.
27. WEITZMANN, M.N., C. ROGGIA, G. TORALDO, *et al.* 2002. Increased production of IL-7 uncouples bone formation from bone resorption during estrogen deficiency. J. Clin. Invest. **110:** 1643–1650.
28. TAKAYANAGI, H., K. OGASAWARA, S. HIDA, *et al.* 2000. T-cell-mediated regulation of osteoclastogenesis by signalling cross-talk between RANKL and IFN-gamma. Nature **408:** 600–605.
29. VERMEIRE, K., H. HEREMANS, M. VANDEPUTTE, *et al.* 1997. Accelerated collagen-induced arthritis in IFN-gamma receptor-deficient mice. J. Immunol. **158:** 5507–5513.
30. SATO, K., T. SATOH, K. SHIZUME, *et al.* 1992. Prolonged decrease of serum calcium concentration by murine gamma-interferon in hypercalcemic, human tumor (EC-GI)-bearing nude mice. Cancer Res. **52:** 444–449.
31. TOHKIN, M., S. KAKUDO, H. KASAI & H. ARITA. 1994. Comparative study of inhibitory effects by murine interferon gamma and a new bisphosphonate (alendronate) in hypercalcemic, nude mice bearing human tumor (LJC-1-JCK). Cancer Immunol. Immunother. **39:** 155–160.
32. BAKER, P.J., M. DIXON, R.T. EVANS, *et al.* 1999. CD4(+) T cells and the proinflammatory cytokines gamma interferon and interleukin-6 contribute to alveolar bone loss in mice. Infect. Immun. **67:** 2804–2809.

33. ARNOLDI, J., J. GERDES & H.D. FLAD. 1990. Immunohistologic assessment of cytokine production of infiltrating cells in various forms of leprosy. Am. J. Pathol. **137:** 749–753.
34. FIRESTEIN, G.S., J.M. ALVARO-GRACIA, R. MAKI & J.M. ALVARO-GARCIA. 1990. Quantitative analysis of cytokine gene expression in rheumatoid arthritis. J. Immunol. **144:** 3347–3353.
35. CANNON, G.W., S.H. PINCUS, R.D. EMKEY, *et al.* 1989. Double-blind trial of recombinant gamma-interferon versus placebo in the treatment of rheumatoid arthritis. Arthritis Rheum. **32:** 964–973.
36. VEYS, E.M., C.J. MENKES & P. EMERY. 1997. A randomized, double-blind study comparing twenty-four-week treatment with recombinant interferon-gamma versus placebo in the treatment of rheumatoid arthritis. Arthritis Rheum. **40:** 62–68.
37. MANN, G.N., T.W. JACOBS, F.J. BUCHINSKY, *et al.* 1994. Interferon-gamma causes loss of bone volume in vivo and fails to ameliorate cyclosporin A-induced osteopenia. Endocrinology **135:** 1077–1083.
38. KEY, L.L. JR., R.M. RODRIGUIZ, S.M. WILLI, *et al.* 1995. Long-term treatment of osteopetrosis with recombinant human interferon gamma. N. Engl. J. Med. **332:** 1594–1599.
39. RODRIGUIZ, R.M., L.L. KEY JR. & W.L. RIES. 1993. Combination macrophage-colony stimulating factor and interferon-gamma administration ameliorates the osteopetrotic condition in microphthalmic (mi/mi) mice. Pediatr. Res. **33:** 384–389.
40. HUANG, W., R.J. O'KEEFE & E.M. SCHWARZ. 2003. Exposure to receptor-activator of NFkappaB ligand renders pre-osteoclasts resistant to IFN-gamma by inducing terminal differentiation. Arthritis Res. Ther. **5:** R49–R59.
41. ATTISANO, L., J.L. WRANA, F. LOPEZ-CASILLAS & J. MASSAGUE. 1994. TGF-beta receptors and actions. Biochim. Biophys. Acta. **1222:** 71–80.
42. MASSAGUE, J. 1990. The transforming growth factor-beta family. Annu. Rev. Cell Biol. **6:** 597–641.
43. SHI, Y. & J. MASSAGUE. 2003. Mechanisms of TGF-beta signaling from cell membrane to the nucleus. Cell **113:** 685–700.
44. PELTON, R.W., S. NOMURA, H.L. MOSES, *et al.* 1989. Expression of transforming growth factor beta 2 RNA during murine embryogenesis. Development **106:** 759–767.
45. MILLAN, F.A., F. DENHEZ, P. KONDAIAH & R.J. AKHURST. 1991. Embryonic gene expression patterns of TGF beta 1, beta 2 and beta 3 suggest different developmental functions in vivo. Development **111:** 131–143.
46. SCHMID, P., D. COX, G. BILBE, *et al.* 1991. Differential expression of TGF beta 1, beta 2 and beta 3 genes during mouse embryogenesis. Development **111:** 117–130.
47. PRUD'HOMME, G.J. & C.A. PICCIRILLO. 2000. The inhibitory effects of transforming growth factor-beta-1 (TGF-beta1) in autoimmune diseases. J. Autoimmun. **14:** 23–42.
48. DERYNCK, R. & X.H. FENG. 1997. TGF-beta receptor signaling. Biochim. Biophys. Acta. **1333:** F105–F150.
49. FAGENHOLZ, P.J., S.M. WARREN, J.A. GREENWALD, *et al.* 2001. Osteoblast gene expression is differentially regulated by TGF-beta isoforms. J. Craniofac. Surg. **12:** 183–190.

50. Breen, E.C., R.A. Ignotz, L. McCabe, *et al.* 1994. TGF beta alters growth and differentiation related gene expression in proliferating osteoblasts in vitro, preventing development of the mature bone phenotype. J. Cell Physiol. **160:** 323–335.
51. Spinella-Jaegle, S., S. Roman-Roman, C. Faucheu, *et al.* 2001. Opposite effects of bone morphogenetic protein-2 and transforming growth factor-beta1 on osteoblast differentiation. Bone **29:** 323–330.
52. Yan, T., B.L. Riggs, W.J. Boyle & S. Khosla. 2001. Regulation of osteoclastogenesis and RANK expression by TGF-beta1. J. Cell Biochem. **83:** 320–325.
53. Erlebacher, A. & R. Derynck. 1996. Increased expression of TGF-beta 2 in osteoblasts results in an osteoporosis-like phenotype. J. Cell Biol. **132:** 195–210.
54. Kalu, D.N., E. Salerno, Y. Higami, *et al.* 1993. In vivo effects of transforming growth factor-beta 2 in ovariectomized rats. Bone Miner. **22:** 209–220.
55. Beaudreuil, J., G. Mbalaviele, M. Cohen-Solal, *et al.* 1995. Short-term local injections of transforming growth factor-beta 1 decrease ovariectomy-stimulated osteoclastic resorption in vivo in rats. J. Bone Min. Res. **10:** 971–977.
56. Gray, T.K., B. Lipes, T. Linkhart, *et al.* 1989. Transforming growth factor beta mediates the estrogen induced inhibition of UMR106 cell growth. Connect Tissue Res. **20:** 23–32.
57. Finkelman, R.D., N.H. Bell, D.D. Strong, *et al.* 1992. Ovariectomy selectively reduces the concentration of transforming growth factor beta in rat bone: implications for estrogen deficiency-associated bone loss. Proc. Natl. Acad. Sci. USA **89:** 12190–12193.
58. Bord, S., S. Beavan, D. Ireland, *et al.* 2001. Mechanisms by which high-dose estrogen therapy produces anabolic skeletal effects in postmenopausal women: role of locally produced growth factors. Bone **29:** 216–222.
59. Manolagas, S.C., S. Kousteni & R.L. Jilka. 2002. Sex steroids and bone. Recent Prog. Horm. Res. **57:** 385–409.
60. Pfeilschifter, J., R. Koditz, M. Pfohl, *et al.* 2002. Changes in proinflammatory cytokine activity after menopause. Endocr. Rev. **23:** 90–119.
61. Hofbauer, L.C., S. Khosla, C.R. Dunstan, *et al.* 2000. The roles of osteoprotegerin and osteoprotegerin ligand in the paracrine regulation of bone resorption. J. Bone Miner. Res. **15:** 2–12.
62. Suda, T., N. Takahashi, N. Udagawa, *et al.* 1999. Modulation of osteoclast differentiation and function by the new members of the tumor necrosis factor receptor and ligand families. Endocr. Rev. **20:** 345–357.
63. Roggia, C., Y. Gao, S. Cenci, *et al.* 2001. Up-regulation of TNF-producing T cells in the bone marrow: a key mechanism by which estrogen deficiency induces bone loss in vivo. Proc. Natl. Acad. Sci. USA **98:** 13960–13965.
64. Ammann, P., R. Rizzoli, J.P. Bonjour, *et al.* 1997. Transgenic mice expressing soluble tumor necrosis factor-receptor are protected against bone loss caused by estrogen deficiency. J. Clin. Invest. **99:** 1699–1703.
65. Kimble, R., S. Bain & R. Pacifici. 1997. The functional block of TNF but not of IL-6 prevents bone loss in ovariectomized mice. J. Bone Miner. Res. **12:** 935–941.
66. Pacifici, R., C. Brown, E. Puscheck, *et al.* 1991. Effect of surgical menopause and estrogen replacement on cytokine release from human blood mononuclear cells. Proc. Natl. Acad. Sci. USA **88:** 5134–5138.

67. RALSTON, S.H. 1994. Analysis of gene expression in human bone biopsies by polymerase chain reaction: evidence for enhanced cytokine expression in postmenopausal osteoporosis. J. Bone Miner. Res. **9:** 883–890.
68. KITAZAWA, R., R.B. KIMBLE, J.L. VANNICE, *et al.* 1994. Interleukin-1 receptor antagonist and tumor necrosis factor binding protein decrease osteoclast formation and bone resorption in ovariectomized mice. J. Clin. Invest. **94:** 2397–2406.
69. KIMBLE, R.B., S. SRIVASTAVA, F.P. ROSS, *et al.* 1996. Estrogen deficiency increases the ability of stromal cells to support murine osteoclastogenesis via an interleukin-1 and tumor necrosis factor-mediated stimulation of macrophage colony-stimulating factor production. J. Biol. Chem. **271:** 28890–28897.
70. RALSTON, S.H., S.J. GALLACHER, U. PATEL, *et al.* 1990. Use of bisphosphonates in hypercalcaemia due to malignancy [letter]. Lancet **335:** 737.
71. SHANKER, G., M. SORCI-THOMAS & M.R. ADAMS. 1994. Estrogen modulates the expression of tumor necrosis factor alpha mRNA in phorbol ester-stimulated human monocytic THP-1 cells. Lymphokine. Cytokine. Res. **13:** 377–382.
72. CHAUDHARY, L.R., T.C. SPELSBERG & B.L. RIGGS. 1992. Production of various cytokines by normal human osteoblast-like cells in response to interleukin-1β and tumor necrosis factor-α: lack of regulation by 17β-estradiol. Endocrinology **130:** 2528–2534.
73. ABRAHAMSEN, B., V. SHALHOUB, E.K. LARSON, *et al.* 2000. Cytokine RNA levels in transiliac bone biopsies from healthy early postmenopausal women. Bone **26:** 137–145.
74. ABRAHAMSEN, B., K. BENDTZEN & H. BECK-NIELSEN. 1997. Cytokines and T-lymphocyte subsets in healthy post-menopausal women: estrogen retards bone loss without affecting the release of IL-1 or IL-1ra. Bone **20:** 251–258.
75. GRCEVIC, D., S.K. LEE, A. MARUSIC & J.A. LORENZO. 2000. Depletion of CD4 and CD8 T lymphocytes in mice In vivo enhances 1, 25- dihydroxyvitamin D(3)-stimulated osteoclast-like cell formation In vitro by a mechanism that is dependent on prostaglandin synthesis. J. Immunol. **165:** 4231–4238.
76. KONG, Y.Y., U. FEIGE, I. SAROSI, *et al.* 1999. Activated T cells regulate bone loss and joint destruction in adjuvant arthritis through osteoprotegerin ligand. Nature **402:** 304–309.
77. TENG, Y.T., H. NGUYEN, X. GAO, *et al.* 2000. Functional human T-cell immunity and osteoprotegerin ligand control alveolar bone destruction in periodontal infection. J. Clin. Invest. **106:** R59–R67.
78. CENCI, S., G. TORALDO, M.N. WEITZMANN, *et al.* 2003. Estrogen deficiency induces bone loss by increasing T cell proliferation and lifespan through IFN-gamma-induced class II transactivator. Proc. Natl. Acad. Sci. USA **100:** 10405–10410.
79. TAICHMAN, R.S. & P.V. HAUSCHKA. 1992. Effects of interleukin-1 beta and tumor necrosis factor-alpha on osteoblastic expression of osteocalcin and mineralized extracellular matrix in vitro. Inflammation **16:** 587–601.
80. GILBERT, L., X. HE, P. FARMER, *et al.* 2000. Inhibition of osteoblast differentiation by tumor necrosis factor-alpha. Endocrinology **141:** 3956–3964.
81. GILBERT, L., X. HE, P. FARMER, *et al.* 2002. Expression of the osteoblast differentiation factor RUNX2 (Cbfa1/AML3/Pebp2alpha A) is inhibited by tumor necrosis factor-alpha. J. Biol. Chem. **277:** 2695–2701.
82. ROBEY, E.A., F. RAMSDELL, J.W. GORDON, *et al.* 1992. A self-reactive T cell population that is not subject to negative selection. Int. Immunol. **4:** 969–974.

83. TANCHOT, C. & B. ROCHA. 1997. Peripheral selection of T cell repertoires: the role of continuous thymus output. J. Exp. Med. **186:** 1099–1106.
84. GROSSMAN, Z. 1993. Cellular tolerance as a dynamic state of the adaptable lymphocyte. Immunol. Rev. **133:** 45–73.
85. AHMED, R. & D. GRAY. 1996. Immunological memory and protective immunity: understanding their relation. Science. **272:** 54–60.
86. RAMMENSEE, H.G., K. FALK & O. ROTZSCHKE. 1993. Peptides naturally presented by MHC class I molecules. Annu. Rev. Immunol. **11:** 213–244.
87. GROSSMAN, Z. & W.E. PAUL. 2000. Self-tolerance: context dependent tuning of T cell antigen recognition. Semin. Immunol. **12:** 197–203; discussion 257–344.
88. TANCHOT, C., F.A. LEMONNIER, B. PERARNAU, *et al.* 1997. Differential requirements for survival and proliferation of CD8 naive or memory T cells. Science **276:** 2057–2062.
89. BOSS, J.M. & P.E. JENSEN. 2003. Transcriptional regulation of the MHC class II antigen presentation pathway. Curr. Opin. Immunol. **15:** 105–111.
90. NAGARAJAN, U.M., A. BUSHEY & J.M. BOSS. 2002. Modulation of gene expression by the MHC class II transactivator. J. Immunol. **169:** 5078–5088.
91. YANG, N.N., M. VENUGOPALAN, S. HARDIKAR & A. GLASEBROOK. 1996. Identification of an estrogen response element activated by metabolites of 17β-estradiol and raloxifene. Science **273:** 1222–1225.
92. GORELIK, L. & R.A. FLAVELL. 2002. Transforming growth factor-beta in T-cell biology. Nat. Rev. Immunol. **2:** 46–53.
93. BRABLETZ, T., I. PFEUFFER, E. SCHORR, *et al.* 1993. Transforming growth factor beta and cyclosporin A inhibit the inducible activity of the interleukin-2 gene in T cells through a noncanonical octamer-binding site. Mol. Cell Biol. **13:** 1155–1162.
94. KEHRL, J.H., L.M. WAKEFIELD, A.B. ROBERTS, *et al.* 1986. Production of transforming growth factor beta by human T lymphocytes and its potential role in the regulation of T cell growth. J. Exp. Med. **163:** 1037–1050.
95. GORELIK, L., S. CONSTANT & R.A. FLAVELL. 2002. Mechanism of transforming growth factor beta-induced inhibition of T helper type 1 differentiation. J. Exp. Med. **195:** 1499–1505.
96. GAO, Y., W.P. QIAN, K. DARK, *et al.* 2004. Estrogen prevents bone loss through transforming growth factor beta signaling in T cells. Proc. Natl. Acad. Sci. USA **101:** 16618–16623.
97. PISKURICH, J.F., Y. WANG, M.W. LINHOFF, *et al.* 1998. Identification of distinct regions of 5′ flanking DNA that mediate constitutive, IFN-gamma, STAT1, and TGF-beta-regulated expression of the class II transactivator gene. J. Immunol. **160:** 233–240.
98. DELVIG, A.A., J.J. LEE, Z.M. CHRZANOWSKA-LIGHTOWLERS & J.H. ROBINSON. 2002. TGF-beta1 and IFN-gamma cross-regulate antigen presentation to CD4 T cells by macrophages. J. Leukoc. Biol. **72:** 163–166.
99. LEE, Y.J., Y. HAN, H.T. LU, *et al.* 1997. TGF-beta suppresses IFN-gamma induction of class II MHC gene expression by inhibiting class II transactivator messenger RNA expression. J. Immunol. **158:** 2065–2075.
100. DONG, Y., L. TANG, J.J. LETTERIO & E.N. BENVENISTE. 2001. The Smad3 protein is involved in TGF-beta inhibition of class II transactivator and class II MHC expression. J. Immunol. **167:** 311–319.
101. BENVENISTE, E.N., J. KWON, W.J. CHUNG, *et al.* 1994. Differential modulation of astrocyte cytokine gene expression by TGF-beta. J. Immunol. **153:** 5210–5221.

102. VAN ROON, J.A., K.A. GLAUDEMANS, J.W. BIJLSMA & F.P. LAFEBER. 2003. Interleukin 7 stimulates tumour necrosis factor alpha and Th1 cytokine production in joints of patients with rheumatoid arthritis. Ann. Rheum. Dis. **62:** 113–119.
103. HUANG, M., S. SHARMA, L.X. ZHU, *et al.* 2002. IL-7 inhibits fibroblast TGF-beta production and signaling in pulmonary fibrosis. J. Clin. Invest. **109:** 931–937.
104. DUBINETT, S.M., M. HUANG, S. DHANANI, *et al.* 1995. Down-regulation of murine fibrosarcoma transforming growth factor-beta 1 expression by interleukin 7. J. Natl. Cancer Inst. **87:** 593–597.
105. TANG, J., B.L. NUCCIE, I. RITTERMAN, *et al.* 1997. TGF-beta down-regulates stromal IL-7 secretion and inhibits proliferation of human B cell precursors. J. Immunol. **159:** 117–125.

Autoimmunity and Bone

EDWARD M. SCHWARZ, R. JOHN LOONEY, M. HICHAM DRISSI, REGIS J. O'KEEFE, BRENDAN F. BOYCE, LIANPING XING, AND CHRISTOPHER T. RITCHLIN

The Center for Musculoskeletal Research, University of Rochester Medical Center, Rochester, New York 14642, USA

ABSTRACT: Focal erosions of cartilage and bone, which occur in the joints of patients with autoimmune inflammatory arthritis (i.e., rheumatoid arthritis (RA) and psoriatic arthritis [PsA]), represent the most debilitating and irreversible components of the disease. Over the last decade, seminal breakthroughs in our understanding of the cells and signal transduction pathways central to this process have been elucidated. From this information an established paradigm has been developed to explain focal erosions in which osteoclasts responsible for erosions are derived from bone marrow-derived myeloid precursors. Using the tumor necrosis factor (TNF) transgenic mouse model of erosive arthritis and anti-TNF clinical trials with PsA patients, we have demonstrated that systemic TNF induces the migration of CD11b$^+$ osteoclast precursors (OCP) from the bone marrow into peripheral blood. These OCP can then enter the joints in blood vessels, translocate across the receptor activator of NF-κB ligand (RANKL) rich inflamed synovium, and differentiate into active osteoclasts. In direct contrast to this, systemic lupus erythematosus (SLE) patients appear to have an innate resistance to bone resorption. Our hypothesis to explain this phenomenon is that systemic interferon-α (IFN-α) diverts the bone marrow-derived myeloid precursors away from the osteoclast lineage and stimulates their differentiation into dendritic cells (DC). In support of this model, several labs have used microarray analyses to define the IFN-induced transcriptome in peripheral blood mononuclear cells (PBMC) from SLE patients. Here we propose the hypothesis that systemic TNF induces osteoclastic differentiation of PBMC in PsA patients that correlates with their erosive disease, and that the innate immune TNF/IFN axis in patients with autoimmune disease dictates their erosive phenotype. To demonstrate this, we injected wild-type C57B/6 and TNF-Tg mice with poly I:C, which is known to induce systemic IFN responses, and show its dominant effects on increasing the number of circulating CD11b$^+$/CD11c$^+$ precursor dendritic cells (pDC), concomitant with a dramatic reduction in CD11b$^+$/CD11c$^-$ OCP. Thus, systemic factors produced by autoimmunity have a dramatic impact on active myelopoiesis and bone homeostasis.

Address for correspondence: Edward M. Schwarz, Ph.D., Department of Orthopaedics, University of Rochester Medical Center, 601 Elmwood Avenue, Box 665, Rochester, NY 14642. Voice: 585-275-3063; fax: 585-756-4721.

e-mail: Edward_Schwarz@URMC.Rochester.Edu

Ann. N.Y. Acad. Sci. 1068: 275–283 (2006). © 2006 New York Academy of Sciences.
doi: 10.1196/annals.1346.031

Keywords: rheumatoid arthritis; psoriatic arthritis; systemic lupus erythematosus; tumor necrosis factor-α; interferon-α; osteoclast precursors; precursor dendritic cell

TNF IN AUTOIMMUNE DISEASE

It is now firmly established that the proinflammatory cytokine, tumor necrosis factor (TNF), is a central mediator of inflammation and tissue catabolism. Although this article focuses on autoimmunity and bone, no discussion on the subject would be complete without highlighting the remarkable success of anti-TNF therapy in the treatment of rheumatoid arthritis (RA), psoriatic arthritis (PsA), ankylosing spondylitis (AS), psoriasis (PS), and Crohn's disease (CD). For their seminal role in defining TNF at the apex of the proinflammatory cytokine cascade and their pioneering work in anti-TNF clinical trials, Drs. M. Feldmann and R.N. Maini received the 2003 Lasker Clinical Medical Research Award.[1] Beyond these specific achievements, the success of anti-TNF therapy via injection of a recombinant soluble receptor (etanercept, Enbrel,™ Wyeth Pharmaceuticals Inc., Philadelphia, PA) or antibody (infliximab, Remicade,™ Centocor, Inc., Horsham, PA and adalimumab, Humira,™ Abbott Laboratories, Abbott Park, IL) exemplifies the ultimate goal of biotechnology and dramatically underscores how complex autoimmune diseases can be effectively suppressed using a targeted therapeutic approach.

IMMUNE-MEDIATED INFLAMMATORY DISORDERS (IMIDS)

Perhaps the most remarkable feature of anti-TNF therapy is its efficacy in various autoimmune diseases, which have completely different pathologies. In an effort to simplify public awareness, the pharmaceutical industry devised a concept to explain a group of unrelated disease that share the common etiology of dysregulated TNF, and termed these conditions IMIDs.[2] We have attempted to elucidate the common mechanism through which TNF drives IMIDs, and how anti-TNF therapy ameliorates them and have pursued the central hypothesis that all autoimmune diseases lead to systemic TNF expression as a result of end organ pathology. This systemic TNF stimulates myelopoiesis and release of monocytes from the bone marrow into the blood. These peripheral blood mononuclear cells (PBMC) then home to sites of inflammation where they translocate into the parenchymal tissue and differentiate into activated macrophages, dendritic cells (DC), or osteoclasts (OC) depending on the local cytokine milieu to mediate the end organ pathology. Monocyte precursors stimulated with granulocyte macrophage colony-stimulating factor (GM-CSF) plus interleukin (IL)-4 differentiate into DC,[3] but can also form OCs if exposed to receptor activator of NF-κB ligand (RANKL) and monocyte colony-stimulating factor (M-CSF).[4] Treatment of monocytes with M-CSF alone produces tissue macrophages, and interferon-γ (IFN-γ) or IL-6 treatment shifts monocyte precursor differentiation to macrophages rather than DC and OC.[5]

THE EROSIVE VERSUS THE NONEROSIVE PHENOTYPE IN PsA AND SLE

Inflammatory arthritis is one of the most common clinical features in autoimmune disorders manifesting as pain, stiffness redness, and swelling. Radiologic evidence of joint damage (erosions and joint space narrowing) is observed in almost all patients with RA followed for more than 5 years.[6,7] Joint damage is also very common in PsA, an inflammatory joint disease that occurs in 10–15% of psoriasis patients.[8] Gladman and colleagues noted that two-thirds of PsA patients had bone erosions on initial presentation to a rheumatologist.[9] Radiographs often show extensive bone loss manifesting as eccentric erosions, frank tuft resorption, and pencil-in cup deformities.[10] Many PsA patients have aggressive synovitis confirmed histologically as marked synovial hyperplasia, extensive vascular proliferation with a tortuous morphology, and pannus tissue penetrating deep into cartilage and bone.[11] In addition, OCs are prominently situated at the bone–pannus junction and in bone marrow-derived deep resorption cavities traversing the bone matrix,[12] which is consistent with the bi-directional attack originally described by Wooley in severely affected RA joints.[13]

Despite the presence of joint deformities on physical examination, not all forms of inflammatory arthritis degrade cartilage and resorb bone. For example, systemic lupus erythematosus (SLE) is a systemic autoimmune disorder with diverse clinical manifestations, but musculoskeletal symptoms are usually the chief complaint, with 50% of patients reporting articular pain on presentation.[14,15] Lupus patients often develop rheumatoid-like deformities (ulnar deviation, tendinopathies, and subluxation) but only 4–6% of patients display erosive changes on plain radiographs.[16,17] Histopathologic analysis of SLE synovium shows mild-to-moderate synovial hyperplasia, microvascular changes, and perivascular inflammation with mononuclear cells.[18–20] Notably, aggressive synovial tissue-invading cartilage or bone is not described in these studies. Indeed, Bywaters noted the similarity of the SLE joint pattern to that reported by Jaccoud in rheumatic fever. In Jaccoud's arthropathy, joint deformities are manually reducible and radiographs do not depict bone or joint damage. In one large study of 939 SLE patients, Jaccoud's arthropathy was identified in 43% of the patients and was associated with a benign prognosis.[21]

TNF AND IFN ARE THE MASTER CYTOKINES OF INNATE IMMUNITY

Innate immunity is a rapidly mobilized, non-antigen-specific response to pathogens or injury that markedly changes the local tissue to defend against infection and damage, and is effectively terminated when the initiating stimulus is removed. The temporal and spatial demands of this system require that all cells contain a common germline-encoded signaling pathway that will enable them to uniformly respond immediately. Research over the last 30 years has identified two critical cytokine pathwaysthat are indispensable for mammalian innate immunity. The first of these is the TNF pathway, derived from the phylogenetically ancient Toll receptor family, whose evolutionary conservation extends from drosophila to man.[22] Essentially, all mammalian cells express TNF receptors and are capable of responding within seconds to appropriate ligands (TNF and lymphotoxin), primarily through NF-κB, JNK, and p38 signaling.[23] The other potent molecules regulating the innate immune response are the type-I INF, which were originally discovered as the intracellular immune defense against viral infection.[24] Similarly, the IFN pathway relies on a ubiquitously expressed receptor that signals through an immediate biochemical cascade known as the JAK-STAT pathway.[25] More recently it has been recognized that when these innate immune responses are chronically activated, the host becomes highly susceptible to a variety of IMIDs. As described above, overexpression of TNF leads to IMIDs. It has been discovered that dysregulation of IFN-α is a major component of SLE.[3] The most compelling data to support this come from microarray studies of PBMC from SLE patients, which demonstrated a remarkably consistent IFN-induced transcriptome.[26,27] Preclinical studies also support an important role for IFN signaling in the development of lupus nephritis as the susceptibility genes in lupus-prone mice have been mapped to IFN-inducible loci.[28–33] Banchereau and colleagues have proposed a unifying hypothesis to explain the presence and etiologic role of IFN in SLE.[34] In this model, an environmental insult (i.e., viral infection) stimulates plasmacytoid DC differentiation, which processes high levels of IFN. This IFN drives myeloid DC differentiation, which thus provides the antigen-presenting cells for autoantibody production. These autoantibodies then induce B-cell stimulation and plasmacytoid DC differentiation to complete the autoimmune vicious cycle.

Considering the role of IFN in DC differentiation and its critical role as a negative regulator of osteoclastogenesis to maintain bone homeostasis,[35] here we propose that the IFN axis controls the erosive versus nonerosive phenotypes of arthritis in SLE patients. In most of these patients, systemic IFN dominantly biases myelopoiesis toward DC and away from OC such that there are very few osteoclast precursors (OCP) in an inflamed joint that have the potential to form bone-resorbing OC. Here we provide the first experimental evidence for this model by stimulating systemic IFN in wild-type C57Bl/6 and TNF-transgenic (TNF-Tg) mice.

METHODS

Animal Studies and Collection of PBMC

Under an approved protocol by the University of Rochester Committee for Animal Resources, we induce systemic IFN in five 12-week-old female

C57Bl/6 (Jackson Labs, Bar Harbor, ME) or TNF-Tg mice,[36] via administration of acute poly (I:C) (Sigma, St. Louis, MO), as described previously.[37] Briefly, PBS placebo or 100 mg of polyI:C was injected into the mice via intraperitoneal (i.p.) injection once daily for 3 days. Afterward, the mice were euthanized, and pooled PBMC were collected via a ficoll gradient, as we described previously.[38]

$CD11b^+/CD11c^+$ pDC and $CD11b^+/CD11c^-$ OCP Frequency and TRAP Osteoclastogenesis Assays

OCP and precursor dendritic cells (pDC) frequency were determined simultaneously, immediately after monocyte purification of PBMC, by two-color FACS, as we described previously.[38] The murine-specific antibodies used were CD11b-PE (clone M1/70), CD11c-APC (clone HL3), (BD Biosciences, San Jose, CA). The percentage of $CD11b^+/CD11c^+$ pDC and $CD11b^+/CD11c^-$ OCP was determined by gating on the populations that contained a fluorescence signal greater than 99% of the isotype controls.

TRAP Staining and Quantitation of Osteoclasts

Tartrate-resistant acid phosphatase (TRAP) osteoclastogenesis was performed, as we described previously.[38] Briefly, PBMC from TNF-Tg mice and their wild-type littermates were cultured in α-modified essential medium (GIBCO BRL, Grand Island, NY) with 10% fetal calf serum (FCS, Hyclone Laboratories, Logan, UT), RANKL (100 ng/mL), and M-CSF (10 ng/mL) for 4 and 6 days, respectively. Cells were fixed and stained for TRAP using the Diagnostics Acid Phosphatase Kit (Sigma) to identify OCs. The OCs were quantified by manual tracing of the photographed OC area, as we described previously.[39]

RESULTS AND DISCUSSION

Systemic IFN Increases the Circulating pDC Frequency and Reduces OCP Frequency

In order to test our hypothesis that systemic IFN mediates a decrease in erosions by biasing myelopoiesis away from OC and toward DC differentiation, we used the well-established polyI:C injection model, which induces an IFN antiviral innate immune response. Wild-type C57B/6 mice and their TNF-Tg littermates were given X injections of poly I:C for 3 days and then PBMC were harvested for analysis by fluorescence-activated cell sorting (FACS) to

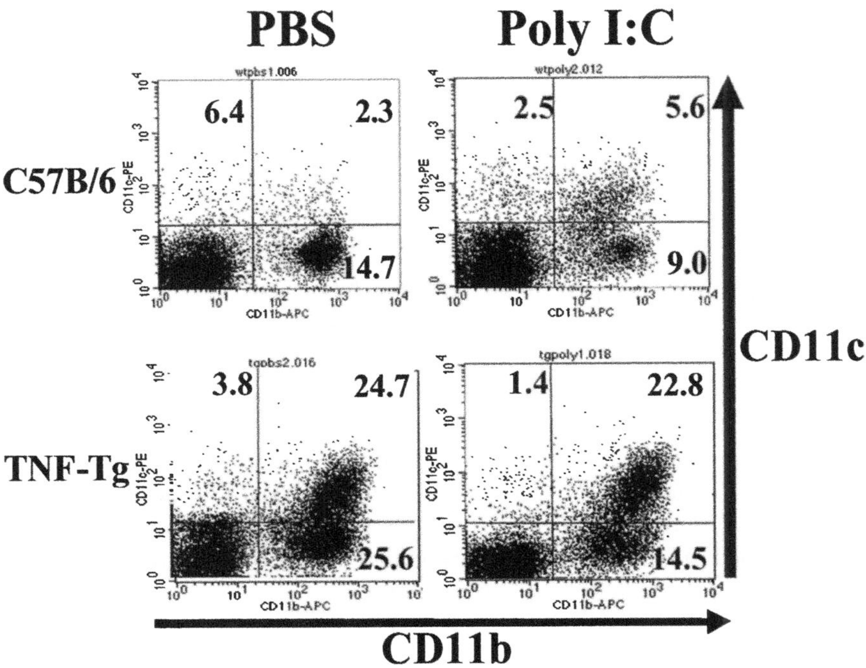

FIGURE 1. PolyI:C treatment alters pDC and OCP frequency in PBMC. Wild-type C57Bl/6 mice and their TNF-Tg littermates ($n = 5$) were treated with PBS placebo or polyI:C (100 mg i.p./day for 3 days) and PBMC from each strain were isolated and pooled together. The PMBC were labeled by antibodies against CD11b and CD11c, and analyzed by two-color FACS, as described in "Methods" section. The percentage of single and double positive cells in each quadrant is indicated.

determine the CD11b$^+$/CD11c$^+$ pDC and CD11b$^+$/CD11c$^-$ OCP frequency. Consistent with our previous reports,[38,40] FIGURE 1 shows that TNF-Tg mice with erosive arthritis have a significant increase in circulating CD11b$^+$ PBMC compared to their nontransgenic littermates. Here, we see that this is due to an increase in both CD11b$^+$/CD11c$^+$ pDC and CD11b$^+$/CD11c$^-$ OCP. Also evident are the effects of the polyI:C treatment, which resulted in a 2.4-fold increase in CD11b$^+$/CD11c$^+$ pDC in wild-type mice. This increase was associated with a 38.8% decrease in CD11b$^+$/CD11c$^-$ OCP in wild-type mice and a 43.4% decrease in TNF-Tg mice.

To confirm these findings, the PBMC were cultured in M-CSF and RANKL to induce osteoclastogenesis, which was quantified by TRAP staining. Consistent with our previous findings,[38,40] osteoclastogenesis in the TNF-Tg PBMC cultures peaked on day 4, while osteoclastogenesis in the wild-type PBMC cultures peaked on day 6. Quantification of the TRAP$^+$ OC in these cultures demonstrated that the polyI:C treatment significantly reduced the number of OCP in both strains (FIG. 2). This finding suggests that systemic IFN can bias

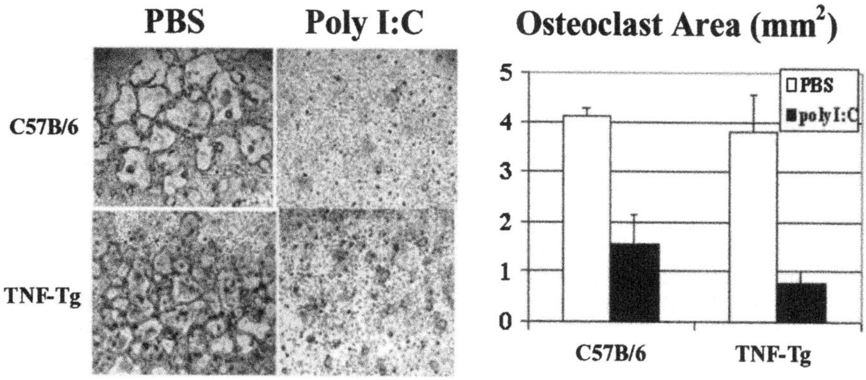

FIGURE 2. PolyI:C treatment dramatically reduces the osteoclastogenic potential of PBMC. The PBMC from C57B/6 and TNF-Tg mice treated with PBS placebo or polyI:C described in FIGURE 1 were cultured in M-CSF and RANKL until the peak in osteoclastogenesis (day 6 and day 4, respectively), and then were fixed and stained for TRAP activity. Representative microphotographs of the cultures ($n = 4$) are shown together with the quantified mean osteoclast area ± SEM. The osteoclast area in the cultures of PBMC from polyI:C-treated mice was significantly less than that of the PBS controls ($P < 0.05$).

myelopoiesis toward DC differentiation at the expense of OCP, even under conditions that favor osteoclastogenesis. Although future studies are needed to determine the precise stage of myelopoiesis that is affected by IFN and whether the IFN has a direct effect on the inhibition of erosive arthritis, these findings support our hypothesis that high IFN levels in SLE patients pushes myeloid differentiation away from osteoclastogenesis.

ACKNOWLEDGMENTS

The authors are supported by Public Health Service grants from NIH (ES011854, AR46545, AR478167, AR46545, AR48697, AR43510, and AR48149).

REFERENCES

1. FELDMANN, M. & R.N. MAINI. 2003. Lasker Clinical Medical Research Award. TNF defined as a therapeutic target for rheumatoid arthritis and other autoimmune diseases. Nat. Med. **9:** 1245–1250.
2. WILLIAMS, J.P. & J.A. MEYERS. 2002. Immune-mediated inflammatory disorders (I.M.I.D.s): the economic and clinical costs. Am. J. Manag. Care. **8:** S664.
3. PASCUAL, V., J. BANCHEREAU & A.K. PALUCKA. 2003. The central role of dendritic cells and interferon-alpha in SLE. Curr. Opin. Rheumatol. **15:** 548–556.

4. BOYLE, W.J., W.S. SIMONET & D.L. LACEY. 2003. Osteoclast differentiation and activation. Nature **423:** 337–342.
5. DELNESTE, Y., P. CHARBONNIER, N. HERBAULT, *et al.* 2003. Interferon-gamma switches monocyte differentiation from dendritic cells to macrophages. Blood **101:** 143–150.
6. KAARELA, K., R. LUUKKAINEN & S. KOSKIMIES. 1993. How often is seropositive rheumatoid arthritis an erosive disease? A 17-year followup study. J. Rheumatol. **20:** 1670–1673.
7. KAARELA, K. & S. SARNA. 1993. Correlations between clinical facets of outcome in rheumatoid arthritis. Clin. Exp. Rheumatol. **11:** 643–644.
8. MEASE, P.J. 2002. Tumour necrosis factor (TNF) in psoriatic arthritis: pathophysiology and treatment with TNF inhibitors. Ann. Rheum. Dis. **61:** 298–304.
9. GLADMAN, D.D., V.T. FAREWELL & C. NADEAU. 1995. Clinical indicators of progression in psoriatic arthritis: multivariate relative risk model. J. Rheumatol. **22:** 675–679.
10. RESNICK, D. & G. NIWAYAMA. 1989. Psoriatic arthritis. *In* Bone and Joint Imaging. RESNICK, Ed.: 320. WB Saunders. Philadelphia.
11. FEARON, U., K. GRIOSIOS, A. FRASER, *et al.* 2003. Angiopoietins, growth factors, and vascular morphology in early arthritis. J. Rheumatol. **30:** 260–268.
12. RITCHLIN, C.T., S.A. HAAS-SMITH, P. LI, *et al.* 2003. Mechanisms of TNF-alpha- and RANKL-mediated osteoclastogenesis and bone resorption in psoriatic arthritis. J. Clin. Invest. **111:** 821–831.
13. BROMLEY, M., H. BERTFIELD, J.M. EVANSON, *et al.* 1985. Bidirectional erosion of cartilage in the rheumatoid knee joint. Ann. Rheum. Dis. **44:** 676–681.
14. KAPOSI, M. 1872. New reports on knowledge of systemic lupus erythematosus. Arch. Dermat. u. Syph. **4:** 36–78.
15. STEVENS, M. 1983. The Clinical Management of Systemic Lupus Erythematosus. *In* P.H. SHUR, Ed.: 63–84. Grune & Stratton. New York.
16. ESDAILE, J.M., D. DANOFF, L. ROSENTHALL & A. GUTKOWSKI. 1981. Deforming arthritis in systemic lupus erythematosus. Ann. Rheum. Dis. **40:** 124–126.
17. GRIGOR, R., J. EDMONDS, R. LEWKONIA, *et al.* 1978. Systemic lupus erythematosus. A prospective analysis. Ann. Rheum. Dis. **37:** 121–128.
18. GOLDENBERG, D.L. & A.S. COHEN. 1978. Synovial membrane histopathology in the differential diagnosis of rheumatoid arthritis, gout, pseudogout, systemic lupus erythematosus, infectious arthritis and degenerative joint disease. Medicine (Baltimore) **57:** 239–252.
19. LABOWITZ, R. & H.R. SCHUMACHER, JR. 1971. Articular manifestations of systemic lupus erythematosus. Ann. Intern. Med. **74:** 911–921.
20. NATOUR, J., L.C. MONTEZZO, L.A. MOURA & E. ATRA. 1991. A study of synovial membrane of patients with systemic lupus erythematosus (SLE). Clin. Exp. Rheumatol. **9:** 221–225.
21. MOLINA, J.F., G. CITERA, D. ROSLER, *et al.* 1995. Coexistence of human immunodeficiency virus infection and systemic lupus erythematosus. J. Rheumatol. **22:** 347–350.
22. LOCKSLEY, R.M., N. KILLEEN & M.J. LENARDO. 2001. The TNF and TNF receptor superfamilies: integrating mammalian biology. Cell **104:** 487–501.
23. CHEN, G. & D.V. GOEDDEL. 2002. TNF-R1 signaling: a beautiful pathway. Science **296:** 1634–1635.
24. AREND, W.P. 2001. The innate immune system in rheumatoid arthritis. Arthritis. Rheum. **44:** 2224–2234.

25. AARONSON, D.S. & C.M. HORVATH. 2002. A road map for those who don't know JAK-STAT. Science **296:** 1653–1655.
26. BENNETT, L., A.K. PALUCKA, E. ARCE, *et al.* 2003. Interferon and granulopoiesis signatures in systemic lupus erythematosus blood. J. Exp. Med. **197:** 711–723.
27. BAECHLER, E.C., F.M. BATLIWALLA, G. KARYPIS, *et al.* 2003. Interferon-inducible gene expression signature in peripheral blood cells of patients with severe lupus. Proc. Natl. Acad. Sci. USA **100:** 2610–2615.
28. DRAKE, C.G., S.J. ROZZO, H.F. HIRSCHFELD, *et al.* 1995. Analysis of the New Zealand black contribution to lupus-like renal disease. Multiple genes that operate in a threshold manner. J. Immunol. **154:** 2441–2447.
29. KOTZIN, B.L. 1996. Systemic lupus erythematosus. Cell **85:** 303–306.
30. ROZZO, S.J., T.J. VYSE, C.G. DRAKE & B.L. KOTZIN. 1996. Effect of genetic background on the contribution of New Zealand black loci to autoimmune lupus nephritis. Proc. Natl. Acad. Sci. USA **93:** 15164–15168.
31. ROZZO, S.J., T.J. VYSE, K. MENZE, *et al.* 2000. Enhanced susceptibility to lupus contributed from the nonautoimmune C57BL/10, but not C57BL/6, genome. J. Immunol. **164:** 5515–5521.
32. VYSE, T.J., S.J. ROZZO, C.G. DRAKE, *et al.* 1997. Control of multiple autoantibodies linked with a lupus nephritis susceptibility locus in New Zealand black mice. J. Immunol. **158:** 5566–5574.
33. VYSE, T.J., S.J. ROZZO, C.G. DRAKE, *et al.* 1998. Contributions of Ea(z) and Eb(z) MHC genes to lupus susceptibility in New Zealand mice. J. Immunol. **160:** 2757–2766.
34. BANCHEREAU, J., V. PASCUAL & A.K. PALUCKA. 2004. Autoimmunity through cytokine-induced dendritic cell activation. Immunity **20:** 539–550.
35. TAKAYANAGI, H., S. KIM, K. MATSUO, *et al.* 2002. RANKL maintains bone homeostasis through c-Fos-dependent induction of interferon-beta. Nature **416:** 744–749.
36. KEFFER, J., L. PROBERT, H. CAZLARIS, *et al.* 1991. Transgenic mice expressing human tumour necrosis factor: a predictive genetic model of arthritis. Embo J. **10:** 4025–4031.
37. SANTIAGO-RABER, M.L., R. BACCALA, K.M. HARALDSSON, *et al.* 2003. Type-I interferon receptor deficiency reduces lupus-like disease in NZB mice. J. Exp. Med. **197:** 777–788.
38. LI, P., E.M. SCHWARZ, R.J. O'KEEFE, *et al.* 2004. Systemic tumor necrosis factor alpha mediates an increase in peripheral CD11bhigh osteoclast precursors in tumor necrosis factor alpha-transgenic mice. Arthritis. Rheum. **50:** 265–276.
39. HUANG, W., R.J. O'KEEFE & E.M. SCHWARZ. 2003. Exposure to receptor-activator of NFkappaB ligand renders pre-osteoclasts resistant to IFN-gamma by inducing terminal differentiation. Arthritis. Res. Ther. **5:** R49–R59.
40. LI, P., E.M. SCHWARZ, R.J. O'KEEFE, *et al.* 2004. RANK signaling is not required for TNFalpha-mediated increase in CD11(hi) osteoclast precursors but is essential for mature osteoclast formation in TNFalpha-mediated inflammatory arthritis. J. Bone. Miner. Res. **19:** 207–213.

Nonsteroid Immune Modulators and Bone Disease

RONALD TAMLER[a] AND SOLOMON EPSTEIN[a,b]

[a]*Mount Sinai School of Medicine, New York, New York 10029, USA*

[b]*Doylestown Hospital, Doylestown, Pennsylvania 18901, USA*

ABSTRACT: Glucocorticoids have been the main agents for preventing organ rejection, but unfortunately they possess serious side effects. Newer immunosuppressive agents have therefore been introduced to overcome these effects and have had a dramatic impact on reducing the incidence of organ rejection, enhancing donor organ acceptance, and hence patient survival posttransplantation. However, calcineurin inhibitors (CIs), such as cyclosporine and tacrolimus, also have serious effects causing rapid and severe bone loss in animal models and humans. The mechanism accounting for this action is unclear at present, but the role of T lymphocyte action *via* RANKL seems to be of essence in triggering bone loss. The mechanism is complex and *in vitro* studies often produce results that are opposite to those seen *in vivo*. In addition to acute, rapid, and severe bone loss (ARSBL), the clinical picture shows an extremely high incidence of fractures at all sites, and depends upon the organ transplanted, preexisting bone disease, interval before transplantation, and the dose and duration of multiple immunosuppressive drugs. Other immune-modifying drugs, such as azathioprine, mycophenolate mofetil, and sirolimus, which are used in conjunction with glucocorticoids and CIs have not been shown to promote bone loss experimentally or clinically. With the exception of glucocorticoids, all of the agents discussed here demand further investigation with regard to their effects on bone health in the clinical setting.

KEYWORDS: organ transplantation; bone loss; fracture; immunosuppressive agents; calcineurin inhibitors; cyclosporine; tacrolimus

IMMUNE MODULATOR DRUGS

Immune modulators are a relatively new class of drug that was developed after the identification of T and B lymphocytes and their subsets as mediators of immune function in organ transplantation and virtually every disease process including infection, cancer, and cardiovascular disease. Calcineurin inhibitors

Address for correspondence: Solomon Epstein, M.D., Doylestown Hospital, 595 W. State Street, Doylestown, PA 18901. Voice: 215-345-2867; fax: 215-345-2532.
e-mail: bonedocsol@aol.com

Ann. N.Y. Acad. Sci. 1068: 284–296 (2006). © 2006 New York Academy of Sciences.
doi: 10.1196/annals.1346.032

(CIs) and noncalcineurin immune modulators can either enhance or suppress the immune reaction depending on the type of modulation required to affect the disease outcome. In organ transplantation, the main requirement is to prevent donor organ rejection by the recipient's immune system. In autoimmune diseases, these drugs must inhibit the recognition of self-destructive autoantibodies. Immune modulators have had an enormous impact on prolonging the life span of these patient populations. However, this progress is not without its drawbacks, one of which is the deleterious effect of CIs on bone. This overview will focus on the effect of immunomodulatory drugs other than glucocorticoids on bone. These drugs include the CIs, cyclosporine and tacrolimus, and the non-CIs, rapamycin, mycophenolate mofetil, methotrexate, and azathioprine. While more immunosuppressive drugs, such as CTLA4Ig (Abatacept) and LEA29Y (Belatacept),[1] continue to evolve, data on the effects on bone are lacking or have not yet been studied.

THE CLINICAL SCENARIO

The impact of immune modulators, such as CIs, on humans is difficult to study. First, multidrug regimens frequently employed in the treated disease states tend to compound each other's effects. For instance, it is a common practice to concomitantly administer three drugs—one common combination consists of glucocorticoids (rapamycin can act as a substitute), cyclosporin or tacrolimus, and azathioprine (mycophenalate mofetil is frequently substituted for azathioprine). This makes it difficult, if not impossible, to assess the contribution of each individual drug to bone loss. Additionally, even if the glucocorticoid dose is decreased or discontinued, and the patient continues on two drugs, it still would be very unclear which drug is to be blamed for the initial *versus* ongoing bone loss. The initial rapid bone loss is mainly the result of a glucocorticoid effect[2] and occurs typically in the first 3 to 6 months.

For optimal data, the effects of one drug need to be compared against the other in a head-to-head study. However, immunosuppressive regimens and drug dosages differ from one center to another. Small studies directly implicate the CIs in bone loss following organ transplantation,[3,4] which confirms the experimental evidence. However, given the small number of transplant recipients at any given center, these studies need to be validated in large multicenter trials.

Finally, the role of secondary hyperparathyroidism is unclear. The etiology most commonly cited is impaired renal function posttransplantation.[5–7] While secondary hyperparathyroidism may also be correlated with glucocorticoid dosage, CIs, and bone loss,[7] no such relationship has been demonstrated in other studies,[8,9] and the role of increased parathyroid hormone (PTH) levels in glucocorticoid-induced bone loss has been called into question.[10]

CIs AND THEIR EFFECTS ON BONE

Cyclosporine (CsA) and tacrolimus (FK506) are inhibitors of calcineurin. In a first step essential for immunosuppression, both bind to intracellular proteins, either cyclophilins[11–13] or FK-binding proteins,[14,15] respectively. These proteins, called immunophilins, are peptidylprolyl-*cis*-trans isomerase enzymes.[16] In turn, the complexes inhibit the intracellular phosphatase enzyme calcineurin,[11,12,17] which prevents transcription of T lymphocyte cytokine genes and genes that control membrane molecules such as CD40 ligand.[18,19] Calcineurin, a serine/threonine phosphatase, is uniquely regulated by Ca^{2+} and calmodulin.[17] Calcineurin interacts with nuclear factors of activated T cells (NF-AT),[20] which constitute a family of transcription factors necessary for activation of genes involved in the inflammatory process and immune system. Thus, by inhibiting calcineurin,[21–25] CsA and FK506 prevent activation of NF-AT with the inhibition of cell growth and differentiation.

Both CsA and FK506 cause acute, rapid, and severe bone loss (ARSBL).[26,27] This was extensively demonstrated in the rat model where administration of immunosuppressive doses to normal, young or old, male or female oophorectomized rats not only produced significant loss of trabecular, but also cortical bone within weeks. Bone loss was dose dependent and reversible after discontinuing the drug.[28] Histomorphometry characterized the phenomenon as extremely high-turnover bone loss with increases in resorption and formation markers. Interestingly, serum 1,25-dihydroxyvitamin D in rats was increased to a higher level with CsA than with FK506 due to stimulation of the 1-hydroxylase enzyme in the kidney.[29] This increase in vitamin D metabolite was not seen in humans treated with CsA.

Certain features of this bone loss are peculiar. The phenomenon is seen only *in vivo* because *in vitro,* CsA inhibits bone resorption by isolated osteoclasts. Furthermore, the T lymphocyte is critical for bone loss,[30] as the occurrence cannot be reproduced in T lymphocyte-deficient nude rats. Thus, the challenge is to establish an experimental model *in vitro* and in gene-modified mice to study the CIs that promote bone loss so dramatically and significantly *in vivo.*

Rat studies revealed that osteoblasts contain mRNA and protein for all isoforms of calcineurin A and B.[31] TAT-assisted transduction of a fusion protein, TAT-calcineurin Aα, into osteoblasts resulted in the enhanced expression of the osteoblast differentiation markers *Runx*-2, alkaline phosphatase, bone sialoprotein (BSP), and osteocalcin. This was associated with a dramatic enhancement of bone formation in intact calvarial cultures. Calcineurin Aα-/- mice displayed severe osteoporosis, markedly reduced mineral apposition rates, and attenuated colony formation in 10-day *ex vivo* stromal cell cultures (CFU-Fs). The latter was associated with significant reductions in *Runx2,* BSP, and osteocalcin expression paralleled by similar decreases in response to FK506. Together, the gain- and loss-of-function experiments indicate that calcineurin regulates bone formation through an effect on osteoblast differentiation.[31]

The Ca^{2+}/calmodulin-sensitive phosphatase calcineurin was found to be a necessary downstream mediator for osteoclast differentiation. [32] Using quantitative polymerase chain reaction (qPCR), we detected all calcineurin isoforms, namely Aα, Aβ, Aγ (catalytic), as well as B1 and B2 (regulatory) in osteoclast precursor RAW-C3 cells. We found that while the expression of these isoforms remained relatively unchanged during osteoclast differentiation, the expression of their primary substrate for calcineurin, NFATc1, increased profoundly. For gain-of-function studies, we incubated osteoclast precursors for 10 min with a calcineurin fusion protein, TAT-calcineurin Aα—this resulted in its receptor-less influx into >90% of the precursor cells. After that, the expression of the osteoclast differentiation markers tartrate-resistant acid phosphatase (TRAP) and integrin β_3 increased markedly. In addition, the expression of NFATc1, as well as the alternative substrate for calcineurin, IκBα, was enhanced significantly. Similarly, transfection with constitutively active NFAT ($NFAT^{ca}$) resulted in an increased expression of both TRAP and β_3. In parallel loss-of-function studies, transfection with dominant-negative NFAT ($NFAT^{dn}$) not only inhibited osteoclast formation, but also reversed the induction of NFATc1, TRAP, and β_3 by TAT-calcineurin Aα. The expression of these markers was also inhibited by calcineurin Aα U1 snRNA. Consistent with these observations, we also noticed a reduction in osteoclastogenesis in calcineurin Aα-/- cells, as well as in osteoclast precursors treated with the calcineurin inhibitors, CsA and FK506.

Together the gain- and loss-of-function experiments establish that calcineurin Aα is necessary for osteoclast formation from its precursor, and that this occurs *via* an NFATc1-dependent mechanism. Thus these studies confirm the critical role of calcineurin in osteoblast and osteoclast differentiation, which may help understand their clinical roles.

Cyclosporine

The first study that linked CsA to bone abnormalities was published in 1988.[33] In this study, histology in renal transplant patients revealed a picture that was unexpected and showed high bone turnover, unlike the results seen with glucocorticoid administration or secondary hyperparathyroidism.[8] In fact, the results resembled those seen in experimental studies in rats. A clinical study in heart transplant recipients also attributed the bone loss seen after cardiac transplantation to CsA; again, the biochemical findings resembled a high turnover osteoporosis.[3] These studies, as mentioned earlier, have all been confounded by triple immunosuppression, and the role of any single drug as the culprit cannot be ascertained. Studies demonstrated that bone mineral density (BMD) in the lumbar spine did not decrease more with CsA monotherapy than with other CsA-containing regimens after 12 months; at times, it even increased after 18 months.[34–37]

The most convincing study was a comparison of CsA monotherapy with a non-CsA-containing regimen of prednisone and azathioprine to isolate the effects of CsA alone.[8] This study used BMD as well as bone histomorphometry. The results showed that both regimens decreased bone at the distal radius and less significantly at the lumbar spine with no differences in the degree of bone loss. Histopathological analysis showed increased osteoclast numbers and a surprising decrease in osteoblast numbers, mineral apposition, and bone formation rates, again with no differences between the groups. Neither biochemical nor histomorphological findings supported a role for PTH, although this factor has been considered by some investigators to be pivotal for the bone loss after transplantation. The surprising decrease in bone formation in this study may be related to the time that had elapsed after transplantation, as the subjects were seen 140 $\pm$ 75 months after the procedure.

FK506

The investigation of the effects of FK506 faces challenges similar to those found with CsA: Although animal studies demonstrated bone loss with a high-turnover remodeling state that was similar in nature and even worse than seen with CsA,[38] the role of FK506 has also been hard to separate from other immunosuppressants used to prevent rejection. Studies attempting to resolve the role of FK506 were performed with patients receiving low-dose prednisone and FK506 compared to normal dose steroids and CsA.[9] It was shown that when the cumulative dose of steroids was adjusted for both groups, bone loss in patients receiving FK506 equaled that seen in the group receiving CsA and low-dose prednisone. No relationship to PTH levels was established.

A prospective, longitudinal, randomized, double-blind study was undertaken over 2 years to assess the effects of FK506, glucocorticoids, azathioprine, or mycophenalate mofetil after heart transplant with or without 1,25(OH)2D.[39] Results showed that overall BMD decreased in both groups after transplantation compared to normal age-matched subjects. While BMD increased in the lumbar spine by 14% in the group receiving vitamin D analog, a slight rise in the placebo group made the differences between the two insignificant. BMD was maintained in the femoral neck region when treated with calcitriol, but the placebo group lost bone significantly. Again, no between-group significance could be shown. The study concluded that FK506 is associated with rapid bone loss comparable to CsA, which can be modified by low-dose calcitriol therapy over 2 years. A more convincing study would be to compare FK506 to CsA in comparable posttransplant patients not receiving glucocorticoids to see if a difference between the two CIs does exist.

Other immune-modifying drugs such as rapamycin (sirolimus) are also used to prevent organ rejection after transplantation. Rapamycin is a macrolide and product of *Streptomyces hygroscopicus,* which does not inhibit the production of interleukins resulting from antigen T cell activation but does inhibit the cellular proliferation stimulated by growth factor signal transduction in response to alloantigens.[40,41] Rapamycin binds to the intracellular immunophilin FKBP12 to form a complex. However, this product does not target calcineurin, but mTor (mammalian target of rapamycin), which inhibits translation of mRNAs that encode for cell-cycle regulators and T cell proliferation.[39] *In vitro* studies have shown that in bone marrow stromal cells, rapamycin, like CsA and FK506, decreases osteoprotegerin (OPG) mRNA and protein levels and increases receptor activator of NF-κB (RANKL), which potentially can induce bone loss.[42] Rapamycin may also act via transforming growth factor (TGF)-β to enhance osteoclastogenesis by inducing monocyte–macrophage cell differentiation into osteoclasts.[43] Conversely, rapamycin does not cause bone loss in rats but may interfere with longitudinal bone growth and, at high doses, may decrease cortical bone in young, rapidly growing animals.[42] An effect on gonadal function has also been described with rapamycin.[42] In human subjects, no studies on the effect of rapamycin therapy on bone without concomitant administration of glucocorticoids or CIs have been reported, and they may not be feasible in a population of transplant recipients. However, the bone loss seen with CIs may be mitigated by combining rapamycin with low-dose cyclosporin. This strategy has been shown effective in rats and does not compromise immune suppression.[44] Everolimus, a derivative of rapamycin, has a similar mechanism of action, but its effects on bone have not been studied.[45]

Mycophenolate Mofetil

Mycophenolate mofetil is another addition to the armamentarium of immune-modifying drugs. Together with CsA or FK506 and glucocorticoids, mycophenolate mofetil is now increasingly replacing azathioprine to prevent organ rejection.

Mycophenolate mofetil is converted *in vivo* to mycophenolic acid, a natural product of *penicillium fungi* that selectively inhibits the proliferation of T and B cells as well as arterial wall smooth-muscle cells. Most other tissues are resistant to the action of mycophenolate mofetil because of alternate nucleotide synthesis. Mycophenolate mofetil also causes less bone marrow suppression than azathioprine. No evidence of alteration in bone metabolism and no loss of bone volume have been noted experimentally *in vivo*.[46] However, recent reports suggest that prednisone and mycophenolate mofetil, in the absence of CsA, may also be associated with a highturnover bone loss on bone histomorphometry.[47] This demonstrates the need for further study.

Azathioprine

Azathioprine has long been an essential part of therapy in combination with CsA and glucocorticoids to prevent organ rejection. It is a purine antagonist and therefore inhibits rapidly proliferating cells, which include T and B lymphocytes, as well as hematopoietic cells in the bone marrow.[48] In rats receiving azathioprine, albeit for a short period, no effect on bone except on bone formation markers was seen.[48] How this would translate with long-term use into clinically relevant outcomes is at present unknown.

CHEMOTHERAPEUTIC AGENTS

The effects of chemotherapeutic agents on bone have largely been neglected, because they are commonly used in combination with other toxic agents and patients receiving these drugs often have diseases that affect bone loss irrespective of the drugs.

In children with acute lymphoblastic leukemia, radiological lucencies, lytic and sclerotic lesions, periosteal elevations, osteopenia, and fractures have been described.[49–51] In a study of 40 children treated with a "cocktail" of prednisone, vincristine, l-asparaginase, methotrexate, and 6-mercaptopurine, 39% of the patients developed fractures during treatment over 24 months.[52] However, it is hard to determine each drug's contribution to these results. Both methotrexate and adriamycin have been reported to cause low-turnover bone loss in animals and humans[53] after a relatively brief administration of 14 days. In children treated for leukemia with methotrexate, an increased incidence of fracture has been reported.[52] Despite its widespread use, no long-term effects of methotrexate on bone have been described until recently when it was shown that rats given two 5-day courses of methotrexate, displayed depressed cancellous and longitudinal bone growth at 80 and 170 days after treatment. In addition, they displayed decreased osteoblast activity, bone volume, and formation as well as increased osteoclast activity.[54] This demonstrates the adverse long-term effect of methotrexate long after cessation of treatment. Despite these remarkable findings, there was no difference on biomechanical testing between treated animals and controls. Thus, the cause of increased fracture incidence in these patients is still unclear.

T CELL SUBSET-SPECIFIC ANTIGEN RECEPTOR BLOCKADE ANTIBODIES

Various antibodies against lymphocytes have been used as adjunct or rescue therapy in patients with acute rejection or as prophylaxis against rejection.

Muromonab-CD3 (OKT3) is a monoclonal antibody against the CD3 component of the T cell antigen receptor. Recently, the CD40:CD40 ligand pathway has been recognized as having a major role in T cell activation pathways.[8,55] The CD40 ligand is expressed on activated CD4 T cells. Stimulation of CD40 is also important in providing signals for antibody production by B cells, which underscores the cross reactivity between T and B cells. A recombinant fusion protein consisting of the extracellular domain of CTLA-4 (a molecule that binds ligands on activated antigen-presenting cells) linked to the constant region of immunoglobulin (Ig)G1 has been shown to be useful in preventing an immune response to transplanted pancreatic islets and has prolonged the survival of cardiac allografts in animals. Given that CsA inhibits CD40 ligand expression in T lymphocytes from transplant populations, it is important to study a role of these T cell subset antigen-blocking antibodies in bone metabolism. At present, no studies have been known to deal with this matter.

Malononitrilamides

Another class of compounds under development are the malononitrilamides, which represent a low-molecular-weight immunosuppressive agent. Malononitrilamides are derivatives of leflunomide. These compounds can block both B and T cell proliferation and suppress IgG and IgM antibody production.[56] The effect on bone is yet unknown and may need to be studied. The real potential of these agents, if indeed they are found to be "safe" for bone or bone sparing, is that their use may allow decreased doses or even elimination of immunosuppressants known to induce bone loss while not compromising the organ transplant.

HYPOGONADISM

Chemotherapeutic Agents Causing Hypogonadism

Although it is stretching the boundaries of this article, another important cause of bone loss needs to be mentioned. These are agents, which although not classified as immune modulators, do have an impact on the immune precursor cells present in the bone marrow. The use of these bone marrow suppressants to treat malignant diseases either of the bone marrow itself or resulting from metastatic spread of the primary tumor, results in bone loss and sometimes an increased incidence of fracture (e.g., in breast and prostate cancer chemotherapy with adjunct therapy).[57–59] Bone loss in these patients may well be the result of sex steroid deficiency owing to either chemical or surgical ablation. The results are similar to what is generally seen with acute loss of sex steroid production (e.g., after acute ovarian sex steroid production inhibition). However, data from long-term trials examining the consequences of these drugs on bone loss and fracture incidence in a randomized controlled manner are lacking, and the effect of the co-variants (e.g., immobilization, polypharmacy, poor nutrition, etc.) needs to be taken into account.

Hormonal Antineoplastic Therapy

Aromatase inhibitors are a class of drugs, which are now frequently used in the treatment of breast cancer and produce a state of estrogen deficiency.

A recent study demonstrated that anastrozole not only significantly prolonged disease-free survival and time to recurrence, but also reduced distant metastases and occurrence of contralateral breast cancer.[60] While fewer side effects were reported than with tamoxifen, an increased risk of fractures was noted. Anastrozole is now recommended as the preferred initial treatment for postmenopausal women with localized estrogen-receptor positive breast cancer.

Men receiving antineoplastic therapy targeting sex hormones should also be concerned about bone health. Fracture risk with androgen-deprivation treatment (ADT) for prostate cancer was studied using records of 50,613 men ≥66 years of age.[61] Excluding events that occurred in the first 12 months after diagnosis, fractures were recorded in 19.4% of those receiving ADT compared to 12.6% for those not receiving such treatment. Patients were further stratified into how many doses of GnRH agonist therapy they received. Those who received ≥9 doses and those who had an orchiectomy had a drastically higher fracture risk than those receiving fewer doses of ADT or no treatment. With an annual incidence of prostate cancer of 220,000, and at least 40% of these men receiving ADT, approximately 3,000 excess fractures are expected to occur yearly as a result of androgen-deprivation treatment.

CONCLUSION

The nonsteroidal immunosuppressants of the CI family have been shown experimentally and clinically to produce severe and rapid high-turnover bone loss. These drugs are often used with glucocorticoids, which are known to compound that effect. However, while the glucocorticoids lead to low-turnover bone loss, the bone biopsy histomorphometry in patients taking CIs reveals a high turnover state. The result of combining these classes of drugs is rapid and severe bone loss with a very high rate of fracture. Immunosuppressants such as sirolimus, azathioprine, and mycophenolate mofetil have as of yet not been clearly demonstrated to adversely affect the bony skeleton. Even fewer data exist for other, newer immune modulators. Separating the contribution of an individual drug as the culprit for producing unfavorable skeletal effects remains a formidable challenge given confounding variables such as disease states and multidrug regimens commonly found in the population receiving transplants.

The clinical picture of bone loss after transplantation and its treatment has been extensively reviewed elsewhere.[62–65]

REFERENCES

1. INGELFINGER, J.R., R.S. SHWARTZ. 2005. Immunosuppression—the promise of specificity. N. Engl. J. Med. **353:** 836–839.

2. MIKULS, T.R., B.A. JULIAN, A. BARTOLUCCI, *et al.* 2003. Bone mineral density changes within six months of renal transplantation. Transplantation **75:** 49–54.
3. THIEBAUD, D., M.A. KRIEG, D. GILLARD-BERGUER, *et al.* 1996. Cyclosporine induces high bone turnover and may contribute to bone loss after heart transplantation. Eur. J. Clin. Invest. **26:** 549–555.
4. RICH, G.M., G.H. MUDGE, G.L. LAFFEL, *et al.* 1992. Cyclosporine A and prednisone-associated osteoporosis in heart transplant recipients. J. Heart Lung Transplant **11:** 950–958.
5. HEAF, J., E. TVEDEGAARD, I.L. KANSTRUP, *et al.* 2000. Clinical bone loss after renal transplantation: role of hyperparathyroidism, acidosis, cyclosporine and systemic disease. Transplantation **14:** 457–463.
6. CAYCO, A.V., J. WYSOLMERSKI, C. SIMPSON, *et al.* 2000. Posttransplant bone disease: evidence for a high bone resorption state. Transplantation **70:** 1722–1728.
7. GUO, C.Y., A. JOHNSON, T.J. LOCKE, *et al.* 1998. Mechanisms of bone loss after cardiac transplantation. Bone **22:** 267–271.
8. CUETO-MANZANO, A.M., S. KONEL & V. CROWLEY. 2003. Bone histopathology and densitometry comparison between cyclosporine a monotherapy and prednisolone plus azathioprine dual immunosuppression in renal transplant patients. Transplantation **75:** 2053–2058.
9. GOFFIN, E., J.P. DEVOGELAER, G. DEPRESSEUX, *et al.* 2003. Evaluation of bone mineral density after renal transplantation under a tacrolimus-based immunosuppression: a pilot study. Clin. Nephrol. **59:** 190–195.
10. RUBIN, M.R. & J.P. BILEZIKIAN. 2002. Clinical review 151: the role of parathyroid hormone in the pathogenesis of glucocorticoid-induced osteoporosis: a reexamination of the evidence. J. Clin. Endocrinol. Metab. **87:** 4033–4041.
11. WALSH, C.T., L.D. ZYDOWSKY & F.D. MCKEON. 1992. Cyclosporine A, the cyclophilin class of peptidylprolyl isomerases, and blockade of T cell signal transduction. J. Biol. Chem. **267:** 13115–13118.
12. HANDSCHUMACHER, R.E., M.W. HARDING, J. RICE, *et al.* 1984. Cyclophilin: a specific cytosolic binding protein for cyclosporin A. Science **226:** 544–547.
13. CARDENAS, M.E., R.S. MUIR, T. BREUDER, *et al.* 1995. Targets of immunophilin–immunosuppressant complexes are distinct highly conserved regions of calcineurin A. EMBO J. **14:** 2772–2783.
14. DOLINSKI, K., S. MUIR, M. CARDENAS, *et al.* 1997. All cyclophilins and FK506 binding proteins are, individually and collectively, dispensable for viability in Saccharomyces cerevisiae. Proc. Natl. Acad. Sci. USA **94:** 13093–13098.
15. LIU, J., M.W. ALBERS, T.J. WANDLESS, *et al.* 1992. Inhibition of T cell signaling by immunophilin–ligand complexes correlates with loss of calcineurin phosphatase activity. Biochemistry **31:** 3896–3901.
16. GOTHEL, S.F. & M.A. MARAHIEL. 1999. Peptidyl-prolyl cis-trans isomerases, a superfamily of ubiquitous folding catalysts. Cell. Mol. Life Sci. **55:** 423–436.
17. SIEKIERKA, J.J. & N.H. SIGAL. 1992. FK-506 and cyclosporin A: immunosuppressive mechanism of action and beyond. Curr. Opin. Immunol. **4:** 548–552.
18. FULEIHAN, R., N. RAMESH, A. HORNER, *et al.* 1994. Cyclosporin A inhibits CD40 ligand expression in T lymphocytes. J. Clin. Invest. **93:** 1315–1320.
19. SHAW, K.T., A.M. HO, A. RAGHAVAN, *et al.* 1995. Immunosuppressive drugs prevent a rapid dephosphorylation of transcription factor NFAT1 in stimulated immune cells. Proc. Natl. Acad. Sci. USA **92:** 11205–11209.
20. MAGES, H.W., R. BAAG, B. STEINER, *et al.* 1998. Utilization of an NF-ATp binding promoter element for EGR3 expression in T cells but not fibroblasts provides

a molecular model for the lymphoid cell specific effect of cyclosporin A. Mol. Cell. Biol. **18:** 7157–7165.
21. ZHANG, B.W., G. ZIMMER, J. CHEN, *et al.* 1996. T cell responses in calcineurin A alpha-deficient mice. J. Exp. Med. **183:** 413–420.
22. HODGE, M.R., A.M. RANGER, F. CHARLES DE LA BROUSSE, *et al.* 1996. Hyperproliferation and dysregulation of IL-4 expression in NF-ATp-deficient mice. Immunity **4:** 397–405.
23. XANTHOUDAKIS, S., J.P. VIOLA, K.T. SHAW, *et al.* 1996. An enhanced immune response in mice lacking the transcription factor NFAT1. Science **272:** 892–895.
24. YOSHIDA, H., H. NISHINA, H. TAKIMOTO, *et al.* 1998. The transcription factor NF-ATc1 regulates lymphocyte proliferation and Th2 cytokine production. Immunity **8:** 115–124.
25. RANGER, A.M., M.R. HODGE, E.M. GRAVALLESE, *et al.* 1998. Delayed lymphoid repopulation with defects in IL-4-driven responses produced by inactivation of NF-ATc. Immunity **8:** 125–134.
26. MOVSOWITZ, C., S. EPSTEIN, M. FALLON, *et al.* 1988. Cyclosporin-A in vivo produces severe osteopenia in the rat: effect of dose and duration of administration. Endocrinology **123:** 2571–2577.
27. MOVSOWITZ, C., S. EPSTEIN, F. ISMAIL, *et al.* 1989. Cyclosporin A in the oophorectomized rat: unexpected severe bone resorption. J. Bone Miner. Res. **4:** 393–398.
28. SCHLOSBERG, M., C. MOVSOWITZ, S. EPSTEIN, *et al.* 1989. The effect of cyclosporin A administration and its withdrawal on bone mineral metabolism in the rat. Endocrinology **124:** 2179–2184.
29. STEIN, B., B.P. HALLORAN, T. REINHARDT, *et al.* 1991. Cyclosporin-A increases synthesis of 1,25-dihydroxyvitamin D3 in the rat and mouse. Endocrinology **128:** 1369–1373.
30. BUCHINSKY, F.J., Y. MA, G.N. MANN, *et al.* 1996. T lymphocytes play a critical role in the development of cyclosporin A induced osteopenia. Endocrinology **137:** 2278–2285.
31. SUN, L., H.C. BLAIR, Y. PENG, *et al.* 2005. Calcineurin regulates bone formation by the osteoblast. Proc. Natl. Acad. Sci. USA **47:** 17130–17135.
32. SUN, L., Y. PENG, N. ZAIDI, *et al.* 2005. Evidence that calcineurin is required for the genesis of bone-resorbing osteoclasts. Am. J. Physiol. Submitted.
33. AUBIA, J., J. MASRAMON, S. SERRANO, *et al.* 1988. Bone histology in renal transplant patients receiving cyclosporin [letter]. Lancet **1**(8593): 1048–1049.
34. PONTICELLI, C. & A. AROLDI. 2001. Osteoporosis after organ transplantation. Lancet **357:** 1623.
35. GROTZ, W., A. MUNDINGER, B. GUGEL, *et al.* 1994. Missing impact of cyclosporine on osteoporosis in renal transplant recipients. Transplant. Proc. **26:** 2652–2653.
36. MCINTYRE, H.D., B. MENZIES, R. RIGBY, *et al.* 1995. Long-term bone loss after renal transplantation: comparison of immunosuppressive regimens. Clin. Transplant. **9:** 20–24.
37. MORENO, A., J.V. TORREGROSA, F. PONS, *et al.* 1999. Bone mineral density after renal transplantation: long-term follow-up. Transplant. Proc. **31:** 2322–2323.
38. CVETKOVIC, M., G.N. MANN, D.F. ROMERO, *et al.* 1994. The deleterious effects of long term cyclosporin A, cyclosporin G and FK506 on bone mineral metabolism in vivo. Transplantation **57:** 1231–1237.
39. STEMPFLE, H.U., C. WERNER, U. SIEBERT, *et al.* 2002. The role of tacrolimus (FK506)-based immunosuppression on bone mineral density and bone turnover

after cardiac transplantation: a prospective, longitudinal, randomized, double-blind trial with calcitriol. Transplantation **73:** 547–552.
40. KUO, C.J., J. CHUNG, D.F. FIORENTINO, *et al.* 1992. Rapamycin selectively inhibits interleukin-2 activation of p70 S6 kinase. Nature **358:** 70–73.
41. HOFBAUER, L.C., C. SHUI, B.L. RIGGS, *et al.* 2001. Effects of immunosuppressants on receptor activator of NF-kappaB ligand and osteoprotegerin production by human osteoblastic and coronary artery smooth muscle cells. Biochem. Biophys. Res. Commun. **280:** 334–339.
42. ROMERO, D.F., F.J. BUCHINSKY, B. RUCINSKI, *et al.* 1995. Rapamycin: a bone sparing immunosuppressant? J. Bone Miner. Res. **10:** 760–768.
43. SHUI, C., B.L. RIGGS & S. KHOSLA. 2002. The immunosuppressant rapamycin, alone or with transforming growth factorbeta, enhances osteoclast differentiation of RAW264.7 monocyte-macrophage cells in the presence of RANK-ligand. Calcif. Tissue Int. **71:** 437–446.
44. GOODMAN, G.R., I.R. DISSANAYAKE, B.R. SODAM, *et al.* 2001. Immunosuppressant use without bone loss—implications for bone loss after transplantation. J. Bone Miner. Res. **16:** 72–78.
45. EISEN, H.J., E.M. TUZCU, R. DORENT, *et al.* 2003. RAD B253 Study Group. Everolimus for the prevention of allograft rejection and vasculopathy in cardiac-transplant recipients. N. Engl. J. Med. **349:** 847–858.
46. DISSANAYAKE, I.R., G.R. GOODMAN, A.R. BOWMAN, *et al.* 1998. Mycophenolate mofetil: a promising new immunosuppressant that does not cause bone loss in the rat. Transplantation **65:** 275–278.
47. HAMDY, N.A.T., M.J.K. MALLAT, N. BRAVENBOER, *et al.* 2003. Factors determining the prevalence of osteoporosis and fractures a year after successful kidney transplantation. Paper presented at the 25th annual meeting of ASBMR September 19–23, Minneapolis, MN, S174.
48. BRYER, H.P., J.A. ISSEROW, E.C. ARMSTRONG, *et al.* 1995. Azathioprine alone is bone sparing and does not alter cyclosporin A-induced osteopenia in the rat. J. Bone Miner. Res. **10:** 132–138.
49. NESBIT, M., W. KRIVIT, R. HEYN, *et al.* 1976. Acute and chronic effects of methotrexate on hepatic, pulmonary, and skeletal systems. Cancer **37:** 1048–1057.
50. SCHWARTZ, A.M. & J.C. LEONIDAS. 1984. Methotrexate osteopathy. Skeletal Radiol. **11:** 13–16.
51. RAGAB, A.H., R.S. FRECH & T.J. VIETTI. 1970. Osteoporotic fractures secondary to methotrexate therapy of acute leukemia in remission. Cancer **25:** 580–585.
52. ATKINSON, S.A., L. FRAHER, C.M. GUNDBERG, *et al.* 1989. Mineral homeostasis and bone mass in children treated for acute lymphoblastic leukemia. J. Pediatr. **114:** 793–800.
53. FRIEDLAENDER, G.E., R.B. TROSS, A.C. DOGANIS, *et al.* 1984. Effects of chemotherapeutic agents on bone. I. Short-term methotrexate and doxorubicin (adriamycin) treatment in a rat model. J. Bone Joint. Surg. **66:** 602–607.
54. WHEELER, D.L., R.A. VANDER GRIEND, T.J. WRONSKI, *et al.* 1995. The short- and long-term effects of methotrexate on the rat skeleton. Bone **16:** 215–221.
55. WEITZMANN, M., S. CENCI, L. RIFAS, *et al.* 2001. T cell activation induces human osteoclast formation via receptor activator of nuclear factor B ligand-dependent and -independent mechanisms. J. Bone Miner. Res. **16:** 328–337.
56. BILOLO, K.K., J. OUYANG, X. WANG, *et al.* 2003. Synergistic effects of malononitrilamides (FK778, FK779) with tacrolimus (FK506) in prevention of acute heart

and kidney allograft rejection and reversal of ongoing heart allograft rejection in the rat. Transplantation **75:** 1881–1887.
57. HADJI, P., C. MASKOW, M. GOTTSCHALK, *et al.* 2003. Influence of chemotherapy (AC) on bone mineral density (BMD) and bone ultrasonometry (QUS) in women with breast cancer. Paper presented at the 25th annual meeting of ASBMR September 19–23, Minneapolis, MN, S198.
58. EASTELL, R. 2003. Effect of anastrozole on bone mineral density: 2-year results of the "Arimidex" (anastrozole), tamoxifen, alone or in combination (ATAC) trial. 25th Annual Meeting ASBMR September 19–23, Minneapolis, MN, S312.
59. CHEN, Z., M. MARICIC, T.L. BASSFORD, *et al.* 2003. Increased fracture risk among breast cancer survivors—results from the women's health initiative presentation. Paper presented at the 25th annual meeting of ASBMR September 19–23, Minneapolis, MN, S22.
60. HOWELL, A., J. CUZICK, M. BAUM, *et al.* 2005. Results of the ATAC (arimidex, tamoxifen, alone or in combination) trial after completion of 5 years' adjuvant treatment for breast cancer. Lancet **365:** 60–62.
61. SHAHINIAN, V.B., K. YONG-FANG, J.L. FREEMAN, *et al.* 2005. Risk of fracture after androgen deprivation for prostate cancer. N. Engl. J. Med. **352:** 154–164.
62. EPSTEIN, S. 1996. Post-transplantation bone disease: the role of immunosuppressive agents and the skeleton. J. Bone Miner. Res. **11:** 1–7.
63. EPSTEIN, S. & A. INZERILLO. 2004. Nonsteroid immunosuppressive agents and bone. Clin. Rev. Bone Miner. Metab. **2:** 91–102.
64. SHANE, E. & S. EPSTEIN. 2001. Transplantation osteoporosis. Transplant. Rev. **15:** 11–32.
65. COHEN, A. & E. SHANE. 2003. Osteoporosis after solid organ and bone marrow transplantation. Osteoporos Int. **14:** 617–630.

Pathogenesis of Osteopenia/Osteoporosis Induced by Highly Active Anti-Retroviral Therapy for AIDS

GEORGE PAN,[a] ZHEN YANG,[a] SCOTT W. BALLINGER,[a] AND JAY M. MCDONALD[a,b]

[a]*Department of Pathology, The University of Alabama at Birmingham, Birmingham, Alabama 35294, USA*

[b]*Veterans Administration Medical Center, Birmingham, Alabama 35233, USA*

ABSTRACT: The advent of highly active anti-retroviral therapy (HAART) has dramatically decreased the rate of AIDS-related mortality and significantly extended the life span of patients with AIDS. A variety of metabolic side effects are associated with these therapies, one of which is metabolic bone disease. A higher prevalence of osteopenia and osteoporosis in HIV-infected patients receiving anti-retroviral therapy than in patients not on therapy has now been reported in several studies. Several factors have been demonstrated to influence HIV-associated decreases in bone mineral density (BMD), including administration of nucleoside reverse transcriptase inhibitors (NRTIs). In this article, discussion will focus on the molecular pathogenesis and treatment of HAART-associated osteopenia and osteoporosis.

KEYWORDS: HIV; osteopenia/osteoporosis; HAART; NRTI; osteoclastogenesis; T cell cytokines; mitochondrial toxicity; RANK/RANKL/OPG

INTRODUCTION

Bone remodeling depends on the tightly integrated activity of two major cell types, osteoblasts, which make new bone (bone formation), and osteoclasts, which destroy old bone (bone resorption). Therefore, the balance between the number and activity of osteoclasts and osteoblasts is crucial in determining bone mass, which is directly related to bone fragility and fracture risk. Two important molecules: macrophage colony-stimulating factor (M-CSF) and receptor for activation of nuclear factor-kappa B ligand (RANKL) produced from osteoblasts/stromal cells regulate the differentiation, function, and survival of

Address for correspondence: Jay M. McDonald, M.D., Department of Pathology, The University of Alabama at Birmingham, 701 19 Street S., LHR 504 Birmingham, AL 35294. Voice: 205-934-6666; fax: 205-975-9927.
e-mail: mcdonald@uab.edu

Ann. N.Y. Acad. Sci. 1068: 297–308 (2006). © 2006 New York Academy of Sciences.
doi: 10.1196/annals.1346.057

osteoclasts.[1] Two transcription factors, Runx2 and Osterix, have been reported to regulate osteoblast differentiation.

HIV Infection and Bone Disease

Bone metabolism in HIV-infected individuals has been studied since the late 1980s. Before the widespread use of highly active anti-retroviral therapy (HAART), studies indicated that bone mineral metabolism was only minimally affected in HIV-infected patients. Serrano *et al.* assessed histomorphometry in HIV-positive patients and found that many parameters of histomorphometry were significantly lower in patients than in controls.[2] Paton *et al.* reported that 45 HIV-infected patients had marginally lower bone mineral density (BMD) at the lumbar spine. None of the patients had reduced BMD to levels associated with a diagnosis of osteoporosis.[3] Recently, Amiel *et al.* analyzed BMD in 48 HIV-infected treatment-naïve patients, 49 HIV-infected patients on protein inhibitors, 51 HIV-infected patients on no-protein inhibitors, and 81 HIV-uninfected control subjects. The results showed a significant decrease of BMD in all HIV-infected patients compared to the control subjects.[4]

The increasing frequency of osteopenia and osteoporosis being reported in HIV-infected patients receiving anti-retroviral therapy suggests an independent role of anti-retroviral drugs in altering bone metabolism.

HAART and Bone Diseases

HAART is a complex therapeutic regimen consisting of up to 15 anti-retroviral agents. In general, HAART includes two major therapeutic regimens: a protease inhibitor (PI)-based regimen and a nucleoside reverse transcriptase inhibitor (NRTI)-based regimen. The PI-based regimen uses one or two PIs combined with two NRTIs, whereas the NRTI-based regimen uses two NRTIs combined with one non-nucleoside reverse transcriptase inhibitor (NNRTI). With more effective therapies as a result of HAART, the prevalence of HAART-associated bone diseases has increased.

A high incidence of osteopenia/osteoporosis has been associated with both PI and NRTI use. Tebas *et al.* measured BMD in AIDS patients receiving a PI and found that 50% of the patients had osteopenia and 21% had osteoporosis.[5] This incidence is significantly increased compared to patients without therapy or normal controls. Moore *et al.* confirmed that 71% of HIV-infected patients on PI therapy have reduced BMD.[6] Jain *et al.* compared the effect of various PIs on bone resorption and found that some PIs, but not all, increase bone resorption.[7] Similarly, Carr *et al.* reported that 3% of 44 HIV-infected patients receiving NRTIs developed osteoporosis and 22% developed osteopenia.[8] Tsekes *et al.* determined BMD and whole body fat by dual-energy X ray absorbance (DEXA) of HIV-infected patients on Zidovudine (AZT) and other NRTIs and found significant decreases in both body fat and BMD.[9]

It has also been noted that HIV-infected patients have an increased risk for osteonecrosis of the hip.[10] Keruly *et al.* reported 15 cases of avascular hip necrosis in HIV-infected patients and suggested that the incidence of osteonecrosis in HIV-infected patients was higher than the general HIV-negative population.[11] Allison *et al.* summarized the epidemiology, etiology, and clinical management of osteonecrosis.[10] It is not known whether osteonecrosis of the hip is attributable to HIV infection itself, HAART, or other HIV-associated complications. Belmonte *et al.* suggested a possible association between the presence of anti-phospholipid antibodies and osteonecrosis.[12]

MOLECULAR PATHOGENESIS OF HAART-ASSOCIATED BONE LOSS

At the cellular level, the regulation of osteoclast differentiation by PIs has been investigated *in vitro* by measuring the expression of RANK/RANKL/OPG. The results showed that some PIs inhibit osteogenesis and expression of osteoprotegerin (OPG) leading to increased osteoclastogenesis and bone resorption.[7] Wang *et al.* reported that the PI, Indinavir, attenuates the function and recruitment of osteoblasts increasing bone loss, whereas another PI, Ritonavir, blocks osteoclastogenesis decreasing bone loss.[13] To study the pathogenesis of NRTI-associated bone mineral loss, we examined the contributions of NRTIs on osteoclastogenesis, T cell cytokines, and mitochondrial toxicity.

Increased Osteoclastogenesis

We reported the osteoclastogenic effects of AZT *in vitro* and *in vivo.*[14] AZT enhances osteoclastogenesis in RAW264.7 cells and primary mouse osteoclast precursors in the presence of RANKL. This osteoclastogenic effect of AZT is associated with increased activities of the tartrate-resistant acid phosphatase (TRAP) promotor and the NF-κB transcription factor.[14] The osteoclastogenic effect of AZT was concentration dependent (FIG. 1A), and the expression of TRAP and calcitonin receptor (CTR), but not RANK, were markedly increased by AZT plus RANKL (FIG. 1B). Note that in the experiment shown in FIGURE 1A, the incubation time is only 3 days, which explains the low value in the presence of RANKL alone. Maximum stimulation of osteoclastogenesis by RANKL alone requires 5–6 days.[14]

We recently extended our studies to include two other NRTI compounds, didanosine (ddI) and lamivudine (3TC), alone and in combination. In the absence of RANKL, these NRTIs failed to stimulate the differentiation of RAW264.7 cells into TRAP-positive osteoclasts, whereas in the presence of RANKL treatment both ddI and 3TC, like AZT, significantly increased

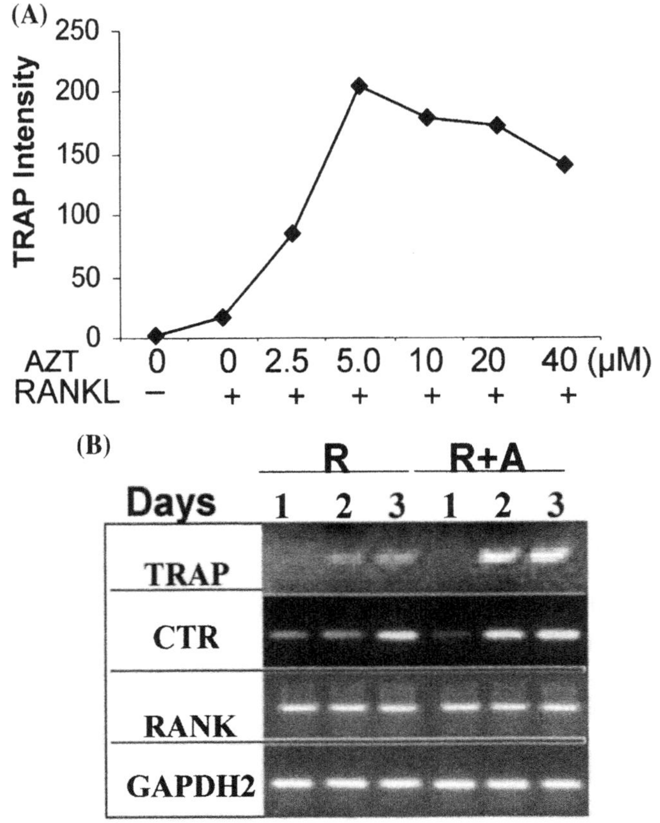

FIGURE 1. The osteoclastogenic effect of AZT in RAW264.7 cells (**A**) RAW264.7 cells were treated with AZT at indicated concentrations in the presence of RANKL (50 ng/mL) for 3 days. Osteoclasts were measured by TRAP staining assay. The Y-axis represents the TRAP intensity per well. (**B**) RAW264.7 cells were treated with RANKL (50 ng/mL, R) or RANKL plus AZT (25 μM, R + A) for 1–3 days as indicated and total mRNAs were isolated. RT-PCR for tartrate-resistant acid phosphatase (TRAP), calcitonin receptor (CTR), RANK, and GAPDH2 genes was performed.

TRAP-positive osteoclasts. Furthermore, there were neither additive nor synergistic effects on osteoclastogenesis when cells are treated with various combinations of NRTIs.[15]

Although bone marrow is usually used as a source of osteoclast precursors, functional osteoclasts can be differentiated *in vitro* from human peripheral blood mononuclear cells (PBMCs). Susa *et al.* described an *in vitro* protocol for the production of functional osteoclasts in the presence of RANKL and M-CSF using both fresh and frozen PBMCs.[16] We therefore determined the osteoclastogenic effect of AZT in PBMCs.[15] Here we compared the effect of HIV infection on osteoclastogenesis efficacy of AZT using PBMCs from

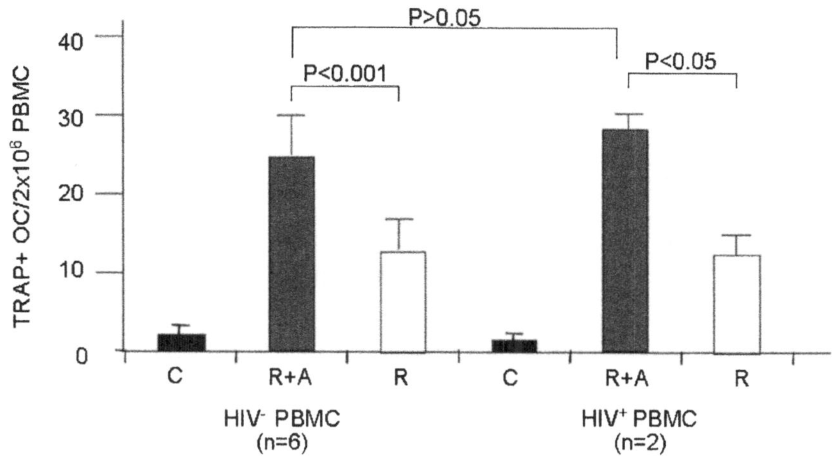

FIGURE 2. The osteoclastogenic effect of AZT in PBMCs. PBMCs (2×10^6 PBMCs per well) from two HIV-positive, HAART-naïve patients and six age-, sex-matched HIV-negative individuals were cultured with RANKL alone (R), or RANKL plus AZT (R+A) in the presence of M-CSF for 2 weeks. Treatment with M-CSF alone was a control (C). The number of multinucleated TRAP positive osteoclasts were counted. The results of the Student's *t*-test are indicated.

HIV-negative and HIV-positive HAART-naïve patients. The results show that AZT plus RANKL profoundly increased osteoclastogenesis in HIV-positive HAART-naïve PBMCs and a similar osteoclastogenic efficacy of AZT in HIV-negative PBMCs (FIG. 2). These results support the conclusion that HIV infection does not itself accelerate osteoclastogenesis or the effect of AZT *in vitro*.

The effects of AZT on bone remodeling were also demonstrated *in vivo*.[14] Compared to control mice, BMD in mice treated with AZT was significantly reduced by 6.15 ± 2.02% ($P < 0.05$, FIG. 3A). Histomorphometric analysis was performed on spine sections. Compared to control mice, both the number (OcN, $P < 0.05$) and surface (OcS, $P < 0.001$) of osteoclasts in AZT-treated mice were markedly increased (FIG. 3B). The number (ObN) and surface (ObS) of osteoblasts were not significantly changed in the NRTI-treated animals. These results indicate that decreased BMD in AZT-treated mice is mediated by increased osteoclastogenesis resulting in increased bone resorption.

T Cell Cytokines

There is a close relationship between the function of T cells and bone remodeling.[17,18] The immune system affects normal bone growth and restructuring. In many inflammatory conditions, including autoimmune diseases, allergy, infection, and injury, bone metabolism is disrupted. Various inflammatory cytokines released from T cells or macrophages, such as

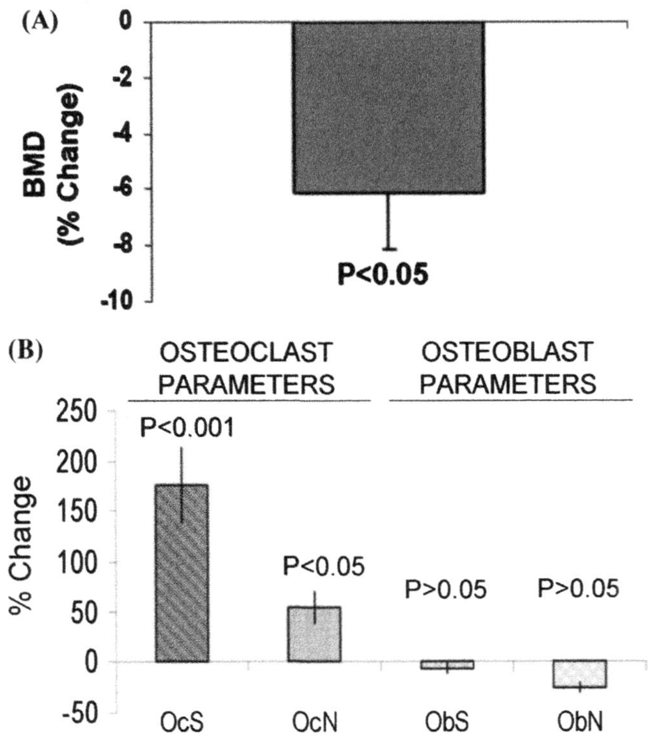

FIGURE 3. Effects of AZT on BMD and bone histomorphometry in mice. Mice, 5 months old, were orally fed PBS ($n = 6$, control) or AZT ($n = 5$, AZT 100 mg/kg/day) for 30 days. (**A**) Bone mineral density (BMD) was measured by DEXA. (**B**) Spines were fixed, decalcificated, and stained with trichrome. Histomorphometry was analyzed in three vertebrae of each animal. Osteoclast parameters included number (OcN) and surface (OcS) of osteoclasts. Osteoblast parameters included number (ObN) and surface (ObS) of osteoblasts. All parameters were compared to the control values, and reported as percentage change induced by AZT. The results were mean ± SE. *P* values are compared to controls.

interferon-α (IFN-α), IFN-β, and tumor necrosis factor-α (TNF-α), various interleukins, CD40L, granulocyte-macrophage colony-stimulating factor (GM-CSF), and macrophage inflammation protein-1 (MIP-1), have been reported to modulate osteoclast formation and function.[17,18] Moreover, activated T cells release RANKL that can trigger osteoclastogenesis and bone loss.[17,19,20] In addition, a crucial counter-regulatory mechanism has been discovered to inhibit the RANKL-induced maturation and activation of osteoclasts. Activated T cells secrete IFN-γ that blocks osteoclastogenesis *in vitro*.[17,21,22] In HIV infection, an overexpression of RANKL has been documented.[23,24] Elevated RANKL alone or in combination with TNF-α in HIV infection augments HIV replication forming a positive feedback loop.[24] In addition, various cytokines triggering osteoclastogenesis have been reported to be changed in HIV infection and in HAART-treated AIDS patients.[25,26] However, our *in vitro* studies using RAW264.7 cells indicate that TNF-α neither promoted osteoclastogenesis when used alone nor enhanced RANKL-induced osteoclastogenesis.[15]

Mitochondrial Toxicity

Mitochondria perform numerous important cell functions, including energy production, redox signaling, calcium storage, and the regulation of cell growth and death. In this regard, it has been shown that mitochondrial damage and dysfunction is involved in the pathogenesis of numerous metabolic degenerative disorders, such as diabetes, lipodystrophy, and heart diseases. Lactic acidosis is a common feature among certain forms of mitochondrial disease, which has also been linked to the observed decrease in BMD associated with NRTI therapy. Carr *et al.* suggested that these effects are caused by nucleoside analogue effects on osteoblast mitochondria.[8] However, other reports failed to show a significant correlation between NRTI-associated mitochondrial toxicity and osteopenia/osteoporosis.[27]

Mitochondria have their own DNA (mtDNA), which is distinct from the nuclear DNA (nDNA). The mtDNA is replicated by a γ-polymerase, which has a relatively high error rate compared to the α-DNA polymerase which is important for nDNA replication. Whereas the success of NRTIs is based upon their ability to interfere with HIV reverse transcriptase and thus inhibit viral replication,[28] NRTIs also inhibit γ-polymerase and mtDNA synthesis. The mtDNA depletion in mitochondria results in mitochondrial damage and dysfunction. Interestingly, RANKL appears to prevent mtDNA damage associated with NRTI treatment. FIGURE 4A shows that AZT induces mtDNA damage while RANKL treatment completely reverses mtDNA damage associated with AZT treatment.

Mitochondrial superoxide dismutase (SOD) is an important cellular antioxidant that converts the superoxide to H_2O_2.[29] Mitochondrial aconitase is widely recognized as a citric acid cycle enzyme that is highly sensitive to oxidative inactivation by superoxide radicals, and consequently, its activity has been used as a measure of oxidant load associated with superoxide levels. Because of their importance in cellular energy and oxidant regulation, mitochondrial SOD and aconitase are important markers of mitochondrial function and oxidant stress. Consequently, we quantified mitochondrial SOD protein levels and aconitase activity in RAW264.7 cells treated with AZT for 3 days with AZT, RANKL, or AZT + RANKL via Western blot and spectrophotometric assays, respectively. Aconitase activity in RAW264.7 cells treated with AZT for 3 days was decreased (FIG. 4B) to 41% of the activity in the control suggesting that AZT treatment resulted in inactivation of aconitase. In contrast, aconitase activity

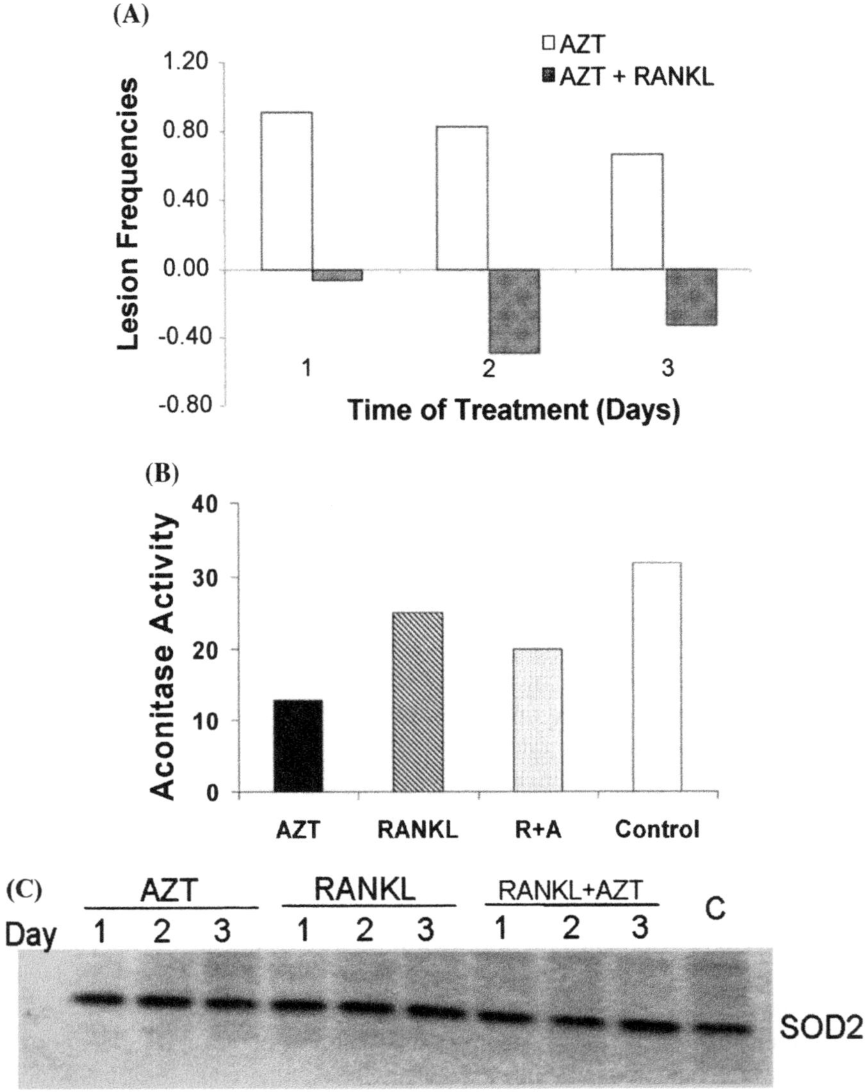

FIGURE 4. Mitochondrial toxicity of AZT. (**A**) RAW264.7 cells (2×10^6) were plated in 100-mm plates and treated with AZT (25 μM), RANKL (100 ng/mL), and AZT plus RANKL for 1–3 days. Cells were washed with cold PBS twice and then cell pellets were collected and stored in –80°C. Mitochondrial DNA damage was determined by qPCR. The frequency of mtDNA damage induced by AZT and AZT+RANKL was expressed relative to the frequency of mtDNA damage in RANKL-treated cells. (**B**) RAW264.7 cells (2×10^6) were treated as indicated for 3 days. Aconitase activity of cell lysates from RAW264.7 cells treated as indicated [AZT, RANKL, RANKL + AZT (R+A)] for 3 days, and (**C**) Western blot for mitochondrial SOD in RAW264.7 cells treated as indicated [AZT, RANKL, RANKL + AZT no treatment (c)] for 1–3 days.

was ameliorated by addition of RANKL, to approximately 78% of the control. Western blot analysis of mitochondrial SOD showed that protein levels were not significantly altered (FIG. 4C). Overall, these data suggest that AZT induces mtDNA damage and increased oxidant stress (aconitase inactivation) that is prevented in the presence of RANKL. Because it has been reported that RANKL expression is elevated by HIV infection,[23,24] it is conceivable that NRTI therapy on a background of increased RANKL levels minimizes the deleterious effects of NRTIs on osteoclast mitochondria, thereby promoting osteopenia/osteoporosis.

CLINICAL TREATMENT OF HAART-ASSOCIATED BONE LOSS

Currently, the management of osteopenia/osteoporosis in HIV-infected patients is focused on reducing the risks for bone loss and includes intake of calcium and vitamin D, weight-bearing exercise, recombinant growth hormone, or other anabolic agents.[30,31] Bisphosphonates are also widely used as a principal drug.[32,33]

It has been reported that anabolic steroids increase spinal bone density after treatment but no effect on the rate of fractures has been documented.[34] However, a study of the steroid, oxandrolone, had no effect on BMD among a small group of HIV-infected individuals.[35] On the other hand, bisphosphonates, such as alendronate (ALN) have been used to treat or prevent osteopenia/osteoporosis in AIDS patients on HAART.[32,33] Inhibition of osteoclast activity and increase of apoptosis of osteoclasts by bisphosphonates are considered the key mechanisms responsible for their efficacy in treating osteopenia and osteoporosis.[36] Their efficacy in treating HIV-1-infected patients is consistent with the increased osteoclastogenesis induced by AZT in our studies.

New therapeutic agents for treating bone disease include OPG and RANK-Fc fusion proteins. The OPG fusion protein is being evaluated for treatment of various metabolic bone and joint disorders caused by elevation of activated osteoclasts and for diminishing advanced bone pain induced by cancer metastasis.[37] RANK-Fc is a recombinant soluble form of RANK, in which the extracellular domain is expressed as a fusion protein with human IgG-Fc. Studies *in vitro* have shown that RANK-Fc functions as an antagonist to RANK-mediated signaling and acts to sequester endogenously produced RANKL.[38] The potent antiresorptive effects of RANK-Fc *in vivo* highlight the potential utility of disrupting RANK signaling as a novel therapeutic approach in humoral hypercalcemia of malignancy and possibly multiple myeloma and skeletal metastases associated with osteolysis.[39] An anti-osteoclastogenic effect of RANK-Fc in HIV-infected cells *in vitro* has been reported implying that RANK-Fc may provide a therapeutic option for HIV-associated osteopenia and osteoporosis.[23] However, no clinical trial treating HIV-associated osteopenia/osteoporosis has been reported.

CONCLUSION

Although AZT and other NRTIs have been beneficial in slowing disease progression of AIDS, osteopenia and osteoporosis are important complications affecting these patients. With the number of AIDS patients receiving HAART increasing, the incidence of osteopenia/osteoporosis is expected to increase. However, there are many controversial issues regarding the pathogenesis and mechanisms responsible for bone loss in HIV-infected patients on HAART. Increased osteoclastogenesis induced by AZT is likely to play a critical role in the pathogenesis of osteopenia and osteoporosis in these patients. Because the osteoclastogenic effect of AZT requires the presence of RANKL, interrupting RANKL by OPG, RANK-Fc, or other compounds may be potent therapeutic options, possibly better than bisphosphonates alone. Carefully controlled clinical prospective trails and additional basic science studies are still required to determine the incidence, etiology, risk factors, and most appropriate treatment for the bone complications in HIV-infected patients.

REFERENCES

1. Teitelbaum, S.L. 2000. Bone resorption by osteoclasts. Science **289:** 1504–1508.
2. Serrano, S. *et al.* 1995. Bone remodelling in human immunodeficiency virus-1-infected patients. A histomorphometric study. Bone **16:** 185–191.
3. Paton, N.I. *et al.* 1997. Bone mineral density in patients with human immunodeficiency virus infection. Calcif. Tissue Int. **61:** 30–32.
4. Amiel, C. *et al.* 2004. BMD is reduced in HIV-infected men irrespective of treatment. J. Bone Miner. Res. **19:** 402–409.
5. Tebas, P. *et al.* 2000. Accelerated bone mineral loss in HIV-infected patients receiving potent anti-retroviral therapy. AIDS **14:** F63–F67.
6. Moore, A.L. *et al.* 2001. Reduced bone mineral density in HIV-positive individuals. AIDS **15:** 1731–1733.
7. Jain, R.G. & J.M. Lenhard. 2002. Select HIV protease inhibitors alter bone and fat metabolism ex vivo. J. Biol. Chem. **277:** 19247–19250.
8. Carr, A. *et al.* 2001. Osteopenia in HIV-infected men: association with asymptomatic lactic acidemia and lower weight pre-anti-retroviral therapy. AIDS **15:** 703–709.
9. Tsekes, G. *et al.* 2002. Body composition changes in protease inhibitor-naive HIV-infected patients treated with two nucleoside reverse transcriptase inhibitors. HIV Med. **3:** 85–90.
10. Allison, G.T. *et al.* 2003. Osteonecrosis in HIV disease: epidemiology, etiologies, and clinical management. AIDS **17:** 1–9.
11. Keruly, J.C. *et al.* 2001. Increasing incidence of avascular necrosis of the hip in HIV-infected patients. J. Acquir. Immune. Defic. Syndr. **28:** 101–102.
12. Belmonte, M.A. *et al.* 1993. Avascular necrosis of bone in human immunodeficiency virus infection and antiphospholipid antibodies. J. Rheumatol. **20:** 1425–1428.

13. WANG, M.W. *et al.* 2004. The HIV protease inhibitor ritonavir blocks osteoclastogenesis and function by impairing RANKL-induced signaling. J. Clin. Invest. **114:** 206–213.
14. PAN, G. *et al.* 2004. AZT enhances osteoclastogenesis and bone loss. AIDS Res. Hum. Retroviruses **20:** 608–620.
15. PAN, G. *et al.* 2006. Modulation of osteoclastogenesis induced by nucleoside reverse transcriptase inhibitors. AIDS Res. Hum. Retroviruses (Submitted). **2(6):** 1–13.
16. SUSA, M. *et al.* 2004. Human primary osteoclasts: in vitro generation and applications as pharmacological and clinical assay. J. Transl. Med. **2(6):** 1–13.
17. TAKAYANAGI, H. *et al.* 2000. T-cell-mediated regulation of osteoclastogenesis by signalling cross-talk between RANKL and IFN-gamma. Nature **408:** 600–605.
18. WEI, S. *et al.* 2002. Interleukin-4 reversibly inhibits osteoclastogenesis via inhibition of NF-kappa B and mitogen-activated protein kinase signaling. J. Biol. Chem. **277:** 6622–6630.
19. WEI, S. *et al.* 2005. IL-1 mediates TNF-induced osteoclastogenesis. J. Clin. Invest. **115:** 282–290.
20. WONG, T.Y. & M.K. SO. 2001. Primary human immunodeficiency virus infection: heightened awareness needed. Hong Kong Med. J. **7:** 205–208.
21. MANOURY-SCHWARTZ, B. *et al.* 1997. High susceptibility to collagen-induced arthritis in mice lacking IFN-gamma receptors. J. Immunol. **158:** 5501–5506.
22. VERMEIRE, K. *et al.* 1997. Accelerated collagen-induced arthritis in IFN-gamma receptor-deficient mice. J. Immunol. **158:** 5507–5513.
23. FAKRUDDIN, J.M. & J. LAURENCE. 2003. HIV envelope gp120-mediated regulation of osteoclastogenesis via receptor activator of nuclear factor kappa B ligand (RANKL) secretion and its modulation by certain HIV protease inhibitors through interferon-gamma/RANKL cross-talk. J. Biol. Chem. **278:** 48251–48258.
24. FAKRUDDIN, J.M. & J. LAURENCE. 2004. Interactions among human immunodeficiency virus (HIV)-1, interferon-gamma and receptor of activated NF-kappa B ligand (RANKL): implications for HIV pathogenesis. Clin. Exp. Immunol. **137:** 538–545.
25. AUKRUST, P. *et al.* 1999. Decreased bone formative and enhanced resorptive markers in human immunodeficiency virus infection: indication of normalization of the bone-remodeling process during highly active anti-retroviral therapy. J. Clin. Endocrinol. Metab. **84:** 145–150.
26. BISWAS, D.K. *et al.* 1998. Inhibition of HIV-1 replication by combination of a novel inhibitor of TNF-alpha with AZT. J. Acquir. Immune Defic. Syndr. Hum. Retrovirol. **18:** 426–434.
27. COSSARIZZA, A. & G. MOYLE. 2004. Anti-retroviral nucleoside and nucleotide analogues and mitochondria. AIDS **18:** 137–151.
28. PAPADOPULOS-ELEOPULOS, E. *et al.* 1999. A critical analysis of the pharmacology of AZT and its use in AIDS. Curr. Med. Res. Opin. **15:** S1–S45.
29. BENOV, L. & I. FRIDOVICH. 1998. Growth in iron-enriched medium partially compensates Escherichia coli for the lack of manganese and iron superoxide dismutase. J. Biol. Chem. **273:** 10313–10316.
30. TAI, V.W. *et al.* 2002. Effects of recombinant human growth hormone on fat distribution in patients with human immunodeficiency virus-associated wasting. Clin. Infect. Dis. **35:** 1258–1262.

31. GRINSPOON, S. & K. MULLIGAN. 2003. Weight loss and wasting in patients infected with human immunodeficiency virus. Clin. Infect. Dis. **36:** S69–S78.
32. SCHAMBELAN, M. *et al.* 2002. Management of metabolic complications associated with anti-retroviral therapy for HIV-1 infection: recommendations of an International AIDS Society-USA panel. J. Acquir. Immune. Defic. Syndr. **31:** 257–275.
33. MONDY, K. & P. TEBAS. 2003. Emerging bone problems in patients infected with human immunodeficiency virus. Clin. Infect. Dis. **36:** S101–S105.
34. EIKEN, P. *et al.* 1997. Effects on bone mass after eight years of hormonal replacement therapy. Br. J. Obstet. Gynaecol. **104:** 702–707.
35. LAWAL, A. *et al.* 2001. Effect of oxandrolone upon bone mineral density content in malnourished HIV+ patients. (Chicago, IL) no.636. Eighth Conference on Retroviruses and Opportunistic Infections. Feb. 4–8, 2001.
36. EVIO, S. *et al.* 2004. Effects of alendronate and hormone replacement therapy, alone and in combination, on bone mass and markers of bone turnover in elderly women with osteoporosis. J. Clin. Endocrinol. Metab. **89:** 626–631.
37. LUGER, N.M. *et al.* 2001. Osteoprotegerin diminishes advanced bone cancer pain. Cancer Res. **61:** 4038–4047.
38. ANDERSON, D.M. *et al.* 1997. A homologue of the TNF receptor and its ligand enhance T-cell growth and dendritic-cell function. Nature **390:** 175–179.
39. OYAJOBI, B.O. *et al.* 2001. Therapeutic efficacy of a soluble receptor activator of nuclear factor kappaB-IgG Fc fusion protein in suppressing bone resorption and hypercalcemia in a model of humoral hypercalcemia of malignancy. Cancer Res. **61:** 2572–2578.

TSH and Bone Loss

LI SUN,[a] TERRY F. DAVIES,[a] HARRY C. BLAIR,[b] ETSUKO ABE,[a] AND MONE ZAIDI[a]

[a]*Mount Sinai Bone Program and Department of Medicine, Mount Sinai School of Medicine, New York, New York 10029*

[b]*University of Pittsburgh and VA Medical Center, Pittsburgh 15260, USA*

ABSTRACT: We have recently challenged the view that the bone loss associated with hyperthyroidism is solely due to elevated thyroid hormone levels. We find that thyroid-stimulating hormone (TSH), derived from the anterior pituitary gland, inhibits bone resorption by the osteoclast. Mice haploinsufficient in the TSH receptor show reduced bone density and evidence of enhanced bone resorption in the face of normal thyroid function. In humans, TSH inhibits markers of bone resorption with a single administration, and low TSH levels correlate with increased fracture risk. The evidence that low TSH levels predispose to osteoporosis in hyperthyroidism is discussed in view of the emerging role of pituitary hormones in bone biology.

KEYWORDS: osteoclast; osteoporosis; thyroid; TSH

INTRODUCTION

The glycoprotein hormone, thyroid-stimulating hormone (TSH), interacts with a seven transmembrane, glycosylated G protein–coupled TSH receptor (TSHR) on thyroid follicular cells to regulate the synthesis and secretion of thyroid hormone. Other putative functions of TSH have been speculated over the years, though none have been proven. TSHRs have nevertheless been documented on the thymus, testes, kidney, brain, adipose tissue, lymphocytes, and fibroblasts (review: Davies *et al.*, 2002).[1] TSH is thought to play a role in recruiting intestinal epithelial lymphocyte subsets to the gut.[2] Adipocytic TSHRs are believed to invoke lipolysis.[3] TSHRs in anterior pituitary folliculo-stellate cells have been implicated in the ultra-short loop regulation of TSH production,[4] whereas receptors in hypothalamic, hippocampal, and pyriform and postcingulate cortices are thought to regulate the hypothalamic–pituitary axis.[5] Tsai *et al.*[6] identified TSHRs in cultures of primary bone marrow-derived osteoblasts, but concluded that the effect was not relevant physiologically.

Address for correspondence: Mone Zaidi, M.D., Ph.D., PO Box 1055, Endocrinology, Mount Sinai School of Medicine, New York, NY 10029. Voice: 212-241-8798; fax: 212-426-8312.
e-mail: mone.zaidi@mssm.edu

Ann. N.Y. Acad. Sci. 1068: 309–318 (2006). © 2006 New York Academy of Sciences.
doi: 10.1196/annals.1346.033

We recently reported direct effects of TSH on bone remodeling by demonstrating that TSHR haploinsufficient mice with normal circulating thyroid hormone levels have osteoporosis and focal osteosclerosis.[7] We also found an abundant expression of TSHRs on both osteoclast and osteoblast precursors, and showed that recombinant TSH negatively regulated both osteoclastic bone resorption and osteoblastic bone formation *via* distinct intracellular signals.[7] We therefore suggest that TSH is a single necessary molecular switch that negatively regulates both components of skeletal remodeling.

THE CLINICAL CONUNDRUM

An association between the thyroid gland and bone disease dates back to the time of von Recklinghausen. Osteoporosis is seen in a variety of hyperthyroid states, notably in multinodular goiter, Graves' disease, subclinical hyperthyroidism, and in patients receiving suppressive doses of thyroxine.[8,9] In all these conditions, serum TSH is reduced often to undetectable levels, whereas thyroid hormones, T3 and T4, can be high to normal depending upon the clinical situation. Specific dilemma has surrounded the notion that subclinical hyperthyroidism, without elevated thyroid hormones is associated with substantial bone loss. It has therefore been suggested that there could be a relationship between low TSH levels and bone disease, although the traditional attribution has been to an elevation in thyroid hormones.[10] Consistent with this, the high bone mass and increased cortical width noted in hypothyroid patients has been thought to result from low thyroid hormone levels, rather than from accompanying elevations in serum TSH.

There is a substantial body of early literature, reviewed elsewhere, on the *in vitro* effects of thyroid hormones on bone remodeling.[11] However, it was shown only recently that bone remodeling is not uncoupled in mice in which the α_1 and β thyroid hormone receptors (TRs) are disrupted genetically.[12] Instead, these mice displayed skeletal morphogenesis and growth plate abnormalities.[12] The observation that $\tilde{\alpha}_1/\beta$TR deficiency produced no bone remodeling phenotype, interpreted in light of the strong clinical association between bone loss and low TSH levels, prompted us to speculate that the osteoporosis of hyperthyroidism arose from accompanying alterations in serum TSH.

SKELETAL DEFECTS IN THE TSHR$^{-/-}$ MOUSE

To explore the role of TSH in bone remodeling, we deleted exon 1 of the *TSHR* gene and replaced it with a green fluorescent protein (GFP) cassette.[13] Heterozygote mice, haploinsufficient in the TSHR, were euthyroid, had a normal growth pattern, and unchanged circulating thyroid hormone and TSH levels. Homozygotic TSHR null mice, in contrast, had runted growth, low

(A) (B)

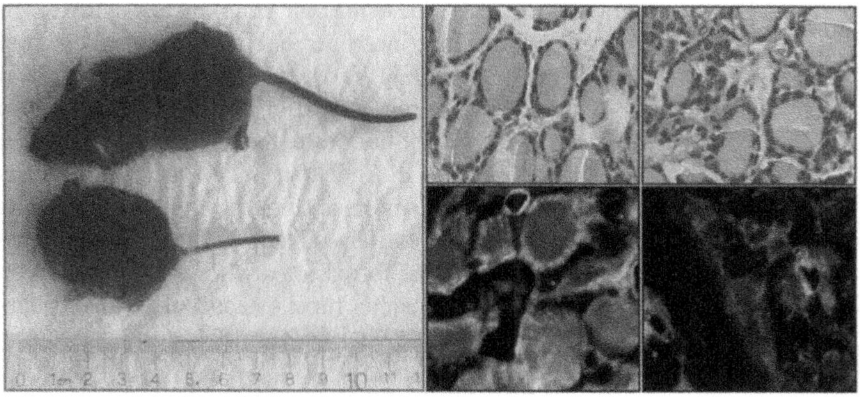

FIGURE 1. $TSHR^{+/-}$ and $TSHR^{-/-}$ mice and histological sections of their thyroid glands under light (**B**: upper panels) and epifluorescence microscopy to examine follicular structure and GFP expression (**B**: lower panels), respectively. Note the significant runting of hypothyroid homozygote mice that had a disrupted follicular structure and a normal follicular pattern in haploinsufficient heterozygote mice that were euthyroid (**A**). (Photomicrographs reproduced from Marians *et al.*)

thyroid hormone levels, a high circulating TSH, and required thyroid extracts for growth and survival.[7,13]

Examination of the thyroid follicles, shown in FIGURE 1, showed two features. First, GFP expression was noted both in the heterozygote and homozygote thyroid follicles indicating that the TSHR had been effectively deleted. Second, we found that thyroid follicles in the heterozygotes were normal, expected from a biochemically euthyroid state. In contrast, the thyroid follicular pattern in the homozygotes was disorganized. Thus, we concluded that a ~50% reduction in TSHR levels in heterozygote mice neither affected thyroid follicular growth or development, nor the secretion of thyroid hormones.[13]

On this background of normal thyroid function, the heterozygotes showed evidence of reduced bone mineral density (BMD) at all sites, indicating that the bone phenotype was due directly to a decrease in the TSHR concentration.[7] That the reduced BMD in homozygote mice was not reversed by thyroid extract replacement, an intervention that normalized the growth defect, further substantiated our conclusion for a direct action of TSH on bone remodeling. In addition to the generalized osteopenia, there was also surprising evidence for focal osteosclerosis. This comprised areas reminiscent of human Paget's bone disease, which upon polarized micrography, showed intense remodeling activity, consisting both of increases in osteoclastic bone resorption and osteoblastic bone formation.

This dual phenotype consisting of high-turnover osteoporosis and focal sclerosis is unusual. Normally, after gonadal steroid withdrawal, when the primary cellular defect is osteoclast activation, resorption precedes and exceeds the enhanced bone formation.[14] However, due to the tight coupling between resorption and formation, bone formation also increases, but lags behind resorption. The overall result of such "high remodeling" is osteoporosis, but without focal sclerosis. In addition to temporal coupling, formation and resorption are also spatially coupled. Each resorptive cavity therefore must be refilled with new bone to maintain skeletal integrity. TSHR knockout mice had focal sclerosis in addition to the high remodeling osteoporosis, suggesting that osteoblastic activity was pronounced, autonomous, and dissociated spatially from resorption. There was therefore "ectopic" bone formation at sites that had not been resorbed previously. We do not know whether focal osteosclerosis also occurs in human hyperthyroidism because of the lack of histological information in these patients.

DIRECT EFFECTS OF TSH ON OSTEOCLASTS AND OSTEOBLASTS

Through the combined use of polymerase chain reaction, immunostaining, and binding studies, we showed that TSH receptors were present in both osteoclast and osteoblast precursors.[7] We also found that TSH inhibits osteoclastogenesis and osteoblastogenesis through distinct and unrelated mechanisms. TSH inhibits osteoclast formation induced by RANKL by causing the attenuation of two of the several components of RANKL signaling, namely IκBα and *Janus* N-terminal kinase (JNK) phosphorylation.[7] This attenuation was associated with the reduced nuclear translocation of the respective downstream transcription factors, c-*jun* and p65, the latter being one of the three NF-κB subunits.

The profound inhibitory effects of TSH on osteoclast formation are likely exerted via the concurrent inhibition both the JNK/c-*jun* and NF-κB pathways. Although we have not tested this directly, reduced c-*jun* levels, even in the face of unaltered c-*fos* levels, can result in reduced homo- and heterodimer formation, and hence, diminish the transcription of critical osteoclast target genes bearing AP-1 sites, such as TRAP and cathepsin K.[15] An identical mechanism transduces the anti-osteoclastogenic effect of estrogen.[16,17] In contrast to estrogen, however, TSH additionally inhibits IκBα phosphorylation making dephosphorylated IκBα available to prevent NF-κB nuclear translocation. Consistent with the inhibitory effects of TSH, the increased osteoclastogenesis *ex vivo* in the $TSHR^{-/-}$ mice was associated with the enhanced basal phosphorylation of IκBα and JNK.

The increased osteoclastogenesis accompanying TSHR deficiency also resulted from increased tumor necrosis factor-α (TNF-α) production.[7] A blocking antibody to TNF-α attenuated the increased osteoclast formation in the TSHR null mouse, substantiating a causal role for elevated TNF-α expression in enhancing osteoclastogenesis. Even more significant, our preliminary studies indicate that crossing $TSHR^{+/-}$ mice with TNF-α null mice reversed the osteoporosis due to TSHR deficiency. Notably, therefore, we found that the depletion of TNF-α, but not its haploinsufficiency reversed the decline in BMD. It is interesting to note that TNF-α levels are also elevated in hyperthyroidism.[18]

In addition to the effect of TSH on osteoclast formation, the peptide also reduced osteoclastic bone resorption in the pit assay in cells cultured from both wild-type and heterozygote mice, but expectedly not in the TSHR-deficient cells.[7] TSH also reduced the survival of both mature osteoclasts and osteoclast precursors. These effects were associated with elevated annexin V staining.[7] TSH also inhibited osteoblast formation by inhibiting the expression of both the LRP-5 and Flk-1 (VEGF) receptors, but without effects on *Runx-2* or osterix levels. Consistent with this, TSHR-deficient mice showed evidence of elevated LRP-5 and VEGF receptor expression in osteoblast precursors. It is of note that mutations of the LRP-5 gene in humans or its targeted disruption in the mouse result in osteopenia.[19] In these conditions, *Runx-2* expression remains normal. Thus, it is not surprising that TSH exerts an effect by inhibiting LRP-5 independently of *Runx-2*. We are nevertheless unclear on the mechanism through which Flk-1 mediates TSH effects on osteoblastogenesis. There is evidence, however, that VEGF can synergize the actions of BMP-4 by activating Flk-1 to enhance mesenchymal cell proliferation.[20]

Overall, therefore, the effects of TSH in suppressing both osteoclasts and osteoblasts appear consistent with the phenotype observed in the $TSHR^{-/-}$ mouse, and defines a new role for the peptide as a bone-suppressing hormone.[21]

BACK TO HUMANS

There are several lines of growing evidence suggesting that TSHRs are critically important in human skeletal homeostasis. First, is the elegant demonstration by Onigata and colleagues that thyroxine-supplemented, euthyroid individuals with TSHR mutations have decreased BMD and elevated serum osteocalcin and N-telopeptide levels, indicative of high remodeling osteoporosis.[22] Consistent with previous studies on TSHR mutations,[23] the G498S and R450H, and to a lesser extent, the R519C mutations display this phenotype, whereas patients with thyroid hormone receptor mutations have normal bone density. Further studies using larger databases are required to substantiate these conclusions.

More recently, recombinant human TSH was administered in a single dose to postmenopausal women to examine the effects on bone turnover.[24] Serum cross-laps fell to premenopausal levels within 2 days of administering a single TSH dose subcutaneously (FIG. 2). Preclinical studies in mice are ongoing to evaluate whether TSH can be administered to humans using a dose or schedule

TABLE 1. High risk of fracture in women with low TSH levels ($n = 686$)

TSH Level	<0.1 mU/L	>0.1–<0.5 mU/L	>0.5–0.5 mU/L
HIP	3.6 (1–12.9)	1.9 (0.7–4.8)	1.0
Nonvertebral	2.3 (0.8–6.8)	2.0 (0.9–4.5)	1.0
Vertebral	4.5 (1.3–15.6)	2.8 (1.0–8.5)	

where maximal effects are seen on the skeleton, with minimal or no actions on the thyroid. Such a preliminary study has been reported in rats, where substantial increases in BMD were noted.[25]

Furthermore, in an elegant epidemiological study of over 600 women, Bauer *et al.*[26] concluded that there was a strong correlation between vertebral, nonvertebral, and hip fractures and serum TSH levels in patients with hyperthyroidism (TABLE 1). In fact, serum TSH levels <0.1 mU/L caused a substantial 4.5-fold increase in the risk of vertebral fracture and a 2.8-fold increase in the risk of hip fracture. The effects were much less pronounced with TSH levels between 0.1 and 0.5 mU/L. The notion arising from this epidemiological study has more recently been tested further in a small clinical trial. It has been shown that a levothyroxine regimen that causes a mild suppression of TSH (that does not adversely affect BMD) is equally effective in reducing nodule size as a regimen that fully suppresses TSH to undetectable levels.[27] This means that low TSH levels, and more importantly levels that fall below detection limit, are key determinants for fracture risk and bone loss in humans. One is therefore tempted to use these values as cut-offs for the empirical use of bisphosphonates to prevent fracture in patients with hyperthyroidism.

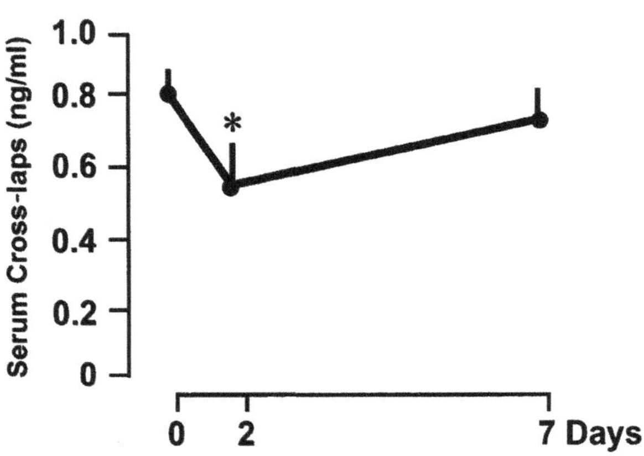

FIGURE 2. The effects of administering human recombinant TSH to postmenopausal women on serum cross-laps (C-telopeptides). Mean ± SEM (Adapted from Mazziotti *et al.*)

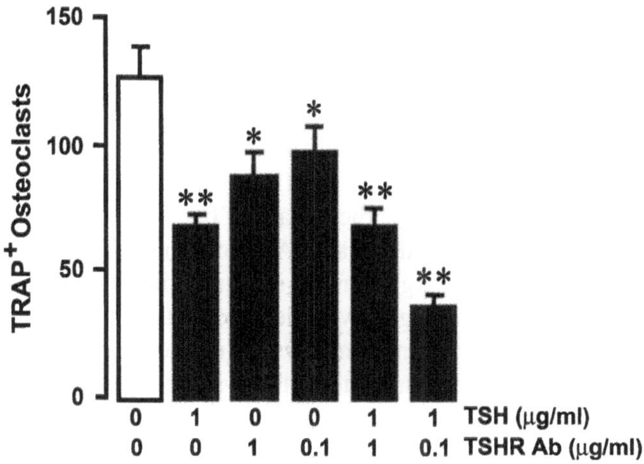

FIGURE 3. Modest inhibitory effect of a monoclonal TSH receptor (TSHR) agonist antibody developed by Ando *et al.*[28] (concentrations as indicated) on osteoclast formation triggered by M-CSF and RANKL in primary bone marrow cell cultures. $^{*}P < 0.05$.

A more fundamental pathophysiological question arises, however, as to how patients with Graves' disease who have a circulating TSHR agonist antibody lose bone. In other words, does the agonist antibody protect against the loss of TSH in these patients? We therefore tested the human monoclonal antibody[28] *in vitro* and found evidence for only weak effects in suppressing osteoclast formation (FIG. 3). This discrepancy between the effects of the antibody in thyroid and bone cells likely reflects the tissue-specific activation properties of the TSHR. We speculate therefore that the mild protection afforded by the antibody is insufficient to antagonize the marked effects of lowering serum TSH. However, this paradigm needs to be proven clinically.

INTEGRATIVE HYPOTHALAMIC CONTROL OF BONE REMODELING

That the skeletal loss in hyperthyroidism may be due to low TSH levels as opposed solely to high T3 and T4 levels would mean revisiting the role of TSH in physiology. Undoubtedly, there is substantive early literature on the direct stimulatory effects of thyroid hormone on osteoblasts and osteoclasts. On the other hand, $TR\alpha1^{-/-}\beta^{-/-}$ mice do not exhibit any bone remodeling defects; instead, they have abnormalities in skeletal maturation characterized by severe runting and growth plate defects.[12] These features are recapitulated only in hypothyroid $TSHR^{-/-}$ mice, not in euthyroid $TSHR^{+/-}$ mice with a profound remodeling phenotype. Importantly, thyroid replacement reverses only the maturation defect, not the remodeling defect in $TSHR^{-/-}$ mice. The studies thus make a compelling argument for the regulation of bone remodeling by TSH and the TSHR, and of skeletal morphogenesis by thyroid hormones.

An integrative model for skeletal surveillance by the hypothalamus has been proposed recently.[29] Its first arm is amplitude-modulated (AM) and regulates bone mass through changes in the circulating levels of pituitary hormones, of which TSH is the one established so far.[7] These hormones either regulate bone cells directly, or indeed, stimulate the release of bone-active master hormones from the thyroid, adrenals, and gonads. A second neurogenic arm regulates bone mass intermittently through changes in stimulation frequency (frequency modulation or FM), functioning to override the effects of the AM arm. Signals originating from the hypothalamic mainly leptinergic neurons are relayed *via* sympathetic and other neurotransmitter pathways directly to bone cells.[30] From a systems standpoint, the FM arm may modify the effects of the AM arm. For example, in hyperthyroidism, the excessive sympathetic discharge from the FM arm may result in even greater bone loss. These paradigms are likely to be tested by crossing mice lacking specific SNS components, such as Dbh and Adrb2, and those lacking or overexpressing pituitary hormones.

ACKNOWLEDGMENTS

This work was supported by grants to MZ, TFD, EA, and HCB from the National Institutes of Health. MZ, EA, and HCB also acknowledge the Department of Veterans Affairs (Merit Review Awards and GRECC).

REFERENCES

1. Davies, T., R. Marians & R. Latif. 2002. The TSH receptor reveals itself [review]. J. Clin. Invest. **110:** 161–164.
2. Wang, J. & J.R. Klein. 1995. Hormonal regulation of extrathymic gut T cell development: involvement of thyroid stimulating hormone. Cell. Immunol. **16:** 299–302.
3. Valyasevi, M.A., D.E. Maddox & J.T. Li. 1999. Differentiation of human orbital preadipocyte fibroblasts induces expression of functional thyrotropin receptor. J. Clin. Endocrinol. Metab. **84:** 2557–2562.
4. Kohn, L.D., H. Shimura, Y. Shimura, *et al.* 1995. The thyrotropin receptor [review]. Vitam. Horm. **50:** 287–384.
5. Prummel, M.F., L.J. Brokken, G. Meduri, *et al.* 2000. Expression of the thyroid-stimulating hormone receptor in the folliculo-stellate cells of the human anterior pituitary. J. Clin. Endocrinol. Metab. **85:** 4347–4353.
6. Tsai, J.A., A. Janson, E. Bucht, *et al.* 2004. Weak evidence of thyrotropin receptors in primary cultures of human osteoblast-like cells. Calcif. Tissue. Int. **74:** 486–491.
7. Abe, E., R.C. Marians, W. Yu, *et al.* 2003. TSH is a negative regulator of skeletal remodeling. Cell **115:** 151–162.

8. ROSS, D.S. 1994. Hyperthyroidism, thyroid hormone therapy, and bone [review]. Thyroid **4:** 319–326.
9. BARAN, D.T. 1994. Thyroid hormone and bone mass: the clinician's dilemma. Thyroid **4:** 143–144.
10. LAKATOS, P. 2003. Thyroid hormones: beneficial or deleterious for bone [review]? Review Calcif. Tissue Int. **73:** 205–209.
11. WENZEL, K.W. 1994. Thyroid hormone and bone. Thyroid **4:** 508–509.
12. GOTHE, S., Z. WANG, L. NG, *et al.* 1999. Mice devoid of all known thyroid hormone receptors are viable but exhibit disorders of the pituitary-thyroid axis, growth, and bone maturation. Genes Dev. **13:** 1329–1341.
13. MARIANS, R.C., L. NG, H.C. BLAIR, *et al.* 2002. Defining TSH-dependent and TSH-independent steps of thyroid hormone synthesis using thyrotropin receptor-null mice. Proc. Natl. Acad. Sci. USA **99:** 15776–15781.
14. RIGGS, B.L., S. KHOSLA & L.J. MELTON III. 2002. Sex steroids and the construction and conservation of the adult skeleton [review]. Endocr. Rev. **23:** 279–302.
15. KAWAIDA, R., T. OHTSUKA, J. OKUTSU, *et al.* 2003. Jun dimerization protein 2 (JDP2), a member of the AP-1 family of transcription factor, mediates osteoclast differentiation induced by RANKL. J. Exp. Med. **197:** 1029–1035.
16. SHEVDE, N.K., A.C. BENDIXEN, K.M. DIENGER & J.W. PIKE. 2000. Estrogens suppress RANK ligand-induced osteoclast differentiation via a stromal cell independent mechanism involving c-Jun repression. Proc. Natl. Acad. Sci. USA **97:** 7829–7834.
17. SRIVASTAVA, S., G. TORALDO, M.N. WEITZMANN, *et al.* 2001. Estrogen decreases osteoclast formation by down-regulating receptor activator of NF-kappa B ligand (RANKL)-induced JNK activation. J. Biol. Chem. **276:** 8836–8840.
18. SIMSEK, G., Y KARTER, S. AYDIN & H. UZUN. 2003. Osteoporotic cytokines and bone metabolism on rats with induced hyperthyroidism; changes as a result of reversal to euthyroidism. Chin. J. Physiol. **46:** 181–186.
19. KATO, M., M.S. PATEL, R. LEVASSEUR, *et al.* 2002. Cbfa1-independent decrease in osteoblast proliferation, osteopenia, and persistent embryonic eye vascularization in mice deficient in Lrp5, a Wnt coreceptor. J. Cell. Biol. **157:** 303–314.
20. PENG, H., V. WRIGHT, A. USAS, *et al.* 2002. Synergistic enhancement of bone formation and healing by stem cell-expressed VEGF and bone morphogenetic protein-4. J. Clin. Invest. **110:** 751–759.
21. NOVACK, D. 2003. TSH, the bone suppressing hormone. Cell **115:** 129–130.
22. ZAIDI, M., L. SUN, K. ONIGATA, *et al.* 2005. Direct actions of pituitary hormones on bone. Proceedings of 196th Meeting of the Society for Endocrinology, London. Endocrine Abstracts. **10:** S10.
23. NAGASHIMA, T., M. MURAKAMI, K. ONIGATA, *et al.* 2001. Novel inactivating missense mutations in the thyrotropin receptor gene in Japanese children with resistance to thyrotropin. Thyroid **11:** 551–559.
24. MAZZIOTTI, G., F. SORVILLO, M. PISCOPO, *et al.* 2005. Recombinant human TSH modulates in vivo C-telopeptides of type-1 collagen and bone alkaline phosphatase, but not osteoprotegerin production in postmenopausal women monitored for differentiated thyroid carcinoma. J. Bone Miner. Res. **20:** 480–486.
25. SAMPATH, K.T., P. SIMIC, R. SENDAK, *et al.* 2005. Systemic administration of thyroid stimulating hormone (TSH) prevents and restores bone loss in rats following ovariectomy. J. Bone Miner. Res. **20:** S415

26. BAUER, D.C., B. ETTINGER, M.C. NEVITT & K.L. STONE. 2001. Study of Osteoporotic Fractures Research Group. Risk for fracture in women with low serum levels of thyroid-stimulating hormone. Ann. Intern. Med. **134:** 561–568.
27. APPETECCHIA, M. 2005. Effects on bone mineral density by treatment of benign nodular goiter with mildly suppressive doses of L-thyroxine in a cohort women study. Horm. Res. **64:** 293–298.
28. ANDO, T., R. LATIF, A. PRITSKER, *et al.* 2002. A monoclonal thyroid-stimulating antibody. J. Clin. Invest. **110:** 1667–1674.
29. ZAIDI, M. 2005. Neural surveillance of skeletal homeostasis. Cell. Metab. **1:** 219–221.
30. ELEFTERIOU, F. 2005. Neuronal signaling and the regulation of bone remodeling. Cell. Mol. Life Sci. **62:** 2339–2349.

Bone Destruction in Multiple Myeloma

TOSHIO MATSUMOTO AND MASAHIRO ABE

Department of Medicine and Bioregulatory Sciences, University of Tokushima Graduate School of Health Biosciences, Tokushima 770-8503, Japan

ABSTRACT: Multiple myeloma (MM) is characterized by accumulation of monoclonal plasma cells in the bone marrow and progression of lytic bone lesions. MM cells enhance bone resorption by triggering a coordinated increase in RANK ligand and decrease in osteoprotegerin in the bone marrow. Macrophage inflammatory protein (MIP)-1α and (MIP)-1β are secreted by MM cells, and play a major role in the enhancement of bone resorption by MM cells. Furthermore, the growth and survival of MM cells are enhanced by contact with osteoclasts (OCs) suggesting the presence of a vicious cycle between OCs and MM cells. OCs also enhance angiogenesis in concert with MM cells largely through the cooperative actions of osteopontin from OCs and VEGF from MM cells. The angiogenic effect may further facilitate the vicious cycle between bone destruction and MM cell expansion. In addition, MM cells secrete soluble factor(s) to suppress bone formation. Secreted Frizzled-related protein (sFRP)-2, an inhibitor of Wingless type (Wnt) binding to Frizzled, is produced by most MM cells, and immunodepletion of sFRP-2 abrogates the inhibition of bone formation. Thus, MM cells enhance bone resorption and suppress bone formation to cause destructive bone lesions. Further elucidation of the mechanism of bone destruction by MM may lead to a novel therapeutic approach to prevent bone destruction and tumor growth.

KEYWORDS: bone resorption; osteoclast; receptor-activator of NF-κB (RANK) ligand; osteoprotegerin; macrophage inflammatory protein-1; osteopontin; vascular endothelial growth factor (VEGF); interleukin-6; vascular cell adhesion molecule (VCAM)-1; angiogenesis; bone formation; osteoblast; Wingless-type (Wnt); secreted Frizzled-related protein (sFRP); dickkopf (Dkk)

INTRODUCTION

Multiple myeloma (MM) almost exclusively develops and expands in the bone marrow, and generates devastating bone destruction by osteoclasts (OCs). The bone destruction causes debilitating clinical symptoms including

Address for correspondence: Toshio Matsumoto, Department of Medicine and Bioregulatory Sciences, University of Tokushima Graduate School of Health Biosciences, 3-18-15 Kuramoto-cho, Tokushima 770-8503, Japan. Voice: +81-88-633-7119; fax: +81-88-633-7407.
e-mail: toshimat@clin.med.tokushima-u.ac.jp

Ann. N.Y. Acad. Sci. 1068: 319–326 (2006). © 2006 New York Academy of Sciences.
doi: 10.1196/annals.1346.035

intractable bone pain, disabling multiple fractures, and hypercalcemia. The severity of bone disease correlates with the tumor burden, and is one of the major parameters in Durie and Salmon's clinical staging system. Furthermore, the aggressive features of MM bone lesions have significantly contributed to its poor prognosis despite the recent development of intensive chemotherapeutic regimens.[1,2] Therefore, elucidation of the molecular mechanism of bone destruction and tumor progression is essential for the development of effective therapies to improve survival as well as quality of life of MM patients.

ENHANCEMENT OF BONE RESORPTION

Interactions between receptor activator of nuclear factor-κB (RANK) expressed on the surface of cells of the OC lineage and RANK ligand expressed on stromal cells play a key role in the formation and activation of OCs, while osteoprotegerin (OPG), a decoy receptor for RANK ligand secreted from various types of cells, inhibits RANK ligand-RANK signaling. MM cells stimulate OC formation by triggering a coordinated increase in RANK ligand and decrease in OPG in the bone marrow. As shown in FIGURE 1, secretion of OPG from a stromal cell line, KM102, is markedly suppressed by co-culturing with a myeloma cell line, ARH77. Addition of conditioned medium from ARH77 cell culture partially suppressed OPG secretion from KM102 cells, suggesting a partial involvement of humoral mechanism in the suppression of OPG secretion from stromal cells by myeloma cells (FIG. 1).

MM cells reside in the proximity of stromal cells and activate OCs in bone destructive lesions. We and others have reported that osteoclastogenic C-C chemokines, macrophage inflammatory protein (MIP)-1α and (MIP)-1β are secreted by most of primary MM cells from patients with multiple osteolytic lesions, and potently induce osteoclastogenesis.[3,4] The ability of MM cells to secrete these chemokines correlated well with the extent of MM bone lesions as well as levels of bone resorption markers in patients with MM, suggesting a causal role for these chemokines in the development of lytic bone lesions.[5] In osteoclastogenic cultures with bone cells on dentine slices, MM cells enhance both formation and function of OCs. These effects were mostly abrogated by neutralizing antibodies against MIP-1α and β in combination, suggesting critical roles for these chemokines in the enhancement of bone resorption. Because the osteoclastogenic activity of (MIP)-1α and (MIP)-1β is completely inhibited by OPG, and because these chemokines induce RANK ligand expression by a stromal cell line, ST-2, in the presence of a physiological concentration of 1,25-dihydroxyvitamin D_3, the effects of MM cells on bone resorption are dependent on the stimulation of RANK ligand expression in stromal cells by MIP-1.[3]

MM cells express CCR1 and CCR5, cognate receptors for MIP-1, and MIP-1 acts on MM cells in an autocrine/paracrine fashion to enhance adhesion of MM

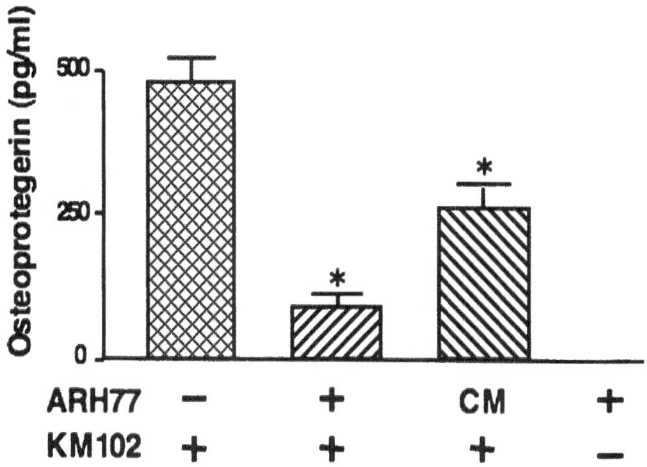

FIGURE 1. OPG production by bone marrow stromal cells. A human bone marrow-derived stromal cell line, KM102, (1×10^5/mL) was cultured in a 24-culture plate in the presence or absence of ARH77 myeloma cells (2×10^5/mL) or their conditioned media (20%) for 2 days. Culture supernatants were harvested and their OPG levels were measured by ELISA. *Significant suppression (<0.05) compared to KM102 alone.

cells to stromal cells via the interaction of VLA-4 with VCAM-1. MM cell adhesion through VLA-4/VCAM-1 interaction further enhances osteoclastogenesis as well as the secretion of MIP-1 by MM cells (unpublished data), creating a microenvironment with a close cell–cell interaction and a high concentration of these chemokines around them.

A portion of MM cells also express RANK ligand on their surface, and concomitantly produce vascular endothelial growth factor (VEGF), which can substitute for M-CSF in the induction of osteoclastogenesis by RANK ligand-RANK interaction. Thus, MM cells potently enhance osteoclastogenesis in the bone marrow microenvironment mainly via the interaction with stromal cells to cause extensive bone resorption and destructive bone lesions. In addition, there may also be a direct interaction of MM cells with OCs and their precursors to directly enhance osteoclastogenesis and bone resorption (FIG. 2).

ENHANCEMENT OF MYELOMA CELL GROWTH AND SURVIVAL BY OSTEOCLASTS

The fact that MM cells almost exclusively grow and expand in the bone marrow suggests the importance of the bone marrow microenvironment in supporting MM cell growth and survival. Among cell components in the bone marrow microenvironment, roles of stromal cells have been extensively studied.[6,7] However, MM cells also adhere to OCs, and we have demonstrated that there is a close interaction between MM cells and OCs as well. The growth

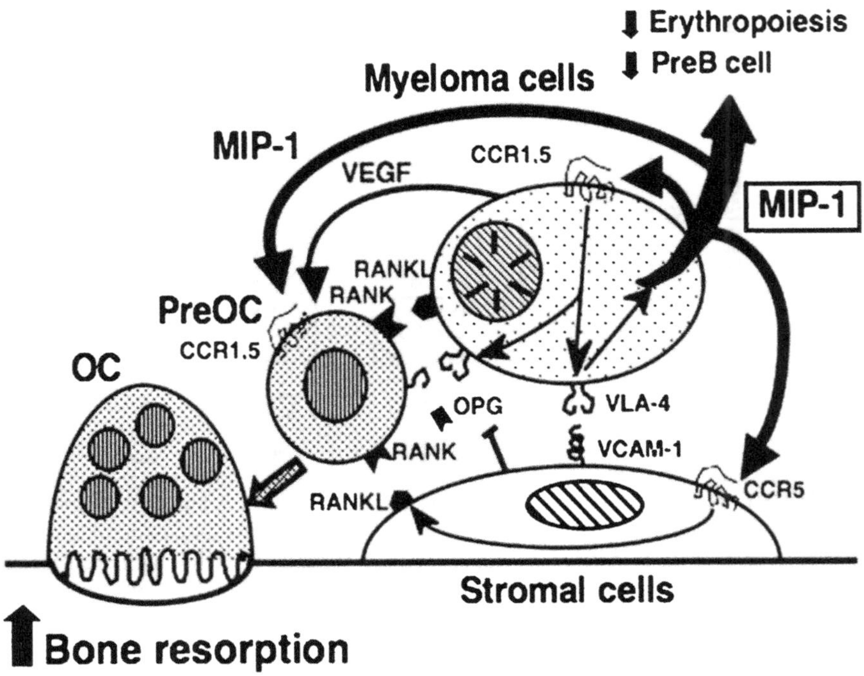

FIGURE 2. Mechanisms of osteolysis enhanced by multiple myeloma. MIP-1α and MIP-1β are constitutively expressed in most of myeloma cells, and activate integrins to enhance adhesion of myeloma cells to stromal cells. Myeloma cell-stromal cell interaction via VLA-4 binding to VCAM-1 further upregulates MIP-1 production by myeloma cells, and stimulates RANK ligand expression and concomitantly suppresses OPG production in stromal cells. MIP-1 also acts on monocytic progenitor cells to induce osteoclastic differentiation (data not shown). Furthermore, a portion of myeloma cells constitutively express RANK ligand as well as VEGF, which can substitute for M-CSF, leading to osteoclastogenesis without the presence of stromal cells. Thus, MM cells potently enhance osteoclastogenesis in the bone marrow microenvironment to cause extensive bone resorption.

and survival of MM cells are potently enhanced by a cell–cell interaction with OCs, and the effects of OCs are only partially inhibited by an anti-human interleukin (IL)-6 neutralizing antibody, despite increased production of IL-6 by OCs via the interaction with MM cells.[8] In addition, the growth of human MM cells is enhanced even under interactions with mouse or rabbit OCs. Because mouse and rabbit IL-6 cannot act on human cells, the stimulatory effect of OCs on MM cell growth cannot be explained by the enhancement of IL-6 secretion. Furthermore, prevention of cellular contact between MM cells and OCs completely abolishes the effect of OCs on MM cell growth.[8]

Because MM cells become refractory to chemotherapeutic agents in advanced stages when systemic bone destruction develops, there is a possibility that the interaction with OCs may have a protective role against the cytotoxic effects of anticancer agents. Our previous study demonstrates that cell–cell contact with OCs not only enhances the growth of doxorubicin-sensitive MM cells but also causes marked resistance to cytotoxic effects of doxorubicin.[8] These observations are in agreement with clinical findings that MM at advanced stages with extensive osteolytic lesions shows refractoriness to chemotherapies, and suggest that increased osteoclast number and/or activity contributes to aggressiveness or drug resistance of MM cells. Taken together, there is a vicious cycle between MM cells and OCs, and OCs, which are generated and activated by MM cells, not only resorb and destroy bone but also enhance MM cell growth and survival largely via cell–cell interactions with MM cells (FIG. 3).

ENHANCEMENT OF ANGIOGENESIS BY MYELOMA CELLS

Angiogenesis is enhanced along with osteoclastogenesis in the bone marrow of patients with MM, which has drawn considerable attention as a potential therapeutic target. As mentioned earlier, MM cells secrete high levels of VEGF. In addition, our preliminary results demonstrate that OCs constitutively secrete high levels of an angiogenic noncollagenous matrix protein, osteopontin (OPN), which is known to cooperatively act with VEGF in angiogenesis. Vascular tubule formation is enhanced by an addition of conditioned media from OCs as potently as those from MM cells, which is further enhanced by conditioned media from co-cultures of OCs and MM cells. These results suggest that there are cooperative interactions between OCs and MM cells in stimulating angiogenesis. When antibodies against OPN or VEGF are added alone, they partially inhibit the enhanced angiogenesis by conditioned media from co-cultures of MM cells and OCs, whereas vascular tubule formation is almost completely abrogated when these antibodies are added together (unpublished data). Thus, OCs enhance angiogenesis in concert with MM cells largely through cooperative actions of OPN from OCs and VEGF from MM cells. The angiogenic effect elaborated by the interactions of MM cells with OCs may further facilitate the vicious cycle between bone destruction and MM cell expansion.

SUPPRESSION OF BONE FORMATION

In typical destructive bone lesions of MM, enhanced bone resorption is accompanied by impaired bone formation and mineralization, which causes the well-known "punched-out" lesions on X rays. Analyses of bone turnover by biochemical bone markers also suggest an imbalance of bone turnover with enhanced bone resorption and suppressed bone formation.[5] However, little has been known about the mechanism(s) responsible for the impairment of

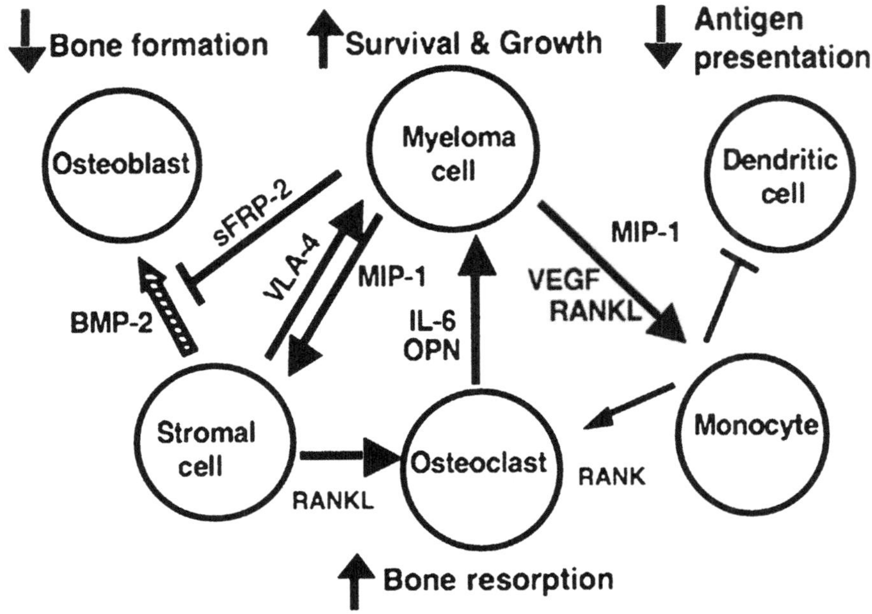

FIGURE 3. Cellular interactions in the bone marrow microenvironment around myeloma cells. Osteoclastogenesis is enhanced by myeloma cells as mentioned in FIGURE 2. The increased osteoclastogenesis in turn enhances myeloma cell growth and survival via myeloma cell–osteoclast interaction. Thus, there is a vicious cycle between myeloma cells and osteoclasts to enhance osteoclastic bone resorption and myeloma growth. Although cell–cell contact between myeloma cells and osteoclasts enhances IL-6 and osteopontin production from osteoclasts, the enhancement of myeloma cell growth and survival is only partially dependent upon osteoclast-derived IL-6 and osteopontin. While MIP-1 enhances osteoclastogenesis from monocytic progenitor cells, it also suppresses the formation of myeloid dendritic cells from common progenitor cells, causing suppressed antigen presentation with resultant impairment in immune responses (data not shown). Furthermore, most myeloma cells produce a soluble Wnt antagonist, sFRP-2, which suppresses osteoblast differentiation induced by BMP-2. sFRP-2 appears to play an important role in the defective bone formation caused by myeloma. Therefore, myeloma cells closely interact with bone marrow microenvironment to cause enhanced bone resorption, immune suppression, and impaired bone formation, forming the clinical features with destructive bone lesions and myeloma expansion.

bone formation. A canonical Wingless-type (Wnt) signaling pathway has recently been shown to play a critical role in osteoblast differentiation.[9,10] Wnt proteins are secreted by cystein-rich glycoproteins and are known as regulators of the differentiation of hematopoietic and mesenchymal cells as well as embryonic development. Because conditioned media from MM cells suppress osteoblast differentiation and mineralization, and because several secreted Frizzled-related protein (sFRP) and dickkopf (DKK) family members are known as soluble Wnt antagonists, the expression of sFRP-1, -2, and -3 as well as DKK-1 is examined in MM cells. Various MM cell lines including U266, RPMI8226, and ARH77 secrete only sFRP-2 at protein levels, although sFRP-3 is also expressed at mRNA levels in all these cell lines. Although DKK-1, an inhibitor of an LDL receptor-related protein (LRP)5/6 Wnt co-receptor, was reported to play an important role in the impairment of bone formation by MM,[11] DKK-1 mRNA can be detected only in U266 cells. More importantly, sFRP-2 mRNA and protein can be detected in most MM cells from patients with advanced or terminal stages of MM. In an effort to clarify the role of sFRP-2 in the suppression of bone formation by MM, recombinant sFRP-2 is added to MC3T3-E1 osteoblastic cell cultures together with BMP-2. Exogenous sFRP-2 partially suppresses alkaline phosphatase activity and almost completely blocks mineralized nodule formation enhanced by BMP-2 (submitted for publication). Furthermore, immunodepletion of sFRP-2 restored mineralized nodule formation by MC3T3-E1 cells in the presence of MM cell-conditioned media. These results are consistent with the notion that MM cell-derived sFRP-2 is among the predominant factors responsible for the impairment of bone formation by MM. In contrast to sFRP-2, DKK-1 is preferentially expressed in mature types of MM cells, but barely expressed in plasmablastic or immature types of MM cells.[11] Nevertheless, because MM cells have been shown to secrete factors other than sFRP-2 that are implicated as inhibitors of osteoblast differentiation, including DKK-1, insulin-like growth factor-binding protein 4 and IL-3, sFRP-2 may act in concert with other factors to cause a potent suppression of bone formation in destructive bone lesions. Further elucidation of the mechanism of the development of destructive bone lesions by MM should lead to novel therapeutic approaches to prevent devastating bone destruction and MM cell expansion.

REFERENCES

1. Bataille, R. & J.L. Harousseau. 1997. Multiple myeloma. N. Engl. J. Med. **336:** 1657–1664.
2. Myeloma Trialists' Collaborative Group. 1998. Combination chemotherapy versus melphalan plus prednisone as treatment for multiple myeloma: an overview of 6,633 patients from 27 randomized trials. J. Clin. Oncol. **16:** 3832–3842.
3. Abe, M., K. Hiura, J. Wilde, *et al.* 2002. Role for macrophage inflammatory protein (MIP)-1alpha and MIP-1beta in the development of osteolytic lesions in multiple myeloma. Blood **100:** 2195–2202.
4. Choi, S.J., J.C. Cruz, F. Craig, *et al.* 2000. Macrophage inflammatory protein 1-alpha is a potential osteoclast stimulatory factor in multiple myeloma. Blood 96: 671–675.
5. Hashimoto, T., M. Abe, T. Oshima, *et al.* 2004. Ability of myeloma cells to secrete macrophage inflammatory protein (MIP)-1alpha and MIP-1beta correlates with lytic bone lesions in patients with multiple myeloma. Br. J. Haematol. **125:** 38–41.

6. HIDESHIMA, T., D. CHAUHAN, K. PODAR, *et al.* 2001. Novel therapies targeting the myeloma cell and its bone marrow microenvironment. Semin. Oncol. **28:** 607–612.
7. DAMIANO, J.S., A.E. CRESS, L.A. HAZLEHURST, *et al.* 1999. Cell adhesion mediated drug resistance (CAM-DR): role of integrins and resistance to apoptosis in human myeloma cell lines. Blood **93:** 1658–1667.
8. ABE, M., K. HIURA, J. WILDE, *et al.* 2004. Osteoclasts enhance myeloma cell growth and survival via cell-cell contact: a vicious cycle between bone destruction and myeloma expansion. Blood **104:** 2484–2491.
9. GONG, Y., R.B. SLEE, N. FUKAI, *et al.* 2001. LDL receptor-related protein 5 (LRP5) affects bone accrual and eye development. Cell **107:** 513–523.
10. BOYDEN, L.M., J. MAO, J. BELSKY, *et al.* 2002. High bone density due to a mutation in LDL-receptor-related protein 5. N. Engl. J. Med. **346:** 1513–1521.
11. TIAN, E., F. ZHAN, R. WALKER, *et al.* 2003. The role of the Wnt-signaling antagonist DKK1 in the development of osteolytic lesions in multiple myeloma. N. Engl. J. Med. **349:** 2483–2494.

Osteogenic Regulation of Vascular Calcification

DWIGHT A. TOWLER, JIAN-SU SHAO, SU-LI CHENG, JOYCE M. PINGSTERHAUS, AND ARLEEN P. LOEWY

Department of Internal Medicine, Division of Bone and Mineral Diseases, Washington University School of Medicine, Barnes-Jewish Hospital, St. Louis, Missouri 63110, USA

ABSTRACT: **Vascular calcification increasingly afflicts our aging and dysmetabolic population, predisposing patients to cardiovascular mortality and lower extremity amputation. Active osteogenic processes are evident in most histoanatomic variants, including elaboration of BMP2-Msx2 signals required for craniofacial bone formation. We developed an animal model of diet-induced diabetes, dyslipidemia, and vascular calcification. High-fat diets promote vascular calcification in male low-density lipoprotein receptor (LDLR)-deficient mice, with concomitant upregulation of aortic BMP2 and Msx2 gene expression. We wished to test if Msx2 exerts pro-calcific actions during vascular calcification, as it does in craniofacial bone. We studied CMV-Msx2Tg+;LDLR+ transgenic mice (C57Bl/6), a model previously demonstrated to recapitulate features of Msx2 signaling during craniosynostosis. After 16 weeks of fatty diets, vascular calcification was studied in CMV-Msx2Tg+ versus nontransgenic sibs. Only CMV-Msx2Tg+ mice fed high-fat diets exhibited vascular calcium accumulation by alizarin red staining, noted in the tunica media of coronary arteries and the aorta. Gene expression studies revealed that while Msx2 was expressed primarily in adventitial cells, alkaline phosphatase (ALP) expression and calcification occurred primarily in the tunica media. Msx2 promotes the elaboration of a pro-osteogenic milieu by upregulating expression of Wingless type (Wnt) ligands while downregulating the canonical antagonist, Dickkopf (Dkk1). Msx2 upregulates aortic Wnt signaling *in vivo*, revealed by the analysis of TOPGAL+ (Wnt reporter) versus CMV-Msx2Tg+; TOPGAL+ mice. Aortic Msx2 exerts pro-osteogenic signaling *in vivo* and *in vitro*, mediated in part via the enhancement of paracrine Wnt signaling. Strategies that selectively inhibit aortic Msx2-Wnt cascades may help diminish the initiation and progression of diabetic vascular disease.**

KEYWORDS: **diabetes BMP2; Msx2; Wnt; Dkk1; β-catenin; alkaline phosphatase; calcium; PTH/PTHrP receptor; tunica media; adventitia; osterix; myofibroblast; osteoblast**

Address for correspondence: Dwight A. Towler, M.D., Ph.D., Department of Internal Medicine, Division of Bone and Mineral Diseases, Washington University School of Medicine, Barnes-Jewish Hospital North Campus Box 8301, 660 South Euclid Ave., St. Louis, MO 63110. Voice: 314-454-7434; fax: 314-454-8434.
e-mail: dtowler@im.wustl.edu

Ann. N.Y. Acad. Sci. 1068: 327–333 (2006).
doi: 10.1196/annals.1346.036

For almost two centuries, it has been appreciated that calcified tissue formation is a component of vascular disease processes.[1] Virchow in fact coined the terms arteritis deformans and atherosclerosis to highlight histopathological features of vascular calcification that resembled bone formation in the setting of tissue inflammatory responses.[2] Using electron microscopy coupled with histochemistry, two decades ago Anderson and colleagues demonstrated alkaline phosphatase (ALP)-containing matrix vesicles as key features of both atherosclerotic and medial artery calcification, providing ultrastructural evidence that vascular calcification is in part an osteogenic process.[3] However, it was seminal studies by Demer and colleagues that first revealed the molecular fingerprints of ectopic osteogenesis on the pathobiology of vascular calcification.[4] They demonstrated that the powerful bone morphogen BMP2 was in fact expressed in calcifying atherosclerotic plaques.[5] Moreover, they showed that inflammatory oxysterols and tumor necrosis factor (TNF)-α[6] enhance osteogenic mineralization via a subset of vascular smooth muscle cells called calcifying vascular cells (CVCs).[4,7] Paracrine interactions arising with cytokine-activated macrophages markedly upregulated CVC ALP activity.[6] Recently, it has become apparent that with advancing age and uremia, there is a reciprocal relationship between vascular mineral content and bone mineral content.[8] Indeed, several groups have identified that osteotropic hormones can reciprocally regulate bone and vascular content, indicating that endocrine axes participate in vascular mineral metabolism.[9,10] Thus, while disruption of key tissue mineralization inhibitors clearly augment vascular calcium load, vascular calcification is subject to active osteogenic regulation in response to morphogenetic, metabolic, mechanical, inflammatory, and endocrine cues.[1]

Vascular calcification can be usefully classified into four histoanatomic variants[1]: (*a*) intimally oriented atherosclerotic calcification; (*b*) medial artery calcification; (*c*) cardiac valve calcification; and (*d*) vascular soft-tissue calciphylaxis. In addition, myocardial tissue, like skeletal muscle, can undergo postinjury myositis ossificans. Our research interests have centered on medial artery calcification, a nonendochondral mineralization process which resembles that of membranous bone formation and dentinogenesis. Medial artery calcification is a highly characteristic feature of diabetes[11] and end-stage renal disease.[12] Once thought benign as compared to the prothrombotic atherosclerotic plaque, epidemiological studies have clearly shown that medial artery calcification increases the risk of cardiovascular morbidity and mortality in patients with diabetes and uremia.[8,11,12] The excess risk for lower extremity amputation and cardiovascular mortality[8,11,12] may arise from a type of vascular "diastolic" dysfunction that arises with reduced vascular compliance of elastic arteries. During systole, potential energy is stored within large elastic arteries such as the aorta.[13] Known as Windkessel physiology,[13] kinetic energy is subsequently released during the relaxation phase of the cardiac cycle, providing diastolic perfusion of the myocardium and sustained perfusion of distal vascular beds. With vessel stiffening, elevated pulse pressure and highly pulsatile flow kinetics interact with elevated systolic blood pressure and increased myocardial oxygen consumption to increase workload and decrease distal tissue perfusion.[13] Compromised elastic artery Windkessel function is likely a major contributor to the increased risk for lower extremity amputation of patients with type II diabetes (T2DM).[11]

The molecular mechanisms that regulate vascular calcification in T2DM are poorly understood. In 1998, we identified that high-fat diets induce obesity, insulin-resistant diabetes, dyslipidemia, and vascular calcification in male low-density lipoprotein receptor (LDLR)-deficient mice.[14] In response to this physiologically relevant model of T2DM, an aortic adventitial BMP2-Msx2 signaling cascade was activated.[14,15] Previous work by others had demonstrated that BMP-Msx signaling is crucial during craniofacial membranous bone formation and dentinogenesis.[16–18] Consistent with this, our cell culture studies demonstrated that Msx2 enhanced osteogenic differentiation (ALP induction, calcification) of aortic myofibroblasts via osterix-dependent signals—but concomitantly suppressed adipogenic potential (e.g., lipid accumulation) of these multipotent mesenchymal progenitors.[7,15,19] Cell autonomous actions of Msx2 were shown to contribute via inhibition of C/EBP-α and PPAR-γ gene expression.[15,19] However, analysis of conditioned media from Msx2-expressing 10T1/2 mesenchymal cells revealed the induction of an adipostatic, pro-osteogenic activity[20]—highly reminiscent of the newly described osteogenic canonical Wnt, Wnt10b.[21] Canonical Wnts signal via the heteromeric LDLR-related protein receptors LRP5 and LRP6 to activate osteogenic gene expression via nuclear β-catenin-dependent transcription.[22,23] We demonstrated that Msx2-expressing cells express a factor that enhances nuclear accumulation of β-catenin, and upregulates activity of the β-catenin-dependent transcription driven by a TCF/LEF optimal promotor (TOP)-reporter construct.[20] Pro-osteogenic and adipostatic activities of Msx2 were both reversed by treatment with recombinant purified Dkk1, an inhibitory ligand of canonical Wnt signaling. Moreover, we demonstrated that Msx2 suppresses mesenchymal expression of Dkk1 while upregulating the expression of genes encoding canonical Wnt ligands Wnt3a and Wnt7a.[20] We had previously shown that PTH1R agonist teriparatide could suppress vascular calcification and aortic Msx2 gene expression while promoting orthotopic bone formation.[9] Consistent with the working hypothesis, teriparatide suppressed Wnt3a/Wnt7a expression and antagonized canonical Wnt3a signaling in vascular smooth muscle cells.[20] To test whether osteogenic Msx2-Wnt signaling contributed to vascular calcification *in vivo*, we generated and evaluated CMV-Msx2 transgenic mice. This model had been previously validated in studies of Msx2-dependent ectopic bone formation during craniosynostosis.[24] Quantitative RT-PCR demonstrated threefold upregulation of aortic Msx2 in our CMV-Msx2 transgenic mice versus nontransgenic siblings. Wnt3a and Wnt7a were concomitantly upregulated with the suppression of aortic Dkk1.[20] Immunohistochemistry demonstrated the expression of Msx2 in the aortic adventitia, but localized the induced expression of ALP in the tunica media. Calcium deposition revealed by alizarin red staining coincides

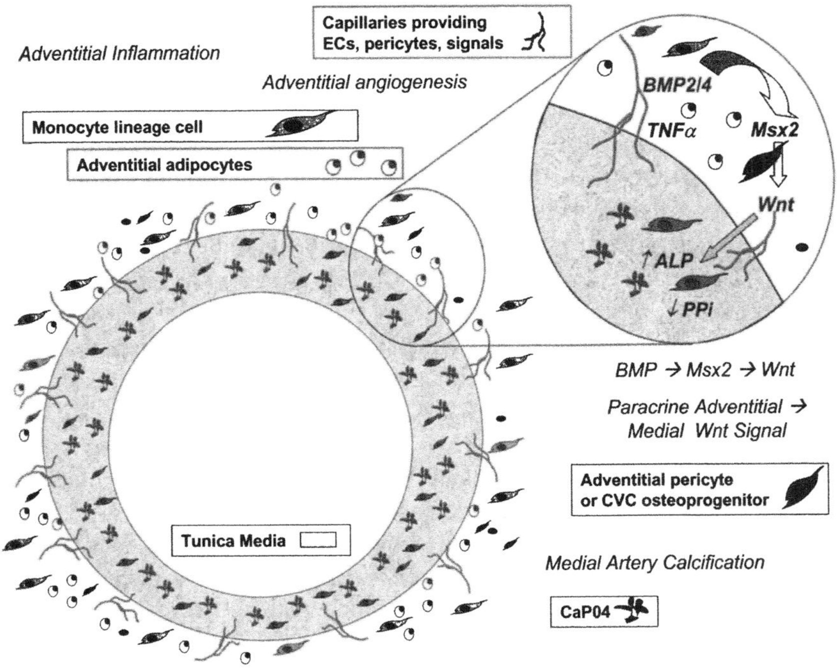

FIGURE 1. Diabetic medial artery calcification: A working model—Diabetes is a global metabolic disorder, characterized by hyperglycemia, dyslipidemia, low-grade inflammation, and obesity (in T2DM) .[1,11,34] Hyperglycemia and dyslipidemia activate the adipogenic differentiation of adventitial pericytes,[15] and generates pro-inflammatory oxidative stress that activates endothelial cells.[34] Neoangiogenesis and monocyte infiltration occurs. Reactive oxygen species and inflammatory cytokines such as TNF-α upregulate endothelial cell (EC) BMP production.[33] With activation of adventitial angiogenesis and capillary sprouting, the pool of multipotent pericytic smooth muscle cells expands, with a subset upregulating Msx2 expression in response to BMPs.[15,16] As an osteogenic transcription factor, Msx2 prevents the adipogenic differentiation of adventitial pericytes, downregulates Dkk1 expression, and enhances the elaboration of canonical Wnt ligands such as Wnt3a and Wnt7a.[20] Adventitial capillaries of the expanded vasa vasorum arising in response to diabetes and dyslipidemia perfuse the tunica media, providing the conduit for paracrine delivery of the Wnt-enriched osteogenic milieu. CVCs arising either from mural smooth muscle cell populations or derived from activated adventitial progenitors are recruited to the osteogenic lineage, increasing the expression of ALP and matrix vesicle production.[3,4,29] ALP cleaves inorganic pyrophosphate (PPi), which rapidly depletes this important inhibitor of epitaxial mineralization.[26,27] Circumferential matrix vesicle-mediated calcium phosphate deposition occurs within the tunica media.[1,11,12] Whether osteogenic CVCs arise solely from transdifferentiation of multipotent pericytic progenitors[7,15,30] or also from circulating osteoprogenitors[31,36] is currently unknown. See text for details.

with ALP expression.[20] The seminal studies of Heistad had shown that the vector of mural microvascular flow is concentric, from the tunica adventitia to the tunica media,[25] therefore, we posited that paracrine Wnt signals were elaborated by Msx2 expressing cells of the tunica adventitia that programmed osteogenic differentiation and mineralization of CVCs within the tunica media (FIG. 1). Consistent with this, the Msx2 transgene significantly upregulated LacZ reporter activity in the aortic tunica media of TOPGAL (TCF/LEF optimal promotor—galactosidase reporter) mice[20]; little if any LacZ induction was observed in the tunica adventitia. Thus, a working model emerges in which a paracrine BMP-Msx2-Wnt signaling cascade controls the osteogenic differentiation and mineralization of vascular progenitors via nonendochondral processes. The concentric medial calcification of diabetes and uremia is posited to arise in part from the anatomic relationship between (*a*) the Msx2-expressing cells of the periaortic adventitia that elaborate a Wnt-laden osteogenic milieu and (*b*) CVCs of Demer in the tunica media that undergo osteogenic differentiation in response to these paracrine cues. Wnt-induced ALP activity degrades extracellular pyrophosphate, an important determinant of vascular mineralization and differentiation,[26,27] exacerbating the deleterious consequences of deficiency in passive inhibitors of tissue mineralization such as fetuin.[28,29] As such, diabetic medial artery calcification can be viewed in part as a vascular "Wntopathy" that upregulates medial artery ALP activity and promotes ectopic osteogenic matrix vesicle-mediated calcification.[20,27,29] The precise origins of the Msx2-expressing cells and CVCs are not known; however, a Sca1+ stem cell population has recently been shown to reside within the aortic adventitia.[30] Whether these cells arise from circulating progenitors[31] or aortic mesoangioblasts[32] is also unclear. However, an integrative model can be drawn in which paracrine interactions between endothelial cells, macrophages, and pericytic myofibroblasts—CVCs—form the vascular osteogenic triad. Intriguingly, adventitial oxidative stress—a stimulus for BMP production by endothelial cells[33] has been shown to be markedly increased in diabetes.[34] Thus, paracrine endothelial–mesenchymal interactions,[35] reminiscent of the epithelial–mesenchymal interactions that control craniofacial skeletogenesis,[16] are now emerging in the pathogenesis of vascular calcification. As the tools of developmental biology, endocrinology, vascular physiology, and molecular genetics are brought to bear, novel therapeutic strategies can be developed to address the unmet clinical need—to ameliorate the devastating cardiovascular consequences of medial artery calcification in diabetes and end-stage renal disease (ESRD).[11,12]

ACKNOWLEDGMENTS

This research has been supported by NIH grants HL69229 and AR43731, and the American Diabetes Association.

REFERENCES

1. VATTIKUTI, R. & D.A. TOWLER. 2004. Osteogenic regulation of vascular calcification: an early perspective. Am. J. Physiol. Endocrinol. Metab. **286:** E686–E696.

2. VIRCHOW, R. 1989. Cellular pathology. As based upon physiological and pathological histology. Lecture XVI—Atheromatous affection of arteries. 1858. Nutr. Rev. **47:** 23–25.
3. TANIMURA, A., D.H. MCGREGOR & H.C. ANDERSON. 1983. Matrix vesicles in atherosclerotic calcification. Proc. Soc. Exp. Biol. Med. **172:** 173–177.
4. DEMER, L.L. & Y. TINTUT. 2005. Return to ectopia: stem cells in the artery wall. Arterioscler. Thromb. Vasc. Biol. **25:** 1307–1308.
5. BOSTROM, K., K.E. WATSON, S. HORN, *et al.* Bone morphogenetic protein expression in human atherosclerotic lesions. J. Clin. Invest. **91:** 1800–1809.
6. TINTUT, Y., J. PATEL, M. TERRITO, *et al.* Monocyte/macrophage regulation of vascular calcification *in vitro*. Circulation **105:** 650–655.
7. TINTUT, Y., Z. ALFONSO, T. SAINI, *et al.* 2003. Multilineage potential of cells from the artery wall. Circulation **108:** 2505–2510.
8. LONDON, G.M., C. MARTY, S.J. MARCHAIS, *et al.* 2004. Arterial calcifications and bone histomorphometry in end-stage renal disease. J. Am. Soc. Nephrol. **15:** 1943–1951.
9. SHAO, J.S., S.L. CHENG, N. CHARLTON-KACHIGIAN, *et al.* 2003. Teriparatide (human parathyroid hormone (1-34)) inhibits osteogenic vascular calcification in diabetic low density lipoprotein receptor-deficient mice. J. Biol. Chem. **278:** 50195–50202.
10. DAVIES, M.R., R.J. LUND, S. MATHEW, *et al.* 2005. Low turnover osteodystrophy and vascular calcification are amenable to skeletal anabolism in an animal model of chronic kidney disease and the metabolic syndrome. J. Am. Soc. Nephrol. **16:** 917–928.
11. LEHTO, S., L. NISKANEN, M. SUHONEN, *et al.* 1996. Medial artery calcification. A neglected harbinger of cardiovascular complications in non-insulin-dependent diabetes mellitus. Arterioscler. Thromb. Vasc. Biol. **16:** 978–983.
12. LONDON, G.M., A.P. GUERIN, S.J. MARCHAIS, *et al.* 2003. Arterial media calcification in end-stage renal disease: impact on all-cause and cardiovascular mortality. Nephrol. Dial. Transplant. **18:** 1731–1740.
13. SAFAR, M.E. & H.S. BOUDIER. 2005. Vascular development, pulse pressure, and the mechanisms of hypertension. Hypertension **46:** 205–209.
14. TOWLER, D.A., M. BIDDER, T. LATIFI, *et al.* 1998. Diet-induced diabetes activates an osteogenic gene regulatory program in the aortas of low density lipoprotein receptor-deficient mice. J. Biol. Chem. **273:** 30427–30434.
15. CHENG, S.L., J.S. SHAO, N. CHARLTON-KACHIGIAN, *et al.* 2003. MSX2 promotes osteogenesis and suppresses adipogenic differentiation of multipotent mesenchymal progenitors. J. Biol. Chem. **278:** 45969–45977.
16. JOWETT, A.K., S. VAINIO, M.W. FERGUSON, *et al.* 1993. Epithelial-mesenchymal interactions are required for msx 1 and msx 2 gene expression in the developing murine molar tooth. Development **117:** 461–470.
17. BARLOW, A.J. & P.H. FRANCIS-WEST. 1997. Ectopic application of recombinant BMP-2 and BMP-4 can change patterning of developing chick facial primordia. Development **124:** 391–398.
18. KIM, H.J., D.P. RICE, P.J. KETTUNEN, *et al.* 1998. FGF-, BMP- and Shh-mediated signalling pathways in the regulation of cranial suture morphogenesis and calvarial bone development. Development **125:** 1241–1251.
19. ICHIDA, F., R. NISHIMURA, K. HATA, *et al.* 2004. Reciprocal roles of MSX2 in regulation of osteoblast and adipocyte differentiation. J. Biol. Chem. **279:** 34015–34022.

20. SHAO, J.S., S.L. CHENG, J.M. PINGSTERHAUS, *et al.* 2005. Msx2 promotes cardiovascular calcification by activating paracrine Wnt signals. J. Clin. Invest. **115:** 1210–1220.
21. ROSS, S.E., N. HEMATI, K.A. LONGO, *et al.* 2000. Inhibition of adipogenesis by Wnt signaling. Science **289:** 950–953.
22. JOHNSON, M.L., K. HARNISH, R. NUSSE, *et al.* 2004. LRP5 and Wnt signaling: a union made for bone. J. Bone Miner. Res. **19:** 1749–1757.
23. HU, H., M.J. HILTON, X. TU, *et al.* 2005. Sequential roles of Hedgehog and Wnt signaling in osteoblast development. Development **132:** 49–60.
24. LIU, Y.H., R. KUNDU, L. WU, *et al.* 1995. Premature suture closure and ectopic cranial bone in mice expressing Msx2 transgenes in the developing skull. Proc. Natl. Acad. Sci. USA **92:** 6137–6141.
25. HEISTAD, D.D., M.L. MARCUS, G.E. LARSEN, *et al.* 1981. Role of vasa vasorum in nourishment of the aortic wall. Am. J. Physiol. **240:** H781–H787.
26. JOHNSON, K., M. POLEWSKI, D. VAN ETTEN, *et al.* 2005. Chondrogenesis mediated by PPi depletion promotes spontaneous aortic calcification in NPP1-/- mice. Arterioscler. Thromb. Vasc. Biol. **25:** 686–691.
27. TOWLER, D.A. 2005. Inorganic pyrophosphate: a paracrine regulator of vascular calcification and smooth muscle phenotype. Arterioscler. Thromb. Vasc. Biol. **25:** 651–654.
28. MERX, M.W., C. SCHAFER, R. WESTENFELD, *et al.* 2005. Myocardial stiffness, cardiac remodeling, and diastolic dysfunction in calcification-prone fetuin-A-deficient mice. J. Am. Soc. Nephrol. **16:** 3357–3364.
29. REYNOLDS, J.L., A.J. JOANNIDES, J.N. SKEPPER, *et al.* 2004. Human vascular smooth muscle cells undergo vesicle-mediated calcification in response to changes in extracellular calcium and phosphate concentrations: a potential mechanism for accelerated vascular calcification in ESRD. J. Am. Soc. Nephrol. **15:** 2857–2867.
30. HU, Y., Z. ZHANG, E. TORSNEY, *et al.* 2004. Abundant progenitor cells in the adventitia contribute to atherosclerosis of vein grafts in ApoE-deficient mice. J. Clin. Invest. **113:** 1258–1265.
31. EGHBALI-FATOURECHI, G.Z., J. LAMSAM, D. FRASER, *et al.* 2005. Circulating osteoblast-lineage cells in humans. N. Engl. J. Med. **352:** 1959–1966.
32. BRUNELLI, S., E. TAGLIAFICO, F.G. DE ANGELIS, *et al.* 2004. Msx2 and necdin combined activities are required for smooth muscle differentiation in mesoangioblast stem cells. Circ. Res. **94:** 1571–1578.
33. SORESCU, G.P., H. SONG, S.L. TRESSEL, *et al.* 2004. Bone morphogenic protein 4 produced in endothelial cells by oscillatory shear stress induces monocyte adhesion by stimulating reactive oxygen species production from a nox1-based NADPH oxidase. Circ. Res. **95:** 773–779.
34. ZHANG, L., A. ZALEWSKI, Y. LIU, *et al.* 2003. Diabetes-induced oxidative stress and low-grade inflammation in porcine coronary arteries. Circulation **108:** 472–478.
35. ARMULIK, A., A. ABRAMSSON & C. BETSHOLTZ. 2005. Endothelial/pericyte interactions. Circ. Res. **97:** 512–523.
36. SATA, M., K. TANAKA & R. NAGAI. 2005. Circulating osteoblast-lineage cells; author reply 737–738. N. Engl. J. Med. **353:** 737–738.

The Role of OPG/TRAIL Complex in Multiple Myeloma

The OPG/TRAIL Complex in an *In Vitro* Osteoclastogenesis Model Derived From Human Multiple Myeloma-Bone Disease

GIACOMINA BRUNETTI,[a] SILVIA COLUCCI,[a] RITA RIZZI,[b] GIORGIO MORI,[a] GRAZIANA COLAIANNI,[a] ANGELA ORANGER,[a] ALBERTA ZALLONE,[a] VINCENZO LISO,[b] AND MARIA GRANO[a]

[a]*Department of Human Anatomy and Histology, University of Bari Medical School, 70124 Bari, Italy*

[b]*Hematology Section, Department of Internal Medicine and Public Medicine, University of Bari Medical School, 70124 Bari, Italy*

ABSTRACT: Multiple myeloma (MM) is often associated with an increased osteoclast (OC) activity. Using an *in vitro* osteoclastogenesis model consisting of MM unstimulated and unfractionated peripheral blood mononuclear cells (PBMCs), we showed that T cells support OC formation and survival. Differently, in T cell-depleted MM PBMC cultures, exogenous macrophage-colony stimulating factor (M-CSF) and receptor activated of nuclear factor-κB ligand (RANKL) were necessary for osteoclastogenesis. We found RANKL, OPG, and TRAIL overexpression by fresh MM T cells. Despite high osteoprotegerin (OPG) levels, the persistence of osteoclastogenesis can be related to the formation of the OPG/TRAIL complex. Our results highlight that MM T cells support OC formation and survival, possibly involving OPG/TRAIL interaction.

KEYWORDS: **multiple myeloma; osteoclast; T cells; TRAIL; OPG**

INTRODUCTION

Multiple myeloma (MM) is a B cell neoplasm characterized by clonal expansion of malignant plasma cells in the bone marrow with frequent occurrence of lytic bone disease.[1] It results from an unbalanced bone turnover with enhanced resorption related to increased osteoclast (OC) recruitment and activity

Address for correspondence: Maria Grano, Department of Human Anatomy and Histology, University of Bari Medical School, Piazza Giulio Cesare, 11, 70124 Bari, Italy. Voice: +39-080-547-8306; fax: +39-080-547-8308.
e-mail: m.grano@anatomia.uniba.it

**Ann. N.Y. Acad. Sci. 1068: 334–340 (2006). © 2006 New York Academy of Sciences.
doi: 10.1196/annals.1346.049**

and low bone formation.[1] The OCs are derived from cells of the monocyte-macrophage lineage, which then fuse to form active resorbing cells in the presence of macrophage-colony stimulating factor (M-CSF) and RANKL.[2] Osteoprotegerin (OPG) is a recently identified cytokine that acts as a decoy receptor for RANKL.[3] Furthermore, it has recently been shown that TRAIL may regulate OPG activity through the OPG/TRAIL interaction.[3] The aim of our study was to investigate the potential involvement of MM T cells in osteoclastogenesis.

MATERIALS AND METHODS

Patients and Cell cultures

The samples included peripheral blood from 32 patients diagnosed as having MM and from 32 patients with non-neoplastic disease without any skeletal involvement, matched for age and sex. Peripheral blood mononuclear cells (PBMCs) were cultured in the presence or absence of 25 ng/mL rh-M-CSF and 30 ng/mL RANKL (R&D Systems Inc., Minneapolis, MN, USA). Mature OCs were identified as TRAP+ multinucleated cells. T cell depletion was performed by using anti-CD2 Ab-coated immunomagnetic Dynabeads (Dynal, Lake Success, NY, USA). For some experiments, PBMCs were cultured in the presence of different concentrations of anti-TRAIL monoclonal antibodies (mAb; R&D Systems Inc.; 10–500 ng/mL) with or without RANKL

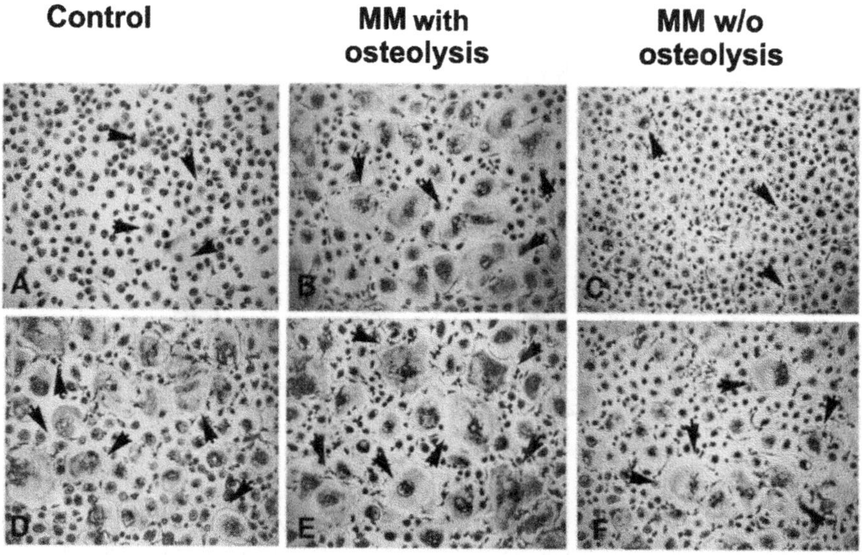

FIGURE 1. TRAP+ OCs obtained from unfractionated PBMCs of MM patients with or without osteolysis (B,E,C,F) and controls (A–D).

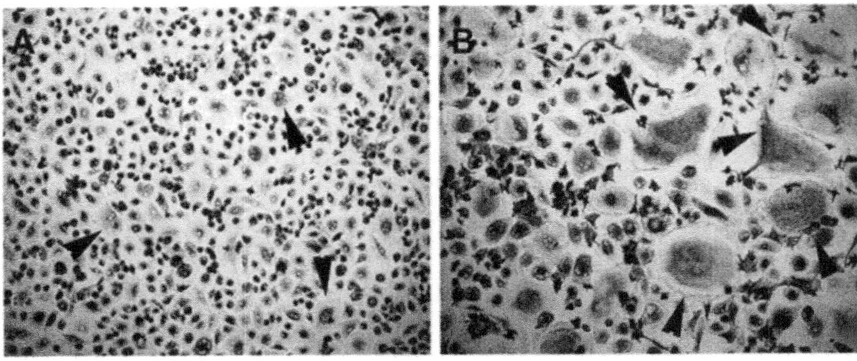

FIGURE 2. Osteoclastogenesis in T cell-depleted MM PBMCs cultured in the absence (A) or in the presence (B) of M-CSF and RANKL (B).

(30 ng/mL). Cell viability was measured by the 3-(4,5-dimethylthiazol-2-yl)-2,5-diphenyltetrazolium bromide (MTT) assay.

RNA Isolation and Reverse-Transcriptase (RT)-PCR Amplification

The freshly isolated T cells and OCs obtained from PBMC cultures of the MM patients and controls were subjected to mRNA extraction using spin

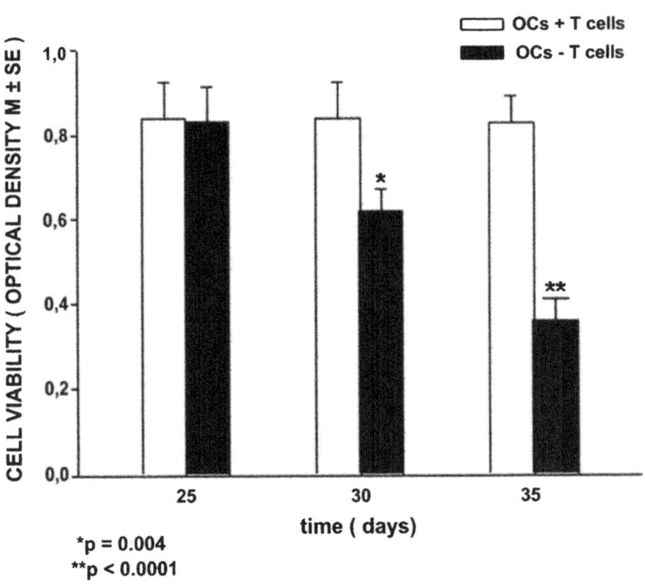

FIGURE 3. Cell viability was measured by MTT assay. MM PBMCs from patients with osteolysis were cultured in the presence (OCs + T cells) or in the absence (OCs–T cells) of T cells for 25, 30 and 35 days. M-CSF and RANKL were added to the T cell-depleted cultures in order to allow OC formation.

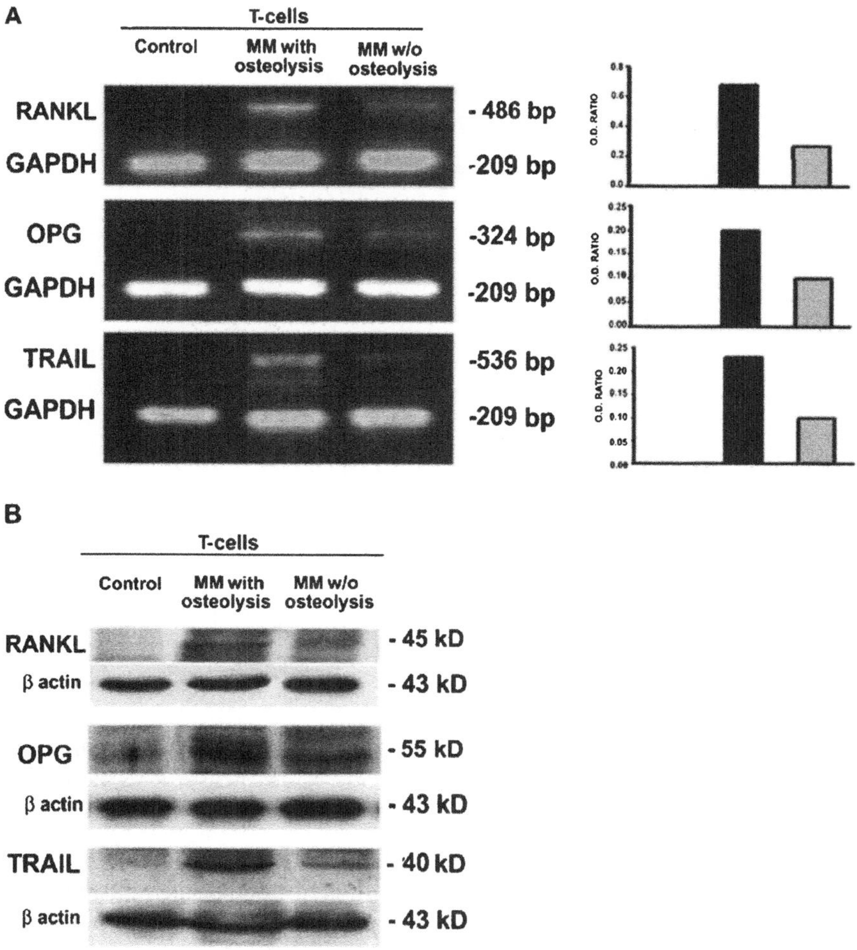

FIGURE 4. RANKL, OPG and TRAIL expression by fresh T cells purified from PBMCs of MM patients with or without osteolysis and controls by RT-PCR (A) and western-blotting (B).

columns (Qiagen, Hilden, Germany). The first-strand cDNA synthesis was prepared according to the manufacturer's instructions (Invitrogen, Carlsbard, CA, USA). Amplification reactions were carried out using platinum *taq*/polymerase (Invitrogen).

Western Blot Analysis

The fresly isolated T cells and OCs obtained from MM and control PBMC cultures were solubilized with lysis buffer. Proteins were subjected to SDS-PAGE gel and transferred to nitrocellulose membrane and were probed for

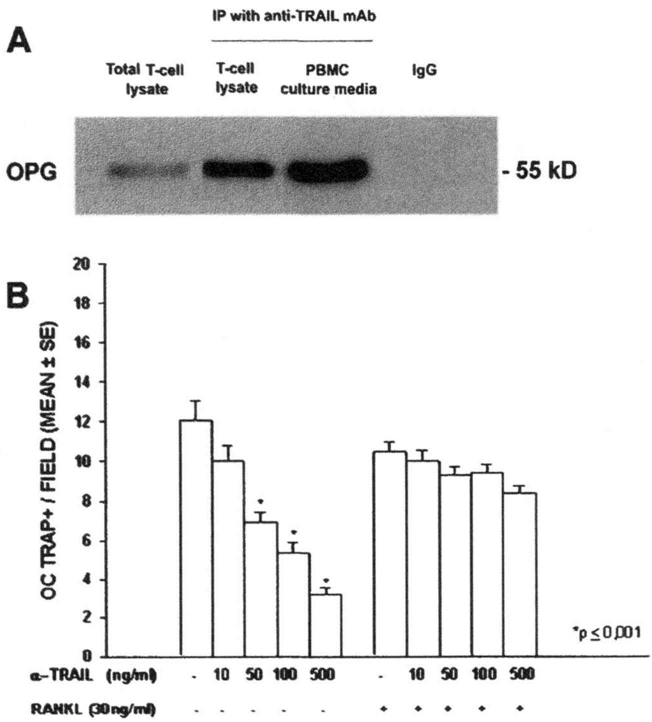

FIGURE 5. OPG/TRAIL complex in fresh T cells and culture media from PBMCs of MM patients (A), and anti-TRAIL mAb effect on osteoclastogenesis ± RANKL (B).

various protein (RANKL, OPG, TRAIL, DR4, DcR2, and β-actin). To detect the TRAIL/OPG complex, immunoprecipitation of T cell lysates and the media collected from PBMC cultures was performed using anti-TRAIL mAb.

RESULTS AND DISCUSSION

Numerous TRAP+ OCs were identified in the unstimulated PBMC cultures from MM patients with osteolysis (FIG. 1B), whereas fewer OCs appeared in the PBMC cultures from the controls and the MM patients without osteolysis(FIG. 1A–C). Moreover, the addition of M-CSF and RANKL to the PBMC cultures from MM patients with osteolysis did not significantly change the OC number (FIG. 1E) compared to the parallel unstimulated cultures (FIG. 1B). However, the above cytokines were essential to trigger and sustain osteoclastogenesis of PBMCs from MM patients without osteolysis and from the controls (FIG. 1D–F). Thus, T cell-depleted PBMC cultures derived from patients with MM were established, since literature data demonstrated that in a pathological condition T cells support osteoclastogenesis.[4] The unstimulated

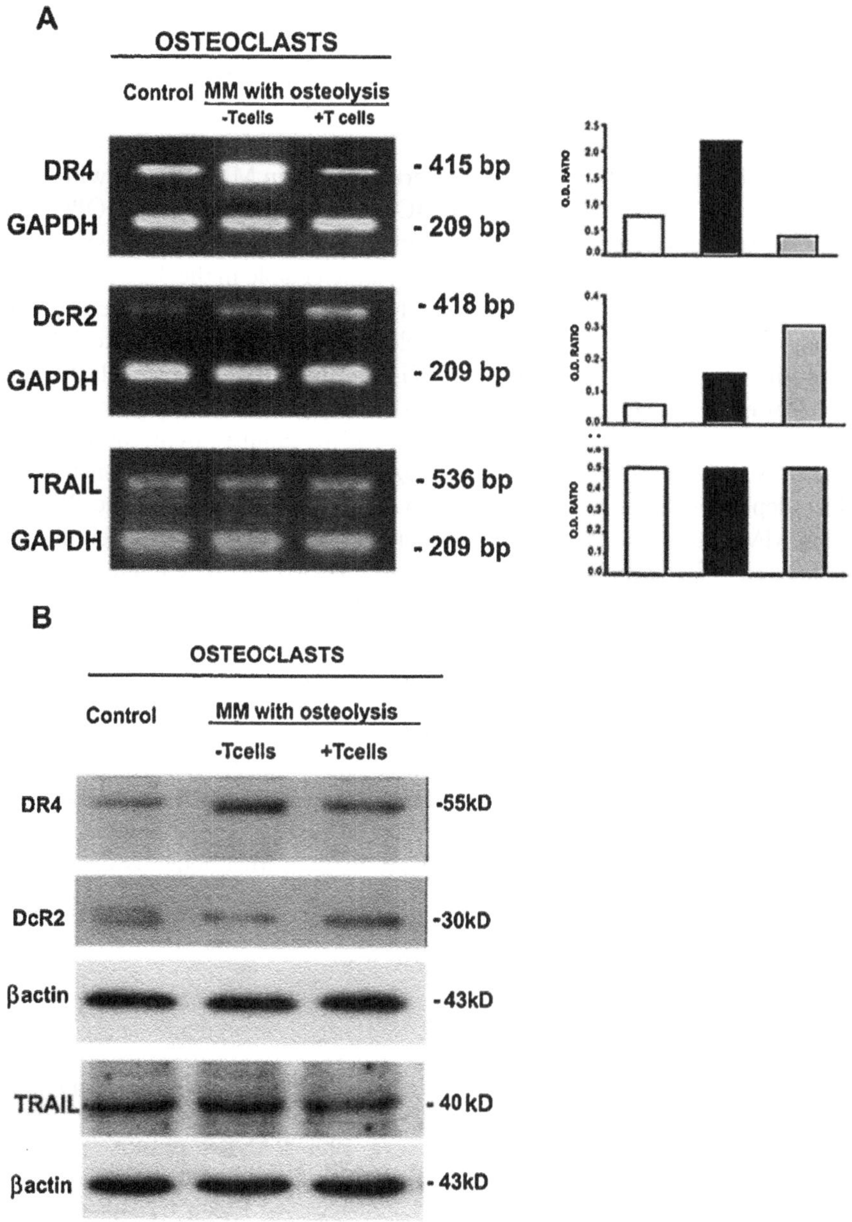

FIGURE 6. OC expression of TRAIL and its death (DR4) and decoy (DcR2) receptors was analyzed by RT-PCR (A) and western blotting (B).

cultures of T cell-depleted PBMCs from MM patients with osteolysis resulted in the development of few small-sized OCs (FIG. 2A). In a different manner, the addition of M-CSF and RANKL to these cultures induced the formation of numerous TRAP+ OCs (FIG. 2B), which exhibited a significantly lower survival than those formed in MM-unfractionated PBMCs (FIG. 3). Moreover, we demonstrated by RT-PCR and Western blot that in MM patients with osteolysis, fresh T cells purified from PBMCs overexpressed RANKL, OPG, and TRAIL, whereas these were expressed in a lower fashion in the fresh T cells from MM patients without osteolysis, and undetectable in the T cells from the controls (FIG. 4A,B). The persistence of osteoclastogenesis despite the high levels of OPG in our system could be explained by the OPG binding to TRAIL. To demonstrate that TRAIL-bound OPG, inhibiting the anti-osteoclastogenic OPG effect, we detected the OPG/TRAIL complex in T cell lysates and in the culture media collected from the MM PBMC cultures by immunoprecipitation with an anti-TRAIL mAb (FIG. 5A). Furthermore, we demonstrated a dose-dependent inhibition of spontaneous osteoclastogenesis in the PBMCs from MM patients with osteolysis cultured in the presence of different concentrations of a neutralizing anti-TRAIL mAb (FIG. 5B). The longer survival, exhibited exclusively by the OCs, generated in the presence of MM T cells, prompted a T cell control of OC survival. In these OCs, we investigated the TRAIL receptor expression by RT-PCR and Western blot and we demonstrated upregulated decoy receptor DcR2 and downregulated death receptor DR4 expression, whereas DR4 was found overexpressed and DcR2 at low levels on the OCs from T cell-depleted cultures (FIG. 6A,B). In conclusion, our results highlight the T cell regulation of OC formation and survival through RANKL, OPG, and TRAIL overexpression, and provide the basis for a potential role for the novel OPG/TRAIL interaction in the biology of MM bone disease.

REFERENCES

1. BARILLÉ-NION, S. & R. BATAILLE. 2003. New insights in myeloma-induced osteolysis. Leuk. Lymphoma. **44:** 1463–1467.
2. BOYLE, W.J., W.S. SIMONET & D.L. LACEY. 2003. Osteoclast differentiation and activation. Nature **423:** 337–342.
3. THEILL, L.E., W.J. BOYLE & J.M. PENNINGER. 2002. RANK-L and RANK: T cell, bone loss, and mammalian evolution. Ann. Rev. Immunol. **20:** 795–823.
4. KONG, Y.Y., U. FEIGE, I. SAROSI, *et al.* 1999. Activated T cells regulate bone loss and joint destruction in adjuvant arthritis through osteoprotegerin ligand. Nature **402:** 304–309.

Restoration of Bone Mass in *Hpg* Mouse by Preoptic Area Grafting

GOPALAN RAJENDREN, HANG ZHOU, BALJIT S. MOONGA, MONE ZAIDI, AND LI SUN

Department of Medicine, The Mount Sinai Bone Program, and the Bronx VA Medical Center, New York, New York 10029, USA

ABSTRACT: Hereditary hypogonadism in the *hpg* mouse, caused by a deletion mutation in the gonadotropin-releasing hormone (GnRH) gene, is associated with sterility, absent ovarian development, and undetectable circulating sex steroids. Eight-month-old female *hpg* mice had a significantly reduced bone mineral density (BMD) at the lumbar spine, femur, and tibia. In addition, the mice showed significant reductions in liver and kidney weight, with virtually nonexistent ovaries. Successfully transplanted *hpg* mice with preoptic area grafts contained GnRH-positive neurons, consistent with our previous experience, and the host median eminence was innervated by GnRH immunoreactive fibers. A return of reproductive function was evident from increased ovarian weight and vaginal cornification. Of note was that grafted *hpg* mice showed a complete reversal to baseline of their BMD measured at all three sites. This establishes that the low bone mass that occurs in old *hpg* mice can be fully and rapidly ameliorated by preoptic area grafting.

KEYWORDS: osteoporosis; hypogonadism; gonadotropin–releasing hormone

INTRODUCTION

The release of the gonadotropins, LH and FSH, is controlled by gonadotropin-releasing hormone (GnRH) which is secreted from the hypothalamus in a pulsatile manner. A deletion mutation of at least 33.5 kb encompassing the distal half of the gene for the common precursor of GnRH and the GnRH-associated peptide (GAP) thus results in hypogonadism in the *hpg* mouse.[1,2] The homozygotic females are sterile, lack ovarian and uterine development, have extremely low pituitary and plasma gonadotropin levels, and undetectable hypothalamic GnRH content.[3] Similarly, human hypogonadotrophic

Address for correspondence: Li Sun, M.D., Ph.D., Endocrinology 1055, Mount Sinai School of Medicine, One Gustave L. Levy Place, New York, NY 10029. Voice: 212-241-8797; fax: 212-426-8312.
e-mail: mone.zaidi@mssm.edu

Ann. N.Y. Acad. Sci. 1068: 341–347 (2006). © 2006 New York Academy of Sciences.
doi: 10.1196/annals.1346.050

hypogonadism is characterized by delayed puberty, infertility, and low serum gonadotropins.

To normalize reproduction in the *hpg* mice, attempts have been made to reinnervate the hypothalamus with GnRH-containing neurons from wild-type mice. Thus, when brain grafts of normal fetal preoptic area (POA) containing GnRH cells were implanted in the third ventricle of adult *hpg* mice, both pituitary and plasma gonadotropin levels increased[4] and the deficits in fertility was corrected to a great extent.[5] Grafted GnRH neurons were found to innervate the median eminence of the host, and even low levels of innervation, support pulsatile LH secretion, gonadal development, and sexual function.[6]

In our preliminary studies, we have shown that as both male and female *hpg* mice age, they develop a low-turnover form of osteoporosis.[7] In this article, we have attempted to rescue the osteopenia by POA grafting in female *hpg* mice.

MATERIALS AND METHODS

GnRH-deficient (hpg) *Mice*

GnRH-deficient *hpg* mice have been bred and maintained in-house for several years. Colonies of heterozygotes that are sexually and hormonally intact, are crossed to produce *hpg* mice that are genotyped by Southern blotting of tail DNA. Female *hpg* mice have infantile ovaries, underdeveloped secondary sexual tissue, and low circulating levels of LH and FSH.

POA Grafting

Grafting of GnRH cell-containing POA was performed in 8-month-old female *hpg* mice as described before.[5] Fetal POA obtained from 17- to 19-day-old wild-type mouse fetuses were stereotaxically implanted into the third cerebral ventricle (1.0 mm posterior to bregma at midline and 5.2 mm ventral to the dura) of *hpg* mice under chloral hydrate anesthesia (380 mg/kg bw). The success of the graft and gonadal development was apparent in 3–4 weeks, as indicated by vaginal opening and the presence of cornified epithelial cells in the vaginal smear.[5]

Bone Densitometry

Six weeks after the graft surgery, bone mineral density (BMD) measurements were performed using a Lunar *Piximus* with a precision of <1.5%. After determination of the whole body weight, mice were anesthetized by the intraperitoneal administration of chloral hydrate (380 mg/kg). Following whole body BMD measurements, with the cranium excluded, a region of interest was selected manually for measurements at specific sites, notably the spine (L4–L6), right and left femur, and right and left tibia. The instrument was calibrated each time before use by employing a phantom *per* manufacturer's recommendation.

Immunocytochemistry

Under deep chloral hydrate anesthesia (760 mg/kg), the mice were perfused transcardially with normal saline (15 mL) followed by 4% paraformaldehyde (60 mL) immediately after the BMD measurements. The brains were removed and stored in 4% paraformaldehyde overnight. At this time the wet weight of the liver, kidney, and ovaries was determined and recorded. Brains were sectioned at 50 μM using a Vibratone. A mouse monoclonal anti-GnRH antibody (QED Advanced Research, San Diego, CA), which has been used in earlier studies,[8] was employed to label GnRH cells. Briefly, the sections were incubated with anti-GnRH antibody (1:5,000) for 5 days followed by an anti-mouse IgG-rhodamine conjugate (1:200) for 2 h. They were then mounted onto gelatin-coated slides and observed under an Olympus AX-70 fluorescent microscope (Olympus America Inc., Melville, NY) equipped with an appropriate filter.

RESULTS AND DISCUSSION

Although idiopathic hypogonadotrophic hypogonadism in humans is associated with osteoporosis of a low-turnover variety,[9] the precise cellular and molecular dysfunction is unclear. The osteoporosis is not ameliorated by sex steroid replacement, particularly when administered during skeletal immaturity.[9] Here, we show that 8-month-old congenitally GnRH-deficient female mice have lowered bone density at the lumbar spine, tibia, and femur (FIG. 1). Notable is that in *hpg* mice, bone loss is both cortical and trabecular, whereas predominantly trabecular osteopenia is seen following gonadal steroid withdrawal.[10] Similarly, old *hpg* males have osteopenia, again, at both cortical and trabecular sites.[7] This resembles senile osteoporosis in animals and humans of both sexes, which is characterized with both cortical and trabecular bone loss.[11]

The most significant finding, however, is that the low bone mass in the 8-month-old *hpg* female is ameliorated fully by the successful implantation of a POA graft containing GnRH-positive neurons. Thus, 6 weeks following POA grafting, BMD at all three sites normalized to levels that were no different from that of wild-type littermates (FIG.1). In addition, the reproductive function of the *hpg* mice, which were otherwise sexually immature with grossly underdeveloped ovaries, was expectedly restored.[12] Namely, 5/5 mice achieved

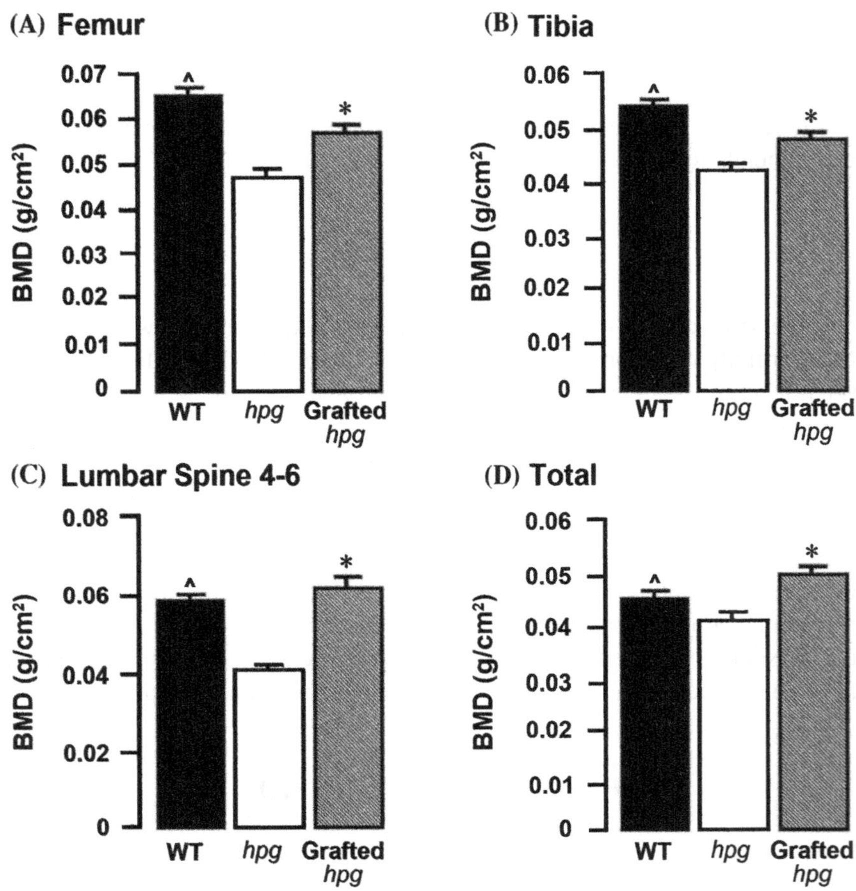

FIGURE 1. Effect of POA grafting on BMD (g/cm^2) in 8-month-old *hpg* mice measured by a Lunar *Piximus* at the femur, tibia, lumbar spine (L4–L6), and total body. Note that the significant decrements in BMD are reversed fully following surgery. Statistics: unpaired Student's *t*-test, $P < 0.05$, ∗grafted *hpg* mice versus control *hpg* mice; ^control *hpg* mice versus normal mice.

vaginal cornification that was nonexistent in *hpg* females. Furthermore, ovarian weight increased by ~40-fold (FIG. 2). Together, the findings indicated an intact hypothalmic–pituitary axis in the grafted mice.

Furthermore, immunostaining of the hypothalamus shown in FIGURE 3 provided unambiguous evidence for re-innervation of the median eminence with GnRH-positive neurons. The latter appears to be a prerequisite in the return of gonadal function in the *hpg* mouse. It has been shown, however, that innervation of the median eminence even by less than 10 GnRH-positive neurons can restore gonadal function.[6] It has also been demonstrated that such innervation is associated with a full return of pulsatile LH secretion, intact reproductive function, and full sexual function.[5,6]

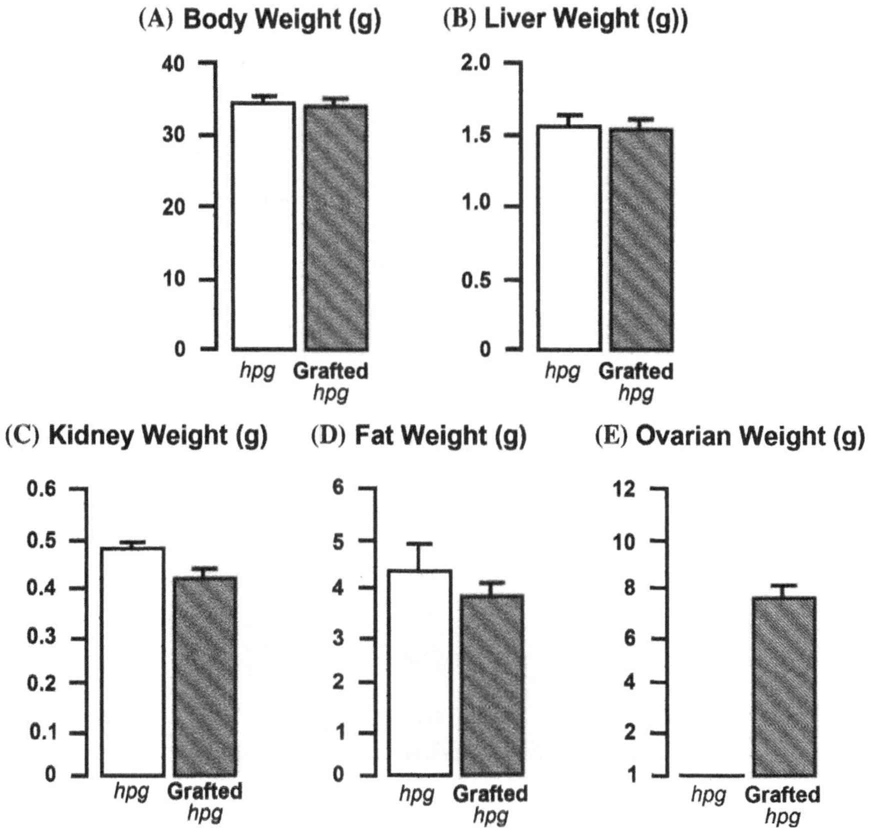

FIGURE 2. POA grafting in *hpg* mice did not affect total body weight, or wet weight of liver, kidney or fat, whereas ovarian weight was dramatically increased. Statistics: unpaired Student's *t*-test, $P < 0.001$ for ovarian weight comparing control with grafted *hpg* mice. Other parameters were not significant.

Interestingly, female *hpg* mice had reduced liver and kidney weights compared with wild-type littermates. The same has been shown for male *hpg* mice in other studies.[13] The liver and spleen did not, however, achieve a normal size upon POA grafting. Finally, we found no difference in body weight between normal mice, *hpg* mice, and POA grafted mice.

Thus, while restoration of the GnRH-pituitary-gonadal steroid axis cures the osteopenia in the aging female *hpg* mouse, its precise cellular and molecular mechanism remains unclear. There are some peculiarities. The osteoporosis does not mimic that seen with gonadal steroid withdrawal; yet, restoration of the hypothalamus–pituitary axis with POA grafting ameliorates the bone loss. It is possible that while restoration of sex steroids may contribute to this effect, other pituitary hormones, FSH and LH, or even GnRH itself, may have direct actions on the skeleton.

FIGURE 3. Photomicrograph of brain section showing a POA graft containing GnRH immunoreactivity (red) in the third ventricle of the *hpg* mouse. The arrows indicate the GnRH-positive cells within the graft median eminence (ME). Scale bar: 50 μM.

ACKNOWLEDGMENTS

MZ acknowledges support from the National Institutes of Health (RO1 AG14197, DK70526, and AG23176). MZ and BSM also acknowledge support of the Department of Veterans Affairs (Merit Review). LS was recipient of the 2005 Award for Outstanding Research in the Pathophysiology of Osteoporosis.

REFERENCES

1. Riggs, B.L., S. Khosla & L.J. Melton III. 2002. Sex steroids and the construction and conservation of the adult skeleton. Endocr. Rev. **3:** 279–302.
2. Mason, A.J., J.S. Hayflick, R.T. Zoeller, *et al.* 1986. A deletion truncating the gonadotropins-releasing hormone gene is responsible for hypogonadism in the *hpg* mouse. Science **234:** 1366–1371.

3. CATTANACH, B.M., C.A. IDDON, H.M. CHARLTON, *et al.* 1977. Gonadotropin releasing hormone deficiency in a mutant mouse with hypogonadism. Nature **269:** 338–340.
4. KRIEGER, D.T., M.J. PERLOW, M.J. GIBSON, *et al.* 1982. Brain grafts reverse hypogonadism of gonadotropin releasing hormone deficiency. Nature **298:** 468–471.
5. GIBSON, M.J., D.T. KRIEGER, H.M. CHARLTON, *et al.* 1984. Mating and pregnancy can occur in genetically hypogonadal mice with preoptic area brain grafts. Science **225:** 949–951.
6. GIBSON, M.J., G.M. MILLER & A.J. SILVERMAN. 1991. Pulsatile luteinizing hormone secretion in normal female mice and in hypogonadal female mice with preoptic area implants. Endocrinology **128:** 965–971.
7. WU, X.B., G. RAJENDREN, L. SUN, *et al.* 2002. Amelioration of low-turnover osteoporosis in the GnRH-deficient hypogonadal hyperleptinaemic *(hpg)* male mouse by estrogen. J. Bone Miner. Res. **16:** S522.
8. RAJENDREN, G. & M.J. GIBSON. 1998. Comparison of afferent and efferent competence of transplanted GnRH cells in the brains of hypogonadal female mice. J. Neuroendocrinol. **10:** 249–257.
9. FINKELSTEIN, J.S., A. KLIBANSKI, R.M. NEER, *et al.* 1989. Increases in bone density during treatment of men with idiopathic hypogonadotrophic hypogonadism. J. Clin. Endocrinol. Metabol. **69:** 776–783.
10. KIMBLE, R.B., S. BAIN & R. PACIFICI. 1997. The functional block of TNF but not of IL-6 prevents bone loss in ovariectomized mice. J. Bone Miner. Res. **12:** 935–941.
11. RIGGS, B.L., S. KHOSLA & L.J. MELTON III. 1998. A unitary model for involutional osteoporosis: estrogen deficiency causes both type I and type II osteoporosis in postmenopausal women and contributes to bone loss in aging men. J. Bone Miner. Res. **13:** 763–773.
12. RAJENDREN, G. & M.J. GIBSON. 2001. Increased synaptic input to gonadotropin releasing hormone cells in preoptic area grafts that support reproductive development in female hypogonadal mice. J. Neuroendocrinol. **13:** 270–274.
13. CRAWFORD, B.A., J. SINGH, J.M. SIMPSON & D.J. HANDELSMAN. 1993. Androgen regulation of circulating insulin-like growth factor-I during puberty in male hypogonadal mice. J. Endocrinol. **139:** 57–63.

IL-7 Drives T Cell-Mediated Bone Loss Following Ovariectomy

MICHAELA ROBBIE-RYAN, ROBERTO PACIFICI, AND M. NEALE WEITZMANN

Division of Endocrinology & Metabolism & Lipids, Emory University School of Medicine, Atlanta, Georgia 30322, USA

ABSTRACT: Ovariectomy-induced bone loss stems in large measure from a realignment of adaptive immune responses leading to the activation and expansion of tumor necrosis factor (TNF)-producing T cells. The mechanisms driving this T cell expansion are complex but we have recently reported that the pro-osteoclastogenic cytokine interleukin (IL)-7 plays a critical role in this process. The mechanisms of IL-7 action are intricate and poorly defined. We present herein an overview of our current understanding of IL-7 action on bone turnover and the role of IL-7 in ovariectomy-induced bone loss.

KEYWORDS: IL-7; T cells; estrogen; osteoclast; osteoblast

IL-7 AND OSTEOCLASTOGENESIS

IL-7 has long been known to stimulate osteoclastic bone loss when injected into mice *in vivo.*[1] However, the mechanisms of IL-7 action *in vivo* are complex and both the B cell and T cell lineages are strongly impacted by this cytokine. As IL-7 administration *in vivo* mimics both the bone destruction and the expansion of B lineage cells observed during ovariectomy, it was originally proposed that IL-7 may stimulate bone destruction by a mechanism involving B cells.[1] However, in an *in vitro* model of human osteoclastogenesis, we discovered that IL-7 stimulated osteoclast formation in the complete absence of B lineage cells. In contrast, T cells were an absolute requirement. Additional investigations determined that IL-7 stimulated osteoclastogenesis *in vitro* by inducing the production of RANKL by T lymphocytes.[2] To investigate the role of T cells in IL-7 mediated bone loss *in vivo*, we injected IL-7 into T cell-replete heterozygous nude mice and T cell-deficient nude mice. While the injection of IL-7 into T cell-replete mice induced significant bone

Address for correspondence: M. Neale Weitzmann, Ph.D., Division of Endocrinology & Metabolism & Lipids, Emory University School of Medicine, 101 Woodruff Circle, 1305 WMRB, Atlanta, GA 30322-0001. Voice: 404-727-1389; fax: 404-727-1300.
e-mail: mweitzm@emory.edu

Ann. N.Y. Acad. Sci. 1068: 348–351 (2006). © 2006 New York Academy of Sciences.
doi: 10.1196/annals.1346.051

loss, T cell-deficient nude mice were completely protected. Reconstitution of T cells into nude mice by means of adoptive transfer rescued the capacity of IL-7 to induce bone destruction. Furthermore, equal numbers of splenocytes derived from T cell-replete mice were found to secrete both RANKL and tumor necrosis factor (TNF)-α *ex vivo* following IL-7 administration *in vivo*. In contrast, T cell-deficient splenocytes from nude mice failed to produce RANKL or TNF-α but T cell reconstitution of nude mice again rescued the capacity of IL-7 to elicit cytokine production from splenocytes.[3]

IL-7 may in fact have two different competing actions on osteoclastogenesis. Addition of IL-7 to bone marrow cultures stimulated by exogenous RANKL and M-CSF inhibits osteoclast formation.[4] The mechanism of this direct suppressive effect of IL-7 is not clear; however, it is reported that early B cell precursors of the $B220^+$ lineage, which populate the bone marrow and are known to expand during estrogen deficiency,[5] may be capable of acting as an additional pool of osteoclast precursors *in vitro* as these precursor cells have the capacity to differentiate into bone-resorbing osteoclasts.[3,4,6] IL-7 may suppress the differentiation of this osteoclast precursor population, possibly by stimulating their differentiation toward mature B cells. While the relevance of $B220^+$ cells in ovariectomy-induced bone destruction *in vivo* remains to be investigated, the net effect of IL-7 *in vivo* appears to be pro-osteoclastogenic.

IL-7 AND BONE LOSS IN ESTROGEN DEFICIENCY

To address the issue of whether IL-7 plays a role in the bone destruction stimulated by ovariectomy, we measured IL-7 levels in sham operated and ovariectomized mice.[7] Significant elevations in IL-7 protein and mRNA were detected in bone marrow from ovariectomized mice. Importantly, injection of a neutralizing antibody against IL-7 completely prevented bone destruction following ovariectomy. Indeed, IL-7 has potent actions on the entire T cell lineage, and preliminary studies now suggest that IL-7 acts during ovariectomy to induce the expansion of hematopoietic stem cells in the bone marrow giving rise to enhanced numbers of T cell precursors. In addition, IL-7 appears to stimulate the process of thymic export as well as play a role in peripheral T cell expansion.[8] IL-7 consequently acts at multiple sites to expand T cell numbers and to stimulate T cell osteoclastogenic cytokine production.

Further investigations revealed that IL-7 neutralization *in vivo* prevented the rise in biochemical indices of bone turnover while causing a significant upregulation in indices of bone formation. This data suggested that IL-7 not only stimulates bone resorption but also suppresses bone formation. To verify the suppressive effect of IL-7 on bone formation, we treated neonatal mouse calvaria with IL-7 in organ cultures. IL-7 suppressed new bone formation in calvarial cultures, in addition to suppressing bone morphogenetic protein

(BMP)-2 stimulated new bone formation. Finally, IL-7 administration in mice dramatically suppressed bone formation *in vivo* as quantitated by double calcein labeling. To investigate the mechanism of IL-7 action on bone formation we evaluated the effects of IL-7 on the endogenous transcription of Runx2 (Cbfa-1) using a heterologous Runx2-driven promotor transfected into ROS 17/2.8 osteoblastic cells. In addition, we quantitated runx2 transcription in response to IL-7 using a reporter driven by the distal osteoblast-specific runx2 promotor. In both cases IL-7 suppressed reporter transcription by approximately 50%.

These data suggested that in ovariectomy conditions elevated concentrations of IL-7 not only upregulate bone resorption but simultaneously reduce the magnitude of the natural compensatory increase in bone formation, thus

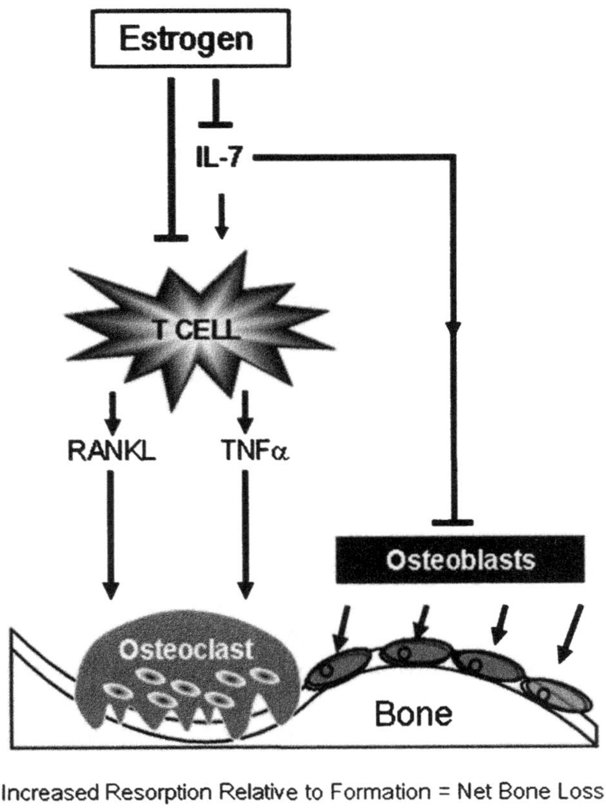

FIGURE 1. Mechanism of IL-7 action in estrogen deficiency bone loss. During estrogen deficiency, upregulated levels of IL-7 stimulate expansion of T cell populations leading to enhanced TNF- and RANKL-driven osteoclastogenesis. In addition, IL-7 suppresses the differentiation of osteoblasts limiting the magnitude of the compensatory increase in bone formation exacerbating bone loss.

preventing the reacquisition of bone homeostasis. The overall effects of IL-7 on bone mass are summarized diagrammatically in FIGURE 1.

CONCLUSION

Taken together, a growing body of evidence now suggests that the suppression of IL-7 by estrogen is a key mechanism by which estrogen protects bone mass, and that ovariectomy leads to an IL-7-driven wave of bone destruction.

REFERENCES

1. MIYAURA, C. *et al.* 1997. Increased B-lymphopoiesis by interleukin 7 induces bone loss in mice with intact ovarian function: similarity to estrogen deficiency. Proc. Natl. Acad. Sci. USA **94:** 9360–9365.
2. WEITZMANN, M.N. *et al.* 2000. Interleukin-7 stimulates osteoclast formation by up-regulating the T- cell production of soluble osteoclastogenic cytokines. Blood **96:** 1873–1878.
3. TORALDO, G. *et al.* 2003. IL-7 induces bone loss in vivo by induction of receptor activator of nuclear factor kappa B ligand and tumor necrosis factor alpha from T cells. Proc. Natl. Acad. Sci. USA **100:** 125–130.
4. LEE, S.K. *et al.* 2003. Interleukin-7 is a direct inhibitor of in vitro osteoclastogenesis. Endocrinology **144:** 3524–3531.
5. MASUZAWA, T. *et al.* 1994. Estrogen deficiency stimulates B lymphopoiesis in mouse bone marrow. J. Clin. Invest. **94:** 1090–1097.
6. SATO, T. *et al.* 2001. Generation of bone-resorbing osteoclasts from B220+ cells: its role in accelerated osteoclastogenesis due to estrogen deficiency. J. Bone Miner. Res. **16:** 2215–2221.
7. WEITZMANN, M.N. *et al.* 2002. Increased production of IL-7 uncouples bone formation from bone resorption during estrogen deficiency. J. Clin. Invest. **110:** 1643–1650.
8. RYAN, M.R. *et al.* 2005. An IL-7-dependent rebound in thymic T cell output contributes to the bone loss induced by estrogen deficiency. Proc Natl. Acad. Sci. USA **102:** 16735–16740.

Stem Cells and Tissue Engineering: Past, Present, and Future

JULIA M. POLAK AND ANNE E. BISHOP

Tissue Engineering & Regenerative Medicine Centre, Imperial College, Chelsea & Westminster Campus, London SW10 9NH, UK

ABSTRACT: Tissue engineering is an interdisciplinary field that brings together the principles of the life sciences and medicine with those of engineering. The increase in its development over the past decade has resulted from a variety of factors; advances in genomics and proteomics, the advent of new biomaterials as potential templates for tissue growth, improvements in bioreactor design, and increased understanding of healing processes. Possibly the greatest contribution has come from our increased knowledge and understanding of stem cell biology, which is paving the way for the generation of unlimited cells of specific phenotypes for incorporation into engineered tissue constructs. Thus, tissue engineering approaches for expanding and engrafting the differentiated progeny of embryonic, fetal, or adult stem cells have major potential for tissue repair and will make a major contribution to medicine in the 21st century.

KEYWORDS: stem cells; tissue engineering; differentiation; bioactive materials

PAST

Tissue engineering has been broadly defined as the design and construction in the laboratory of living, functional components that can be used for the maintenance, regeneration, or replacement of malfunctioning tissues. Although it is widely assumed to be a relatively new field, the first report of tissue engineering was published in 1933 when tumor cells were implanted into a pig wrapped in a polymer membrane to protect them from immune attack.[1] However, the modern era of tissue engineering could be said to have begun at the start of the 1980s with the development and clinical use of skin replacements.[2,3] The field continues to expand and, although there have been some hold-ups on the industrial development side, including problems in devising cost-efficient scaleable processes, guaranteeing product viability, and

Address for correspondence: Julia M. Polak, Tissue Engineering & Regenerative Medicine Centre, Imperial College, Chelsea & Westminster Campus, Fulham Road, London SW10 9NH, UK. Voice: +44-208-237-2569; fax: +44-9746-5619.
e-mail: julia.polak@imperial.ac.uk

**Ann. N.Y. Acad. Sci. 1068: 352–366 (2006). © 2006 New York Academy of Sciences.
doi: 10.1196/annals.1346.001**

satisfying regulators, the basic research is rapidly evolving. Without doubt, a lot of the momentum that has driven the development of tissue engineering as a discipline in its own right has come from the quaternary of Vacanti brothers, Charles, Joseph, Martin, and Frank.

Tissue engineering is highly interdisciplinary and amalgamates the principles of the life sciences and medicine with those of engineering. The basic components of tissue engineering are cells, scaffolds, and signals. Tissue engineers have turned to virtually every cell type over the past two decades to try to define the best cell source for each particular tissue being constructed. A major advance has been the recognition of the huge advantages offered by stem cells; these include primarily their ability to self-renew, providing a theoretically inexhaustible supply, and to differentiate to any cell type when grown under the appropriate conditions.

This article provides a resumé of stem cell biology and tissue engineering, the current status quo, and we will also attempt to predict some of the key areas where stem cells are likely to be used successfully in future tissue engineering strategies.

PRESENT

Tissue Engineering

The challenge for tissue engineers is to optimize cell isolation, proliferation and, in the case of stem cells, differentiation and to design scaffolds and/or delivery systems that promote the coordinated growth of three-dimensional tissues. A widely used basic strategy has been to harvest cells from a patient, expand them in cell culture, and seed them onto a scaffold that provides a biomechanical environment that drives formation of the required tissue until the cells produce their own extracellular matrix. The tissue can be grown on a scaffold that will completely resorb as the new tissue grows, so that only the new tissue will be implanted; or a biocomposite on a permanent scaffold can be implanted. After implantation, the tissue engineered construct must then be able to survive, restore normal function, and integrate with the surrounding tissues.

Cell Sources

The single most important element to the success of tissue engineering is the ability to generate sufficient numbers of cells that maintain the appropriate phenotype and perform the required biological functions. For example, cells must produce extracellular matrix in the correct organization, secrete cytokines and other signaling molecules, and interact with neighboring cells/tissues. Immediately, this raises a number of potential problems, not least of which is obtaining appropriate cell numbers to promote repair.

Primary Cells

Primary cells are mature cells specific to tissue type that are harvested from explant material removed by surgical procedure. An example is primary human osteoblasts that are harvested from the femoral heads removed during total hip replacement operations. Primary cells are the most desirable with regard to immunological compatibility but in general they are differentiated and post-mitotic. This can also be compounded by the tendency of some cell types to de-differentiate during *ex vivo* cultivation and express an inappropriate phenotype (e.g., articular chondrocytes in culture often produce fibrocartilage as opposed to hyaline cartilage). Although primary, especially autologous, cells are still used in tissue engineering, yields and proliferation rates tend to be low and for some phenotypes, e.g., spinal cord neurons, harvesting from a patient or donor is not an option. These limitations have stimulated studies to find and develop alternative cell sources for tissue engineering strategies and stem cells are already providing solutions to some of the problems encountered using primary cells released from explanted tissues.

Stem Cells

Stem cells can be defined as undifferentiated cells that can proliferate and have the capacity both to self-renew and to differentiate to one or more types of specialized cells. However, there has been some reconsideration of this definition recently in view of the observation of de-differentiation and trans-differentiation of certain mature cells.[4–6] In view of this, it has been suggested that it should be made broader and applicable to a biological function that can be induced in a range of cell types, including differentiated cells, rather than a single entity.[7] Stem cells can be isolated from embryos, fetuses, or from adult tissue, but the range of cell types to which they can differentiate varies. For tissue engineering, stem cells can provide a virtually inexhaustible cell source. Current research is focused on promoting stem cell differentiation to required lineages, purification of consequent cells, confirmation that there is no residual carcinogenic potential in the cell population, and implantation in a form that will replace, or augment the function of, diseased or injured tissues. An initial step is the selection of the most appropriate stem cell to form the required tissue.

Embryonic. Embryonic stem cells (ESC) are the most plastic cell source available to tissue engineers.[8–10] They were first described more than two decades ago, when they were isolated from the inner cell mass of the developing murine blastocyst and grown in the laboratory.[11,12] ESC have since been shown to be totipotent, differentiating to all lineages, including the germ line and trophoblast.[13–15] *In vitro*, murine ESC proliferate indefinitely in the undifferentiated state and retain the capacity to differentiate to all mature somatic phenotypes when they receive the appropriate signals. Maintenance of their undifferentiated state in culture depends on the presence of leukemia inhibitory factor (LIF) and serum or murine feeder layer cells. Study of the control of murine ESC self-renewal has recently shown that overexpression of the anti-apoptotic protein Bcl-2, acting in concert with LIF, removes the need for serum or feeders.[16]

The initial isolation of murine ESC lines gave developmental biologists a major tool as it provided a simple model system to study the processes of early embryonic development and cellular differentiation. However, it also paved the way for tissue engineering applications if similar totipotent cells could be extracted from human blastocysts. Human ESC were duly derived in 1998,[17,18] providing a tremendous boost for tissue engineering.

Human ESC show several distinct differences from murine ESC; they tend to proliferate more slowly and mostly form flat, rather than spherical, colonies. They are also generally easier to dissociate into single cells than their murine counterparts.[19] Human ESC are unresponsive to LIF and most lines need to be grown on feeder layers in order to sustain self-renewal (FIG. 1). Much recent research has centered on the identification of the pathways controlling

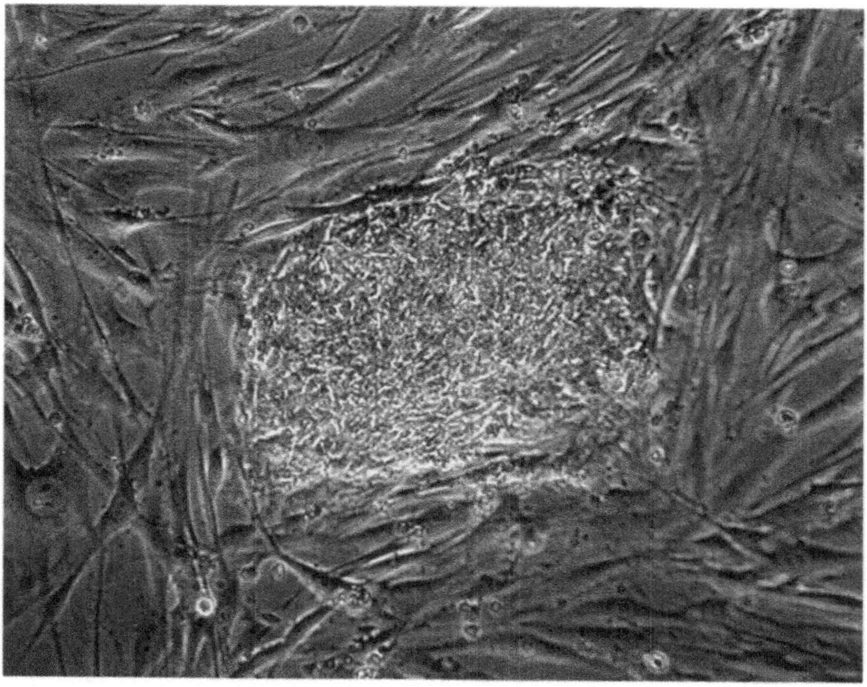

FIGURE 1. A colony of human ESC (H1) growing on a feeder layer of mitotically inactivated murine embryonic fibroblasts. (Inverted microscope image of living cells.)

human ESC renewal. For tissue engineers, one of the main reasons for this has been the need to grow human ESC without any contamination by animal cells or proteins; a problem highlighted by a report in January 2005 that showed the expression of a nonhuman protein by human ESC lines grown on animal feeder layers.[20] Various means have been tried, including addition of basic fibroblast growth factor to the culture medium,[19,21] growth on matrigel or laminin in feeder cell-conditioned medium[22] or in the presence of GSK-3-specific inhibitors such as 6-bromoindirubin-3′-oxime (BIO),[23] and growth on fibroblasts derived from human ESC.[24,25] Although the adapted culture methods remove the need for animal feeder cells, none of the protocols is entirely free of animal products and so the problem remains.

Different cultures and religious groups have their own ethical stance on the use of human ESC.[9] In the UK, human ESC lines can only be derived legally under licence from the Human Fertilization and Embryology Authority and the lines must be placed in the Medical Research Council's Stem Cell Bank. At present, the UK is the only country that allows cloning of human ESC and this is restricted to therapeutic nuclear transfer, to create autologous cells or to produce diseased cells for *in vitro* pathogenetic or drug discovery investigations, with reproductive cloning being illegal.

Fetal. The recent identification of human fetal mesenchymal stem cells (MSC) has raised the possibility of using autologous cells for *in utero* treatments.[26] The MSC population extracted from fetal blood contains adherent cells that divide in culture for 20 to 40 passages and can differentiate into mesenchymal lineages including bone and cartilage, but also have the ability to form oligodendrocytes and hematopoietic cells. These cells, which can be found circulating only during the first trimester, are similar to hematopoietic populations in fetal liver and bone marrow and have been shown in a xenogeneic sheep model to engraft into multiple organs and undergo site-specific tissue differentiation.

Adult. An adult (or postnatal somatic) stem cell is an undifferentiated cell found among differentiated cells in a tissue or organ that can renew itself. There is an extensive repository of these stem cells located in various tissue niches throughout the body, including bone marrow, brain, liver, and skin as well as in the circulation (FIG. 2).[27] Their main role is to maintain and repair the tissue in which they are found. Originally, these cells were considered to have only very limited differentiation potential but there have been many reports that they can show a considerable degree of plasticity although there is some debate about quite how much.[28–32] In theory, therefore, these cells could be removed from a patient, incorporated into a tissue construct, and put back into the same individual when repair becomes necessary, thereby removing the need for immunosuppression. Clearly, adult-derived progenitor cells need to be investigated and their clinical usefulness established. However, as with

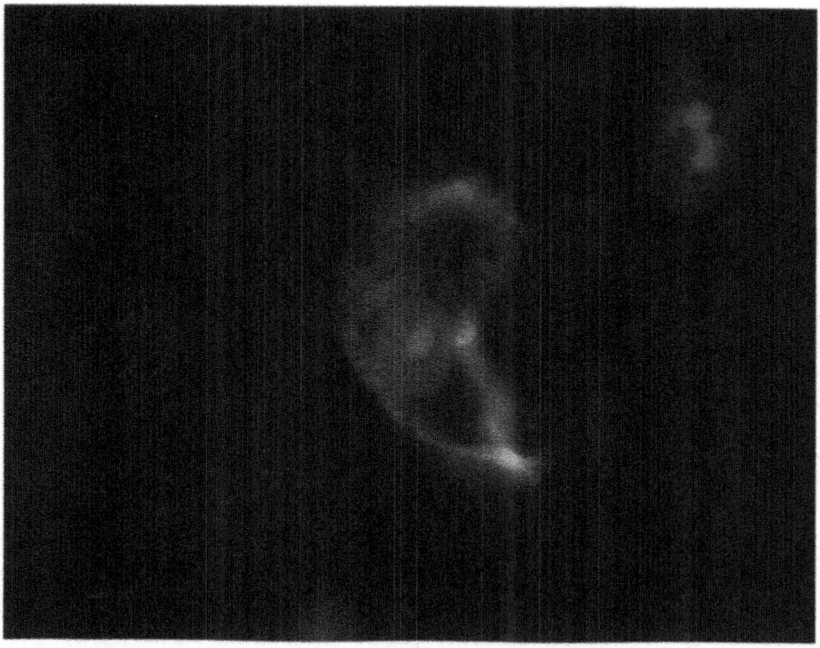

FIGURE 2. A single CD34–human cord blood stem cell fixed and immunostained for α-feto-protein, a differentiation marker for endodermal cells. (Green: indirect immunofluorescence for α-fetoprotein; red: propdium iodide as nuclear counterstain.)

primary mature cells, for some stem cell types, problems with accessibility, low frequency (e.g., there is roughly 1 stem cell per 100,000 bone marrow cells), restricted differentiation potential, and poor growth tends to limit their usefulness for tissue engineering. For a long time, adult stem cells have been considered to be a safer option for clinical applications than ESC because they have not been shown to form teratomas. However, new observations of the spontaneous transformation of human MSC during prolonged culture (4–5 months) argue for reassessment of the biosafety aspects of these cells.[33] The situation is compounded for human MSC that have been transduced with the human telomerase reverse transcriptase gene (hTERT) in order to extend their life span and provide a large number of cells for therapeutic applications.[34] These telomerized cells also undergo spontaneous genetic alteration leading to tumorigenicity in immunodeficient mice.[35] Newly identified sources of stem cells include fat,[36] placenta,[37] and spleen.[38] These sites can be an abundant source of stem cells; for example, a single gram of human adipose tissue has been reported to yield more than 70,000 pluripotent adipose-derived adult stem cells after 1 day in culture.[36]

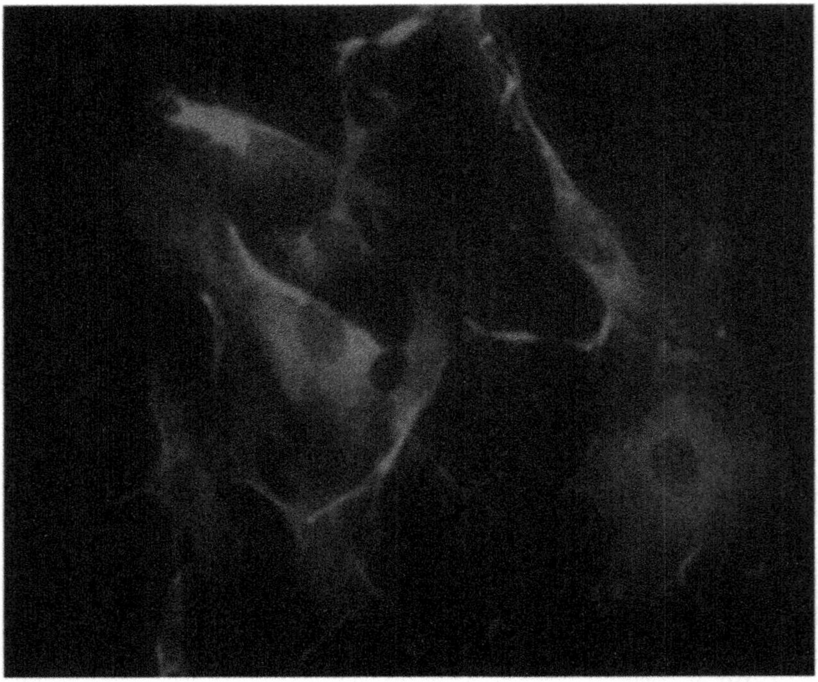

FIGURE 3. Type I pneumocytes, identified by their distinct morphology and immunoreactivity for aquaporin 5, differentiated from murine ESC by exposure to an extract of murine pneumocytes. (Red: indirect immunofluorescence for aquaporin 5.)

Directed differentiation of stem cells. Stem cells can be encouraged to differentiate to the required phenotype by manipulating the culture conditions under which they are maintained. In this way, it is possible to control or restrict the available differentiation pathways and to generate selectively cultures enriched with a particular phenotype. Such manipulations include stimulation of cells with particular cytokines, growth factors, amino acids, other proteins and active ions and co-culture with a relevant cell/tissue type (see Microenvironment). A new approach is to use cell extract-based re-programming, where permeabilized stem cells are exposed to extracts of the target phenotype (FIG. 3).[39] Often, cell sorting techniques such as fluorescence-activated cell sorting (FACS) are used to purify the target cell type further. Using these approaches, virtually every cell type in the body has been derived from stem cells in the laboratory.

Stem cells, in particular ESC, are also amenable to genetic manipulation and have been instrumental in the creation of transgenic and gene knockout animals that provide the means for more detailed investigation of the genome and the specific functions of a particular gene. This genetic tractability also offers the potential to introduce genes to promote lineage-restricted differentiation and provides a basis for gene therapy to introduce therapeutic genes and potentially to modulate the immune response allowing implantation of "non-self" cells/tissues.

SUPPORT MATERIALS

The prime requirements for any scaffold in tissue engineering are biocompatibility, the capacity to sustain and/or promote the growth of the relevant cells/tissue, and provision of a template for tissue growth in three dimensions. Biological materials have frequently been used in tissue engineering; for example, the submucosa of the small bowel has been used in the engineering of blood vessels for 40 years.[40,41] The tissues are decellularized and not only allogeneic but also xenogeneic sources have been used. Naturally occurring collagen has also been used in a variety of forms. There are obvious limitations associated with biological materials, including lack of consistency and structure malleability. The developments in materials science that have occurred in recent years, particularly in the generation of bioactive materials, have provided a major advance in tissue engineering.[42–44] The new materials offer many advantages as they can be designed to meet specific spatial and strength requirements and their rate of degradation can be precisely controlled. Biocompatible, nonbiological materials fall into three classes: bioinert, resorbable, and bioactive. No material can be said to be totally inert when implanted but the group known as bioinert materials only provoke formation of scar tissue. The stainless steel used in artificial hips is an example of a widely used bioinert material. The resorbable materials eventually dissolve when implanted with the generation of harmless dissolution products. There is a wide range of such material already in clinical use such as the polymers (polyglycolic, PGA; polylactic acids, PLLA) used for suturing. These polymers have proved popular with tissue engineers as, in addition to already being used in the clinic, they are easy to shape and are tough with high tensile properties. Bioactive materials are those that stimulate a biological response from the body and are classified for bone tissue engineering as osteoconductive and osteoproductive. Osteoconductive materials (e.g., synthetic hydroxyapatite ceramics) bond to bone and stimulate growth at the edge of the graft while osteoproductive materials (e.g., bioactive glasses) stimulate the growth of new bone within the material.[45]

MICROENVIRONMENT

To maximize the chances of success, tissue engineers try to replicate as far as possible the natural microenvironment in which the cell/tissue would normally grow and function within the body. This comprises a range physical and chemical factors, broadly divided into physical (insoluble macromolecules, e.g., collagen), chemical (soluble macromolecules, e.g., cytokines), and cell–cell interactions (proteins on adjacent cells).[46] Thus, tissue engineers have provided extracellular matrix and soluble proteins to cells in constructs and, for load-bearing musculoskeletal tissues, mechanical stimulation may also be required. This approach is being refined by investigation of the specific cues cells receive in their local tissue milieu so that the crucial cells can be created in the tissue construct.[47] For stem cells, co-culture with mature cells or tissues is also being used increasingly as a means to drive their differentiation toward required lineages.[48–54]

Among the latest developments in this area is the use of synthetic biomaterials to create microenvironments that mimic natural extracellular matrix, with the particular aim of directing stem cell differentiation and morphogenesis of the target tissue.[46,55] This approach has been employed, for example, to construct a scaffold that enhances differentiation of neural stem cells while inhibiting the development of glial cells.[56]

STEM CELL APPLICATIONS IN TISSUE ENGINEERING

In aiming to get engineered tissues into the clinic, there is a variety of considerations in addition to identifying appropriate cell sources. Key among these are:

- *Complex tissues:* The final construct for implantation needs to replicate the architecture and complex cellular interdependence found in the corresponding normal tissue. Most current constructs involve only one or two cell types grown and their clinical viability is, therefore, limited.
- *Vascularization:* Rapid angiogenic stimulus is needed on implantation to ensure maintenance of the nutrient supply to the tissue construct as it integrates *in situ*.
- *Interface stability:* The host tissue–graft interface must permit/promote integration of the construct with the native tissue.
- *Function:* It must be possible to demonstrate that constructs will have the required level of normal activity *in vivo*.
- *Storage:* The viability of cell-based products needs to be maintained during storage and transport.
- *Sterility:* Sterility must be maintained during the production of each construct until implantation is complete. GMP compliance is essential.
- *Cost:* The manufacturing costs have limited the translation of many tissue engineering strategies to clinical applications. The background technology in tissue engineering continues to grow but there remains an urgent need to develop cost-efficient, scaleable processes, and rapid quality control tools.

At present, four main therapeutic approaches using stem cells have been defined: direct administration of adult stem cells, regeneration mediated by stimulation of endogenous stem cells, implantation of differentiated stem cells, and tissue engineering. Thus, stem cells are being used in the clinic, but most applications are based on the application of adult or fetal stem cells and involve cell delivery rather than tissue engineering per se. However, broader tissue engineering strategies, including those using ESC are being developed and tested. For example, engineered skin has been used for some time to treat burns and ulcers and most techniques expand basal keratinocytes, comprising stem, transient amplifying and committed cells, and fibroblasts from biopsies and grow them on structural or matrix proteins or on a polymer scaffold.[57] For the skeleton, interoperative orthopedic procedures using adult stem cells are being developed, particularly for the treatment of patients with nonunion fractures. In these procedures adult osteogenic precursor cells, harvested from bone marrow, are grown on a scaffold and have the advantage over adult cells of laying down new extracellular matrix without the need for removal of old matrix, as happens with adult bone autografts.[58,59] The first clinical application of bone engineered from stem cells was the successful grafting of the distal phalanx of the thumb comprising osteogenic cells from the distal radius periosteum.[60] The recent demonstration that human ESC can be encouraged to form osteoblasts *in vitro* and *in vivo* provides a source for bone tissue engineering as well as a model system for studying bone formation.[61] The fact that cartilage has no blood supply means that it has relatively little capacity for repair, although a progenitor cell population has recently been demonstrated within the surface of articular cartilage.[62] Until now, tissue engineering of cartilage has been based mainly on the harvesting and expansion *in vitro* of autologous chondrocytes from articular surfaces on resorbable scaffolds, such as PGA.[63–65] Human MSC[66] and ESC[67] yield chondrocytes *in vitro* and recent work growing rat MSC on a polyethylene glycol-based hydrogel has shown the formation of an entire chondryle, with chondral and osseous components, which suggests the potential of engineering total joints for replacement.[68]

Promising results are being obtained for the cardiovascular system. In the clinic, there has been success in the treatment of myocardial infarct by stem cell delivery using autologous bone marrow, although the clinical trials at present are mostly small and without satisfactory controls.[69] ESC differentiated to cardiomyocytes were first tested as a means to repair the heart in an animal model in 1996.[70] Although sufficient cardiomyocytes were generated, they did not become fully incorporated into the cardiac muscle *in situ*. More recently, implantation of undifferentiated murine ESC contained in a liquid matrix (Matrigel) into infarcted heart significantly improved contractility.[71] Human ESC have been differentiated into endothelial cells that were isolated with antibodies to platelet endothelial cell adhesion molecule-1.[72] When cultured in matrigel *in vitro*, the cells formed tube-like structures and, when implanted into severe combined immune-deficient (SCID) mice on biodegradable polymer scaffolds, the cells appeared to form microvessels containing mouse blood cells. Tissue constructs, except for cartilage, need a microvascular network and attempts have been made to encourage vasculogenesis on scaffolds, for example, by seeding with endothelial progenitor cells (EPC) isolated from human cord blood.[73] Interestingly, the EPC maintained their phenotype on PGA-PLLA scaffolds but only formed distinct capillary-like structures when they were seeded in combination with smooth muscle cells.[74]

FUTURE

The major medical challenges of the 21st century are likely to be of a very different variety from those of today. Whereas fighting infectious disease has long been a preoccupation of medicine, in the future, dealing with the consequences of a predominantly aging population is likely to take priority. There is much to do if the full promise of stem cells and tissue engineering is to be realized. Our understanding of stem cell biology continues to increase but we must be able to not only control but also optimize the differentiation of stem cells. This includes determination of the signals that trigger differentiation, including chemical stimuli, matrix/substrate-associated cues, and the role of physical factors, with the latter being particularly unexplored.

Beyond the science, as stem cell biology is translated into clinical application, there are other issues and challenges that will need to be faced. The use of ESC remains contentious for some governments and religious groups and this position is unlikely to alter in the short term. However, governments change and, once clear clinical benefits are demonstrated, some current, opposing stances may become reconciled to the use of ESC. A pragmatic issue is the move from small-scale laboratory experiments to large-scale production of cells in GMP facilities. Clearly, innovative bioreactor technology will be needed. Standardization of the type and level of process quality controls will also be required to meet the conditions of the appropriate regulatory bodies.

If any kind of off-the-shelf availability of tissue and organs is to be achieved for a particular clinical application, then allogeneic cells will need to be employed. This raises the issue of the immunogencity of the cells. Everything that we now know suggests that a differentiated cell, one derived from a stem cell, will exhibit the normal immunogenic characteristics of that particular type of differentiated cell. If immune acceptance is an issue, there will then need to be a strategy for overcoming this. One means could be tissue typing, as is currently done for donor organs; and a recent publication suggests that as few as 150 ESC lines deposited in a bank may be sufficient to alllow a satisfactory rate of matching.[75] A possibility is therapeutic nuclear transfer and licences have been granted in the UK by the Human Embryology and Fertilization Authority for the cloning of human stem cells in order to make them immunocompatible.

In closing, finding the most effective ways of using stem cells, from adult, fetal, and embryonic sources, and triggering their differentiation in a controlled manner will provide cell banks for the *in vitro* growth of tissue and for cell replacement therapy. Developing these concepts from bench to bedside will be crucial in meeting healthcare needs in the new century.

REFERENCES

1. BISCEGLIE, V. 1933. Uber die antineoplastische immunitat; heterologe Einpflanzung von Tumoren in Huhner-embryonen. Ztschr. Krebsforsch **40:** 122–140.
2. BELL, E., P. EHRLICH, D.J. BUTTLE, *et al.* 1981. Living tissue formed in vitro and accepted as skin-equivalent of full thickness. Science **221:** 1052–1054.
3. BURKE, J.F., I.V. YANNAS, W.C. QUIMBY, *et al.* 1981. Successful use of a physiologically acceptable artificial skin in the treatment of extensive burn injury. Ann. Surg. **194:** 413–448.
4. BJORNSON, C.B., R.L. RIETZE, B.A. REYNOLDS, *et al.* 1999. Turning brain into blood: a haematopoietic fate adopted by adult neural stem cells in vivo. Science **283:** 534–537.
5. MEZEY, E., K.J. CHANDROSS, G. HARTA, *et al.* 2000. Turning blood into brain: cells bearing neuronal antigens generated in vivo from bone marrow. Science **290:** 1779–1782.
6. TOMA, J., M. AKHARVAN, K.J.L. FERNANDES, *et al.* 2001. Isolation of multipotent adult stem cells from the dermis of mammalian skin. Nat. Cell Biol. **3:** 778–784.
7. BLAU, H.M., T.R. BRAZELTON & J.M. WEIMANN. 2001. The evolving concept of stem cells: entity or function? Cell **105:** 829–841.
8. RIPPON, H.J. & A.E. BISHOP. 2004. Embryonic stem cells. Cell Prolif. **37:** 23–34.
9. MUMMERY, C. 2004. Stem cell research: immortality or a healthy old age? Eur. J. Endocrinol. **151:** U7–U12.
10. WOBUS, A.M. & K.R. BOHELER. 2005. Embryonic stem cells: prospects for developmental biology and cell therapy. Physiol. Rev. **85:** 635–678.
11. EVANS, M.J. & M.H. KAUFMAN. 1981. Establishment in culture of pluripotential cells from mouse embryos. Nature **292:** 154–156.
12. MARTIN, G.R. 1981. Isolation of a pluripotent cell line from early mouse embryos cultured in medium conditioned by teratocarcinoma stem cells. Proc. Natl. Acad. Sci. USA **78:** 7634–7638.
13. NAGY, A., E. GOCZA, E.M. DIAZ, *et al.* 1990. Embryonic stem cells alone are able to support fetal development in the mouse. Development **110:** 815–821.
14. BRADLEY, A., M. EVANS, M.H. KAUFMAN, *et al.* 1984. Formation of germ-line chimaeras from embryo-derived teratocarcinoma cells lines. Nature **309:** 255–256.
15. XU, R.H., X. CHEN, D.S. LI, *et al.* 2002. BMP4 initiates human embryonic stem cells differentiation to trophoblast. Nat. Biotechnol. **20:** 1261–1264.
16. YAMANE, T., S.J. DYLLA, M. MUIJTJENS, *et al.* 2005. Enforced Bcl-2 expression overrides serum and feeder cell requirements for mouse embryonic stem cell self-renewal. Proc. Natl. Acad. Sci. USA **102:** 3312–3317.
17. THOMSON, J.A., J. ITSKOVITZ-ELDOR, S.S. SHAPIRO, *et al.* 1998. Embryonic stem cell lines derived from human blastocysts. Science **282:** 1145–1147.
18. REUBINOFF, B.E., M.F. PERA, C.Y. FONG, *et al.* 2000. Embryonic stem cell lines from human blastocysts: somatic differentiation in vitro. Nat. Biotechnol. **18:** 399–404.

19. LASLETT, A.L., A.A. FILIPCZYK & M.F. PERA. 2003. Characterization and culture of human embryonic stem cells. Trends Cardiovasc. Med. **13:** 295–301.
20. MARTIN, M.J., A. MUOTRI, F. GAGE, *et al.* 2005. Human embryonic stem cells express an immunogenic nonhuman sialic acid. Nat. Med. **11:** 228–232.
21. XU, R.H., R.M. PECK, D.S. LI, *et al.* 2005. Basic FGF and suppression of BMP signaling sustain undifferentiated proliferation of human ES cells. Nat. Methods **2:** 185–190.
22. XU, C., M.S. INOKUMA, J. DENHAM, *et al.* 2001. Feeder-free growth of undifferentiated human embryonic stem cells. Nat. Biotechnol. **19:** 971–974.
23. SATO, N., L. MEIJER, L. SKALTSOUNIS, *et al.* 2004. Maintenance of pluripotency in human and mouse embryonic stem cells through activation of Wnt signaling by a pharmacological GSK-3-specific inhibitor. Nat. Med. **10:** 55–63.
24. XU, C., J. JIANG, V. SOTTILE, *et al.* 2004. Immortalized fibroblast-like cells derived from human embryonic stem cells support undifferentiated cell growth. Stem Cells **22:** 972–980.
25. STOJKOVIC, P., M. LAKO, R. STEWART, *et al.* 2005. An autogeneic feeder cell system that efficiently supports growth of undifferentiated human embryonic stem cells. Stem Cells **23:** 306–314.
26. CAMPAGNOLI, C., I.A. ROBERTS, S. KUMAR, *et al.* 2001. Identification of mesenchymal stem/progenitor cells in human first-trimester fetal blood, liver, and bone marrow. Blood **98:** 2396–2402.
27. PRESNELL, S.C., B. PETERSEN & M. HEIDARAN. 2002. Stem cells in adult tissues. Cell & Dev. Biol. **13:** 369–376.
28. HAWLEY, R.G. & D.A. SOBIESKI. 2002. Somatic stem cell plasticity: to be or not to be. Stem Cells **20:** 195–197.
29. HOLDEN, C. & G. VOGEL. 2002. Stem cells. Plasticity: time for a reappraisal? Science **296:** 2126–2129.
30. VERFAILLIE, C.M., M.F. PERA & P.M. LANDSDORP. 2002. Stem cells: hype and reality. Hematology (Am. Soc. Hematol. Edu. Program) **2002:** 369–391.
31. POULSOM, R., M.R. ALISON, S.J. FORBES, *et al.* 2002. Adult stem cell plasticity. J. Pathol. **197:** 441–456.
32. RAFF, M. 2003. Adult stem cell plasticity: fact or artefact. Annu. Rev. Cell Dev. Biol. **19:** 1–22.
33. RUBIO, D., J. GARCIA-CASTRO, M.C. MARTIN, *et al.* 2005. Spontaneous human adult stem cell transformation. Cancer Res. **65:** 3035–3039.
34. SIMONSEN, J.L., C.ROSADA, N.SERAKINCI, *et al.* 2002. Telomerase expression extends the proliferative life-span and maintains the osteogenic potential of human bone marrow stromal cells. Nat. Biotech. **20:** 592–596.
35. BURNS, J.S., B.M. ABDALLAH, P. GULDBERG, *et al.* 2005. Tumorigenic heterogeneity in cancer stem cells evolved from long-term cultures of telomerase-immortalized human mesenchymal stem cells. Cancer Res. **65:** 3126–3135.
36. WICKHAM, M.Q., G.R. ERICKSON, J.M. GIMBLE, *et al.* 2003. Multipotent human adult stem cells derived from infrapatellar fat pad of the knee. Clin. Orthop. **412:** 196–212.
37. YEN, B.L., H.I. HUANG, C.C. CHIEN, *et al.* 2005. Isolation of multipotent cells from human term placenta. Stem Cells **23:** 3–9.
38. KODAMA, S., M. DAVIS & D.L. FAUSTMAN. 2005. Diabetes and stem cell researchers turn to the lowly spleen. Sci. Aging Knowledge Environ. **2005:** pe2.
39. QIN, M.D., G.P. TAI, P. COLLAS, *et al.* 2005. Cell extract-derived differentiation of embryonic stem cells. Stem Cells **23:** 712–718.

40. MATSUMOTO, T., R.H. HOLMES, C.O. BURDICK, *et al.* 1965. The fate of the inverted segment of small bowel used for the replacement of major veins. Surgery **60:** 739–743.
41. JERNIGAN, T.W., M.A. CROCE, C. CAGIANNOS, *et al.* 2004. Small intestinal submucosa for vascular reconstruction in the presence of gastrointestinal contamination. Ann. Surg. **239:** 733–738.
42. HENCH, L.L. & J.M. POLAK. 2002. Third-generation biomedical materials. Science **295:** 1014–1017.
43. LANGER, R. & D.A. TIRRELL. 2004. Designing materials for biology and medicine. Nature **428:** 487–492.
44. LUTOLF, M.P. & J.A. HUBBELL. 2005. Synthetic biomaterials as instructive extracellular microenvironments for morphogenesis in tissue engineering. Nat. Biotechnol. **23:** 47–55.
45. JONES, J.R. & L.L. HENCH. 2003. Regeneration of trabecular bone using porous ceramics. Curr. Opin. Solid State Mat. Sci. **7:** 301–307.
46. KLEINMAN, H.K., D. PHILIP & M.P. HOFFMAN. 2003. Role of the extracellular matrix in morphogenesis. Curr. Opin. Biotechnol. **14:** 526–532.
47. REDDI, A.H. 2000. Morphogenesis and tissue engineering of bone and cartilage: inductive signals, stem cells, and biomimetic biomaterials. Tissue Eng. **6:** 351–359.
48. BUTTERY, L.D.K., S. BOURNE, J.D. XYNOS, *et al.* 2001. Differentiation of osteoblasts and in vitro bone formation from murine embryonic stem cells. Tissue Eng. **7:** 89–94.
49. MUMMERY, C., D. WARD, C.E. VAN DEN BRINK, *et al.* 2002. Cardiomyocyte differentiation of murine and human embryonic stem cells. J. Anat. **200:** 233–242.
50. ALEXANIAN, A.R. & M. SIEBER-BLUM. 2003. Differentiating adult hippocampal stem cells into neural crest phenotypes. Neuroscience **118:** 1–5.
51. FAIR, J.H., B.A. CAIRNS, M. LAPAGLIA, *et al.* 2003. Induction of hepatic differentiation in embryonic stem cells by co-culture with embryonic cardiac mesoderm. Surgery **134:** 189–196.
52. BALL, S.G., A.C. SHUTTLEWORTH & C.M. KIELTY. 2004. Direct cell contact influences bone marrow mesenchymal stem cell fate. Int. J. Biochem. Cell Biol. **36:** 714–727.
53. HWANG, J.H., S.H. YUK, J.H. LEE, *et al.* 2004. Differentiation of stem cells isolated from rat smooth muscle. Mol. Cells **17:** 57–61.
54. VAN VRANKEN, B.E., H.M. ROMANSKA, J.M. POLAK, *et al.* 2005. Co-culture of embryonic stem cells with pulmonary mesenchyme: a microenvironment that promotes differentiation of pulmonary epithelium. Tissue Eng. **11:** 1177–1187.
55. ZHANG, S. 2003. Fabrication of novel biomaterials through molecular self-assembly. Nat. Biotechnol. **21:** 1171–1178.
56. SILVA, G.A., C. CZEISLER, K.L. NIECE, *et al.* 2004. Selective differentiation of neural progenitor cells by high epitope-density nanofibers. Science **303:** 1352–1355.
57. MARGULIS, A., W. ZHANG & J.A. GARLICK. 2005. In vitro fabrication of engineered human skin. Methods Mol. Biol. **289:** 61–70.
58. MUSCHLER, G. & R.J. MIDURA. 2002. Connective tissue progenitors: practical concepts for clinical applications. Clin. Orthop. **395:** 66–80.
59. MUSCHLER, G.F., H. NITTO & Y. MATSUKURA. 2003. Spine fusion using cell matrix composites enriched in bone marrow-derived cells. Clin. Orthop. **407:** 102–118.

60. Vacanti, C.A., L.J. Bonassar, M.P. Vacanti, *et al.* 2001. Replacement of an avulsed phalanx with tissue-engineered bone. N. Engl. J. Med. **344:** 1511–1514.
61. Bielby, R.C., A.R. Boccaccini, L.D.K. Buttery, *et al.* 2004. In vitro differentiation and in vivo mineralization of osteogenic cells derived from human embryonic stem cells. Tissue Eng. **10:** 1518–1525.
62. Dowthwaite, G.P., J.C. Bishop, S.N. Redman, *et al.* 2004. The surface of articular cartilage contains a progenitor cell population. J. Cell Sci. **117:** 889–897.
63. Vacanti, C.A., L.G. Cima & D. Ratowski. 1992. Tissue-engineered growth of new cartilage in the shape of a human ear using synthetic polymers seeded with chondrocytes. Mat. Res. Sec. Symp. Proc. **252:** 367–373.
64. Puelacker, W.C., J. Wisser, C.A. Vacanti, *et al.* 1994. Temperomandibular joint disc replacement made by tissue-engineered growth of cartilage. J. Oral Maxillofac. Surg. **52:** 1172–1178.
65. Ibarra, C., C. Janetta, C.A. Vacanti, *et al.* 1997. Tissue-engineered meniscus transplantation. Transplant. Proc. **29:** 986–988.
66. Pittenger, M.F., A.M. Mackay, S.C. Beck, *et al.* 1999. Multilineage potential of human adult mesenchymal stem cells. Science **284:** 143–147.
67. Vats, A., R. Bielby, N. Tolley, *et al.* 2006. Chondrogenic differentiation of human embryonic stem cells: the effect of the micro-environment. Stem Cells Tiss. Eng. In press.
68. Alhadlaq, A. & J.J. Mao. 2005. Tissue-engineered osteochondral constructs in the shape of an articular condyle. J. Bone Joint Surg. Am. **87:** 936–944.
69. Wollert, K.C., G.P. Meyer, J. Lotz, *et al.* 2004. Intracoronary autologous bone-marrow cell transfer after myocardial infarction: the BOOST randomised controlled clinical trial. Lancet **364:** 141–148.
70. Klug, M.G., M.H. Soonpaa, G.Y. Koh, *et al.* 1996. Genetically selected cardiomyocytes from differentiating embryonic stem cells form stable intracardiac grafts. J. Clin. Invest. **98:** 216–224.
71. Kofidis, T., J.L. de Bruin, G. Hoyt, *et al.* 2004. Injectable bioartificial myocardial tissue for large-scale intramural cell transfer and functional recovery of injured heart muscle. J. Thorac. Cardiovasc. Surg. **128:** 571–578.
72. Levenberg, S., J.S. Golub, M. Amit, *et al.* 2002. Endothelial cells derived from human embryonic stem cells. Proc. Natl. Acad. Sci. USA **99:** 4391–4396.
73. Kaushal, S., G.E. Amiel, K.J. Guleserian, *et al.* 2001. Functional small-diameter neovessels created using endothelial progenitor cells expanded ex vivo. Nat. Med. **7:** 1035–1040.
74. Wu, X., E. Rabkin-Aikawa, K.J. Guleserian, *et al.* 2004. Tissue-engineered microvessels on three-dimensional biodegradable scaffolds using human endothelial progenitor cells. Am. J. Physiol. Heart Circ. Physiol. **287:** H480–H487.
75. Taylor, C.J., E. M. Bolton, S. Pocock, *et al.* 2005. Banking on human embryonic stem cells: estimating the number of donor cell lines needed for HLA matching. Lancet **366:** 2019–2025.

Bisphosphonates

From Bench to Bedside

R. GRAHAM G. RUSSELL

Botnar Research Centre and Oxford University Institute of Musculoskeletal Sciences, Oxford OX3 7LD, UK

ABSTRACT: The discovery and development of the bisphosphonates (BPs) as a major class of drugs for the treatment of bone diseases has been a fascinating journey that is still not over. In clinical medicine, several BPs are established as the treatments of choice for various diseases of excessive bone resorption, including Paget's disease of bone, myeloma and bone metastases, and osteoporosis. Bisphosphonates are chemically stable analogues of inorganic pyrophosphate, and are resistant to breakdown by enzymatic hydrolysis. Bisphosphonates inhibit bone resorption by being selectively taken up and adsorbed to mineral surfaces in bone, where they interfere with the action of the bone-resorbing osteoclasts. Bisphosphonates are internalized by osteoclasts and interfere with specific biochemical processes. Bisphosphonates can be classified into at least two groups with different molecular modes of action. The simpler non-nitrogen-containing bisphosphonates (such as clodronate and etidronate) can be metabolically incorporated into nonhydrolyzable analogues of adenosine triphosphate (ATP) that may inhibit ATP-dependent intracellular enzymes. The more potent, nitrogen-containing bisphosphonates (such as pamidronate, alendronate, risedronate, ibandronate, and zoledronate) are not metabolized in this way but can inhibit enzymes of the mevalonate pathway, thereby preventing the biosynthesis of isoprenoid compounds that are essential for the posttranslational modification of small GTP-binding proteins (which are also GTPases) such as rab, rho, and rac. The inhibition of protein prenylation and the disruption of the function of these key regulatory proteins explain the loss of osteoclast activity and induction of apoptosis. The key target for bisphosphonates is farnesyl pyrophosphate synthase (FPPS) within osteoclasts, and the recently elucidated crystal structure of this enzyme reveals how BPs bind to and inhibit at the active site via their critical N atoms. In conclusion, bisphosphonates are now established as an important class of drugs for the treatment of many bone diseases, and their mode of action is being unraveled. As a result their full therapeutic potential is gradually being realized.

Address for correspondence: Prof. Graham Russell, The Botnar Research Centre, Nuffield Department of Orthopaedic Surgery, University of Oxford, Headington, Oxford, OX3 7LD, UK. Voice: +44-(0)-1865-22-7388; fax: +44-(0)-1865-22-7966.
e-mail: graham.russell@ndos.ox.ac.uk

Ann. N.Y. Acad. Sci. 1068: 367–401 (2006). © 2006 New York Academy of Sciences.
doi: 10.1196/annals.1346.041

KEYWORDS: bone resorption; osteoclasts; bisphosphonates; protein prenylation; bone metastases; myeloma; osteoporosis

INTRODUCTION

The bisphosphonates (BPs) are stable analogues of a naturally occurring pyrophosphate (PPi) compound. BPs were first synthesized in the 1800s,[1] but have only been used progressively in medicine since the 1960s.

Etidronate, the first BP to be used in humans in Paget's disease,[2] was synthesized just over 100 years ago.[3] The early uses of BPs were industrial, mainly as corrosion inhibitors, also as complexing agents in the textile, fertilizer, and oil industries, as well as for many other industrial processes.[4] Their use as "water softeners" was based on their ability to inhibit calcium carbonate precipitation, as do polyphosphates, and has been applied in the prevention of scaling in domestic and industrial water installations. There are many books and review articles available that describe the chemistry, pharmacology, and clinical applications of BPs.[5–18]

EARLY STUDIES ON PPi AND BPs

In the early 1960s, William Neuman and Herbert Fleisch[19] were studying mechanisms of calcification induced by collagen, and showed that body fluids, such as plasma and urine, contained inhibitors of calcification. Since it had been known since the 1930s that trace amounts of polyphosphates were capable of acting as water softeners by inhibiting the crystallization of calcium salts, such as calcium carbonate, they proposed that compounds of this type might be natural regulators of calcification under physiological conditions. Fleisch and his colleagues showed that inorganic PPi, a naturally occurring polyphosphate and a known byproduct of many biosynthetic reactions in the body, was present in serum and urine and could prevent calcification by binding to newly forming crystals of HAP.[20,21] It was therefore postulated that PPi might be the body's own "water softener" that normally prevents calcification of soft tissues, and regulates bone mineralization. The concentrations of PPi in body fluids would be expected to be regulated by hydrolytic enzymes.

At this time, I was completing my Ph.D. in the UK and was studying kidney stone formation and other disorders of calcification. It seemed possible that some of these pathologic disorders might be linked to disturbances in PPi metabolism.[22] Prominent among these was the rare but fascinating inherited disorder, hypophosphatasia, in which lack of alkaline phosphatase is associated with mineralization defects of the skeleton. My studies showed that PPi levels were elevated in urine,[23] and indicated that alkaline phosphatase was probably the key extracellular enzyme responsible for hydrolyzing PPi. We later showed elevated concentrations of PPi in plasma,[24] further supporting the notion that the activity of alkaline phosphatase regulates circulating amounts of PPi to below the critical levels that would otherwise prevent normal physiological calcification processes.

Attempts to exploit these concepts by using PPi and polyphosphates to inhibit ectopic calcification in blood vessels, skin, and kidneys in laboratory animals were successful only when the compounds were injected.[25] Orally administered PPi and polyphosphates were inactive, due to the hydrolysis of PPi in the gastrointestinal tract, probably by mucosal brush border phosphatases. During the search for more stable analogues of PPi that might also have the antimineralization properties of PPi but that would be resistant to hydrolysis, several different chemical classes were studied, including P-N-P and P-C-C-P compounds. The BPs (at that time called diphosphonates) characterized by P-C-P motifs were among these.[26] These early studies with BPs were the result of a very successful collaboration with Marion (Dave) Francis of the Procter and Gamble company and the group working in Switzerland.[27,28]

Like PPi, BPs had high affinity for bone mineral[29] and were found to prevent the formation and aggregation of calcium phosphate crystals. BPs had high affinity for bone mineral and were found to prevent calcification both *in vitro* and *in vivo*, but, unlike PPi, were also able to prevent experimentally induced pathological calcification when given orally to rats *in vivo.*[30] This property of being active by mouth was key to their future use in man.

In these early studies BPs were shown to not only prevent the experimentally induced calcification of many soft tissues, including skin, kidneys, and blood vessels *in vivo*, but with some of the compounds, for example, etidronate to also inhibit mineralization of ectopic bone as well of normal calcified tissues such as bone and cartilage.[31] BPs appear to prevent calcification by physicochemical mechanisms producing direct impairment of the calcification process by acting as crystal poisons after adsorption to mineral surfaces, rather than by effects on the deposition of matrix.

Perhaps the most important step toward the future use of BPs occurred when we found that BPs, like we had already shown for PPi,[32] also had the novel property of being able to inhibit the dissolution of HAP crystals.[33] This led to studies to determine whether they might also inhibit bone resorption.

Many studies using a variety of experimental systems showed that BPs inhibit osteoclast-mediated bone resorption, not only in organ cultures of bone *in vitro*, but also both in normal animals and in those with experimentally increased resorption. The first experimental model studied was thyroparathyroidectomized rats treated with parathyroid hormone to stimulate bone resorption *in vivo.* BPs also suppress resorption induced by many other agents such as calcitriol, vitamin D, and retinoids. The effect on retinoid-induced hypercalcemia has been used to develop a powerful and rapid-screening assay for new compounds.[34,35] In growing intact rats, the BPs block the removal of both bone and cartilage, thus retarding the remodeling of the metaphysis, which becomes club-shaped and radiologically denser than normal.[36] This effect is the basis of the 'Schenk' model also used to compare the potency of new compounds. The inhibition of endogenous bone resorption can also be monitored by kinetic studies[37] using radio-calcium (^{45}Ca), and by using biochemical markers of bone resorption.

The BPs are also effective in preventing bone destruction in a number of animal models of human disease. One of the first to be studied was the sciatic nerve section as a model of immobilization osteoporosis,[38] as well as bone loss due to spinal cord section. Other commonly used models of osteoporosis include the prevention of bone loss associated with ovariectomy. Less commonly used models involve orchidectomy, lactation, low calcium diets, or the administration of agents such as heparin or corticosteroids. If not given in excess, BPs maintain or improve the biomechanical properties of bone both in normal animals and in experimental models of osteoporosis.[39]

In general there is a good correlation between potency and structure–activity relationships *in vitro* and *in vivo.*[40] In the presence of BPs isolated osteoclasts form fewer and smaller erosion cavities on various mineralized matrices *in vitro*.[41]

THE PHARMACOLOGICAL DEVELOPMENT OF BPs

Once the potential clinical value of BPs had been appreciated, research efforts were devoted to the development of compounds with a more powerful antiresorptive activity, but without a corresponding ability to inhibit mineralization. With compounds such as etidronate there was only a 10- to 100-fold difference between doses that inhibit mineralization compared with doses that reduce bone resorption. Enhancing this window was readily achieved and many hundreds of BPs have been synthesized, and more than a dozen have been used in humans. With the development of BPs that were more potent inhibitors of bone resorption, these dose differences widened to several orders of magnitude, which meant that inhibition of skeletal mineralization ceased to be a major clinical concern. The gradation of potency evaluated in the animal models corresponded quite well with that found in humans, although the differences in potency are much smaller in humans.

Since BPs accumulate in bone, it is important to know what happens during long-term administration. It is reassuring from a clinical point of view that the inhibition of bone resorption reaches a new steady-state level, rather than becoming progressively lower, even when the compounds are given continuously.[42] The level of suppression depends on the administered dose, and has also been observed in humans.[43] These results show that there is no progression of the antiresorptive effect with time and suggest that the bisphosphonate buried in the bone is inactive at least as long as it remains buried there. They

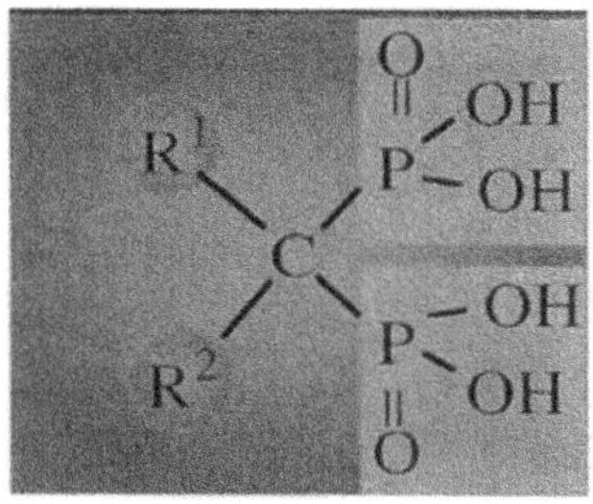

FIGURE 1. The generic structure of bisphosphonates and their functional domains.

also show that, within the therapeutic dosage range, there is little risk of a continuous decrease in bone turnover in the long run, which might lead to an increase in bone fragility. An additional important pharmacological property of BPs is that the total dose administered is a major determinant of their effects. This has been well studied for ibandronate[44] and zoledronate.[45] In both cases the same inhibition of bone resorption has been documented whether the BP is given in small frequent (e.g., daily) doses compared with larger doses given less frequently. This is the basis for the development of intermittent dosing regimens in humans.

DEFINING STRUCTURE–ACTIVITY RELATIONSHIPS

The features of the bisphosphonate molecule necessary for biological activity were well defined in the early studies. The P-C-P moiety is responsible for the strong affinity of the BPs for binding to hydroxyapatite (HAP) and allows for a number of variations in structure based on substitution in the R1 and R2 positions on the carbon atom (FIG. 1). The ability of the BPs to bind to HAP crystals, and to prevent both crystal growth and dissolution, was enhanced when the R1 side chain (attached to the geminal carbon atom of the P-C-P group) was a hydroxyl group (as in etidronate) rather than a halogen atom such as chlorine (as in clodronate). The presence of a hydroxyl group at the R1 position increases the affinity for calcium (and thus bone mineral) due to the ability of BPs to chelate calcium ions by tridentate rather than bidentate binding.[46]

The ability of BPs to inhibit bone resorption *in vitro* and *in vivo* also requires the P-C-P structure. Monophosphonates, for example, pentane monophosphonate, or P-C-C-P or P-N-P compounds, are ineffective as inhibitors of bone resorption. Furthermore, the antiresorptive effect cannot be accounted for simply by adsorption of BPs to bone mineral and prevention

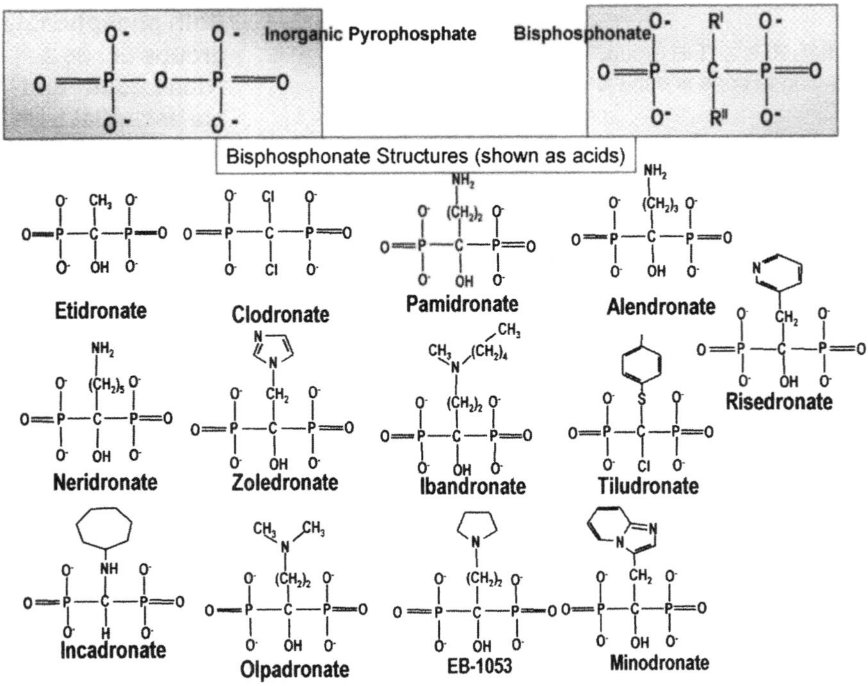

FIGURE 2. Structures of the bisphosphonates used in clinical studies and under clinical development.

of HAP dissolution. It became clear that BPs must inhibit bone resorption by cellular effects on osteoclasts, rather than simply by physicochemical mechanisms.

Following the successful clinical use of clodronate and etidronate in the 1970s and 1980s, more potent antiresorptive BPs were studied, which had different R2 side chains, but in which R1 was unaltered. In particular, BPs containing a basic primary nitrogen atom in an alkyl chain (as in pamidronate and alendronate) were found to be 10- to 100-fold more potent than etidronate and clodronate. After this in the 1980s, there was a phase in which synthesis of novel compounds took place specifically to determine their possible effects on calcium metabolism, with the result that compounds highly effective as inhibitors of bone resorption were identified and studied (FIG. 2).

These compounds, especially those that contain a tertiary nitrogen, such as ibandronate[47] and olpadronate,[48] were even more potent at inhibiting bone resorption. Among this generation of compounds that were synthesized to optimize their antiresorptive effects, the most potent antiresorptive BPs were those containing a nitrogen atom within a heterocyclic ring (as in risedronate[49] and zoledronate[50]), which are up to 10,000-fold more potent than etidronate in some experimental systems.

The analysis of structure–activity relationships allowed the spatial features of the active pharmacophore to be defined in considerable detail, even before the molecular mechanism of action was fully elucidated. For maximal potency, the nitrogen atom in the R2 side chain must be a critical distance away from the P-C-P group, and in a specific spatial configuration.[51] This was used successfully for predicting the features required in the chemical design of new and more active compounds.

Although the structure of the R2 side chain is the major determinant of antiresorptive potency, both phosphonate groups are also required for the drugs to be pharmacologically active. Alterations to one or both phosphonate groups reduces the affinity for bone mineral and this may be one reason why such bisphosphonate analogues are less active. For example, replacement of one of the phosphonate hydroxyl groups with a methyl group (to form a phosphonophosphinate) markedly reduces both bone affinity and antiresorptive potency. Methylation of both phosphonate groups to form a bisphosphinate leads to loss of bone affinity and loss of antiresorptive activity *in vivo*. However, bisphosphonate analogues (e.g., a phosphonophosphinate and a phosphonocarboxylate) with similar affinity for bone can have very different antiresorptive potencies. This suggests that the two phosphonate groups (or alternatively, the combination of a phosphonate and a carboxylate group) are required both for targeting the bone and for the molecular mechanism of antiresorptive action, presumably because BPs mimic naturally occurring PPi-containing compounds.

In summary, studies of the relationships between bisphosphonate structure and antiresorptive potency suggested that the ability of BPs to inhibit bone resorption depended on two separate properties of the bisphosphonate molecule. The two phosphonate groups, together with a hydroxyl group at the R1 position,[52] impart high affinity for bone mineral and act as a "bone hook," allowing rapid and efficient targeting of BPs to bone mineral surfaces (FIG. 3). Once localized within bone, the structure and three-dimensional conformation of the R2 side chain (as well as the phosphonate groups in the molecule) determined the biological activity of the molecule and influenced the ability of the drugs to interact with specific molecular targets. Our understanding of what these molecular targets might be has become much clearer as a result of recent work.

CLINICAL APPLICATIONS OF BPs

After it was shown that BPs inhibited experimentally induced calcification and bone resorption, their potential application to clinical disorders was obvious but it took many years for them to become well established. The most impressive clinical application of BPs was as inhibitors of bone resorption, often for diseases where no effective treatment existed previously. Thus BPs became the treatment of choice for a variety of bone diseases in which excessive

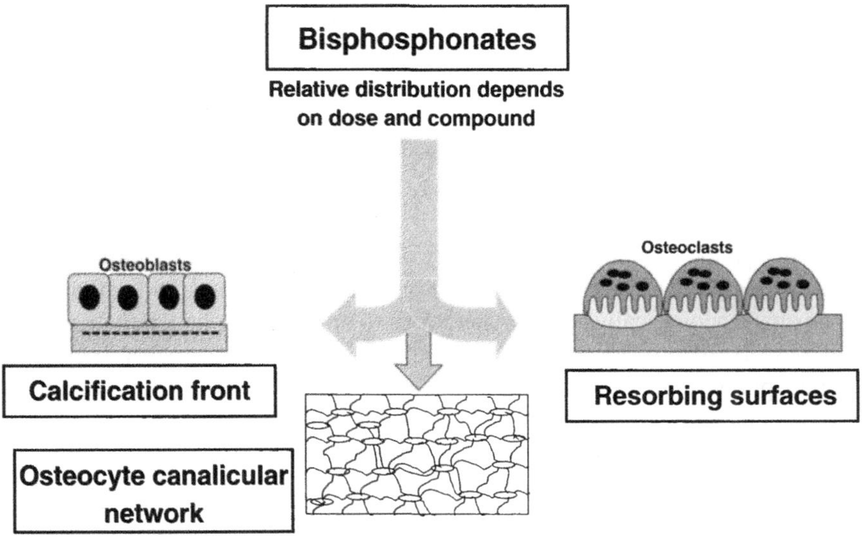

FIGURE 3. Uptake and distribution of BPs into different compartments of bone. This distribution may be affected by both the dose and the compound given.

osteoclast activity is an important pathological feature, including Paget's disease of bone, metastatic and osteolytic bone disease, and hypercalcemia of malignancy, as well as osteoporosis.

BPs and Inhibition of Calcification

Exploration of BPs as inhibitors of calcification showed some promise and early applications of etidronate included use in fibrodysplasis ossicans progressiva (FOP, formerly known as myositis ossificans) and in patients who had undergone total hip replacement surgery to prevent subsequent heterotopic ossification and to improve mobility.[53] Etidronate has also been used to prevent ectopic calcification and ossification, after spinal cord injury, and as topical applications in toothpastes to prevent dental calculus.

It should be emphasized that these effects required very high doses of etidronate, and that inhibition of skeletal mineralization is not a significant clinical problem when etidronate is used at the low doses recommended in the treatment of osteoporosis.

BPs for Radio-Nuclide Imaging of Bone

One of the earliest clinical uses of BPs was as agents for bone imaging, "bone scanning," for which they still remain outstandingly useful for detecting bone metastases and other bone lesions. The application of PPi and simple BPs as bone-scanning agents depends on their strong affinity for bone mineral, particularly at sites of increased bone turnover, and their ability to be linked to a gamma-emitting technetium isotope.[54,55]

BPs in Paget's Disease

Paget's disease was the first clinical disorder in which a dose-dependent inhibition of bone resorption could be demonstrated using BPs in human.[56,57]

The central feature of Paget's disease is the osteoclast, since the pathological characteristics of the disease (such as bone pain, fractures, and skeletal deformities) are the result of increased numbers of osteoclasts and increased osteoclast activity. The pathogenesis of Paget's disease is gradually being elucidated. The reason why large multinucleate osteoclasts accumulate may be because they do not undergo apoptosis in the normal way. This may have a genetic basis[58] in many patients associated with mutations in the sequestosome 1 (SQSTM1) gene, which encodes for a scaffold protein in the NF-kappaB-signaling pathway. A viral etiology may also contribute.[59] If defective apoptosis of osteoclasts contributes to the pathogenesis of Paget's disease, the BPs may be viewed as bone-selective drugs that specifically induce apoptosis in the affected osteoclasts.

BPs have become the most important drugs used in the treatment of Paget's disease.[60] For many years pamidronate given by intravenous infusion was extensively used,[61] but the newer and more potent BPs can produce even more profound suppression of disease activity than was possible with the BPs available in former years.[62,63] The latest advance is with zoledronate,[64] which when given as a single 5-mg infusion produced a greater and longer-lasting suppression of excess bone turnover than even oral risedronate given at 30 mg/day over 2 months, hitherto one of the most effective treatments.

BPs in Oncology

Many cancers in humans are associated with hypercalcemia (raised blood calcium) and/or increased bone destruction. This may be due to the release from tumors of factors that increase bone resorption, such as parathyroid hormone-related protein (PTH-rP), or bone-resorbing cytokines such as interleukin 6.

BPs can prevent the increase in bone resorption associated with experimental tumors, particularly those that localize in or metastasize to bone.[65] In view of clinical results, it is interesting that BPs may not only reduce metastases in bone but reduce the overall tumor burden,[66] although in some models soft tissue tumor mass may increase. The reasons for changes in tumor burden induced by BPs are still uncertain, but are of considerable potential clinical significance. These effects may be due to changes in the release of growth factors, which are present in bone matrix and which may stimulate tumor cell growth during bone resorption.[67,68] In addition, there may be direct effects of BPs on tumor cells themselves, for example, by altering cell attachment and inducing apoptosis.[69] Many of these appear to be due to inhibition of the mevalonate pathway, and there is evidence that synergistic antitumor effects can be achieved in the presence of other chemotherapeutic agents.

BPs are remarkably effective in the treatment of bone problems associated with malignancy[70] and are now the drugs of choice.[71–73] Clinical trials investigating the benefit of bisphosphonate therapy utilize a composite end point defined as a skeletal-related event (SRE) or bone event, which typically includes pathologic fracture, spinal cord compression, radiation, or surgery to bone, and hypercalcemia of malignancy. BPs significantly reduced the incidence of these events in myeloma[74] and in patients with breast cancer metastases,[75,76] and in metastatic prostate cancer,[77] lung cancer, renal cell carcinoma, and other solid tumors. The goals of treatment for bone metastases are also to prevent disease-related skeletal complications, palliate pain, and maintain quality of life. Zoledronate,[78] pamidronate, clodronate, and ibandronate[79,80] have demonstrated efficacy compared with placebo.

There is an important possibility that the survival of patients may be prolonged[81–83] in some groups of patients. Recently, osteonecrosis of the jaw (ONJ)[84] has been identified as a potential complication of high-dose BP therapy in malignant diseases.

BPs in Osteoporosis

Osteoporosis is acknowledged to be a major health problem.[85] There have been impressive advances in understanding the epidemiology and pathogenesis of osteoporosis[86] and its associated fractures, and in the application of physical and biochemical methods to its diagnosis and evaluation. Up until the 1990s, there were few treatments for osteoporosis, but in the past few years, there have been remarkable advances in the therapeutic approaches to prevention and treatment of postmenopausal and other forms of osteoporosis.

As a drug class, the BPs have emerged as the leading effective treatments for postmenopausal and other forms of osteoporosis. Etidronate was the first of these,[87–89] followed by alendronate,[90–92] and then risedronate.[93,94] All have been approved as therapies in many countries, and can increase bone mass and reduce fracture rates in the spine by 30–50%, and also other sites in postmenopausal women.[95] The reduction in fractures may be related not only to the increase in bone mass arising from the inhibition of bone resorption and reduced activation frequency of bone remodeling units, but also to enhanced osteon mineralization.[96] These BPs also prevent bone loss associated with glucocorticosteroid administration.[97,98]

Pamidronate has proved remarkably effective in increasing bone in children with the inherited "brittle bone" disorder, osteogenesis imperfecta.[99,100]

Among the newer BPs, ibandronate[101] has been recently introduced as a once monthly tablet. In addition to formulations to be taken by mouth weekly or monthly, new routes of administration are being studied, especially periodic (e.g., 3 monthly) injections with ibandronate, and once yearly treatment with zoledronate.[102] This has the great attraction of delivering a defined dose without the variability associated with oral administration as well as avoiding potential gastrointestinal intolerance. If these approaches are accompanied by greater compliance and convenience, they may become popular methods of treatment.

CLINICAL PHARMACOLOGY AND USE OF BPs

The clinical pharmacology of BPs is characterized by low intestinal absorption (~1–4%), but highly selective localization and retention in bone. Significant side effects of BPs are minimal.[103–105] Although there are more similarities than differences among individual compounds and each bisphosphonate is potentially capable of treating any of the disorders of bone resorption in which they are used, in practice different compounds have come to be favored for the treatment of different diseases. Currently, there are at least 10 BPs (etidronate, clodronate, tiludronate, pamidronate, alendronate, risedronate, zoledronate, and ibandronate, and also to a limited extent olpadronate and neridronate) that have been registered for various clinical applications in various countries. To a major extent, the diseases in which they are used reflect the history of their clinical development and the degree of commercial interest in and sponsorship of the relevant clinical trials. The one most used in Paget's disease was formerly pamidronate given parenterally, which is now being superseded by zoledronate. In cancer, i.v. pamidronate or clodronate given orally has been extensively used, but is being replaced by zoledronate, which is superior in trials. In osteoporosis, the major current drugs are risedronate and alendronate.

Other clinical issues under consideration with BPs include the choice of therapeutic regimen, for example, the use of intermittent dosing rather than continuous, intravenous versus oral therapy, the optimal duration of therapy, the combination with other drugs such as teraparatide, and their extended use in related indications, for example, glucocorticosteroid-associated osteoporosis, male osteoporosis, childhood osteopenic disorders, arthritis, and other disorders. There is therefore much that needs to be done to improve the way in which existing drugs can be used, as well as enable new drugs to be introduced.

UNDERSTANDING THE MECHANISMS OF ACTION OF BPs AT A CELLULAR LEVEL

The pronounced selectivity of BPs for bone rather than other tissues is the basis for their value in clinical practice. Their preferential uptake by and adsorption to mineral surfaces in bone brings them into close contact with osteoclasts. During bone resorption, BPs are probably internalized by endocytosis, along with other products of resorption. Many studies have shown that BPs can affect osteoclast-mediated bone resorption in a variety of ways that include effects on osteoclast recruitment, differentiation, and resorptive activity, and may induce apoptosis.

Since mature, multinucleated osteoclasts are formed by the fusion of mononuclear precursors of hematopoietic origin, BPs could also inhibit bone resorption by preventing osteoclast formation, in addition to affecting mature osteoclasts. *In vitro*, BPs can inhibit in a dose-dependent manner the formation of osteoclast-like cells in long-term cultures of human bone marrow.[106] In organ culture also, some BPs can inhibit the generation of mature osteoclasts, possibly by preventing the fusion of osteoclast precursors.[107,108]

It is likely that BPs are selectively internalized by osteoclasts rather than other cell types because of their accumulation in bone and the endocytic activity of osteoclasts. During the process of bone resorption, the subcellular space beneath the osteoclast is acidified by the action of vacuolar-type proton pumps in the ruffled border of the osteoclast membrane.[109] The acidic pH of this microenvironment causes dissolution of the HAP bone mineral, while the breakdown of the extracellular bone matrix is brought about by the action of proteolytic enzymes. Since BPs adsorb to bone mineral, especially at sites of bone resorption where the mineral is most exposed,[110,111] osteoclasts are the cell type in bone most likely to be exposed to the highest concentrations of free, non-mineral-bound bisphosphonate, as a result of the release of the bisphosphonate from bone mineral in the low pH environment beneath osteoclasts. It has been estimated that pharmacological doses of alendronate that inhibit bone resorption *in vivo* could give rise to local concentrations as high as 1 mM alendronate in the resorption space beneath an osteoclast. This is much higher than the concentrations of BPs required to affect osteoclast morphology and cause osteoclast apoptosis *in vitro*.[112]

Another observation that remains unexplained is that BPs may act at nanomolar concentrations to stimulate osteoblasts to produce an osteoclast inhibitory factor.[113–116]

Despite reports that some BPs do not have toxic effects on osteoclasts,[117] it is clear from many studies that BPs can reduce osteoclast number[118] and can induce apoptotic cell death in osteoclasts.[119] These effects occur both with the nitrogen-containing bisphosphonates (N-BPs), as well as the simpler BPs, such as clodronate.[120]

In addition, apoptosis triggered by exposure to BPs is not restricted to osteoclasts, since macrophages, such as the murine cell line J774,[121,122] and human myeloma cell lines[123,124] also undergo apoptosis after treatment with several (N-BPs) *in vitro*.

UNDERSTANDING THE MECHANISMS OF ACTION OF BPs AT A BIOCHEMICAL LEVEL

Over the years there have been many efforts to explain how BPs work on cells, especially via inhibitory effects on enzymes, for example, by direct or indirect inhibition of the osteoclast proton-pumping H^+ATPase,[125–127] phosphatases, or lysosomal enzymes.[128,129]

Because osteoclasts are highly endocytic, bisphosphonate present in the resorption space is likely to be internalized by endocytosis, and thereby affect osteoclasts directly.[130] The uptake of BPs by osteoclasts *in vivo* has been confirmed using radiolabeled[131] and fluorescently labeled alendronate, which was internalized into intracellular vacuoles. Following cellular uptake, a characteristic morphological feature of bisphosphonate-treated osteoclasts is the lack of a ruffled border, the region of invaginated plasma membrane facing the resorption cavity. BPs also disrupt the cytoskeleton of the osteoclast.[132] Early explanations for these effects invoked the inhibition of protein kinases or phosphatases that regulate cytoskeletal structure, such as protein tyrosine phosphatases.[133–136] However, a more likely mechanism by which the cytoskeleton may be affected involves loss of function of small GTPases, such as Rho and Rac.

Since the early 1990s there has been a systematic effort to elucidate the molecular mechanisms of action of BPs. Our work in this area has been led by Michael Rogers,[137] and we have proposed that BPs can be classified into at least two major groups with different modes of action (FIG. 4). The first group comprises the non-nitrogen BPs that perhaps most closely resemble PPi, such as clodronate and etidronate, and these can be metabolically incorporated into nonhydrolyzable analogues of adenosine triphosphate (ATP). It is likely that intracellular accumulation of these metabolites within osteoclasts inhibits their function and may cause osteoclast cell death. In contrast, the second group contains the more potent, (N-BPs), such as alendronate, risedronate, and zoledronate. Members of this group interfere with other metabolic reactions, notably in the mevalonate biosynthetic pathway, and affect cellular activity and cell survival by interfering with protein prenylation and therefore the signaling functions of key regulatory proteins. These mechanisms are discussed in greater detail below.

The ATP BP Metabolite Mechanism

This work has its origins in the study of the inhibitory effects of BPs on the growth of the amebas of the slime mold *Dictyostelium discoideum.*[138,139] In the first of what appear to be two major but distinct molecular mechanisms by which BPs affect osteoclasts, we found that some, but not all, BPs could be metabolically incorporated by the amebas into analogues of adenosine

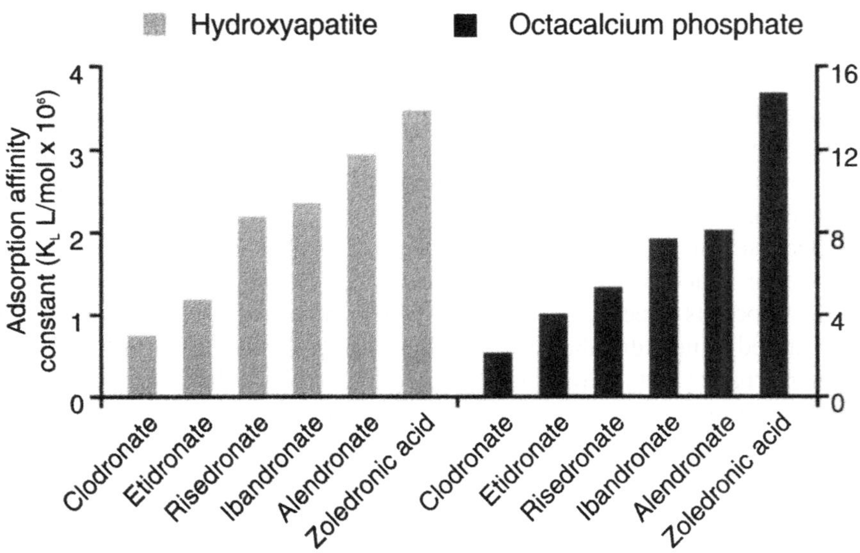

FIGURE 4. Differential binding of BPs to 2 calcium phosphate minerals found in bone, hydroxyapatite and octacalcium phosphate. There are unexpected differences between the BPs indicating that not only the P-C-P structure but also the R2 side chains contribute to mineral binding. Nancollas GH, et al. Bone 2006 in press. doi:10.1016/j.bone.2005.05.003

triphosphate (ATP or Appp).[140,141] The resulting metabolites contained the P-C-P moiety in place of the β,γ-phosphate groups of ATP, thus resulting in nonhydrolyzable (AppCp) nucleotides.

The BPs that were metabolized by *Dictyostelium discoideum* all contained short R1 and R2 side chains, with the exception of tiludronate, and were relatively weak inhibitors of bone resorption. A similar classification into metabolizable and nonmetabolizable BPs was obtained with cell-free lysates from mammalian cells. We found that the incorporation of BPs into these AppCp nucleotide analogues is brought about by members of the family of aminoacyl-tRNA synthetases,[142] which catalyze a reversible reaction in which an amino acid condenses with ATP to form an aminoacyladenylate, together with the release of PPi (reaction 1, shown below). Since this reaction is reversible, it appears that BPs with short R1 and R2 side chains (which most resemble PPi in structure) can replace PPi in the back reaction (reaction 2). This results in the condensation of a bisphosphonate (pCp) with an aminoacyladenylate (amino acid-AMP), to form an analogue of ATP (AppCp).

1. Enzyme + amino acid + ATP ⇔ amino-acyl-AMP + PPi
2. Amino-acyl-AMP + pCp ⇔ amino acid + AppCp

The aminoacyl-tRNA synthetases that can use a bisphosphonate in place of PPi all belong to the type II subclass of enzymes (e.g., Asn-, Asp-, Gly-, His-, Lys-, Phe-, Ser-aminoacyl-tRNA synthetases), which differ from the type

I subclass in the structure of the catalytic site. Thus, it appears that BPs with short side chains, but also rather surprisingly tiludronate, can replace PPi and be accommodated into the active site of type II aminoacyl-tRNA synthetases. In contrast, the more potent BPs that contain a nitrogen atom in the R2 side chain are not metabolized, presumably since the different, and in some cases bulkier, structure of the R2 side chain prevents these BPs from binding at the active site of these aminoacyl-tRNA synthetase enzymes.

Although the formation of AppCp-type bisphosphonate metabolites was first demonstrated in slime mold amebas (which also produced diadenosine tetraphosphate metabolites, AppCppA) and with cell-free lysates,[143] it also occurs in intact mammalian cells *in vitro* (J774 macrophage-like cells and MG63 osteosarcoma cells), which can also metabolize clodronate to an analogue of ATP (AppCCl2p),[144] as confirmed by mass spectrometric analysis of cell lysates from clodronate-treated cells,[145] including purified rabbit osteoclasts *in vitro.*[146]

The aminoacyl-tRNA synthetases are cytoplasmic enzymes, and the metabolism of BPs is dependent on cellular uptake.[147] As a result of the accumulation in the cell cytoplasm of these nonhydrolyzable AppCp analogues of ATP, they are likely to inhibit intracellular enzymes, thus having adverse effects on cell metabolism, function, and survival.

The AppCp-type metabolites of BPs are cytotoxic when internalized and cause similar changes in morphology to those observed in clodronate-treated cells, possibly by interference with mitochondrial ATP translocases.[148] Overall, this group of BPs therefore appears to act as prodrugs, being converted to active metabolites following intracellular uptake by osteoclasts *in vivo*.

The Mevalonate Pathway Mechanism

The potent, (N-BPs) are apparently not metabolized to AppCp-type metabolites as described above. A major step forward has been the demonstration that the (N-BPs) used as inhibitors of bone resorption all appear to be inhibitors of the mevalonate pathway. This is a biosynthetic pathway responsible for the production of cholesterol, other sterols, and isoprenoid lipids such as isopentenyldiphosphate (also known as isopentenylpyrophosphate IPP), as well as farnesyldiphosphate (FPP) and geranylgeranyldiphosphate (GGPP) (FIG. 5). FPP and GGPP are required for the posttranslational modification (prenylation) of small GTPases such as Ras, Rab, Rho, and Rac, which are prenylated at a cysteine residue in characteristic C-terminal motifs.[149] Small GTPases are important signaling proteins, which regulate a variety of cell processes important for osteoclast function, including cell morphology, cytoskeletal arrangement, membrane ruffling, trafficking of vesicles, and apoptosis.[150–153] Prenylation is required for the correct function of these proteins, since the lipid prenyl group serves to anchor the proteins in cell membranes and may also participate in protein–protein interactions.[154]

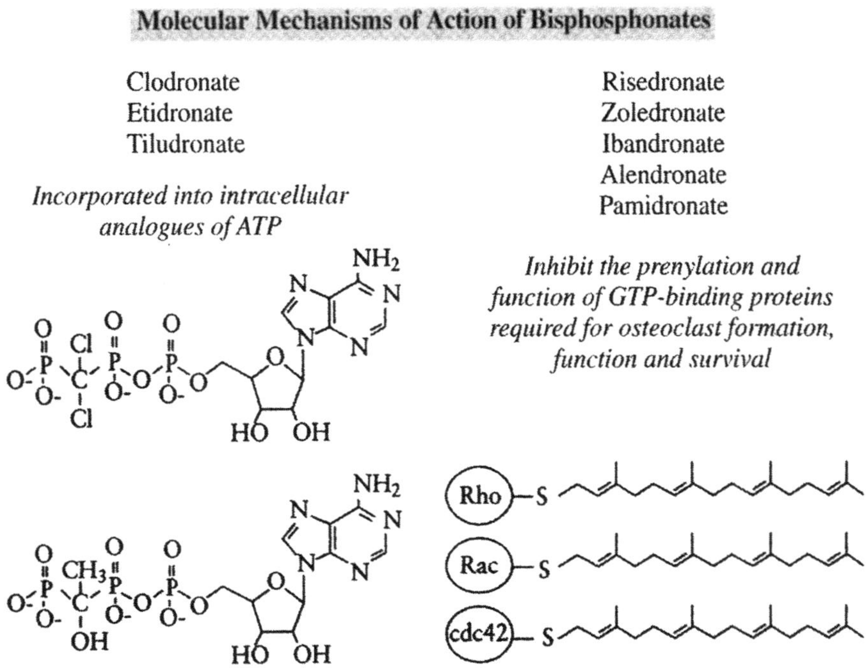

FIGURE 5. Bisphosphonates can be divided into two classes (those that do or do not contain nitrogen in the R2 side chain) according to their intracellular actions. Protein prenylation involves the transfer of a farnesyl or geranylgeranyl group onto cysteine residues near the C-terminus of small GTPases such as Rho, Rac, Rab and Ras.

There are now many observations that point to the importance of the mevalonate pathway for osteoclast function, and strengthen the proposal that the (N-BPs) inhibit osteoclastic bone resorption predominantly by inhibition of this pathway. These BPs inhibit the synthesis of mevalonate metabolites including FPP and GGPP, and thereby impair the prenylation of proteins,[155] and cause loss of function of small GTPases. There is a strong structure–activity relationship so that changes to the structure of the nitrogen-containing R2 side chain or to the phosphonate groups, which altered antiresorptive potency and also influenced the ability to inhibit protein prenylation to a corresponding degree.[156] An important verification of the critical importance of this pathway has come from showing that the addition of intermediates of the mevalonate pathway (such as FPP and GGPP) could overcome bisphosphonate-induced apoptosis and other events in many cell systems. A further prediction was that if inhibition of the mevalonate pathway could account for the antiresorptive effects of BPs, then the statin drugs should also inhibit bone resorption. Statins are inhibitors of HMG-CoA reductase, one of the first steps in the mevalonate pathway. They proved to be even more potent than BPs at inhibiting osteoclast formation and bone resorption *in vitro,*[157,158] an effect that could also be overcome by the addition of geranylgeraniol (which can be used for protein geranylgeranylation) but not farnesol (which is utilized for protein farnesylation). Hence, it appears that, although (N-BPs) can prevent both farnesylation and geranylgeranylation of proteins (probably by inhibiting enzymes required for synthesis of FPP and GGPP), loss of geranylgeranylated proteins in osteoclasts is of greater consequence than loss of farnesylated proteins. This is consistent with the known role of geranylgeranylated proteins, such as Rho, Rac, and Rab in processes that are fundamental to osteoclast formation and function (e.g., cytoskeletal rearrangement, membrane ruffling, and vesicular trafficking[159]), and further work has confirmed this, particularly the importance of Rab proteins.

The comparison between BPs and statins is interesting. The statins are widely used as cholesterol-lowering drugs, through their ability to lower cholesterol biosynthesis by inhibiting HMG-CoA reductase. Despite several studies there is no substantial evidence that statins have effects on bone when used clinically, perhaps because they are selectively taken up by liver rather than bone, which is the converse of the case for BPs. This is therefore an excellent example of how drug specificity is achieved by highly selective tissue targeting (FIG. 6).

The exact enzymes of the mevalonate pathway that are inhibited by individual BPs have been partially elucidated. Several enzymes of the pathway utilize isoprenoid diphosphates as a substrate (IPP isomerase, FPP synthase, GGPP synthase, squalene synthase) and thus are likely to have similar substrate binding sites. Thus, if (N-BPs) act as substrate analogues of an isoprenoid diphosphate, it is possible that these BPs will inhibit more than one of the enzymes of the mevalonate pathway. Early studies revealed that incadronate and ibandronate, but not other BPs, are inhibitors of squalene synthase, an enzyme in the mevalonate pathway required for cholesterol biosynthesis.[160,161] Inhibition of squalene synthase would not, however, lead to inhibition of protein prenylation.

However, it is now clear that farnesyl PPi synthase (FDPS or FPPS) is a major site of action of the N-BPs.[162] FPPS catalyzes the successive condensation of isopentenyl PPi with dimethylallyl PPi and geranyl PPi (FIG. 7). There is a strong relationship among individual BPs between inhibition of bone resorption and inhibition of FPPS with the most potent BPs having IC50s in the nM range.[163] Modeling studies provide a molecular rationale for BP binding to FPPS.[164] Our recent studies using protein crystallography, enzyme kinetics, and isothermal titration calorimetry have led to the first high-resolution X-ray structures of the human enzyme in complexes with risedronate and zoledronate.[165] These agents bind to the dimethylallyl/geranyl PPi ligand pocket and induce a conformational change. The interactions of the N-BPs cyclic nitrogen with Thr201 and Lys200 suggest that these inhibitors achieve potency by positioning their nitrogen in a proposed carbocation binding site.[166] This explains why the nitrogen moiety is so important to the potency of these BPs (FIG. 8). Kinetic analyses reveal that inhibition is competitive with geranyl

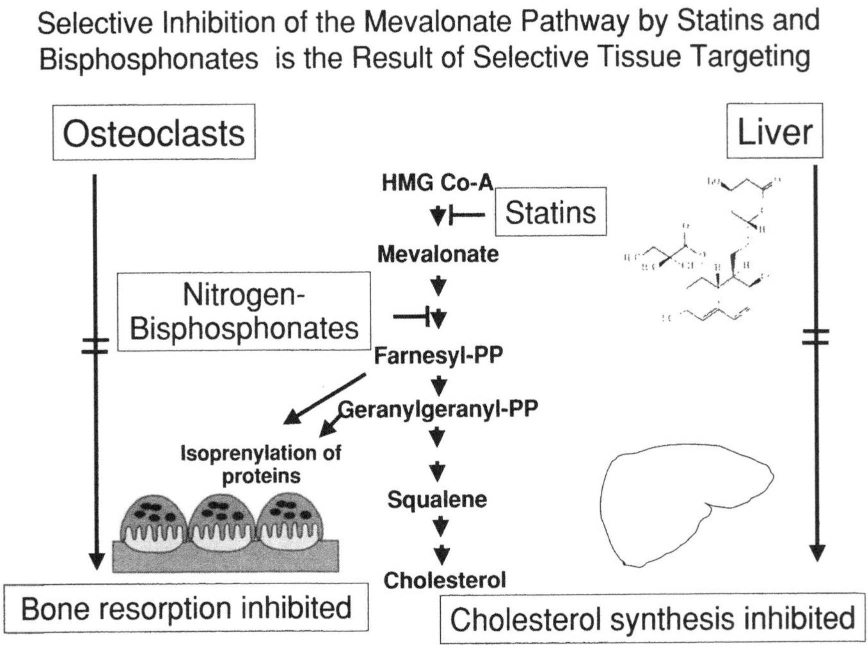

FIGURE 6. The differential effects of statins and BPs in the mevalonate pathway, showing how tissue selectivity of uptake determines their pharmacological specificity.

PPi and is of a slow, tight-binding character, indicating that isomerization of an initial enzyme-inhibitor complex occurs upon binding of the N-BP.

Taken together, these observations clearly indicate that BPs can be grouped into two classes: those that can be metabolized into nonhydrolyzable analogues of ATP (the least potent BPs) and those that are not metabolized but that can inhibit protein prenylation (the potent, N-BPs). The identification of two such classes may help to explain some of the other pharmacologic differences between the two classes.

CURRENT ISSUES AND NEW LEADS WITH BPs

Update on PPi

Since the early studies that indicated that PPi was a potential endogenous regulator of mineralization, there have been significant advances in understanding the metabolism of PPi and in identifying clinical disorders in which alterations in PPi may have a pathogenic role.

Much of the PPi in the extracellular compartment may be generated at the cell surface by the action of nucleoside triphosphate pyrophosphohydrolases

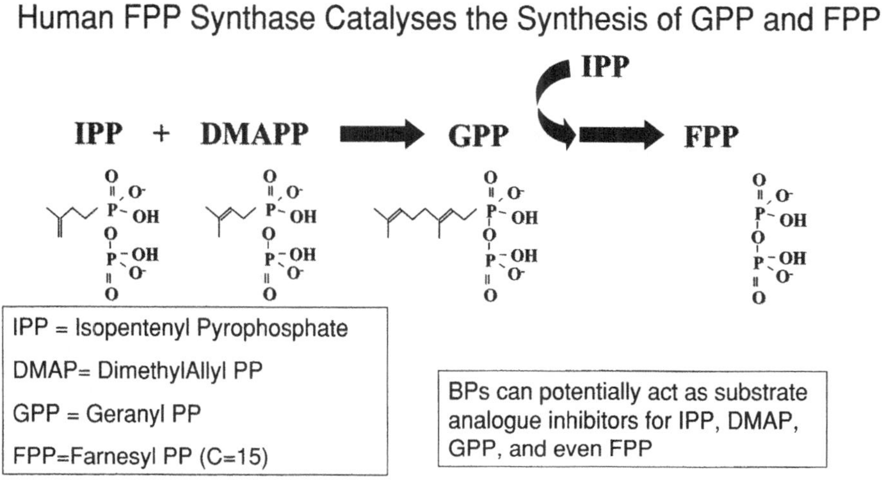

FIGURE 7. The enzyme reactions catalysed by Farnesyl Pyrophosphate Synthase. N-BPs act as substrate analogues for GPP.

(NTP-PPases), which liberate PPi from NTPs, such as ATP. The major enzyme involved in removing PPi is alkaline phosphospatase or tissue nonspecific alkaline phosphatase (TNAP), as has been known for many years. TNAP is also located at cell sufaces and its tissue distribution is restricted, particularly to liver, cartilage, and bone. A third regulator of extracellular PPi has been postulated to be a membrane transporter of PPi called ANK, which is thought to extrude PPi from within cells.

Genetic mutations of all three of these regulatory proteins are associated with disturbances in PPI metabolism and disordered calcification.[167–169] Skeletal mineralization is defective when PPi is high, e.g., in hypophosphatasia due to inactivating mutations in TNAP.[170,171] Conversely, excessive mineralization and bone formation may occur when NTP-PPase (PC-1) is defective and PPi levels are low as in juvenile vascular calcification[172] and another rare condition called ossification of the posterior longitudinal ligament (OPLL)[173] of the spine and osteoarthritis, which occurs particularly in Japanese populations.

Mutations of the ANK gene in mice produces a skeletal phenotype of progressive ankylosis and aberrant calcification,[174] while in humans mutations of ANKH are associated with familial chondrocalcinosis, a condition in which calcium PPi crystals deposit in articular cartilage and other sites.[175–177] Mutations of the ANKH gene are also somewhat unexpectedly associated with craniometaphyseal dysplasia (CMD).[178,179]

So far there is no evidence that BPs used clinically interfere with the endogenous metabolism of extracellular PPi.

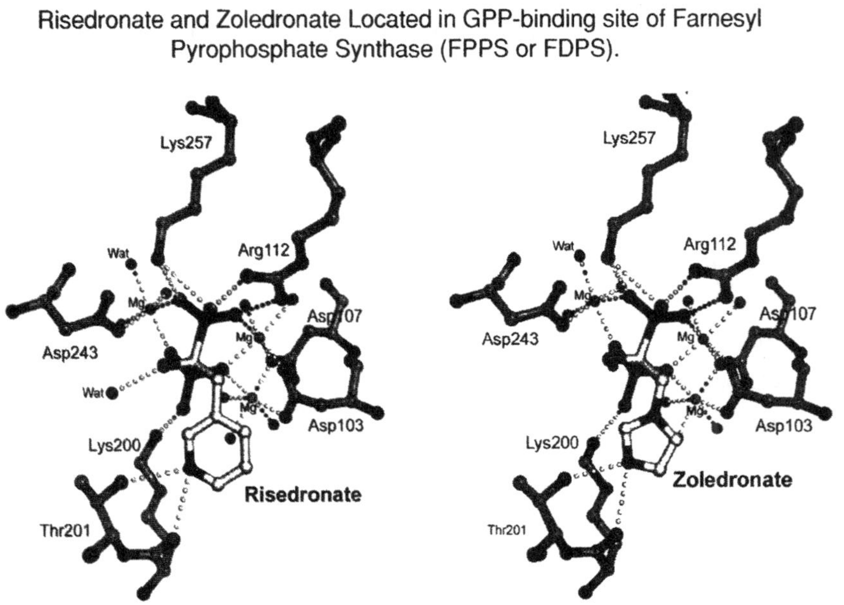

FIGURE 8. Risedronate and Zoledronate located in the GPP binding site of Farnesyl Pyrophosphate Synthase (FDPS or FPPS). See Kavanagh K., Guo K, Dunford J, Wu X, Knapp S., Ebetino FH, Rogers MJ, Russell RGG, Oppermann U. Human Farnesyl Diphosphate Synthase (FDPS): Crystal structure and molecular interactions with nitrogen-containing bisphosphonates. PNAS in press 2006 & J Bone Min Res Sa409 2005

The Effects of BP on Bone Architecture, Structure and Strength, and on Bone Healing and Fracture Repair

Many experimental and clinical studies show that BPs conserve bone architecture and strength.[180–184] However, there have been concerns about whether the use of prolonged high doses of BPs may impair bone turnover to such an extent that bone strength is impaired. High doses in animals are associated with increased microdamage[185,186] and even fractures.[187] It has been suggested that BPs might prevent naturally occurring microscopic cracks in bone from healing. There are isolated reports of adynamic bone associated with BP usage[188] but the long-term use of the BPs in the therapy of osteoporosis appears to be safe.[189] Case reports of induction of ostepetrosis-like lesions in children treated with excessive doses of pamidronate have been reported.[190]

A question often asked is whether BPs inhibit fracture repair. By reducing bone turnover one might expect BPs to interfere with fracture healing. However, a recent long-term study in a beagle dog model that simulated fracture repair has demonstrated that ibandronate treatment did not adversely affect normal bone healing.[191] Studies of repair processes after creating drill hole defects in dogs also showed no impairment with ibandronate.[192]

Several other recent studies raise the intriguing possibility that BPs may enhance fracture repair and related processes.[193] In studies of the osseointegration of metal implants in OVX rats, treatment with ibandronate resulted in improved osseointegration, rather than impairment of the healing process.[194] Potential applications of BPs in orthopedics include protection against loosening of prostheses,[195] better integration of biomaterials and implants, improved healing in distraction osteogenesis,[196] and conserving bone architecture after osteonecrosis[197,198] and in Perthes disease.[199]

How do BPs Work? Explaining the Long Duration of Action. Recycling Within Bone

BPs are well accepted as the main class of antiresorptive agents and have many clinical applications. There are potentially important differences between clinically useful BPs regarding their potency and duration of action. Efficacy is closely related to affinity for bone mineral and ability to inhibit FPP synthase. Recent studies (see FIG. 9) have shown that there are marked differences among BPs in binding to HAP,[200] may explain the variations in retention and persistence of effect observed in animal and clinical studies. In the case of zoledronic acid in particular, the remarkable magnitude of effect and prolonged duration of action can be explained in part by these new observations. In explaining the long duration of action, it has been proposed that there is continuous recycling of BP off and back onto the bone surface. This notion is supported by observations that BPs can be found in plasma and urine many months after dosing.

How do BPs Work? Actions on Osteocytes

In contrast to their ability to induce apoptosis in osteoclasts, which contributes to the inhibition of resorptive activity, some experimental studies suggest that BPs may protect osteocytes and osteoblasts from apoptosis induced by glucocorticoids.[201] Recent evidence suggests that the inhibition of osteocyte apoptosis by BPs is mediated through the opening of connexion43 hemichannels and activation of extracellular signal-regulated kinases.[202] The possibility that BPs used clinically may get access to osteocytes differentially depending on their mineral-binding affinities and inherent structural properties needs to be studied.

The "Acute Phase Response"

A well-recognized side effect of the (N-BPs) is to cause an acute phase response *in vivo,*[203–205] which can lead to induction of fever and "flu"-like

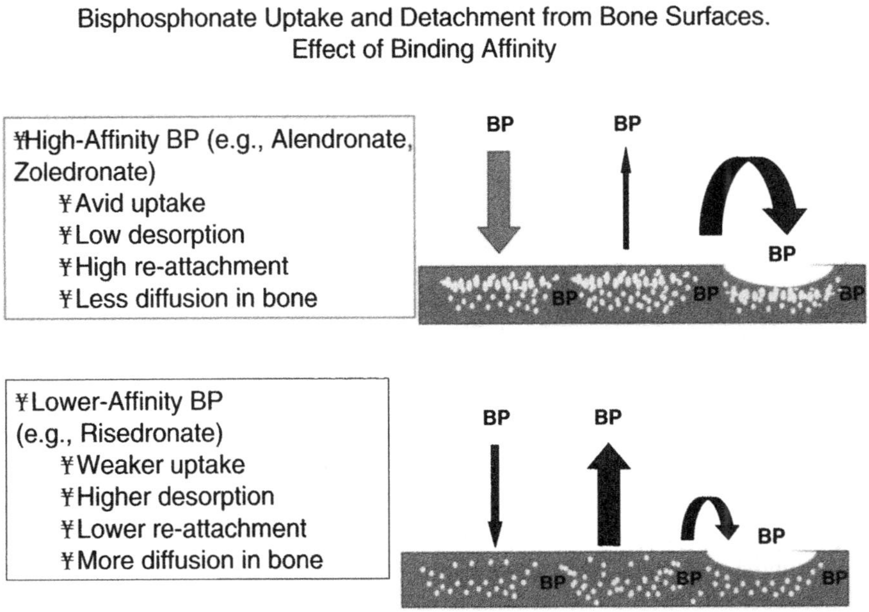

FIGURE 9. Effect of Binding Affinity of Bisphosphonates on their Uptake and Detachment from bone surfaces (Adapted from Nancollas GH, et al. Bone 2006 in press.)

symptoms in patients. These effects are transient and occur predominantly on first exposure to the drug, especially with i.v. administration. The mechanism has been attributed to release of pro-inflammatory cytokines, and the mechanism has been further unraveled by showing that it involves selective receptor-mediated activation of γ, Δ T cells leading to their proliferation and activation.[206] The BP effect involves the mevalonate pathway *in vitro* and can be overcome by using statins.[207]

Nonskeletal Effects of BPs

There are numerous examples of BPs having effects on cells and tissues outside the skeleton. The effects on osteoclast precursors, tumor cells, macrophages, and γ Δ T cells are examples, and in all cases are probably explained by sufficient BPs entering cells to inhibit the mevalonate pathway.

A particularly interesting aspect of these nonskeletal effects is the observations made on protozoan parasites, the growth of which can be inhibited by BPs acting on FPPS.[208,209] The therapeutic potential is enticing given the importance of these diseases. The range of affected protozoa include Entamoeba,[210] Plasmodia,[211] Trypanosomes,[212] Toxoplasma,[213] Cryptosporidia,[214] and Leishmania spp.[215]

SUMMARY AND FUTURE PROSPECTS

It has taken over 30 years since the discovery of the profound effects of the BPs on calcium metabolism for them to become well established as clinically successful antiresorptive agents, and their availability has enabled new approaches to the therapy of bone diseases.

There have now been many years of mostly favorable experience with the use of BPs in diseases such as Paget's disease of bone, myeloma, and bone metastases. BPs represent an important class of drugs for the treatment of these bone diseases.

Their application in osteoporosis is more recent and was spurred on by the development of techniques to measure bone mass with precision, the increased awareness of osteoporosis as a major socioeconomic problem, and the willingness of the larger pharmaceutical companies to invest in clinical studies on the scale necessary to demonstrate their effects on fractures.

The difficulties of bringing these drugs to the market are illustrated by those that fall by the wayside, such as oral pamidronate and tiludronate. There are important lessons to be learned from the need to do good dose-response studies during phase II development and making appropriate choices of doses.

However, despite the enormous potential for developing "better" BPs based on current knowledge of their structure–activity properties, it is uncertain, given the high cost of development, that further agents will be developed unless they offer distinct advantages over currently available BPs.

Other clinical indications ripe for future study include the prevention of bone loss and erosions in rheumatoid arthritis, possible applications in other joint diseases, and the reduction of bone loss associated with periodontal disease, and loosening of joint prostheses.

The recent elucidation of the likely mode of action of BPs within cells opens up the possibility of exploiting the subtle and potentially important differences between classes of BPs and individual compounds.

REFERENCES

1. MENSCHUTKIN, N. 1865. Ueber die Einwirkung des Chloracetyls auf phosphorige Saure. Ann. Chem. Pharm. **133:** 317–320.
2. SMITH, R., R.G.G. RUSSELL & M. BISHOP. 1971. Diphosphonates and Paget's disease of bone. Lancet **1:** 945–947.
3. VON BAEYER, H. & K.A. HOFMANN. 1897. Acetodiphosphorige Saure. Berichte Dtsch. Chem. Ges. **20:** 1973–1978.
4. BLOMEN, L.J.M.J. 1995. Discovery and history of the non-medical uses of bisphosphonates. Chapter 7. *In* Bisphosphonates on Bone. O. Bijvoet, *et al.* Eds.: 111–124. Elsevier Science B.V. Amsterdam.
5. RUSSELL, R.G. & M.J. ROGERS. 1999. Bisphosphonates: from the laboratory to the clinic and back again [review]. Bone **25:** 97–106.

6. Bijvoet, O., H. Fleisch, R.E. Canfield & R.G.G. Russell, Eds. 1995. Bisphosphonates on Bone. Elsevier Science B.V. Amsterdam.
7. Fleisch, H. 2000. Bisphosphonates in Bone Disease. From the Laboratory to the Patient, 4th ed. Academic Press. New York.
8. Fleisch, H. 1998. Bisphosphonates: mechanisms of action. Endocr. Rev. **19:** 80–100.
9. Ebetino, F.H., M.D. Francis, M.J. Rogers, *et al.* 1998. Etidronate. Mechanisms of action of etidronate and other bisphosphonates. Revs. Contemp. Pharma **9:** 233–243.
10. Green, J.R. 2004. Bisphosphonates: preclinical review. Oncologist. **9**(Suppl 4): 3–13.
11. Russell, R.G.G. & M.J. Rogers. 1997. Introduction to bisphosphonates and the clinical pharmacology of alendronate. Br. J. Rheumatol. **36**(S): 10–14.
12. Johnson, S. & F.N. Johnson, Eds. 1998. Etidronate in Osteoporosis, **Vol 9**. Rev in Contemp Pharm. Marius Press. Lancashire, UK.
13. Geddes, A.D., S.M. D'Souza, F.H. Ebetino, *et al.* 1994. Bisphosphonates: structure-activity relationships and therapeutic implications. Bone Miner. Res. **8:** 265–306.
14. Sietsema, W.K., F.H. Ebetino, A.M. Salvagno, *et al.* 1989. Antiresorptive dose-response relationships across three generations of bisphosphonates. Drugs Exp. Clin. Res. **15:** 389–396.
15. Rogers, M.J. 2004. From molds and macrophages to mevalonate: a decade of progress in understanding the molecular mode of action of bisphosphonates [review]. Calcif. Tissue Int. **75:** 451–461.
16. Papapoulos, S.E. 1995. Pharmacodynamics of bisphosphonates in man. Chapter 15. *In* Bisphosphonates on Bone. O. Bijvoet, *et al.* Eds.: Elsevier Science B.V. Amsterdam.
17. Rondeau, J.M., F. Bitsch, E. Boirgier, *et al.* 2006. Structural basis for the exceptional *in vivo* efficacy of bisphosphonate drugs. Chem. Med. Chem. **1:** 267–273.
18. Michaelson, M.D. & M.R. Smith. 2005 Bisphosphonates for treatment and prevention of bone metastases [review]. J. Clin. Oncol. **23:** 8219–8224.
19. Fleisch, H. & W.F. Neuman. 1961. Mechanisms of calcification: role of collagen, polyphosphates, and phosphatase. Am. J. Physiol. **200:** 1296–1300.
20. Fleisch, H. & S. Bisaz. 1962. Isolation from urine of pyrophosphate, a calcification inhibitor. Am. J. Physiol. **203:** 671–675.
21. Fleisch, H., R.G.G. Russell & F. Straumann. 1966. Effect of pyrophosphate on hydroxyapatite and its implications in calcium homeostasis. Nature **212:** 901–903.
22. Russell, R.G.G. & A. Hodgkinson. 1966. The urinary excretion of inorganic pyrophosphate by normal subjects and patients with renal calculus. Clin. Sci. **31:** 51–62.
23. Russell, R.G.G. 1965. Excretion of inorganic pyrophosphate in hypophosphatasia. Lancet **10:** 461–464.
24. Russell, R.G.G., S. Bisaz, A. Donath, *et al.* 1971. Inorganic pyrophosphate in plasma in normal persons and in patients with hypophosphatasia, osteogenesis imperfecta and other disorders of bone. J. Clin. Invest. **50:** 961–969.
25. Schibler, D., R.G.G. Russell & H. Fleisch. 1968. Inhibition by pyrophosphate of aortic calcification induced by Vitamin D3 in rats. Clin. Sci. **35:** 363–372.

26. Fleisch, H., R.G.G. Russell, S. Bisaz, *et al.* 1968. The influence of pyrophosphate analogues (diphosphonates) on the precipitation and dissolution of calcium phosphate in vitro and in vivo. Calcif. Tissue Res. **2:** 10–10A.
27. Francis, M.D., R.G.G. Russell & H. Fleisch. 1969. Diphosphonates inhibit formation of calcium phosphate crystals *in vitro* and pathological calcification *in vivo*. Science **165:** 1264–1266.
28. Fleisch, H., R.G.G. Russell & M.D. Francis. 1969. Diphosphonates inhibit hydroxyapatite dissolution *in vitro* and bone resorption in tissue culture and *in vivo*. Science **165:** 1262–1264.
29. Jung, A., S. Bisaz & H. Fleisch. 1973. The binding of pyrophosphate and two diphosphonates by hydroxyapatite crystals. Calcif. Tissue Res. **11:** 269–280.
30. Fleisch, H.A., R.G.G. Russell, S. Bisaz, *et al.* 1970. The inhibitory effect of phosphonates on the formation of calcium phosphate crystals *in vitro* and on aortic and kidney calcification *in vivo*. Eur. J. Clin. Invest. **1:** 12–18.
31. Schenk, R., W.A. Merz, R. Muehlbauer, *et al.* 1973. Effect of ethane-1-hydroxy-1,1-diphosphonate (EHDP) and dichloromethylene diphosphonate (Cl2MDP) on the calcification and resorption of cartilage and bone in the tibial epiphysis and metaphysis of rats. Calcif. Tissue Res. **11:** 196–214.
32. Fleisch, H., J. Maerki & R.G.G. Russell. 1966. Effect of pyrophosphate on dissolution of hydroxyapatite and its possible importance in calcium homeostasis. Proc. Soc. Exp. Biol. **122:** 317–320.
33. Russell, R.G.G., R.C. Muhlbauer, S. Bisaz, *et al.* 1970. The influence of pyrophosphate, condensed phosphates, phosphonates and other phosphate compounds on the dissolution of hydroxyapatite in vitro and on bone resorption induced by parathyroid hormone in tissue culture and in thyroparathyroidectomised rats. Calcif. Tissue Res. **6:** 183–196.
34. Trechsel, U., A. Stutzer & H. Fleisch. 1987. Hypercalcaemia induced with an arotinoid in thyroparathyroidectomised rats. A new model to study bone resorption in vivo. J. Clin. Invest. **80:** 1679–1686.
35. Stutzer, A., H. Fleisch & U. Trechsel. 1988. Short and long term effects of a single dose of bisphosphonates on retinoid-induced bone resorption in thyroparathyroidectomised rats. Calcif. Tissue Int. **43:** 294–299.
36. Schenk, R., P. Eggli, H. Fleisch, *et al.* 1986. Quantitative morphometric evaluation of the inhibitory activity of new aminobisphosphonates on bone resorption in the rat. Calcif. Tiss. Int. **38:** 342–349.
37. Gasser, A.B., D.B. Morgan, H.A. Fleisch, *et al.* 1972. The influence of two diphosphonates on calcium metabolism in the rat. Clin. Sci. **43:** 31–45.
38. Muehlbauer, R.C., R.G.G. Russell, D.A. Williams, *et al.* 1971. The effects of diphosphonates, polyphosphates and calcitonin on "immobilisation osteoporosis" in rats. Eur. J. Clin. Invest. **1:** 336–344.
39. Ferretti, J.L. 1995. Effects of bisphosphonates on bone biomechanics. Chapter 31. *In* Bisphosphonates on Bone. O. Bijvoet, *et al.* Eds.: Elsevier Science B.V. Amsterdam.
40. Shinoda, H., G. Adamek & R. Felix, *et al.* 1983. Structure-activity relationships of various bisphosphonates. Calcif. Tissue Int. **35:** 87–99.
41. Flanagan, A.M. & T.J. Chambers. 1989. Dichloromethylenebisphosphonate (Cl2MBP) inhibits bone resorption through injury to osteoclasts that resorb Cl2MBP-coated bone. Bone Miner. **6:** 33–43.

42. REITSMA, P.H., O.L.M. BIJVOET, H. VERLINDEN-OOMS, *et al.* 1980. Kinetic studies of bone and mineral metabolism during treatment with (3-amino-1-hydroxypropylidene)-1,1-bisphosphonate (APD) in rats. Calcif. Tissue Int. **32:** 145–157.
43. GARNERO, P., W.J. SHIH, E. GINEYTS, *et al.* 1994. Comparison of new biochemical markers of bone turnover in late postmenopausal osteoporotic women in response to alendronate treatment. J. Clin. Endocrinol. Metab. **79:** 1693–1700.
44. BAUSS, F. & R.G.G. RUSSELL. 2004. Ibandronate in osteoporosis: preclinical data and rationale for intermittent dosing. Osteoporos. Int. **15:** 423–433.
45. GASSER, J.A. & J.R. GREEN. 2002. Long-term protective effect of a single IV injection of zoledronic acid on cancellous bone structure and cortical bone in ovariectomized rats. Bone. **30**(Suppl.): 41S.
46. EBRAHIMPOUR, A. & M.D. FRANCIS. 1995. Bisphosphonate therapy in acute and chronic bone loss: physical chemical considerations in bisphosphonate-related therapies. *In* Bisphosphonate on bones. O. Bijvoet, *et al.* Eds.: Elsevier Science B.V Amsterdam.
47. MUEHLBAUER, R.C., F. BAUSS, R. SCHENK, *et al.* 1991. BM 21.0955, a potent new bisphosphonate to inhibit bone resorption. J. Bone Miner. Res. **6:** 1003–1011.
48. PAPAPOULOS, S.E., K. HOEKMAN, C.W.G.M. LOWIK, *et al.* 1988. Application of an *in vitro* model and a clinical protocol in the assessment of the potency of a new bisphosphonate. J. Bone Miner. Res. **4:** 775–781.
49. GOA, K.L. & J.A. BALFOUR. 1998. Risedronate. Drugs Aging **13:** 83–91.
50. GREEN, J.R., K. MUELLER & K.A. JAEGGI. 1994. Preclinical pharmacology of CGP 42′446, a new, potent, heterocyclic bisphosphonate compound. J. Bone Miner. Res. **9:** 745–751.
51. EBETINO, F.H., A.V. BAYLESS, J. AMBURGEY, *et al.* 1996. Elucidation of a pharmacore for the bisphosphonate mechanism of bone antiresorptive activity. Phosphorus Sulfur Silicon Relat. Elem. **109/110:** 217–220.
52. VAN BEEK, E., C. LOWIK, I. QUE, *et al.* 1996. Dissociation of binding and antiresorptive properties of hydroxybisphosphonates by substitution of the hydroxyl with an amino group. J. Bone Miner. Res. **11:** 1492–1499.
53. SMITH, R. 1995. Ectopic calcification and ossification. Chapter 31. *In* Bisphosphonate on Bones. O. Bijvoet, *et al.* Eds.: Elsevier Science B.V. Amsterdam.
54. FOGELMAN, I., R.G. BESSENT, J.F. TURNER, *et al.* 1978. The use of whole-body retention of Tc-99m diphosphonate in the diagnosis of metabolic bone disease. J. Nucl. Med. **19:** 270–275.
55. FRANCIS, M.D. & I. FOGELMAN. 1987. 99mTc- diphosphonate uptake mechanisms on bone. *In* Bone Scanning in Clinical Practice. I. Fogelman, Ed.: 1–6. Springer Verlag. London.
56. RUSSELL, R.G.G., R. SMITH, C. PRESTON, *et al.* 1974. Diphosphonates in Paget's disease. Lancet **1:** 894–898.
57. DELMAS, P.D. & P.J. MEUNIER. 1997. The management of Paget's disease of bone. N. Engl. J. Med. **336:** 558–566.
58. DAROSZEWSKA, A. & S.H. RALSTON. 2005. Genetics of Paget's disease of bone. Clin. Sci. (Lond). **109:** 257–263.
59. MEE, A.P. 1999. Paramyxoviruses and Paget's disease: affirmative view. Bone **24:** 19S–21S.
60. DOUGLAS, D.L., R.G.G. RUSSELL, C.J. PRESTON, *et al.* 1980. Effect of dichloromethylene diphosphonate in Paget's disease of bone and in hypercalcaemia due to primary hyperparathyroidism or malignant disease. Lancet **1:** 1043–1047.

61. Coukell, A.J. & A. Markham. 1998. Pamidronate. A review of its use in the management of osteolytic bone metastases, tumour-induced hypercalcaemia and Paget's disease of bone. Drugs Aging **12:** 149–168.
62. Singer, F.R., T.L. Clemens & R.E. Eusebio. 1998. Risedronate, a highly effective oral agent in the treatment of patients with severe Paget's disease. J. Clin. Endocrinol Metab. **83:** 1906–1910.
63. Siris, E.S., A.A. Chines, R.D. Altman, *et al.* 1998. Risedronate in the treatment of Paget's disease of bone: an open label, multicenter study. J. Bone Miner. Res. **13:** 1032–1038.
64. Reid, I.R., P. Miller, K. Lyles, *et al.* 2005. Comparison of a single infusion of zoledronic acid with risedronate for Paget's disease. N. Engl. J. Med. **353:** 898–908.
65. Martodam, R.R., K.S. Thornton, D.A. Sica, *et al.* 1983. The effects of dichloromethylene diphosphonate on hypercalcemia and other parameters of the humoral hypercalcemia of malignancy in the rat Leydig cell tumor. Calcif. Tissue Int. **35:** 512–519.
66. Sasaki, A., B.F. Boyce, B. Story, *et al.* 1995. Bisphosphonate risedronate reduces metastatic human breast cancer burden in bone in nude mice. Cancer Res. **55:** 3551–3557.
67. Mundy, G.R. & T. Yoneda. 1995. Facilitation and suppression of bone metastasis. Clin. Orthop. Relat. Res. **312:** 34–44.
68. Mundy, G.R. 1998. Bisphosphonates as anticancer drugs. N. Engl. J. Med. **339:** 398–400.
69. Clezardin, P., F.H. Ebetino & P.G. Fournier. 2005. Bisphosphonates and cancer-induced bone disease: beyond their antiresorptive activity [review]. Cancer Res. **65:** 4971–4974.
70. Ross, J.R., Y. Saunders, P.M. Edmonds, *et al.* 2004. A systematic review of the role of bisphosphonates in metastatic disease [review]. Health Technol. Assess. **8:** 1–176.
71. Paterson, A.D., J.A. Kanis, E.C. Cameron, *et al.* 1983. The use of dichloromethylene diphosphonate for the management of hypercalcaemia in multiple myeloma. Br. J. Haematol. **54:** 121–131.
72. Ralston, S.H., S.J. Gallacher, U. Patel, *et al.* 1989. Comparison of three bisphosphonates in cancer-associated hypercalcaemia. Lancet **ii:** 1180–1182.
73. Body, J.J. 2004. Hypercalcemia of malignancy [review]. Semin. Nephrol. **24:** 48–54.
74. Sirohi, B. & R. Powles. 2004. Multiple myeloma [review]. Lancet **363:** 875–887.
75. Pavlakis, N., R. Schmidt & M. Stockler. 2005. Bisphosphonates for breast cancer. Cochrane Database Syst Rev. CD003474.
76. Coleman, R.E. 2005. Bisphosphonates in breast cancer [review]. Ann. Oncol. **16:** 687–695.
77. Parker, C.C. 2005. The role of bisphosphonates in the treatment of prostate cancer [review]. BJU Int. **95:** 935–938.
78. Smith, M.R. 2005. Zoledronic acid to prevent skeletal complications in cancer: corroborating the evidence. Cancer Treat. Rev. **31**(Suppl 3): 19–25.
79. Bell, R. 2005. Efficacy of ibandronate in metastatic bone disease: review of clinical data. Oncologist **10**(Suppl 1): 8–13.
80. Body, J.J., I.J. Diel, M. Lichinitzer, *et al.* 2004. Oral ibandronate reduces the risk of skeletal complications in breast cancer patients with metastatic bone disease: results from two randomised, placebo-controlled phase III studies. Br. J. Cancer **90:** 1133–1137.

81. McCloskey, E.V., I.C.M. MacLennan, M. Drayson, *et al.* 1998. A randomized trial of the effect of clodronate on skeletal morbidity in multiple myeloma. MRC working party on leukaemia in adults. Br. J. Haematol. **100:** 317–325.
82. Berenson, J.R., A. Lichtenstein, L. Porter, *et al.* 1996. Efficacy of pamidronate in reducing skeletal events in patients with advanced multiple myeloma. N. Engl. J. of Med. **334:** 488–493.
83. Diel, I.J., E-F. Solomayer, S.D. Costa, *et al.* 1998. Reduction in new metastases in breast cancer with adjuvant clodronate treatment. N. Engl. J. Med. **339:** 357–363.
84. Maerevoet, M., C. Martin & L. Duck. 2005. Osteonecrosis of the jaw and bisphosphonates. N. Engl. J. Med. **353:** 99–102.
85. 2004. Bone Health and Osteoporosis: A Report of the Surgeon General. U.S. Department of Health and Human Services, Office of the Surgeon General. Rockville, MD. www.surgeongeneral.gov/library.
86. Russell, R.G.G. 2003. Pathogenesis of osteoporosis. *In* Rheumatology, 3rd ed. M.C. Hochberg *et al.* Eds.: 2075–2147. publ. Mosby.
87. Watts, N.B., S.T. Harris, H.K. Genant, *et al.* 1990. Intermittent cyclical etidronate treatment of postmenopausal osteoporosis. N. Engl. J. Med. **323:** 73–79.
88. Storm, T., G. Thamsborg, T. Steiniche, *et al.* 1990. Effect of intermittent cyclical etidronate therapy on bone mass and fracture rate in women with postmenopausal osteoporosis. N. Engl. J. Med. **322:** 1265–1271.
89. van Staa, T.P., L. Abenhain & C. Cooper. 1998. Use of cyclical etidronate and prevention of non-vertebral fractures. Br. J. Rheumatol. **37:** 87–94.
90. Liberman, U.A., S.R. Weiss, J. Broll, *et al.* 1995. Effect of oral alendronate on bone mineral density and the incidence of fractures in postmenopausal osteoporosis. N. Engl. J. Med. **333:** 1437–1443.
91. Black, D.M., S.R. Cummings, D.B. Karpf, *et al.* 1996. Randomised trial of effect of alendronate on risk of fracture in women with existing vertebral fractures. Fracture intervention trial research group. Lancet **348:** 1535–1541.
92. Bone, H.G., D. Hosking, J.-P. Devogelaer, *et al.* for the Alendronate Phase III Osteoporosis Treatment Study Group. 2004. Ten years' experience with alendronate for osteoporosis in postmenopausal women. N. Engl. J. Med. **350:** 1189–1199.
93. Reginster, J., H.W. Minne, O.H. Sorensen, *et al.* 2000. Randomized trial of the effects of risedronate on vertebral fractures in women with established postmenopausal osteoporosis. Vertebral Efficacy with Risedronate Therapy (VERT) Study Group. Osteoporos. Int. **11:** 83–91.
94. McClung, M.R., P. Geusens, P.D. Miller, *et al.* and Hip Intervention Program Study Group. 2001. Effect of risedronate on the risk of hip fracture in elderly women. N. Engl. J. Med. **344:** 333–340.
95. Boonen, S., R.F. Laan, I.P. Barton, *et al.* 2005. Effect of osteoporosis treatments on risk of non-vertebral fractures: review and meta-analysis of intention-to-treat studies. Osteoporos. Int. **16:** 1291–1298.
96. Chavassieux, P.M., M.E. Arlot, C. Reda, *et al.* 1997. Histomorphometric assessment of the long term effects of alendronate on bone quality and remodeling in patients with osteoporosis. J. Clin. Invest. **100:** 1475–1480.
97. Adachi, J.D., W.G. Bensen, J. Brown, *et al.* 1997. Intermittent etidronate therapy to prevent corticosteroid-induced osteoporosis. N. Engl. J. Med. **337:** 381–387.

98. SAAG, K.G., K. EMKEY, T.J. SCHNITZER, *et al.* 1998. Alendronate for the prevention and treatment of glucocorticoid-induced osteoporosis. N. Engl. J. Med. **339:** 292–229.
99. GLORIEUX, F.H., N.J. BISHOP, H. PLOTKIN, *et al.* 1998. Cyclic administration of pamidronate in children with severe osteogenesis imperfecta. N. Engl. J. Med. **33:** 947–952.
100. RAUCH, F. & F.H. GLORIEUX. 2005. Osteogenesis imperfecta, current and future medical treatment. Am. J. Med. Genet. C. Semin. Med. Genet. **139:** 31–37.
101. CHESNUT, C.H., A. SKAG, C. CHRISTIANSEN, *et al.* 2004. Effects of oral ibandronate administered daily or intermittently on fracture risk in postmenopausal osteoporosis. J. Bone Miner. Res. **19:** 1241–1249.
102. REID, I.R., J.P. BROWN, P. BURCKHARDT, *et al.* 2002. Intravenous zoledronic acid in postmenopausal women with low bone mineral density. N. Engl. J. Med. **346:** 653–661.
103. ADAMI, S. & N. ZAMBERLAN. 1996. Adverse effects of bisphosphonates. Drug Saf. **14:** 158–170.
104. VAN STAA, T.P. 1998. Post marketing survey with ICT-etidronate therapy. Rev. Contemp. Pharm. **9:** 277–286.
105. DEGROEN, P.C., D.F. LUBBE, L.J. HIRSCH, *et al.* 1996. Esophagitis associated with the use of alendronate. N. Engl. J. Med. **355:** 1016–1021.
106. HUGHES, D.E., B.R. MACDONALD, R.G.G. RUSSELL, *et al.* 1989. Inhibition of osteoclast-like cell formation by bisphosphonates in long-term cultures of human bone marrow. J. Clin. Invest. **83:** 1930–1935.
107. BOONEKAMP, P.M., L.J.A. VAN DER WEE-PALS, M.L.L. VAN WIJK-VAN LENNEP, *et al.* 1986. Two modes of action of bisphosphonates on osteoclastic resorption of mineralized matrix. Bone Miner. **1:** 27–39.
108. LÖWIK, C.W.G.M., G. VAN DER PLUIJM, L.J.A. VAN DER WEE-PALS, *et al.* 1988. Migration and phenotypic transformation of osteoclast precursors into mature osteoclasts: the effect of a bisphosphonate. J. Bone Miner. Res. **3:** 185–192.
109. SUDA, T., I. NAKAMURA, E. JIMI, *et al.* 1997. Regulation of osteoclast function. J. Bone Miner. Res. **12:** 869–879.
110. MASARACHIA, P., M. WEINREB, R. BALENA, *et al.* 1996. Comparison of the distribution of 3H-alendronate and 3H-etidronate in rat and mouse bones. Bone **19:** 281–290.
111. AZUMA, Y., H. SATO, Y. OUE, *et al.* 1995. Alendronate distributed on bone surfaces inhibits osteoclastic bone resorption in vitro and in experimental hypercalcemia models. Bone **16:** 235–245.
112. SATO, M. & W. GRASSER. 1990. Effects of bisphosphonates on isolated rat osteoclasts as examined by reflected light microscopy. J. Bone Miner. Res. **5:** 31–40.
113. NISKIKAWA, M., T. AKATSU, Y. KATAYAMA, *et al.* 1996. Bisphosphonates act on osteoblastic cells and inhibit osteoclast formation in mouse marrow cultures. Bone **18:** 9–14.
114. YU, X., J. SCHOLLER & N.T. FOGED. 1996. Interaction between effects of parathyroid hormone and bisphosphonate on regulation of osteoclast activity by the osteoblast-like cell line UMR-106. Bone **19:** 339–345.
115. SAHNI, M., H. GUENTHER, H. FLEISCH, *et al.* 1993. Bisphosphonates act on rat bone resorption through the mediation of osteoblasts. J. Clin. Invest. **91:** 2004–2011.

116. VITTE, C., H. FLEISCH & H.L. GUENTHER. 1996. Bisphosphonates induce osteoblasts to secrete an inhibitor of osteoclast-mediated resorption. Endocrinology **137:** 2324–2333.
117. ITO, M., M. CHOKKI, Y. OGINO, *et al.* 1998. Comparison of the cytotoxic effects of bisphosphnates *in vitro* and *in vivo*. Calcif. Tissue Int. **63:** 143–147.
118. BREUIL, V., F. COSMAN, L. STEIN, *et al.* 1998. Human osteoclast formation and activity *in vitro*: effects of alendronate. J. Bone Miner. Res. **13:** 1721–1729.
119. HUGHES, D.E., K.R. WRIGHT, H.L. UY, *et al.* 1995. Bisphosphonates promote apoptosis in murine osteoclasts *in vitro* and *in vivo*. J. Bone Miner. Res. **10:** 1478–1487.
120. SELANDER, K.S., J. MÖNKKÖNEN, E.K. KARHUKORPI, *et al.* 1996. Characteristics of clodronate-induced apoptosis in osteoclasts and macrophages. Mol. Pharmacol. **50:** 1127–1138.
121. ROGERS, M.J., K.M. CHILTON, F. COXON, *et al.* 1996. Bisphosphonates induce apoptosis in mouse macrophage-like cells by a nitric-oxide-independent mechanism. J. Bone Miner. Res. **11:** 1482–1491.
122. COXON, F.P., H.L. BENFORD, R.G.G. RUSSELL, *et al.* 1998. Protein synthesis is required for caspase activation and induction of apoptosis by bisphosphonate drugs. Mol. Pharmacol. **54:** 631–638.
123. SHIPMAN, C.M., M.J. ROGERS, J.F. APPERLEY, *et al.* 1997. Bisphosphonates induce apoptosis in human myeloma cells; a novel anti-tumour activity. Br. J. Haematol. **98:** 665–672.
124. SHIPMAN, C.M., P.I. CROUCHER, R.G.G. RUSSELL, *et al.* 1998. The bisphosphonate incadronate (YM175) causes apoptosis of human myeloma cells in vitro by inhibiting the mevalonate pathway. Cancer Res. **58:** 5294–5297.
125. DAVID, P., H. NGUYEN, A. BARBIER, *et al.* 1996. The bisphosphonate tiludronate is a potent inhibitor of the osteoclast vacuolar H^+ ATPase. J. Bone Miner. Res. **11:** 1498–1507.
126. CARANO, A., S.A. TEITELBAUM, J.D. KONSEK, *et al.* 1990. Bisphosphonates directly inhibit the bone resorption activity of isolated avian osteoclasts in vitro. J. Clin. Invest. **85:** 456–461.
127. ZIMOLO, Z., G. WESOLOWSKI & G.A. RODAN. 1995. Acid extrusion is induced by osteoclast attachment to bone. J. Clin. Invest. **96:** 2277–2283.
128. FELIX, R., R.G.G. RUSSELL & H. FLEISCH. 1976. The effect of several diphosphonates on acid phosphohydrolases and other lysosomal enzymes. Biochim. Biophys. Acta **429:** 429–438.
129. LERNER, U.H. & A. LARSSON. 1987. Effects of four bisphosphonates on bone resorption, lysosomal enzyme release, protein synthesis and mitotic activities in mouse calvarial bones *in vitro*. Bone **8:** 179–189.
130. SATO, M., W. GRASSER, N. ENDO, *et al.* 1991. Bisphosphonate action. Alendronate localisation in rat bone and effects on osteoclast ultrastructure. J. Clin. Invest. **88:** 2095–2105.
131. MURAKAMI, H., N. TAKAHASHI, T. SASAKI, *et al.* 1995. A possible mechanism of the specific action of bisphosphonates on osteoclasts: tiludronate preferentially affects polarized osteoclasts having ruffled borders. Bone **17:** 137–144.
132. SCHMIDT, A., S.J. RUTLEDGE, N. ENDO, *et al.* 1995. Alendronate inhibition of protein tyrosine phosphatase activity. Proc. Natl. Acad. Sci. USA **93:** 3068–3073.
133. ENDO, N., S.J. RUTLEDGE, E.E. OPAS, *et al.* 1996. Human protein tyrosine phosphatases: alternative splicing and inhibition by bisphosphonates. J. Bone Miner. Res. **11:** 535–543.

134. OLPAS, E.E., S.J. RUTLEDGE, E. GOLUB, *et al.* 1997. Alendronate inhibition of protein-tyrosine-phosphatase-Meg1. Biochem. Pharmacol. **54:** 721–727.
135. MURAKAMI, H., N. TAKAHASHI, S. TANAKA, *et al.* 1997. Tiludronate inhibits protein tyrosine phosphatase activity in osteoclasts. Bone **20:** 399–404.
136. ROGERS, M.J. 2004. From molds and macrophages to mevalonate: a decade of progress in understanding the molecular mode of action of bisphosphonates [review]. Calcif. Tissue Int. **75:** 451–461.
137. ROGERS, M.J., D.J. WATTS, R.G.G. RUSSELL, *et al.* 1994. Inhibitory effects of bisphosphonates on growth of amoebae of the cellular slime mould *Dictyostelium discoideum*. J. Bone Miner. Res. **9:** 1029–1039.
138. ROGERS, M.J., X. XIONG, R.J. BROWN, *et al.* 1995. Structure-activity relationships of new heterocycle-containing bisphosphonates as inhibitors of bone resorption and as inhibitors of growth of *Dictyostelium discoideum* amoebae. Mol. Pharmacol. **47:** 398–402.
139. ROGERS, M.J., X. JI, R.G.G. RUSSELL, *et al.* 1994. Incorporation of bisphosphonates into adenine nucleotides by amoebae of the cellular slime mould *Dictyostelium discoideum*. Biochem. J. **303:** 303–311.
140. ROGERS, M., R.G.G. RUSSELL, G.M. BLACKBURN, *et al.* 1992. Metabolism of halogenated bisphosphonates by the cellular slime mould *Dictyostelium discoideum*. Biochem. Biophys. Res. Commun. **189:** 414–423.
141. ROGERS, M.J., R.J. BROWN, V. HODKIN, *et al.* 1996. Bisphosphonates are incorporated into adenine nucleotides by human aminoacyl-tRNA synthetase enzymes. Biochem. Biophys. Res. Commun. **224:** 863–869.
142. PELORGEAS, S., J.P. MARTIN & M. SATRE. 1992. Cytotoxicity of dichloromethane diphosphonate and of 1-hydroxyethane-1,1-diphosphonate in the amoebae of the slime mould *Dictyostelium discoideum*. Biochem. Pharmacol. **44:** 2157–2163.
143. FRITH, J.C., J. MÖNKKÖNEN, G.M. BLACKBURN, *et al.* 1996. Clodronate and liposome-encapsulated clodronate are metabolized to a toxic ATP analog, adenosine 5′-(β,γ-dichloromethylene) triphosphate, by mammalian cells *in vitro*. J. Bone Miner. Res. **12:** 1358–1367.
144. AURIOLA, S., J. FRITH, M.J. ROGERS, *et al.* 1997. Identification of adenine nucleotide-containing metabolites of bisphosphonate drugs using ion-pair liquid chromatography-electrospray mass spectrometry. J. Chrom. B. **704:** 187–195.
145. FRITH, J.C., J. MONKKONEN, S. AURIOLA, *et al.* 2001. The molecular mechanism of action of the antiresorptive and antiinflammatory drug clodronate: evidence for the formation in vivo of a metabolite that inhibits bone resorption and causes osteoclast and macrophage apoptosis. Arthritis Rheum. **44:** 2201–2210.
146. ROGERS, M.J., X. XIONG, X. JI, *et al.* 1997. Inhibition of growth of *Dictyostelium discoideum* amoebae by bisphosphonates is dependent on cellular uptake. Pharm. Res. **14:** 625–630.
147. LEHENKARI, P.P., M. KELLINSALMI, J.P. NAPANKANGAS, *et al.* 2002. Further insight into mechanism of action of clodronate: inhibition of mitochondrial ADP/ATP translocase by a nonhydrolyzable, adenine-containing metabolite. Mol. Pharmacol. **61:** 1255–1262.
148. ZHANG, F.L. & P.J. CASEY. 1996. Protein prenylation: molecular mechanisms and functional consequences. Annu. Rev. Biochem. **65:** 241–269.
149. RIDLEY, A.J. & A. HALL. 1992. The small GTP-binding protein Rho regulates the assembly of focal adhesions and actin stress fibers in response to growth factors. Cell **70:** 389–399.

150. RIDLEY, A.J., H.F. PATERSON, C.L. JOHNSTON, *et al.* 1992. The small GTP-binding protein, rac, regulates growth factor-induced membrane ruffling. Cell **70:** 401–410.
151. ZERIAL, M. & H. STENMARK. 1993. Rab GTPases in vesicular transport. Curr. Opin. Cell Biol. **5:** 613–620.
152. ZHANG, D., N. UDAGAWA, I. NAKAMURA, *et al.* 1995. The small GTP-binding protein, Rho p21, is involved in bone resorption by regulating cytoskeletal organization in osteoclasts. J. Cell Sci. **108:** 2285–2292.
153. MARSHALL, C. 1993. Protein prenylation: a mediator of protein-protein interactions. Science **259:** 1865–1866.
154. LUCKMAN, S.P., D.E. HUGHES, F.P. COXON, *et al.* 1998. Nitrogen-containing bisphosphonates inhibit the mevalonate pathway and prevent post-translational prenylation of GTP-binding proteins, including Ras. J. Bone Miner. Res. **13:** 581–589.
155. LUCKMAN, S.P., F.P. COXON, F.H. EBETINO, *et al.* 1998. Heterocycle-containing bisphosphonates cause apoptosis and inhibit bone resorption by preventing protein prenylation: evidence from structure-activity relationships in J774 macrophages. J. Bone Miner. Res. **13:** 1668–1678.
156. STAAL, A., J.C. FRITH, M.H. FRENCH, *et al.* 2003. The ability of statins to inhibit bone resorption is directly related to their inhibitory effect on HMG-CoA reductase activity. J. Bone Miner. Res. **18:** 88–96.
157. FISHER, J.E., M.J. ROGERS, J.M. HALASY, *et al.* 1999. Mechanism of action of alendronate: geranylgeraniol, an intermediate of the mevalonate pathway, prevents inhibition of osteoclast formation, bone resorption and kinase activation in vitro. Proc. Natl. Acad. Sci. USA **96:** 133–138.
158. HALL, A. 1998. Rho GTPases and the actin cytoskeleton. Science **279:** 509–514.
159. AMIN, D., S.A. CORNELL, M.H. PERRONE, *et al.* 1996. 1-hydroxy-3-(methylpentylamino)-propylidene-1,1-bisphosphonic acid as a potent inhibitor of squalene synthase. Drug Res. **46:** 759–762.
160. AMIN, D., S.A. CORNELL, S.K. GUSTAFSON, *et al.* 1992. Bisphosphonates used for the treatment of bone disorders inhibit squalene synthase and cholesterol biosynthesis. J. Lipid. Res. **33:** 1657–1663.
161. VAN BEEK, E., E. PIETERMAN, L. COHEN, *et al.* Nitrogen-containing bisphosphonates inhibit isopentenyl pyrophosphate isomerase/farnesyl pyrophosphate synthase activity with relative potencies corresponding to their antiresorptive potencies *in vitro* and *in vivo*. Biochem. Biophys. Res. Commun. **255:** 491–494.
162. DUNFORD, J.E., K. THOMPSON, F.P. COXON, *et al.* 2001. Structure-activity relationships for inhibition of farnesyl diphosphate synthase in vitro and inhibition of bone resorption in vivo by nitrogen-containing bisphosphonates. J. Pharmacol. Exp. Ther. **296:** 235–242.
163. EBETINO, F.H., C.N. ROZÉ, C.E. MCKENNA, *et al.* 2005. Molecular interactions of nitrogen-containing bisphosphonates within farnesyl diphosphate synthase. J. Organomet. Chem. **690:** 2679–2687.
164. KAVANAGH, K., K. GUO, J. DUNFORD, *et al.* 2006. Human Farnesyl Diphosphate Synthase (FDPS): crystal structure and molecular interactions with nitrogen-containing bisphosphonates. Proc. Natl. Acad. Sci. USA In press.
165. MARTIN, M.B., W. ARNOLD, H.T. HEATH, III, *et al.* 1999. Nitrogen-containing bisphosphonates as carbocation transition state analogs for isoprenoid biosynthesis. Biochem. Biophys. Res. Commun. **263:** 754–758.

166. TERKELTAUB, R.A. 2001. Inorganic pyrophosphate generation and disposition in pathophysiology [review]. Am. J. Physiol. Cell Physiol. **281:** C1–C11.
167. ANDERSON, H.C., D. HARMEY, N.P. CAMACHO, *et al.* 2005. Sustained osteomalacia of long bones despite major improvement in other hypophosphatasia-related mineral deficits in tissue nonspecific alkaline phosphatase/nucleotide pyrophosphatase phosphodiesterase 1 double-deficient mice. Am. J. Pathol. **166:** 1711–1720.
168. TIMMS, A., Y. ZHANG, R.G.G. RUSSELL, *et al.* 2002. Genetic studies of disorders of calcium crystal deposition. Rheumatology (Oxford) **41:** 725–729.
169. MUMM, S., J. JONES, P. FINNEGAN, *et al.* 2001. Hypophosphatasia: molecular diagnosis of Rathbun's original case. J. Bone Miner. Res. **16:** 1724–1727.
170. FEDDE, K.N., L. BLAIR, J. SILVERSTEIN, *et al.* 1999. Alkaline phosphatase knock-out mice recapitulate the metabolic and skeletal defects of infantile hypophosphatasia. J. Bone Miner. Res. **14:** 2015–2026.
171. RUF, N., B. UHLENBERG, R. TERKELTAUB, *et al.* 2005. The mutational spectrum of ENPP1 as arising after the analysis of 23 unrelated patients with generalized arterial calcification of infancy (GACI). Hum. Mutat. **26:** 495–496.
172. NAKAMURA, I., S. IKEGAWA, A. OKAWA, *et al.* 1999. Association of the human NPPS gene with ossification of the posterior longitudinal ligament of the spine (OPLL). Hum. Genet. **104:** 492–497.
173. HO, A.M., M.D. JOHNSON & D.M. KINGSLEY. 2000. Role of the mouse ank gene in control of tissue calcification and arthritis. Science **289:** 265–270.
174. WILLIAMS, C.J., Y. ZHANG, A. TIMMS, *et al.* 2002. Autosomal dominant familial calcium pyrophosphate dihydrate deposition disease is caused by mutation in the transmembrane protein ANKH. Am. J. Hum. Genet. **71:** 985–991.
175. ZHANG, Y., K. JOHNSON, R.G. RUSSELL, *et al.* 2005. Association of sporadic chondrocalcinosis with a -4-basepair G-to-A transition in the 5'-untranslated region of ANKH that promotes enhanced expression of ANKH protein and excess generation of extracellular inorganic pyrophosphate. Arthritis Rheum. **52:** 1110–1117.
176. PENDLETON, A., M.D. JOHNSON, A. HUGHES, *et al.* 2002. Mutations in ANKH cause chondrocalcinosis. Am. J. Hum. Genet. **71:** 933–940.
177. REICHENBERGER, E., V. TIZIANI, S. WATANABE, *et al.* 2001. Autosomal dominant craniometaphyseal dysplasia is caused mutations in the transmembrane protein ANK. Am. J. Hum. Genet. **68:** 1321–1326.
178. NUERNBERG, P., H. THIELE, D. CHANDLER, *et al.* 2001. Heterozygous mutations in ANKH, the human ortholog of the mouse progressive ankylosis gene, result in craniometaphyseal dysplasia. Nat. Genet. **28:** 37–41.
179. BORAH, B., E.L. RITMAN, T.E. DUFRESNE, *et al.* 2005. The effect of risedronate on bone mineralization as measured by micro-computed tomography with synchrotron radiation: correlation to histomorphometric indices of turnover. Bone **37:** 1–9.
180. BORAH, B., T.E. DUFRESNE, P.A. CHMIELEWSKI, *et al.* 2002. Risedronate preserves trabecular architecture and increases bone strength in vertebra of ovariectomized minipigs as measured by three-dimensional microcomputed tomography. J. Bone Miner. Res. **17:** 1139–1147.
181. BORAH, B., T.E. DUFRESNE, P.A. CHMIELEWSKI, *et al.* 2004. Risedronate preserves bone architecture in postmenopausal women with osteoporosis as measured by three-dimensional microcomputed tomography. Bone **34:** 736–746.

182. LALLA, S., L.A. HOTHORN, N. HAAG, *et al.* 1998. Lifelong administration of high doses of ibandronate increases bone mass and maintains bone quality of lumbar vertebrae in rats. Osteoporos. Int. **8:** 97–103.
183. BAUSS, F., S. LALLA, R. ENDELE, *et al.* 2002. Effects of treatment with ibandronate on bone mass, architecture, biomechanical properties, and bone concentration of ibandronate in ovariectomized aged rats. J. Rheumatol. **29:** 2200–2208.
184. MASHIBA, T., S. MORI, D.B. BURR, *et al.* 2005. The effects of suppressed bone remodeling by bisphosphonates on microdamage accumulation and degree of mineralization in the cortical bone of dog rib. J. Bone Miner. Metab. **23**(Suppl): 36–42.
185. MASHIBA, T., T. HIRANO, C.H. TURNER, *et al.* 2000. Suppressed bone turnover by bisphosphonates increases microdamage accumulation and reduces some biomechanical properties in dog rib. J. Bone Miner. Res. **15:** 613–620.
186. FLORA, L., G.S. HASSING, G.G. CLOYD, *et al.* 1981. The long-term skeletal effects of EHDP in dogs. Metab. Bone Dis. Relat. Res. **3:** 289–300.
187. ODVINA, C.V., J.E. ZERWEKH, D.S. RAO, *et al.* 2005. Severely suppressed bone turnover: a potential complication of alendronate therapy. J. Clin. Endocrinol. Metab. **90:** 1294–1301.
188. OTT, S.M. 2005. Long-term safety of bisphosphonates. J. Clin. Endocrinol. Metab. **90:** 1897–1899.
189. WHYTE, M.P., D. WENKERT, K.L. CLEMENTS, *et al.* 2003. Bisphosphonate-induced osteopetrosis. N. Engl. J. Med. **349:** 457–463.
190. MONIER-FAUGERE, M.C., Z. GENG, E.P. PASCHALIS, *et al.* 1999. Intermittent and continuous administration of the bisphosphonate ibandronate in ovariohysterectomized beagle dogs: effects on bone morphometry and mineral properties. J. Bone Miner. Res. **14:** 1768–1778.
191. BAUSS, F., R.K. SCHENK, S. HORT, *et al.* 2004. New model for simulation of fracture repair in full-grown beagle dogs: model characterization and results from a long-term study with ibandronate. J. Pharmacol. Toxicol. Methods **50:** 25–34.
192. LITTLE, D.G., M. MCDONALD, R. BRANSFORD, *et al.* 2005. Manipulation of the anabolic and catabolic responses with OP-1 and zoledronic acid in a rat critical defect model. J. Bone Miner. Res. **20:** 2044–2052.
193. BOBYN, J.D., S.A. HACKING, J.J. KRYGIER, *et al.* 2005. Zoledronic acid causes enhancement of bone growth into porous implants. J. Bone Joint Surg. Br. **87:** 416–420.
194. KURTH, A.H., C. EBERHARDT, S. MÜLLER, *et al.* 2005. The bisphosphonate ibandronate improves implant integration in osteopenic ovariectomized rats. Bone **37:** 204–210.
195. WILKINSON, J.M., A.C. EAGLETON, I. STOCKLEY, *et al.* 2005. Effect of pamidronate on bone turnover and implant migration after total hip arthroplasty: a randomized trial. J. Orthop Res. **23:** 1–8.
196. LITTLE, D.G., N.C. SMITH, P.R. WILLIAMS, *et al.* 2003. Zoledronic acid prevents osteopenia and increases bone strength in a rabbit model of distraction osteogenesis. J. Bone Miner. Res. **18:** 1300–1307.
197. LITTLE, D.G., R.A. PEAT & A. MCEVOY. 2003. Zoledronic acid treatment results in retention of femoral head structure after traumatic osteonecrosis in young Wistar rats. J. Bone Miner. Res. **18:** 2016–2022.
198. LAI, K.A., W.J. SHEN, C.Y. YANG, *et al.* 2005. The use of alendronate to prevent early collapse of the femoral head in patients with nontraumatic osteonecrosis. A randomized clinical study. J. Bone Joint Surg. Am. **87:** 2155–2159.

199. LITTLE, D.G., M. MCDONALD, I.T. SHARPE, *et al.* 2005. Zoledronic acid improves femoral head sphericity in a rat model of Perthes disease. J. Orthop. Res. **23:** 862–868.
200. NANCOLLAS, G.H., R. TANG, R.J. PHIPPS, *et al.* 2005. Novel insights into actions of bisphosphonates on bone: differences in interactions with hydroxyapatite. Bone [Epub ahead of print].
201. PLOTKIN, L.I., R.S. WEINSTEIN, A.M. PARFITT, *et al.* 1999. Prevention of osteocyte and osteoblast apoptosis by bisphosphonates and calcitonin. J. Clin. Invest. **104:** 1363–1374.
202. PLOTKIN, L.I., J.I. AGUIRRE, S. KOUSTENI, *et al.* 2005. Bisphosphonates and estrogens inhibit osteocyte apoptosis via distinct molecular mechanisms downstream of extracellular signal-regulated kinase activation. J. Biol. Chem. **280:** 7317–7325.
203. SAUTY, A., M. PECHERSTORFER, I. ZIMMERROTH, *et al.* 1996. Interleukin-6 and tumor necrosis factor alpha levels after bisphosphonate treatment *in vitro* and in patients with malignancy. Bone **18:** 133–139.
204. SCHWEITZER, D.H., M. OOSTENDORP-VAN DE RUIT, G. VAN DER PLUIJM, *et al.* 1995. Interleukin-6 and the acute phase response during treatment of patients with Paget's disease with the nitrogen-containing bisphosphonate dimethylamino-hydroxypropylidene bisphosphonate. J. Bone Miner. Res. **10:** 956–962.
205. ADAMI, S., A.K. BHALLA, R. DORIZZI, *et al.* 1987. The acute-phase response after bisphosphonate administration. Calcif. Tissue Int. **41:** 326–331.
206. SANDERS, J.M., S. GHOSH, J.M. CHAN, *et al.* 2004. Quantitative structure-activity relationships for gammadelta T cell activation by bisphosphonates. J. Med. Chem. **47:** 375–384.
207. THOMPSON, K. & M.J. ROGERS. 2004. Statins prevent bisphosphonate-induced gamma,delta-T-cell proliferation and activation *in vitro*. J. Bone Miner. Res. **19:** 278–288.
208. GABELLI, S.B., J.S. MCLELLAN, A. MONTALVETTI, *et al.* 2005. Structure and mechanism of the farnesyl diphosphate synthase from *Trypanosoma cruzi*: implications for drug design. Proteins **62:** 80–88 [Epub ahead of print].
209. SZABO, C.M., Y. MATSUMURA, S. FUKURA, *et al.* 2002. Inhibition of geranylgeranyl diphosphate synthase by bisphosphonates and diphosphates: a potential route to new bone antiresorption and antiparasitic agents. J. Med. Chem. **45:** 2185–2196.
210. GHOSH, S., J.M. CHAN, C.R. LEA, *et al.* 2004. Effects of bisphosphonates on the growth of Entamoeba histolytica and Plasmodium species in vitro and in vivo. J. Med. Chem. **47:** 175–187.
211. MARTIN, M.B., J.S. GRIMLEY, J.C. LEWIS, *et al.* 2001. Bisphosphonates inhibit the growth of *Trypanosoma brucei*, *Trypanosoma cruzi*, *Leishmania donovani*, *Toxoplasma gondii*, and *Plasmodium falciparum*: a potential route to chemotherapy. J. Med. Chem. **44:** 909–916.
212. LING, Y., G. SAHOTA, S. ODEH, *et al.* 2005. Bisphosphonate inhibitors of Toxoplasma gondi growth: in vitro, QSAR, and in vivo investigations. J. Med. Chem. **48:** 3130–3140.
213. MORENO, B., B.N. BAILEY, S. LUO, *et al.* 2001. (31)P NMR of apicomplexans and the effects of risedronate on *Cryptosporidium parvum* growth. Biochem. Biophys. Res. Commun. **284:** 632–637.
214. RODRIGUEZ, N., B.N. BAILEY, M.B. MARTIN, *et al.* 2002. Radical cure of experimental cutaneous leishmaniasis by the bisphosphonate pamidronate. J. Infect. Dis. **186:** 138–140.

Novel Targets and Therapeutics for Bone Loss

RONALD N. MARGOLIS[a] AND SUNIL J. WIMALAWANSA[b]

[a]*National Institute of Diabetes and Digestive and Kidney Diseases, NIH, Democracy 2, Room 693, Bethesda, Maryland 20892-5460, USA*

[b]*Division of Endocrinology, Department of Medicine, UMDNJ-Robert Wood Johnson Medical School, New Brunswick, New Jersey 08903-0019, USA*

ABSTRACT: Advances in the treatment of osteoporosis over the past decade have resulted in the generation of novel therapeutic agents aimed at providing both anticatabolic and anabolic effects in bone. In-depth understanding of the biology of key factors regulating bone metabolism has begun to reveal new approaches to treating this costly and debilitating disease. During the next decade we will observe the development and evolution of several new classes of therapeutic targets and agents to combat this disease.

KEYWORDS: osteoporosis; novel therapies; RANKL; OPG; Wnt; leptin; prostaglandins; SERMs

INTRODUCTION

The human skeleton is a finely tuned, carefully constructed, and yet extremely labile framework for the support of the vertebrate organism.[1] It has the ability to change over time, through the early developmental stage to the adult. The skeleton ultimately reaches a state of dynamic stasis, followed by a decline with age. These transitions represent several regulated schema that can be subject to change with activity, diet, hormonal status, and disease. Our ability to understand pathways of endocrine, paracrine, and autocrine regulation involved in day-to-day maintenance, as well as growth and decay of the skeleton opens up many venues for possible interventions to stabilize peak bone mass, or restore the age- or disease-depleted skeleton. At an NIH-sponsored workshop held in 1995,[2] the state of the science in the field suggested several factors as potential anabolic factors or targets. Since that time significant progress has led to a better understanding of the regulation of bone formation and resorption. The work presented at the "Novel Targets and Therapies

Address for correspondence: Ronald N. Margolis, Ph.D., Division of Diabetes, Endocrinology and Metabolic Diseases, National Institute of Diabetes and Digestive and Kidney Diseases, NIH, 6707 Democracy Blvd., Room 693, Bethesda, MD 20892-5460. Voice: 301-594-8819; fax: 301-435-6047.
e-mail: rm76f@nih.gov

Ann. N.Y. Acad. Sci. 1068: 402–409 (2006). © 2006 New York Academy of Sciences.
doi: 10.1196/annals.1346.040

for Bone Loss" session of this NYAS symposium expanded upon this earlier theme and provided new insights. Bone formation and resorption are tightly coupled and most therapies to date have focused solely on the resorption side of the equation. Newer approaches have focused on anabolic factors in bone, some including hormonal therapies such as parathyroid hormone (PTH).

NEW TARGETS

Stems Cells

Mesenchymal stem cells (MSCs), found in bone marrow constitute among other possible lineages the undifferentiated precursors for both adipocytes and osteoblasts. Divergent signaling pathways lead to final commitment of one or the other differentiated cell type. Better understanding of the signaling path at early uncommitted stages can be useful for therapeutic strategies designed to increase flow of osteoblast precursors to the places they are needed: bone-forming units.[3] The Wingless type (Wnt) signaling pathway has been implicated in both the early commitment stage[4] and the regulation of the late, mature, bone cell (see below). In the MSC, in marrow, Wnt signaling is crucial to osteoblastogenesis at least partly through suppression of PPARγ expression and subsequent adipogenesis.[5] Recent evidence suggests that osteoblast-lineage cells circulate in humans, thus providing the basis for therapeutic strategies aimed at identifying, amplifying, and targeting these cells to the appropriate place of need.[4,5] Therefore, interventions aimed at the Wnt signaling pathway in the undifferentiated MSC population both in marrow and the circulation could represent a novel target for therapeutics designed to stimulate the production and targeting of new osteoblasts.

Osteoprotegerin/RANKL

While studies have been carried out to investigate the combined use of antiresorptive agents with an anabolic agent (e.g., bisphosphonates/PTH 1-34), some data suggest that such combination therapy may not be the most efficacious way to give long-lasting increase in bone strength and quality.[1] Functional coupling between bone resorption and formation is now well recognized, with the bone resorption pits created by the osteoclast ultimately leading to recruitment and activation of osteoblasts and subsequent bone formation. Indeed, high turnover of bone in postmenopausal osteoporosis is a hallmark of the net loss of bone that occurs as bone formation is unable to keep up with high levels of bone resorption. The result is increased incidence of fracture.

Since the osteoblast produces many of the factors (e.g., RANKL and osteoprotegerin [OPG]) that are responsible for the regulation of osteoclastogenesis and osteoclast activation, intervention at the level of blocking or modulating production and/or activity of these paracrine factors could produce a net anabolic stimulus.[6] Thus, activation of the bone anabolic response via intermittent PTH administration together with anticatabolic modulation of osteoclast activation may help to maximize the anabolic effects of PTH. Since resorption and formation are tightly coupled, a complete block of osteoclast formation/activation would not be desired.

Bone turnover is a cooperative process with many players involved beyond the interactions of RANKL and OPG. For example, recently discovered factors such as the secreted frizzled-related protein-1 (sFRP-1) produced by osteoblasts have been shown to block osteoclast activity by binding to RANKL, much as does OPG.[7] Again, the desired end point may be *modulation* of RANKL levels, rather than complete abrogation of its effects on the osteoclast. The goal would be to maintain a proper balance of osteoblasts/osteoclasts, with a slight tilt toward formation over resorption. Interestingly, sFRP-1 has been shown to be a decoy receptor for the Wnt signaling pathway, through the Ftz receptor, though sFRP-1 also binds to RANKL. As a result the ability to sequester RANKL may also be important. In homeostasis, high levels of expression of sFRP-1 by the osteoblast may represent a compensatory mechanism designed to limit osteoclastogenesis.

Wnt Pathways

The role of the Wnt pathway in bone formation has been amply demonstrated with the discovery of human mutations in the Wnt co-receptor LRP5, in which gain of function mutations equates to higher bone densities and loss of function leads to severe osteoporosis.[8] Thus, in addition to the role of Wnts at the level of the MSC, the Wnt/LRP5 pathway represents an important potential target for therapeutic intervention directed at bone formation, per se.[9] That Wnts are implicated in osteoblastogenesis comes from evidence that Wnt10b is essential for MSCs to commit to the bone phenotype, at the expense of adipogenesis.[5] Aging-related declines in osteoblastogenesis may be related to differential switching between PPARγ-mediated adipogenesis and Wnt10b-mediated osteoblastogenesis. Initially, the Wnt/LRP5 pathway leading to bone formation was thought to involve actions specific to the osteoblast.[4] Indeed, during osteoblastogenesis transcription factors required for osteoblast development are turned on, but in the mature osteoblast recent findings suggest that an additional mechanism is present. One of the downstream targets of Wnt signaling in mature, fully differentiated osteoblasts is OPG, and thus blockade of osteoclastogenesis may be the focal point of Wnt signaling in the mature osteoblast. The net result would be increased bone mass.[9] Clearly, various components of the Wnt pathway represent important potential targets for development of anabolic agents that could increase bone mass in the adult distinct from effects directed at MSC commitment early in bone development.

Nuclear Hormone Superfamily

Considerable progress in understanding potential pharmacological interventions in women with postmenopausal osteoporosis suggests major roles for the nuclear hormone receptor superfamily in the regulation of bone turnover.[10] Studies that have focused on the roles of estrogen, androgens, vitamin D, and other nuclear receptors (NRs), have begun to identify possible new therapeutic targets to address low bone mass. Emerging evidence of roles in metabolism for several of the orphan NRs (ONRs) in the NR superfamily have led to some intriguing suggestions that the ONRs, such as estrogen-related receptor (ERR)-α, can play important roles in osteoblast biology.[11] As with many ONRs, the key role for ERR-α may be during early bone development, though in mature obsteoblasts, and blocking ERRalpha expression causes defects in osteoblastogenesis, making it also a potential new target for the development of new anabolic therapeutics for bone.

GPCRs

That PTH and parathyroid hormone-related protein (PTHrP) function through 7tm G-protein-coupled receptors (GPCRs) is well known.[12] While it is also well known that intermittent PTH has an anabolic effect on bone, with use in the clinic now established for severe osteopenia associated with postmenopausal osteoporosis, it is less clear how signaling through GPCRs couple PTH to the end effect. This signaling is rapidly attenuated by the action of a family of protein kinases, the G-protein-coupled receptor kinase (GRK).[13] Recent evidence suggests that inhibitors of a specific member of the GRK family in osteoblasts lead to age-dependent effects on bone mass. In murine models, GRK actions at earlier time points were implicated in increased bone mass, while in the adult little effect was seen, unless PTH was also present. PTH appears to unmask an anabolic effect mediated through GRK action that leads to increased bone mass, particularly in trabecular bone. Therefore, it is possible to consider one form of combination therapy that would involve intermittent PTH combined with a specific inhibitor of GRK that would allow the anabolic effects of PTH in bone to be fully expressed. PTHrP has also been found to have anabolic effects on bone through actions directed at bone formation, and while the mode of action has not been fully understood (likely involving GPCRs), this too represents a potentially valuable anabolic target.[14]

Prostaglandins

The role of prostaglandins (PGs), particularly PGE2, in bone formation has been well documented[15]; however, there are many other PGs that are produced and have local effects on tissues and cells. Recently, PGD2 has been shown to bind two receptors in bone and potentially contribute to anabolic effects in bone. Both receptors are members of the large 7tm GPCR superfamily, one of which (DP) is coupled to cAMP production, while the second (CRTH2) is coupled to intracellular Ca fluxes. It is important to note that activation of the latter suppresses RANKL production, and activation of the former suppresses OPG production. Both are expressed in human osteoblasts. Stimulation of PGD2 production by numerous hormones and cytokines, including PTH, contributes to an overall anabolic effect. In addition, stimulation of CRTH2 produces a strong chemotactic signal for osteoblasts, resulting in increased recruitment of osteoblasts to resorption pits. Thus, drugs targeted at the modulation of local PGD2 production could have significant effects on osteoblasts and hence on bone formation.

NEW THERAPEUTICS

Calcimimetics

From studies and clinical trials with calcimimetics, such as the NPS compound NPS R-568, it has been possible to learn how the calcium-activated receptor (CaR) functions in physiology and pathophysiology.[16] By modulating the activity of the CaR in parathyroid cells it has been possible to control parathyroid hyperplasia and consequent bone lysis associated with secondary hyperparathyroidism (HPT) stemming from end-stage renal disease. Therefore, compounds that can modulate endogenous PTH secretion (or actions) without the necessity of injecting PTH could prove efficacious in treating osteoporosis. Daily administered, but orally active compounds that could modulate the activity of the CaR on the parathyroid cell could conceivably give the same effect as intermittent PTH injection.

Leptin

The novel and controversial finding that leptin, a hormone originally discovered in adipose cells that regulates satiety, lipid metabolism, and reproduction, also regulates bone resorption through a central nervous system mechanism[17] has opened the path to consideration for how leptin, or compounds like it could be effective in the clinic. Signaling through the sympathetic nervous system appears to be essential and might be exploited to positively impact bone formation.

Selective Estrogen Receptor Modulators

The deleterious effects of estrogen loss on the skeleton in postmenopausal women have driven research designed to develop replacement therapies.

Until recently this meant hormone replacement therapy (HRT) for postmenopausal women, but the development of selective estrogen receptor modulators (SERMs) as alternatives with fewer side effects has resulted in the development of first- (tamoxifen) and second- (raloxifene) generation SERMs.[18] One of the unexpected findings from early trials of SERMs, such as tamoxifen, for the treatment of estrogen receptor-positive breast cancer, was the observation that women in the trial also had increased bone mass.[19] Further development of SERMs, extended also to include the novel concept of a selective estrogen receptor downregulator or SERD activity,[19] has provided novel and innovative approaches to the development of new therapeutics for low bone mass.

Bone loss resulting from deficiency in ovarian estrogen due to premature ovarian failure, oophorectomy, or postmenopausal status was attributed to the loss of the effects of estrogen on bone. Receptors for the estrogen receptor (ER) α form of the receptor were found in osteoblasts and osteoclasts with effects that appeared to be at the level of inhibition of NF-κB signaling cascades in osteoclasts, making estrogen antiresorptive, with SERMs in effect being anticatabolic agents in bone. These findings were reinforced through the NIH-funded WHI study. Although the estrogen-progestin arm of the trial was shut down early due to deleterious effects of estrogen on breast cancer and stroke risk, bone density was increased.[20] This has resulted in concerns over long-term HRT use. New generations of SERMs may be able to express tissue-specific effects, including in bone.[21] Nuclear hormone receptors, like the ER, respond to ligand by change in conformation leading to recruitment of co-activators, thus the focus of therapeutic development may need to shift to compounds that alter receptor conformation in tissue-specific contexts to permit recruitment of co-regulators to enhance the anticatabolic effects on bone without the harmful effects often seen elsewhere.

Novel Approaches

Gene therapy has long been touted as an important and novel approach to the restoration of lost function, correction of the effects of naturally occurring mutations, and introduction of specific cell-based therapies. Despite considerable animal experimentation and human trials, the promise has yet to be fulfilled. Nonetheless, intriguing possibilities remain such as in studies on allografts to repair bone damage in experimental animal models.[22] These allografts often fail to thrive and become completely integrated to the area requiring repair. When allograft materials were coated with adeno-associated virus (AAV) vectors that drive the expression of vascular endothelial growth factor (VEGF) and RANKL the graft survived and began to undergo the remodeling necessary for successful integration and repair. By using factors that stimulate vascularization and bone remodeling new bone can be formed to replace the inert implanted material. The use of factors such as RANKL that initiate bone resorption as part of an anabolic cascade is a key to initiating a remodeling sequence with increased bone formation. Thus, in this scenario, RANKL appears to function as an anticatabolic factor.[23]

Finally, the intriguing possibility that compounds that modulate tissue nitric oxide (NO) levels can also have positive effects on bone mass in postmenopausal women opens the door to consideration of additional, nontraditional approaches as potential alternatives to HRT.[24]

FUTURE DIRECTIONS

Basic research has long served as an important platform for the development of the means to achieve restoration of bone mass. In particular, the development of targeted therapeutic approaches to achieve anabolic actions on bone represents an important advance that has been translated to the clinic. Additional research to fully elucidate the signaling pathways regulating osteoblast and osteoclast generation, function, and interaction will help to identify additional potential targets for the development of new and improved therapeutic agents. The desired result is to reverse the devastating effects on the skeleton that occur during aging- and disease-dependent bone loss.

REFERENCES

1. Martin, T.J. & N.A. Sims. 2005. Osteoclast-derived activity in the coupling of bone formation to resorption. Trends Mol. Med. **11:** 76–81.
2. Margolis, R.N., E. Canalis & N.C. Partridge. 1996. Invited review of a workshop: anabolic hormone in bone: basic research and therapeutic potential. J. Clin. Endocrinol. Metab. **81:** 872–877.
3. Eghbali-Fatourechi, G.Z., J. Lamsam, D. Fraser, *et al.* Circulating osteoblast-lineage cells in humans. N. Engl. J. Med. **352:** 1959–1966.
4. Kennell, J.A. & O.A. MacDougald. 2005. Wnt signaling inhibits adipogenesis through B-catenin-dependent and -independent mechanisms. J. Biol. Chem. **280:** 24004–24010.
5. Bennett, C.N., K.A. Longo & O.A. MacDougald. 2005. Regulation of osteoblastogenesis and bone mass by Wnt10b. Proc. Natl. Acad. Sci. USA **102:** 3324–3329.
6. Glass, D.A., P. Bialek, J.D. Ahn, *et al.* 2005. Canonical Wnt signaling in differentiated osteoblasts controls osteoblast differentiation. Dev. Cell **8:** 751–764.
7. Hausler, K.D., N.J. Horwood, Y. Chuman, *et al.* 2004. Secreted frizzled-related protein-1 inhibits RANKL-dependent osteoclast formation. J. Bone Miner. Res. **19:** 1873–1881.
8. Ferrari, S.L., S. Deitsch & S.E. Antonarakis. 2005. Pathogenic mutations and polymorphisms in the lipoprotein receptor-related protein 5 reveal a new biological pathway for the control of bone mass. Curr. Opin. Lipid. **16:** 207–214.

9. HOLMEN, S.L., C.R. ZYLSTRA, A. MUKHERJEE, *et al.* 2005. Essential role of B-catenin in postnatal bone acquisition. J. Biol. Chem. **280:** 21162–21168.
10. REGINSTER, J.-Y. 2004. Prevention of postmenopausal osteoporosis with pharmacological therapy: practice and possibilities. J. Intern. Med. **255:** 615–628.
11. BONNELYE, E. & J.E. AUBIN. 2005. Review: estrogen receptor-related receptor alpha: a mediator of estrogen response in bone. J. Biol. Chem. **280:** 3104–3111.
12. QIN, L., L.D. QUARLES & R.F. SPURNEY. 2004. Unmasking the osteoinductive effects of a G-protein-coupled receptor (GPCR) kinase (GRK) inhibitor by treatment with PTH (1-34). J. Bone Miner. Res. **19:** 1661–1670.
13. HORWITZ, M.J., M.B. TEDESCO, C. GUNDBERG, *et al.* 2003. Short-term, high-dose parathyroid hormone-related protein as a skeletal anabolic agent for the treatment of postmenopausal osteoporosis. J. Clin. Endocrinol. Metab. **88:** 569–575.
14. GALLANT, M.A., R. SAMADFAM, J.A. HACKETT, *et al.* 2005. Production of prostaglandin D2 by human osteoblasts and modulation of osteoprotegerin, RANKL, and cellular migration by DP and CRTH2 receptors. J. Bone Miner. Res. **20:** 672–681.
15. NAGANO, N. & E.F. NEMETH. 2005. Functional proteins involved in regulation of intracellular $Ca2^+$ for drug development: the extracellular calcium receptor and an innovative medical approach to control secondary hyperparathyroidism by calcimimetics. J. Pharmacol. Sci. **97:** 335–360.
16. ELEFTERIOU, F., J.D. AHN, S. TAKEDA, *et al.* 2005. Leptin regulation of bone resorption by the sympathetic nervous system and CART. Nature **434:** 514–521.
17. MCDONNELL, D.P. 2005. The molecular pharmacology of estrogen receptor modulators: implications for the treatment of breast cancer. Clin. Cancer Res. **11:** 871–877.
18. AGNUSDEI, D. & N. IORI. 2000. Raloxifene: results from the MORE study. J. Musculoskelet. Neuronal Interact. **1:** 127–132.
19. ROSSOUW, J.E., G.L. ANDERSON, R.L. PRENTICE, *et al.* 2002. Risks and benefits of estrogen plus progestin in healthy postmenopausal women: principal results from the Women's health initiative randomized controlled trial. JAMA **288:** 321–333.
20. KOMM, B.S., Y.P. KHARODE, P.V.N. BODINE, *et al.* 2005. Bazedoxifene acetate: a selective estrogen receptor modulator with improved selectivity. Endocrinology **146:** 3999–4008.
21. ITO, H.A., C.P. MILLER & C.R. LYTTLE. 2005. Remodeling of cortical bone allografts mediated by adherent rAAV-RANKL and VEGF gene therapy. Nat. Med. **11:** 291–297.
22. BOYCE, B.F., L. XING, W. SHAKESPEARE, *et al.* 2003. Regulation of bone remodeling and emerging breakthrough drugs for osteoporosis and osteolytic bone metastases. Kidney Int. **85** [Suppl.] S2–S5.
23. WIMALAWANSA, S.J. 2000. Nitroglycerin therapy is as efficacious as standard estrogen replacement therapy (Premarin) in prevention of oopohorectomy-induced bone loss: a human pilot clinical study. J. Bone Miner. Res. **15:** 2240–2244.

Advanced Imaging Assessment of Bone Quality

HARRY K. GENANT AND YEBIN JIANG

Osteoporosis and Arthritis Research Group, University of California, San Francisco, California 94143, USA

ABSTRACT: Noninvasive and/or nondestructive techniques can provide structural information about bone, beyond simple bone densitometry. While the latter provides important information about osteoporotic fracture risk, many studies indicate that bone mineral density (BMD) only partly explains bone strength. Quantitative assessment of macrostructural characteristics, such as geometry, and microstructural features, such as relative trabecular volume, trabecular spacing, and connectivity, may improve our ability to estimate bone strength. Methods for quantitatively assessing macrostructure include (besides conventional radiographs) dual X ray absorptiometry (DXA) and computed tomography (CT), particularly volumetric quantitative computed tomography (vQCT). Methods for assessing microstructure of trabecular bone noninvasively and/or nondestructively include high-resolution computed tomography (hrCT), microcomputed tomography (micro-CT), high-resolution magnetic resonance (hrMR), and micromagnetic resonance (micro-MR). vQCT, hrCT, and hrMR are generally applicable *in vivo*; micro-CT and micro-MR are principally applicable *in vitro*. Despite progress, problems remain. The important balances between spatial resolution and sampling size, or between signal-to-noise and radiation dose or acquisition time, need further consideration, as do the complexity and expense of the methods versus their availability and accessibility. Clinically, the challenges for bone imaging include balancing the advantages of simple bone densitometry versus the more complex architectural features of bone, or the deeper research requirements versus the broader clinical needs. The biological differences between the peripheral appendicular skeleton and the central axial skeleton must be further addressed. Finally, the relative merits of these sophisticated imaging techniques must be weighed with respect to their applications as diagnostic procedures, requiring high accuracy or reliability, versus their monitoring applications, requiring high precision or reproducibility.

KEYWORDS: osteoporosis; micro-imaging; micro-CT; micro-MRI; bone microarchitecture; bone quality

Address for correspondence: Harry K. Genant, University of California, San Francisco, San Francisco, CA 94143.
e-mail: harry.genant@ucsf.edu

Ann. N.Y. Acad. Sci. 1068: 410–428 (2006). © 2006 New York Academy of Sciences.
doi: 10.1196/annals.1346.038

INTRODUCTION

More than standard bone densitometry,[1] noninvasive and/or nondestructive techniques are capable of providing macro- or microstructural information about bone.[2] While the bone densitometry provides important information about osteoporotic fracture risk, numerous studies indicate that bone strength is only partially explained by bone mineral density (BMD). Quantitative assessment of macrostructural characteristics, such as geometry, and microstructural features, such as relative trabecular volume, trabecular spacing, and connectivity, may improve our ability to estimate bone strength.

The methods available for quantitatively assessing macrostructure include computed tomography (CT) and particularly volumetric quantitative computed tomography (vQCT). Noninvasive and/or nondestructive methods for assessing microstructure of trabecular bone include high-resolution computed tomography (hrCT), microcomputed tomography (micro-CT), high-resolution magnetic resonance (hrMR), and micromagnetic resonance (micro-MR). vQCT, hrCT, and hrMR are generally applicable *in vivo*, micro-CT and micro-MR are principally applicable *in vitro*.

VOLUMETRIC CT

The use of standard QCT has centered on a two-dimensional characterization of vertebral trabecular bone, but there is also interest in developing three-dimensional, or vQCT techniques to improve spinal measurements and to extend QCT assessments to the proximal femur (FIG. 1, 2). These three-dimensional techniques encompass the entire object of interest with stacked slices or spiral CT scans, and can use anatomic landmarks to automatically define coordinate systems for reformatting CT data into anatomically relevant projections.

Volumetric CT can determine bone mineral content (BMC) or BMD of the entire bone or subregion, such as a vertebral body or femoral neck, as well as provide separate analysis of the trabecular or cortical components. Since a true and highly accurate volumetric rendering is provided, important geometrical and biomechanically relevant assessments, such as cross-sectional moment of inertia and finite-element analyses can be derived.[2–8] Highly accurate assessment of bone size and density, independent of the artifacts of projectional radiographic and densitometric techniques (such as absorptiometry), can also be derived for epidemiologic studies and for studies of nutritional, racial, and genetic influences on bone size and density.[9]

Because of the complex anatomy of the proximal femur and its dramatic three-dimensional variations in bone density, vQCT has particularly important ramifications for both research and clinical applications at this biologically relevant site. vQCT and finite-element analysis modeling have been used by Lotz

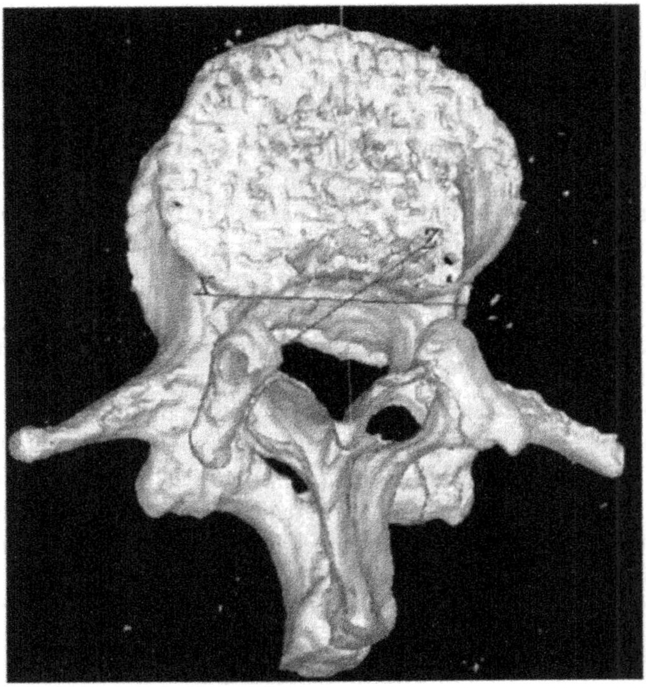

FIGURE 1. Volumetric QCT of the spine may be used to analyze bone mineral density in bone compartments and to accurately measure vertebral geometry. (Courtesy of Thomas Lang.)

et al.[7] and by Keyak *et al.*[4] to improve estimations of proximal femoral strength over global projectional densitometry. *In vitro* studies by Lang *et al.*[5,6] have shown enhanced prediction of *in vitro* fracture load using subregional vQCT of the hip. QCT three-dimensional finite-element model of the proximal femur shows moderate reproducibility in postmenopausal women.[10] QCT-based finite-element models of the hip in 51 women aged 74 ($\pm$5) years showed different risk factors for hip fracture during single-limb stance and falls, which agree with epidemiologic findings of different risk factors for cervical and trochanteric fractures.[11]

Finite-element models derived from QCT scans may improve the prediction of vertebral strength because they mechanically integrate all the geometrical and material property data within the scans. QCT BMD values of each bone voxel can be converted into elastic modulus values using predetermined correlation between the elastic modulus and QCT-derived BMD. Finite-element models integrate mechanically all of the anisotropic, inhomogeneous, and complex geometry of the bone structure examined. It has been demonstrated that voxel-based finite-element model-derived estimates of strength are better predictors of *in vitro* vertebral compressive strength than clinical measures of bone density derived from QCT with or without geometry.[12] Though imaging

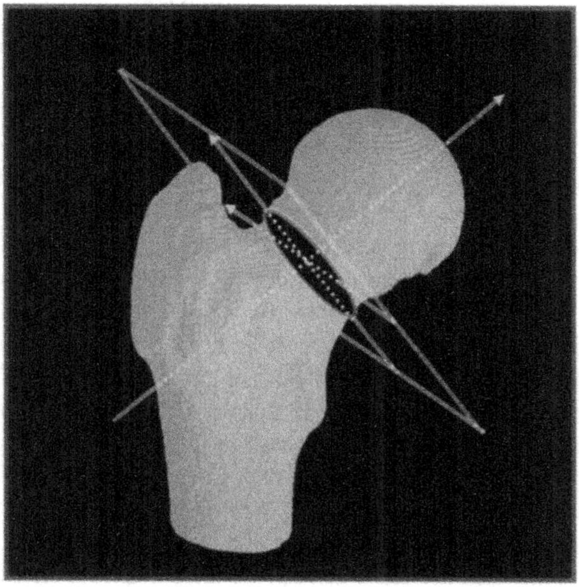

FIGURE 2. vQCT of the hip may be used to analyze BMD in bone compartments and to accurately measure proximal femoral geometry, as well as, to provide a basis for finite-element analyses. (Courtesy of Klaus Engelke.)

resolution is not critical in cross-sectional studies using clinical CT scanners, longitudinal studies that seek to track more subtle changes in stiffness over time should account for the small but highly significant effects of voxel size.[13] However, these data were generated in the laboratory with excised bone specimen without soft tissue: this remains to be confirmed from clinical studies and from epidemiology studies. In addition, the compressive loading experiments used in the laboratory differ from *in vivo* conditions.

HIGH-RESOLUTION COMPUTED TOMOGRAPHY

There is much research under way in the areas of hrCT (micro-CT). The spatial resolution of clinical CT scanners (typically >0.4 mm) is inadequate for highly accurate cortical measurements and for the analysis of discrete trabecular morphological parameters, and new CT developments address this issue. There are two main approaches: the development of new image acquisition and analysis protocols using state-of-the-art clinical CT scanners; and the development of new hrCT scanners for *in vivo* investigations of the peripheral skeleton, or of new micro-CT scanners for *in vitro* two- or three-dimensional structural analysis of very small bone samples (typically <2 cm^3).

State-of-the-art spiral CT scanners use a relatively high resolution ($\approx$0.4 mm $\times$ 0.4 mm), thin slice (1–1.5 mm) to provide images of the spine and hip that clearly display structural information. However, it requires a higher radiation exposure than is employed for standard QCT. Also, the extraction of the quantitative structural information is difficult and the results vary substantially according to the threshold and image processing techniques used. This is due to substantial partial volume effects at this resolution relative to the typical dimensions of trabeculae (100–400 μm) and trabecular spaces (200–2000 μm). hrCT has been employed to measure a feature called the trabecular fragmentation index (length of the trabecular network divided by the number of discontinuities) in an effort to separate osteoporotic subjects from normal subjects,[14] and a similar trabecular textural analysis approach has been reported by Ito *et al.*[15] Gordon *et al.*[16] reported on a hrCT technique that extracted a texture parameter reflecting trabecular hole area, analogous to star volume that appears to enhance vertebral fracture discrimination relative to BMD (FIG. 3). One *in vitro* study showed that a combination of BMD and trabecular structural parameters of bone cubes examined by pQCT improved prediction of bone biomechanical properties.[17] Recently, Link' group has provided a comprehensive analysis of multislice hrCT with hrMR applied to the peripheral skeleton.[18]

High-resolution spinal CT with postprocessing steps used to assess trabecular structure from CT image (A). The structure is segmented by defining the boundary between cortical and trabecular bone (B). The trabecular network is reduced to a binary image (C), which is then thinned to produce a representation of the trabecular form (D).

Higher-resolution CT scanners for peripheral skeletal measurements *in vivo* have been developed and evaluated by Rüegsegger, Durand, and Müller.[19–21] The images show trabecular structure in the radius, with a spatial resolution of 170 μm isotropic. The images can be used for quantitative trabecular structure analysis and also for a separate assessment of cortical and trabecular BMD. Note, however, that these state-of-the-art scanners approach the limits of spatial resolution achievable *in vivo* with acceptable radiation exposure.[22]

More recent studies of high-resolution and volumetric CT have further documented the unique capabilities of these techniques. They have shown that trabecular structural analysis from multi-detector row CT images can better discriminate postmenopausal women with vertebral fracture than dual X ray absorptiometry (DXA).[23] High-resolution spiral CT assessment of the trabecular structure of the vertebral body in combination with BMD improves prediction of biomechanical properties.[24] In elderly men, there is an independent association of sex steroid levels with cortical and trabecular area and their QCT volumetric BMD, but lack such association in young men.[25]

MICROCOMPUTED TOMOGRAPHY

To achieve very high-spatial resolution images, Feldkamp *et al.*[26,27] constructed a unique micro-CT system for three-dimensional *in vitro* analyses

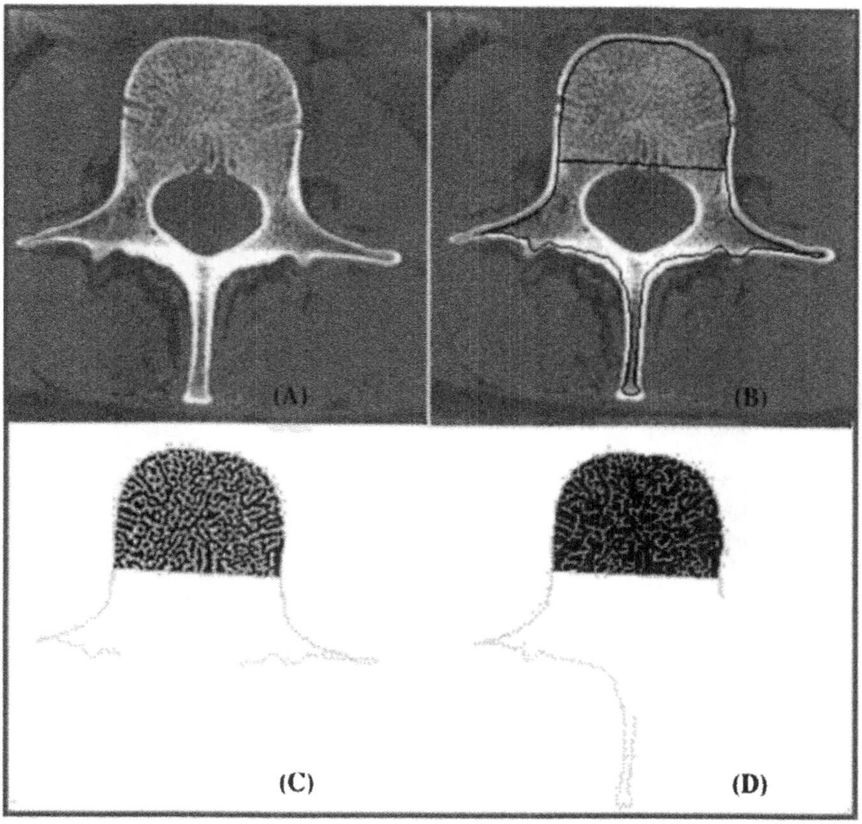

FIGURE 3. High-resolution spinal CT with post-processing steps used to access trabecular structure from CT image (**A**). The structure is segmented by defining the boundary between cortical and trabecular bone (**B**). The trabecular network is reduced to a binary image (**C**), which is then thinned to produce a representation of the trabecular from (**D**). (Courtesy of Christopher Gordon.)

of small bone samples. The system employed cone-beam geometry and a three-dimensional reconstruction algorithm. The spatial resolution of $\approx$60 μm clearly visualized individual trabeculae, allowing a three-dimensional analysis of trabecular network. Goulet *et al.*[28] used images of bone cubes generated by this system to examine standard histomorphometric parameters as well as additional parameters, such as Euler number, an index of connectivity, and mean intercept length, a means of determining anisotrophy. Goulet also related these image-based parameters to Young's modulus, a measure of elasticity of bone. Based on data sets from Feldkamp's micro-CT, Engelke *et al.*[29,30] developed a three-dimensional digital model of trabecular bone for comparing two- and three-dimensional structural analysis methods, and to investigate the effects of spatial resolution and image processing techniques on the extraction of structural parameters (FIG. 4). Three-dimensional data sets from these micro-CT

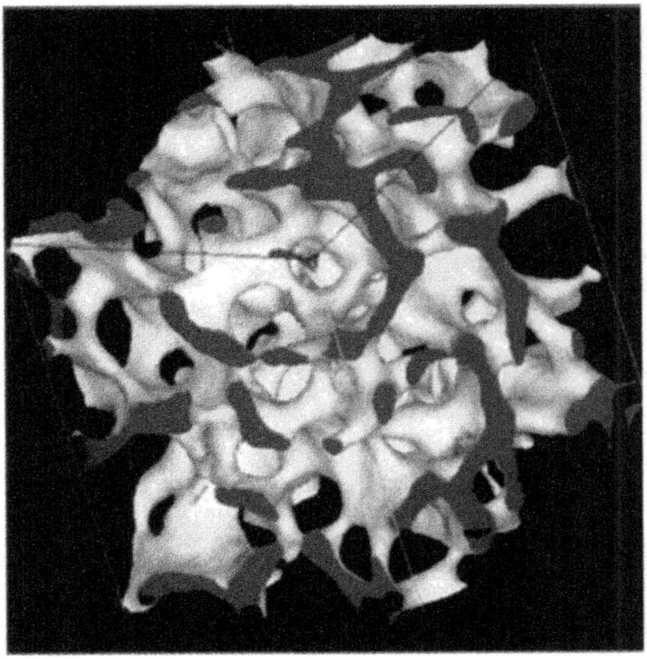

FIGURE 4. Three-dimensional digital model of trabecular bone based on micro-CT image at approximately 60 μm^3 resolution. (Courtesy of Klaus Engelke.)

systems can be used for calculating classical histomorphometric parameters like trabecular thickness and separation[31–33] as well as for determining topological measurements like the Euler number and connectivity.

Another *in vitro* micro-CT scanner with a spatial resolution of 15–20 μm^3 was developed by Rüeggsegger *et al.*[34,35] and has been used extensively in laboratory investigations. Its high accuracy in relation to standard two-dimensional histomorphometry as well as to serial grindings and their derived three-dimensional parameters has been reported.[16] The relationship of these parameters to *in vitro* measures of strength and their application to micro-finite-element modeling has been shown.[36,37] More recently, additional special purpose ultra-high-resolution micro-CT systems have been developed for imaging bone microstructure at resolutions approaching 10 μm or better.[38] These various micro-CT systems have found wide application in both preclinical animal studies and clinical research settings.[39] Similarly, in animal studies, micro-CT has recently found applications in the assessment of skeletal phenotype in gene knockout or in mice,[40–43] in osteoporotic[44] or arthritic rodents.[45]

In a human study by Jiang *et al.*, the rapid deterioration in trabecular microarchitecture in women experiencing menopause was documented by paired iliac crest biopsies before and approximately 5 years after the menopause (FIG. 5). Prominent thinning of trabeculae and conversion of plate-like to rod-like trabecular structure were observed.[39]

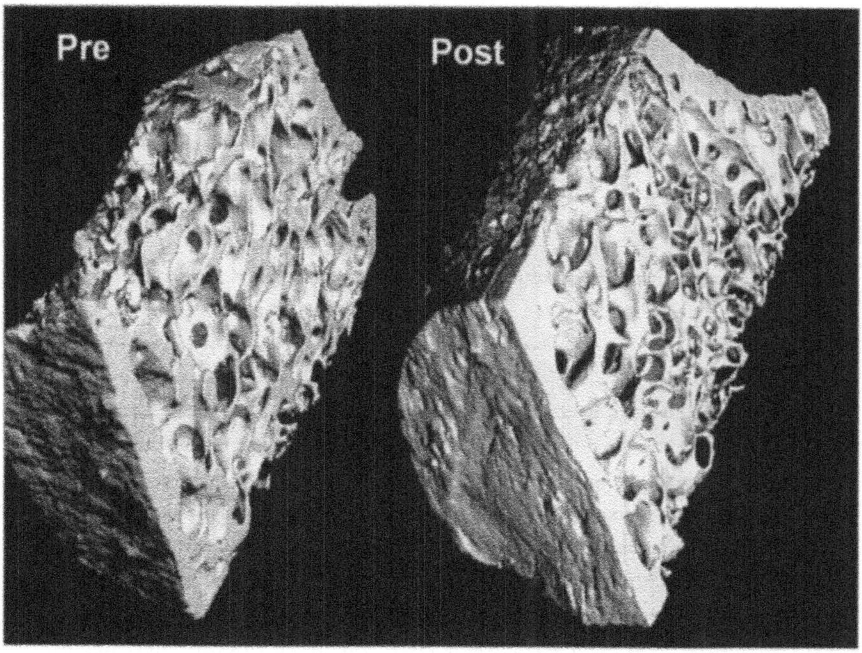

FIGURE 5. Micro-CT image at ~20 μm^3 resolution of trabecular structure of serial iliac crest biopsies from a postmenopausal woman before and 2 years after estrogen replacement therapy. (Courtesy of Yebin Jiang.)

Jiang *et al.* also used micro-CT with three-dimensional analyses compared to standard two-dimensional histomorphometry to study the longitudinal impact of teriparatide (PTH 1–34) treatment versus placebo on the skeleton of postmenopausal women (FIG. 6). In this analysis the changes of the more simple two-dimensional indices pertaining to cancellous bone structure, such as trabecular number, thickness, and spacing did not reach significance after PTH treatment. However, more stereologically correct indices, like marrow star volume and micro-CT-based three-dimensional indices, revealed significant changes, further corroborating the superiority of these techniques for structural analysis of small samples, like bone biopsies. The root mean square CV as reproducibility of micro-CT examination after rescanned and reanalyzed 20 human biopsy specimens was 2–6% for trabecular structural parameters.[46,47]

While the micro-CT scanners described above use an X ray tube as radiation source, other investigators have explored the potential of high-intensity, tight collimation synchrotron radiation, which allows either faster scanning or higher spatial resolution in imaging bone specimens. These systems have been referred to as X ray tomographic microscopy (XTM) and can achieve spatial resolution of 10 μm or better. Bonse, Graeff, and Engelke[18,48–50] were among the first to apply this approach to imaging of bone specimens. Kinney and Lane *et al.*[51,52] have applied the XTM approach to imaging the rat tibia at ultra

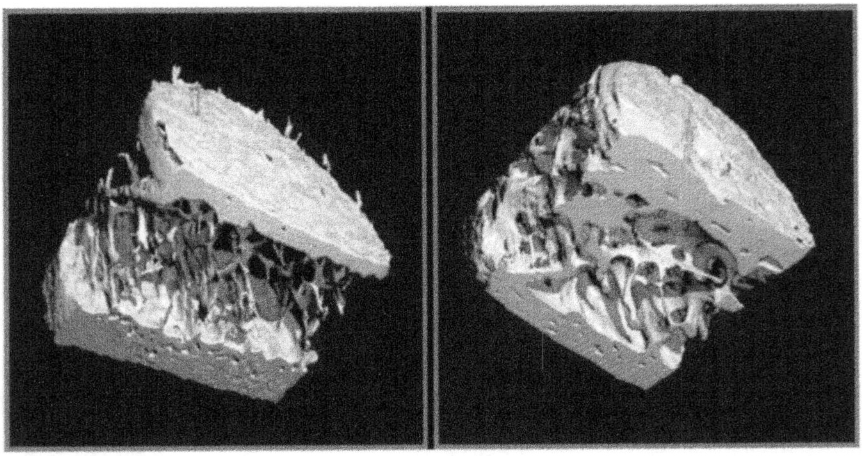

FIGURE 6. Paired biopsy sample was obtained from a 65-year-old woman treated with PTH. Compared to the baseline biopsy (**A**), PTH treatment (**B**) increased trabecular bone volume, trabecular connectivity, and cortical thickness, and shift trabecular morphology from a rod-like structure to a more plate-like pattern. (Courtesy of Yebin Jiang.)

high resolution, both *in vitro* and *in vivo*, and have documented the impact of oophorectomy and PTH treatment on two- and three-dimensionally derived trabecular bone indices. Ritman and Peyrin *et al.*[53] have used synchrotron-based XTM to image trabecular bone ultrastructure at resolutions approaching 1–2 μm, thereby providing the capability to assess additional features, such as resorption cavities. In recently reported studies using synchrotron radiation, micro-CT examination of sequential iliac biopsies showed that treatment with a bisphosphonate did not cause significant hypermineralization but did increase the mineralization at the tissue level.[54] Also, standard micro-CT and histomorphometric assessments of serial iliac crest bone biopsies from postmenopausal women treated with another bisphosphonate for 10 years showed normal microarchitecture.[55]

At recent international congresses there have been many additional studies using micro-CT, including technical developments and various applications. An *in vivo* human micro-CT scanner developed in Europe has been used to examine distal radius or distal tibia with isotropic resolution of about 100 μm.[56] Its application in United States and other places remains to be seen. Advanced micro-finite analyses with models based on these three-dimensional peripheral micro-CT systems (isotropic voxel resolution of 165 μm) have predicted the failure load of the human radius Colles fractures better than by DXA or bone morphology and geometry measurements.[57] Micro-CT (and micro-MRI, see below) replicate the complex trabecular architecture on a macroscopic scale for visual or biomechanical analysis. A complete set of three-dimensional image data provides a basis for finite-element modeling for virtual biomechanics to predict mechanical properties.[58]

MR MICROSCOPY

High-resolution MR and micro-MR—referred to collectively as MR microscopy—have received considerable attention as research tools and as potential clinical tools for assessment of trabecular bone architecture. Magnetic resonance is a complex technology based on the application of high-magnetic fields, transmission of radiofrequency (RF) waves, and detection of RF signals from excited hydrogen protons. A noninvasive, nonionizing radiation technique, MR can provide three-dimensional images in arbitrary orientations and can depict trabecular bone as a negative image by virtue of the strong signal generated by the abundant fat and water protons in the surrounding marrow tissue. The appearance of the MR image is affected by many factors beyond spatial resolution, including the field strength and specific pulse sequence used, the echo time, and the signal-to-noise achieved.[59–61] Analysis and interpretation of MR images are more complicated than for the X ray-based images of CT. Nevertheless, MR microscopy holds much promise for improved quantitative assessment of trabecular structure both *in vivo* and *in vitro*.

Because of the relation of signal-to-field strength in MR, special purpose, small bore, high-field magnets have been employed to obtain very high resolution or micro-MR images of bone specimens *in vitro*. Wehrli and colleagues[62–65] obtained 78-μm isotopic resolution of human and bovine bone cubes using three-dimensional imaging at 9.4 Tesla, and derived anisotropy ellipsoids from the analysis of mean intercept length. They also found good correlations between MR-derived parameters and standard histomorphometric measures. Antich *et al.*[66] have conducted similar experiments and found changes in accordance with histomorphometry measures. Kapadia *et al.*[67] extended the *in vitro* techniques to obtain images in an ovariectomized rat model and were able to measure changes in trabecular structure following ovariectomy. Hipp and colleague[68] examined bovine cubes in a small bore microimaging spectrometer at 60-μm^3 resolution and found three-dimensional results heavily dependent upon the threshold and image processing algorithm. Majumdar *et al.*[69] examined human cadaver specimens using a standard clinical MR scanner at 1.5 Tesla and a spatial resolution of $117 \times 117 \times 300$ μm, and compared these images with XTM images and with serial grindings to determine the impact of in-plane resolution and slice thickness on both two- and three-dimensional structural and textural parameters. Considerable resolution dependence was observed for traditional stereological parameters, some of which could be modulated by appropriate thresholds and image processing techniques.

Limitations of signal-to noise, spatial resolution, and total imaging time, prevent resolution of smaller individual trabeculae *in vivo* at clinical field strengths, but the images show the larger trabeculae and the texture of the trabecular network. The trabecular structure can still be quantified using standard techniques of stereology as well as textural parameters, such as fractal analysis.

In an early study by Majumdar *et al.*[60] establishing the feasibility of using such images to quantify trabecular structure, MR images of the distal radius were obtained using a modified gradient echo sequence, a 1.5 Tesla imager, and a spatial resolution of 156 μm^2 and slice thickness of 0.7 mm. Representative axial sections from normal and osteoporotic subjects clearly depicted the loss of the integrity of the trabecular network with the development of osteoporosis (FIG. 7). Similar images of the calcaneus of normal subjects showed that the orientation of the trabeculae is significantly different in various anatomic regions. Ellipses, representing the mean intercept length, showed a preferred orientation and hence mapped the anisotropy of trabecular structure. In preliminary *in vivo* studies of the calcaneus, *gray* scale reference values from fat, muscle, and tendon were used to calculate reproducible threshold values.[70] This approach gave a midterm, *in vivo* precision of ≈3.5% CV for trabecular width and spacing.

Both Wehrli and colleagues,[71] and Glüer and colleagues,[72–74] have used clinical imagers at 1.5 Tesla with special RF coil designs and have measured trabecular and cortical bone in the phalanges, a convenient anatomic site particularly suitable for obtaining high signal-to-noise and high-spatial resolution images *in vivo*. Resolution of 78–150 μm and slice thickness of 300 μm have been achieved in the phalanges. Stampa *et al.*[74] used these phalangeal images to derive quantitative three-dimensional parameters based on an algorithm and model for defining trabecular rods and plates. Others, including Link *et al.*,[75] Majumdar *et al.*[76] and Wehrli *et al.*[77–79] have shown the ability to discriminate spine and/or hip fractures using trabecular structure or textural parameters from *in vivo* MR images of the radius or calcaneus. Recently, Engelke has provided a comprehensive analysis of threshold effects in hrMR of the calcaneus when compared with ultra-high-resolution anatomic sectioning *in vitro*.[80] Newitt has reported promising results on the application of micro-finite-element analyses based on hrMR images of the distal radius *in vivo*.[81] Structure parameters determined in high-resolution MR images of the proximal femur specimen correlated significantly with bone strength, with the highest correlations obtained combining DXA BMD and structure measures.[82]

The resolution achievable *in vivo* by MRI is not sufficient to depict precisely individual trabeculae and, thus, does not permit the quantification of the "true" trabecular bone morphology and topology. Trabecular samples of the distal radius imaged using MRI at 156 × 156 × 300 μ correlate well with micro-CT at 34 × 34 × 34 μ, with r^2 ranging from 0.57 to 0.64 for morphological measurements.[83] Trabecular bone structure parameters assessed in the distal radius on high-resolution MR and multislice CT images are significantly correlated with those determined on contact radiographs of the corresponding specimen sections. For the MR imaging, the threshold algorithm used for binarizing the images substantially affected these correlations.[84]

In an animal study by Jiang *et al.*, MRI microscopy showed that ovariectomy induces deterioration of trabecular microstructure and of the biomechanical

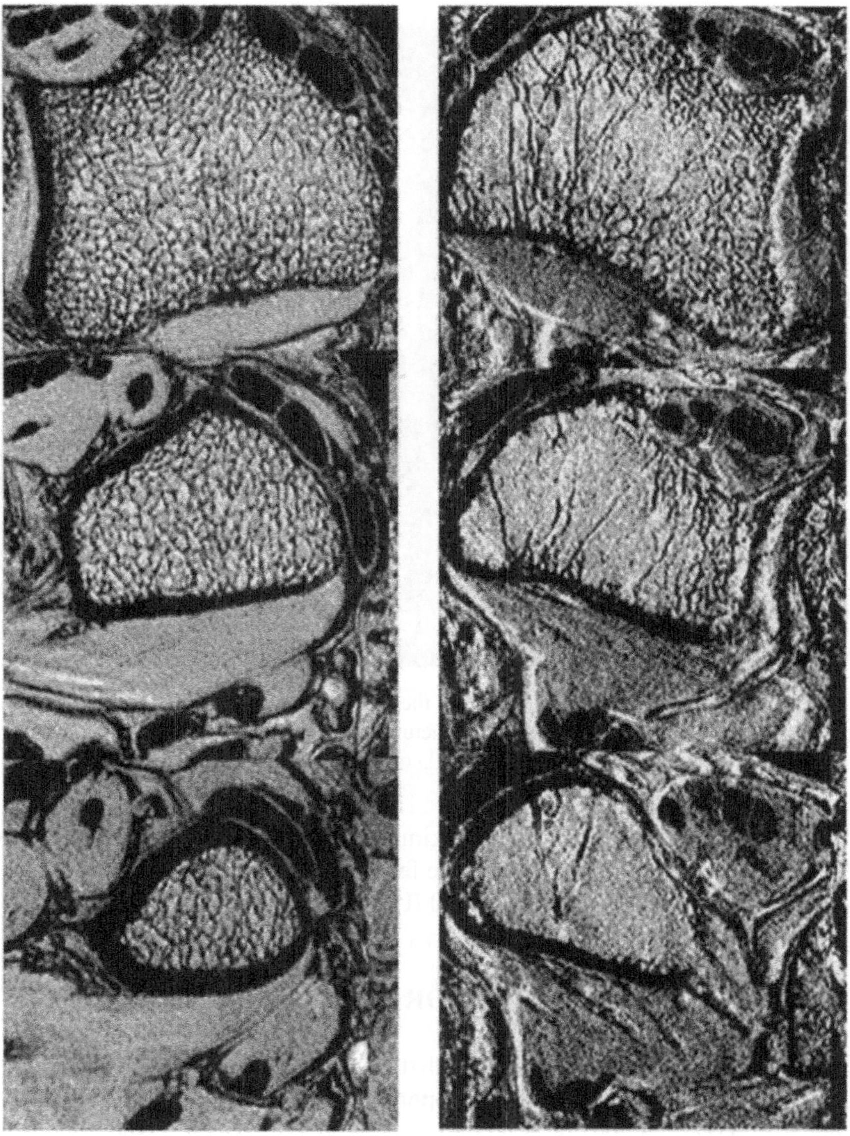

FIGURE 7. High-resolution (~150 × 150 × 500 μm) axial gradient echo MR images of the distal radius of a young woman (**L**) and an elderly osteoporotic woman (**R**) clearly show the deterioration in trabecular and cortical structure. (Courtesy of Sharmila Majumdar.)

properties in the femoral neck of ewes (FIG. 8). Calcitonin treatment prevented ovariectomy (OVX) induced changes in a dose-dependent manner. The femoral neck trabecular microstructure significantly correlates with biomechanical properties, and its combination with BMD further improved the prediction of bone quality.[85]

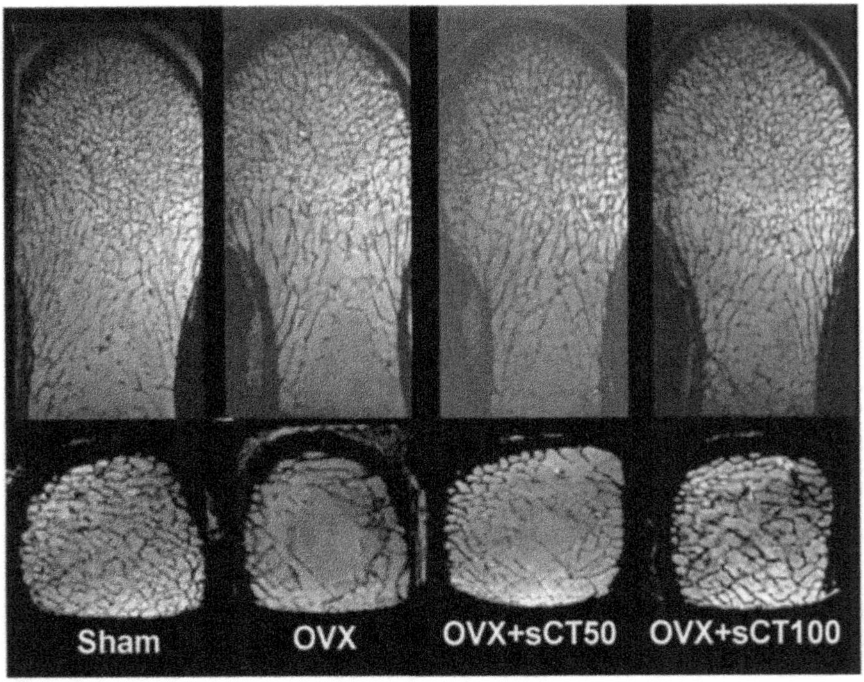

FIGURE 8. MR microscopic images of the proximal femur show that the ovariectomy (OVX)-induced loss in trabecular microstructure in the femoral neck is prevented by the treatment of salmon calcitonin (sCT) at 50 IU or 100 IU.

MR microscopic images of the proximal femur show that the OVX-induced loss in trabecular microstructure in the femoral neck is prevented by the treatment of salmon calcitonin (sCT) at 50 IU or 100 IU.

CHALLENGES FOR BONE IMAGING

Despite the considerable progress that has been made over the past decade in advanced bone imaging for osteoporosis assessment, a number of challenges remains. Technically, the challenges reflect the balances and trade-offs between spatial resolution, sample size, signal-to noise, radiation exposure and acquisition time, or between the complexity and expense of the imaging technologies versus their availability and accessibility. Clinically, the challenges for bone imaging include balancing the advantages of standard BMD information versus the more complex architectural features of bone, or the deeper research requirements of the laboratory versus the broader needs of clinical practice. The biological differences between the peripheral appendicular skeleton and the central axial skeleton and their impact on the relevant bone imaging methods must be further clarified. Finally, the relative merits of these sophisticated imaging techniques must be weighed with respect to their applications as diagnostic procedures, requiring high accuracy or reliability, versus their applications as monitoring procedures, requiring high precision or reproducibility.

ACKNOWLEDGMENTS

This research work was adapted in part from Genant HK, Gordon C, Jiang Y, Lang TF, Link TM, Majumdar S. 1999. Advanced imaging of bone macro- and microstructure. Bone **25:** 149–152.

REFERENCES

1. GENANT, H.K., K. ENGELKE, T. FUERST, *et al.* 1996. Noninvasive assessment of bone mineral and structure: state of the art. J. Bone Miner. Res. **11:** 707–730.
2. FAULKNER, K.G., C.E. CANN & B.H. HASEGAWA. 1990. CT-derived finite element model to determine vertebral cortex strength. *In* Medical Imaging IC: Image Processing, Vol. 1233. M.H. Loew, Ed.: 194–202. SPIE Newport Beach, California.
3. LANG, T.F., G. GUGLIELMI, C. VAN KUIJK, *et al.* 2002. Measurement of bone mineral density at the spine and proximal femur by volumetric quantitative computed tomography and dual-energy X-ray absorptiometry in elderly women with and without vertebral fractures. Bone **30:** 247–250.
4. KEYAK, J.H., S.A. ROSSI, K.A. JONES & H.B. SKINNER. 1998. Prediction of femoral fracture load using automated finite element modeling. J. Biomech. **31:** 125–133.
5. LANG, T.F., J. LI, S.T. HARRIS & H.K. GENANT. 1999. Assessment of vertebral bone mineral density using volumetric quantitative computed tomography. J. Comp. Assist. Tomogr. **23:** 130–137.
6. LANG, T.F., J.H. KEYAK, M.W. HEITZ, *et al.* 1997. Volumetric quantitative computed tomography of the proximal femur: precision and relation to bone strength. Bone **21:** 101–108.
7. LOTZ, J.C., T.N. GERHART & W.C. HAYES. 1990. Mechanical properties of trabecular bone from the proximal femur: a quantitative CT study. J. Comp. Assist. Tomogr. **14:** 107–114.
8. MCBROOM, R.J., W.C. HAYES, W.T. EDWARDS, *et al.* 1985. Prediction of vertebral body compressive fracture using quantitative computed tomography. J. Bone Joint Surg. Am. **67:** 1206–1214.
9. JERGAS, M., M. BREITENSEHER, C.C. GLÜER, *et al.* 1995. Estimates of volumetric bone density from projectional measurements improve the discriminatory capability of dual X-ray absorptiometry. J. Bone Miner. Res. **10:** 1101–1110.
10. SODE, M., J. KEYAK, M. BOUXSEIN & T. LANG. 2004. Assessment of femoral neck torsional strength indices. J. Bone Miner. Res. **19:** S238.
11. LANG, T.F., J.H. KEYAK, A. YU, *et al.* 2003. Determinants of proximal femoral strength in elderly women. J. Bone Miner. Res. **18:** S266.
12. CRAWFORD, R.P., C.E. CANN & T.M. KEAVENY. 2003. Finite element models predict in vitro vertebral body compressive strength better than quantitative computed tomography. Bone **33:** 744–750.

13. CRAWFORD, R.P., W.S. ROSENBERG & T.M. KEAVENY. 2003. Quantitative computed tomography-based finite element models of the human lumbar vertebral body: effect of element size on stiffness, damage, and fracture strength predictions. J. Biomech. Eng. **125:** 434–438.
14. CHEVALIER, F., A.M. LAVAL-JEANTET, M. LAVAL-JEANTET, *et al.* 1992. CT image analysis of the vertebral trabecular network in vivo. Calcif. Tissue Int. **51:** 8–13.
15. ITO, M., M. OHKI, K. HAYASHI, *et al.* 1995. Trabecular texture analysis of CT images in the relationship with spinal fracture. Radiology **194:** 55–59.
16. GORDON, C.L., T.F. LANG, P. AUGAT, *et al.* 1998. Image-based assessment of spinal trabecular bone structure from high-resolution CT images. Osteoporos. Int. **8:** 317–325.
17. JIANG, Y., J. ZHAO, P. AUGAT, *et al.* 1998. Trabecular bone mineral and calculated structure of human bone specimens scanned by peripheral quantitative computed tomography: relation to biomechanical properties. J. Bone Miner. Res. **13:** 1783–1790.
18. LINK, T.M., V. VIETH, C. STEHLING, *et al.* 2003. High-resolution MRI vs multi-slice spiral CT: which technique depicts the trabecular bone structure best? Eur. Radiol. **13:** 663–671.
19. DURAND, E.P. & P. RÜEGSEGGER. 1992. High-contrast resolution of CT images for bone structure analysis. Med. Phys. **19:** 569–573.
20. MÜLLER, R., M. HAHN, M. VOGEL, *et al.* 1996. Morphometric analysis of noninvasively assessed bone biopsies: comparison of high-resolution computed tomography and histologic sections. Bone **18:** 215–220.
21. MÜLLER, R., T. HILDEBRAND, H.J. HAUSELMANN & P. RUEGSEGGER. 1996. In vivo reproducibility of three-dimensional structural properties of noninvasive bone biopsies using 3D-pQCT. J. Bone Miner. Res. **11:** 1745–1750.
22. ENGELKE, K., W. GRAEFF, L. MEISS, *et al.* 1993. High spatial resolution imaging of bone mineral using computed microtomography. Comparison with microradiography and undecalcified histologic sections. Invest. Radiol. **28:** 341–349.
23. TAKADA, M., K. KIKUCHI, S. UNAU & K. MURATA. 2004. Three-dimensional analysis of trabecular bone structure of human vertebra in vivo using image data from multi-detector row computed tomography—correlation with bone mineral density and ability to discriminate women with vertebral fracture. J Bone Miner. Res. **19:** S371.
24. BAUER, J.S., D. MUELER, M. FISCHBECK, *et al.* 2004. High resolution spiral-CT for the assessment of osteoporosis: which site of the spine and region of the vertebra is best suited to obtain trabecular bone structural parameters? J. Bone Miner. Res. **19:** S169.
25. KHOSLA, S., L.J. MELTON, E.J. ATKINSON, *et al.* 2004. Relationship of volumetric density, geometry and bone structure at different skeletal sites to sex steroid levels in men. J. Bone Miner. Res. **19:** S88.
26. FELDKAMP, L.A., S.A. GOLDSTEIN, A.M. PARFITT, *et al.* 1989. The direct examination of three-dimensional bone architecture in vitro by computed tomography. J. Bone Miner. Res. **4:** 3–11.
27. KUHN, J.L., S.A. GOLDSTEIN, L.A. FELDKAMP, *et al.* 1990. Evaluation of a micro-computed tomography system to study trabecular bone structure. J. Orthop. Res. **8:** 833–842.
28. GOULET, R.W., S.A. GOLDSTEIN, M.J. CIARELLI, *et al.* 1994. The relationship between the structural and orthogonal compressive properties of trabecular bone. J. Biomech. **27:** 375–389.

29. ENGELKE, K., C. KLIFA, B. MUNCH, *et al.* 1994. Morphological analysis of the trabecular network: the influence of image processing technique on structural parameters. 10th International Bone Densitometry Workshop, Venice Italy. J. Bone Miner. Res. **25:** S8.
30. ENGELKE, K., S.M. SONG, C.C. GLÜER & H.K. GENANT. 1996. A digital model of trabecular bone. J. Bone Miner. Res. **11:** 480–489.
31. PARFITT, A.M., C. MATTHEWS, A. VILLANUEVA. 1983. Relationships between surface, volume and thickness of iliac trabecular bone in aging and in osteoporosis. J. Clin. Invest. **72:** 1396–1409.
32. PARFITT, A.M. 1983. The stereologic basis of bone histomorphometry. Theory of quantitative microscopy and reconstruction of the third dimension. *In* Bone Histomorphometry: Techniques and Interpretations. R. Recker, Ed.: 53–87. CRC Press. Boca Raton.
33. ODGAARD, A. & H.J.G. GUNDERSEN. 1993. Quantification of connectivity in cancellous bone, with special emphasis on 3-D reconstructions. Bone **14:** 173–182.
34. RÜEGSEGGER, P., B. KOLLER & R. MULLER. 1996. A microtomographic system for the nondestructive evaluation of bone architecture. Calcif. Tissue Int. **58:** 24–29.
35. MÜLLER, R., F. BAUSS, S.Y. SMITH & M.K. HANNAN. 2001. Mechano-structure relationships in normal, ovariectomized and Ibandronate treated aged macaques as assessed by micro-tomographic imaging and biomechanical testing. Trans. Orthop. Res. Soc. **26:** 66.
36. MÜLLER, R. & P. RUEGSEGGER. 1996. Analysis of mechanical properties of cancellous bone under conditions of simulated bone atrophy. J. Biomech. **29:** 1053–1060.
37. VAN RIETBERGEN, B., H. WEINANS, R. HUISKES & A. ODGAARD. 1995. A new method to determine trabecular bone elastic properties and loading using micromechanical finite-element models. J. Biomech. **28:** 69–81.
38. ENGELKE, K., M. KAROLCZAK, S. SCHALLER, *et al.* 1998. A cone beam micro-computed tomography (μCT) system for imaging of 3D trabecular bone structure. Presented at the 13th International Bone Densitometry Workshop, 4–8 October, Wisconsin, USA.
39. JIANG, Y., P. ZHAO, E.Y. LIAO, *et al.* 2005. Application of micro CT assessment of 3D bone microstructure in preclinical and clinical studies. J. Bone Miner. Metab. (In press).
40. SOHASKEY, M.L., Y. JIANG, J. ZHAO, *et al.* 2004. Insertional mutagenesis of osteopotentia, a novel transmembrane protein essential for skeletal integrity. J. Bone Miner. Res. **19:** S21.
41. SOROCÉANU, M.A., D. MIAO, Y. JIANG, *et al.* 2004. *Pthrp* haploinsufficiency impairs bone formation but potentiates the bone anabolic effects of PTH (1-34). J. Bone Miner. Res. **19:** S97.
42. TAKESHITA, S., N. NAMBA, J. ZHAO, *et al.* 2002. SHIP-deficient mice are severely osteoporotic due to increased numbers of hyper-resorptive osteoclasts. Nat. Med. **8:** 943–949 (Online published on Aug. 5, 2002).
43. BERGO, M.O., B. GAVINO, J. ROSS, *et al.* 2002. Zmpste24 deficiency in mice causes spontaneous bone fractures, muscle weakness, and a prelamin A processing defect. Proc. Natl. Acad. Sci. USA **99:** 13049–13054.
44. LANE, N.E., M. BALOOCH, J. ZHAO, *et al.* 2004. Glucocorticoids induce changes around the osteocyte lacunae that reduces bone strength and bone mineral content independent of apoptosis: preliminary data from a glucocorticoid-induced bone loss model in male mice. J. Bone Miner. Res. **19:** S434–S435.

45. JIANG, Y., J.J. ZHAO, R. MANGADU, *et al.* 2004a. Assessment of 3D cortical and trabecular bone microstructure and erosion on micro CT images of a murine model of arthritis. J. Bone Miner. Res. **19:** S474.
46. JIANG, Y., J. ZHAO, B.H. MITLAK, *et al.* 2003. Recombinant human parathyroid hormone (1–34) (teriparatide) improves both cortical and cancellous bone structure. J. Bone Miner. Res. **18:** 1932–1941.
47. JIANG, Y., J. ZHAO, E.F. ERIKSEN & H.K. GENANT. 2003b. Reproducibility of micro CT quantification of 3D microarchitecture of the trabecular and cortical bone in the iliac crest of postmenopausal osteoporotic women and their treatment with teriparatide [rhPTH(1-34)]. RSNA **03:** 571.
48. BONSE, U., F. BUSCH, O. GUNNEWIG, *et al.* 1994. 3D computed x-ray tomography of human cancellous bone at 8 μm spatial and 10–4 energy resolution. Bone Miner. **25:** 25–38.
49. ENGELKE, K., W. DIX, W. GRAEFF, *et al.* 1991. Quantitative microtomography and microradiography of bones using synchrotron radiation. Presented at the 8th International Workshop on Bone Densitometry. Bad Reichenhall, Germany, Osteoporos. Int.
50. GRAEFF, W. & K. ENGELKE. 1991. Microradiography and microtomography. *In* Handbook on Synchrotron Radiation. S. Ebashi, M. Koch, E. Rubenstein, Eds.: 361–405. North-Holland, Amsterdam.
51. KINNEY, J.H., N.E. LANE & D.L. HAUPT. 1995. In vivo, three-dimensional microscopy of trabecular bone. J. Bone Miner. Res. **10:** 264–270.
52. LANE, N.E., J.M. THOMPSON, G.J. STREWLER & J.H. KINNEY. 1995. Intermittent treatment with human parathyroid hormone (hPTH[1–34]) increased trabecular bone volume but not connectivity in osteopenic rats. J. Bone Miner. Res. **10:** 1470–1477.
53. PEYRIN, F., M. SALOME, P. CLOETENS, *et al.* 1998. What do micro-CT examinations reveal at various resolutions: a study of the same trabecular bone samples at the 14, 7, and 2 micron level. Presented at the Symposium on Bone Architecture and the Competence of Bone. Ittingen, Switzerland, 3–5 July.
54. BORAH, B., E.L. RITMAN, T.E. DUFRESNE, *et al.* 2004. Five year risedronate therapynormalizes mineralization: synchrotron radiation micro-computed tomography study of sequential triple biopsies. J. Bone Miner. Res. **19:** S308.
55. RECKER, R., K. ENSRUD, S. DIEM, *et al.* 2004. Normal bone histomorphometry and 3D microarchitecture after 10 years alendronate treatment of postmenopausal women. J. Bone Miner. Res. **19:** S45.
56. NEFF, M., M. DAMBACHER, S. HAEMMERLE, *et al.* 2004. 3D evaluation of bone microarchitecture in humans using high resolution pQCT; a new in vivo, non invasive and time saving procedure. J. Bone Miner. Res. **19:** S236.
57. PISTOIA, W., B. VAN RIETBERGEN, E.M. LOCHMULLER, *et al.* 2002. Estimation of distal radius failure load with micro-finite element analysis models based on three-dimensional peripheral quantitative computed tomography images. Bone **30:** 842–848.
58. BORAH, B., G.J. GROSS, T.E. DUFRESNE, *et al.* 2001. Three-dimensional microimaging (MRmicroI and microCT), finite element modeling, and rapid prototyping provide unique insights into bone architecture in osteoporosis. Anat. Rec. **15:** 101–110.
59. MAJUMDAR, S., H. GENANT, A. GIES & G. GUGLIELMI. 1993. Regional variations in trabecular structure in the calcaneus assessed using high resolution magnetic resonance images and quantitative image analysis. J. Bone Miner. Res. **8S:** 351.

60. MAJUMDAR, S., H. GENANT, S. GRAMPP, *et al.* 1994. Analysis of trabecular bone structure in the distal radius using high resolution MRI. Eur. Radiol. **4:** 517–524.
61. MAJUMDAR, S., D.C. NEWITT, M. JERGAS, *et al.* 1995. Evaluation of technical factors affecting the quantification of trabecular bone structure using magnetic resonance imaging. Bone **17:** 417–430.
62. CHUNG, H., F.W. WEHRLI, J.L. WILLIAMS & S.D. KUGELMASS. 1993. Relationship between NMR transverse relaxation, trabecular bone architecture, and strength. Proc. Natl. Acad. Sci. USA **90:** 10250–10254.
63. CHUNG, H.W., F.W. WEHRLI, J.L. WILLIAMS, *et al.* 1995. Quantitative analysis of trabecular microstructure by 400 MHz nuclear magnetic resonance imaging. J. Bone Miner. Res. **10:** 803–811.
64. CHUNG, H.W., F.W. WEHRLI, J.L. WILLIAMS & S.L. WEHRLI. 1995. Three-dimensional nuclear magnetic resonance microimaging of trabecular bone. J. Bone Miner. Res. **10:** 1452–1461.
65. HWANG, S.N., F.W. WEHRLI & J.L. WILLIAMS. 1997. Probability-based structural parameters from three-dimensional nuclear magnetic resonance images as predictors of trabecular bone strength. Med. Phys. **24:** 1255–1261.
66. ANTICH, P., R. MASON, R. MCCOLL, *et al.* 1994. Trabecular architecture studies by 3D MRI microscopy in bone biopsies. J. Bone Miner. Res. **9S1:** 327.
67. KAPADIA, R.D., W. HIGH, D. BERTOLINI & S.K. SARKAR. 1993. MR microscopy: a novel diagnostic tool in osteoporosis research. *In* 4th International Symposium on Osteoporosis & Consensus Development Conference. C. Christiansen, Ed.: 28. Hong Kong.
68. SIMMONS, C.A. & J.A. HIPP. 1997. Method-based differences in the automated analysis of the three-dimensional morphology of trabecular bone. J. Bone Miner. Res. **12:** 942–947.
69. MAJUMDAR, S., D. NEWITT, A. MATHUR, *et al.* 1996. Magnetic resonance imaging of trabecular bone structure in the distal radius: relationship with X-ray tomographic microscopy and biomechanics. Osteoporos. Int. **6:** 376–385.
70. OUYANG, X., K. SELBY, P. LANG, *et al.* 1997. High resolution magnetic resonance imaging of the calcaneus: age-related changes in trabecular structure and comparison with dual X-ray absorptiometry measurements. Calcif. Tissue Int. **60:** 139–147.
71. JARA, H., F.W. WEHRLI, H. CHUNG & J.C. FORD. 1993. High-resolution variable flip angle 3D MR imaging of trabecular microstructure in vivo. Magn. Reson. Med. **29:** 528–539.
72. KÜHN, B., B. STAMPA & C.C. GLÜER. 1997. Hochauflösende Darstellung und Quantifierung der trabekulären Knochenstruktur der Fingerphalangen mit der Magnetresonanztomographie. Z. Med. Phys. **7:** 162–168.
73. STAMPA, B., B. KÜHN, M. HELLER, C.C. GLÜER. 1998. Rods or plates: a new algorithm to characterize bone structure using 3D magnetic resonance imaging. Presented at the 13th International Bone Densitometry Workshop, 4–8 October. Wisconsin, USA.
74. STAMPA, B., B. KUHN, C. LIESS, *et al.* 2002. Characterization of the integrity of three-dimensional trabecular bone microstructure by connectivity and shape analysis using high-resolution magnetic resonance imaging in vivo. Top. Magn. Reson. Imaging **13:** 357–363.
75. LINK, T., S. MAJUMDAR, P. AUGAT, *et al.* 1998. In vivo high resolution MRI of the calcaneus: differences in trabecular structure in osteoporotic patients. J. Bone Miner. Res. **13:** 1175–1182.

76. MAJUMDAR, S., H.K. GENANT, S. GRAMPP, *et al.* 1997. Correlation of trabecular bone structure with age, bone mineral density and osteoporotic status: in vivo studies in the distal radius using high resolution magnetic resonance imaging. J. Bone Miner. Res. **12:** 111–118.
77. WEHRLI, F.W., S.N. HWANG, J. MA, *et al.* 1998. Cancellous bone volume and structure in the forearm: noninvasive assessment with MR microimaging and image processing. Radiology **206:** 347–357.
78. WEHRLI, F.W., B.R. GOMBERG, P.K. SAHA, *et al.* 2001. Digital topological analysis of in vivo magnetic resonance microimages of trabecular bone reveals structural implications of osteoporosis. J. Bone Miner. Res. **16:** 1520–1531.
79. WEHRLI, F.W., P.K. SAHA, B.R. GOMBERG, *et al.* 2002. Role of magnetic resonance for assessing structure and function of trabecular bone. Top. Magn. Reson. Imaging **13:** 335–355.
80. ENGELKE, K., M. HAHN, M. TAKADA, *et al.* 2001. Structural analysis of high resolution in vitro MR images compared to stained grindings. Calcif. Tissue Int. **68:** 163–171.
81. NEWITT, D.C., S. MAJUMDAR, B. VAN RIETBERGEN, *et al.* 2002. In vivo assessment of architecture and micro-finite element analysis derived indices of mechanical properties of trabecular bone in the radius. Osteoporos. Int. **13:** 6–17.
82. LINK, T.M., V. VIETH, R. LANGENBERG, *et al.* 2003. Structure analysis of high resolution magnetic resonance imaging of the proximal femur: in vitro correlation with biomechanical strength and BMD. Calcif. Tissue Int. **72:** 156–165.
83. POTHUAUD, L., A. LAIB, P. LEVITZ, *et al.* 2002. Three-dimensional-line skeleton graph analysis of high-resolution magnetic resonance images: a validation study from 34-microm-resolution microcomputed tomography. J. Bone Miner. Res. **17:** 1883–1895.
84. LINK, T.M., V. VIETH, C. STEHLING, *et al.* 2003. High-resolution MRI vs multi-slice spiral CT: which technique depicts the trabecular bone structure best? Eur. Radiol. **13:** 663–671.
85. JIANG, Y., J. ZHAO, P. GEUSENS, *et al.* 2005. Femoral neck trabecular microstructure in ovariectomized ewes treated with calcitonin: MRI microscopic evaluation. J. Bone Miner. Res. **20:** 125–130.

Bone Strength: Current Concepts

CHARLES H. TURNER

Orthopaedic Research Laboratories and Biomechanics and Biomaterials Research Center, Indiana University Purdue University at Indianapolis, Indianapolis, Indiana 46202, USA

ABSTRACT: Bones serve several mechanical functions, including acoustic amplification in the middle ear, shielding vital organs from trauma, and serving as levers for muscles to contract against. Bone is a multiphase material made up of a tough collagenous matrix intermingled with rigid mineral crystals. The mineral gives bone its stiffness. Without sufficient mineralization, bones will plastically deform under load. Collagen provides toughness to bone making it less brittle so that it better resists fracture. Bone adapts to mechanical stresses largely by changing its size and shape, which are major determinants of its resistance to fracture. Tissue is added in regions of high mechanical stress providing an efficient means for improving bone strength. Experiments have shown that small additions of bone mineral density (BMD) (5–8%) caused by mechanical loading can improve bone strength by over 60% and extend bone fatigue life by 100-fold. Consequently, it is clear that bone tissue possesses a mechanosensing apparatus that directs osteogenesis to where it is most needed for improving bone strength. The biological processes involved in bone mechanotransduction are poorly understood and further investigation of the molecular mechanisms involved might uncover drug targets for osteoporosis. Several pathways are emerging from current research, including membrane ion channels, ATP signaling, second messengers, such as prostaglandins and nitric oxide, insulin-like growth factors, and Wnt signaling.

KEYWORDS: biomechanics; fracture; osteoporosis; mechanotransduction

BACKGROUND

The skeleton serves several functions.[1] Probably most importantly, it provides a ready source of calcium ions to maintain appropriate serum calcium levels and facilitate proper muscle function. Bones also provide acoustic amplification and impedance matching in the middle ear. But here we will focus

Address for correspondence: Charles H. Turner, Orthopaedic Research Laboratories and Biomechanics and Biomaterials Research Center, Indiana University Purdue University at Indianapolis, 1120 South Drive, FH 115, Indianapolis, IN 46202. Voice: 317-274-3226; fax: 317-274-3702.
e-mail: turnerch@iupui.edu

Ann. N.Y. Acad. Sci. 1068: 429–446 (2006). © 2006 New York Academy of Sciences.
doi: 10.1196/annals.1346.039

on the skeleton as a structure. There are at least two key biomechanical roles for bones. First, bones shield vital organs from trauma. In the skull, mechanical energy resulting from blunt trauma concentrates mainly in the spongy bone separating the bony plates. Consequently, very little energy is transferred to the innermost bony plate and the cranial vault is preserved. In addition, bones serve as levers for muscles to contract against. To better distribute joint forces, the ends of long bones are broadened to reduce stress (load/area) and that stress is carried by trabecular bone beneath the joint surface to distribute the force into the long bone cortex.

BASIC BIOMECHANICS

There are a number of biomechanical parameters that can be used to characterize the integrity of bone. The key relationship is that between load applied to a structure and displacement in response to the load-displacement curve (FIG. 1). The slope of the elastic region of the load-displacement curve represents the extrinsic stiffness or rigidity of the structure (S). Besides stiffness, several other biomechanical properties can be derived, including ultimate force (F_u), work to failure (area under the load-displacement curve, U), and ultimate displacement (d_u). Each of these measured parameters reflects a different property of the bone: ultimate force reflects the general integrity of the bone structure; stiffness is closely related to the mineralization of the bone; work to failure is the amount of energy necessary to break the bone; and ultimate

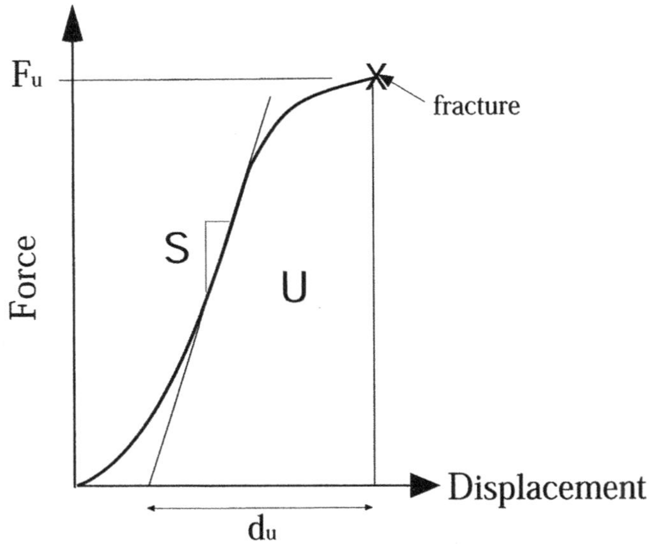

FIGURE 1. Load-displacement curve for bone tissue. The slope of the linear region of the curve represents the extrinsic stiffness or rigidity of the structure (S); the height of the curve denoted ultimate force (F_u); area under the curve is work to failure (U); and total displacement to fracture is ultimate displacement (d_u).

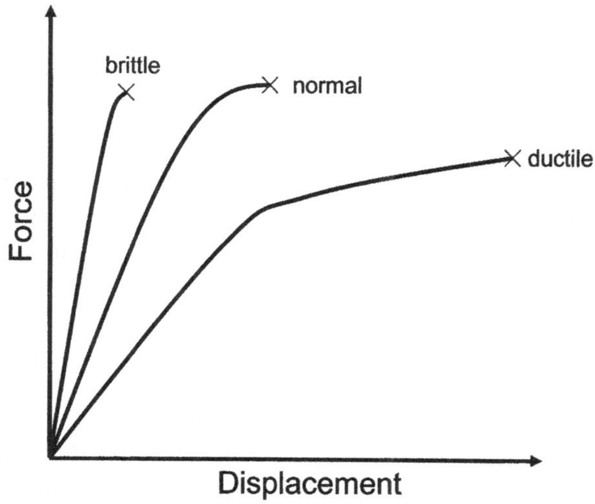

FIGURE 2. Load-displacement curves for different bone conditions. Osteopetrotic bone is brittle and thus displays reduced work to failure. On the other hand, a bone from a young child is ductile with larger ultimate displacement, resulting in increased work to failure.

displacement is inversely related to the brittleness of the bone. The biomechanical status of bone may be poorly described by just one of these properties. For instance, a bone from an osteopetrotic patient will tend be very stiff, but also very brittle, resulting in reduced work to failure and increased risk of fracture (FIG. 2). On the other hand, a bone from a young child will tend to be poorly mineralized and weak, but very ductile (large ultimate displacement), resulting in increased work to failure. Because of these properties, "greenstick" fractures, in which the bone undergoes large deformation but does not completely break, are sometimes observed in children.

When load is converted to stress and deformation converted to strain by engineering formulas, the relationship between stress and strain in bone follows a curve called the stress–strain curve. The slope of the stress–strain curve within the elastic region is called the elastic or Young's modulus (E). The Young's modulus is a measure of the intrinsic stiffness of the material. The area under the stress–strain curve is a measure of the amount of energy needed to cause material failure. This property of a material is called energy absorption or modulus of toughness or just toughness. The maximum stress and strain the bone can sustain are called the ultimate strength and ultimate strain, respectively. It should be noted that strength, as it is defined by the stress–strain curve, is an intrinsic property of bone. That is, these strength values are independent of the size and shape of the bone. The force required to break the bone is different from the intrinsic strength, because ultimate load will vary with bone size. It is important to keep this distinction in mind because intrinsic strength and ultimate load can show different trends in drug or genetic studies, especially if the drug or gene affects the size of the bone. Strength measures that are not presented in units of stress do not represent the intrinsic strength of the material but are influenced by extrinsic factors like specimen size and shape.

The elastic strain region and the plastic strain region of the stress-strain curve are separated by the yield point (FIG. 3). The yield point represents a gradual transition, above which stresses begin to cause permanent damage to the bone structure. Post-yield strains (i.e., strains beyond the yield point) represent permanent deformations of bone structure caused by slip at cement lines, trabecular microfracture, crack growth, or combinations of these. The yield point is seldom well defined when testing bone specimens. Several methods have been proposed to determine the yield point. For instance, the yield point is often defined as the point where the stress-strain curve begins to become nonlinear.[2] Other techniques include offset methods where a line parallel with the linear portion of the stress-strain curve and offset by 0.03% to 0.2% strain is constructed.[3] The point where this line intersects the stress-strain curve is arbitrarily called the yield point.

WHY BONES BREAK

Strength and stiffness are typically used to define the "health" of a bone, but they are not as clearly related to risk of fracture as is the amount of energy

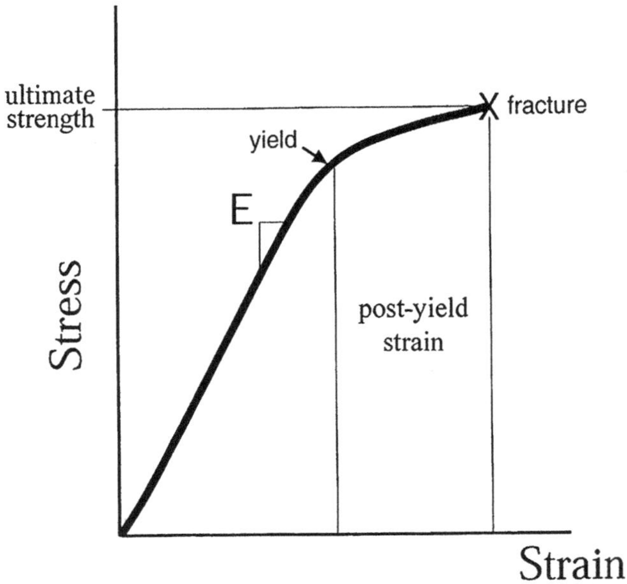

FIGURE 3. Stress-strain curve for bone. The slope of the curve is called Young's modulus (*E*). The height of the curve is the ultimate strength. The yield point represents a transition, above which strains begin to cause permanent damage to the bone structure. Post-yield strain is inversely proportional to the brittleness of the bone.

required to cause fracture. Trauma transfers energy into the bone and, if that energy exceeds what the bone can absorb, the bone will break. A bone that is highly mineralized is also stiff and brittle and will require much less energy to fracture (the area under the curve) than a bone that is more compliant. Bone mineral density (BMD) is highly correlated with strength and stiffness, but there is an inverse relationship between bone stiffness (or Young's modulus) and ultimate strain (FIG. 4). Deer antler, which is not highly mineralized, has a very high ultimate strain (ductility) but low Young's modulus. This combination of properties makes an antler exceedingly difficult to break.[4] On the other hand, the tympanic bulla of a fin whale has high mineral content, high Young's modulus, and is very brittle. The bulla is not as strong as antler or bone due to its brittleness. However, it is not designed for strength, but for acoustic properties. Like the middle ear ossicles, the bulla is highly mineralized to provide proper acoustic impedance. Long bone tissue has intermediate mechanical properties.

Cracks forming in bone tissue provide a means by which energy is released. Energy absorbed by the bone during loading builds up and is released as damage accumulates or when the bone fractures. Bone subjected to high-energy trauma, such as a gunshot wound, will form many cracks because energy accumulates quickly and must be released. High-energy trauma causes

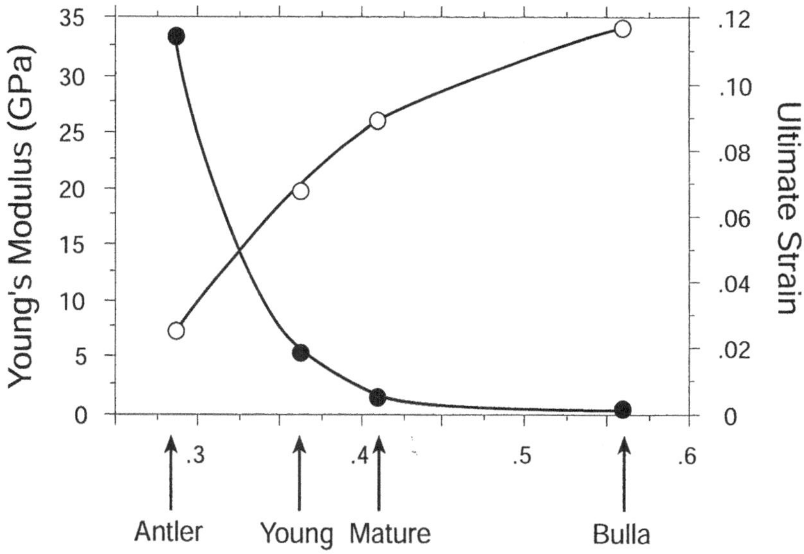

FIGURE 4. Biomechanical properties of mineralized tissues (from Currey 1990). As mineral volume fraction (x-axis) increases, Young's modulus (*open circles*) is improved but ultimate strain (*closed circles*) decreases. Deer antler is less stiff (lower Young's modulus) and less brittle (higher ultimate strain), whereas the tympanic bulla from a fin whale is very brittle and stiff. Bone tissue from young (1-year-old) cows is less stiff and brittle than tissue taken from mature (9-year-old) cows. (Reprinted from Bone Formation, F. Bronner and M.C. Farach-Carson, eds., copyright 2004, with permission from Springer-Verlag.)

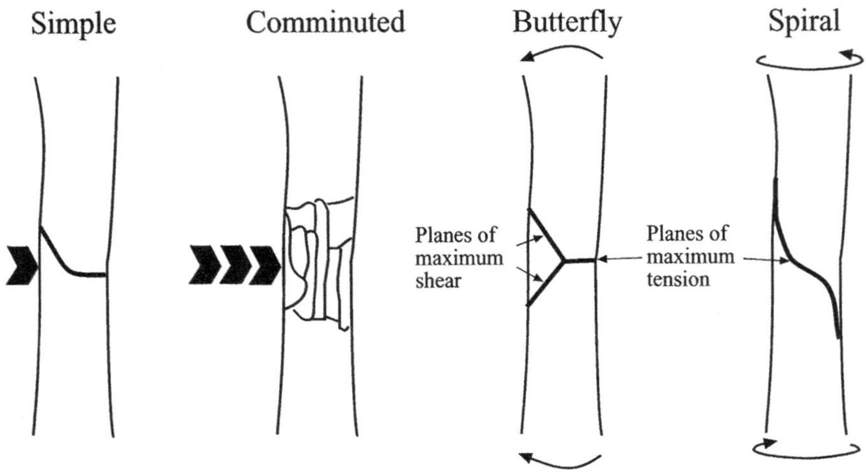

FIGURE 5. Types of bone fractures. A low-energy impact causes a simple fracture, whereas high energy causes fragmented or comminuted fracture. Bending tends to cause a butterfly fracture with fracture planes following maximum shear and tension lines. Torsion causes a spiral fracture with the fracture plane following a maximum tension plane. (Adapted from *Orthopaedic Basic Science*, 2nd Edition, American Academy of Orthopaedic Surgeons with permission.)

the bone to break into fragments. Lower energy trauma causes simple fractures, without fragmentation (FIG. 5). Bone tissue is weakest in tension or shear and strongest in compression, thus cracks tend to propagate along tension or shear planes within the bone tissue. Shear planes run at 45° angles from compressive stresses and because of this bending forces can create "butterfly" fractures. Fracture lines follow shear planes to create a fragment that resembles a butterfly wing. Similarly, a plane of maximum tension falls 45° from the shear plane, so torsional (shear) loads often cause spiral fractures that propagate along a helical plane of maximum tension.

MATRIX COMPOSITION AND BONE BIOMECHANICS

Bone tissue is a two-phase porous composite material composed primarily of collagen and mineral, with mechanical properties determined primarily by the amounts, arrangement, and molecular structure of these primary constituents. The mineral component confers strength and stiffness to the tissue. The collagen phase is tough and improves bone's work to failure or toughness.

The ratio of mineral to collagen in bone affects both bone's strength and brittleness.[5] Excessive mineral content, or a change in the stoichiometry or quality of the mineral increases brittleness and is detrimental. Collagen has a small influence on the strength and stiffness of bone, but mostly improves bone's toughness.[6] The most obvious clinical example of the mechanical effects of a collagen defect is osteogenesis imperfecta (OI). OI is a family of heritable disorders that involve mutations in the type I procollagen genes. People with OI have a markedly increased risk of fracture, and often present with multiple fractures at a young age. This is likely due to a combination of poor bone tissue properties, together with low bone volume and thin long bone cortices.

BONE GEOMETRY AND STRONGER BONES

Long bones are for the most part thick-walled tubes. This geometry allows bones to carry loads effectively but remain relatively light. To understand the principle behind this, first consider that long bones are loaded mainly in bending.[7] Deflection of a beam in bending is given by $ML^2/8EI$, where M is the bending moment, L is length, E is the Young's modulus, and I is the second moment of area. For a given value of M, beam deflection can be reduced by shortening the beam (decrease L), stiffening the beam material (increase E), or increasing I. Anabolic therapies for osteoporosis take advantage of the latter option. For a tubular bone, the second moment of area I equals $\pi/4\ (r_p{}^4 - r_e{}^4)$, where r_p is the periosteal radius and r_e is the endosteal radius. For mammalian long bones, r_p is about 1.8 times r_e making $I = 0.71 r_p{}^4$.[8] This equation demonstrates the importance of the periosteal radius for bone structural rigidity. An 8% increase in r_p will cause the bone rigidity to increase by 36%. If we incorporate the effects of aging by increasing the endosteal radius by 35% to simulate bone resorption, the cross-sectional area of the cortex is decreased by 12% yet the bone rigidity still increases by 17% due to the modest 8% increase in periosteal radius (FIG. 6). Consequently, adding bone tissue to the periosteal surface will substantially strengthen the bone. Mechanical loading applied to skeleton preferentially increases periosteal bone formation in regions where mechanical stresses are highest causing large increases in mechanical strength.

STRUCTURAL OPTIMIZATION

Efficiency of structures is enhanced if mass can be reduced as much as possible without compromising strength or rigidity. Structural engineers have adapted several strategies to achieve optimum structure. One effective approach involves the reshaping of a structure to equalize the stresses as much as possible. In theory, one should eliminate structural elements that carry no stress, since these are superfluous and can be removed to reduce mass, and augment elements that carry high levels of stress as these are at risk of failure. Mattheck[9] illustrated this principle with the analysis of a shaft with a rectangular aperture (FIG. 7). This mechanical part developed a stress concentration in the fillet when loaded in bending. They employed an algorithm that allowed

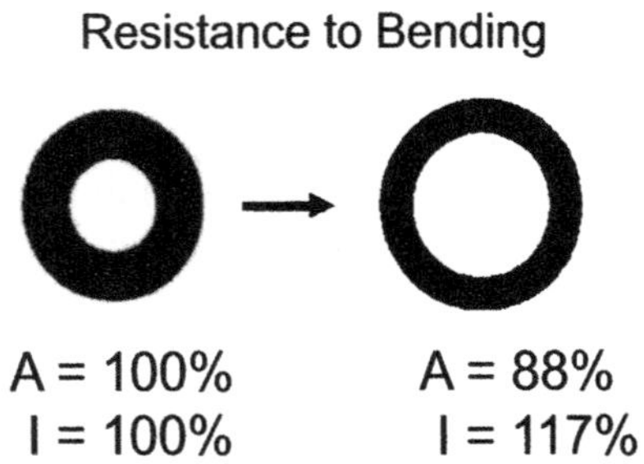

FIGURE 6. Long bones resemble hollow tubes. Under bending loads, the rigidity of a long bone is determined largely by the second moment of area (I). The effects of aging were simulated in the right hand cross-section by increasing the endosteal radius by 35% to simulate bone resorption. The cross-sectional area (A) of the cortex is decreased by 12% yet the bone rigidity still increased by 17% due to the modest 8% increase in periosteal radius.

the fillet to "grow" in areas of high stress, thus changing the shape in the fillet slightly. The stress concentration was eliminated in the optimized part and mechanical tests showed that the improved design sustained over 40-fold more loading cycles before failing.

The structural optimization algorithm favored by Mattheck dictates that bone should be formed only where stresses are highest. Bone should not be formed in regions of low stress, as this adds little structural benefit and increases the risk of inappropriate bone formation that might impinge on nerves or other adjacent tissues. Recent studies[10,11] show how long bones apply this algorithm in response to loading. Cyclic mechanical loads were applied axially along the ulna of adult rats. The rat ulna has a natural curvature in the medial-lateral direction, so axial loads induce bending of the bone. Under load, the medial periosteal surface of the bone was subject to compressive stresses (or strains) and the lateral periosteal surface was under tension. The ratio of compressive to tensile strain magnitude in the loaded adult rat ulna is about 1.5, indicating that the highest strains in the ulna occur at the periosteal medial surface in compression (FIG. 8). The pattern of bone formation induced by loading

FIGURE 7. Structural optimization applied to a shaft with a rectangular aperture. The original design for the part caused a stress concentration in the fillet (*detail A*). To reduce the high stress, material was added to the fillet. Although the amount of new material added was modest (*middle left panel*), the mises stress at the fillet was reduced by about 70% (*middle right panel*). The original design failed after only 200,000 use cycles, whereas the new, optimized design lasted for over 8,000,000 cycles. This represents over 40-fold improvement in performance of the part. (Reprinted from Mattheck C. Design in Nature. Learning from Trees. Springer, Berlin. 1997, with permission.)

Shaft with Rectangular Aperture

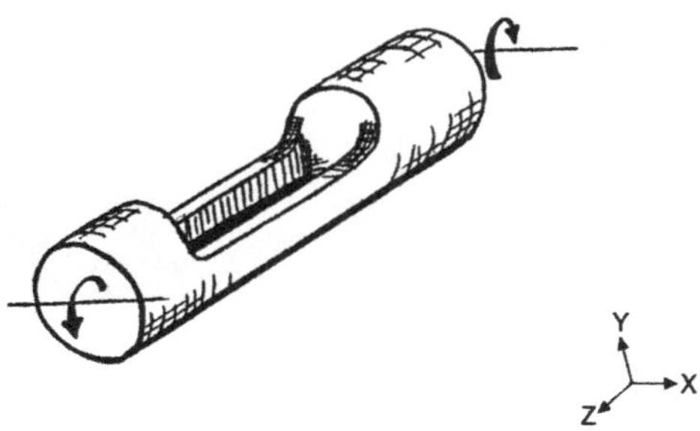

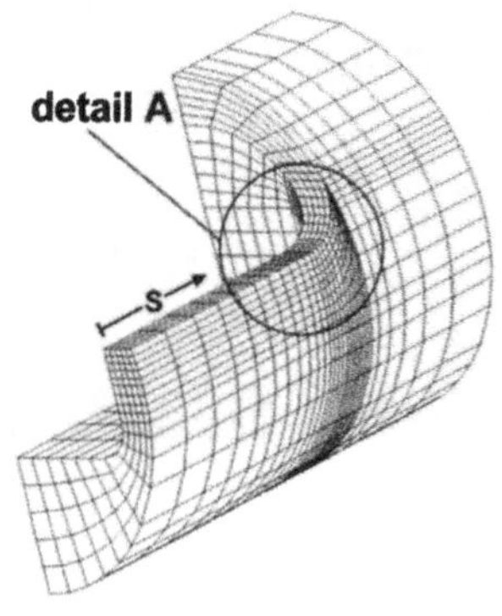

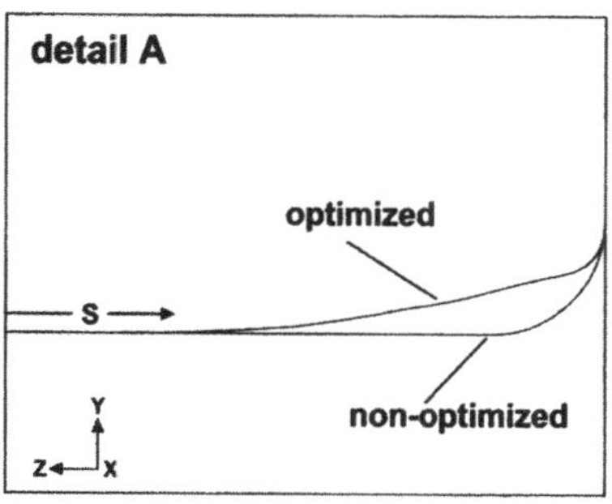

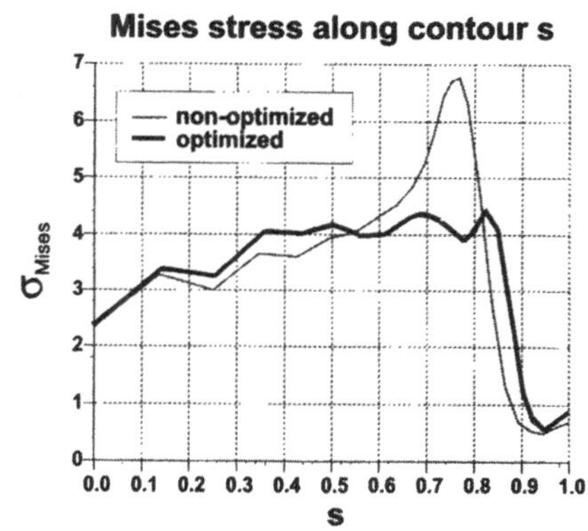

Failure after 200,000 cycles

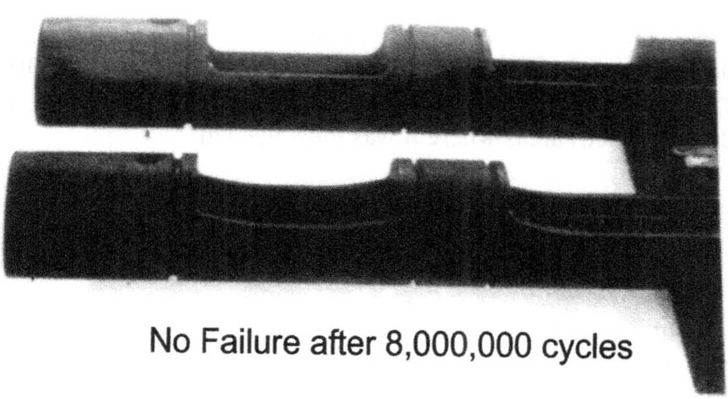

No Failure after 8,000,000 cycles

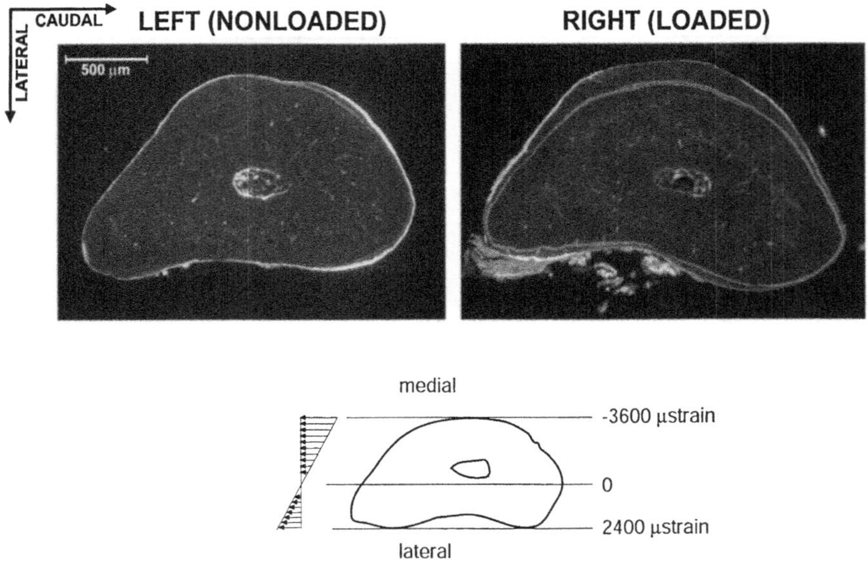

FIGURE 8. The rat ulna is strained more on the medial (*top*) surface when loaded. The bottom figure shows the strain profile across the loaded ulna. The strains are designated in units of microstrain. Positive values are tensile strain and negative values are compressive strain. Bone formation is shown in the right panel. The bright lines within the bone show labels at the beginning of loading. Little new bone was formed in the left (control) ulna, but the loaded ulna showed increased bone formation mostly in regions where the strains were of greatest magnitude. (Reprinted from Ref. 10. Used with permission from the American Society for Bone and Mineral Research.)

resembles the strain distribution, with more bone formation where the strains are highest. The improvement in bone structure is evidenced by a 64% increase in bone strength and a 100-fold increase in fatigue life, yet the improvement in areal BMD was only a modest 5–8%. Therefore, loading caused dramatic improvements in bone biomechanical properties, even with small changes in BMD. The structural efficiency of the ulna was improved by bone formation preferentially on the periosteal surface, with more bone formation in highly strained (or stressed) areas where it was most needed.

THE CHARACTERISTICS OF THE SKELETAL MECHANOTRANSDUCER

Mechanotransduction in bone involves several cell types. The cells that ultimately form or resorb bone may not be the same that transduce and respond to mechanical signals. Mechanotransduction might involve signaling through mechanically activated ion channels in the cell membrane, focal adhesions of the cytoskeleton, or a G protein-coupled mechanoreceptor. In cultured osteoblastic cells, fluid flow increases intracellular calcium within minutes and this response is suppressed by gadolinium, a blocker of the stretch-activated calcium channel.[12] In addition, the L-type voltage-operated calcium channel probably plays a role in bone cell mechanotransduction. Studies using bone explants showed that gadolinium abolished loading-related responses in osteocytes, while a blocker of L-type calcium channels inhibited loading-related responses in osteoblasts.[13] In addition, two blockers of L-type calcium channels, verapamil and nifedipine, strongly suppress mechanically induced bone formation in rats.[14,15]

An increase in intracellular calcium is observed in osteoblastic cells seconds after a mechanical stimulus. The inositol 1,4,5-triphosphate pathway plays a key role in intracellular calcium release after a mechanical stimulus[16] and intracellular calcium signaling appears to be requisite for expression of bone matrix proteins. Intracellular calcium mobilization triggers a mitogen-activated protein kinase (MAPK) signaling pathway, which is linked to the expression of osteopontin.[17]

Osteoblastic cells attach to the bone matrix through the integrin-cytoskeleton complex. Integrins are heterodimeric transmembrane proteins that bind to the extracellular matrix (ECM) on the outside of cells, and are linked to the actin cytoskeleton via the short cytoplasmic domain of the β subunit on the inside of cells at specialized sites known as focal adhesions. Several lines of evidence obtained with various cell types, including fibroblasts, epithelial cells, endothelial cells, neutrophils, as well as osteoblasts, indicate that a key molecule in mediating linkage of actin filaments to integrin cytoplasmic domains is the protein α-actinin.[18] Fluid flow or mechanical stretch applied to osteoblastic cells induces recruitment of integrins to focal adhesions and causes the actin filaments in the cell to reorganize into large bundles of actin filaments called stress fibers.[19,20] Microinjection into osteoblasts of a 53-kDa proteolytic fragment of α-actinin, which contains the integrin-binding domain, but not the actin-binding domain, causes the competitive displacement of the endogenous α-actinin from focal adhesions and blocks fluid flow-induced gene expression.[19]

AUTOCRINE AND PARACRINE MESSENGERS FOR SKELETAL MECHANOTRANSDUCTION

Within minutes after a mechanical stimulus, a number of biochemical signaling pathways are set in motion. The interacting effects of paracrine and autocrine signaling pathways on cell to cell communication or osteoblastic activity are poorly understood. For instance, release of prostaglandins is consistently observed after loading of bone explants or the application of a mechanical stimulus to osteoblasts in culture.

Prostaglandins and nitric oxide are released from bone cells within minutes after dynamic mechanical loading.[21–24] Blockade of prostaglandin synthesis using nonsteroidal anti-inflammatory drugs (NSAID) suppresses mechanically induced bone formation *in vivo*,[25–27] as does the nitric oxide synthesis inhibitor L-NAME.[28] Nitric oxide release from bone cells appears to be involved in cellular mechanotransduction, but it is not clear how nitric oxide affects intracellular signaling pathways in bone cells. In other cell types, nitric oxide binds to soluble guanylyl cyclase, thus stimulating the enzyme and increasing intracellular cyclic GMP and may cause similar effects in bone cells. Cyclic GMP has been suggested as a mediator of mechanical loading in osteoblastic cells.[29] Nitric oxide may play an important role as a mediator of the suppressive effects of mechanical loading on osteoclasts. Nitric oxide is known to be a strong inhibitor of osteoclast activity[30–32] and has been shown to decrease expression of receptor activator of NF-κB ligand (RANKL, a.k.a., osteoclast differentiation factor) and increases expression of osteoprotegerin (OPG, an inhibitor of osteoclast differentiation), which in turn leads to decreased recruitment of osteoclasts.[33] Mechanical stimuli applied to marrow stromal cells increases reduced expression of RANKL, which in turn decreases osteoclast number.[34,35] Consequently, cells of osteoblastic lineage appear to be mediators of the suppressive effects of mechanical stimuli on bone resorption, perhaps by releasing nitric oxide.

The two most active prostaglandins in bone cells are PGE_2 and PGI_2. Both are released from osteoblasts or osteocytes shortly after mechanical loading and have numerous effects on bone including the recruitment of osteoblasts from marrow stroma.[36] Exogenous PGE_2 administered in rats is strongly osteogenic and results in increased recruitment of osteoblasts and accelerated osteoblastic activity.[37] The anabolic effects of PGE_2 are mediated mainly through the EP_4 prostaglandin receptor,[38] but the EP_2 receptor may also be important in bone matrix synthesis.[39] There are two isoforms of prostaglandin synthase (cyclooxygenase), constitutive (COX-1) and inducible (COX-2). Selective inhibition of COX-2 using NS-398 is considerably more effective in blocking loading-induced bone formation *in vivo* than is indomethacin, which blocks both isoforms of cyclooxygenase.[26,27] Administration of NS-398 3 h before mechanical loading suppressed bone formation by 67% in the rat tibia, whereas administration of the drug 30 min after loading had no significant effect.[27] These findings demonstrate that prostaglandin synthesis is most important prior to loading, suggesting that prostaglandins must be available at the time of loading to potentiate the osteogenic response.

Adenosine triphosphate (ATP) signaling plays a role in skeletal mechanotransduction. Osteoblastic cells can communicate through autocrine or paracrine activity of secreted ATP on $P2Y_2$ purinergic receptors[40] and $P2Y_2$ signaling appears to be mechanosensitive.[41] A local mechanical stimulus initiates intercellular calcium signaling, mediated by ATP receptors, which rapidly propagates from cell to cell. In addition, the P2X family of receptors is probably an important target for mechanically derived signals. $P2X_7$ receptor knockout mice have an osteopenic phenotype that resembles the skeleton of animals subjected to chronic disuse. $P2X_7$ signaling is important for promoting osteoblastic activity and bone formation, whereas $P2X_7$ signaling suppresses osteoclastic bone resorption.[42] Mice with a null mutation of $P2X_7$ are 60–70% less responsive to mechanical loading of the skeleton, indicating that $P2X_7$ signaling is critical for proper skeletal mechanotransduction.[43]

Wingless-type (Wnt) signaling through the Lrp5 receptor is important in osteoblast mechanotransduction. Mice with nonfunctional Lrp5 receptors respond poorly to mechanical loading of the skeleton, with 88–99% reduced bone formation compared to controls.[44] In addition, osteoblasts cultured from Lrp5-deficient mice do not express osteopontin after a mechanical stimulus. However, early mechanotransduction events are normal in Lrp5-deficient osteoblasts. In response to a mechanical stimulus, these cells release ATP and PGE_2 and extracellular-regulated kinase (ERK)1/2 (a hallmark of osteoblast mechanotransduction[17]) is activated normally. Consequently, it appears that Wnt signaling through Lrp5 is necessary to couple early mechanotransduction events that occur within 15 min after a mechanical stimulus with synthesis of matrix proteins, which occurs many hours after a mechanical stimulus.

The insulin-like growth factors (IGF) play a role in anabolic mechanical loading responses although that role has not been clearly defined. IGF-1 interacts synergistically with mechanical loading in stimulating bone formation in mice[45] yet others have shown IGF-II, not IGF-I, to mediate mechanical loading responses.[46] The diminished bone formation seen with disuse is due in part to resistance to the effects of IGF-I on bone formation[47] and IGF-I expression is increased in osteocytes and bone-lining cells within 6 h of mechanical loading.[48] Therefore, IGF-I appears to be a necessary mediator of mechanically induced bone formation yet the molecular mechanisms that link IGF-I to mechanical loading are poorly understood.

INTERACTIONS BETWEEN HORMONES AND SKELETAL MECHANOTRANSDUCTION

There are examples of hormones that might amplify the effects of mechanical loading; these include parathyroid hormone (PTH) or the 1-34 PTH fragment and estrogen. PTH(1-34) acts synergistically with mechanical loading to enhance periosteal bone formation[49] and PTH(1-34) has been shown to enhance the anabolic effect of mechanical loading on endocortical and trabecular surfaces in rats.[50] In addition, the anabolic effect of mechanical loading is abolished if the parathyroid gland is removed.[50] In cultured osteoblasts, PTH(1-34) sensitizes cells to mechanical forces possibly by enhancing the mobilization of intracellular Ca^{2+}.[15,51] PTH(1-34) differs from mechanical loading in that the former stimulates bone resorption whereas the latter suppresses it. This discrepancy was addressed by Bakker *et al.*,[52] who applied mechanical stress to primary osteoblasts in culture and observed that both nitric oxide and PGE_2

production were elevated twofold, whereas PTH(1-34) increased PGE_2 production with no effect on nitric oxide release. Thus there are opposite effects of PTH(1-34) and mechanical loading on nitric oxide production, which may explain the different actions these two stimuli have on bone resorption.

Estrogen is another hormone that may interact with mechanical loading pathways, but the nature of that interaction remains unclear. Mechanical loading increases estrogen receptor alpha phosphorylation in osteoblasts through activation of ERK 1.[53] In addition, mice deficient in estrogen receptor alpha

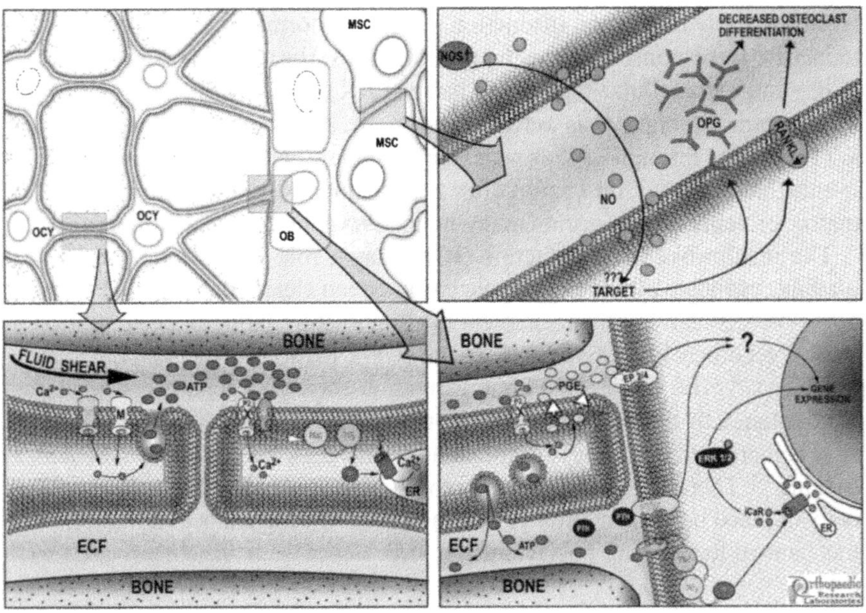

FIGURE 9. Model for mechanotransduction in bone. Fluid shear on osteocytes (OCY) induces an influx of extracellular Ca^{2+} via voltage-sensitive (V) and perhaps mechanosensitive (M) channels. Shear stress also enhances ATP release, which binds to the purinergic receptors P2X (ionotropic) and P2Y (metabotropic). Signaling through P2Y is required for Ca^{2+} release from intracellular stores via a G_q – PLC – PIP_2 – IP_3 pathway. ERK 1/2 is activated by intracellular Ca^{2+} and activated gene expression. ATP release causes PGE_2 release through signaling downstream of the $P2X_7$ receptor. PGE_2 binds and signals through one of the EP receptors, probably EP2 and/or EP4, and ultimately results in enhanced bone formation. PTH signaling also appears to be required for mechanotransduction to occur, but the intracellular pathways involved are not well understood (question mark). Wnt signaling through the Lrp5 receptor also appears to be important in mechanically induced bone formation. Pressure in the marrow cavity and/or fluid shear forces on marrow stromal cells (MSC) may stimulate nitric oxide synthase (NOS) activity and nitric oxide (NO) release. NO is a strong inhibitor of bone resorption and probably acts by inhibiting RANKL expression, while increasing osteoprotegerin (OPG) production (RANKL enhances osteoclast differentiation, while OPG suppresses this process). OCY = osteocyte; OB = osteoblast; MSC = marrow stromal cell. (Copyright by Alexander G. Robling 2004.)

expression show suppressed osteogenic responsiveness to mechanical loading.[54] In contrast, others have shown that estrogen suppresses the anabolic effect of mechanical loading.[55] These observations might be reconciled by considering that the effects of estrogen on the skeleton are site specific. Estrogen suppresses bone resorption on trabecular and endocortical bone surfaces, thus preserving bone mass. Conversely, estrogen suppresses bone formation on the periosteal surfaces.[56] Consequently, the interaction between estrogen and mechanical loading may depend upon which bone surfaces are involved.

CLOSING REMARKS

Bone strength is affected primarily by matrix composition and geometry. Increased mineralization of bone has the combined effects of stiffening the tissue while making it more brittle. This has implications for osteoporosis therapies that change bone turnover. A therapy that reduces bone turnover will increase mineralization and stiffness but at the same time bone becomes more brittle, whereas a therapy that increases bone turnover will have opposite effects. The most effective way to strengthen bone is by adding new bone tissue where bone stresses are greatest. This occurs when bone adapts to mechanical loading. Bone cells act as mechanotransducers that detect high stresses and signal locally for an anabolic response. There are a number of pathways involved in bone cell mechanotransduction (FIG. 9). Targeting these pathways using pharmacotherapy could effectively strengthen bones.

REFERENCES

1. MARTIN, R.B., D.B. BURR & N.A. SHARKEY. 1998. Skeletal Tissue Mechanics. Springer-Verlag. New York.
2. HVID, I. & J. JENSEN. 1984. Cancellous bone strength at the proximal human tibia. Eng. Med. **13:** 21–25.
3. TURNER, C.H. 1989. Yield behavior of cancellous bone. J. Biomech. Eng. **111:** 1–5.
4. CURREY, J.D. 1990. Physical characteristics affecting the tensile failure properties of compact bone. J. Biomech. **23:** 837–844.
5. CURREY, J. 2002. Bones: structure and Mechanics. Princeton University Press. Princeton, NJ.
6. WANG, X. *et al.* 2002. Age-related changes in the collagen network and the toughness of bone. Bone **31:** 1–7.
7. RUBIN, C.T. & L.E. LANYON. 1982. Limb mechanics as a function of speed and gait: a study of functional strains in the radius and tibia of horse and dog. J. Exp. Biol. **101:** 187–211.
8. CURREY, J. 1984. The Mechanical Adaptations of Bones. Princeton University Press. Princeton, NJ.
9. MATTHECK, C. 1997. Design in Nature. Learning from Trees. Springer. Berlin.

10. ROBLING, A.G. *et al.* 2002. Improved bone structure and strength after long-term mechanical loading is greatest if loading is separated into short bouts. J. Bone Miner. Res. **17:** 1545–1554.
11. WARDEN, S.J. *et al.* 2005. Bone adaptation to a mechanical loading program significantly increases skeletal fatigue resistance. J. Bone Miner. Res. **20:** 809–816.
12. HUNG, C.T. *et al.* 1996. Intracellular Ca2+ stores and extracellular Ca2+ are required in the real-time Ca2+ response of bone cells experiencing fluid flow. J. Biomech. **29:** 1411–1417.
13. RAWLINSON, S.C., A.A. PITSILLIDES & L.E. LANYON. 1996. Involvement of different ion channels in osteoblasts' and osteocytes' early responses to mechanical strain. Bone **19:** 609–614.
14. LI, J. *et al.* 2002. L-type calcium channels mediate mechanically induced bone formation in vivo. J. Bone Miner. Res. **17:** 1795–1800.
15. LI, J. *et al.* 2003. Parathyroid hormone enhances mechanically induced bone formation, possibly involving L-type voltage-sensitive calcium channels. Endocrinology **144:** 1226–1233.
16. CHEN, N.X. *et al.* 2000. Ca(2+) regulates fluid shear-induced cytoskeletal reorganization and gene expression in osteoblasts. Am. J. Physiol. Cell Physiol. **278:** C989–C997.
17. YOU, J. *et al.* 2001. Osteopontin gene regulation by oscillary fluid flow via intracellular calcium mobilization and activation of mitogen-activated protein kinase in MC3T3-E1 osteoblasts. J. Biol. Chem. **276:** 13365–13371.
18. PAVALKO, F.M. & K. BURRIDGE K. 1991. Disruption of the actin cytoskeleton after microinjection of proteolytic fragments of alpha-actinin. J. Cell Biol. **114:** 481–491.
19. PAVALKO, F.M. *et al.* 1998. Fluid shear-induced mechanical signaling in MC3T3-E1 osteoblasts requires cytoskeleton-integrin interactions. Am. J. Physiol. Cell Physiol. **275:** C1591–C1601.
20. MEAZZINI, M.C. *et al.* 1998. Osteoblast cytoskeletal modulation in response to mechanical strain in vivo. J. Orthop. Res. **16:** 170–180.
21. REICH, K.M., C.V. GAY & J.A. FRANGOS. 1990. Fluid shear stress as a mediator of osteoblast cyclic adenosine monophosphate production. J. Cell Physiol. **143:** 100–104.
22. KLEIN-NULEND, J. *et al.* 1997. Pulsating fluid flow stimulates prostaglandin release and inducible prostaglandin G/H synthase mRNA expression in primary mouse bone cells. J. Bone Miner. Res. **12:** 45–51.
23. JOHNSON, D.L., T.N. MCALLISTER & J.A. FRANGOS. 1996. Fluid flow stimulates rapid and continuous release of nitric oxide in osteoblasts. Am. J. Physiol. Endocrinol. Metab. **271:** E205–E208.
24. PITSILLIDES, A.A. *et al.* 1995. Mechanical strain-induced NO production by bone cells: a possible role in adaptive bone (re)modeling? FASEB J. **9:** 1614–1622.
25. CHOW, J.W. *et al.* 1999. Role of nitric oxide and prostaglandins in mechanically induced bone formation. J. Bone Miner. Res. **13:** 1039–1044.
26. FORWOOD, M.R. 1996. Inducible cyclo-oxygenase (COX-2) mediates the induction of bone formation by mechanical loading in vivo. J. Bone Miner. Res. **11:** 1688–1693.
27. LI, J., D.B. BURR & C.H. TURNER. 2002. Suppression of prostaglandin synthesis with NS-398 has different effects on endocortical and periosteal bone formation induced by mechanical loading. Calcif. Tissue Int. **70:** 320–329.

28. TURNER, C.H. *et al.* 1996. Nitric oxide inhibitor L-NAME suppresses mechanically induced bone formation in rats. Am. J. Physiol. Endocrinol. Metab. **270:** E634–E639.
29. RODAN, G.A. *et al.* 1975. Cyclic AMP and cyclic GMP: mediators of the mechanical effects on bone remodeling. Science **189:** 467–469.
30. MACINTYRE, I. *et al.* 1991. Osteoclastic inhibition: an action of nitric oxide not mediated by cyclic GMP. Proc. Natl. Acad. Sci. USA **88:** 2936–2940.
31. LOWIK, C.W. *et al.* 1994. Inducible production of nitric oxide in osteoblast-like cells and in fetal mouse bone explants is associated with suppression of osteoclastic bone resorption. J. Clin. Invest. **93:** 1465–1472.
32. KASTEN, T.P. *et al.* 1994. Potentiation of osteoclast bone-resorption activity by inhibition of nitric oxide synthase. Proc. Natl. Acad. Sci. USA **91:** 3569–3573.
33. FAN, X. *et al.* 2004. Nitric oxide regulates RANKL and OPG expression in bone marrow stromal cells. Endocrinology **145:** 751–759.
34. RUBIN, J. *et al.* 1997. Pressure regulated osteoclast formation and MCSF expression in marrow cultures. J. Cell Physiol. **170:** 81–87.
35. RUBIN, J. *et al.* 2000. Mechanical strain inhibits expression of osteoclast differentiation factor by murine stromal cells. Am. J. Physiol. Cell. Physiol. **278:** C1126–C1132.
36. KEILA, S., A. KELNER & M. WEINREB. 2001. Systemic prostaglandin E2 increases cancellous bone formation and mass in aging rats and stimulates their bone marrow osteogenic capacity in vivo and in vitro. J. Endocrin. **168:** 131–139.
37. LI, X.J. *et al.* 1990. Transient effects of subcutaneously administered prostaglandin E2 on cancellous and cortical bone in young adult dogs. Bone **11:** 353–364.
38. MACHWATE, M. *et al.* 2001. Prostaglandin receptor EP4 mediates the bone anabolic effects of PGE2. Mol. Pharmacol. **60:** 36–41.
39. PARALKAR, V.M. *et al.* 2003. An EP2 receptor-selective prostaglandin E2 agonist induces bone healing. Proc. Natl. Acad. Sci. USA **100:** 6736–6740.
40. JØRGENSEN, N.R. *et al.* 1997. ATP- and gap junction-dependent intercellular calcium signaling in osteoblastic cells. J. Cell Biol. **139:** 497–506.
41. YOU, J. *et al.* 2002. P2Y purinoceptors are responsible for oscillatory fluid flow-induced intracellular calcium mobilization in osteoblastic cells. J. Biol. Chem. **277:** 48724–48729.
42. KE, H.Z. *et al.* 2003. Deletion of the P2x7 nucleotide receptor reveals its regulatory roles in bone formation and resorption. Mol. Endocrinol. **17:** 1356–1367.
43. LI, J. *et al.* 2005. The P2x7 nucleotide receptor mediates skeletal mechanotransduction. J. Biol. Chem. **280**: 42952–42959.
44. SAWAKAMI, K. *et al.* 2004. Site-specific osteopenia and decreased mechanoreactivity in Lrp5-mutant mice. J. Bone Miner. Res. **19**(Suppl 1): S38.
45. GROSS, T.S. *et al.* 2002. Noninvasive loading of the murine tibia: an in vivo model for the study of mechanotransduction. J. Bone Miner. Res. **17:** 493–501.
46. CHENG, M. *et al.* 1999. Mechanical strain stimulates ROS cell proliferation through IGF-II and estrogen through IGF-I. J. Bone Miner. Res. **14:** 1742–1750.
47. SAKATA, T. *et al.* 2003. Skeletal unloading induces resistance to insulin-like growth factor I on bone formation. Bone **32:** 669–680.
48. LEAN, J.M. *et al.* 1995. Increased insulin-like growth factor I mRNA expression in rat osteocytes in response to mechanical stimulation. Am. J. Physiol. Endocrinol. Metab. **268:** E318–E327.

49. MA, Y. *et al.* 1999. Parathyroid hormone and mechanical usage have a synergistic effect in rat tibial diaphyseal bone. J. Bone Miner. Res. **14:** 439–448.
50. CHOW, J.W. *et al.* 1998. Role for parathyroid hormone in mechanical responsiveness of rat bone. Am. J. Physiol. Endocrinol. Metab. **274:** E146–E154.
51. MIYAUCHI, A. *et al.* 2000. Parathyroid hormone-activated volume-sensitive calcium influx pathways in mechanically loaded osteocytes. J. Biol. Chem. **275:** 3335–3342.
52. BAKKER, A.D. *et al.* 2003. Interactive effects of PTH and mechanical stress on nitric oxide and PGE2 production by primary mouse osteoblastic cells. Am. J. Physiol. Endocrinol. Metab. **285:** E608–E613.
53. JESSOP, H.L. *et al.* 2001. Mechanical strain and estrogen activate estrogen receptor alpha in bone cells. J. Bone Miner. Res. **16:** 1045–1055.
54. LEE, K. *et al.* 2003. Bone adaptation requires oestrogen receptor-alpha. Nature **424:** 389.
55. JARVINEN, T.L. *et al.* 2003. Estrogen deposits extra mineral into bones of female rats in puberty, but simultaneously seems to suppress the responsiveness of female skeleton to mechanical loading. Bone **32:** 642–651.
56. KIM, B.T. *et al.* 2003. The structural and hormonal basis of sex differences in peak appendicular bone strength in rats. J. Bone Miner. Res. **18:** 150–155.

Future Anti-Catabolic Therapeutic Targets in Bone Disease

BRENDAN F. BOYCE,[a] LIANPING XING,[a] ZHENQIANG YAO,[a]
WILLIAM C. SHAKESPEARE,[b] YIHAN WANG,[b]
CHESTER A. METCALF III,[b] RAJI SUNDARAMOORTHI,[b]
DAVID C. DALGARNO,[b] JOHN D. IULIUCCI,[b]
AND TOMI K. SAWYER[b]

[a]*University of Rochester Medical Center, Rochester, New York 14642, USA*

[b]*ARIAD Pharmaceuticals, Cambridge, Massachusetts, USA*

ABSTRACT: Understanding of the regulation of bone catabolism has advanced significantly over the past two decades with the identification of key enzymes that regulate osteoclast formation, activation, and survival following their knockout in mice or recognition of mutations in humans. This led to the discovery of specific inhibitors of some of these key enzymes as proof-of-concept lead compounds or potential clinical candidates for the prevention of osteoporosis and other diseases associated with increased bone resorption. Bisphosphonates have been the major therapeutic agents prescribed for the prevention of bone loss in a variety of pathologic conditions for over 30 years. More potent amino bisphosphonates have increased efficacy than earlier drugs, but side effects such as upper gastrointestinal symptoms and the requirement to take them at least 2 h before food have limited patient compliance. This, coupled with the growing knowledge of the pathways regulating osteoclast function, has driven efforts to identify small molecular lead compounds that could be developed into new therapeutic agents with efficacy that matches or supersedes that of bisphosphonates for the prevention of bone loss. In this article, we review briefly the effects of specific inhibitors of bone resorption that have been developed to date and highlight in a variety of models of increased bone resorption the effects of Src kinase inhibitors that have been targeted to bone to limit potential unwanted side effects on other cells.

KEYWORDS: bone resorption; osteoclasts; cathepsin K; vacuolar ATPase; mTOR; Src

Address for correspondence: Brendan F. Boyce, M.D., Department of Pathology and Laboratory Medicine, University of Rochester Medical Center, 601 Elmwood Avenue, Box 626, Rochester, NY, 14642. Voice: 585-275-5837; fax: 585-273-3637.
e-mail: brendan_boyce@urmc.rochester.edu

**Ann. N.Y. Acad. Sci. 1068: 447–457 (2006). © 2006 New York Academy of Sciences.
doi: 10.1196/annals.1346.042**

POTENTIAL ANTI-CATABOLIC TARGETS IN OSTEOCLASTS

Understanding of the regulation of osteoclast formation, activation, and survival has increased dramatically in the past 15 years since the identification initially of a role for Src tyrosine kinase in osteoclast activation[1,2] and later for RANKL/RANK signaling through NF-κB in all these processes[3–6] and of the enzymes that mediate the secretion of protons[7] and chloride ions[8] for the dissolution of the mineral and degradation of the matrix of bone.[9,10] Furthermore, additional pathways activated by RANKL/RANK signaling in osteoclasts have been identified along with the enzymes that mediate this signaling (reviewed by Boyle *et al.*[11]). These studies led to the development or identification from screens of compound libraries of a number of molecules that specifically inhibit these enzymes, including cathepsin K, vacuolar ATPase (V-ATPase), and rapamycin, whose effects on bone resorption will be described below.

In addition to these attempts to develop specific anti-catabolic agents, many investigators have attempted to determine the active ingredients in a variety of natural product-related herbal remedies or foods that have been reported to have osteoprotective effects. These include tanshinone, the main active diterpene quinone of a herbal medicine used to treat coronary heart disease and also present in Chinese green tea,[12] the water extract of *Dioscorea spongiosa*,[13] and the soybean isoflavones, genistein and daidzein.[14,15] However, the osteoprotective effects of the phytoestrogens remains controversial[15] and double-blind controlled clinical trials will be required to prove the efficacy of all herbal compounds in humans.

CATHEPSIN K INHIBITORS

Cathepsin K is a cysteine protease, which is selectively and highly expressed by osteoclasts and degrades several key matrix components of bone, including type I collagen.[9,16] Inactivating mutations in the gene encoding cathepsin K are responsible for the rare human skeletal disorder, pycnodysostosis, which is characterized by osteopetrosis, short stature, and craniofacial abnormalities.[17,18] Cathepsin K knockout mice develop osteopetrosis because, while their osteoclasts can demineralize bone, they cannot degrade the matrix effectively.[9,10]

Following the identification of cathepsin K as the key matrix-degrading enzyme in osteoclasts, several small molecule inhibitors of cathepsin K that inhibit bone resorption *in vitro* have been developed.[19–22] However, *in vivo* efficacy was difficult to demonstrate for such compounds in ovariectomized rats because the human and rat cathepsins differ in key residues in their active site resulting in differences in their potency between the human and rat enzymes, the latter being lower.[19,23] SB 331750, a potent rat and murine cathepsin K inhibitor, partially prevented parathyroid hormone-induced hypercalcemia in an acute rat model of bone resorption and when administered for 4 weeks intraperitoneally (10, or 30 mg/kg/day), it ameliorated bone loss in the ovariectomized rat.[21] This azepanone-based cathepsin K inhibitor was modified to make it orally active in rats and monkeys[22] and it displayed some efficacy in ovariectomized monkeys, reducing urinary NT*x* levels by 32% over the first 24 h, and serum CT*x* levels by 32% after 1.5 h relative to vehicle controls. Further studies are ongoing to improve the efficacy of such cathepsin inhibitors in order to develop potential clinical candidates.

V-ATPASE INHIBITORS

V-ATPases are ATP-dependent proton pumps, which like chloride ion pumps are highly expressed in the ruffled border membranes of osteoclasts and function to acidify the fluid milieu in resorption lacunae.[7,24] V-ATPase is one of three major classes of ATP-driven cation pumps (V-, P-, and F-ATPase), and V-ATPase antisense oligonucleotides inhibit bone resorption *in vitro*.[25] Although it is not clear if there is an osteoclast-specific isoform,[26,27] V-ATPase a3 subunit knockout mice[7] and humans with a defect in the TCIRG1 subunit of V-ATPase[28] have severe osteopetrosis because they lack osteoclast-mediated extracellular acidification, but they can acidify lysosomes and transport protons in renal microsomes.[7] However, V-ATPase is also present in the plasma membranes of renal intercalated cells and macrophages.[29] Thus, any V-ATPase inhibitor should have specificity for osteoclasts. For example, Bafilomycin A1, a potent and specific inhibitor of V-ATPase, equally inhibits all three V-ATPases, but it has undesirable side effects when administered to animals.[30] Thus, attempts were made to discover compounds that inhibited V-ATPases without affecting either P- or F-ATPase.

Oral administration of SB 242784 (10 mg/kg/day) inhibited retinoid-induced hypercalcemia in thyroparathyroidectomized rats and prevented OVX-induced bone loss in the distal femoral metaphysis and proximal tibia of 8-month-old rats when administered over a 6-month period.[31] However, the effects of a lower dose (5 mg/kg/day) were not significantly different from those of the controls, suggesting that although SB 242784 has efficacy *in vivo*, it may have a narrow therapeutic range. A more recent report described the effects of a novel specific V-ATPase inhibitor, FR167356, preventing extrusion of acid from acid-loaded osteoclasts *in vitro*.[32] It also inhibited PTH-induced resorption of murine calvariae *in vitro* and resorption pit formation without altering osteoclast numbers. In these latter experiments, inhibition occurred at 1 μM concentration of the compound, which the authors indicate was accompanied by no change in osteoclast numbers. However, there was no dose response, and they did not exclude a toxic effect of FR167356 on the cells. Furthermore, FR167356 inhibited macrophage V-ATPase and kidney V-ATPase with a similar potency for the inhibition of osteoclast V-ATPase. Thus, the search for an effective, osteoclast-specific V-ATPase inhibitor remains ongoing.

MAMMALIAN TARGET OF RAPAMYCIN INHIBITORS

Mammalian target of rapamycin (mTOR) is a serine/threonine protein kinase member of the phosphoinositol kinase-related kinase (PIKK) family involved in the regulation of cell growth through the initiation of gene translation in response to nutrients, such as amino acids, growth factors, insulin, and mitogens.[33,34] mTOR initiates translation by activating the ribosomal p70S6k protein kinase (S6K1) and by inhibiting the eIF4E inhibitor, 4E-BP1. It is thought to be involved in several other cellular functions, including actin organization, membrane trafficking, secretion, protein degradation, protein kinase C signaling, ribosome biogenesis, and tRNA synthesis,[34] and thus could have important regulatory roles in osteoclasts. Rapamycin, a natural product macrolide, binds to the immunophilin FK-506 binding protein, FKBP12 peptide prolyl cis/trans isomerase, and interaction of this FKBP12-rapamycin complex with mTOR inhibits its function.[35] Rapamycin inhibits proliferation of synovial fibroblasts from rheumatoid arthritis patients[36] and induces osteoclast apoptosis *in vitro*.[37] Thus, it could be effective in rheumatoid arthritis as an anti-inflammatory and anti-catabolic agent, but its effects on osteoclasts are controversial. A recent report indicates that everolimus, an orally bioavailable derivative of the natural product macrolide rapamycin, inhibits mouse and human osteoclasts *in vitro*,[38] similar to a previous report in mouse primary cells,[37] and that it reduces ovariectomized (OVX)-induced loss of cancellous bone by 60%, mainly by decreasing osteoclast-mediated bone resorption. However, rapamycin has been reported to stimulate osteoclast differentiation in the mouse macrophage cell line, RAW 264.7.[39] This discrepant finding may reflect differences between osteoclast formation from primary osteoclast precursors versus formation from a macrophage/monocyte cell line, and further studies will be required to confirm that rapamycin is an effective anti-catabolic agent *in vivo*.

SRC TYROSINE KINASE INHIBITORS

A role for Src, a protein tyrosine kinase, in osteoclasts was identified unexpectedly when *Src-/-* mice were generated.[1] Despite the ubiquitous expression of Src, the only cells that appear to have a required function in the knockout mice were osteoclasts. *Src-/-* mice formed increased numbers of osteoclasts, but the cells failed to resorb bone because they did not form ruffled borders,[2] which severely limited their ability to resorb bone matrix. Thus the mice were osteopetrotic and their teeth did not erupt. Src is activated in osteoclasts following integrin binding when the cells attach to bone matrix to initiate bone resorption[40] and it mediates the complex intracellular cytoskeletal reorganization that is required for polarization of the cytoplasm, including ruffled border formation. It is activated also in response to RANKL/RANK interaction in osteoclasts following the recruitment of TNF receptor activated factor 6 (TRAF6) to the intracellular domain of RANK to mediate downstream signaling.[41] Src binds to TRAF6 and recruits a number of signaling proteins, including Cbl,[42] Pyk-2[43], and cortactin,[44] which mediate polarization of the cell, actin ring, and ruffled border formation in an as yet incompletely understood process.

In addition to these osteoclast-activating actions, Src also appears to mediate osteoclast survival in response to RANKL *in vitro*. In studies in which RANKL and macrophage colony-stimulating factor (M-CSF) were withdrawn from *Src-/-* or wt osteoclasts the wt cells survived in the presence of RANKL alone, while *Src-/-* osteoclasts died by apoptosis.[41,45] Src appears to mediate this protective effect by recruiting Akt to the intracellular domain of TRAF6 and activating the PI3-kinase/mTOR pathway, which promotes osteoclast survival by phosphorylating the apoptosis inducer, caspase 3, and thus inactivating it.[46] However, recent studies have questioned this role for Akt.[47] Therefore, Src expression in osteoclasts could influence bone mass by regulating not only osteoclast activation, but also osteoclast life span. In contrast, the *Src-/-* cells survived, albeit in slightly reduced numbers, when they were treated with M-CSF alone, indicating that Src does not mediate M-CSF survival signaling. Interestingly, osteoclast apoptosis is not increased in *Src-/-* mice *in vivo*, and indeed osteoclast numbers are actually higher in *Src-/-* mice than in wt controls.[45] Thus, these *in vitro* findings reflect a phenomenon that does not apply *in vivo* or M-CSF or perhaps other Src family members substitute for Src *in vivo* in the Src-/- mice to enhance their survival. Further studies will be required to determine which of these is correct.

There are three major structural domains in Src that could mediate its functions in osteoclasts: the kinase domain, which phosphorylates tyrosine residues in molecules that bind to the SH2 domain, and the SH3 domain, which binds proline-rich portions of interacting molecules. Studies designed to determine which domain(s) of the molecule mediates its resorptive function suggested surprisingly that Src kinase activity is not essential for bone resorption *in vivo*.[46] In these studies, the TRAP promotor was used to drive expression of wild-type and various mutated *src* genes in *Src-/-* mice to generate *Src-/-* transgenic mice. As expected, expression of wild-type Src in the *src-/-* transgenic mice rescued the defect in ruffled border formation and resorption in the mice, which had normal bone volumes and tooth eruption. However, a surprising observation was that in 25% of the *Src-/-* mice expressing the kinase-inactive mutant, K295M, in which the ATP-binding site of Src is mutated to prevent phosphorylation of tyrosine residues on interacting molecules, bone volumes were normal, and the mice also had normal tooth eruption and ruffled border formation. Equally surprising was the observation that the osteopetrosis became more pronounced in 25% of the mice, which had increased osteoclast apoptosis, while no phenotypic change was observed in the remaining 50% of the mice.[46]

We found it difficult to explain these divergent findings definitively, but on the basis of our previous experience that expression levels in osteoclasts of SV-40 large T antigen driven by this same TRAP promotor in transgenic mice correlated with the number of apoptotic osteoclasts and the severity of osteopetrosis,[48] we speculated that K295M expression levels varied in the *src-/-* transgenic mice. We further speculated that at high levels of expression Src K295M acted in a dominant-negative fashion to inhibit recruitment of interacting molecules, while optimal expression levels rescued the defect, and that there was low expression in the majority of mice, which thus had no change in the intensity of their osteopetrosis. Unfortunately, we were unable to assess the levels of K295M protein expression in these mice with the antibodies available at the time. Nevertheless, our interpretation was supported by our observation that the percentage of apoptotic osteoclasts correlated with the severity of the osteopetrosis in the Src K295M transgenic mice. Furthermore, expression of the SH2 and SH3 domains of the molecule without the kinase domain also induced more pronounced osteopetrosis in *src-/-* transgenic mice as well as in wild-type mice due to increased osteoclast apoptosis.[45] Osteoclasts from these mice had reduced phospho-Akt and PI3 kinase, consistent with a failure of Src activation of the survival pathway.

Our conclusion that Src kinase activity is not essential for osteoclastic resorption *in vivo* has been challenged recently.[49] These authors overexpressed the Src K295M mutant and other Src mutant molecules in osteoclasts *in vitro* using adenoviral infection and concluded that kinase activity was essential for resorption, based on profoundly reduced resorptive activity by osteoclasts expressing Src K295M. This reduced resorptive activity was not rescued by concomitant expression of the constitutively active Src Y527F src mutant, which rescued the defective resorption of the kinase-deficient mutant, Src Y416M. Although these are interesting and provocative findings, we believe that, on the basis of their reported findings, there is an alternative explanation to the authors' conclusions. Specifically, the level of expression of the Src K295M mutant assessed by Western blot analysis was slightly higher in the *Src-/-* osteoclasts than that of native Src in the control *Src-/-* mice. Thus, we propose that in these experiments, the authors had achieved a level of expression of Src K295M in osteoclasts *in vitro* that was equivalent to the levels we had achieved *in vivo* in osteoclasts in our Src K295M mice, which had a slight increase in the degree of their osteopetrosis. The osteoclasts infected with the K295M retroviral construct did not have increased apoptosis and thus the level of expression of this construct is likely to have been lower than that which induces detectably increased osteoclast apoptosis *in vivo*. Further studies will be required to determine the role of the kinase domain of Src in osteoclasts for their resorptive activity, but until these are done our *in vivo* data in our opinion remain more compelling than these *in vitro* observations.

In addition to its role in osteoclasts, Src tyrosine kinase also has important functions in malignant cells, including regulation of cell division, growth factor signaling, and movement.[50] Its activity is upregulated in certain cancers. Thus, in patients with metastatic bone disease, Src inhibitors could potentially have negative effects not only on osteoclasts, but also on tumor cells and their interactions with osteoclasts. Because of these roles for Src in osteoclasts and tumor cells, several pharmaceutical companies have advanced Src inhibitors as potential therapeutic agents for use in osteoporosis, osteolytic bone metastasis, and related diseases with increased resorption. Examples of such Src inhibitors include the ATP-based lead compounds NVP-AAK980[51] and AP23451 *(vide infra)*. Several other noteworthy small molecule classes of Src kinase inhibitors have been advanced and these include the pyrazolopyrimidines PP1 and PP2,[52] the pyrrolopyrimidines CGP-76775 and CGP-76030,[51,53] the pyridopyrimidinones PD166326 and PD180970,[54,55] the quinazoline AZM475271,[56] the quinoline SKI-606,[57] and the indolinone SU6656.[58] Hereafter, we will briefly describe the *in vitro* and *in vivo* effects of a Src kinase inhibitor, AP23451, developed by ARIAD Pharmaceuticals (Cambridge, MA).

AP23451 is a purine-based Src tyrosine kinase inhibitor with a diphosphonate group attached to it to target it to bone in an attempt to avoid unwanted

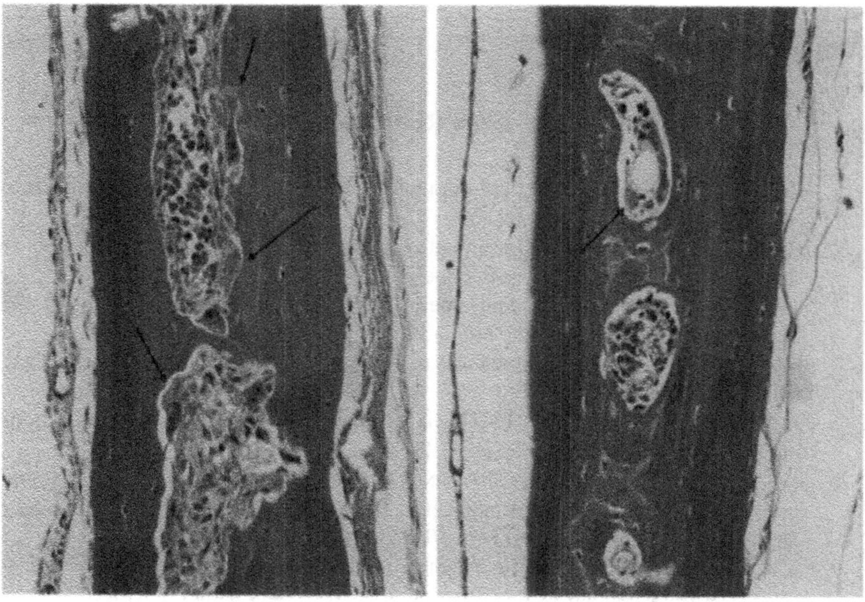

FIGURE 1. Inhibitory effect of Src inhibitor, AP23451, on PTH-induced bone resorption. PTH (20 μg/kg, given subcutaneously 4× daily for 3 days) stimulated increased osteoclast formation and activity (arrows) *in vivo* in mouse calvaria (left panel), an effect prevented by AP23451 (10 mg/kg) given subcutaneously 2× daily for 2 days before the start of the PTH injections and continued for 3 days (right panel).

possible side effects on other tissues. It exemplifies a key lead compound among a series of bone-targeted Src kinase inhibitors of varying ATP-related templates.[59–61] AP23451 inhibited osteoclast formation and induced osteoclast apoptosis *in vitro* in the 0.1- to 1-μM range. The diphosphonate group does not possess significant antiresorptive activity and the effects of AP23451 to induce osteoclast apoptosis were not prevented by the addition of geranylgeraniol, which did prevent alendronate-induced apoptosis. AP23451 dose dependently prevented parathyroid hormone (PTH)-induced bone resorption (FIG. 1) and hypercalcemia, and ovariectomy-induced bone loss.[62] When given to mice inoculated with MDA-231 breast cancer cells into the left cardiac ventricle, AP23451 prevented metastasis-induced osteolysis similar to zolendronic acid in the same experiment,[63] but it also significantly reduced the volume of tumor cells inside the bone marrow cavities of the mice. This inhibitory effect on tumor cell volume was not observed in mice treated with zolendronic acid, suggesting that the Src inhibitor may have had an inhibitory effect on the tumor cells distinct from the effect on osteoclasts. Inhibition of resorption reduces the efflux of bone-derived growth factors that participate in a vicious cycle to stimulate tumor cell growth or release of osteoclast-stimulating factors, such as parathyroid hormone related protein (PTHrP).[64] As mentioned previously, Src is expressed by many tumor cells and is thus a target for inhibitors along with osteoclasts. These findings suggest that Src inhibitors could have a dual action in metastatic bone disease to prevent the growth of metastatic lesions and their progression.

REFERENCES

1. SORIANO, P. *et al.* 1991. Targeted disruption of the c-src proto-oncogene leads to osteopetrosis in mice. Cell **64:** 693–702.
2. BOYCE, B.F. *et al.* 1992. Requirement of pp60^{c-src} expression for osteoclasts to form ruffled borders and resorb bone in mice. J. Clin. Invest. **90:** 1622–1627.
3. FRANZOSO, G. *et al.* 1997. Requirement for NF-kappaB in osteoclast and B-cell development. Genes Dev. **11:** 3482–3496.
4. IOTSOVA, V. *et al.* 1997. Osteopetrosis in mice lacking NF-κB1 and NF-κB2. Nat. Med. **3:** 1285–1289.
5. DOUGALL, W.C. *et al.* 1999. RANK is essential for osteoclast and lymph node development. Genes Dev. **13:** 2412–2424.
6. KONG, Y.Y., W.J. BOYLE & J.M. PENNINGER. 1999. Osteoprotegerin ligand: a common link between osteoclastogenesis, lymph node formation and lymphocyte development. Immunol. Cell Biol. **77:** 188–193.
7. LI, Y.P. *et al.* 1999. Atp6i-deficient mice exhibit severe osteopetrosis due to loss of osteoclast-mediated extracellular acidification. Nat. Genet. **23:** 447–451.
8. KORNAK, U. *et al.* 2001. Loss of the ClC-7 chloride channel leads to osteopetrosis in mice and man. Cell **104:** 205–215.
9. GOWEN, M. *et al.* 1999. Cathepsin K knockout mice develop osteopetrosis due to a deficit in matrix degradation but not demineralization. J. Bone Miner. Res. **14:** 1654–1663.

10. HOFBAUER, L.C. & A.E. HEUFELDER. 1999. Osteopetrosis in cathepsin K-deficient mice. Eur. J. Endocrinol. **140:** 376–377.
11. BOYLE, W.J., W.S. SIMONET, & D.L. LACEY. 2003. Osteoclast differentiation and activation. Nature **423:** 337–342.
12. CUI, L. *et al.* 2004. Tanshinone prevents cancellous bone loss induced by ovariectomy in rats. Acta Pharmacol. Sin. **25:** 678–684.
13. YIN, J. *et al.* 2004. Antiosteoporotic activity of the water extract of Dioscorea spongiosa. Biol. Pharm. Bull. **27:** 583–586.
14. PICHERIT, C. *et al.* 2000. Daidzein is more efficient than genistein in preventing ovariectomy-induced bone loss in rats. J. Nutr. **130:** 1675–1681.
15. WEAVER, C.M. & J.M. CHEONG. 2005. Soy isoflavones and bone health: the relationship is still unclear. J. Nutr. **135:** 1243–1247.
16. ATLEY, L.M. *et al.* 2000. Proteolysis of human bone collagen by cathepsin K: characterization of the cleavage sites generating by cross-linked N-telopeptide neoepitope. Bone **26:** 241–247.
17. GELB, B.D. *et al.* 1996. Pycnodysostosis, a lysosomal disease caused by cathepsin K deficiency. Science **273:** 1236–1238.
18. NISHI, Y. *et al.* 1999. Determination of bone markers in pycnodysostosis: effects of cathepsin K deficiency on bone matrix degradation. J. Bone Miner. Res. **14:** 1902–1908.
19. VOTTA, B.J. *et al.* 1997. Peptide aldehyde inhibitors of cathepsin K inhibit bone resorption both *in vitro* and *in vivo*. J. Bone Miner. Res. **12:** 1396–1406.
20. YASUMA, T. *et al.* 1998. Synthesis of peptide aldehyde derivatives as selective inhibitors of human cathepsin L and their inhibitory effect on bone resorption. J. Med. Chem. **41:** 4301–4308.
21. LARK, M.W. *et al.* 2002. A potent small molecule, nonpeptide inhibitor of cathepsin K (SB 3317.50) prevents bone matrix resorption in the ovariectomized rat. Bone **30:** 746–753.
22. GABRIELSSON, J. & D. WEINER. 1997. Pharmacokinetic and Pharmacodynamic Data Analysis: concepts and Applications. Stockholm: Apotekarsocieteten.
23. MARQUIS, R.W. *et al.* 2004. An azepanone-based inhibitor of human cathepsin K with improved oral bioavailability in the rat and the monkey. Mol. Pharm. **1:** 97–100.
24. BLAIR, H.C. *et al.* 1989. Osteoclastic bone resorption by a polarized vacuolar proton pump. Science **245:** 855–857.
25. LAITALA, T. & H.K. VAANANEN. 1994. Inhibition of bone resorption in vitro by antisense RNA and DNA molecules targeted against carbonic anhydrase II or two subunits of vacuolar H(+)-ATPase. J. Clin. Invest. **93:** 2311–2318.
26. CHATTERJEE, D. *et al.* 1992. The osteoclast proton pump differs in its pharmacology and catalytic subunits from other vacuolar H(+)-ATPases. J. Exp. Biol. **172:** 193–204.
27. LI, Y.P., W. CHEN & P. STASHENKO. 1996. Molecular cloning and characterization of a putative novel human osteoclast-specific 116-kDa vacuolar proton pump subunit. Biochem. Biophys. Res. Commun. **218:** 813–821.
28. FRATTINI, A. *et al.* 2000. Defects in TCIRG1 subunit of the vacuolar proton pump are responsible for a subset of human autosomal recessive osteopetrosis. Nat. Genet. **25:** 343–346.
29. BOWMANN, E.J., A. SIEBERS & K. ALTENDORF. 1988. Bafilomycin: a class of inhibitors of membrane ATPases from microorganisms, animal cells, and plant cells. Proc. Natl. Acad. Sci. USA **85:** 7972–7976.

30. KEELING, D.J. *et al.* 1998. Tissue-selective inhibition of vacuolar acid pumps. Acta Physiol. Scand. **643:** 195–201.
31. VISENTIN, L. *et al.* 2000. A selective inhibitor of the osteoclastic V-H(+)-ATPase prevents bone loss in both thyroparathyroidectomized and ovariectomized rats. J. Clin. Invest. **106:** 309–318.
32. NIIKURA, K., M. TAKANO & M. SAWADA. 2004. A novel inhibitor of vacuolar ATPase, FR1673.56, which can discriminate between osteoclast vacuolar ATPase and lysosomal vacuolar ATPase. Br. J. Pharmacol. **142:** 558–566.
33. MARTIN, K.A. & J. BLENIS. 2002. Coordinate regulation of translation by the PI 3-kinase and mTOR pathways. Adv. Cancer Res. **86:** 1–39.
34. RICHARDSON, C.J. *et al.* 2004. PI3-kinase and TOR: PIKTORing cell growth. Sem. Cell Develop. Biol. **15:** 147–159.
35. SEHGAL, S.N. 2003. Sirolimus: its discovery, biological properties, and mechanism of action. Transplant Proc. 35: S7–S14.
36. MIGITA, K. *et al.* 1996. The effects of the immunosuppressant rapamycin on the growth of rheumatoid arthritis (RA) synovial fibroblast. Clin. Exp. Immunol. **104:** 86–91.
37. GLANTSCHNIG, H. *et al.* 2003. M-CSF, TNF alpha and RANK ligand promote osteoclast survival by signaling through mTOR/S6 kinase. Cell Death Differ. **10:** 1165–1177.
38. KNEISSEL, M. *et al.* 2004. Everolimus suppresses cancellous bone loss, bone resorption, and cathepsin K expression by osteoclasts. Bone **35:** 1144–1156.
39. SHUI, C., B.L. RIGGS & S. KHOSLA. 2002. The immunosuppressant rapamycin, alone or with transforming growth factor-beta, enhances osteoclast differentiation of RAW264.7 monocyte-macrophage cells in the presence of RANK-ligand. Calcif. Tissue Int. **71:** 437–446.
40. INSOGNA, K.L. *et al.* 1997. Colony-stimulating factor-1 induces cytoskeletal reorganization and c-src-dependent tyrosine phosphorylation of selected cellular proteins in rodent osteoclasts. J. Clin. Invest. **100:** 2476–2485.
41. WONG, B.R. *et al.* 1999. TRANCE, a TNF family member, activates Akt/PKB through a signaling complex involving TRAF6 and c-Src. Mol. Cell **4:** 1041–1049.
42. TANAKA, S. *et al.* 1996. c-Cbl is downstream of c-Src in a signaling pathway necessary for bone resorption. Nature **383:** 528–531.
43. SANJAY, A. *et al.* 2001. Cbl associates with Pyk2 and Src to regulate Src kinase activity, alpha(v)beta(3) integrin-mediated signaling, cell adhesion, and osteoclast motility. J. Cell Biol. **152:** 181–195.
44. TRUBERT, C.L. *et al.* 1997. PP60c-src expression in osteoclasts from osteopetrotic children and in giant tumor cells. Eur. J. Histochem. **41:** 169–176.
45. XING, L. *et al.* 2001. Genetic evidence for a role for Src family kinases in TNF family receptor signaling and cell survival. Genes Dev. **15:** 241–253.
46. SCHWARTZBERG, P.L. *et al.* 1997. Rescue of osteoclast function by transgenic expression of kinase-deficient Src in src-/- mutant mice. Genes Dev. **11:** 2835–2844.
47. SUGATANI, T. & K.A. HRUSKA. 2005. Akt1/Akt2 and mammalian target of rapamycin/Bim play critical roles in osteoclast differentiation and survival, respectively, whereas Akt is dispensable for cell survival in isolated osteoclast precursors. J. Biol. Chem. **280:** 3583–3589.
48. BOYCE, B.F. *et al.* 1995. Targeting simian virus 40 T antigen to the osteoclast in transgenic mice causes osteoclast tumors and transformation and apoptosis of osteoclasts. Endocrinology **136:** 5751–5759.

49. MIYAZAKI, T. *et al.* 2004. Src kinase activity is essential for osteoclast function. J. Biol Chem. **279:** 17660–17666.
50. SCHWARTZBERG, P.L. 1998. The many faces of Src: multiple functions of a prototypical tyrosine kinase. Oncogene **17:** 1463–1468.
51. MISSBACH, M., E. ATTMANN & M. SUSA. 2000. Tyrosine kinase inhibition in bone metabolism. Curr. Opin. Drug Disc. **3:** 541–548.
52. HANKE, J.H. *et al.* 1996. Discovery of a novel, potent, and Src family-selective tyrosine kinase inhibitor. J. Biol. Chem. **271:** 695–701.
53. MISSBACH, M. *et al.* 1999. A novel inhibitor of the tyrosine kinase Src suppresses phosphorylation of its major cellular substrates and reduces bone resorption in vitro and in rodent models in vivo. Bone **24:** 437–449.
54. HURON, D.R. *et al.* 2003. A novel pyridopyrimidine inhibitor of Abl kinase is a picomolar inhibitor of Bcr-Abl-driven K562 cells and is effective against STI571-resistant Bcr-Abl mutants. Clin. Cancer Res. **9:** 1267–1273.
55. KRAKER, A.J. *et al.* 2000. Biochemical and cellular effects of c-Src kinase-selective pyrido[2,3-d]pyrimidine tyrosine kinase inhibitors. Biochem. Pharmacol. **60:** 885–898.
56. PLE, P.A. *et al.* 2004. Discovery of a new class of anilinoquinazoline inhibitors with high affinity and specificity for the tyrosine kinase domain of c-Src. J. Med. Chem. **47:** 871–887.
57. BOSCHELLI, D.H. 2002. 4-Anilino-3-quinolinecarbonitriles: an emerging class of kinase inhibitors. Curr. Topics Med. Chem. **2:** 1051–1063.
58. BLAKE, R.A. *et al.* 2000. SU6656, a selective Src family kinase inhibitor used to probe growth factor signaling. Molec. Cell. Biol. **20:** 9018–9027.
59. WANG, Y. *et al.* 2003. Bone-Targeted 2,6,9-Trisubstituted Purines: novel inhibitors of Src tyrosine kinase for the treatment of bone diseases. Med. Chem. Lett. **13:** 3067–3070.
60. SUNDARAMOORTHI, R. *et al.* 2003. Bone-targeted Src kinase inhibitors: novel pyrrolo- and pyrazolopyrimidine analogs. Bioorg. Med. Chem. Lett. **13:** 3063–3066.
61. VU, C. *et al.* 2003. Bone-targeted pyrido[2,3-d]pyrimidin-7-ones: potent inhibitors of Src tyrosine kinase as novel antiresorptive agents. Bioorg. Med. Chem. Lett. **13:** 3071–3074.
62. BOYCE, B.F. *et al.* 2002. Development of a novel bone-targeted Src tyrosine kinase inhibitor AP2345.1 having potent *in vitro* and *in vivo* anti-resorptive properties. J. Bone Miner. Res. **17:** S159.
63. BOYCE, B.F. *et al.* 2003. Regulation of bone remodeling and emerging breakthrough drugs for osteoporosis and osteolytic bone metastases. Kidney Int. **85** (Suppl.): S2–S5.
64. YIN, J.J. *et al.* 1999. TGF-beta signaling blockade inhibits PTHrP secretion by breast cancer cells and bone metastases development. J. Clin. Invest. **103:** 197–206.

Mechanisms Involved in Skeletal Anabolic Therapies

T.J. MARTIN,[a] J.M.W. QUINN,[a] M.T. GILLESPIE,[a] K.W. NG,[a] M.A. KARSDAL,[c] AND N.A. SIMS,[a,b]

[a]*St. Vincent's Institute of Medical Research, Fitzroy Victoria 3065, Australia*

[b]*Department of Medicine, University of Melbourne, St. Vincent's Hospital, Fitzroy Victoria 3063, Australia*

[c]*Nordic Bioscience A/S, Herlevhovedgade 207, DK-2730 Herlev, Denmark*

ABSTRACT: Since parathyroid hormone (PTH) is the only proven anabolic therapy for bone, it becomes the benchmark by which new treatments will be evaluated. The anabolic effect of PTH is dependent upon intermittent administration, but when an elevated PTH level is maintained even for a few hours it initiates processes leading to new osteoclast formation, and the consequent resorption overrides the effects of activating genes that direct bone formation. Identification of PTH-related protein (PTHrP) production by cells early in the osteoblast lineage, and its action through the PTH1R upon more mature osteoblastic cells, together with the observation that PTHrP± mice are osteoporotic, all raise the possibility that PTHrP is a crucial paracrine regulator of bone formation. The finding that concurrent treatment with bisphosphonates impairs the anabolic response to PTH, adds to other clues that osteoclast activity is necessary to complement the direct effect that PTH has in promoting differentiation of committed osteoblast precursors. This might involve the generation of a coupling factor from osteoclasts that are transiently activated by receptor activator of nuclear factor-κB ligand (RANKL) in response to PTH.

New approaches to anabolic therapies may come from the discovery that an activating mutation in the LRP5 gene is responsible for an inherited high bone mass syndrome, and the fact that this can be recapitulated in transgenic mice, whereas inactivating mutations result in severe bone loss. This has focused attention on the Wnt/frizzled/β-catenin pathway as being important in bone formation, and proof of the concept has been obtained in experimental models.

KEYWORDS: osteoblast; osteoclast; parathyroid hormone; Wnt signaling

Address for correspondence: T. John Martin, St. Vincent's Institute of Medical Research, 9 Princes Street, Fitzroy Vic 3065, Australia. Voice: 61-3-9288-2480; fax: 61-3-9416-2676.
e-mail: jmartin@svi.edu.au

Ann. N.Y. Acad. Sci. 1068: 458–470 (2006). © 2006 New York Academy of Sciences.
doi: 10.1196/annals.1346.043

FORMATION OF BONE

In the adult human skeleton, approximately 5–10% of the existing bone is replaced every year through the process of bone remodeling. The remodeling process, which continues throughout adult life, is an integral part of the calcium homeostatic system and provides a mechanism for self-repair and adaptation to physical stress. The cellular sequence is always initiated by osteoclastic bone resorption to be followed by osteoblastic new bone formation. This sequence of events is initiated asynchronously throughout the skeleton, at sites that are geographically and chronologically separated from each other. Both bone formation and resorption occur at the same place so that there is no change in the shape of the bone.

The maintenance of a normal, healthy skeletal mass depends on information transfer taking place among osteoblasts, osteoclasts, immune cells, and constituents of the bone matrix. Remodeling thus maintains the mechanical integrity of the skeleton by replacing old bone with new bone.[1–5] The maintenance of adequate trabecular and cortical bone requires that bone formation and resorption should be balanced, such that a high or low level of resorption is usually associated with a similar change in the level of bone formation. The theory that resorption is followed by an equal amount of formation has come to be known as "coupling." However, during life, the effects of growth and aging, including changes in mechanical stress mean that this theory of equal bone replacement rarely holds true. During growth there is a positive balance, with the amount of bone replaced at individual basic multicellular units (BMUs) exceeding that lost,[3] and with aging there is a negative balance at individual BMUs,[6] with gradual attrition of bone. In common metabolic states, such as postmenopausal osteoporosis, while coupling exists and both bone formation and resorption are occurring at a higher level than normal, the amount of bone formed is not equal to that resorbed and bone density is therefore reduced.

Until the early 1980s it was understood that bone metabolism was regulated by circulating hormones. The discoveries in subsequent years revealed that, although circulating hormones are important controlling factors, the key influences are locally generated cytokines that influence bone cell function and communication in complex ways, and often are themselves regulated by the hormones. Any approach to treatment of bone disease requires an understanding of these relationships. Until recently the established means of osteoporosis prevention and treatment have been limited entirely to drugs that inhibit bone resorption. Now that the first bone-forming treatment has been developed, and others are being investigated as new prospects, it is essential that the local communication networks in bone be taken into consideration. The processes central to successful anabolic skeletal therapies are those that enhance osteoblast differentiation, most importantly doing so in ways that preserve normal bone shape and structure. Thus, the integrity of the processes involved in the normal control of bone turnover needs to be maintained in the course of any anabolic treatment. This article focuses on the cellular and molecular mechanisms that are relevant to current and future approaches to anabolic therapies.

TRANSCRIPTIONAL CONTROLS

Cbfa1 (Runx2) is an essential and early transcriptional regulator of osteoblast differentiation.[7] It also controls bone cell function by maintaining the differentiated phenotype of the osteoblast in maturity. Transgenic overexpression of a dominant negative form of runx2 postnatally in mice led to decreased production of runx2, as well as diminished expression of the genes required for the differentiated osteoblast.[8] Runx2 is central to the replenishment of osteoblasts after bone loss, a key requirement in restoring bone.

The search for transcription factors other than runx2 led to the identification of a novel zinc finger containing transcription factor, osterix (Osx), which is specifically expressed in developing bones and not in other tissues. Mice rendered null for the *osx* gene did not develop mineralized bone, but, like the *runx2-/-* mice, had an entirely cartilaginous skeleton and died at birth.[9] In *Osx-/-* mice, runx2 mRNA was expressed at the same level as in the wild type, but the mutant mice had no Osx mRNA. This suggests that Osx is an important transcription factor in osteoblast differentiation, which functions downstream from runx2 and perhaps acts cooperatively with it. The bone morphogenetic proteins (BMPs) are important in osteoblast differentiation through their paracrine role in bringing into play signaling mechanisms that are crucial for the process,[10] including sonic and indian hedgehog, and Wingless-type (Wnt) signaling.

However important runx2 is in osteoblast differentiation and bone formation, it is clearly subject to regulatory mechanisms mediated by other transcription factors, growth factors, and hormones.[7] These include parathyroid hormone (PTH), which activates the *runx2* promoter, increasing production of mRNA and protein for runx2 in osteoblasts, raising the question of whether this might contribute to the anabolic action of PTH.[11] This regulatory pathway contains a series of molecular steps that very likely can serve as targets for the development of drugs to influence bone formation.

THE WNT SIGNALING PATHWAY AND BONE FORMATION

Intriguing insights into the control of bone mass come from discoveries of mutations in a gene associated with the osteoporosis pseudoglioma syndrome, and with a high bone mass syndrome. The genetic abnormalities consist of inactivating mutations in the gene for low-density lipoprotein receptor-related protein 5 (LRP5), resulting in impaired bone mass and severe osteoporosis, with heterozygous carriers also having reduced bone mass and an increased incidence of osteoporotic fractures.[12] On the other hand, an activating mutation of the same gene in another kindred is associated with greatly increased mass of bone that is nevertheless of normal shape and architecture.[13,14] *In vitro* studies showed that the normal inhibition of Wnt signaling by another protein, the Wnt antagonist, Dickkopf (Dkk 1) was defective in the presence of the mutant LRP5, providing a molecular explanation for the increased activity of the Wnt signaling pathway.

LRP5 is a single-pass membrane receptor that forms part of a complex necessary for activation in the Wnt signaling pathway. LRP5 interacts with the Wnt-frizzled ligand–receptor complex, resulting in the inhibition of β-catenin phosphorylation by glycogen synthase kinase-3β (GSK-3β). Since GSK-3β activity facilitates ubiquitin-mediated breakdown of β-catenin, the LRP5 effect is to prevent this, allowing accumulation of β-catenin and hence its translocation to the nucleus, where it interacts with TCF/LEF transcription factors to activate gene transcription. LRP5 and LRP6 are co-receptors for Wnt proteins, with LRP5 inhibited by the Wnt antagonists, Dkk-1 or Dkk-2, and activated by Dkk-3 or Dkk-4. A potential link between the PTH and Wnt signaling pathways arises from *in vivo* gene array studies in rats treated with PTH, in which several components of the Wnt–receptor complex were upregulated in PTH-treated rat bone, and furthermore, PTH treatment of osteoblastic cells *in vitro* activated the TCF/LEF transcriptional pathway.[15] Notably, activating the Wnt signaling pathway in rats by treatment with a GSK-3β inhibitor resulted in increased bone formation.[16]

These findings together have revealed a previously unrecognized control pathway for osteoblast differentiation and function and bone growth, a pathway studied extensively in the development in many organisms, as well as in cancer. The control mechanisms for this pathway are complex, providing several points that are the focus of attention toward the development of drugs anabolic for bone. Its very complexity, the number of participants, and the likely involvement of this pathway in cell proliferation control and in neoplastic processes in some tissues, provide major challenges.

STIMULATION OF BONE FORMATION BY PTH

The only means currently available for skeletal anabolic therapy is intermittent treatment with PTH(1-34), approved in the United States as "Forteo" for the treatment of osteoporosis.[17] Other active forms of PTH, including PTH(1-84) are under study, including being in clinical trial. The efficacy of PTH in promoting bone formation is such that it sets a standard for any new treatments that are to be developed. It is therefore appropriate to consider its actions in further detail.

The ability of intermittent injections of PTH to increase bone formation was already recognized in the 1930s. Recent clinical studies established that

PTH(1-34) has a powerful anabolic effect, revealing it as the first therapy capable of restoring lost bone.[17] Not surprisingly, PTH-related protein (PTHrP) exerts similar actions, with PTHrP (1-34), (1-36), and (1-74) having all been shown to be anabolic in rodents.[18,19] Clinical data from the measurement of bone markers indicate that PTHrP(1-36) is also anabolic in human subjects.[20]

Growth Factors as Mediators of PTH Anabolic Response

The two general mechanisms proposed for the PTH anabolic effect are promotion by PTH of differentiation of committed osteoblast precursors,[21] and inhibition of osteoblast apoptosis by PTH.[22] PTH administered intermittently *in vivo* enhances the production of proteins associated with bone formation, including runx2, osteocalcin, and type I collagen. There is also evidence from a number of approaches that the PTH anabolic response could be mediated at least in part, through increased production and/or activation of growth factors known to be capable of enhancing bone formation.[23] Insulin-like growth factor-1 (IGF-1) and transforming growth factor (TGF)-β are two of the major growth factors of bone. IGF-1 is provided in a latent form in matrix because of the regulated production of several specific binding proteins that sequester it.[24] The bone matrix stores large amounts of latent TGF-β, which can be activated through the acid conditions of bone resorption[25] or by proteases produced by osteoblasts.[26] PTH treatment of osteoblasts has been shown to activate TGF-β[26] as well as to enhance IGF-1 production by osteoblasts and by bone in organ culture. Such mechanisms could contribute to the PTH anabolic effect. Consistent with this possibility is the finding that the anabolic response to PTH is lost in mice rendered null for *IGF-1*.[27]

Anabolic Effect of PTH Requires Transient Exposure to Increased Hormone

The ability of PTH to promote bone formation is dependent upon the hormone being administered intermittently in a way that yields a peak blood level that is not maintained.[17,28] In that circumstance, processes are initiated in bone, which result in anabolic effects, presumably as a result of the activation of genes responding specifically to a transiently activated signaling system that requires a rapid increase in PTH or PTHrP, with a rapid decline to preexisting levels. On the other hand, if PTH or PTHrP is infused, or administered in such a way that elevated plasma levels are maintained, the dominant effect is stimulation of osteoclast formation and bone resorption, to the extent that these override any anabolic response. Studies in the rat *in vivo* support this view. Infusion of PTH into rats caused a robust and sustained increase in RANKL and decrease in osteoprotegerin (OPG) production in bone, which preceded hypercalcemia and enhanced osteoclast formation. In these conditions also, sustained elevated levels of PTH resulted in decreased expression of genes associated with bone formation.[29] These included cbfa1, osteocalcin, bone sialoprotein, and type 1 collagen. In contrast, repeated single injections of PTH triggered a rapid but transient increase in the RANKL/OPG ratio, with a peak 1 h after PTH injection and a rapid return to control levels. This was associated with increased bone formation and enhanced expression of the genes associated with bone formation.[30] Finally, and most importantly, Holtrop *et al.*[31] showed that intravenous injection of PTH in young rats resulted in transient activation of osteoclasts *in vivo*, evident within 30 min, and followed only some hours later at high PTH doses by increased osteoclast number. In light of current knowledge, it is most likely that rapid, transient activation of osteoclasts following PTH injection is the result of rapid induction of RANKL expression in cells of the osteoblast lineage.

The results of these experiments raised the question: "Is an effect of PTH upon osteoclasts necessary for the anabolic effect?" A number of lines of evidence are consistent with this possibility.

Osteoclast Activity and Bone Formation Responses

When sheep were treated with PTH for 3 months the anabolic effect was significantly reduced when a bisphosphonate (tiludronate) was co-administered.[32] In ovariectomized rats treated with PTH, bone formation parameters (assessed by histomorphometry) were reduced when co-treatment was provided either with estrogen or with the bisphosphonate, risedronate.[33] A similar conclusion, that osteoclast activity is necessary for the anabolic response to PTH, came from studies in mice rendered null for the *c-fos* gene, which are osteopetrotic because they cannot develop osteoclasts, and furthermore fail to show an anabolic response to PTH.[34] This evidence from animals is supported by recent clinical studies. Treatment of patients with osteoporosis concomitantly with PTH and a bisphosphonate resulted in significant early blunting of the anabolic response to PTH,[35,36] with the obvious implication that combining the anabolic PTH with antiresorptive bisphosphonate would be contraindicated.[35–38]

Observations made in both human and mouse genetics suggest that the osteoclast could be a source of an activity that contributes to bone formation. In individuals with the osteopetrotic syndrome, ADOII, due to inactivating mutations in the chloride-7 channel (ClC-7), bone resorption is deficient because of the failure of the osteoclast acidification process. These patients have increased osteoclast numbers, but bone formation is nevertheless normal, rather than diminished as might be expected because of the greatly impaired resorption.[39] Furthermore, in mice deficient in either c-*src*,[40] ClC-7,[41] or tyrosine phosphatase epsilon,[42] bone resorption is inhibited without inhibition of formation. In these three knockout mouse lines osteoclast resorption is greatly reduced by the mutation, although osteoclast numbers are not reduced. Indeed, osteoclast numbers are actually increased in vacuolar ATPase (V-ATPase)-deficient mice because of reduced osteoclast apoptosis. A possibility is that these osteoclasts, although unable to resorb bone, are nevertheless capable of generating a factor (or factors) required for bone formation. An illustration of this concept is provided in studies of reparative bone growth around calcium phosphate implants, in which inhibition of bone resorption by the V-ATPase inhibitor, bafilomycin, led to increased nonresorbing osteoclasts surrounding the neo-implants, resulting in a large increase in bone formation.[43] Thus the bone formation increase was not dependent on resorption by the osteoclasts.

The cytokines that signal through glycoprotein (gp)130 play an important role in intercellular communication processes in bone, with evidence indicating that they can be involved in the regulation of both bone resorption and formation.[44–46] We addressed this by studying mice in which each of the two gp130-dependent signaling pathways was specifically attenuated, and found that inactivation of the SHP2/ras/MAPK signaling pathway yielded mice with greater osteoclast numbers and bone resorption, as well as greater bone formation than wild-type mice. The net effect was a decreased amount of bone in these mice. When these mice were crossed with IL-6 null mice, the resulting mutants had similarly high osteoclast numbers and increased bone resorption, but without any increase in bone formation and therefore with extremely low bone mass. Thus resorption alone is insufficient to promote the coupled bone formation, but the active osteoclasts are the likely source of an activity that is IL-6 dependent, but not necessarily IL-6 itself.[46,47]

Our interpretation of the foregoing and other data is that what is needed for the full expression of the anabolic action of PTH, in addition to its direct effects on committed preosteoblasts, is a transient effect on the osteoclast, achieved by promoting *activation,* rather than *formation,* of osteoclasts.[47] This is an important distinction, and the precise way in which the osteoclast is involved in the anabolic process needs to be clearly understood because of the implications for sequential or combined use of therapeutic resorption inhibitors and anabolic agents, and for the development of new anabolic agents. FIGURE 1 illustrates this schematically, indicating the known effect of PTH to increase osteoclast activity indirectly by acting on osteoblastic cells to increase RANKL. The RANKL-activated osteoclasts are depicted as releasing a biological activity that enhances osteoblast maturation, either independently or cooperatively with PTH, or with the locally important factors promoting bone formation. The impaired anabolic response to PTH when there is a complete blockade of osteoclasts, either through their impaired formation, as in *c-fos* deficient mice, or resulting from prolonged bisphosphonate treatment, implies that the full expression of the anabolic effect might require a contribution from the active osteoclast. The genetic data implies further that inhibition of osteoclastic resorption, e.g., by preventing acidification, might nevertheless provide osteoclasts that are capable of producing this anabolic stimulus. An interesting further implication is that resorption inhibitors based

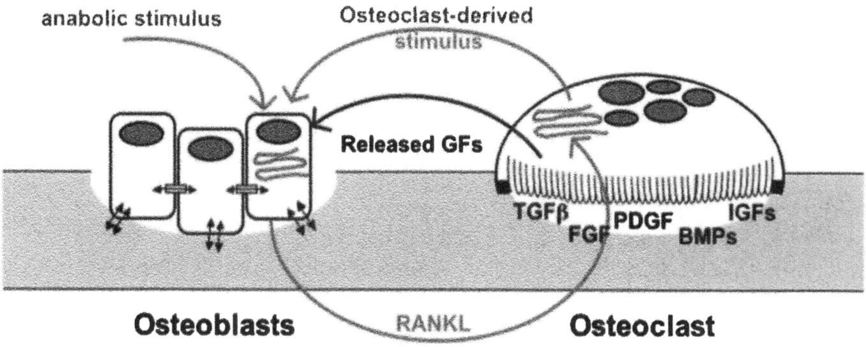

FIGURE 1. Osteoclasts can contribute to bone formation either by their promotion of resorption, with release of stored growth factors from matrix, or by the release from transiently activated osteoclasts of activity that contributes to anabolic effects on bone. (Modified from Martin, T.J. & N.A.Sims[47] with permission).

on inhibition of acidification of *c-Src* might not impair the anabolic response to PTH.

Significance of PTHrP Production in Bone

The discovery of PTHrP production in bone[48,49] raised the intriguing possibility that it has important local actions in bone, not only those reflecting the anabolic action of PTH, but also possibly producing other effects through actions of differently processed forms of the molecule. The cells producing PTHrP in bone are likely distinct from those responding, at least through the PTH1R. In isolated cells from rat calvariae, for example, cells of early enzyme digests were those most abundantly producing PTHrP, whereas PTH responses through the common PTH/PTHrP receptor (PTH1R) were enriched in the osteoblast-rich populations of later digests (TABLE 1). Furthermore, in studying differentiation of mouse stromal osteoblast precursors, decline in PTHrP mRNA production was associated with increase in PTH1R receptor mRNA and PTH- induced cAMP response.[50]

One function that has been studied extensively is the role of PTHrP in endochondral bone formation. Targeted disruption of the genes for PTHrP or PTH1R in mice resulted in death in the perinatal period with gross skeletal abnormalities consistent with chondrodysplasia.[51,52] Histological studies suggest a central role of PTHrP in fetal endochondral bone formation through its actions in maintaining a pool of proliferating chondrocytes, inhibition of terminal chondrocyte differentiation, retardation of cartilage matrix mineralization, and differentiation of periosteal mesenchymal precursors into cells of the chondrocytic or osteoblastic lineages.

It appears that PTHrP might be involved in intramembranous bone formation also. In an experimental model of this process in the rabbit bone, formation

TABLE 1. Alkaline Phosphatase Activity, PTHrP Production and cAMP Response in Sequentially Digested Rat Calvarial Cells

	Alk Phos (nmol/min/mg protein)	PTHrP (fmol/10^6 cells)	cAMP (response to 10 nM PTHrP)
1	32 ± 1.9	20.9 ± 2.1	0.16
2	43 ± 4.1	9.8 ± 0.5	0
3	63 ± 7.0	3.9 ± 0.1	0.11
4	214 ± 15	6.7 ± 0.2	7.5
5	291 ± 16	7.5 ± 0.3	6.5
6	714 ± 24	4.2 ± 0.1	23

cAMP data are expressed as the fold response, with triplicate assays for control and treatment with 10 nM PTHrP. (Used with permission from Suda *et al.*[48])

begins with transformation of primitive marrow mesenchymal cells into trabecular bone without any cartilage intermediate, and the synchronized appearance of hemopoietic marrow elements follows the onset of matrix mineralization. In this model, cells of the osteoblast lineage consistently expressed PTHrP mRNA and protein throughout the bone formation sequence, with prominent production by cuboidal, actively synthesizing osteoblasts, and weaker expression in lining cells on the mineralized trabeculae.[49] These observations, together with those of Suda *et al.*,[48] are consistent with a role for PTHrP in the differentiation of mesenchymal precursors to the osteogenic lineage.

A most important observation made in the course of the genetic experiments was that heterozygous *PTHrP* null mice were shown to be phenotypically normal at birth, but by 3 months of age they demonstrated a reduction in trabecular bone volume, and an increase in adipocytes was observed in the bone marrow.[53] The absolute requirement of PTHrP for normal bone integrity not only in the fetus but also postnatally, together with the finding that PTHrP haploinsufficiency is associated with less bone and with the preferential differentiation of stromal mesenchymal cells into adipocytes rather than osteocytes, all point to PTHrP as a critical molecule in the development and maintenance of bone. Evidence for a crucial role of locally generated PTHrP in bone remodeling was published while this article was in press. Miao *et al.* established this in mice in which *PTHrP* gene was specifically ablated in osteoblasts.[54]

What are the ways in which PTHrP can act as a paracrine/autocrine factor in bone? The pharmacologic effects of intermittent versus sustained PTH/PTHrP treatment are striking and very different. If the behavior of osteoblasts in response to stimulation through PTHR1 requires this type of variation in the delivery of the relevant ligand, can PTH, as a circulating peptide hormone, achieve this? That is doubtful. On the other hand, the regulated, local production of PTHrP could fulfill this role, with its regulation being the result of hormonal, cytokine, or neural control. In the case of PTHrP there is an added possibility that biological activities within the remainder of the molecule could influence local events, either through independent processes or by modifying actions through the PTH1R.

In summary, the efficacy of PTH in promoting bone formation provides a benchmark for the evaluation of new anabolic therapies. New approaches to anabolic treatment might also be expected from increased understanding of the molecular and cellular basis of this action of PTH/PTHrP, and by answering the question whether the primary physiological regulator of bone formation is the paracrine factor, PTHrP, rather than the hormone, PTH. Other pathways also offer targets for the development of skeletal anabolics, with the pathway of most current interest the Wnt signaling pathway, which may even interact with PTH/PTHrP.

ACKNOWLEDGMENTS

Work from the authors' laboratories was supported by a Program Grant from the NHMRC (Australia).

REFERENCES

1. FROST, H.M. 1964. Dynamics of bone remodeling. *In* Bone Biodynamics. H.M. Frost, Ed.: 315–333. Little, Brown & Co. Boston.
2. HARRIS, W.H. & R.P. HEANEY. 1969. Skeletal renewal and metabolic bone disease. N. Engl. J. Med. **280:** 193–202.
3. PARFITT, A.M. 1982. The coupling of bone formation to bone resorption: a critical analysis of the concept and of its relevance to the pathogenesis of osteoporosis. Metab. Bone Dis. Relat. Res. **4:** 1–6.
4. ERIKSEN, E.F. 1986. Normal and pathological remodeling of human trabecular bone: three dimensional reconstruction of the remodeling sequence in normals and in metabolic bone disease. Endocr. Rev. **7:** 379–408.
5. MARTIN, T.J. & G.A. RODAN. 2001. Coupling of bone resorption and formation during bone remodeling. *In* Osteoporosis. 2nd ed. R. Marcus, *et al.* Eds.: 361–372. Academic Press. San Diego.
6. LIPS, P., P. COURPRON & P.J. MEUNIER. 1978. Mean wall thickness of trabecular bone packets in the human iliac crest: changes with age. Calcif. Tissue Res. **26:** 13–17.
7. KARSENTY, G. 2001. Minireview: transcriptional control of osteoblast differentiation. Endocrinology **142:** 2731–2733.
8. DUCY, P., M. STARBUCK, M. PRIEMEL, *et al.* 1999. A Cbfa1-dependent genetic pathway controls bone formation beyond embryonic development. Genes Devel. **13:** 1025–1036.
9. NAKASIMA, K., X. ZHOU, G. KUNKEL, *et al.* 2002. The novel zinc finger-containing transcription factor osterix is required for osteoblast differentiation and bone formation. Cell **108:** 17–29.
10. YAMAGUCHI, A., T. KOMORI & T. SUDA. 2000. Regulation of osteoblast differentiation mediated by bone morphogenetic proteins, hedgehogs and cbfa1. Endocrine Rev. **21:** 393–411.

11. Krishnan, V., T.L. Moore, Y.L. Ma, *et al.* 2003. Parathyroid hormone bone anabolic action requires Cbfa1/Runx2-dependent signalling. Mol. Endocrinol. **17:** 423–435.
12. Gong, Y. *et al.* 2001. LDL-receptor-related protein (LRP5) affects bone accrual and eye development. Cell **197:** 513–523.
13. Little, R.D. *et al.* 2002. A mutation in the LDL receptor-related protein 5 gene result in the autosomal dominant high bone mass trait. Am. J. Hum. Genet. **70:** 11–19.
14. Boyden, L.M., J. Mao, J. Belsky, *et al.* 2002. High bone density due to a mutation in LDL-receptor-related protein 5. N. Engl. J. Med. **346:** 1513–1521.
15. Kulkarni, N.H., D.L. Halladay, R.R. Miles, *et al.* 2005. Effect of parathyroid hormone on Wnt Signaling in bone. J. Cell Biochem. **95:** 1178–1190.
16. Kulkarni, N.H., M. Liu, D.L. Halladay, *et al.* 2004. Orally available GSK3 α/ß dual inhibitor increases markers of cellular differentiation *in vitro* and bone mass *in vivo* [abstract]. J. Bone Miner. Res. **19:** S57.
17. Neer, R.M., C.D. Arnaud, J.R. Zanchetta, *et al.* 2001. Effect of parathyroid hormone (1-34) on fractures and bone mineral density in postmenopausal women with osteoporosis. N. Engl. J. Med. **344:** 1434–1441.
18. Hock, J.M., J. Fonseca, M. Gunness-Hey, *et al.* 1989. Comparison of the anabolic effects of synthetic parathyroid hormone-related protein (PTHrP) 1-34 and PTH 1-34 on bone in rats. Endocrinology **125:** 2022–2027.
19. Weir, E.C., G. Terwilliger, L. Sartori, *et al.* 1992. Synthetic parathyroid hormone-like protein (1-74) is anabolic for bone in vivo. Calcif. Tissue Int. **51:** 30–34.
20. Everhart-Caye, M., S.E. Inzucchi, J. Gunness-Hey, *et al.* 1996. Parathyroid hormone (PTH)-related protein (1-36) is equipotent to PTH(1-34) in humans. J. Clin. Endocrinol. Metab. **81:** 199–220.
21. Dobnig, H. & R.T. Turner. 1995. Evidence that intermittent treatment withparathyroid hormone increases bone formation in adult rats by activation of bone lining cells. Endocrinology **136:** 3632–3638.
22. Manolagas, S.C. 2000. Birth and death of bone cells: basic regulatory mechanisms and implications for the pathogenesis and treatment of osteoporosis. Endocr. Rev. **21:** 115–137.
23. Canalis, E., T.L. McCarthy, & M. Centrella. 1989a. The role of growth factors in skeletal remodelling. Endocrin. Metab. Clin. N. Am. **19:** 903–918.
24. Canalis, E.R., M. Centrella, W. Burch, *et al.* 1989b. Insulin-like growth factor 1 mediates selective anabolic effects of parathyroid hormone in bone cultures. J. Clin. Invest. **83:** 60–65.
25. Dallas, S.L., J.L. Rosser, G.R. Mundy, *et al.* 2002. Proteolysis of latent transforming growth factor-beta (TGF-beta)-binding protein-1 by osteoclasts. A cellular mechanism for release of TGF-beta from bone matrix. J. Biol. Chem. **277:** 21352–21360.
26. Yee, J.A., L. Yan, J.C. Dominguez, *et al.* 1993. Plasminogen-dependent activation of latent transforming growth factor beta by growing cultures of osteoblast-like cells. J. Cell Physiol. **157:** 528–534.
27. Bikle, D.D., T. Sakata, C. Leary, *et al.* 2002. Insulin-like growth factor I is required for the anabolic actions of parathyroid hormone on mouse bone. J. Bone Miner. Res. **17:** 1570–1578.
28. Frolik, C.A., R.L. Cain, M. Sato, *et al.* 2003. Anabolic and catabolic bone effects of human parathyroid hormone(1-34) are predicted by duration of hormone exposure. Bone **33:** 372–379.

29. MA, Y.L., R.L. CAIN, D.L. HALLADAY, *et al.* 2001. Catabolic effects of continuous human PTH(1-38) in vivo is associated with sustained stimulation of RANKL and inhibition of osteoprotegerin and gene-associated bone formation. Endocrinology **142:** 4047–4054.
30. ONYIA, J.E., J. BIDWELL, J. HERRING, *et al.* 1995. In vivo, human parathyroid hormone fragment (hPTH 1-34) transiently stimulates immediate early response gene expression, but not proliferation, in trabecular bone cells of young rats. Bone **17:** 479–484.
31. HOLTROP, M.E., G.J. KING, K.A. COX, *et al.* 1979. Time-related changes in the ultrastructure of osteoclasts after injection of parathyroid hormone in young rats. Calcif. Tissue Int. **27:** 129–135.
32. DELMAS, P.D., P. VERGNAUD, M.E. ARLOT, *et al.* 1995. The anabolic effect of human PTH(1-34) on bone formation is blunted when bone resorption is inhibited by the bisphosphonate tiludronate—is activate resorption a prerequisite for the in vivo effect of PTH on formation in a remodelling system? Bone **16:** 603–610.
33. WRONSKI, T.J., C.-F. YEN, H. QI, *et al.* 1993. Parathyroid hormone is more effective than estrogen or bisphosphonates for restoration of lost bone mass in ovariectomized rats. Endocrinology **132:** 823–831.
34. DEMIRALP, B., H.L. CHEN, A.J. KOH, *et al.* 2002. Anabolic actions of parathyroid hormone during bone growth are dependent on c-fos. Endocrinology **143:** 4038–4047.
35. BLACK, D.M., S.L. GREENSPAN, K.E. ENSRUD, *et al.* 2003. The effects of parathyroid hormone and alendronate alone or in combination in postmenopausal osteoporosis. N. Engl. J. Med. **349:** 1207–1215.
36. FINKELSTEIN, J.S., A. HAYES, J.L. HUNZELMAN, *et al.* 2003. The effects of parathyroid hormone, alendronate, or both in men with osteoporosis. N. Engl. J. Med. **349:** 1216–1226.
37. KHOSLA, S. 2003. Parathyroid hormone plus alendronate–a combination that does not add up. N. Engl. J. Med. **349:** 1277–1279.
38. MARTIN, T.J. 2004. Does bone resorption inhibition affect the anabolic response to parathyroid hormone? Trends Endocrinol. Metab. **15:** 49–50.
39. CLEIREN, E., O. BENICHOU, E. VAN HUL, *et al.* 2001. Albers-Schonberg disease (autosomal dominant osteopetrosis, type II) results from mutations in the ClCN7 chloride channel gene. Hum. Mol. Genet. **10:** 2861–2867.
40. MARZIA, M., N.A. SIMS, S. VOIT, *et al.* 2000. Decreased c-Src expression enhances osteoblast differentiation and bone formation. J. Cell Biol. **151:** 311–320.
41. KARSDAL, M.A., K. HENRIKSEN, M.G. SORENSEN, *et al.* 2005. Acidification of the osteoclastic resorption lacunae controls the coupling of bone resorption to bone formation. Am. J. Pathol. **166:** 467–476.
42. CHIUSAROLI, R., H. KNOBLER, C. LUXENBURG, *et al.* 2004. Tyrosine phosphatase epsilon is a positive regulator of osteoclast function in vitro and in vivo. Mol. Biol. Cell. **15:** 234–244.
43. RZESZUTEK, K., F. SARRAF & J.E. DAVIES. 2003. Proton pump inhibitors control osteoclastic resorption of calcium phosphate implants and stimulate increased local reparative bone growth. J. Craniofac. Surg. **14:** 301–307.
44. ROMAS, E., N. UDAGAWA, H. ZHOU, *et al.* 1996. The role of gp130-mediated signals in osteoclast development: regulation of interleukin 11 production by osteoblasts and distribution of its receptor in bone marrow cultures. J. Exp. Med. **183:** 2581–2591

45. HEYMANN, D. & A.V. ROUSSELLE. 2000. gp130 Cytokine family and bone cells. Cytokine **12:** 1455–1468.
46. SIMS, N.A., B.J. JENKINS, J.M. QUINN, *et al.* 2004. Glycoprotein 130 regulates bone turnover and bone size by distinct downstream signaling pathways. J. Clin. Invest. **113:** 379–389.
47. MARTIN, T.J. & N.A. SIMS. 2005. On the coupling of bone formation to resorption. Trends Molec. Med. **11:** 76–81.
48. SUDA, N., M.T. GILLESPIE, K. TRAIANEDES, *et al.* 1996. Expression of parathyroid hormone-related protein in cells of osteoblast lineage. J. Cell Physiol. **166:** 94–104.
49. KARTSOGIANNIS, V., J. MOSELEY, B. MCKELVIE, *et al.* 1997. Temporal expression of PTHrP during endochondral bone formation in mouse and intramembranous bone formation in an *in vivo* rabbit model. Bone **21:** 385–391.
50. ALLAN, E.H., P.W. HO, A. UMEZAWA, *et al.* 2003. Differentiation potential of a mouse bone marrow stromal cell line. J. Cell. Biochem. **90:** 158–169.
51. AMIZUKA, N., H. WARSHAWSKY, J.E. HENDERSON, *et al.* 1994. Parathyroid hormone-related peptide-depleted mice show abnormal epiphyseal cartilage development and altered endochondral bone formation. J. Cell Biol. **126:** 1611–1623.
52. KARAPLIS, A.C. & H.M. KRONENBERG. 1996. Physiological roles for parathyroid hormone-related protein: lessons from gene knockout mice. Vitam. Horm. **52:** 177–193.
53. AMIZUKA, N., A.C. KARAPLIS, J.E. HENDERSON, *et al.* 1996. Haploinsufficiency of parathyroid hormone-related peptide (PTHrP) results in abnormal postnatal bone development. Dev. Biol. **175:** 166–176.
54. MIAO, D., B. HE, Y. JIANG, *et al.* 2005. Osteoblast-derived PTHrP is a potent endogenous bone anabolic agent that modifies the therapeutic efficacy of administered PTH(1-34). J. Clin. Invest. **115:** 2402–2411.

Misconceptions About Calcimimetics

E.F. NEMETH

NPS Pharmaceuticals, Inc., Toronto, Ontario M5G 1L8, Canada

ABSTRACT: Calcimimetics are ligands that activate the calcium receptor. Some are small molecules and of these, the most extensively studied are phenylalkylamines like cinacalcet. This compound is a positive allosteric modulator that selectively targets the parathyroid calcium receptor to inhibit the secretion of parathyroid hormone. Cinacalcet is the first calcimimetic compound to attain regulatory approval for the treatment of hyperparathyroidism resulting from end-stage renal disease. The discovery of calcimimetics and the receptor they act on are considered with the intent of extracting lessons relevant to medical research and the discovery of new drugs.

KEYWORDS: calcimimetics; calcium receptor; allosteric modulators; drug discovery

INTRODUCTION

Extracellular calcium is the primary physiological regulator of the parathyroid cell and most, if not all its effects are mediated by a cell surface calcium receptor. The calcium receptor is a G protein–coupled receptor that stands apart from other receptors in this superfamily because its physiological ligand is an inorganic ion.[1] Calcimimetics are inorganic or organic ligands that act on the calcium receptor to mimic or potentiate the effects of extracellular Ca^{2+}.[2] The term encompasses at least two classes of ligands that are distinguished by their molecular mechanism of action and not by their structure or by their binding sites within the receptor. Type I calcimimetics act as orthosteric ligands and are conventional agonists; they depress parathyroid hormone (PTH) secretion even in the absence of extracellular Ca^{2+}. Type II calcimimetics act as positive allosteric modulators to increase the sensitivity of the calcium receptor to activation by extracellular Ca^{2+}; they do not affect PTH secretion under extreme hypo- or hypercalcemic conditions.

The calcimimetic compound, cinacalcet, has a number of distinctions: it is the first drug to act on the calcium receptor, it is a first-in-class therapy for secondary hyperparathyroidism in patients on dialysis and, as a type II

Address for correspondence: E.F. Nemeth, NPS Pharmaceuticals, Inc., 101 College Street, South Tower, Suite 800, Toronto, ON M5G 1L8 Canada. Voice: 416-849-5605; fax: 416-929-0610.
e-mail: enemeth@npsp.com

Ann. N.Y. Acad. Sci. 1068: 471–476 (2006). © 2006 New York Academy of Sciences.
doi: 10.1196/annals.1346.044

calcimimetic, it is the first allosteric modulator of *any* G protein–coupled receptor to achieve regulatory approval. Cinacalcet is also distinguished by a widespread misunderstanding of how it, and the receptor it acts on, came to be.

This brief commentary is a first attempt at articulating two of these misconceptions. Placing the discovery of the calcium receptor and calcimimetics into a historically accurate context reveals some lessons relevant to the conduct of scientific research in general and drug discovery in particular.

MISCONCEPTION 1—THE CALCIUM RECEPTOR WAS "AN UNANTICIPATED DISCOVERY" IN 1993

This is a popular misunderstanding and it is a pity that such inaccurate views make it into high-profile journals.[3,4] Some receptors are discovered by molecular cloning but the calcium receptor is not one of them. Pharmacologists have been studying receptors and creating drugs that act on them long before one was ever cloned. The existence of some calcium receptor-like mechanism on parathyroid cells was revealed by attempts to understand how extracellular Ca^{2+} regulated PTH secretion. The hypothesis that parathyroid cells express a cell surface calcium receptor that enables them to detect and respond to small fluctuations in the level of extracellular Ca^{2+} was formulated almost a decade before 1993 when several independent groups, using physiological and pharmacological techniques, all came to the same conclusion. A biophysicist was probably the first to clearly articulate the hypothesis: "These findings suggest that the interaction of divalent cations with parathyroid cells is not based on their different permeability through the membrane and point toward the possible existence of a divalent cation receptor in the external surface of these cells."[5] The group in Sweden studied Ca^{2+} fluxes in rat and in bovine parathyroid cells and concluded: "The results suggest the existence of a novel Ca^{2+} permeability physiologically activated by binding of Ca^{2+} to an external receptor."[6] Their report appeared just 2 months (and in the same Journal) after we reported results showing that extracellular Ca^{2+} acts at the cell surface to cause the rapid mobilization of intracellular Ca^{2+}, and this led to the proposal "...that there exists a receptor on the surface of parathyroid cells which is promiscuous and responds to a variety of divalent cations."[7] Biochemical evidence consistent with the mobilization of intracellular Ca^{2+} soon followed and all these findings were collected for discussion in a workshop at the 1989 ASBMR meeting in Montreal.[8] So the idea that parathyroid cells express a cell surface calcium receptor was gaining widespread acceptance prior to 1993. This idea in its most nascent form can even be detected in Franklin McLean's opening address at the First Symposium on Advances in Parathyroid Research in 1960: "It appears then that the parathyroid glands are themselves the receptors of the stimuli, in the form of circulating calcium ions, that determine the rate of output of the parathyroid hormone."[9]

By 1990, several characteristics of the calcium receptor were known: this putative receptor was promiscuous and could be activated by a variety of inorganic and organic polycations. The receptor was coupled to the formation of inositol trisphosphate and the mobilization of intracellular Ca^{2+} from a nonmitochondrial pool. There was significant evidence showing that extracellular Ca^{2+} regulated adenylate cyclase in a pertussis toxin-sensitive manner[10] suggesting, already in 1989, that the calcium receptor might be a G protein–coupled receptor. At the 1991 ASBMR meeting in San Diego, results presented by two independent groups showed that functional expression in *Xenopus* oocytes was a viable strategy for cloning the calcium receptor.[11,12] It was this approach that was used by Ed Brown, Steve Hebert, and their colleagues to clone the cDNA encoding the bovine parathyroid calcium receptor.[13] They were probably not all that surprised by the results of homology analysis showing that this cDNA encoded a G protein–coupled receptor. What occurred in 1993 is not the discovery of the calcium receptor *per se* but the unambiguous evidence that it existed. The real and clinically significant discovery that emerged 1993 is found in their companion paper that describes the molecular basis for a number of inherited disorders of systemic calcium homeostasis.[14]

MISCONCEPTION 2—THE CLONING OF THE CALCIUM RECEPTOR ENABLED THE DISCOVERY OF CALCIMIMETICS

This misunderstanding, which appeared early on,[15] permeates the literature and has now become the conventional wisdom.[4] Indeed, it has even been suggested that the "discovery" of the calcium receptor in 1993 "... introduced the possibility that a therapy could be developed ...."[3] This is a rather peculiar statement given the fact that drugs acting on the two classic members of the G protein–coupled receptor superfamily, adrenergic and muscarinic, had been used clinically for nearly 50 years before either receptor was cloned. Similarly, the lack of a cloned calcium receptor did not deter the discovery of calcimimetics.

The *Nature* article[13] appeared in the December 9, 1993 issue but Brown and Hebert had the correct cDNA clone as early as February of that year. A clinical trial with the first calcimimetic, NPS R-568, was initiated just 12 months later; the first human was dosed on February 28, 1994 and the results presented in 1995.[16] A year is typically required just to acquire the preclinical safety and efficacy data needed to produce an Investigational New Drug (IND) application—and that time frame assumes that the clinical candidate has already been selected so that efforts focus on a single compound. Moreover, the first patent on calcimimetic compounds was published in March 1993,[17] and there was at least an 18-month lag time between filing a patent and its publication. Given these time lines, it is difficult to sustain the view reflected in much of the literature. The calcimimetic NPS R-568 was far along in preclinical development before its molecular target was cloned.

Similarly, high-throughput screening was not used to discover calcimimetics.[18] The approach that was used is no secret and is stated plainly in the first sentence of the Results section: "To screen for agonist-like activity on the Ca^{2+} receptor, we assessed the ability of test substances to evoke an increase in $[Ca^{2+}]_i$ in bovine parathyroid cells."[19] In the absence of a parathyroid cell line or a cDNA encoding the calcium receptor, this seemed to be the best, perhaps only, approach. Moreover, most of the evidence supporting the existence of the calcium receptor derived from studies in dissociated bovine parathyroid cells and we knew a lot about the characteristics of the (still putative) calcium receptor in these cells. It was, and remains a tedious and laborious assay and at best permits the screening of 20 to 40 compounds a week—truly ultra-low throughput. By today's standards, a technically unsophisticated assay bordering on the comical. But it was a *functional* assay, with a physiologically meaningful readout, and this, it turns out, was pivotal to the discovery of type II calcimimetics—the allosteric modulators.

Despite all the hype about translational molecular medicine, calcimimetics were discovered in the old fashioned way, by testing compounds on live animal tissue in an assay configured to detect compounds having the desired therapeutic effect. In his Nobel Lecture,[20] Sir James Black explains how this is done and how this approach yielded two (not just one) first-in-class drugs: propranolol and cimetidine. And he did it without a clue about the structure of the receptors these drugs target.

PARFITT'S PLEA

Just what, besides striving for historical accuracy at the risk of perceived self-aggrandizement, is the point of all this rambling? It is to draw attention to and reinforce the view recently articulated Michael Parfitt.[21] In the third installment of his "misconceptions" series, he lamented the sorry state of research as it relates to a specific problem in mineral metabolism. His discontents are twofold. The first is the general lack of interest in consulting the older literature. The second, which emerges from the first, is "...a disdain for physiology, even though a knowledge of physiology and pathophysiology is essential for the interpretation of molecular biology." The pilfering of the misconception titles here is meant to convey my respectful concurrence with his view.

In one respect, the cloning of the calcium receptor has done more harm than good. Because the techniques of molecular biology permit even vanishingly small levels of mRNA to be detected, the calcium receptor can now found in all kinds of cells throughout the body. The literature has become awash with claims that the calcium receptor is involved in all sorts of bodily functions. Not surprisingly, these universally expressed calcium receptors are also claimed as viable drug targets. (Why patients treated with cinacalcet are not keeling over in agony remains a mystery.) I have discussed this elsewhere[22] and argue there that many of these claims are either patently wrong or of such trivial physiological and pharmacological consequence that they can be safely ignored. These misconceptions result from a failure to understand some basic physiological actions of divalent cations on the plasma membrane. There are many actions of extracellular Ca^{2+} on cells, most of them have nothing to do with the calcium receptor, and they are all waiting to be rediscovered in the literature of the 1950s and 1960s.

The penchant to view the function of the body solely through the lens of molecular biology has fostered a purely reductionist approach to drug discovery. This approach supposes that any cloned target is a viable drug target until proven otherwise. Unfortunately, the proof otherwise often occurs only when the compound is well into preclinical development or, even worse, in the clinic. Despite all the “omics” and now, “systems biology”, the attrition rate for drug candidates has not improved; some argue that it is getting worse. It might be that these approaches just need a little time to mature and that their resultant drug candidates have yet to reach the clinic. In any event, the odds today are still about 20:1 that a compound entering phase I trials will fail to win regulatory approval. Yet the case history of cinacalcet flies in the face of this attrition rate: only two calcimimetics entered the clinic and one made it all the way to the market. Sometimes you do get lucky, but calcimimetics beat the odds mostly because the concept was based on physiology and the comparatively simple pathophysiology of hyperparathyroidism. In 1990, when the calcimimetic drug discovery project began in earnest, there were 50 years of physiology dictating what the drug must do—it had to do what extracellular Ca^{2+} does because that is the primary physiological stimulus regulating the secretion of PTH. It was a physiological phenomenon searching for a mechanism. The calcium receptor provided that mechanism, and thereby a target with an intrinsically high probability of success. Most drug discovery programs today start with a molecular mechanism in the hope that it will explain some physiological phenomenon.

There is one other factor that allowed calcimimetics to beat the odds: if you start with a compound that is already a drug, you are likely to end up with a drug. But that is a story for a different audience.

REFERENCES

1. BROWN, E.M., A. CONIGRAVE & N. CHATTOPADHYAY. 2001. Receptors and signaling for calcium ions. *In* The Parathyroids. J.P. Bilezikian, R. Marcus & M.A. Levine, Eds.: 127–142. Academic Press. New York.
2. NEMETH, E.F. 2002. Calcium receptors as novel drug targets. *In* Principles of Bone Biology. J.P. Bilezikian, L.G. Raisz & G.A. Rodan, Eds.: 1339–1359. Academic Press. San Diego, CA.
3. CURHAN, G. 2004. Fooling the parathyroid gland–will there be health benefits? N. Engl. J. Med. **350:** 1565–1567.
4. STEWART, A.F. 2004. Translational implications of the parathyroid calcium receptor. N. Engl. J. Med. **351:** 324–326.

5. LOPEZ-BARNEO, J. & C.M. ARMSTRONG. 1983. Depolarizing response of rat parathyroid cells to divalent cations. J. Gen. Physiol. **82:** 269–294.
6. GYLFE, E., R. LARSSON, H. JOHANSSON, *et al.* 1986. Calcium-activated calcium permeability in parathyroid cells. FEBS Lett. **205:** 132–136.
7. NEMETH, E.F. & A. SCARPA. 1986. Cytosolic Ca^{2+} and the regulation of secretion in parathyroid cells. FEBS Lett. **203:** 15–19.
8. NEMETH, E.F. & E. CARAFOLI. 1990. The role of extracellular calcium in the regulation of intracellular calcium and cell function. Cell Calcium **11:** 319–321.
9. MCLEAN, F.C. 1961. Unsolved problems of parathyroid physiology. *In* The Parathyroids. R.O. Greep & R.V. Talmage, Eds.: 7–18. Charles C. Thomas. Springfield, IL.
10. CHEN, C.J., J.V. BARNETT, D.A. CONGO, *et al.* 1989. Divalent cations suppress 3′-5′-adenosine monophosphate accumulation by stimulating a pertussis toxin-sensitive guanine nucleotide-binding protein in cultured bovine parathyroid cells. Endocrinology **124:** 233–239.
11. SHOBACK, D.M. & T.H. CHEN. 1991. Injection of poly(A)+ RNA from bovine parathyroid tissue into Xenopus oocytes confers sensitivity to extracellular calcium [abstract]. J. Bone Miner. Res. **6:** S135.
12. RACKE, F.K., G.R. DUBYAK & E.F. NEMETH. 1991. Functional expression of the parathyroid cell calcium receptor in Xenopus oocytes [abstract]. J. Bone Miner. Res. **6:** S118.
13. BROWN, E.M., G. GAMBA, D. RICCARDI, *et al.* 1993. Cloning and characterization of an extracellular Ca^{2+}-sensing receptor from bovine parathyroid. Nat. Lett. **366:** 575–580.
14. POLLAK, M.R., E.M. BROWN, Y.H.W. CHOU, *et al.* 1993. Mutations in the human Ca^{2+}-sensing receptor gene cause familial hypocalciuric hypercalcemia and neonatal severe hyperparathyroidism. Cell **75:** 1297–1303.
15. OTT, S.M. 1998. Calcimimetics–new drugs with the potential to control hyperparathyroidism. J. Clin. Endocrinol. Metab. **83:** 1080–1082.
16. HEATH, H. III, E.L. SANGUINETTI, S. OGLESBY & T.B. MARRIOTT. 1995. Inhibition of human parathyroid secretion *in vivo* by NPS-568, a calcimimetic drug that targets the parathyroid cell surface calcium receptor. Bone **16**(Suppl.): 85S.
17. NEMETH, E.F., B.C. VANWAGENEN & M.F. BALANDRIM. 1993. Calcium receptor-active molecules. Chem. Abstr. **119:** P63054s.
18. GOUDET, C., V. BINET, L. PREZEAU, *et al.* 2004. Allosteric modulators of class-C G-protein-coupled receptors open new possibilities for therapeutic application. Drug Discov. Today: Therap. Strat. **1:** 125–133.
19. NEMETH, E.F., M.E. STEFFEY, L.G. HAMMERLAND, *et al.* 1998. Calcimimetics with potent and selective activity on the parathyroid calcium receptor. Proc. Natl. Acad. Sci. USA **95:** 4040–4045.
20. BLACK, J. 1989. Drugs from emasculated hormones: the principle of syntopic antagonism. Biosci. Rep. **9:** 253–272.
21. PARFITT, A.M. 2003. Misconceptions (3): calcium leaves bone only by resorption and enters only by formation. Bone **33:** 259–263.
22. NEMETH, E.F. 2004. Calcimimetic and calcilytic drugs: just for parathyroid cells? Cell Calcium **35:** 283–289.

Osteoblastic Activation in the Hematopoietic Stem Cell Niche

LAURA M. CALVI

Endocrine Division, Department of Medicine, University of Rochester School of Medicine and Dentistry, Rochester, New York 14642, USA

ABSTRACT: Hematopoietic stem cells (HSC) are rare primitive cells capable of reconstituting all blood cell lineages throughout the life of an individual. The microenvironment in which stem cells reside is essential for their survival, self-renewal, and differentiation. This microenvironment, or HSC niche, has been difficult to define in bone and bone marrow, but recent studies from our laboratory and others have shown that osteoblasts, the bone-forming cells, are an essential regulatory component of this complex cellular network. We established that parathyroid hormone (PTH), through activation of the PTH/PTHrP receptor (PTH1R) in osteoblastic cells, could alter the HSC niche resulting in HSC expansion *in vivo* and *in vitro* and improving dramatically the survival of mice receiving bone marrow transplants. These findings are of great clinical appeal, because they suggest that a strategy aimed at modifying supportive cells in a stem cell niche can expand HSC. While a number of molecules have been found to be important for hematopoietic/osteoblastic interactions, we have focused on the Jagged1/Notch signaling pathway, which was necessary for the PTH-dependent HSC expansion. Since the Jagged1/Notch signaling pathway has been implicated in the microenvironmental control of stem cell self-renewal in several organ systems, definition of Jagged1 modulation, which is currently poorly understood, should provide additional molecular targets for stem cell regulation and advance the understanding of stem cell-microenvironmental interactions.

KEYWORDS: stem cell niche; osteoblast; parathyroid hormone; hematopoiesis; Notch; Jagged1; self-renewal

INTRODUCTION

Stem cells are defined as rare cells with extensive proliferative potential, the ability of giving rise to one or more differentiated cell types, and unlimited self-renewal. During mammalian embryonal development, only the fertilized egg

Address for correspondence: Laura M. Calvi, Endocrine Division, Department of Medicine, University of Rochester School of Medicine and Dentistry, 601 Elmwood Avenue, Rochester, NY 14642. Voice: 585-275-2901; fax: 585-273-1288.
e-mail: laura_calvi@urmc.rochester.edu

Ann. N.Y. Acad. Sci. 1068: 477–488 (2006). © 2006 New York Academy of Sciences.
doi: 10.1196/annals.1346.021

and the early cleavage blastomere are totipotent stem cells, capable of generating all cell types. Later in differentiation, the inner cell mass of the blastocyst is composed of stem cells that are pluripotent, or capable of generating all the cell types found in the embryo and adult.[1] In addition to the embryonic stem cells described above, adult stem cells have been hypothesized and identified as rare cells with the ability to indefinitely self-renew. These cells are defined as multipotent, as they are capable of regenerating several but not all cell types present in the adult. Stem cells have to continuously be poised between self-renewal and commitment to differentiation in order to be able to exist throughout the life of an individual, and at the same time function as a reservoir, which can be called upon at times of stress. Since a number of pathologies are caused by destruction or degeneration of tissues and organs, the ability to understand and control adult stem cells at will has enormous therapeutic potential. This stem cell control could be intrinsic of stem cells and studies have begun to define a set of genes that define "stemness."[2–4] Alternatively, or in addition, external regulatory mechanisms could determine stem cell fate decision. Although the existence of a microenvironment or niche was postulated as early as the 1970s,[5] the niche hypothesis was verified *in vivo* only recently, through the work on *Drosophila* ovarian and testis stem cells.[6–8] Interestingly, the demonstration of a hematopoietic stem cell (HSC) niche has been elusive, although much evidence pointed to the bone endosteal surface as an important harbor for the most primitive hematopoietic cells during adult life. Only recently, our laboratory and others have demonstrated that osteoblastic cells are a regulatory component of the HSC niche, and have begun to elucidate the cellular and molecular mechanisms mediating osteoblastic–hematopoietic interactions.

HSC AND THE NICHE HYPOTHESIS

HSC are the best understood stem cells. Surface markers identify oligopotent precursor and differentiated progeny as well as the most primitive cells, allowing for the prospective isolation of HSC, which has been accomplished in both mice and humans.[9–13] These cells can be quantified also through functional assays,[14,15] which rely on stromal/HSC interactions and identify limiting dilution of long-term culture initiating-cells (LTC-ICs). In addition, the ability of these cells to give rise to the full hematopoietic repertoire is demonstrated and can be quantified through competitive transplantation.[16] Finally, not only are these cells well characterized, but they were the first stem cells to be used therapeutically, and they are currently routinely relied upon during bone marrow transplantation for the treatment of hematologic malignancies and bone marrow failure states.[17] Self-renewal, a characteristic of both long-term and short-term HSC, [18] is essential for primitive stem cells to persist for the lifetime of an individual. A number of signaling pathways have been identified that appear to be important for HSC self-renewal, including the Sonic Hedgehog,[19] Wnt,[20–22] and Notch[23–26] signaling pathways. In all of these pathways, interactions of a receptor on the surface of the HSC with either a secreted or a cell-bound ligand lead to physiologic activation of the pathway and self-renewal rather than differentiation. Interestingly, some of these pathways have been found to be important not only for HSC but also for other stem cell systems.

The concept that key control of stem cell fate may not be intrinsic to the stem cell itself but instead could be conferred to the cells by the microenvironment is not a new one. Shofield in 1978 initially proposed that, in order to explain the limited expansion of HSC in transplanted animals, it would be necessary to envision a specialized location, or niche, in which the stem cells would be located and which would regulate their ability to either self-renew or differentiate.[5] Such a niche would be a custom microenvironment, composed of specialized cells capable of supporting stem cells. Stem cells would be physically anchored to support cells, which should be capable to provide the factors necessary for fate determination and self-renewal. While the niche hypothesis was proposed for HSC, proof of its existence was demonstrated initially in the *Drosophila* ovary and testis.[6–8] In the *Drosophila* ovary, somatic cap cells anchor germline stem cells through adherens junctions, stimulate the reception of an essential bone morphogenetic protein (BMP) signal, and are capable of reprogramming cells to become stem cells.[6,27,28] Identification of an HSC had, however, remained elusive, in part because of the complex organization of the bone marrow space, and in part because of the difficulty in identifying and labeling HSC *in vivo*.

OSTEOBLASTS ARE A REGULATORY COMPONENT OF THE HSC NICHE

In spite of the difficulty in physically identifying the HSC niche, mounting evidence had suggested that such a niche would exist. If the hypothesis of a niche as a physical space were to be true for HSC, some architectural organization of the HSC and their progeny would be expected, as was seen in the *Drosophila* gonads. In fact, such spatial organization was already demonstrated by Gong and Lord, who identified more primitive hematopoietic progenitor cells in close proximity to the endosteal surface, while more differentiated cells were seen in the center of the bone marrow space.[29,30] More recent studies confirmed these findings, and showed that infused primitive hematopoietic cells also tend to preferentially home to the endosteal surface.[31]

The ability to assess physiologic exogenous cues has been partly limited by poor definition of stromal constituents of the HSC niche. Stromal cell components have been thought to include fibroblasts, adipocytes, and endothelial cells as well as cells of the osteoblastic lineage. However, in light of the architectural organization of the bone marrow reviewed above, cells on the endosteal surface, and particularly cells of the osteoblastic lineage, would be natural candidates as HSC niche cells. In fact, a number of studies have suggested that osteoblastic cells may play a role in the regulation of the hematopoietic system. Osteoblastic cells can support both terminal granulomatopoiesis and HSC survival *in vitro*,[32] they can stimulate intermediate progenitors (CFU-Cs) and more primitive cells (LTC-ICs),[33] expand HSC two- to fourfold *in vitro*,[34] and produce high levels of factors important for HSC support.[32] In addition, osteoblastic cells engraft during bone marrow transplantation, and their co-transplantation with HSC can increase engraftment rate.[35,36]

Only recently osteoblastic cells have been shown *in vivo* to be a regulatory component in the HSC niche.[37,38] Through the use of genetically altered animal models, specific expansion and/or activation of osteoblastic cells resulted in a specific increase in HSC frequency,[37,38] while osteoblastic destruction resulted in loss of HSC.[39] In the work by Zhang *et al*., mice with conditional inactivation of the BMP receptor IA (BMPRIA) were found to have a characteristic bone phenotype, with additional trabecular bony structures along the femoral endosteal surface.[38] In these mice, the osteoblastic cell pool is expanded, and HSC are phenotypically and functionally increased.[38] The investigators demonstrate that the HSC phenotype of the transgenic mice with conditional inactivation of the BMPRIA is mediated by the microenvironment, given the pattern of expression of the BMPRIA, which is not expressed in HSC, and given the results of reciprocal bone marrow transplant experiments. They then show that HSC are attached to an N-cadherin positive subset of osteoblastic cells. Although visualization of labeled HSC in the intact bone remains a controversial topic, expansion of osteoblastic cells was responsible for the HSC increase, supporting the hypothesis that osteoblastic cells are one component of the HSC niche. Consistent with the importance of N-cadherin in mediating HSC–osteoblastic interactions, angiopoietin and Tie-2 were recently shown to regulate HSC quiescence at least partially through osteoblastic N-cadherin.[40] In addition, c-Myc-deficient HSC have impaired differentiation, increased N-cadherin expression, and are localized to the osteoblastic niche, while c-Myc overexpression in HSC diminishes N-cadherin with loss of HSC self-renewal,[41] once again suggesting the importance of osteoblastic cells for HSC regulation. Finally, the osteoblastic-secreted matrix protein, osteopontin, was recently found to be another regulatory component of the HSC niche capable of negatively regulating HSC self-renewal.[42,43]

PARATHYROID HORMONE MODULATES THE HSC NICHE

In a study simultaneous to the work by Zhang *et al.*,[37] we explored the important relationship between osteoblasts and HSC by studying a genetically altered mouse model in which osteoblast-specific expression of an activated

PTH1R is targeted by the 2.3-kb fragment of the $\alpha 1$(I) collagen promotor (Col1-caPTH1R mice).[44] Since this fragment of this promotor specifically targets cells of the osteoblastic lineage,[45–48] this transgenic model allows for the study of the specific effect of PTH1R activation on osteoblastic cells. The PTH1R, a G-protein-coupled receptor, is linked to the adenylyl cyclase/protein kinase A signaling system and to the phospholipase C/protein kinase C systems.[49,50] In contrast, the mutant PPR used in the Col1-caPPR transgene causes ligand-independent cAMP accumulation, without inositol phosphate production.[51,52] The bones of Col1-caPTH1R mice demonstrated brisk trabecular bone formation, increased trabeculae, and trabecular osteoblastic cells. Osteoblastic cells expressing the constitutively active PTH1R were increased in number, produced high levels of the Notch ligand, Jagged1, and supported an increase in the number of HSC, identified by flow cytometric analysis, functional assays, and competitive transplantation, with evidence of Notch1 activation in stem cells *in vivo*. PTH treatment both *in vivo* and *in vitro* could reproduce the HSC expansion. Interestingly, when myeloablated mice were transplanted with limiting numbers of bone marrow mononuclear cells (BMMC), PTH treatment dramatically improved survival and bone marrow morphology compared with control.[37]

These studies indicate that osteoblastic cells represent a regulatory component of the bone marrow microenvironment that exerts an effect on HSC through Notch signaling. The Notch signaling pathway is activated by direct cell–cell contact of the Notch receptor with one of its ligands. In the absence of Notch signaling, lineage-specific gene expression occurs and cells undergo differentiation. In contrast, when a Notch ligand, such as Jagged or Delta binds the Notch receptor, through γ-secretase enzymatic activity, an activated Notch intracytoplasmic domain (NICD) is released. Nuclear translocation of activated NICD results in transcriptional suppression, with inhibition of differentiation and primitive cell self-renewal.[53] Increased stromal Jagged1 is sufficient to expand HSC,[54] and we identified PTH1R-dependent osteoblastic Jagged1 increase and HSC Notch1 activation. It remains unclear whether PTH1R activation results in increased levels of expression of Jagged1 in all or a particular subpopulation of osteoblastic cells, whether Jagged1 is necessary for PTH-dependent HSC expansion, and whether PTH can directly alter the HSC niche through Notch signaling.

Osteoblasts are the main target of PTH in bone. Progenitors of these cells are largely of mesenchymal origin.[55] Committed osteoprogenitor cells proliferate and then differentiate into later stages. The progression of osteoblastic maturation requires the sequential activation and suppression of genes that encode phenotypic and regulatory proteins, providing a panel of well-established markers that reflect stages of osteoblastic differentiation.[55–58] These stages are recognizable as preosteoblasts, mature osteoblasts, quiescent bone-lining cells, and osteocytes by morphological, histochemical, and molecular criteria.[55,59] The mechanism of action of PTH in these cells appears controversial: PTH

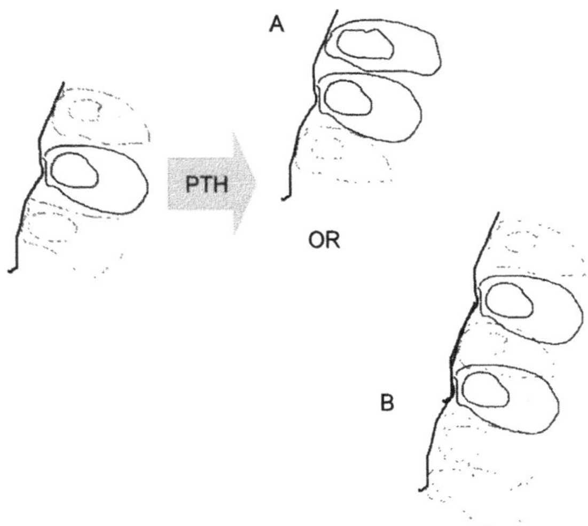

FIGURE 1. PTH may expand HSC by: (**A**) specifically expanding HSC supportive osteoblasts (pink), or (**B**) by simply expanding the total osteoblastic cell population.

either inhibits or stimulates the growth of osteoblastic cells, and exerts either differentiating or de-differentiating effects depending on the *in vitro* system studied.[60–62] In addition, studies have suggested that PTH may have different actions on osteoblastic cells at different stages of differentiation.[63] Therefore, PTH treatment may simply increase total osteoblastic cells and consequently expand the osteoblastic population responsible for HSC support (FIG. 1). Alternatively, PTH could specifically expand or stimulate a population of immature osteoblastic cells, which have increased levels of Jagged1 and are particularly suited to modulate HSC (FIG. 1). As discussed above, *in vivo* evidence provided by Zhang *et al.* also suggests that a particular osteoblastic subpopulation is a key component of the HSC niche.[38] Indeed, we found an expansion of the N-cadherin positive osteoblastic population in the Col1-caPTH1R mice (LMC, in preparation). However, the specific differentiation stage of this osteoblastic subpopulation is poorly defined. Similarly, the degree of differentiation of Jagged1-expressing cells, and whether this is altered by PTH treatment, is unknown. An essential challenge for the understanding of osteoblastic action in the HSC mice will be the precise definition of the HSC-supporting osteoblastic cells, and whether PTH treatment alters their number or function.

ROLE OF THE NOTCH LIGAND JAGGED1 IN OSTEOBLASTIC SUPPORT OF HSC

While it is known that increase of the Notch ligand Jagged1 in the bone marrow stroma is sufficient to expand HSC,[54] that Jagged1 is expressed in osteoblastic cells[64] and increased in clonal osteoblastic cells by PTH,[65] it remains unclear whether Jagged1 expression is necessary to achieve PTH-mediated HSC expansion. The Notch signaling pathway involves membrane-bound ligands and regulates cell-fate specification in a wide variety of systems including HSC self-renewal.[24,25,66–68] By changing the balance of self-renewal versus differentiation, Notch signaling has been shown to increase stem cell numbers without expanding mature cells in a manner reminiscent of the phenotype observed in the Col1-caPPR and in PTH-treated mice.[25,66] Further, Jagged1 has been shown to be expressed by bone marrow-derived stromal cells (BMSCs)[25,54] and murine osteoblasts.[64] Our data showing that γ-secretase activity is necessary *in vitro* to mediate PTH expansion of HSC[37] implicate the Notch signaling pathway, but do not clarify whether Jagged1 is the fundamental Notch ligand mediating this effect. In fact, a recent study questioned the need of Jagged1 for HSC self-renewal.[69] While activated Notch1 levels were increased in HSC in Col1-caPPR mice, it is not known whether Notch1 is necessary to mediate this microenvironmental effect of PTH. Further studies evaluating the effects of PTH on the HSC niche in the absence of osteoblastic Jagged1 will establish the importance of this Notch ligand in the PTH-dependent HSC expansion.

Expression of Jagged1 has been detected in immortalized murine osteoblastic cells,[64,70] and in osteoblastic cells during bone regeneration.[71] Significant controversy exists on the role of Notch signaling in osteoblastic differentiation. Nobta *et al.*[71] suggested that Jagged1 is capable of enhancing bone mineral deposition, and others have shown that activation of Notch signaling either impairs[70,72] or facilitates[71,73] osteoblastic differentiation. However, it is currently unknown whether Jagged1 is regulated during osteoblastic differentiation. In addition, the effects of PTH in altering the osteoblastic pool have been demonstrated both *in vivo* and *in vitro*: in addition to controversial effects of PTH on osteoblastic differentiation discussed above,[60–62] studies have suggested that PTH may increase the proportion of BMSC that commit to the osteoblastic lineage both *in vitro* and *in vivo*,[63,74,75] increase osteoblastic proliferation,[44,76,77] and decrease rates of osteoblastic apoptosis.[44,78] Furthermore, PTH has been shown *in vivo* to expand the osteoblastic cell pool population,[44] an effect that was also present in PTH-treated stromal cell cultures used in LTC-IC assays.[37] It is therefore possible that osteoblastic PTH stimulation, through Jagged1 and Notch1 interactions within the osteoblastic cell population may promote a specific osteoblastic phenotype in the stromal environment, which may better support HSC.

CONCLUSION

While osteoblastic cells have been recently identified as cellular participants in the HSC niche, we are only beginning to define the molecular and cellular mechanisms mediating this very specific hematopoietic–osteoblastic interaction. The data reviewed here suggest that it may be possible to define the complexity of the niche and better understand the systems directing stem cell behavior, with important potential implications for the treatment of disease states. Future challenges of this nascent field include definition of the osteoblastic cell necessary for HSC support, including their differentiation stage and adhesion molecules, and a better understanding of the mechanisms by which PTH improves osteoblastic–HSC interaction. Twenty-five years after the niche hypothesis was proposed for HSC, we are only beginning to shed light on the complex interactions between osteoblasts and hematopoietic cells.

ACKNOWLEDGMENTS

I would like to thank L.A. Milner, C.T. Jordan, J.W. Friedberg, and the members of the Musculoskeletal Research Center at the University of Rochester School of Medicine and Dentistry for helpful discussion. This work is supported by the National Institutes of Health, the Wilmot Cancer Research Fellowship, and the Pew foundation.

REFERENCES

1. Lovell-Badge, R. 2001. The future for stem cell research. Nature **414:** 88–91.
2. Hackney, J.A. *et al.* 2002. A molecular profile of a hematopoietic stem cell niche. Proc. Natl. Acad. Sci. USA **99:** 13061–13066.
3. Ivanova, N.B. *et al.* 2002. A stem cell molecular signature. Science **298:** 601–604.
4. Ramalho-Santos, M. *et al.* 2002. “Stemness”: transcriptional profiling of embryonic and adult stem cells. Science **298:** 597–600.
5. Schofield, R. 1978. The relationship between the spleen colony-forming cell and the haemopoietic stem cell. Blood Cells **4:** 7–25.
6. Xie, T. & A.C. Spradling. 2000. A niche maintaining germ line stem cells in the Drosophila ovary. Science **290:** 328–330.
7. Kiger, A.A. *et al.* 2001. Stem cell self-renewal specified by JAK-STAT activation in response to a support cell cue. Science **294:** 2542–2545.
8. Tulina, N. & E. Matunis. 2001. Control of stem cell self-renewal in Drosophila spermatogenesis by JAK-STAT signaling. Science **294:** 2546–2549.
9. Reya, T. *et al.* 2001. Stem cells, cancer, and cancer stem cells. Nature **414:** 105–111.
10. Spangrude, G.J., S. Heimfeld & I.L. Weissman. 1988. Purification and characterization of mouse hematopoietic stem cells. Science **241:** 58–62.
11. Baum, C.M. *et al.* 1992. Isolation of a candidate human hematopoietic stem-cell population. Proc. Natl. Acad. Sci. USA **89:** 2804–2808.
12. Morrison, S.J. & I.L. Weissman. 1994. The long-term repopulating subset of hematopoietic stem cells is deterministic and isolatable by phenotype. Immunity **1:** 661–673.

13. OSAWA, M. *et al.* 1996. Long-term lymphohematopoietic reconstitution by a single CD34-low/negative hematopoietic stem cell. Science **273:** 242–245.
14. SUTHERLAND, H.J. *et al.* 1989. Characterization and partial purification of human marrow cells capable of initiating long-term hematopoiesis *in vitro.* Blood **74:** 1563–1570.
15. SUTHERLAND, H.J. *et al.* 1990. Functional characterization of individual human hematopoietic stem cells cultured at limiting dilution on supportive marrow stromal layers. Proc. Natl. Acad. Sci. USA **87:** 3584–3588.
16. HARRISON, D.E. 1980. Competitive repopulation: a new assay for long-term stem cell functional capacity. Blood **55:** 77–81.
17. ANTIN, J. &, N. KERNAND 2003. Hematopoietic stem cell transplantation. *In* Blood Principles and Practice of Hematology. R.I. Handin, S.E. Lux & T.P. Stossel, Eds.: 2133–2190. Lippincott Williams and Wilkins. Philadelphia, PA.
18. MORRISON, S.J. *et al.* 1997. Identification of a lineage of multipotent hematopoietic progenitors. Development **124:** 1929–1939.
19. BHARDWAJ, G. *et al.* 2001. Sonic hedgehog induces the proliferation of primitive human hematopoietic cells via BMP regulation. Nat. Immunol. **2:** 172–180.
20. YAMANE, T. *et al.* 2001. Wnt signaling regulates hemopoiesis through stromal cells. J. Immunol. **167:** 765–772.
21. MURDOCH, B. *et al.* 2003. Wnt-5A augments repopulating capacity and primitive hematopoietic development of human blood stem cells in vivo. Proc. Natl. Acad. Sci. USA **100:** 3422–3427.
22. WILLERT, K. *et al.* 2003. Wnt proteins are lipid-modified and can act as stem cell growth factors. Nature **423:** 448–452.
23. VARNUM-FINNEY, B. *et al.* 1998. The Notch ligand, Jagged-1, influences the development of primitive hematopoietic precursor cells. Blood **91:** 4084–4091.
24. VARNUM-FINNEY, B., C. BRASHEM-STEIN & I.D. BERNSTEIN. 2003. Combined effects of Notch signaling and cytokines induce a multiple log increase in precursors with lymphoid and myeloid reconstituting ability. Blood **101:** 1784–1789.
25. KARANU, F.N. *et al.* 2000. The Notch ligand jagged-1 represents a novel growth factor of human hematopoietic stem cells. J. Exp. Med. **192:** 1365–1372.
26. BURNS, C.E. *et al.* 2005. Hematopoietic stem cell fate is established by the Notch-Runx pathway. Genes Dev. **19:** 2331–2342.
27. XIE, T. & A.C. SPRADLING. 1998. Decapentaplegic is essential for the maintenance and division of germline stem cells in the Drosophila ovary. Cell **94:** 251–260.
28. SONG, X. *et al.* 2002. Germline stem cells anchored by adherens junctions in the Drosophila ovary niches. Science **296:** 1855–1857.
29. GONG, J.K. 1978. Endosteal marrow: a rich source of hematopoietic stem cells. Science **199:** 1443–1445.
30. LORD, B.I., N.G. TESTA & J.H. HENDRY. 1975. The relative spatial distributions of CFUs and CFUc in the normal mouse femur. Blood **46:** 65–72.
31. NILSSON, S.K., H.M. JOHNSTON & J.A. COVERDALE. 2001. Spatial localization of transplanted hemopoietic stem cells: inferences for the localization of stem cell niches. Blood **97:** 2293–2299.
32. TAICHMAN, R.S. & S.G. EMERSON. 1994. Human osteoblasts support hematopoiesis through the production of granulocyte colony-stimulating factor. J. Exp. Med. **179:** 1677–1682.
33. TAICHMAN, R.S., M.J. REILLY & S.G. EMERSON. 1996. Human osteoblasts support human hematopoietic progenitor cells in vitro bone marrow cultures. Blood **87:** 518–524.

34. TAICHMAN, R.S., M.J. REILLY & S.G. EMERSON. 2000. The hematopoietic microenvironment: osteoblasts and the hematopoietic microenvironment. Hematology **4:** 421–426.
35. EL-BADRI, N.S. *et al.* 1998. Osteoblasts promote engraftment of allogeneic hematopoietic stem cells. Exp. Hematol. **26:** 110–116.
36. NILSSON, S.K. *et al.* 1999. Cells capable of bone production engraft from whole bone marrow transplants in nonablated mice. J. Exp. Med. **189:** 729–734.
37. CALVI, L.M. *et al.* 2003. Osteoblastic cells regulate the haematopoietic stem cell niche. Nature **425:** 841–846.
38. ZHANG, J. *et al.* 2003. Identification of the haematopoietic stem cell niche and control of the niche size. Nature **425:** 836–841.
39. VISNJIC, D. *et al.* 2004. Hematopoiesis is severely altered in mice with an induced osteoblast deficiency. Blood **103:** 3258–3264.
40. ARAI, F. *et al.* 2004. Tie2/angiopoietin-1 signaling regulates hematopoietic stem cell quiescence in the bone marrow niche. Cell **118:** 149–161.
41. WILSON, A. *et al.* 2004. c-Myc controls the balance between hematopoietic stem cell self-renewal and differentiation. Genes Dev. **18:** 2747–2763.
42. NILSSON, S.K. *et al.* 2005. Osteopontin, a key component of the hematopoietic stem cell niche and regulator of primitive hematopoietic progenitor cells. Blood **106:** 1232–1239.
43. STIER, S. *et al.* 2005. Osteopontin is a hematopoietic stem cell niche component that negatively regulates stem cell pool size. J. Exp. Med. **201:** 1781–1791.
44. CALVI, L.M. *et al.* 2001. Activated parathyroid hormone/parathyroid hormone-related protein receptor in osteoblastic cells differentially affects cortical and trabecular bone. J. Clin. Invest. **107:** 277–286.
45. ROSSERT, J.A. *et al.* 1996. Identification of a minimal sequence of the mouse pro-alpha 1(I) collagen promoter that confers high-level osteoblast expression in transgenic mice and that binds a protein selectively present in osteoblasts. Proc. Natl. Acad. Sci. USA **93:** 1027–1031.
46. ROSSERT, J., & B. DE CROMBRUGGHE. 1996. Type I collagen: structure, synthesis and regulation. *In* Principles of Bone Biology. J.P. Bilezikian, L.G. Raisz, G.A. Rodan, Eds.: 127–142. Academic Press. San Diego.
47. ROSSERT, J., H. EBERSPAECHER & B. DE CROMBRUGGHE. 1995. Separate cis-acting DNA elements of the mouse pro-alpha 1(I) collagen promoter direct expression of reporter genes to different type I collagen-producing cells in transgenic mice. J. Cell Biol. **129:** 1421–1432.
48. PAVLIN, D. *et al.* 1992. Differential utilization of regulatory domains within the alpha 1(I) collagen promoter in osseous and fibroblastic cells. J. Cell Biol. **116:** 227–236.
49. JUPPNER, H. *et al.* 1991. A G protein-linked receptor for parathyroid hormone and parathyroid hormone-related peptide. Science **254:** 1024–1026.
50. ABOU-SAMRA, A.B. *et al.* 1992. Expression cloning of a common receptor for parathyroid hormone and parathyroid hormone-related peptide from rat osteoblast-like cells: a single receptor stimulates intracellular accumulation of both cAMP and inositol trisphosphates and increases intracellular free calcium. Proc. Natl. Acad. Sci. USA **89:** 2732–2736.
51. SCHIPANI, E., K. KRUSE & H. JUPPNER. 1995. A constitutively active mutant PTH-PTHrP receptor in Jansen-type metaphyseal chondrodysplasia. Science **268:** 98–100.

52. SCHIPANI, E. *et al.* 1996. Constitutively activated receptors for parathyroid hormone and parathyroid hormone-related peptide in Jansen's metaphyseal chondrodysplasia [see comments]. N. Engl. J. Med. **335:** 708–714.
53. MILNER, L.A. & A. BIGAS. 1999. Notch as a mediator of cell fate determination in hematopoiesis: evidence and speculation. Blood **93:** 2431–2448.
54. LI, L. *et al.* 1998. The human homolog of rat Jagged1 expressed by marrow stroma inhibits differentiation of 32D cells through interaction with Notch1. Immunity **8:** 43–55.
55. AUBIN, J., & J.T. TRIFFITT. 2002. Mesenchymal stem cells and osteoblast differentiation. *In* Principles of Bone Biology, Vol. 1. J.P. Bilezikian, L.G. Raisz, G.A. Rodan, Eds.: 59–81. Academic Press. San Diego, CA.
56. MALAVAL, L. *et al.* 1994. Cellular expression of bone-related proteins during in vitro osteogenesis in rat bone marrow stromal cell cultures. J. Cell Physiol. **158:** 555–572.
57. OTTO, F. *et al.* 1997. Cbfa1, a candidate gene for cleidocranial dysplasia syndrome, is essential for osteoblast differentiation and bone development [see comments]. Cell **89:** 765–771.
58. BANERJEE, C. *et al.* 1997. Runt homology domain proteins in osteoblast differentiation: AML3/CBFA1 is a major component of a bone-specific complex. J. Cell Biochem. **66:** 1–8.
59. STEIN, G.S. & J.B. LIAN. 1993. Molecular mechanisms mediating proliferation/differentiation interrelationships during progressive development of the osteoblast phenotype. Endocr. Rev. **14:** 424–442.
60. BELLOWS, C.G. *et al.* 1990. Parathyroid hormone reversibly suppresses the differentiation of osteoprogenitor cells into functional osteoblasts. Endocrinology **127:** 3111–3116.
61. ONYIA, J.E. *et al.* 1995. In vivo, human parathyroid hormone fragment (hPTH 1-34) transiently stimulates immediate early response gene expression, but not proliferation, in trabecular bone cells of young rats. Bone **17:** 479–484.
62. CARPIO, L. *et al.* 2001. Induction of osteoblast differentiation indexes by PTHrP in MG-63 cells involves multiple signaling pathways. Am. J. Physiol. Endocrinol. Metab. **281:** E489–E499.
63. ISOGAI, Y. *et al.* 1996. Parathyroid hormone regulates osteoblast differentiation positively or negatively depending on the differentiation stages. J. Bone Miner. Res. **11:** 1384–1393.
64. PEREIRA, R.M. *et al.* 2002. Cortisol regulates the expression of Notch in osteoblasts. J. Cell. Biochem. **85:** 252–258.
65. QIN, L. *et al.* 2003. Gene expression profiles and transcription factors involved in parathyroid hormone signaling in osteoblasts revealed by microarray and bioinformatics. J. Biol. Chem. **278:** 19723–19731.
66. STIER, S. *et al.* 2002. Notch1 activation increases hematopoietic stem cell self-renewal in vivo and favors lymphoid over myeloid lineage outcome. Blood **99:** 2369–2378.
67. VARNUM-FINNEY, B. *et al.* 2000. Pluripotent, cytokine-dependent, hematopoietic stem cells are immortalized by constitutive Notch1 signaling. Nat. Med. **6:** 1278–1281.
68. KARANU, F.N. *et al.* 2001. Human homologues of Delta-1 and Delta-4 function as mitogenic regulators of primitive human hematopoietic cells. Blood **97:** 1960–1967.

69. MANCINI, S.J. *et al.* 2005. Jagged1-dependent Notch signaling is dispensable for hematopoietic stem cell self-renewal and differentiation. Blood **105:** 2340–2342.
70. SCIAUDONE, M. *et al.* 2003. Notch 1 impairs osteoblastic cell differentiation. Endocrinology **144:** 5631–5639.
71. NOBTA, M. *et al.* 2005. Critical regulation of bone morphogenetic protein-induced osteoblastic differentiation by Delta1/Jagged1-activated Notch1 signaling. J. Biol. Chem. **280:** 15842–15848.
72. SHINDO, K. *et al.* 2003. Osteogenic differentiation of the mesenchymal progenitor cells, Kusa is suppressed by Notch signaling. Exp. Cell Res. **290:** 370–380.
73. WATANABE, N. *et al.* 2003. Suppression of differentiation and proliferation of early chondrogenic cells by Notch. J. Bone Miner. Metab. **21:** 344–352.
74. NISHIDA, S. *et al.* 1994. Increased bone formation by intermittent parathyroid hormone administration is due to the stimulation of proliferation and differentiation of osteoprogenitor cells in bone marrow. Bone **15:** 717–723.
75. CHAN, G.K. *et al.* 2003. Parathyroid hormone-related peptide interacts with bone morphogenetic protein 2 to increase osteoblastogenesis and decrease adipogenesis in pluripotent C3H10T 1/2 mesenchymal cells. Endocrinology **144:** 5511–5520.
76. MACDONALD, B.R., J.A. GALLAGHER & R.G. RUSSELL. 1986. Parathyroid hormone stimulates the proliferation of cells derived from human bone. Endocrinology **118:** 2445–2449.
77. MIAO, D. *et al.* 2001. Parathyroid hormone-related peptide stimulates osteogenic cell proliferation through protein kinase C activation of the Ras/mitogen-activated protein kinase signaling pathway. J. Biol. Chem. **276:** 32204–32213.
78. JILKA, R.L. *et al.* 1999. Increased bone formation by prevention of osteoblast apoptosis with parathyroid hormone. J. Clin. Invest. **104:** 439–446.

Circulating Cells with Osteogenic Potential

SUNDEEP KHOSLA AND GUITI Z. EGHBALI-FATOURECHI

Endocrine Research Unit, Mayo Clinic College of Medicine, Rochester, Minnesota 55905, USA

ABSTRACT: While osteoclast lineage cells are clearly present in the peripheral circulation, whether there is a comparable pool of circulating osteoblast lineage cells has remained controversial. Using assays requiring adherence to plastic (as originally described by Friedenstein and colleagues for bone marrow stromal cells over four decades ago), several studies have shown that plastic adherent cells with osteogenic potential are, indeed, present in the circulation of a number of species, but at extremely low concentrations. Work from a number of independent groups over the past decade has also identified a population of nonadherent bone marrow cells with osteogenic potential. Since these nonadherent cells may be much more likely to access the peripheral circulation than plastic adherent cells, we tested for the presence of circulating osteoblast lineage cells in humans using flow cytometry to identify cells in the peripheral blood expressing bone-related proteins. Our findings indicate that these cells are present in the circulation in significant numbers, are markedly increased in the peripheral blood of adolescent boys going through the growth spurt, and may also increase following fractures. These circulating osteogenic cells express bone-related proteins, can mineralize *in vitro*, and form bone *in vivo*. The identification of these osteogenic cells in peripheral blood opens up new questions regarding the possible role of these cells in bone remodeling, in fracture repair, and possibly in vascular calcification.

KEYWORDS: osteoblasts; circulating cells; stromal cells

Pioneering studies by Friedenstein in the 1960s[1] established that the bone marrow stroma contains plastic adherent cells (colony-forming units, fibroblast, CFU-F) that can give rise to a broad spectrum of fully differentiated connective tissues, including cartilage, bone, adipose tissue, fibrous tissue, and myelosupportive stroma. Over the subsequent decades, these studies led to considerable work on the characterization of bone marrow stromal cells in rodent and human systems with osteoblastic potential.[2] While these cells were known to be

Address for correspondence: Sundeep Khosla, M.D., Endocrine Research Unit, Mayo Clinic, 200 First Street SW, 5-194 Joseph, Rochester, MN 55905. Voice: 507-255-6788; fax: 507-255-4828.
e-mail: khosla.sundeep@mayo.edu

**Ann. N.Y. Acad. Sci. 1068: 489–497 (2006). © 2006 New York Academy of Sciences.
doi: 10.1196/annals.1346.022**

present in the bone marrow, the question of whether bone marrow stromal cells or cells with osteogenic potential were also present in the peripheral circulation was of considerable interest. Circulating osteoblast lineage cells would provide, for example, a possible source of cells for bone remodeling in cortical bone at sites not contiguous to red marrow, or perhaps for fracture healing at sites distant from red marrow. Moreover, since osteoclast lineage cells clearly circulated,[3] whether a parallel process existed for osteoblast lineage cells was unknown but certainly plausible.

Early evidence against the presence of circulating osteoblastic cells came from work of Gothlin and Ericsson[4] who used a model of parabiotic rats in which both rats had induction of a femoral fracture. One of the rats was subsequently shielded, whereas the other rat received radiation to destroy at least hematopoietic and possibly mesenchymal precursors (the latter are more radio-resistant). The shielded rat was then given a dose of ^{3}H-thymidine during arrest of cross-circulation, and at various subsequent time points, the fracture sites in both rats were examined for the presence of labeled cells. Since the rat infused with the ^{3}H-thymidine was found to have a label in monocytes, macrophages, endothelial cells, fibroblasts, chondroblasts, osteoblasts, and osteoclasts in the fracture callus, but the contra-lateral, parabiosed rat was found to have a label only in monocytes, macrophages, and osteoclasts, the inevitable conclusion was that osteoblasts did not circulate.

Despite these negative findings, several subsequent studies did demonstrate that plastic adherent cells with bone marrow stromal or osteoblastic characteristics were, indeed, present in the circulation, but at extremely low concentrations.[5,6] Thus, Kuznetsov and colleagues[6] obtained blood from a number of species and demonstrated the presence of plastic adherent colony-forming cells with osteogenic potential in the circulation of mice, rabbits, guinea pigs, and humans. However, as noted above, the frequency of these circulating adherent cells was extremely rare: $\sim$1 in 10^6 in mice and guinea pigs, 1 in 10^7 in rabbits, and < 1 in 10^8 in humans. The isolated cells had variable morphologies, with some colonies showing a fibroblastic appearance, whereas others were composed of more round or cuboidal cells. The cells stained for a number of bone-related proteins, including collagen type I (col I), osteonectin, and osteopontin, but interestingly, were negative for alkaline phosphatase (AP). When placed in hydroxyapatite implants in immunodeficient mice, cells derived from cultures of these circulating cells could form bone. In addition, these cells were also capable of acquiring an adipogenic phenotype when cultured in rabbit serum. Collectively, these findings indicated that circulating cells with osteoblastic and adipogenic potential could be identified in a number of species using assays requiring adherence to plastic, but that the concentration of these cells, particularly in humans, was extremely low.

Concurrent with this extensive body of work on plastic adherent bone marrow stromal cells, Long and colleagues[7–10] identified a plastic nonadherent population of cells in human bone marrow that also had osteogenic potential.

The primary assay they used was cell sorting using antibodies to osteocalcin (OCN), and AP to identify cells expressing these bone-related proteins on the cell surface. By flow cytometry, OCN-positive cells had forward/side scatter characteristics of small, round cells, and when cultured in the presence of TGF-β and accessory bone marrow cells, these cells proliferated and differentiated into mature osteoblastic cells, expressing the expected bone-related genes and demonstrating the ability to mineralize. Since AP is a cell surface protein, it is easy to understand why cell sorting (without having to permeabilize the cells) should work for this protein. It is less clear, however, why cell sorting with an OCN antibody in these and subsequent studies (see below) was able to identify osteogenic cells, since OCN is a secreted protein. Possibilities include the fact that OCN does possess several Gla residues, similar to clotting factor proteins, and these are known to be able to insert into membranes. Thus, in the process of secretion, OCN may become anchored to the cell membrane for long enough for the cell sorting to work. Alternatively, there is evidence for an OCN receptor,[11,12] and it is possible that the cells secreting OCN also have an OCN receptor on the cell surface, allowing the OCN to bind. Clearly, while further work needs to be done to establish just why cell sorting with OCN is feasible, the studies of Long et al.[7–10] and the subsequent studies noted below do establish that this method nonetheless identifies cells producing OCN and having osteogenic potential.

Consistent with the work of Falla *et al.*[13] concomitantly found that whereas normal mouse bone marrow contained 1 in 2.2×10^5 cells capable of forming bone nodules *in vitro*, treatment of the mice *in vivo* with 5-fluorouracil (5-FU) increased this frequency 12-fold (to 1 in 1.75×10^4 cells). 5-FU had previously been commonly used for the purification of hematopoietic stem cells from the bone marrow since it drastically reduces the number of cycling cells by incorporation into DNA during the S-phase of the cell cycle and ultimate cell death of cycling cells, but it spares the quiescent cells of the marrow. Interestingly, similar to Long's observations, these quiescent cells with osteogenic potential were found exclusively in the plastic nonadherent fraction of the bone marrow. These findings provided further support for a nonadherent bone marrow population capable of differentiating into osteoblasts, and also provided a possible explanation for the failure of the parabiotic studies of Gothlin and Ericsson[4] to detect circulating osteogenic cells: thus, if the circulating osteogenic cells came from the bone marrow nonadherent pool of preosteoblastic cells (which seems more likely than originating from plastic adherent bone marrow stromal cells), these cells would not have incorporated the pulse of ^{3}H-thymidine administered as the label (since they are relatively quiescent and 5-FU resistant).

Because of the extensive focus on bone marrow plastic adherent cells as described by Friedenstein *et al.*,[1] however, the work of Long *et al.*[7–10] and Falla *et al.*[13] was largely ignored until further studies by Dominici and colleagues[14] provided additional support for the bone marrow nonadherent cell population as a major source of osteogenic cells. These investigators compared the ability of bone marrow plastic adherent versus nonadherent cells from donor mice to reconstitute hematopoietic and osteogenic compartments of the bone marrow in lethally irradiated host mice. Thus, they first obtained plastic adherent marrow stromal cells from FVB/N mice and labeled them with green fluorescent protein (GFP) using a retroviral vector. These labeled cells were then infused into the host, irradiated mice, and as expected, the adherent bone marrow cells did not contribute to hematopoietic reconstitution of the host marrow. It is to be noted that ~1.5% of osteoblasts or osteocytes in the host were found to be GFP positive. The investigators next isolated bone marrow nonadherent cells from the donor mice, similarly transduced them with the GFP vector, and infused these labeled cells into the host, irradiated mice. They now found that >90% of blood leukocytes, erythrocytes, and platelets were GFP positive and more importantly, that ~18% of osteoblasts/osteocytes in the host were GFP positive (over 10-fold more than when plastic adherent bone marrow cells had been used). Moreover, by analyzing the integration site of the retroviral GFP vector, the investigators were also able to demonstrate that the same clone of donor-transduced cells gave rise to the hematopoietic and osteoblastic cells in the host. Collectively, these data indicated that the nonadherent bone marrow cells had a >10-fold more robust bone-repopulating activity in this assay than plastic adherent cells, and that a unique progenitor cell with both hematopoietic and osteoblastic differentiation potential resided in the nonadherent subset of bone marrow cells.

Further support for a common precursor cell to both hematopoietic and osteoblastic cells came from concurrent studies by Olmsted-Davis and colleagues,[15] who were studying a side population (SP) of bone marrow cells that had previously been shown to display strong hematopoietic reconstituting ability.[16] These SP cells had also been shown to differentiate *in vivo* into vascular endothelial cells.[17] The SP cells had been identified as a small population of bone marrow cells with the ability to extrude the Hoechst red dye, and were positive for the stem cell marker, *c-kit*. These investigators performed similar transplantation studies as described above, this time using donor cells from beta-galactosidase expressing (Rosa26) mice and found that virtually all of the osteoblastic reconstituting ability of whole bone marrow appeared to reside in this bone marrow SP cell fraction that had previously been identified primarily on the basis of providing a pool of hematopoietic stem cells.[15]

Collectively, these series of studies from several different groups indicated that the bone marrow nonadherent population did, in fact, contain a pool of cells with osteogenic potential (perhaps even more than that of adherent bone marrow stromal cells) and that in this pool were also cells with both hematopoietic and mesenchymal (e.g., osteoblastic) reconstituting ability. Based on this growing evidence for osteogenic cells in the bone marrow nonadherent fraction, we recently tested the hypothesis that these nonadherent bone marrow osteoblastic cells did, in fact, access the peripheral circulation and that they were present in peripheral blood in humans in much larger numbers than detected by assays

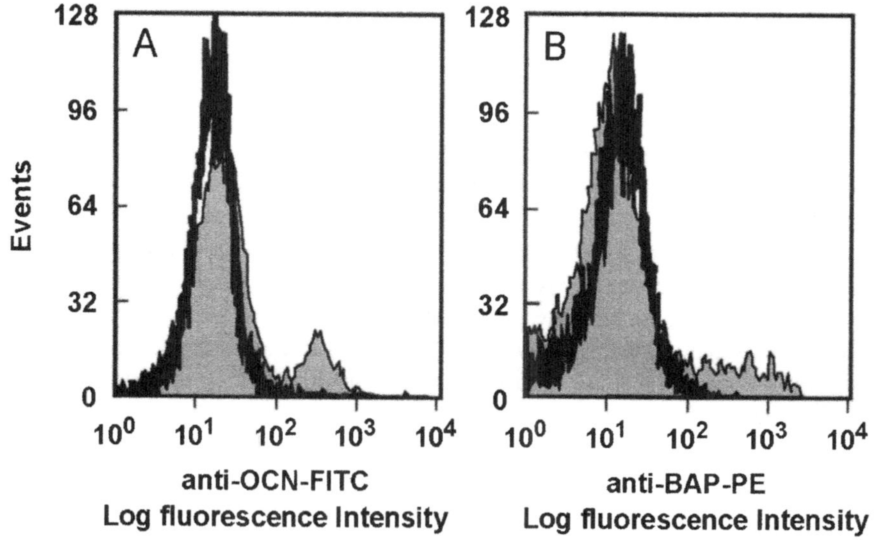

FIGURE 1. Representative histograms of flow cytometry profiles of (**A**) OCN positive (from an adolescent male) and (**B**) AP positive (from an adult male) cells. The histograms for the isotype controls (bold lines) as well as for the specific antibodies are shown.

requiring adherence to plastic.[18] We tested this hypothesis in the context of a clinical study where we compared the percentage of peripheral blood mononuclear cells expressing OCN or AP in 11 adolescent males (age, mean $\pm$ SEM, 14.5 $\pm$ 0.2 years) versus 11 adult males (37.7 $\pm$ 2.3 years). Due to rapid skeletal growth, the adolescent males had markedly elevated serum markers of bone formation ($\sim$sixfold increased) and also had significantly higher insulin-like growth factor-I (IGF-I) and IGF binding protein (IGFBP)-3 levels (both reflecting increased growth hormone secretion).

OCN- and AP-expressing cells were identified by flow cytometry using antibodies to OCN and AP, respectively. FIGURE 1 shows representative flow histograms of isotype antibody controls and OCN and AP positive cells from an adolescent and adult male, respectively. Using these methods, we found that the adolescent males had over a fivefold increase in circulating OCN positive cells as compared to adult males, with no significant difference in circulating AP positive cells between the two groups. Moreover, in contrast to previous studies examining the presence of circulating plastic adherent cells, where circulating osteogenic cells constituted < 1 in 10^8 mononuclear cells, we found a frequency of circulating OCN or AP positive cells on the order of 1 in 100 mononuclear cells. In the two groups combined, circulating OCN positive (but not AP positive) cells also correlated significantly with serum OCN and bone AP levels ($R = 0.72$ and 0.77, respectively, $P < 0.001$) and with serum IGF-I and IGFBP-3 levels ($R = 0.74$ and 0.61, $P < 0.001$ and

< 0.01, respectively), both attesting to the biological validity of our findings and suggesting that increases in growth hormone and/or IGF-I secretion during adolescence may be causally related to the increase in circulating osteoblastic cells.

As part of these studies, we also had the serendipitous opportunity to sequentially assay for circulating OCN and AP positive cells in a subject before and after a severe distal wrist fracture. Thus, one of the adult males in the study sustained a commuted distal forearm fracture 3 weeks after providing his baseline blood sample. We subsequently obtained a second blood sample 3 weeks following the fracture, and found that circulating OCN positive cells were increased over sixfold in the post-fracture sample as compared to baseline with an ~twofold increase in circulating AP positive cells. Although we are in the process of performing a more rigorous study assessing changes in these cell populations following fracture, preliminary data in two additional fracture patients showed elevated values for both circulating OCN and AP positive cells in these patients as compared to our normal adult mean values.

As further validation for our findings, we performed extensive studies excluding the possibility that we were detecting either platelets (which can express at least the OCN mRNA[19]) or granulocytes (which express AP[20]). Thus, flow cytometry of platelet-rich plasma demonstrated that platelets were effectively excluded from our flow cytometry gates, and in addition, we were unable to detect any surface staining for OCN on platelets using our methods, suggesting that while human platelets may express the OCN mRNA, they likely produce little or no OCN protein.[19] Similarly, using dual staining with the anti-CD15 antibody (a granulocyte-specific marker), we established that there was no contamination by granulocytes in our gated regions.

As further validation that we had identified osteoblast lineage cells using our methods, we also demonstrated that OCN positive cells were markedly enriched, by real-time RT-PCR, for mRNA expression of AP and col I, as compared either to unsorted or to OCN negative cells. When expanded in a stem cell medium for 3 weeks followed by osteoblast differentiation medium for 3 weeks, OCN positive cells were also shown to have the capacity to mineralize (assessed by von Kossa staining). Finally, sorted OCN positive cells were shown to form lamellar bone when placed in hydroxyapatite pellets subcutaneously in an immunodeficient mouse in a typical *in vivo* bone formation assay.

Collectively, our findings demonstrated that there are circulating cells in human peripheral blood that express the osteoblastic markers, OCN and AP, that are capable of replication *in vitro*, becoming adherent, and differentiating into osteoblasts, and that are also capable of forming bone *in vivo*. The percentage of these cells also increased markedly in the circulation of adolescent boys going through the growth spurt, and may also increase during fracture healing.

We believe that there are several significant implications for our findings. First, these circulating osteoblastic cells may be a potential source for osteoblasts in cortical bone, at sites distant from red marrow. Second, even in

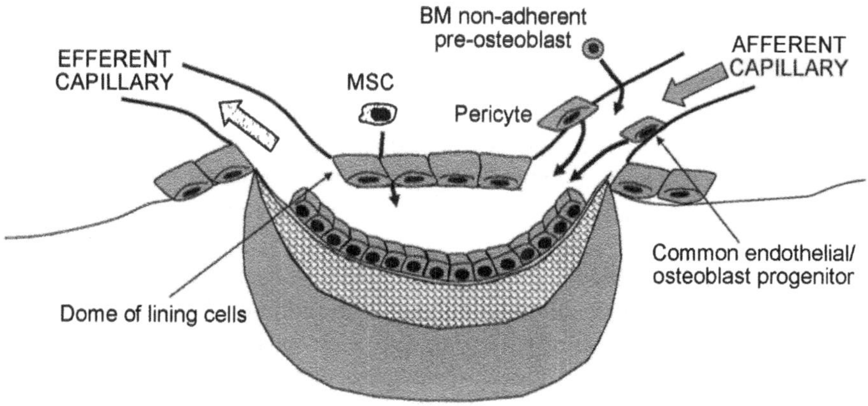

FIGURE 2. Schematic representation of the bone remodeling compartment (BRC). The roof of lining cells penetrated by an afferent marrow capillary, as well as the efferent capillary, is shown. Osteoblasts on the surface of bone could be derived from bone marrow stromal cells (MSC) by diapedesis of these cells through the lining cells; from pericytes surrounding the invading vasculature; from a common endothelial/osteoblast progenitor cell in the vessel wall; or from the bone marrow nonadherent osteogenic cells that have gained access to the marrow microcirculation. These possibilities are not mutually exclusive. Shown is a BRC on a trabecular surface, but a similar scenario may also exist for remodeling sites in cortical bone, with the exception that such sites would not be contiguous to the marrow, and hence there could not be local diapedesis of marrow stromal cells is shown.

trabecular bone, it appears that the basic multicellular unit (BMU) is not in direct contact with the bone marrow but rather is isolated in a closed cavity of flattened, lining cells.[21] The "bone remodeling compartment" or BRC is, however, penetrated by marrow capillary sinusoids (FIG. 2). Thus, osteoblastic cells on trabecular surfaces may come from several potential sources, and these possibilities are by no means mutually exclusive. Bone marrow stromal cells in the contiguous marrow, as they differentiate into osteoblasts, may transit the dome of lining cells by diapedesis. Alternatively, cells in the vascular wall or in the perivascular tissue (i.e., pericytes) may differentiate into osteoblasts as the capillary invades the BRC.[22] Indeed, while there is ongoing controversy on this issue, it appears that CD34 positive (an endothelial/hematopoietic marker) cells can give rise to pericytes[23] and to osteogenic cells.[24] Finally, bone marrow nonadherent cells with osteogenic potential may access the bone marrow capillary circulation and also enter the BRC via this route. These cells also likely exit the BRC and enter the circulation, thus explaining our ability to detect them in peripheral blood.

In addition to providing a source of osteoblasts in trabecular and cortical bone, circulating osteogenic cells may also contribute to the process of fracture repair. Einhorn and colleagues[25] demonstrated over 15 years ago that skeletal injury in one femur induces an increase in histologically assessed bone formation in the contralateral femur, indicating the presence of systemic factors released from the site of skeletal injury that would perhaps mobilize osteogenic cells to the site of injury. Consistent with this, systemically infused adherent bone marrow cells have been shown to localize to sites of fracture in mice,[26] although further studies are needed to assess whether bone marrow nonadherent cells or circulating osteogenic cells also possess this ability and whether, once localized to the fracture site, these circulating cells contribute to fracture healing.

In summary, a growing body of work over the past decade now supports the finding that, in addition to the conventional plastic adherent bone marrow cells identified by Friedenstein[1] over 40 years ago, there exists a bone marrow nonadherent cell population that also has osteogenic potential. Our recent findings would suggest that these nonadherent cells access the peripheral circulation in far greater numbers than do adherent bone marrow stromal cells. The origin of these circulating osteogenic cells (i.e., exclusively from the bone marrow or perhaps from the developing vasculature at the sites of bone remodeling), the precise role these cells play in skeletal remodeling or injury, or in nonskeletal bone formation and calcification (e.g., in vascular calcifications) remain important, unanswered questions.

ACKNOWLEDGMENTS

This work was supported by Grants AG004875 and AG024901 from the National Institutes of Health.

REFERENCES

1. Friedenstein, A.J., R.K. Chailakhyan & U.V. Gerasimov. 1987. Bone marrow osteogenic stem cells: in vitro cultivation and transplantation in diffusion chambers. Cell Tissue Kinet. **20:** 263–272.
2. Bianco, P. & P.G. Robey. 2000. Marrow stromal stem cells. J. Clin. Invest. **105:** 1663–1668.
3. Fujikawa, Y. *et al.* 1996. The human osteoclast precursor circulates in the monocyte fraction. Endocrinology **137:** 4058–4060.
4. Gothlin, G. & J.L.E. Ericsson. 1976. The osteoclast: review of ultrastructure, origin, and structure-function relationship. Clin. Orthop. Rel. Res. **120:** 201–231.
5. Zvaifler, N.J. *et al.* 2000. Mesenchymal precursor cells in the blood of normal individuals. Arthritis Res. **2:** 477–488.
6. Kuznetsov, S.A. *et al.* 2001. Circulating skeletal stem cells. J. Cell Biol. **153:** 1133–1139.
7. Long, M.W., J.L. Williams & K.G. Mann. 1990. Expression of human bone-related proteins in the hematopoietic microenvironment. J. Clin. Invest. **86:** 1387–1395.

8. LONG, M.W. *et al.* 1995. Regulation of human bone marrow-derived osteoprogenitor cells by osteogenic growth factors. J. Clin. Invest. **95:** 881–887.
9. LONG, M.W. *et al.* 1999. Age-related phenotypic alterations in populations of purified human bone precursor cells. J. Gerontol. **54A:** B54–B62.
10. EIPERS, P.G. *et al.* 2000. Bone marrow accessory cells regulate human bone precursor cell development. Exp. Hematol. **28:** 815–825.
11. BODINE, P.V.N. & B.S. KOMM. 1999. Evidence that conditionally immortalized human osteoblasts express an osteocalcin receptor. Bone **25:** 535–543.
12. PI, M. *et al.* 2005. Identification of a novel extracellular cation sensing G-protein coupled receptor. J Biol Chem **280:** 40201–40209.
13. FALLA, N. *et al.* 1993. Characterization of a 5-fluorouracil-enriched osteoprogenitor population of the murine bone marrow. Blood. **82:** 3580–3591.
14. DOMINICI, M. *et al.* 2004. Hematopoietic cells and osteoblasts are derived from a common marrow progenitor after bone marrow transplantation. Proc. Natl. Acad. Sci. USA **101:** 11761–11766.
15. OLMSTED-DAVIS, E.A. *et al.* 2003. Primitive adult hematopoietic stem cells can function as osteoblast precursors. Proc. Natl. Acad. Sci. USA **100:** 15877–15882.
16. GOODELL, M.A. *et al.* 1997. Dye efflux studies suggest that hematopoietic stem cells expressing low or undetectable levels of CD34 antigen exist in multiple species. Nat. Med. **3:** 1337–1345.
17. JACKSON, K.A. *et al.* 2001. Regeneration of ischemic cardiac muscle and vascular endothelium by adult stem cells. J. Clin. Invest. **107:** 1395–1402.
18. EGHBALI-FATOURECHI, G.Z. & S. KHOSLA. 2005. Circulating osteoblast-lineage cells. N. Engl. J. Med. **353:** 737–738.
19. THIEDE, M.A. *et al.* 1994. Presence of messenger ribonucleic acid encoding osteocalcin, a marker of bone turnover, in bone marrow megakaryocytes and peripheral blood platelets. Endocrinology **135:** 929–937.
20. STEWART, C.A. 1974. Leukocyte alkaline phosphatase in myeloid maturation. Pathology **6:** 287–293.
21. HAUGE, E.M. *et al.* 2001. Cancellous bone remodeling occurs in specialized compartments lined by cells expressing osteoblastic markers. J. Bone Miner. Res. **16:** 1575–1582.
22. DOHERTY, M.J. *et al.* 1998. Vascular pericytes express osteogenic potential in vitro and in vivo. J. Bone Miner. Res. **13:** 828–838.
23. HOWSON, K.M. *et al.* 2005. The postnatal rat aorta contains pericyte progenitor cells that form spheroidal colonies in suspension culture. Am. J. Physiol. (Cell Physiol.) **289:** C1396–1407.
24. CHEN, J.L. *et al.* 1997. Osteoblast precursor cells are found in CD34+ cells from human bone marrow. Stem Cells **15:** 368–377.
25. EINHORN, T.A. *et al.* 1990. The osteogenic response to distant skeletal injury. J. Bone Joint Surg. **72-A:** 1374–1378.
26. DEVINE, M.J.M., C.M. Jang, *et al.* 2002. Transplanted bone marrow cells localize to fracture callus in a mouse model. J. Orthop. Res. **20:** 1232–1239.

The Pericellular Matrix as a Transducer of Biomechanical and Biochemical Signals in Articular Cartilage

FARSHID GUILAK,[a] LEONIDAS G. ALEXOPOULOS,[a] MAUREEN L. UPTON,[a] INCHAN YOUN,[a] JAE BONG CHOI,[a] LI CAO,[a] LORI A. SETTON,[a] AND MANSOOR A. HAIDER[b]

[a]*Departments of Surgery and Biomedical Engineering, Duke University Medical Center, Durham, North Carolina 27710, USA*

[b]*Department of Mathematics, North Carolina State University, Raleigh, North Carolina 27695, USA*

ABSTRACT: The pericellular matrix (PCM) is a narrow tissue region surrounding chondrocytes in articular cartilage, which together with the enclosed cell(s) has been termed the "chondron." While the function of this region is not fully understood, it is hypothesized to have important biological and biomechanical functions. In this article, we review a number of studies that have investigated the structure, composition, mechanical properties, and biomechanical role of the chondrocyte PCM. This region has been shown to be rich in proteoglycans (e.g., aggrecan, hyaluronan, and decorin), collagen (types II, VI, and IX), and fibronectin, but is defined primarily by the presence of type VI collagen as compared to the extracellular matrix (ECM). Direct measures of PCM properties via micropipette aspiration of isolated chondrons have shown that the PCM has distinct mechanical properties as compared to the cell or ECM. A number of theoretical and experimental studies suggest that the PCM plays an important role in regulating the microenvironment of the chondrocyte. Parametric studies of cell–matrix interactions suggest that the presence of the PCM significantly affects the micromechanical environment of the chondrocyte in a zone-dependent manner. These findings provide support for a potential biomechanical function of the chondrocyte PCM, and furthermore, suggest that changes in the PCM and ECM properties that occur with osteoarthritis may significantly alter the stress-strain and fluid environments of the chondrocytes. An improved understanding of the structure and function of the PCM may provide new insights into the mechanisms that regulate chondrocyte physiology in health and disease.

KEYWORDS: articular cartilage; chondron; chondrocyte; osteoarthritis; collagen; proteoglycan; collagen VI

Address for correspondence: Farshid Guilak, Ph.D., Orthopaedic Research Laboratories, Department of Surgery, Division of Orthopaedic Surgery, 375 MSRB, Box 3093, Duke University Medical Center, Durham, NC 27710. Voice: 919-684-2521; fax: 919-681-8490.
e-mail: guilak@duke.edu

Ann. N.Y. Acad. Sci. 1068: 498–512 (2006). © 2006 New York Academy of Sciences.
doi: 10.1196/annals.1346.011

INTRODUCTION

Articular cartilage is the connective tissue lining the surfaces of bones in diarthrodial joints, providing a low-friction load-bearing surface that allows joint movement. The extracellular matrix (ECM) of this tissue is maintained by a balance of anabolic and catabolic activities of a single type of cell—the chondrocytes—which are distributed sparsely throughout the tissue in adult cartilage. During normal activities of daily living, articular cartilage is exposed to significant mechanical loads and deformations due to joint loading. Deformation of the cartilage layer results in a complex mechanical environment that is characterized by time- and spatially varying stresses, strains, hydrostatic pressure, interstitial fluid flow, electrokinetic effects, and osmotic pressure.[1–5] At the cellular level, these biophysical factors, in conjunction with biochemical factors (e.g., growth factors, cytokines), play important roles in modulating the physiology of the chondrocytes, and as a consequence, the health of the joint.[6–10]

Within the ECM, chondrocytes are surrounded singly or in groups by a narrow region of tissue termed the pericellular matrix (PCM), which together with the cell has been termed the *chondron*.[11–13] This distinct matrix compartment was originally termed the *chondrone* by Benninghoff, who identified a region of altered structure around chondrocytes, which he described as a "fluid filled bladder."[11] No direct research on the chondron was reported until the observation that a very small number of intact chondrons could be isolated as a by product of cartilage homogenization.[12] With refinement of this technique, the majority of work on the composition and metabolic activity of the chondron was reported in the pioneering work of Poole and co-workers.[14–18] More recently, the development of novel methods for enzymatic[19] or mechanical[20] isolation of intact chondrons have allowed a number of new biological and biomechanical assays for studying the properties and function of the PCM. These studies have provided further evidence that the PCM plays important roles in regulating the biomechanical, biophysical, and biochemical interactions between the cell and the ECM. In this article, we review a number of studies investigating the structure, composition, and mechanical properties of the PCM. Taken together, they provide important new insights into the potential function of this matrix compartment in articular cartilage.

STRUCTURE AND COMPOSITION OF THE PCM

The PCM is a narrow region of matrix immediately surrounding nearly all chondrocytes in adult articular cartilage, and is generally characterized as having a higher concentration of proteoglycans and a fine arrangement of collagen fibers as compared to the ECM (FIG. 1). An excellent review of the structure and composition of the chondron was reported by Poole.[21]

Chondrons have been reported in a variety of shapes, consisting of columns containing multiple cells, spheroidal shapes, or oriented clusters of chondrons (FIG. 2).[22] The size and structure of chondrons can change with age, site, species, and disease state.[17,23,24] For example, horizontally oriented clusters of multiple chondrons are observed in the superficial zone of human ankle, but not knee articular cartilage.[25] In osteoarthritis, the area of the chondron is greatly enlarged and the incidence of chondrocyte proliferation is increased.[17,23]

The PCM contains many of the same molecular constituents as the ECM, although there are some distinct differences in the structure and composition of these regions. This region has been shown to be rich in fibronectin, proteoglycans (e.g., aggrecan, hyaluronan, and decorin), and collagen (types II, VI,

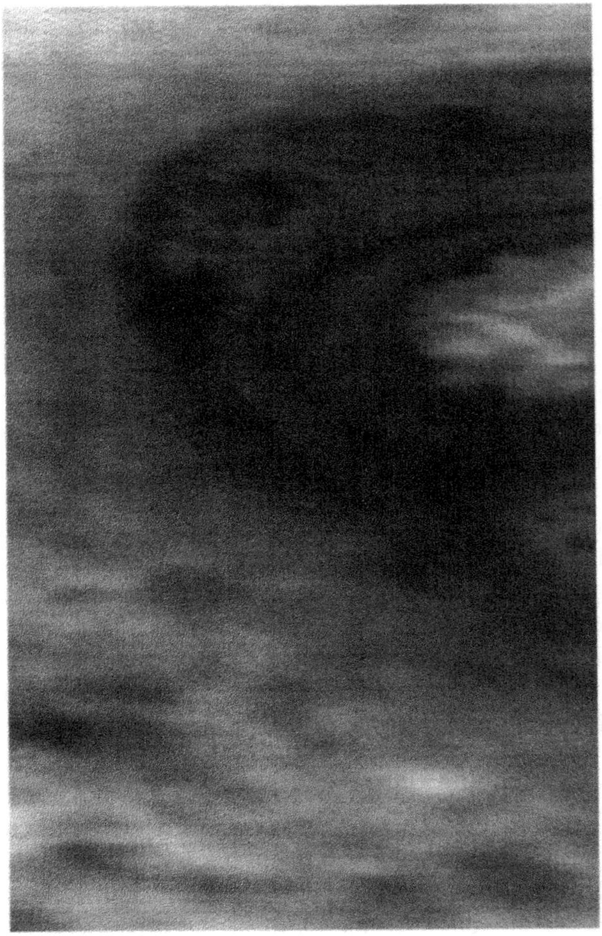

FIGURE 1. Toluidine blue staining of proteoglycan in a section of human articular cartilage indicating the presence of a proteoglycan-rich pericellular region. (From Guilak *et al.*,[74] with permission from Elsevier.)

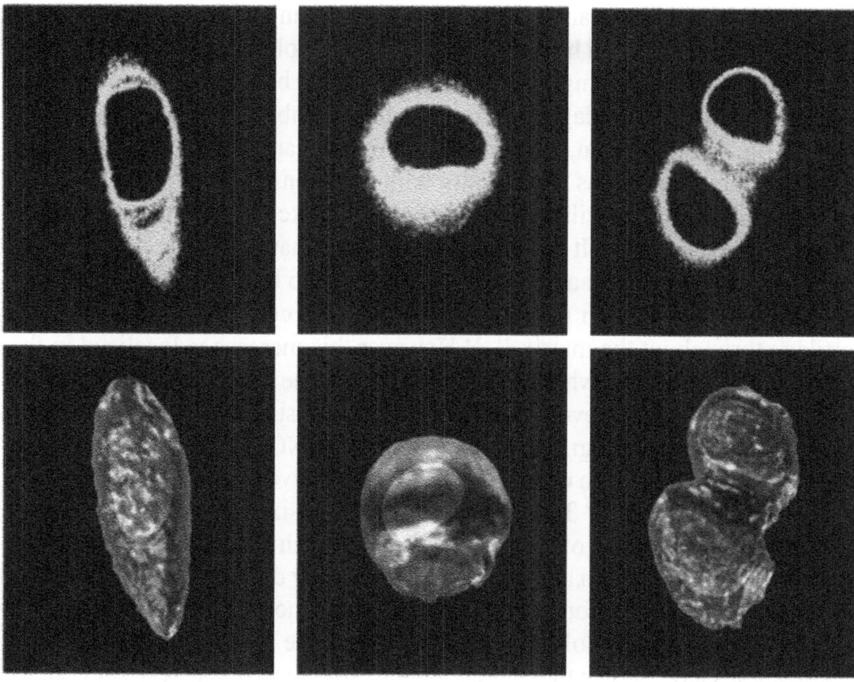

FIGURE 2. Immunofluorescent labeling of type VI collagen (top row) and three-dimensional reconstructions (bottom row) of the chondron in thick sections of articular cartilage from the knee joint of skeletally mature pigs.[22] Significant variations in chondron size and shape were observed depending on the site within the tissue.

and IX),[15,16,18] but is defined primarily by the presence of type VI collagen as compared to the ECM.[14] Many reports suggest that the cell and PCM are surrounded by a "felt-like" capsule, although this aspect of the structure may depend on the site of origin[13] or exist as a byproduct of homogenization since it does not appear in chondrons isolated by enzymatic digestion or microaspiration methods.[19,20] Furthermore, enzymatically isolated chondrons exhibit differences in composition as compared to mechanically isolated chondrons as evidenced by the retention of aggrecan and types II and VI collagen in the PCM, but also by the loss of some specific constituents such as fibronectin and the chondroitin sulfate epitope labeled by the 7D4 antibody.[19]

TYPE VI COLLAGEN AND THE PCM

Type VI collagen is a beaded filament consisting of three different polypeptide chains: $\alpha 1$ (140 kDa), $\alpha 2$ (130 kDa), and $\alpha 3$($\sim$200 kDa). These polypeptide chains, which contain a triple helical domain and an N- and C-terminus domains, are bundled together to form the monomeric unit of type VI collagen, which then forms dimers and tetramers as the basic unit of the beaded filamentous structure. It has been hypothesized that these molecules are secreted from the cells as monomers or higher-order multimers and become incorporated into the ECM.[26] Type VI collagen interacts with a number of ECM components such as biglycan, decorin, fibromodulin, hyaluronan, fibronectin, perlecan, and heparin but it also has low affinity with collagens I and II.[27–29]

In normal articular cartilage collagen type VI is exclusively present in chondrons[13,19,26] (see FIG. 2). It has been hypothesized that type VI collagen forms a network that appears to anchor the chondrocyte to the PCM.[29,30] Biochemical analyses have shown upregulation of type VI collagen in osteoarthritic cartilage throughout the matrix.[31–33] However this increase is localized in the middle and deep zones, whereas the superficial zone shows reduced levels of type VI collagen.[31,33] Several studies have demonstrated cell attachment to collagen VI through integrin receptors or through NG2, a chondroitin sulfate proteoglycan expressed in chondrocytes that is believed to be a cellular receptor for type VI collagen.[34] Thus, it has been hypothesized that type VI collagen is an important regulator of cellular interactions with various matrix proteins.

While type VI collagen appears to be the primary demarcation of the chondron, there is evidence for the presence of other molecules such as type IX collagen or laminin as molecular constituents of the PCM.[18,21] Furthermore, recent studies have shown that knockout mice deficient in the gene for col6a1, which therefore lack collagen VI, still exhibit the formation of a PCM.[35] Therefore other molecules besides collagen VI may be responsible for maintaining the overall structure of the PCM.

BIOLOGICAL FUNCTIONS OF THE PCM

It is well accepted that the interactions between the chondrocyte and the ECM are of critical importance in regulating the development and maintenance of cartilage. For example, interactions between the cell surface and ECM components have a significant effect on proteoglycan metabolism, gene expression, and response to growth factors.[36–38] Because the PCM completely surrounds the chondrocyte, it is likely that any signals that the chondrocyte perceives, either biochemical or biophysical, are influenced by the PCM. Thus any direct interactions between cell surface receptors such as integrins and the tissue matrix are likely to occur at the level of the PCM. Furthermore, molecules that interact with the chondrocyte surface must pass through the pericellular environment. During their passage, soluble mediators such as growth factors may be modified or retained.[39] ECM molecules such as aggrecan may also be modified in this region, as part of the time-dependent maturation of the complex formation of aggrecan with hyaluronan.[40] Retention of the native PCM has been shown to significantly alter the metabolic activity of chondrocytes,[41] suggesting that the PCM may regulate the presentation of biochemical and biophysical factors to the cell and, therefore, play a biological role in controlling cell biosynthesis. Alterations in PCM structure or size, as may occur with aging or osteoarthritis,[17,23,42] may therefore have secondary effects on the cartilage ECM by altering diffusion times of matrix macromolecules as they pass in or out of the chondron.

BIOMECHANICAL FUNCTION OF THE PCM

In addition to the evidence supporting a biological function of the PCM, there has been considerable speculation that an additional function of the chondron is biomechanical in nature,[12] potentially providing a protective role for the chondrocytes during loading through an "adaptive water loss from PCM proteoglycans."[43] A number of early studies have also suggested that the chondron serves as a filter or transducer of mechanical signals,[44,45] potentially through an interaction of type VI collagen with cell surface integrins or hyaluronan.[46–49] These findings are supported by experimental data that show that the presence of a newly formed PCM augments the chondrocytes' metabolic response to static and cyclic loading.[50] Furthermore, mechanical compression alters proteoglycan deposition and matrix deformation around individual cells in cartilage explants.[51] Static compression of cartilage explants was found to stimulate site-directed deposition of proteoglycans in the PCM, superimposed on an overall inhibition of proteoglycan synthesis in the ECM. Conversely, pericellular deposition and overall proteoglycan synthesis was stimulated by dynamic compression. These findings suggest that cell-matrix interactions in the PCM may play an important role in the response of chondrocytes to mechanical compression, possibly by influencing macromolecular transport and local tissue strain around cells.

To examine specific hypotheses on the potential role of the PCM as a transducer of mechanical signals, several studies have developed theoretical models of cell-matrix interactions that incorporate the structure and properties of chondrocytes, the PCM, and the ECM. The primary goal of such approaches has been to characterize the stress-strain, fluid-flow, and osmotic environments of individual cells to provide further insights into the sequence of events through which mechanical loading of cartilaginous tissues is converted to an intracellular response.[52–58] In combination with direct measurements of PCM mechanical properties, these studies provide supporting evidence that the mechanical properties of the PCM, relative to those of the cell and ECM, may significantly influence the mechanical and physicochemical environment of the chondrocyte.[56,59–61]

The results of such theoretical predictions, however, are dependent on the assumed mechanical properties and geometry of the ECM, PCM, and chondrocyte. The mechanical properties of the cartilage ECM have been extensively characterized in various configurations.[62–66] Furthermore, the physical and mechanical properties of chondrocytes have also been determined using several techniques, including gel compression,[67] micropipette aspiration,[68–71] and cytoindentation,[72] all showing the modulus of the cell to be on the order of ~1–4 kPa.

More recently, the zonal variations in the mechanical properties of the PCM from normal and osteoarthritic cartilage were determined in a series of studies on enzymatically or mechanically isolated chondrons using the micropipette technique.[20,73–75] With this technique, a small region on the surface of the PCM is aspirated into a small glass pipette by the application of a controlled pressure, and the ensuing deformation is recorded and quantified using video microscopy (FIG. 3). In this manner, either the equilibrium or transient response of the PCM can be coupled with theoretical formulations of the experimental configuration to determine the elastic or viscoelastic properties of the PCM.

Using various model representations of the chondron such as an elastic half space[20,75] or a biphasic ellipsoid,[73] this approach has been used to determine the Young's modulus, hydraulic permeability, and Poisson's ratio of the PCM from normal and osteoarthritic articular cartilage. The transient behavior of the PCM could be accurately described by a biphasic model, suggesting that the viscoelastic behavior of the PCM is likely due to flow-dependent effects, similar to that of the ECM. The Young's modulus of the PCM was reported to range from ~40 to 70 kPa, approximately 1 order of magnitude less than that of the cartilage ECM, and significantly decreased by approximately 50% with osteoarthritis. The hydraulic permeability of the PCM was 1–2 orders of magnitude lower than that of the ECM, but was more than doubled with osteoarthritis. The Poisson's ratio of the PCM was similar for chondrons isolated from normal or osteoarthritic cartilage ($\nu \sim 0.04$). These findings suggest that the PCM may undergo degenerative processes with osteoarthritis, similar to those occurring in the ECM. Interestingly, no differences were observed in any of these studies between the properties of the PCM of chondrons isolated from the surface zone as compared to those of the middle/deep zone.[20,73,75] These findings are in general agreement with other studies showing little deformation of mechanically isolated chondrons compressed within an agarose matrix, suggesting that the Young's modulus of the chondrons is greater than that of the agarose (25 kPa).[76]

By combining direct experimental measurements of PCM properties with theoretical models of cell-matrix interactions, there is growing evidence that the PCM may play an important biomechanical role in cartilage. For example, a multiscale finite-element model was used to show that simply the presence of a PCM in articular cartilage dramatically changed the stress-strain and fluid-flow environments of the chondrocyte.[60] In a more refined model that accounted for the large zonal differences in ECM properties, it was found that the mismatch between the Young's moduli of PCM and ECM amplifies compressive strains at the cellular level from 15–70% and significantly reduces stress magnitudes in a zone-dependent manner.[59] Furthermore, the lower permeability of PCM

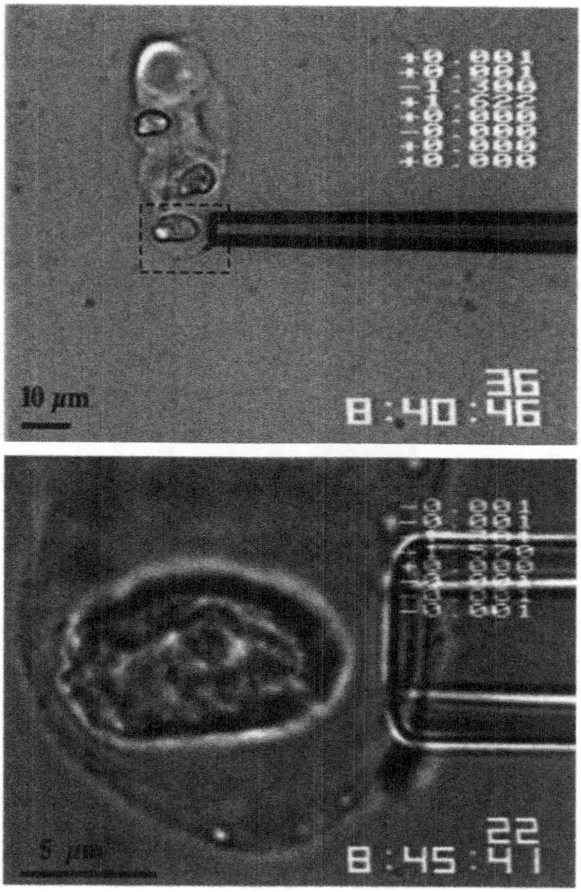

FIGURE 3. Micropipette aspiration of an isolated chondron. Chondrons were isolated by microaspiration and suspended in a specially designed chamber on an inverted optical microscope. The flattened tip of a small glass micropipette was brought in contact with a chondron, and a series of pressure steps was applied to the PCM. The length of PCM aspiration versus applied pressure was used in combination with different theoretical analyses to determine the Young's modulus of the PCM.[20,73] (From Alexopoulos *et al.*,[73] with permission from Elsevier.)

relative to the ECM inhibited fluid flux near the cell by a factor of 30, suggesting that the PCM may have a significant influence on convective transport to and from the chondrocyte. The decreased modulus and increased permeability of both the PCM and ECM with osteoarthritis significantly altered the mechanical environment of the chondrocyte, leading to ~70% higher compressive strains and 40% higher fluid flux near the cell (FIG. 4). These findings are supported by an analytical model of a chondron subjected to cyclic loading, which suggested that the transmission of mechanical deformation to the cell is

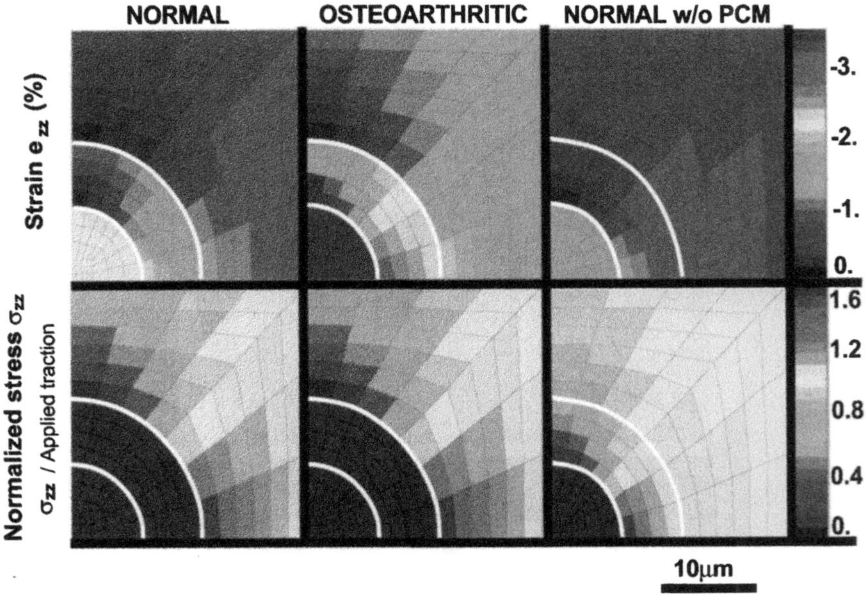

FIGURE 4. The effect of PCM on the local stress and strain environment of the cell determined using a finite-element model. The figure shows the relative magnitude of strain (top) and stress (bottom) in the cell, PCM, and ECM under compression. The presence of a PCM had a significant influence on the magnitude of stress or strain at the cellular level. The presence of a PCM decreases the magnitude of stress at the cellular level, whereas it amplifies the local strains. With osteoarthritis, the decrease in ECM and PCM stiffness results in significantly greater local strains in the tissue under the same loading conditions. (Adapted from Alexopoulos *et al.*,[59] with permission from Elsevier.)

directly dependent on the modulus and permeability of the PCM.[56] Together, these findings provide support for a potential biomechanical role for the chondrocyte PCM, and suggest that changes in the structure or properties of the PCM, as may occur with aging or osteoarthritis, may alter the stress-strain and fluid-flow environment of the chondrocytes and potentially chondrocyte metabolism.

PHYSICOCHEMICAL PROPERTIES OF THE CHONDRON

Several important characteristics of the biomechanical and physicochemical properties of articular cartilage are related to the transport of the interstitial water and ion phases in the tissue.[1] Due to the porous and permeable nature of collagen–proteoglycan matrix, mechanical compression of articular cartilage induces exudation of interstitial water, which leads to an increase in the apparent concentration of proteoglycans and thus, associated changes in interstitial ion concentrations.[77] These transient changes in the fixed-charge density and ion concentrations, and therefore the local osmolality, are believed to contribute to changes in cell shape and volume with tissue compression.[54,55] Thus the physiology and pathology of the chondrocyte may be influenced by the mechano-chemical properties of the PCM. For example, the swelling response of chondrocytes within isolated chondrons in response to an osmotic challenge depends on the method used to isolate the chondrons (i.e., mechanical homogenization versus enzymatic isolation).[78] Furthermore, disruption of the collagen network in osteoarthritis leads to increased water content of the tissue and loss of proteoglycans,[6] and a corresponding decrease in the pericellular osmolality.[79] Thus, a further understanding of the mechano-chemical properties of the PCM will be of great value in determining the relative influence and interactions of mechanical, electrical, and chemical stimuli in regulating chondrocyte activity. In this regard, a recent study has developed a detailed theoretical mechano–chemical model to describe the passive equilibrium response of an isolated chondron under osmotic loading.[61] In this model, the chondrocyte is represented as an ideal osmometer[80] and the PCM model is formulated using triphasic mixture theory that represents the tissue as a continuum mixture of solid, fluid, and ionic phases.[1,81] For the case of a neo-Hookean PCM constitutive law, the model was used to conduct a parametric analysis of cell and chondron deformation under hyper- and hypo-osmotic loading. In particular, model predictions indicated that the passive volumetric swelling response of the chondrocyte under hypo-osmotic loading was significantly diminished, relative to its ideal response in free solution, when the cell is encapsulated by a triphasic PCM. Furthermore, the magnitude of cell swelling was predicted to be highly sensitive to pericellular fixed charge densities that were between roughly 1 to 5 times the reported values for cartilage ECM. These findings suggest that PCM mechanical and chemical properties will have a significant effect on the swelling behavior and osmotic environment of a chondrocyte. In other studies, a similar model has shown that the physicochemical properties of the PCM will significantly affect the electrical, chemical, and mechanical fields around a chondrocyte during compression of articular cartilage.[82] The combination of such theoretical models with various experimental measures on isolated chondrons will aid in determining additional properties of the PCM, such as the tissue-fixed charge density and its relative contribution to PCM mechanical properties.

CONCLUSIONS

These findings provide further support for a potential biomechanical role for the chondrocyte PCM, and furthermore, suggest that changes in the PCM and ECM properties that occur with osteoarthritis may significantly alter the stress-strain and fluid environments of the chondrocytes. While this article focused on the PCM of articular chondrocytes, recent data suggest that other cartilaginous tissues, such as the knee meniscus or intervertebral disc, contain a similar structure surrounding the cells.[83] An improved understanding of the structure and function of the PCM may provide new insights into the mechanisms that regulate chondrocyte physiology in health and disease.

ACKNOWLEDGMENTS

This study was supported by the National Institutes of Health (AG15768, AR48182, AR50245, AR47442, AR45644, GM08555), The Whitaker Foundation (RG-020933), and the NSF (DMS-0211154).

The authors declare that they have no competing financial interests.

REFERENCES

1. Lai, W.M., J.S. Hou & V.C. Mow. 1991. A triphasic theory for the swelling and deformation behaviors of articular cartilage. J. Biomech. Eng. **113:** 245–258.
2. Mow, V.C. & X.E. Guo. 2002. Mechano-electrochemical properties of articular cartilage: their inhomogeneities and anisotropies. Ann. Rev. Biomed. Eng. **4:** 175–209.
3. Mow, V.C., C.C. Wang & C.T. Hung. 1999. The extracellular matrix, interstitial fluid and ions as a mechanical signal transducer in articular cartilage. Osteoarthritis Cartilage **7:** 41–58.
4. Maroudas, A. *et al.* 1991. The effect of osmotic and mechanical pressures on water partitioning in articular cartilage. Biochim. Biophys. Acta **1073:** 285–294.
5. Frank, E.H. & A.J. Grodzinsky. 1987. Cartilage electromechanics—II. A continuum model of cartilage electrokinetics and correlation with experiments. J. Biomech. **20:** 629–639.
6. Guilak, F. *et al.* 1994. Mechanical and biochemical changes in the superficial zone of articular cartilage in canine experimental osteoarthritis. J. Orthop. Res. **12:** 474–484.
7. Guilak, F., R.L. Sah & L.A. Setton. 1997. Physical regulation of cartilage metabolism. *In* Basic Orthopaedic Biomechanics. V.C. Mow & W.C. Hayes, Eds.: 179–207. Lippincott-Raven. Philadelphia.
8. Helminen, H.J. *et al.* 1987. Joint loading effects on articular cartilage: a historical review. *In* Joint Loading: biology and Health of Articular Structures. H.J., Helminen *et al.*, Eds.: 1–46. Wright and Sons. Bristol.
9. Setton, L.A. *et al.* 1994. Mechanical properties of canine articular cartilage are significantly altered following transection of the anterior cruciate ligament. J. Orthop. Res. **12:** 451–463.
10. Stockwell, R.A. 1987. Structure and function of the chondrocyte under mechanical stress. *In* Joint Loading: biology and Health of Articular Structures. H.J. Helminen, *et al.*, Eds.: 126–148. Wright and Sons. Bristol.
11. Benninghoff, A. 1925. Form und bau der Gelenkknorpel in ihren Beziehungen Zur Funktion. Zweiter Teil: der Aufbau des Gelenkknorpels in sienen Bezienhungen zur Funktion. **2:** 783.

12. SZIRMAI, J.A. 1974. The concept of the chondron as a biomechanical unit. *In* Biopolymer und Biomechanik von Bindegewebssystemen. F. Hartmann, Ed.: 87. Academic Press. Berlin.
13. POOLE, C.A., M.H. FLINT & B.W. BEAUMONT. 1987. Chondrons in cartilage: ultrastructural analysis of the pericellular microenvironment in adult human articular cartilages. J. Orthop. Res. **5:** 509–522.
14. POOLE, C.A., S. AYAD & J.R. SCHOFIELD. 1988. Chondrons from articular cartilage: I. Immunolocalization of type VI collagen in the pericellular capsule of isolated canine tibial chondrons. J. Cell Sci. **90:** 635–643.
15. POOLE, C.A. *et al.* 1990. Chondrons from articular cartilage (II): analysis of the glycosaminoglycans in the cellular microenvironment of isolated canine chondrons. Connect. Tissue Res. **24:** 319–330.
16. POOLE, C.A., T.T. GLANT & J.R. SCHOFIELD. 1991. Chondrons from articular cartilage. (IV). Immunolocalization of proteoglycan epitopes in isolated canine tibial chondrons. J. Histochem. Cytochem. **39:** 1175–1187.
17. POOLE, C.A., A. MATSUOKA & J.R. SCHOFIELD. 1991. Chondrons from articular cartilage. III. Morphologic changes in the cellular microenvironment of chondrons isolated from osteoarthritic cartilage. Arthritis Rheum. **34:** 22–35.
18. POOLE, C.A. *et al.* 1997. Immunolocalization of type IX collagen in normal and spontaneously osteoarthritic canine tibial cartilage and isolated chondrons. Osteoarthritis Cartilage **5:** 191–204.
19. LEE, G.M. *et al.* 1997. Isolated chondrons: a viable alternative for studies of chondrocyte metabolism in vitro. Osteoarthritis Cartilage **5:** 261–274.
20. ALEXOPOULOS, L.G. *et al.* 2003. Alterations in the mechanical properties of the human chondrocyte pericellular matrix with osteoarthritis. J. Biomech. Eng. **125:** 323–333.
21. POOLE, C.A. 1997. Articular cartilage chondrons: form, function and failure. J. Anat. **191:** 1–13.
22. YOUN, I. *et al.* 2006. Zonal variations in the 3D morphology of the chondron measured in situ using serial confocal sections. Trans. Orthop. Res. Soc. **30:** 365.
23. LEE, G.M. *et al.* 2000. The incidence of enlarged chondrons in normal and osteoarthritic human cartilage and their relative matrix density. Osteoarthritis Cartilage **8:** 44–52.
24. QUINN, T.M., E.B. HUNZIKER & H.J. HAUSELMANN. 2005. Variation of cell and matrix morphologies in articular cartilage among locations in the adult human knee. Osteoarthritis Cartilage **13:** 672–678.
25. SCHUMACHER, B.L. *et al.* 2002. Horizontally oriented clusters of multiple chondrons in the superficial zone of ankle, but not knee articular cartilage. Anat. Rec. **266:** 241–248.
26. ENGVALL, E., H. HESSLE & G. KLIER. 1986. Molecular assembly, secretion, and matrix deposition of type VI collagen. J. Cell Biol. **102:** 703–710.
27. BIDANSET, D.J. *et al.* 1992. Binding of the proteoglycan decorin to collagen type VI. J. Biol. Chem. **267:** 5250–5256.
28. KIELTY, C.M. *et al.* 1992. Type VI collagen microfibrils: evidence for a structural association with hyaluronan. J. Cell Biol. **118:** 979–990.
29. MARCELINO, J. & C.A. MCDEVITT. 1995. Attachment of articular cartilage chondrocytes to the tissue form of type VI collagen. Biochim. Biophys. Acta **1249:** 180–188.

30. SHERWIN, A.F. *et al.* 1999. The distribution of type VI collagen in the developing tissues of the bovine femoral head. Histochem. J. **31:** 623–632.
31. HAMBACH, L. *et al.* 1998. Severe disturbance of the distribution and expression of type VI collagen chains in osteoarthritic articular cartilage. Arthritis Rheum. **41:** 986–996.
32. MCDEVITT, C.A. *et al.* 1988. Experimental osteoarthritic articular cartilage is enriched in guanidine soluble type VI collagen. Biochem. Biophys. Res. Commun. **157:** 250–255.
33. PULLIG, O., G. WESELOH & B. SWOBODA. 1999. Expression of type VI collagen in normal and osteoarthritic human cartilage. Osteoarthritis Cartilage **7:** 191–202.
34. DOANE, K.J., S.J. HOWELL & D.E. BIRK. 1998. Identification and functional characterization of two type VI collagen receptors, alpha 3 beta 1 integrin and NG2, during avian corneal stromal development. Invest. Ophthalmol. Vis. Sci. **39:** 263–275.
35. ALEXOPOULOS, L.G., P. BONALDO & F. GUILAK. 2005. Accelerated development of osteoarthritis in col6a1 knockout mice: the biomechanical role of the type VI collagen in articular cartilage. Trans. Orthop. Res. Soc. **29:** 393.
36. ADAMS, J.C. & F.W. WATT. 1993. Regulation of development and differentiation by the extracellular matrix. Development **117:** 1183–1198.
37. BOUDREAU, N., C. MYERS & M.J. BISSEL. 1995. From laminin to lamin: regulation of tissue-specific gene expression by the ECM. Trends Cell Biol. **5:** 1–4.
38. LOESER, R.F. 1997. Growth factor regulation of chondrocyte integrins. Differential effects of insulin-like growth factor 1 and transforming growth factor beta on alpha 1 beta 1 integrin expression and chondrocyte adhesion to type VI collagen. Arthritis Rheum. **40:** 270–276.
39. RUOSLAHTI, E. & Y. YAMAGUCHI. 1991. Proteoglycans as modulators of growth factor activities. Cell **64:** 867–869.
40. SANDY, J.D., J.R. O'NEILL & L.C. RATZLAFF. 1989. Acquisition of hyaluronate-binding affinity in vivo by newly synthesized cartilage proteoglycans. Biochem. J. **258:** 875–880.
41. LARSON, C.M. *et al.* 2002. Retention of the native chondrocyte pericellular matrix results in significantly improved matrix production. Matrix Biol. **21:** 349–359.
42. HORIKAWA, O. *et al.* 2004. Distribution of type VI collagen in chondrocyte microenvironment: study of chondrons isolated from human normal and degenerative articular cartilage and cultured chondrocytes. J. Orthop. Sci. **9:** 29–36.
43. POOLE, C.A., M.H. FLINT & B.W. BEAUMONT. 1988. Chondrons extracted from canine tibial cartilage: preliminary report on their isolation and structure. J. Orthop. Res. **6:** 408–419.
44. POOLE, C.A. 1992. Chondrons: the chondrocyte and its pericellular microenvironment. *In* Articular Cartilage and Osteoarthritis. K.E. Kuettner, *et al.*, Eds.: 201–220. Academic Press. London, New York.
45. GRECO, F. *et al.* 1992. Ultrastructural analysis of the adaptation of articular cartilage to mechanical stimulation. Ital. J. Orthop. Traumatol. **18:** 311–321.
46. LEE, V. *et al.* 2000. The roles of matrix molecules in mediating chondrocyte aggregation, attachment, and spreading. J. Cell. Biochem. **79:** 322–333.
47. LOESER, R.F. *et al.* 2000. Integrin expression by primary and immortalized human chondrocytes: evidence of a differential role for alpha1beta1 and alpha2beta1 integrins in mediating chondrocyte adhesion to types II and VI collagen. Osteoarthritis Cartilage **8:** 96–105.

48. KNUDSON, W. & R.F. LOESER. 2002. CD44 and integrin matrix receptors participate in cartilage homeostasis. Cell. Molecul. Life Sci. **59:** 36–44.
49. MCDEVITT, C.A., J. MARCELINO & L. TUCKER. 1991. Interaction of intact type VI collagen with hyaluronan. FEBS Lett. **294:** 167–170.
50. BUSCHMANN, M.D. *et al.* 1995. Mechanical compression modulates matrix biosynthesis in chondrocyte/agarose culture. J. Cell Sci. **108:** 1497–1508.
51. QUINN, T.M. *et al.* 1998. Mechanical compression alters proteoglycan deposition and matrix deformation around individual cells in cartilage explants. J. Cell Sci. **111:** 573–583.
52. BAER, A.E. & L.A. SETTON. 2000. The micromechanical environment of intervertebral disc cells: effect of matrix anisotropy and cell geometry predicted by a linear model. J. Biomech. Eng. **122:** 245–251.
53. BAER, A.E. *et al.* 2003. The micromechanical environment of intervertebral disc cells determined by a finite deformation, anisotropic, and biphasic finite element model. J. Biomech. Eng. **125:** 1–11.
54. GUILAK, F. 1995. Compression-induced changes in the shape and volume of the chondrocyte nucleus. J. Biomech. **28:** 1529–1542.
55. GUILAK, F., A. RATCLIFFE & V.C. MOW. 1995. Chondrocyte deformation and local tissue strain in articular cartilage: a confocal microscopy study. J. Orthop. Res. **13:** 410–421.
56. HAIDER, M.A. 2004. A radial biphasic model for local cell-matrix mechanics in articular cartilage. SIAM J. Appl. Math. **64:** 1588–1608.
57. UPTON, M.L. *et al.* 2006. Finite element modeling predictions of region-specific cell-matrix mechanics in the meniscus. Biomech. Model. Mechanobiol. doi: 10.1007/s10237-006-0031-4.
58. WU, J.Z., W. HERZOG & M. EPSTEIN. 1999. Modelling of location- and time-dependent deformation of chondrocytes during cartilage loading. J. Biomech. **32:** 563–572.
59. ALEXOPOULOS, L.G., L.A. SETTON & F. GUILAK. 2005. The biomechanical role of the chondrocyte pericellular matrix in articular cartilage. Acta. Biomater. **1:** 317–325.
60. GUILAK, F. & V.C. MOW. 2000. The mechanical environment of the chondrocyte: a biphasic finite element model of cell-matrix interactions in articular cartilage. J. Biomech. **33:** 1663–1673.
61. HAIDER, M.A. *et al.* 2006. A mechano-chemical model for the passive swelling response of an isolated chondron under osmotic loading. Biomech. Model. Mechanobiol. doi: 10.1007/s10237-006-0026-1.
62. ATESHIAN, G.A. *et al.* 1997. Finite deformation biphasic material properties of bovine articular cartilage from confined compression experiments. J. Biomech. **30:** 1157–1164.
63. ATHANASIOU, K.A. *et al.* 1991. Interspecies comparisons of in situ intrinsic mechanical properties of distal femoral cartilage. J. Orthop. Res. **9:** 330–340.
64. FRANK, E.H. & A.J. GRODZINSKY. 1987. Cartilage electromechanics—I. Electrokinetic transduction and the effects of electrolyte pH and ionic strength. J. Biomech. **20:** 615–627.
65. MOW, V.C. *et al.* 1980. Biphasic creep and stress relaxation of articular cartilage in compression: theory and experiments. J. Biomech. Eng. **102:** 73–84.
66. ZHU, W. *et al.* 1993. Viscoelastic shear properties of articular cartilage and the effects of glycosidase treatments. J. Orthop. Res. **11:** 771–781.

67. FREEMAN, P.M. *et al.* 1994. Chondrocyte cells respond mechanically to compressive loads. J. Orthop. Res. **12:** 311–320.
68. GUILAK, F. 2000. The deformation behavior and viscoelastic properties of chondrocytes in articular cartilage. Biorheology **37:** 27–44.
69. JONES, W.R. *et al.* 1999. Alterations in the Young's modulus and volumetric properties of chondrocytes isolated from normal and osteoarthritic human cartilage. J. Biomech. **32:** 119–127.
70. TRICKEY, W.R., G.M. LEE & F. GUILAK. 2000. Viscoelastic properties of chondrocytes from normal and osteoarthritic human cartilage. J. Orthop. Res. **18:** 891–898.
71. TRICKEY, W.R. *et al.* 2006. Determination of the Poisson's ratio of the cell: recovery properties of chondrocytes after release from complete micropipette aspiration. J. Biomech. **39:** 78–87.
72. KOAY, E.J., A.C. SHIEH & K.A. ATHANASIOU. 2003. Creep indentation of single cells. J. Biomech. Eng. **125:** 334–341.
73. ALEXOPOULOS, L.G. *et al.* 2005. Osteoarthritic changes in the biphasic mechanical properties of the chondrocyte pericellular matrix in articular cartilage. J. Biomech. **38:** 509–517.
74. GUILAK, F. *et al.* 1999. The deformation behavior and mechanical properties of chondrocytes in articular cartilage. Osteoarthritis Cartilage **7:** 59–70.
75. GUILAK, F. *et al.* 2005. Zonal uniformity in mechanical properties of the chondrocyte pericellular matrix: micropipette aspiration of canine chondrons isolated by cartilage homogenization. Ann. Biomed. Eng. **33:** 1312–1318.
76. KNIGHT, M.M., D.A. LEE & D.L. BADER. 1998. The influence of elaborated pericellular matrix on the deformation of isolated articular chondrocytes cultured in agarose. Biochim. Biophys. Acta. **1405:** 67–77.
77. MAROUDAS, A. 1979. Physicochemical properties of articular cartilage. *In* Adult Articular Cartilage. M. Freeman, Ed.: 215–290. Pitman Medical. Tunbridge Wells.
78. HING, W.A., A.F. SHERWIN & C.A. POOLE. 2002. The influence of the pericellular microenvironment on the chondrocyte response to osmotic challenge. Osteoarthritis Cartilage **10:** 297–307.
79. MAROUDAS, A. *et al.* 1985. Studies of hydration and swelling pressure in normal and osteoarthritic cartilage. Biorheology **22:** 159–169.
80. GUILAK, F., G.R. ERICKSON & H.P. TING-BEALL. 2002. The effects of osmotic stress on the viscoelastic and physical properties of articular chondrocytes. Biophys. J. **82:** 720–727.
81. ATESHIAN, G.A. *et al.* 2004. The correspondence between equilibrium biphasic and triphasic material properties in mixture models of articular cartilage. J. Biomech. **37:** 391–400.
82. LIKHITPANICHKUL, M. *et al.* 2003. The mechano-electrochemical environment of chondrocytes in articular cartilage explants under unconfined compression: emphasis on the cell matrix interactions. Proceedings of the 2003 Summer Bioengineering Conference, Key Biscayne, FL.
83. MCDEVITT, C.A. *et al.* 2001. The isolation of a cell-pericellular matrix complex from meniscus fibrocartilage: the fibrochondron. Trans. Orthop. Res. Soc. **26:** 384.

Clinical Biophysics: The Promotion of Skeletal Repair by Physical Forces

ROY K. AARON,[a] DEBORAH McK. CIOMBOR,[a] SHUO WANG,[a]
AND BRUCE SIMON[b]

[a]*Department of Orthopaedics, Brown Medical School, Providence, Rhode Island 02906, USA*

[b]*EBI, L.P., Parsippany, New Jersey 07054, USA*

ABSTRACT: Skeletal tissues respond to the physical demands of their environment by altering the synthesis and organization of the extracellular matrix. These observations have major implications for how physical environmental demands result in the clinical observations of atrophy and hypertrophy, and how manipulation of the physical environment can be used therapeutically to stimulate repair. Electrical stimulation will be considered as a paradigm of how musculoskeletal tissues respond to physical stimuli. A model of demineralized bone matrix-induced endochondral ossification has been used because it epitomizes the cell biology of endochondral bone formation in a temporally consistent way. We have studied cartilage and bone matrix production, the temporal locus of cell responsiveness, signal dosimetry, and the synthesis of signaling cytokines (TGF-β) using biochemical, immunohistochemical, and molecular techniques. Exposure to certain electrical environments enhances chondrocyte differentiation reflected as a temporal acceleration and quantitative increase of cartilage extracellular matrix, earlier onset of osteogenesis, and more mature trabecular bone. The cell pool competent to respond resides in the mesenchymal stage. The enhancement in chondrogenesis is associated with an increase in TGF-β synthesis mediated at least in part by binding of the transcription factor AP-1 and may be modulated specifically by phosphorylation of JNK. The clinical practice of orthopedics has empirically created a variety of biophysical environments in attempts to optimize skeletal repair. We are beginning to understand the biological effects of biophysical stimulation and are now poised to replace empiricism with treatment paradigms based upon physiologic understandings of dose and biologic response.

KEYWORDS: physical forces; mechanical strain; electrical fields; skeletal repair; fracture healing

Address for correspondence: Roy K. Aaron, M.D., Department of Orthopaedic Surgery, Brown Medical School, 100 Butler Drive, Providence, RI 02906. Voice: 401-274-9660; fax: 401-861-5812.
e-mail: Roy_Aaron@Brown.edu

**Ann. N.Y. Acad. Sci. 1068: 513–531 (2006). © 2006 New York Academy of Sciences.
doi: 10.1196/annals.1346.045**

INTRODUCTION

Physical Forces Regulate Development and Morphogenesis

Skeletal tissues are highly responsive to the physical demands of their environment and respond to changes in physical requirements by altering the synthesis and organization of the extracellular matrix. The physical environment, in other words, exerts regulatory influences on gene expression and the synthesis of structural proteins (including those forming the extracellular matrix) as well as signaling proteins (including growth factors): it therefore has an overall influence on morphogenesis, particularly chondrogenesis and endochondral bone formation. Regulatory pathways involve recognition of physical signals, signal transduction, and regulation of gene expression. The understanding of skeletal morphogenesis, shape, and structure in relation to the physical environment is reflected clinically in observations of atrophy and hypertrophy and has wide-ranging therapeutic implications. It is of considerable interest in microgravity environments; in aging, senescence, and osteoporosis; in athletic training and overuse injuries; and in therapeutic rehabilitation regimens. This understanding has major implications for discussion of how epigenetic physical factors influence gene expression and how manipulation of the physical environment can be used therapeutically to stimulate repair. This article considers how information about the physical environment is carried to cells in the form of signal content, how cells transduce this information, and how physical stimuli regulate gene expression in the context of fracture repair and endochondral ossification. Upregulation of growth factor synthesis by physical agents is advanced as a possible intermediary mechanism of action.

Although genetic factors both permit and constrain the range of expression of skeletal features, these features are also influenced by environmental and physical factors. In the 19th century, Wilhelm Roux proposed that mechanical factors played a central role in the morphogenesis of musculoskeletal tissues and referred to the collective physical processes influencing developmental morphogenesis as "developmental mechanics," or *entwicklungsmechanik.* In modern biochemical and molecular terms, this concept can be defined as genetic potential modified by epigenetic factors. Considerable evidence exists that intermittent physical forces affect cell differentiation, particularly the formation of cartilage. For example, intermittent hydrostatic pressure associated with loading is chondrogenic and maintains the cartilage phenotype. By modulating cartilage growth rates, mechanical loading histories influence morphogenesis. In the absence of appropriate epigenetic cues or in the presence of environmental teratogens (physical or chemical), disturbances in endochondral ossification and morphogenesis can be observed.

Developmental pattern formation and limb morphogenesis are influenced by an integrated three-dimensional signaling network of genetic and epigenetic information. This network consists of genetic potential modifiable by endocrine and other chemical factors and by physical forces, notably mechanical and electric. Alterations in the physical environment have been shown to produce dysmorphogenesis and fetal abnormalities in both animals and humans. An example is the classic experiment of Sokoloff that showed the failure of joint cavitation in paralyzed chick embryos and which demonstrated that muscle contracture and movement of embryonic joints are necessary for normal articular morphogenesis. A correlate in human disease is arthrogryposis, in which restricted joint movement and muscle contractures are associated with abnormalities in joint morphogenesis. In embryogenesis, patterns of ionic currents and voltage gradients accompany morphologic changes and gene expression. Electric fields have been associated with cell migration and differentiation in embryogenesis, including limb pattern formation. Blockage of endogenous electric currents and voltage gradients results in a disturbance of morphogenesis and embryological abnormalities. Hyperthermia is teratogenic in humans depending upon the dose, timing, and gestational age of the exposure. Arthrogryposis and limb reduction defects have been described, as well as a variety of central nervous system and craniofacial abnormalities. Pregnant women who report a fever of 102°F or above for at least 24 h demonstrated a 15.8% incidence of major fetal malformations compared to 4.5% in a control, nonfebrile group. Thus, physical forces that disrupt the informational network of development with intrusive epigenetic information can disrupt normal developmental morphogenesis, including cell differentiation and extracellular matrix synthesis, which appear as dysmorphogenesis. This disruption appears to occur as a consequence of altered regulation of gene expression for important signaling molecules and a consequent alteration in expression in genes for and synthesis of extracellular matrix proteins, prominently proteoglycans and type II collagen.

Empiricism in Orthopedics

Skeletal repair requires the synthesis and organization of an extracellular matrix appropriate to the functional, biomechanical demands of bone, and the failure to produce an appropriate extracellular matrix results in nonunion and loss of weight-bearing capacity (FIG. 1). Failure to synthesize an appropriate extracellular matrix can have catastrophic consequences in terms of loss of function and implant failure. The physical environment exerts regulatory influences on structural extracellular matrix and signaling proteins, which are reflected in the clinical context as the adequacy of fracture repair. The clinical practice of orthopedics empirically creates a variety of biophysical environments in attempts to optimize skeletal repair. These include internal fixation with plates and rods of varying flexibility; external fixators of varying stiffness; dynamized fixation; and a number of weight-bearing, bracing, and rehabilitative regimens. All of these physical environments influence the type of

(A)

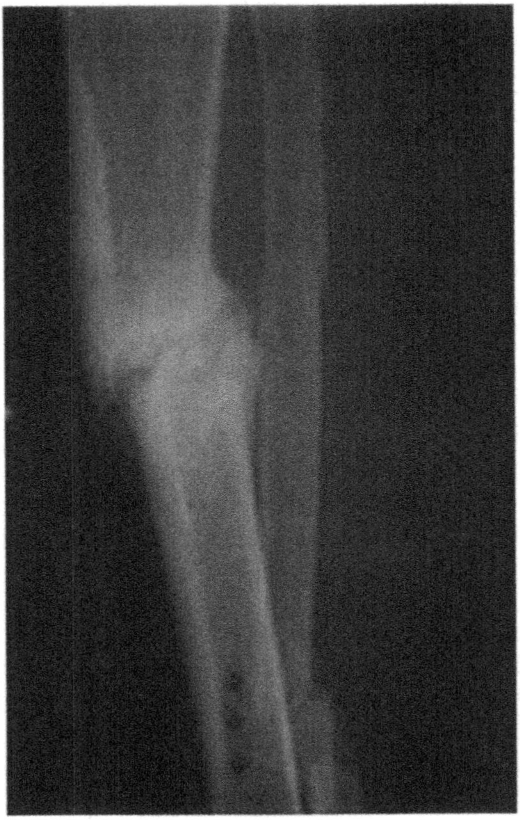

FIGURE 1. Consequences of unsuccessful fracture healing: (**A**) Synthesis and organization of an extracellular matrix inappropriate to the functional biomechanical demands. (**B**) Failure to produce an effective extracellular matrix to restore tissue continuity and function. (**C**) Failure of bone repair, resulting in fracture of fixation plate and gross instability.

extracellular matrix that is formed and, in the clinical context, restore function. FIGURE 2 illustrates fracture healing in two different mechanical environments. The fracture fixed with a rigid plate experiences minimal mechanical strain and is healing by direct osteonal repair, whereas the fracture subject to motion is healing by endochondral ossification. In both cases, extracellular matrix synthesis and gene expression are being regulated by the physical environment. Techniques to augment repair include stimulation with mechanical stress, ultrasound, and electric and electromagnetic fields. With these techniques, the orthopedic surgeon provides physical, epigenetic cues to healing fractures. This environmental cueing alters the timing and extent of gene expression for and protein synthesis of both structural proteins and signaling proteins, notably growth factors. Physical epigenetic cueing provides information transfer on both loading conditions and the composition of extracellular matrix.

(B) (C)

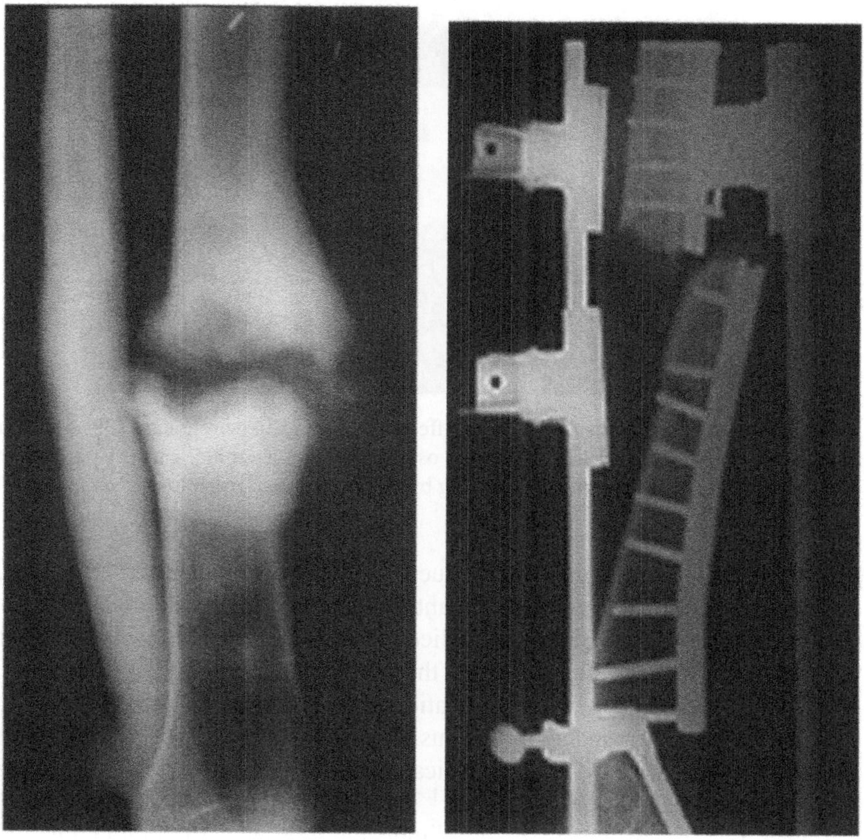

FIGURE 1. *Continued.*

Information transfer is embedded in dosimetric parameters. For essentially all physical agents that regulate morphogenesis, information is carried in the signal characteristics of amplitude, frequency, exposure duration, on-off (duty) cycle, and signal coherence. We are beginning to understand the biological effects, dosimetry, and synergies of biophysical stimulation, and we are now poised at a moment when treatment paradigms based upon physiologic understanding of dose and biological response may be able to replace empiricism.

MECHANICAL STRAIN AND STRAIN-ASSOCIATED EVENTS

Mechanical Stimulation of Fracture Healing

Endochondral bone formation—and in particular fracture healing—can be modified by exposure to mechanical strain. The rate and quality of the formation of extracellular matrix can be regulated by the application of mechanical

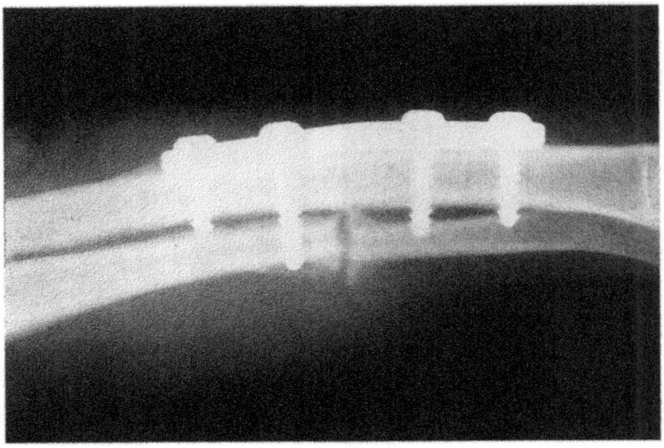

FIGURE 2. Fracture healing in two different mechanical environments. Fixed with a rigid plate, the radius is healing by direct osteonal repair. The ulna, which is in a more flexible mechanical environment, is healing by endochondral ossification.

strains of varying amplitudes and frequencies, and strain magnitude, frequency, and size of fracture gap are determinants of fracture healing. Short-term interfragmentary cyclic micromotion applied at a high strain rate induces a greater amount of periosteal callus than does the same stimulus applied at a low strain rate. Additionally, the time of application of the strain in the fracture healing process influences endochondral callus formation and bone healing. Based on these and other basic studies, mechanical stimulation has been used to enhance healing of tibial diaphyseal fractures.[1] These studies also show dosimetric relationships of fracture gap size, amplitude and frequency of stimulation, and time of exposure in the fracture healing process. The bone response varies with both strain magnitude and cycle number. Typically, as the applied load or strain magnitude decreases, the number of cycles required for activation of bone formation increases.[2] These studies and others demonstrate that the pattern, rate, and effectiveness of bone healing are regulated by mechanical factors.[3,4]

Applied strains <5% and small hydrostatic pressures <0.5 MPa have been shown to stimulate intramembranous bone formation. Strains of 5% to 15% and hydrostatic pressures >0.15 MPa stimulated endochondral bone formation, whereas strain magnitudes >15% resulted in the synthesis of fibrous tissue.[5] This study demonstrated that the amplitude of applied strains regulates the synthesis of extracellular matrix molecules in bone, cartilage, and fibrous tissue, indicating both the regulatory capacity of mechanical forces in a dosimetric fashion and the effects of mechanical factors on fracture healing of clinical consequence.

In a sheep fracture model, osteotomy gap size and interfragmentary motion have a profound influence upon the type of extracellular matrix formed at the

TABLE 1. Mechanical Stimulation of Fracture Repair (Kenwright and Goodship)

Animals:
- Tibial osteotomy followed for 12 weeks
- 1 mm displacement at 0.5 Hz × 17 min/day
- Control = 54% torsional stiffness
- Treated = 83% torsional stiffness

Human:
- Tibial fractures followed for 6 months
- 1 mm displacement at 0.5 Hz × 30 min/day
- Control = 22 weeks to reach stiffness
- Treated = 18 weeks to reach stiffness

fracture site.[6] This study reported the application of interfragmentary cyclic micromovement as a function of strain rate. Strain rates of 400 mm/s produced significantly greater bone mineral content than did strain rates of 2 or 40 mm/s, a change reflected in differential stiffness during gait. Other studies by this group in both animal and human trials have demonstrated that optimum characteristics of mechanical stress can be identified and that fracture healing can be accelerated by the application of optimized mechanical environments (TABLE 1).[7]

Strain-Associated Events

Bone and cartilage cells exist in a complex biomechanical environment defined in large part by considerations of mechanical strain relative to the nature of the extracellular matrix. Although mechanical strain is the most obvious of physical forces, it affects the overall physical environment of tissues, including pH, pO_2, hydrostatic pressure, fluid flow, osmotic pressure, and electrokinetic phenomena (TABLE 2). Gradients in fluid and ionic flow result in electrokinetic events, notably streaming potentials, which result from the movement of mobile charged ions past a stationary charged phase and are proportional to the magnitude of the applied strain. Streaming potentials can be modulated by cyclic loading conditions and by alterations in proteoglycan charge or content, thus constituting a signaling mechanism that reflects both the

TABLE 2. Mechanical Strain Affects the Overall Physical Environment

1. Cell deformation
2. Hydrostatic pressure
3. Fluid flow
   a. Shear stresses
   b. Ion flow
   c. Streaming potentials
4. Osmotic pressure
5. Ion concentration and pH

mechanical environment and the state of the extracellular matrix. One or more of these strain-associated events may transduce information to cells concerning their mechanical environment and the capacity of the extracellular matrix to meet these mechanical demands. Since these other biophysical events occur coincidentally with mechanical strain, it is difficult if not impossible to isolate the effects of the mechanical environment from those of accompanying fluid flow, ion movement, and electrokinetic events. The exogenous application of electric or electromagnetic fields induces an electric field in tissue and permits the isolation of this physical force as a regulator of skeletal function.

Recognition of Electric and Electromagnetic Signals

Studies of signal transduction of electric and electromagnetic fields have focused on two considerations: activation of transmembrane channels and factitious interactions with transmembrane receptors.[8] A recent review has summarized evidence of transmembrane channel activation by mechanical strain and electric and electromagnetic fields, suggesting that an increase in cytosolic calcium and activated calmodulin were common to these physical forces.[9] Other investigators, however, have shown inconsistent responses to electric fields with various transmembrane channels.[10] Dosimetric studies of static and low-frequency fields demonstrated no effect on transmembrane ATP-sensitive potassium channels.[11] Although transmembrane calcium fluxes are well demonstrated, it remains uncertain whether or not transmembrane channels are the mechanisms of signal transduction.

Twenty years ago it was postulated that the effects of electric and electromagnetic fields were mediated at the cell membrane by factitious receptor interactions, particularly with the parathyroid hormone (PTH) receptor.[12,13] These studies suggested that electromagnetic field (EMF) effects on PTH signaling were mediated through conformational changes in the transmembrane portion of the PTH receptor. Other effects of electric and EMF with transmembrane receptors have recently been summarized.[8] Detailed studies have been reported of the interactions between EMF and adenosine A_{2A} receptors.[14] Saturation-binding experiments revealed a significant increase in adenosine A_{2A} receptor density associated with enhanced transmembrane signaling in neutrophils treated with EMF. This effect was dependent on dosimetric parameters and was specific to this receptor population. The mitogen-activated protein (MAP) kinase pathways are thought to mediate the effects of physical forces. Phosphorylated ERK 1/2 mediates most mechanical stimuli.[15] We have found that EMF are mediated primarily by phosphorylated JNK.[16] Several transcription factors have been shown to be upregulated by MAP kinase members as well as protein kinases and calcium/calmodulin-dependent kinases, notably NF-κB, CREB, and AP-1.[16–18] These transcription factors regulate a wide variety of proliferative, differentiative, and adaptive responses. AP-1 in particular regulates the synthesis of transforming growth factor-β (TGF-β), which has a number of effects on chondrogenesis. We describe below the increased synthesis of TGF-β in the context of endochondral ossification and have described elsewhere the association with AP-1 binding to the TGF-β promotor by EMF.[16]

ELECTRIC AND ELECTROMAGNETIC STIMULATION

Electric stimulation of skeletal repair can be achieved by the implantation of DC electrodes, by the external application of electric energy via capacitive coupling, or by the application of EMF. A large number of preclinical and clinical studies have been carried out demonstrating that the rate and quality of extracellular matrix formation in endochondral bone formation can be altered by exposure to exogenous electric and EMF.[19] Healing and gain in mechanical strength of fractures and osteotomies have been shown to be accelerated with these techniques in a number of animal models. The responses of cartilage and bone to electric and electromagnetic stimulation have shown exquisite sensitivity to dosimetry. Discrete frequency and amplitude spectra as well as exposure times have been described that represent optimum exposure parameters in both *in vivo* and *in vitro* models.[9] Studies in our laboratory have demonstrated that extracellular matrix synthesis is dependent upon amplitude and exposure parameters. For a specific therapeutic signal, a peak response has been observed at 0.04–0.1 G with a sustained response thereafter (FIG. 3). Further studies determined the relationships between amplitude and exposure duration, an example of which is shown in FIGURE 4. Similar studies demonstrating dose responses and discrete frequency and amplitude windows were carried out with sine waves.[20] Significant increases in chondrogenesis were observed at 0.07 mT and 0.1 mT, with the peak response occurring at 0.1 mT. A dose response in terms of exposure duration was observed as an increase in chondrogenesis at 7–9 h of daily exposure. At shorter or longer exposure times, the increase in chondrogenesis was not observed, indicating an exposure window of biological response. This field exhibited maximal stimulation of chondrogenesis at 0.1 mT and 7 h of daily exposure. Maximal chondrogenesis was increased 33% over sham-treated controls at these optimal exposure conditions.

REGULATION OF SYNTHESIS OF STRUCTURAL EXTRACELLULAR MATRIX PROTEINS BY ELECTRIC AND ELECTROMAGNETIC FIELDS

The regulation of gene expression for and protein synthesis of extracellular matrix proteins by electrical and is presented as a paradigm of the response of skeletal tissues to physical stimuli in general.

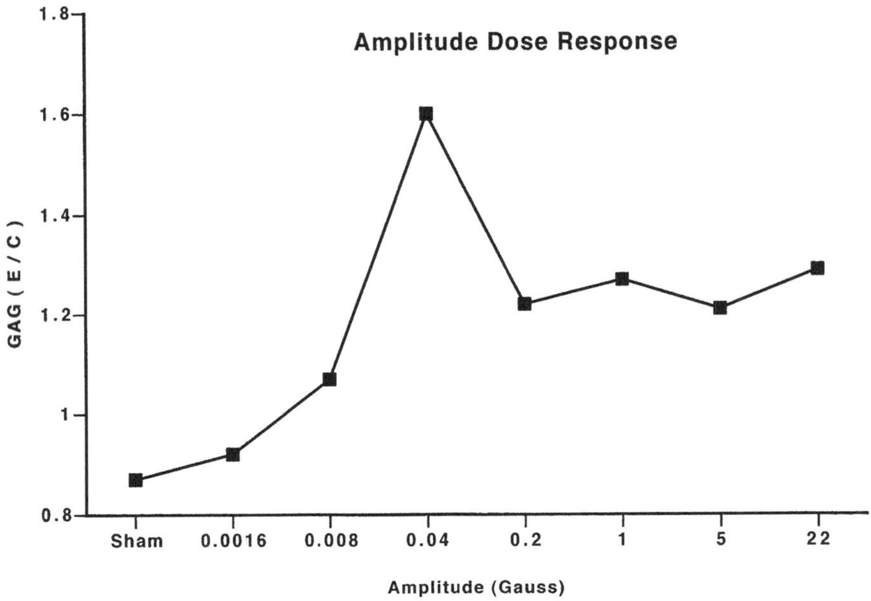

FIGURE 3. The amplitude dose response shows approximately a 20% to 30% enhancement of chondrogenesis at 0.2 G and above. A clear peak at 0.04 to 0.1 G was observed with a 50% to 60% enhancement of chondrogenesis.

The model of demineralized bone matrix (DBM)-induced endochondral ossification has been used to study the responsiveness of chondrogenesis and osteogenesis to EMF because this system recapitulates the cell biology of endochondral ossification in a temporally consistent way. Mesenchymal cells are recruited within the first day after DBM implantation and undergo proliferation. Differentiation of chondrocytes is apparent on day 6, with chondrogenesis becoming maximal on days 8–10. Endochondral calcification is observed on day 10. Progressive calcification occurs, with trabeculae appearing on days 12–14 and maturing by days 18–20. The model has been well characterized in biochemical, molecular, and morphological terms and therefore serves as an excellent test system with which to examine factors that alter the rate or quality of endochondral ossification.

In the following experiment, DBM was prepared from tibial and femoral diaphyseal bone of mature male CD rats, as described by Reddi and Huggins.[19] Twenty-five milligrams of DBM powder were implanted subcutaneously along the thoracic musculature in immature (28- to 35-day-old) male CD rats, and the rats were subjected to EMF stimulation daily beginning on the day following implantation. Control rats were treated identically but without EMF exposure. In our studies, animals were sacrificed and ossicles were harvested periodically to emphasize the mesenchymal and chondrogenic stages of development.

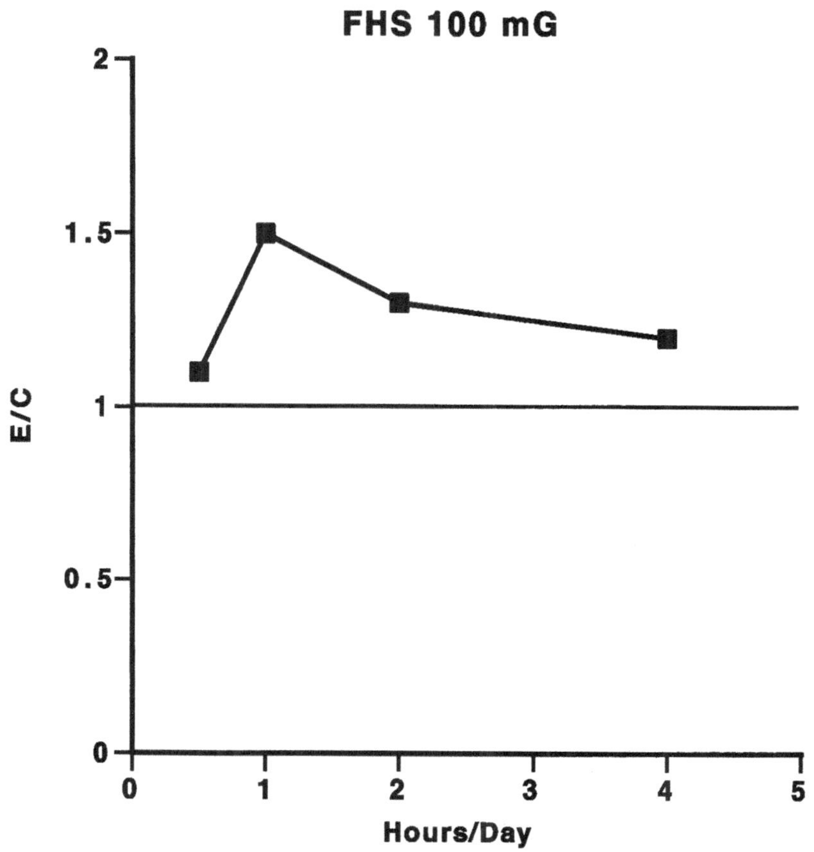

FIGURE 4. Dose response in terms of exposure time per day. Fifty percent enhancement of chondrogenesis was observed for a 100-mG signal at 1 h/day. No enhancement was observed with the same signals at shorter or longer exposure times.

Enhanced Chondrogenic Differentiation

Significant acceleration was observed in the rate of bone formation and in the degree of trabecular maturation in EMF-exposed ossicles compared to controls. Endochondral calcification occurred earlier in stimulated ossicles, and by day 16 of ossicle development the total amount of calcified tissue formed in stimulated ossicles was greater than that in controls. The amount of trabecular bone formed in EMF-stimulated ossicles was significantly increased over the amount in controls, and the trabecular index (trabecular bone/total calcified tissue) demonstrated a greater degree of bone maturation in stimulated ossicles.[19] The DNA content of both control and EMF-treated ossicles rose progressively from day 1 through day 8 of ossicle development. However, no significant differences were observed between the two groups. Similarly, no

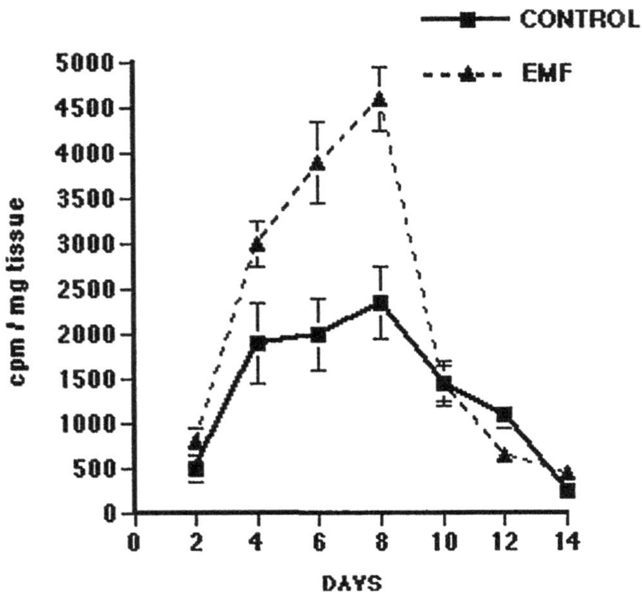

FIGURE 5. Incorporation of radiolabeled sulfate into proteoglycan. Ossicles exposed to electromagnetic field stimulation exhibit a significant increase in proteoglycan synthesis by day 4 of ossicle development, peaking during maximal chondrogenesis on day 8 (P = 0.001) and falling to control levels coincident with the onset of calcification.

differences were observed in ^{3}H thymidine incorporation. These results suggest the absence of enhanced generalized cell proliferation or recruitment by EMF as a mechanism for the acceleration of endochondral ossification.

Chondrogenesis was markedly stimulated by EMF exposure. On day 8 of development during maximal chondrogenesis, ossicles exposed to EMF demonstrated a twofold increase in sulfate incorporation and a 64% increase in glycosaminoglycan content compared to controls (FIG. 5). Both the spatial area of the ossicle composed of cartilage and the number of chondrocytes were increased by EMF exposure. The glycosaminoglycan contents expressed per unit ossicle area or per chondrocyte and the chondrocyte:matrix ratios were the same in control as in exposed ossicles, indicating the production of more chondrocytes in EMF-exposed ossicles rather than an increase of proteoglycan synthesis per chondrocyte (TABLE 3). The DNA content also did not differ between the two groups, indicating that total cell number was unchanged by EMF exposure and that the increased number of chondrocytes makes up a greater fraction of the total cell content in EMF-exposed ossicles compared to controls. The increased deposition of proteoglycan in the extracellular matrix in EMF-exposed ossicles was accompanied by an earlier and quantitatively greater expression of aggrecan and type II collagen genes. A threefold increase in aggrecan expression and a twofold increase in type II collagen

TABLE 3. Chondrogenesis Under EMF Exposure

Index	Control	Exposed	Percent	*P*
mRNA aggrecan	6.1	22.5	269	0.02
mRNA type II collagen	11.8	21.9	86	0.05
$^{35}SO_4$incorporation (cpm/mg)	2166 ± 387	4448 ± 293	105	0.005
GAG content (μg/mg)	1.4 ± 0.2	2.5 ± 0.2	79	0.01
Cartilage area (mm^2)	24 ± 2.1	148 ± 11.7	517	0.001
Chondrocytes (n)	701 ± 227	3582 ± 675	411	0.005
Chondrocyte/cartilage	29.2	24.2		n.s.

expression was observed in EMF-exposed ossicles compared to controls. The proteoglycans synthesized are of normal size distribution and chemical composition, but a greater fraction of proteoglycans synthesized in response to EMF were composed of cartilage-type proteoglycans, which co-elute with bovine nasal cartilage proteoglycan on Sepharose CL-2B columns. Reactivity to the antibodies 3B3 and 5D4 was also observed earlier in EMF-exposed ossicles.

Evidence for enhanced cell differentiation and extracellular matrix synthesis has been presented in morphologic, biochemical, and molecular terms.[21] The evidence for accelerated chondrogenic differentiation in the exposed ossicles is supported by a temporal acceleration and quantitative increase in the expression of mRNA for aggrecan and type II collagen and the earlier appearance and deposition of cartilage-specific proteoglycans compared to control ossicles on days 6 and 8 of development. The spatial area of ossicles composed of cartilage was correspondingly increased, as was the number of chondrocytes, suggesting that chondrogenic differentiation occurred. Maturation was suggested also by the glycosaminoglycan contents expressed per area or per chondrocyte and by the chondrocyte:matrix ratios, which did not change with EMF exposure. Accelerated maturation of cartilage matrix by EMF was also observed morphologically and biochemically. Earlier chondrocyte hypertrophy and matrix calcification were evident. Collectively, these data suggest that chondrogenic differentiation occurs earlier and that cartilage extracellular matrix is synthesized to a greater degree and matures faster in response to EMF exposure.

Locus of the Responsive Cell Pool

Ossicles stimulated during the mesenchymal (days 1–3), chondrogenic (days 4–8), and calcific (days 9–19) phases have been compared on day 20 of development to ossicles stimulated throughout their developmental sequence and to unstimulated controls.[22] Ossicles stimulated during both the mesenchymal and chondrogenic phases (days 1–8) have equivalent calcium content to ossicles stimulated throughout their developmental sequence. Ossicles stimulated

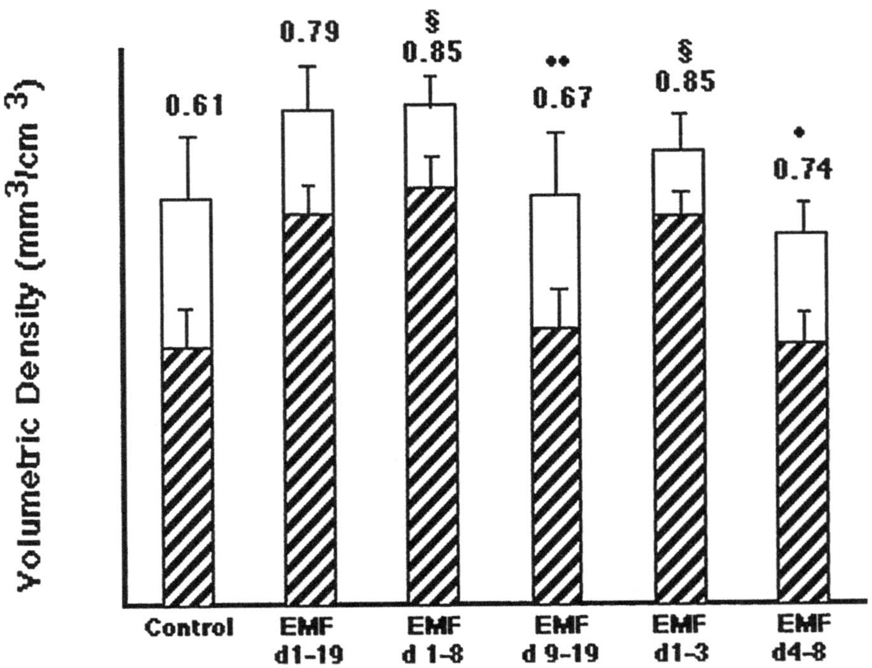

FIGURE 6. The fraction of total calcified tissue composed of trabeculae (trabecular index) is a measure of ossicle maturation. Ossicles exposed during the mesenchymal phase (days 1–3) or mesenchymal and chondrogenic phases (days 1–8) have an amount of calcified tissue and trabecular index equivalent to that of ossicles exposed throughout the full developmental sequence (day 19). Exposure of ossicles to EMF at other developmental stages does not induce maturation as compared with stimulation on days 1–19. § = significant, $^*P < 0.02$, and $^{**}P < 0.01$.

during calcification and bone formation (days 9–19) have calcium contents significantly less than ossicles stimulated for days 1–8 or 1–19, although greater than those of unstimulated controls. Ossicles stimulated during the mesenchymal phase had an amount of calcium equal to those stimulated throughout the entire developmental sequence, whereas those stimulated during chondrogenesis only had an intermediate amount.

Histomorphometry confirmed these results and permitted an examination of trabecular maturation (FIG. 6). The amount of calcified tissue formed in ossicles exposed during the mesenchymal phase was identical to that formed by exposure during the full 20-day developmental sequence. Exposure of ossicles during chondrogenesis only or calcification and bone formation did not result in quantitatively more total calcified tissue than that observed in unstimulated ossicles. Ossicles stimulated during the mesenchymal phase, or mesenchymal and chondrogenic phases, had a trabecular index (0.85 ± 0.02) equal to that seen in ossicles stimulated throughout their developmental sequence

(0.79 ± 0.01), whereas ossicles stimulated in the later phases exhibited trabecular indices equal to those of controls. These data demonstrate accelerated maturation of trabecular bone with EMF stimulation during only the mesenchymal phase of endochondral bone formation. These studies suggest that there are differing sensitivities to stimulation among the developmental stages of endochondral bone formation. This model is most sensitive in the mesenchymal stage during which cell recruitment, proliferation, and differentiation occur. These studies identify a progenitor cell pool in the mesenchymal stage that is competent to respond to EMF stimulation and results in accelerated formation of cartilage and trabecular bone. The competence of the cell pool to respond is diminished by day 4 of ossicle development.

The importance of stimulating early in the developmental process is supported by examination of the effects of strain rate and stimulation timing in the modulation of fracture healing by mechanical forces.[6] Using a strain rate of 400 mm/s applied either directly after an osteotomy or in a delayed application 6 weeks after osteotomy, these studies demonstrated a distinct lag in bone mineral content and walking stiffness index in animals treated after a late start of mechanical fracture stimulation. These observations confirm ours, which suggest that the cell pool responsive to physical stimulation is present in early stages prior to chondrogenesis.

REGULATION OF SYNTHESIS OF SIGNALING PROTEINS BY EMF

Electric and electromagnetic fields increase gene expression for and synthesis of growth factors, and this effect may amplify field effects through autocrine and paracrine signaling.[8] Electric and electromagnetic fields can produce a sustained upregulation of growth factors that enhances but does not disorganize endochondral bone formation.

The initial reports of an enhancement of growth factor synthesis in skeletal tissue in response to electric and electromagnetic fields demonstrated an increase in IGF-2 mRNA and protein and suggested that IGF-2 may in part mediate proliferation of osteoblast-like cells.[23] Further studies with combined magnetic field exposure of both human osteoblast-like and rat fracture callus cells demonstrated increases in IGF-2 levels after 30 min of exposure.[24] These results are similar to those observed in response to mechanical strain and may indicate an intermediary signaling mechanism modulating the effect of biophysical input in general. The stability of the signaling pathways suggests that the mechanism may serve to amplify electric and electromagnetic signaling.[17,23]

EMF exposure stimulates mRNA expression of TGF-β and several bone morphogenic proteins (BMP) (TABLE 4). Exposure of developing chick embryos to an EMF enhances bone formation in calvaria and increases BMP-2

TABLE 4. Regulation of TGF-β/BMP

Study	Technique	Model	Result
Nagai & Ota[25]	IC	Osteoblasts	↑ BMP-2, -4 mRNA
Yajima *et al.*[26]	IC	Osteoblasts	↑ BMP-4, -5, -7 mRNA
Bodamyali *et al.*[27]	IC	Osteoblasts	↑ proliferation, BMP-2, -4 mRNA
Aaron *et al.*[28]	IC	E.O. *in vivo*	↑ differentiation, TGF-β1
Aaron *et al.*[20]	IC	E.O. *in vivo*	↑ differentiation, TGF-β1 mRNA + protein
Lohmann *et al.*[29]	IC	MG63 osteoblasts	↑ differentiation, TGF-β1
Lohmann *et al.*[30]	IC	MLO-Y4 osteocyte-like cells	↑ TGF-β1, PGE_2
Guerkov *et al.*[31]	IC	Human nonunion cells	↑ TGF-β1
Zhuang *et al.*[32]	CC	MC3T3	↑ proliferation, ↑ TGF-β1 mRNA

and -4 mRNA levels 1.6- to 2.7-fold. These results suggest that upregulation of mRNA for BMP-2 and -4 might be related to, and possibly mediate, the bone inductive effect of the EMF.[25] Similar studies were carried out in SV-HFO immortalized human osteoblasts. Under unstimulated, basal conditions, these cells express significant amounts of mRNA for BMP-1, -2, -4, -5, and -7, but not BMP-3 and -6. EMF exposure markedly increased the levels of mRNA for BMP-4, -5, and -7, with a maximal increase observed after 24 h of exposure.[26] Upregulation of BMP-2 and -4 mRNA expression in rat calvarial osteoblasts *in vitro* has also been reported coincident with bone induction.[27]

EMF upregulate mRNA expression for and protein synthesis of TGF-β_1 coincident with increases in extracellular matrix protein and gene expression.[28] Regulation of protein synthesis occurred in a distinct dose-dependent manner in terms of both amplitude and duration of exposure. Importantly, the pattern of TGF-β_1 expression was preserved throughout the developmental sequence, suggesting that EMF exposure enhances but does not disorganize chondrogenesis, endochondral calcification, and the normal physiologic expression pattern of TGF-β_1. Significant sustained increases in TGF-β_1 protein were observed in EMF-stimulated compared to control tissues throughout the developmental sequence (FIG. 7). The normal physiologic peak expression of TGF-β_1 during early chondrogenesis was observed also in EMF-stimulated tissue, although it was quantitatively increased. Increases in TGF-β_1 protein ranged from 9.8 to 16.2 pg/mg. On day 6, during the peak TGF-β_1 production, the increase produced by EMF was 0.08 pg/μg DNA. Overall, EMF exposure resulted in a 32% increase in TGF-β_1 protein and a 2.5-fold increase in mRNA compared to unstimulated tissues. Immunohistochemical studies demonstrated that TGF-β_1 was synthesized by chondrocytes rather than by mesenchymal cells. These data demonstrate that both the transcription of TGF-β_1 mRNA and the

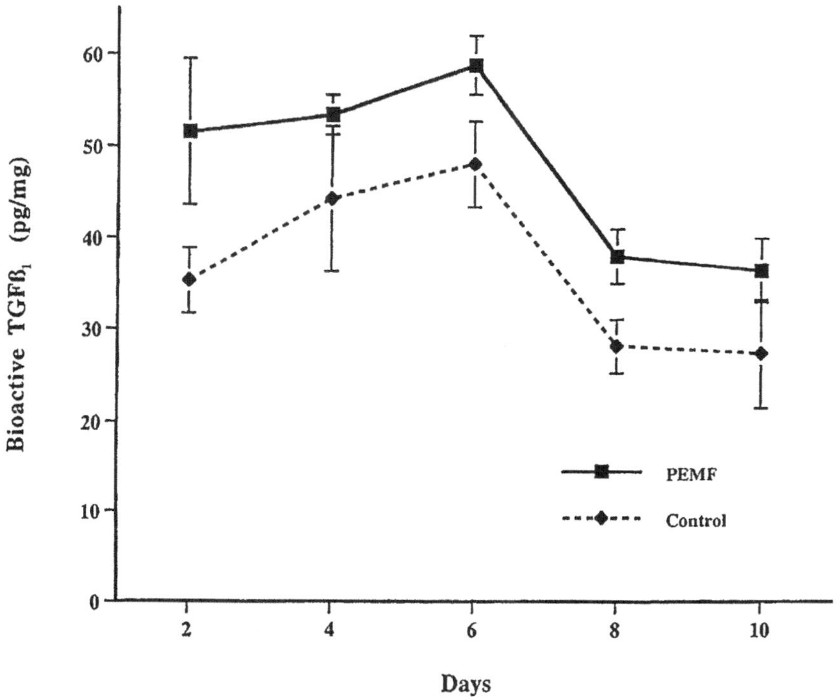

FIGURE 7. Increase in bioactive TGF-β1 protein with EMF stimulation. Significant increases are seen at days 2, 6, and 8 ($P < 0.05$). TGF-β1 protein is increased in prechondrogenic and early chondrogenic tissue (days 2–6) with a decrease on day 8 just prior to matrix calcification. The constitutive pattern of TGF-β1 expression in endochondral bone formation is increased but not disorganized.

accumulation of active TGF-β1 protein are upregulated by EMF coincident with accelerated chondrogenesis, suggesting that these EMF effects may be mediated by TGF-β1.

CONCLUSION

Therapeutic opportunities exist in the understanding of biophysical regulation of connective tissue function. Translational opportunities may be as varied as directly signaling fracture repair, designing optimum rehabilitative and exercise regimens, and designing bioreactors for engineered tissue.

REFERENCES

1. Kenwright, J. & T. Gardner. 1998. Mechanical influences on tibial fracture healing. Clin. Orthop. Relat. Res. **355S:** S179–S190.

2. EINHORN, T.A. 1995. Current concepts review: enhancement of fracture healing. J. Bone Joint Surg. **77A:** 940–956.
3. EVANS, M., J. KENWRIGHT & J.L. CUNNINGHAM. 1988. Design and performance of a fracture monitoring transducer. J. Biomed. Eng. **10:** 64–69.
4. AARON, R.K. & M.E. BOLANDER, Eds. 2005. Physical Regulation of Skeletal Repair. American Academy of Orthopaedic Surgeons. Rosemont, IL.
5. CLAES, L.E. *et al.* 1998. Effects of mechanical factors on the fracture healing process. Clin. Orthop. Relat. Res. **355S:** S132–S147.
6. GOODSHIP, A.E., J. CUNNINGHAM & J. KENWRIGHT. 1998. Strain rate and timing of stimulation in mechanical modulation of fracture healing. Clin. Orthop. Relat. Res. **355S:** S105–S115.
7. GOODSHIP, A.E. 2005. Mechanical regulation of bone healing. *In* Physical Regulation of Skeletal Repair. R.K. Aaron & M.E. Bolander, Eds.: 17–26. American Academy of Orthopaedic Surgeons. Rosemont, IL.
8. AARON, R.K. *et al.* 2004. Stimulation of growth factor synthesis by electric and electromagnetic fields. Clin. Orthop. Relat. Res. **419:** 30–37.
9. BRIGHTON, C. *et al.* 2001. Signal transduction in electrically stimulated bone cells. J. Bone Joint Surg. **83A:** 1514–1523.
10. LACY-JULBERT, A. & J.C. METCALFE. 1998. Biological responses to electromagnetic fields. FASEB J. **12:** 395–420.
11. WANG, K.W. & S.B. HLADKY. 1994. An upper limit for the effect of low frequency magnetic fields on ATP-sensitive potassium channels. Biochim. Biophys. Acta **1195:** 218–222.
12. CAIN, C.D., W.R. ADEY & R.A. LUBEN. 1987. Evidence that pulsed electromagnetic fields inhibit coupling of adenylate cyclase by parathyroid hormone in bone cells. J. Bone Miner. Res. **2:** 437–441.
13. LUBEN, R.A. *et al.* 1982. Effects of electromagnetic stimuli on bone and bone cells in vitro inhibition of responses to parathyroid hormone by low-energy, low-frequency fields. Proc. Natl. Acad. Sci. USA **79:** 4180–4184.
14. VARANI, K. *et al.* 2002. Effect of low frequency electromagnetic fields on A_{2A} adenosine receptors in human neutrophils. Br. J. Pharmacol. **136:** 57–66.
15. IQBAL, J. & M. ZAIDI. 2005. Common cellular signaling mechanisms for mechanotransduction. *In* Physical Regulation of Skeletal Repair. R.K. Aaron & M.E. Bolander, Eds.: 241–245. American Academy of Orthopaedic Surgeons. Rosemont, IL.
16. CIOMBOR, D.M. *et al.* EMF fields activate the JNK pathway resulting in enhanced TGFbeta gene expression. Trans. Orthop. Res. Soc. **29:** 0093.
17. RYABY, J. *et al.* 1994. The role of insulin-like growth factor II in magnetic field regulation of bone formation. Bioelectrochem. Bioenerg. **35:** 87–91.
18. ZHOU, J., G. YAO, J. ZHANG & Z. CHANG. 2002. CREB DNA binding activation by a 50-Hz magnetic field in HL60 cells is dependent on extra- and intracellular Ca^{2+} but not PKA, PKC, ERK, or p38 MAPK. Biochem. Biophys. Res. Comm. **296:** 1013–1018.
19. AARON, R.K., D.M. CIOMBOR & B.J. SIMON. 2004. Treatment of nonunions with electric and electromagnetic fields. Clin. Orthop. Relat. Res. **419:** 21–29.
20. AARON, R.K. *et al.* 1999. Power frequency fields promote cell differentiation coincident with an increase in transforming growth factor-β_1 expression. Bioelectromagnetics **20:** 453–458.

21. CIOMBOR, D.M., G. LESTER, R.K. AARON, *et al.* 2002. Low frequency EMF regulates chondrocyte differentiation and expression of matrix proteins. J. Orthop. Res. **20:** 40–50.
22. AARON, R.K. & D.M. CIOMBOR. 1996. Acceleration of experimental endochondral ossification by biophysical stimulation of the progenitor cell pool. J. Orthop. Res. **14:** 582–589.
23. FITZSIMMONS, R.J. *et al.* 1992. Low-amplitude, low-frequency electric field-stimulated bone cell proliferation may in part be mediated by increased IGF-II release. J. Cell Physiol. **150:** 84–89.
24. FITZSIMMONS, R.J. *et al.* 1995. IGF-II receptor number is increased in TE-85 cells by low-amplitude, low frequency electromagnetic field (EMF) exposure. J. Bone Miner. Res. **10:** 812–819.
25. NAGAI, M. & M. OTA. 1994. Pulsating electromagnetic field stimulates mRNA expression of bone morphogenetic protein-2 and -4. J. Dental Res. **73:** 1601–1605.
26. YAJIMA, A., M. OCHI & Y. HIROSE. 1996. Effects of pulsing electromagnetic fields on gene expression of bone morphogenetic proteins in human osteoblastic cell line in vitro. J. Bone Miner. Res. **11:** 381.
27. BODAMYALI, T. *et al.* 1998. Pulsed electromagnetic fields induce osteogenesis and upregulate transcription of bone morphogenetic proteins 2 and 4 in rat osteoblasts in vitro. Biochem. Biophys. Res. Comm. **250:** 458–461.
28. AARON, R.K., S. WANG & D.M. CIOMBOR. 2002. Upregulation of basal TGFβ_1 levels by EMF coincident with chondrogenesis—implications for skeletal repair and tissue engineering. J. Orthop. Res. **20:** 233–240.
29. LOHMANN, C.H. *et al.* 2000. Pulsed electromagnetic field stimulation of MG63 osteoblast-like cells affects differentiation and local factor production. J. Orthop. Res. **18:** 637–646.
30. LOHMANN, C.H. *et al.* 2003. Pulsed electromagnetic fields affect phenotype and connexin 43 protein expression in MLO-Y4 osteocyte-like cells and ROS 17/2.8 osteoblast-like cells. J. Orthop. Res. **21:** 326–334.
31. GUERKOV, H.H. *et al.* 2001. Pulsed electromagnetic fields increase growth factor release by nonunion cells. Clin. Orthop. Relat. Res. **384:** 265–279.
32. ZHUANG, H. *et al.* 1997. Electrical stimulation induces the level of TGF-β1 mRNA in osteoblastic cells by a mechanism involving calcium/calmodulin pathway. Biochem. Biophys. Res. Comm. **237:** 225–229.

COX-2 has a Critical Role During Incorporation of Structural Bone Allografts

REGIS J. O'KEEFE, PRAROP TIYAPATANAPUTI, CHAO XIE, TIAN FANG LI, CHRIS CLARK, MICHEAL J. ZUSCIK, DI CHEN, HICHAM DRISSI, EDWARD SCHWARZ, AND XINPING ZHANG

Department of Orthopedics, Center for Muscular Skeletal Research, University of Rochester School of Medicine and Dentistry, Rochester, New York 14642, USA

ABSTRACT: **Nonsteroidal anti-inflammatory drugs (NSAIDs), which inhibit cyclooxygenase (COX) activity, reduced pain and are commonly used in patients with skeletal injury. In this article we will also present data to show that selective COX-2 inhibitor delays allograft healing and incorporation. In contrast, local delivery of prostaglandin E2 (PGE2) enhanced bone formation at cortical bone graft junction. A 4-mm mid-diaphyseal segmental femoral defect was created and then repaired by frozen bone allograft of the same size. A 22-gauge metal pin was placed in the intramedullary cavity to stabilize the bone graft. Healing was evaluated weekly by X ray and by a semiquantitative histomorphometric analysis at 5 weeks postsurgery. Celecoxib (25 mg/kg/day) and Ketorolac (4 mg/kg/day) were administered daily for 2 weeks or 5 weeks. PGE2 was infused locally at a dose of 800 nmol/kg per day via osmotic minipump for 4 weeks. Inhibition of cyclooxygenase by daily administration of the Celecoxib or Ketorolac for 5 weeks reduced new bone ingrowth by about 60% ($P < 0.05$). The percentage of bony bridging in both drug-treated groups was significantly decreased at 5 weeks. Temporal administration of Celecoxib for 2 weeks also significantly reduced bone formation by 45% and withdrawal of the Celecoxib only led to slight recovery of bone formation at the graft side. In contrast to the inhibitory effects of NSAIDS, PGE2 infusion at the cortical bone junction increased bone formation by about twofold. These results demonstrated that COX-2 is essential for bone allograft incorporation. Furthermore, our data support the notion that COX-2-dependent PGE2 produced at the early stage of bone healing is prerequisite for efficient skeletal repair.**

KEYWORDS: **cyclooxygenase; PGE2; allografts; COX-2**

Address for correspondence: Dr. Regis J. O'Keefe or Dr. Xinping Zhang, The Center for Musculoskeletal Research, University of Rochester Medical Center, 601 Elmwood Avenue, Rochester, NY 14642. Voice: 716-273-4255; fax: 716-442-3214.

e-mail: Regis_OKeefe@URMC.rochester.edu; Xinping_Zhang@URMC.rochester.edu

Ann. N.Y. Acad. Sci. 1068: 532–542 (2006). © 2006 New York Academy of Sciences.
doi: 10.1196/annals.1346.012

INTRODUCTION

Nonsteroidal anti-inflammatory drugs, also known as NSAIDs, are the most commonly used pain-relieving medications in our society. In 1971, Sir John Vane demonstrated for the first time that the mechanism of action of aspirin and other NSAIDs is via inhibition of cyclooxygenase (COX), which catalyzes the formation of prostaglandins and thromboxanes from arachidonic acid (AA).[1,2] It is now known that at least two isoforms of COX exist, the inducible isoform COX-2 and the constitutive isoform COX-1. While NSAIDs have different selectivity toward these isoforms, inhibition of COX-2 is considered as the basis for the anti-inflammatory and analgesic effects of NSAIDs.[3,4] While selective COX-2 inhibitors have reduced gastrointestinal complications, recent studies have revealed that suppression of COX-2 may perturb normal renal and cardiovascular functions.[5–7]

Several studies show that COX activity is involved in normal bone metabolism and suggest that NSAIDs have a negative impact on bone repair. Ibuprofen, Indomethacin, Ketorolac, Celecoxib, and Rofecoxib have been shown to delay fracture healing in animal models.[8–13] Bone repair in animal models of spine fusion is also impaired by NSAIDs.[14,15] Definitive data in humans are not available since prospective, placebo-controlled clinical trials with bone healing as a primary endpoint have not been performed. However, retrospective studies have suggested that treatment with NSAIDs impairs spinal fusion in humans.[16,17] Retrospective studies of fracture repair in humans have shown mixed results, with some studies showing inhibition and others demonstrating no effect.[18,19] Consistent with an inhibitory effect on bone repair, Indomethacin and Ketorolac have been advocated for preventing heterotopic ossification following joint replacement.[20–23]

The most compelling data implicating COX activity during bone repair comes from genetic models that demonstrate a critical role for COX-2. Simon *et al.*[24] demonstrated that COX-2 -/- mice had impaired fracture healing whereas COX-1 knockout mice had a normal rate of repair.[13] Work in our laboratory also showed diminished bone repair in COX-2 -/- mice following fracture. Although cartilage maturation occurred in COX-2 -/- mice as evidenced by the expressions of *type X collagen* and *MMP-13* genes, completion of endochondral bone formation with the elimination of cartilage was impaired. Moreover, osteoblast differentiation from mesenchymal precursors was markedly impaired, with decreased intramembranous bone formation and reduced osteocalcin gene expression observed in fractures as well as on the calvarial surface following FGF-1 injection. Cell culture studies confirmed the decreased bone formation, since the number of bone nodules was reduced in COX-2 -/- mice. However, the decrease in bone nodule formation in COX-2 -/- mice is reversed by the addition of exogenous PGE2.

These later data suggest that mesenchymal cells remain receptive to PGE2 and bone formation occurs normally once COX-2/PGE2 signals are available.

Several studies using pharmacological inhibition of COX-2 show that after an initial delay, fracture healing progresses and approaches normal rates following either brief or continuous treatments with NSAIDs/COX-2 specific inhibitors.[25] However, animal fracture studies use models where healing is extremely reliable. In contrast, human fractures are heterogeneous and while some fractures, such as the distal radius, essentially always heal, other fractures, such as tibia fractures and spinal fusion have elevated rates of nonunion.[26] Moreover, it has been shown that comorbidity, including various diseases and the associated degree of bone and soft tissue injury impacts on fracture healing.[26] In the current study we used a gain and loss of function approach in order to better understand the role of COX-2 inhibition in a murine segmental femur allograft-healing model.[27,28] This is a more challenging model of bone repair that mimics clinical scenarios where nonunion is more likely to occur. Allografts do not provide cells to the local environment and healing has been shown to depend upon osteoinductive and angiogenic activity from the host at the cortical bone junction.[29,30] In our current study we examined the effects of Celecoxib at a dose of 25 mg/kg and Ketorolac at a dose of 4 mg/kg on bone formation in this model. Furthermore, we infused PGE2 locally at the cortical bone junction via osmotic minipumps and demonstrated that constitutive administration of PGE2 increased allograft healing via increasing host bone formation at the allograft junction.

MATERIALS AND METHODS

Murine Segmental Femoral Graft Model

Ten-week-old C57BL/6 mice were anesthetized, a 7- to 8-mm long incision was made, and the femur was exposed by blunt dissection. A 4-mm mid-diaphyseal segment was removed by osteotomizing the bone using angled wire scissors. Bone grafts were obtained from the femoral shaft of a different strain of mice: 129 (allografts). These 4-mm allografts were sterilized with 70% ethanol and then fresh frozen at -70°C. Prior to re-implantation they were thawed to room temperature and rinsed in saline to remove residual ethanol. The bone graft was secured with a 22-gauge metal pin placed through the segmental graft through the marrow cavity. The incision was closed by silk suture. Graft healing was followed radiographically using a Faxitron X-ray system (Faxitron X-ray Corporation, Wheeling, IL). Mice were sacrificed at 2, 3, 4, and 5 weeks after surgery and samples processed for histological analysis.

Histological and Histomorphometric Analyses

Following sacrifice, grafted femurs were harvested, fixed in 10% neutral-buffered formalin, decalcified, and 3-μm paraffin-embedded sections prepared

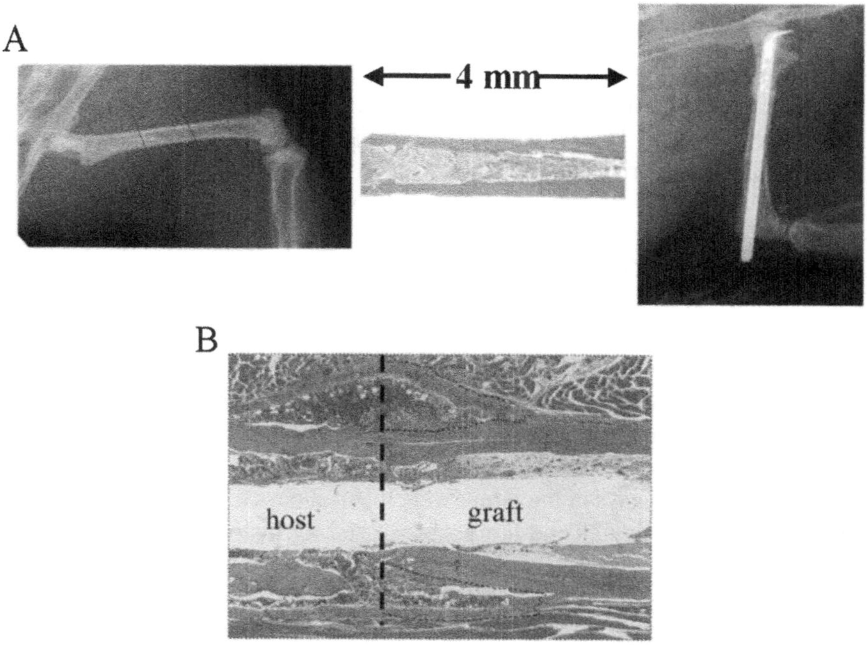

FIGURE 1. Schematic representation of the mouse segmental femoral bone graft model. (**A**) An 4-mm long allograft was used to repair the segmental defect created in a host mouse femur. The graft was stabilized via intramedullary metal pin. (**B**) Illustration of bone histomorphometric measurements for graft incorporation.

and stained with hematoxylin/eosin, or alcian blue/hematoxylin.[28] Histomorphometric analysis was carried out using Osteometrics™ software (Osteometrics, Inc., Decatur, GA) to determine the area of periosteal new bone formation. As illustrated in Figure 1, a line was drawn in the middle of the junction between graft and host bone to separate the new bone formation on the surface of the graft and the host bone. The measurement of periosteal bone formation on the side of the graft reflects the ingrowth of new bone into the graft, and was used to evaluate graft healing in this study.[28] A total 61 allografts were implanted and evaluated. Eleven specimens were excluded from the study due to infection or fracture of the graft. (See Table 1).

TABLE 1. Animals used to evaluate NSAID effects on allograft healing

	Control (5 weeks)	Celecoxib (2 weeks)	Celecoxib (5 weeks)	Ketorolac (5 weeks)
Number of animals used	16	10	20	15
Number of specimens included	16	6	17	11

Drug Administration

Celecoxib (25 mg/kg/day) was administered orally (daily for 2 or 5 weeks) and Ketorolac (4 mg/kg/day) was administered by intramuscular injection (daily for 5 weeks). Drugs therapy was initiated on the day of surgery and all animals were sacrificed at 5 weeks.

PGE2 Delivery via Osmotic Minipumps

Allograft surgery was performed and a polyvinyl catheter was fixed on the bone surface and connected to an Alzet 2002 osmotic minipump (ALZET, Cupertino, CA) implanted in the subcutaneous tissue of the back. The pumps contained PGE2 in ethanol/propylene glycol (40:60, vol/vol). Infusion was performed at a rate of 0.25 μl/h for 4 weeks. PGE2 was infused at a dose of 800 nmol/kg/day. X rays of the femur were taken at days 0, 14, and 28. Animals were sacrificed on day 28 and femurs processed for histology.

Statistical Analysis

Data are expressed as the means ± SEM. Statistical significance was determined by one-way ANOVA. A P value <0.05 was considered statistically significant.

RESULTS

Celecoxib or Ketorolac Treatments Reduce Bone Formation and Allograft Incorporation

The effect of sustained and temporary COX-2 inhibition was examined on allograft healing using the selective COX-2 inhibitor Celecoxib (25 mg/kg/day) and the nonspecific NSAID, Ketorolac (4 mg/kg/day). Since allograft healing relies on the recruitment of mesenchymal stem cell from the host bed, and is limited to bone formation at the host allograft junction, we hypothesized that healing would be highly sensitive to COX-2 inhibition. Continuous COX inhibition for 5 weeks resulted in a 60% decrease in bone formation following both Celecoxib and Ketorolac treatments compared to controls ($P < 0.05$) (FIG. 2). No significant differences in new bone formation were observed between the Celecoxib- and Ketorolac-treated groups. Both groups showed a lower percentage of histologic bony bridging after 5 weeks, 55% ($n = 17, P < 0.05$) and 61% ($n = 11, P < 0.05$) for Celecoxib and Ketorolac, respectively, as compared to 82% ($n = 16$) in the controls. In the group that was treated with

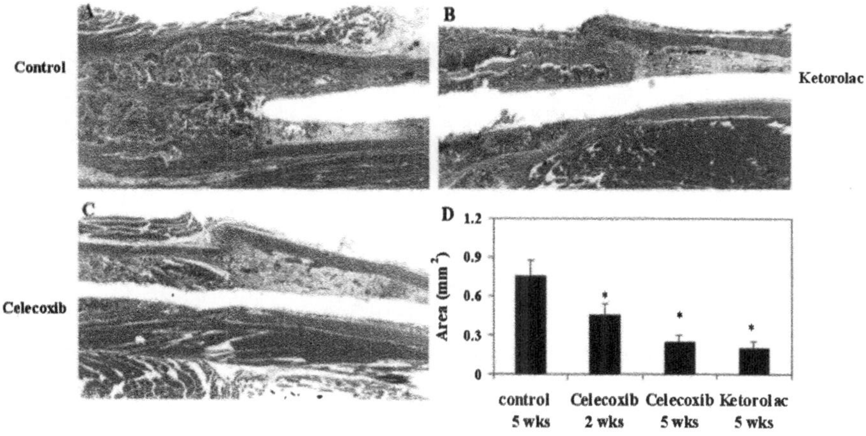

FIGURE 2. Celecoxib or Ketorolac treatment inhibits bone formation on the surface of allografts. Representative histological sections show the allograft incorporation at the host graft junctions at 5 weeks postgrafting in control group (**A**), Ketorolac- (**B**), or Celecoxib-treated mice (**C**). Histomorphometric quantification demonstrated a significant reduction in new bone formation at the graft side in both short-term (2 weeks) and long-term (5 weeks) Celecoxib treatment groups and the group treated with ketorolac for 5 weeks ($^*P < 0.05$).

Celecoxib for only 2 weeks following surgery, a significant reduction in bone formation was also noticed, which was accompanied by a lower rate (66%) of histological bony bridging ($n = 6$). While cessation of Celecoxib at 2 weeks resulted in a slight recovery in bone formation, the recovery was incomplete and a 45% reduction in bone formation remained (FIG. 2D).

PGE2 Infusion Enhances Bone Formation in Allograft Model

To examine COX gain of function, PGE2 was infused into the allograft site at a dose of 800 nmol/kg/day. Weekly radiographs demonstrated increased bone callus formation at the host/allograft junction. Callus was present at day 14 in PGE2-infused mice, while vehicle-treated mice had absent or minimal callus (FIG. 3 C,D). By day 28, callus formation was much greater in PGE2-treated mice compared to the vehicle controls (FIG. 3 E,F). Histological and histomorphometric analyses confirmed increased bone formation following PGE2 infusion (FIG. 4). Quantitative histomorphometry demonstrated about a twofold induction of bone formation on the host side and the allograft side ($n = 3$).

DISCUSSION

A murine allograft-healing model was used to examine the role of COX gain and loss of function in bone repair. Both Ketorolac and Celecoxib

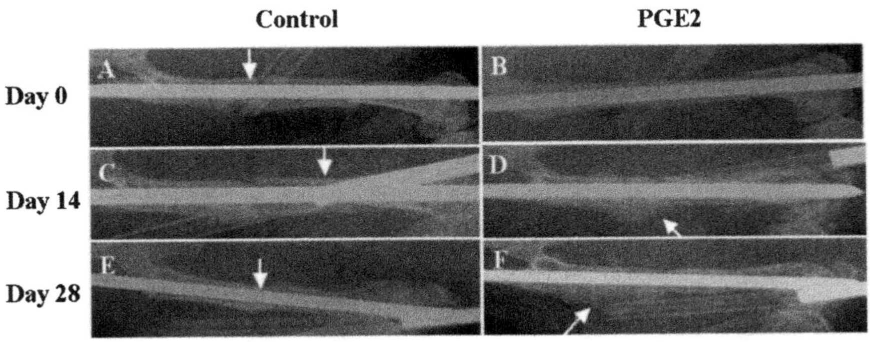

FIGURE 3. Induction of bone callus formation upon *in vivo* PGE2 infusion. An osmotic minipump was inserted subcutaneously into the mouse and an 18G catheter was used to direct vehicle (**A**, **C**, **E**) or PGE2 (**B**, **D**, **F**) to the femoral grafting site. Radiographic healing was monitored by X rays at days 0 (A,B), 14 (C,D), and 28 (E,F) postsurgery. An increase in bone callus formation was observed at the site of the infusion (*arrow*).

markedly inhibited bone repair. Furthermore, short-term administration of Celecoxib for the initial 2 weeks continued to impair bone formation and allograft incorporation at 5 weeks. In contrast, PGE2 infusion stimulated bone formation and healing. Taken together, these results suggest that COX-2-dependent prostaglandins are essential for bone repair and PGE2 generated in the early phase of bone healing is required for efficient bone healing.

It was shown by Gerstenfeld *et al.* in a rat tibia fracture healing model that COX-2 expression is transiently increased during the first 14 days following fracture whereas COX-1 expression does not change over a period of 28 days.[25] The transient induction of COX-2 supports a role during the early phase of bone healing. Indeed, a study by Riew *et al.* suggests that the inhibitory effects of NSAIDs are more significant when administered earlier following spinal fusion.[15] Other studies also suggest that early administration of NSAIDs results in greater inhibition of bone formation.[31–34] However, conflicting data have been presented recently with regard to the effects of short-term inhibition of COX-2 in rat fracture healing models. Gerstenfeld and Einhorn *et al.* report a complete healing recovery in a rat fracture healing model following withdrawal of COX-2 inhibition,[25] whereas O'Connor *et al.* reported a significant healing impairment in a 1- to 14-day treatment regimen.[35] Brown *et al.*, also used a rat model and showed that fracture healing recovers to normal levels, even in the presence of continuous COX inhibition.[36] The current allograft model extends prior observations by examining the role of repair in a more challenging healing environment, similar to the situations commonly encountered in humans. The presence of devitalized allograft involves reduced numbers of bone-forming precursors and decreased vascularity, and therefore may be more sensitive to drug inhibition.

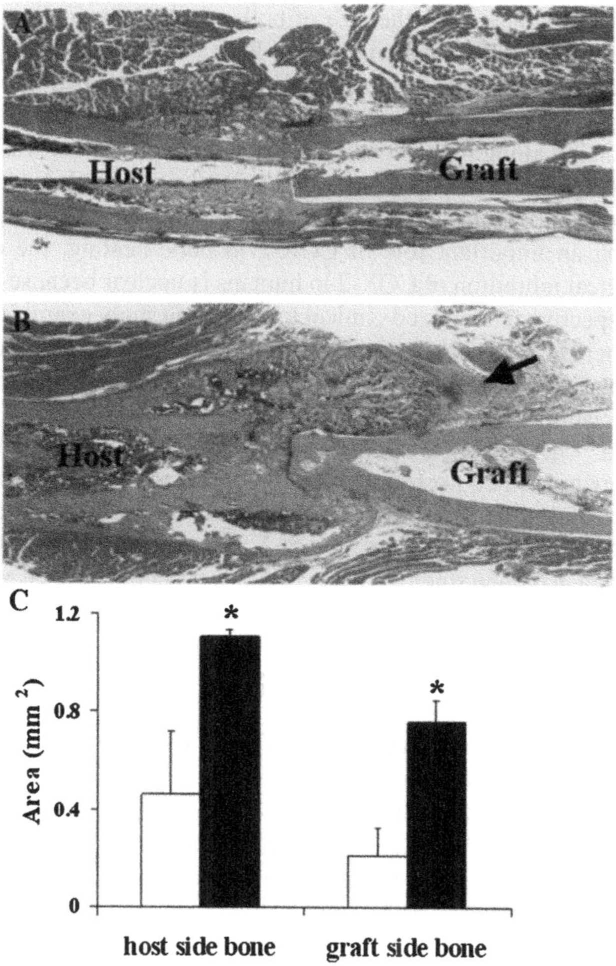

FIGURE 4. Induction of new bone formation upon *in vivo* PGE2 infusion. Representative histology showed the proximal end of bone graft junctions from vehicle (**A**) or PGE2-treated group (**B**) at 4 weeks postgrafting. Histomorphometric analyses demonstrated a significant increase in new bone formation on the surface of host and graft in allografts infused with PGE2 (**C**) (*white bars* = control, *black bars* = PGE2; $n = 3$, $^*P < 0.05$).

For more than two decades it has been known that infusion of PGE2 leads to periosteal hyperostosis in infants.[37] Histological sections of hyperostosis revealed a thickened periosteum and fibrocartilage-like tissue covering an exuberant neocortex of closely aligned woven bone trabeculae.[38] The therapeutic potential of PGE2 in bone healing was not explored until very recently. Studies using both large and small animal models showed that infusion of PGE2, EP2, or EP4 agonists increase periosteum bone formation and improve cortical bone healing.[39–41] A single injection of EP2 agonist CP-533,536 enhances fracture healing by increasing the size of both cartilaginous callus and bone callus. In the current study, local infusion of PGE2 enhanced allograft healing and incorporation. Consistent with prior findings PGE2 infusion stimulated host periosteal bone formation with subsequent creeping bone callus formation that extended along the host–allograft junction. The effect was opposite to the inhibitory effects of Ketorolac or COX-2-selective inhibitor Celecoxib.

In spite of compelling genetic evidence and findings in animal studies demonstrating an important role of COX-2 in bone healing, the effects of pharmacological inhibition of COX-2 in humans is unclear because of the absence of prospective randomized clinical trials.[42] One study examined clinical features associated with nonunion of the femoral shaft and included 32 patients with nonunion and 67 comparable patients with fracture healing.[43] A marked association between nonunion and the use of NSAIDs ($P = 0.000001$) was observed.[43] In contrast, a recent retrospective human study of 80 patients undergoing instrumented posterior spine fusion using autologous bone graft showed that Rofecoxib (50 mg) administered daily for 5 consecutive days did not increase the incidence of nonunion at 1 year postoperation.[44] Given the conflicting data in human studies, COX-2 inhibitors should be used judiciously in following skeletal injury or reconstruction surgery, particularly in patients with a high-risk fracture or a comorbidity that could affect bone repair.

ACKNOWLEDGMENTS

This work is supported by The Orthopedic Research Education Foundation and Musculoskeletal Transplant Foundation and The National Institutes of Health (AR051469, AR46545). We also thank Loralee Gehan and Krista Canary for their assistance with histological work.

REFERENCES

1. VANE, J.R. 1971. Inhibition of prostaglandin synthesis as a mechanism of action for aspirin-like drugs. Nat. New. Biol. **231:** 232–235.
2. FERREIRA, S.H., S. MONCADA & J.R. VANE. 1971. Indomethacin and aspirin abolish prostaglandin release from the spleen. Nat. New Biol. **231:** 237–239.
3. SMITH, C.J. *et al.* 1998. Pharmacological analysis of cyclooxygenase-1 in inflammation. Proc. Natl. Acad. Sci. USA **95:** 13313–13318.
4. SEIBERT, K. *et al.* 1994. Pharmacological and biochemical demonstration of the role of cyclooxygenase 2 in inflammation and pain. Proc. Natl. Acad. Sci. USA **91:** 12013–12017.
5. NUSSMEIER, N.A. *et al.* 2005. Complications of the COX-2 inhibitors parecoxib and valdecoxib after cardiac surgery. N. Engl. J. Med. **352:** 1081–1091.
6. BRESALIER, R.S. *et al.* 2005. Cardiovascular events associated with rofecoxib in a colorectal adenoma chemoprevention trial. N. Engl. J. Med. **352:** 1092–1102.

7. CHENG, H.F. & R.C. HARRIS. 2005. Renal effects of non-steroidal anti-inflammatory drugs and selective cyclooxygenase-2 inhibitors. Curr. Pharm. Des. **11:** 1795–1804.
8. SUDMANN, E. & T. HAGEN. 1976. Indomethacin-induced delayed fracture healing. Arch. Orthop. Unfallchir. **85:** 151–154.
9. BO, J., E. SUDMANN & P.F. MARTON. 1976. Effect of indomethacin on fracture healing in rats. Acta Orthop. Scand. **47:** 588–599.
10. TORNKVIST, H. & T.S. LINDHOLM. 1980. Effect of ibuprofen on mass and composition of fracture callus and bone. An experimental study on adult rat. Scand. J. Rheumatol. **9:** 167–171.
11. SUDMANN, E. & G. BANG. 1979. Indomethacin-induced inhibition of haversian remodelling in rabbits. Acta Orthop. Scand. **50:** 621–627.
12. HO, M.L., J.K. CHANG & G.J. WANG. 1998. Effects of ketorolac on bone repair: a radiographic study in modeled demineralized bone matrix grafted rabbits. Pharmacology **57:** 148–159.
13. ZHANG, X. *et al.* 2002. Cyclooxygenase-2 regulates mesenchymal cell differentiation into the osteoblast lineage and is critically involved in bone repair. J. Clin. Invest. **109:** 1405–1415.
14. DIMAR, J.R.N. *et al.* 1996. The effects of nonsteroidal anti-inflammatory drugs on posterior spinal fusions in the rat. Spine **21:** 1870–1876.
15. RIEW, K.D. *et al.* 2003. Time-dependent inhibitory effects of indomethacin on spinal fusion. J. Bone Joint Surg. Am. **85-A:** 632–634.
16. GLASSMAN, S.D. *et al.* 1998. The effect of postoperative nonsteroidal anti-inflammatory drug administration on spinal fusion. Spine **23:** 834–838.
17. MARTIN, G.J., JR., S.D. BODEN & L. TITUS. 1999. Recombinant human bone morphogenetic protein-2 overcomes the inhibitory effect of ketorolac, a nonsteroidal anti-inflammatory drug (NSAID), on posterolateral lumbar intertransverse process spine fusion. Spine **24:** 2188–2193; discussion 2193–2194.
18. GIANNOUDIS, P.V. *et al.* 2000. Nonunion of the femoral diaphysis. The influence of reaming and non-steroidal anti-inflammatory drugs. J. Bone Joint Surg. **82B:** 655–658.
19. ADOLPHSON, P. *et al.* 1993. No effects of piroxicam on osteopenia and recovery after Colles' fracture. A randomized, double-blind, placebo-controlled, prospective trial. Arch. Orthop. Trauma Surg. **112:** 127–130.
20. RITTER, M.A. & J.M. SIEBER. 1985. Prophylactic indomethacin for the prevention of heterotopic bone formation following total hip arthroplasty. Clin. Orthop. Relat. Res. **196:** 217–225.
21. NEAL, B. *et al.* 2000. Non-steroidal anti-inflammatory drugs for preventing heterotopic bone formation after hip arthroplasty (Cochrane review). Cochrane Database Syst. Rev. **3:** CD001161.
22. MOED, B.R. *et al.* 1998. Heterotopic bone formation in male and female rabbits. J. South Orthop. Assoc. **7:** 86–89.
23. MOED, B.R. *et al.* 1994. Effect of two nonsteroidal antiinflammatory drugs on heterotopic bone formation in a rabbit model. J. Arthroplasty **9:** 81–87.
24. SIMON, A.M., M.B. MANIGRASSO & J.P. O'CONNOR. 2002. Cyclo-oxygenase 2 function is essential for bone fracture healing. J. Bone Miner Res. **17:** 963–976.
25. GERSTENFELD, L.C. *et al.* 2003. Differential inhibition of fracture healing by non-selective and cyclooxygenase-2 selective non-steroidal anti-inflammatory drugs. J. Orthop. Res. **21:** 670–675.

26. Bhandari, M. *et al.* 2003. Predictors of reoperation following operative management of fractures of the tibial shaft. J. Orthop. Trauma **17:** 353–361.
27. Ito, H. *et al.* 2005. Remodeling of cortical bone allografts mediated by adherent rAAV-RANKL and VEGF gene therapy. Nat. Med. **11:** 291–297.
28. Tiyapatanaputi, P. *et al.* 2004. A novel murine segmental femoral graft model. J. Orthop. Res. **22:** 1254–1260.
29. Burchardt, H. 1987. Biology of bone transplantation. Orthop. Clin. N. Am. **18:** 187–196.
30. Goldberg, V.M. & S. Stevenson. 1993. The biology of bone grafts. Semin. Arthroplasty **4:** 58–63.
31. Goodman, S. *et al.* 2002. COX-2 selective NSAID decreases bone ingrowth *in vivo*. J. Orthop. Res. **20:** 1164–1169.
32. Giordano, V. *et al.* 2003. Effect of tenoxicam on fracture healing in rat tibiae. Injury **34:** 85–94.
33. Endo, K. *et al.* 2002. Cyclooxygenase-2 inhibitor inhibits the fracture healing. J. Physiol. Anthropol. Appl. Human Sci. **21:** 235–238.
34. Thaller, J. *et al.* 2005. The effect of nonsteroidal anti-inflammatory agents on spinal fusion. Orthopedics **28:** 299–303; quiz 304–305.
35. Simon, A.M., C.T. Sabatino & J.P. O'Connor. 2001. Effects of cyclooxygenase-2 inhibitors on fracture healing. Trans. Orthop. Res. Soc. **26:** 205.
36. Brown, K.M. *et al.* 2004. Effect of COX-2-specific inhibition on fracture-healing in the rat femur. J. Bone Joint Surg. Am. **86A:** 116–123.
37. Jorgensen, H.R., H. Svanholm & A. Host. 1988. Bone formation induced in an infant by systemic prostaglandin-E2 administration. Acta Orthop. Scand. **59:** 464–466.
38. Faye-Petersen, O.M. *et al.* 1996. Prostaglandin E1-induced hyperostosis: clinicopathologic correlations and possible pathogenetic mechanisms. Pediatr. Pathol. Lab. Med. **16:** 489–507.
39. Yoshida, K. *et al.* 2002. Stimulation of bone formation and prevention of bone loss by prostaglandin E EP4 receptor activation. Proc. Natl. Acad. Sci. USA **99:** 4580–4585.
40. Li, M. *et al.* 2003. A novel, non-prostanoid EP2 receptor-selective prostaglandin E2 agonist stimulates local bone formation and enhances fracture healing. J. Bone Miner. Res. **18:** 2033–2042.
41. Paralkar, V.M. *et al.* 2003. An EP2 receptor-selective prostaglandin E2 agonist induces bone healing. Proc. Natl. Acad. Sci. USA **100:** 6736–6740.
42. Gerstenfeld, L.C. & T.A. Einhorn. 2004. COX inhibitors and their effects on bone healing. Expert Opin. Drug Saf. **3:** 131–136.
43. Giannoudis, P.V. *et al.* 2000. Nonunion of the femoral diaphysis. The influence of reaming and non-steroidal anti-inflammatory drugs. J. Bone Joint Surg. Br. **82:** 655–658.
44. Reuben, S.S. & E.F. Ekman. 2005. The effect of cyclooxygenase-2 inhibition on analgesia and spinal fusion. J. Bone Joint Surg. Am. **87:** 536–542.

The Effect of Anti-TNF-α Therapy on Spinal Bone Mineral Density in Patients with Crohn's Disease

MICHAEL PAZIANAS,[a] ANDREW D. RHIM,[a,b] ANDREW M. WEINBERG,[b] CHINYU SU,[a,b] AND GARY R. LICHTENSTEIN[a,b]

[a]*Department of Medicine, University of Pennsylvania School of Medicine, Hospital of the University of Pennsylvania, Philadelphia, Pennsylvania 19104, USA*

[b]*Division of Gastroenterology, University of Pennsylvania School of Medicine, Hospital of the University of Pennsylvania, Philadelphia, Pennsylvania 19104, USA*

ABSTRACT: Proinflammatory cytokines, including tumor necrosis factor-α (TNF-α) might be, at least partially, responsible for the development of osteopenia or osteoporosis in Crohn's disease. We investigated whether anti-TNF therapy for Crohn's disease could have any skeletal impact. Therefore, we studied the effects of infliximab, a monoclonal antibody against TNF-α with and without bisphosphonates, on spinal bone mineral density (BMD). The effect of corticosteroids was also analyzed. A retrospective cohort analysis was performed on 61 patients with Crohn's disease and low BMD by serial DXA scans. Twenty-three patients were on infliximab and 36 patients were on bisphosphonates. Mean duration between DXA scans was 2.2 ± 0.99 years. After controlling for corticosteroid use, patients with concurrent infliximab and bisphosphonate treatment exhibited a greater increase in BMD compared to those on bisphosphonates alone (+6.7%/year vs. +4.46%/year, $P = 0.045$); corticosteroids inhibited this effect ($P = 0.025$). However, infliximab alone had no effects on BMD. Patients receiving bisphosphonates showed a significant increase in lumbar spine BMD compared to those not on bisphosphonates (+3.97% change in T score/year vs. −3.68%/year, $P < 0.0001$). Concurrent corticosteroid use significantly inhibited this effect (+2.15%/year vs. +4.97%/year, $P = 0.0014$). Concurrent infliximab use may confer an additional benefit to that already documented for bisphosphonate use alone; bisphosphonates are beneficial in the treatment of low BMD in patients with Crohn's disease, though corticosteroids may partially inhibit this effect.

KEYWORDS: Crohn's disease; infliximab; BMD; bisphosphonates; corticosteroids

Address for correspondence: Michael Pazianas, M.D., University of Pennsylvania, 3615 Chestnut Street, Philadelphia, PA 19104. Voice: 215-573-5788; fax: 215-573-8684.
e-mail: pazianas@mail.med.upenn.edu

Ann. N.Y. Acad. Sci. 1068: 543–556 (2006). © 2006 New York Academy of Sciences.
doi: 10.1196/annals.1346.055

INTRODUCTION

Low bone mineral density (BMD) is a relatively common complication of inflammatory bowel disease (IBD). In patients with Crohn's disease, the incidence of osteoporosis is approximately 15%.[1] The issue of bone disease in Crohn's disease has gained increasing attention in recent years, and its pathophysiology remains unclear. While the majority, but not all, studies have shown that total lifetime and concurrent corticosteroid dose have an inverse effect on BMD in patients with Crohn's disease, no clear consensus has emerged from the analyses of other possible modifiable osteoporosis risk factors present in this population, including vitamin D and calcium deficiency and previous intestinal surgery.[1]

Osteoclast-mediated bone resorption is stimulated in response to the release of certain cytokines including tumor necrosis factor-α (TNF-α) and interleukin-6 (IL-6). In Crohn's disease, elevations of these cytokines constitute part of the underlying inflammatory disease activity; therefore, bone resorption could be increased. Indeed, recent studies have shown that the inflammatory state associated with acute and subacute Crohn's disease has deleterious effects on bone composition in laboratory animals.[2,3]

Infliximab (Remicade®, Centocor, Malvern, PA) is a chimeric monoclonal antibody directed against TNF-α. It has a high degree of specificity, affinity, and avidity to TNF-α.[4] It has been shown to be effective in the amelioration of inflammatory disease activity in moderate to severe cases of Crohn's disease.[5,6] Recently, infliximab has been reported to significantly inhibit osteoclast-mediated bone destruction in a rheumatoid arthritis animal model[7] and infliximab, in combination with methotrexate, is indicated for the treatment of moderate-to-severe active rheumatoid arthritis. Finally, recent studies reported improvement of bone turnover markers in Crohn's disease patients treated with infliximab.[8,9] Hence, it is possible that treatment with infliximab may have beneficial effects on BMD.

Bisphosphonates are pyrophosphate analogs that act mainly on the bone-resorbing osteoclasts.[10] The nitrogen-containing bisphosphonates, such as alendronate, risedronate, and ibandronate could inhibit the intracellular mevalonate pathway, a key cholesterol pathway in these cells, resulting in inactivation and apoptosis.[11] Bisphosphonates have been shown to be effective in treating patients with low BMD and preventing fractures associated with postmenopausal osteoporosis.[12,13] In addition, they could prevent/treat glucocorticoid-induced osteoporosis and osteopenia.[14,15] However, data on the effectiveness of bisphosphonate treatment on BMD in patients with Crohn's disease are limited. In a small ($n = 32$) randomized, placebo-controlled study[16] and a larger ($n = 84$) prospective randomized trial,[17] treatment with the bisphosphonate alendronate (10 mg/day for 12 months) and ibandronate (1 mg i.v. every 3 months for 27 months), respectively, both resulted in a 5% increase in BMD in Crohn's patients after 1 year, consistent with trials of bisphosphonate therapy for postmenopausal osteoporosis. The aforementioned studies, however, did not assess the effect of concurrent medications, such as corticosteroids, on the BMD in these patients with Crohn's disease.

We conducted a retrospective cohort analysis to evaluate the effects of infliximab administration on BMD in patients with Crohn's disease and documented low BMD (osteopenia or osteoporosis). We also studied the effects of concurrent bisphosphonate and glucocorticoid treatment.

MATERIALS AND METHODS

Patient Selection

Patients with Crohn's disease receiving care at the Center for IBD at the University of Pennsylvania (UPenn IBD Center) from 1995 to 2002 were identified via computer-generated database. Inclusion criteria were: (*a*) documentation of the diagnosis of Crohn's disease for at least 3 years prior to the study; (*b*) patient had to have received care for their IBD at the UPenn IBD Center for >1 year prior to documentation of osteopenia or osteoporosis; (*c*) documented osteopenia or osteoporosis by DXA scan performed at the UPenn IBD Center (BMD more than one standard deviation [SD] below average for young adults or T score < -1 at the spine); (*d*) documented treatment with bisphosphonates >1 year, if patients were taking bisphosphonates; and (*e*) documented follow-up DXA scan(s) at least 1 year subsequent to the start of bisphosphonate therapy performed at the UPenn IBD Center.

Exclusion criteria included: (*a*) pregnancy or lactation; (*b*) presence of comorbid disease known to significantly affect bone metabolism, such as primary hyperparathyroidism, osteomalacia, and Paget's disease; and (*c*) presence of "anatomical abnormalities" that precluded precise measurements of BMD. The observation period has been defined as the period between first DXA scan result of T score < -1 through the last or most current DXA scan. Institutional review board approval was obtained prior to conducting the research.

Data Collection

All data were extracted from the medical records. Patient data collected included: age at first DXA scan and all subsequent DXA scans, sex, menopausal status, number of pregnancies, history of intestinal resections (including date, length, and location), having any infliximab administration during the observation period, length of bisphosphonate use, side effects of bisphosphonate use, prior or current treatment with estrogen/hormone replacement therapy, calcitonin, calcium, and vitamin D, and smoking status before and during observation period (as defined in exclusion criteria, above), and finally, any glucocorticoid use during the observation period for IBD-related symptoms (or other disorders). Since cumulative corticosteroid dosage could not be accurately calculated for all cases, patients were classified based on documentation of any corticosteroid use (>3 months) during the observation period.

BMD Measurements

All BMD measurements were obtained by DXA scan at the Department of Radiology of the University of Pennsylvania (Hologic QDR; Hologic Inc., Waltham, MA) in order to reduce variability from the use of different densitometers and interpretations from a large number of radiologists.

Measurements of BMD were assessed at the posterior-anterior spine (L1-4); only lumbar spine BMD measurements were included in this report. BMD measurements were expressed as a T score as calculated by the equation: T score = (Measured patient BMD − Mean peak BMD of young adult controls)/(SD of peak BMD of young adults).[18] Osteopenia and osteoporosis were defined according to WHO recommendations[19] as a T score < -1.0 (and up to -2.5 SDs) and -2.5 SDs or lower, respectively.

Outcomes

The primary outcome data were changes in BMD in serial DXA scans over time, as defined as percent (%) change in lumbar spine T score per year. Number of fractures as documented in patient records was also collected as outcome data.

Statistical Methods

All data are reported as mean ± SD. The Mann-Whitney signed rank test was performed on all continuous data sets to determine statistical differences among cohorts. Qualitative data were analyzed by the χ^2 test. A P value of <0.05 was considered significant. The Mann-Whitney signed rank test was performed with StatPlus v2.2 software; all other analyses were performed with Microsoft Excel 2002.

RESULTS

Patient Characteristics

The characteristics of the patients in those with and without bisphosphonate use during the observational period (defined as the period between first

TABLE 1. Patient Characteristics

	+ Bisphos	−Bisphos	+ Infliximab	−Infliximab
n	36	25	23	38
Age (Median, range)	45 (22–77)	42 (22–68)	40* (22–69)	50* (28–77)
Gender (F/M)	20/16	14/11	13/10	20/18
% Premenopausal women	21	17	15	19
Follow-up (year, ±SD)	2.3 ± 1.0	2.2 ± 1.0	2.3 ± 0.9	2.2 ± 1.1
Baseline lumbar spine T score	−2.03 ± 0.78*	−1.26 ± 0.60*	−1.67 ± 0.96	−1.66 ± 0.87

$^*P < 0.001$ for comparison of the +infliximab and –infliximab cohorts.

DXA scan result of T score < -1 through the last or most current DXA scan) and those with and without infliximab use during the observational period are shown in TABLE 1. There were no significant differences between cohorts in gender make-up, number of postmenopausal women, number of women on hormone replacement therapy, follow-up time, or number of patients with recorded corticosteroid use during the observational period by χ^2 test. However, patients who underwent one or more infliximab injections during the observational period were significantly younger than those who did not take infliximab (37.6 vs. 49.8 years; $P = 0.0496$). Prior retrospective studies involving patients taking any dosage of infliximab had a similar average age for the infliximab+ cohort.[20,21] There was no difference in the prevalence of smoking or the degree and number of bowel resections among any of the cohorts (data not shown). Also, there was no difference between the inflix+ and inflix– or the steroid+ and steroid– cohorts in the type and dosage of bisphosphonate and/or corticosteroid taken during the observational period (data not shown). All patients in the bisphos+ cohort were taking oral bisphosphonates, either alendronate or risedronate. The mean duration of the observational period (as defined in the "Methods" section) for all cases was 2.2 ± 1.0 years (range, 1.0–4.5 years). Finally, there was a significant difference in baseline lumbar spine T score between the bisphos+ and bisphos– cohorts; patients who were on bisphosphonates tended to have a lower T score than those who were not on bisphosphonates (−2.03 ± 0.78 vs. −1.26 ± 0.62; $P = 0.037$).

Effect of Infliximab on BMD

There was no statistically significant difference among the infliximab−, bisphos− and infliximab+, bisphos− subgroups: −3.25 ± 4.53% vs. −4.09 ± 3.85% mean change in lumbar spine T score per year; $P > 0.05$. In addition, among patients on bisphosphonates, those on infliximab tended to have greater increases in BMD over time compared to those not on infliximab (+4.38 ± 2.94%/year vs. +3.98 ± 2.35%/year). However, this difference was not significant ($P = 0.068$).

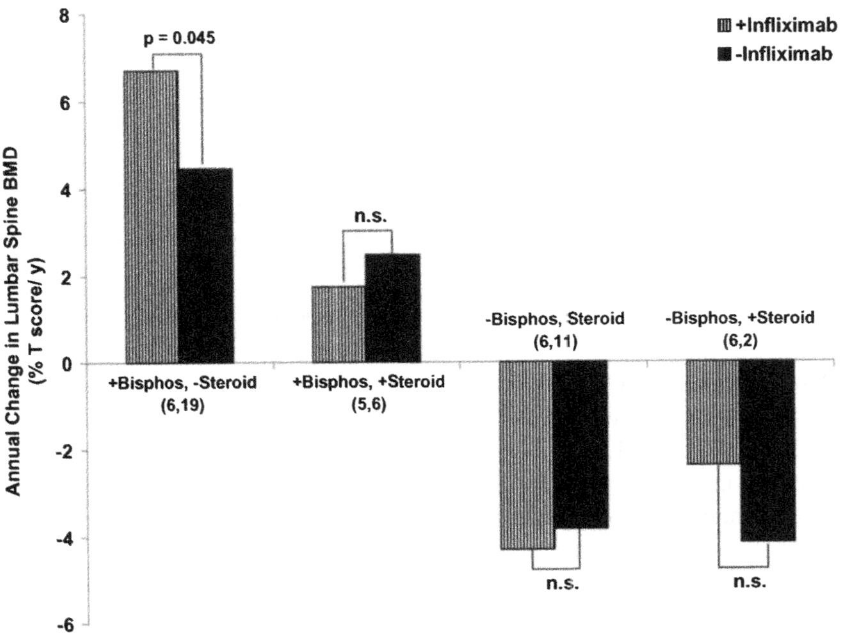

FIGURE 1. The effect of infliximab on mean percent change in lumbar spine T score per year in four subgroups, according to bisphosphonate and corticosteroid usage. Striped bars represent those subgroups that took concurrent infliximab; solid bars represent those that did not have documented infliximab use. Lines and corresponding *P*-values indicate the result of the comparison of the two connected subgroups by Mann-Whitney rank test. *P*-values of >0.05 are listed as n.s. Numbers in parentheses denote the n for those who did and did not take infliximab, respectively. SD given in text.

Further subgroup analysis was completed to assess the effect of infliximab on BMD over time in patients according to bisphosphonate and corticosteroid use (FIG. 1). Infliximab was found to have a positive effect on BMD over time in those patients who took bisphosphonates but not corticosteroids during the observational period (FIG. 1, first pair of bars). Patients in this subgroup (bisphos+, steroid−) who had documented infliximab usage (striped bar) displayed a significantly greater increase in the rate of change of lumbar spine BMD compared to those not on infliximab (solid bar) (6.58 ± 2.06%/ year vs. 4.45 ± 2.30%/year; $P = 0.045$).

Effect of Bisphosphonates on Mean T score Over Time in Crohn's Patients with Low BMD

Data for each individual patient's serial DXA scan results are depicted by cohort in FIGURE 2A, B. As might be expected, patients with low BMD and no documented bisphosphonate use had a significantly greater starting lumbar spine T score compared to those who were started on bisphosphonate treatment

(FIG. 2C) (−1.26 ± 0.63 vs. −2.03 ± 0.78, respectively; $P = 0.0012$). However, at the end of the observation period (as defined in the "Methods" section), the mean T score for the bisphos− cohort was not statistically different from that of the bisphos+ cohort (−1.77 ± 0.69 vs. −1.41 ± 0.94; $P > 0.05$). The mean change in BMD over time was significantly greater in bisphosphonate-treated patients compared to untreated patients (FIG. 2A, B). The average change in lumbar spine T score over time was calculated to be significantly greater in the bisphos+ cohort compared to the bisphos− cohort (+3.97 ± 2.60% vs. −3.68 ± 4.12% change in lumbar spine T score per year; $P < 0.0001$).

Effect of Corticosteroids on BMD in Bisphosphonate-Treated Patients

Subgroup analysis was performed to evaluate the effect of corticosteroid use in the bisphos+ and bisphos− cohorts (FIG. 3). Patients who had documented administration of any dose of intravenous or oral corticosteroid (steroid+, solid bar) in the bisphos+ cohort had a significantly diminished benefit on lumbar spine T score over time compared to those patients in the same cohort with no documented corticosteroid usage (denoted by open bar) (+2.15 ± 1.47% vs. +4.97 ± 2.4% mean change in lumbar spine T score per year; $P = 0.031$). However, these patients in the bisphos+, steroid+ subgroup continued to display significantly greater gains in their lumbar spine BMD compared to the bisphos−, steroid− and bisphos−, steroid+ subgroups (+2.15 ± 1.47 %/year vs. −3.47 ± 3.26 %/year and −4.16 ± 5.94 %/year, respectively; $P < 0.0010$). In other words, all patients who took bisphosphonates, regardless of concurrent corticosteroid use, displayed a significant increase in lumbar spine BMD over time.

DISCUSSION

This is the first attempt to study the effects of infliximab alone or in combination with bisphosphonates, and/or corticosteroids on BMD. We present data to suggest that infliximab may provide an additional benefit in the treatment of low BMD when used in conjunction with bisphosphonates. We could also confirm previous reports[16,17] that the administration of bisphosphonates significantly improves BMD over time in patients with Crohn's disease and osteopenia or osteoporosis. Our data also allowed us to quantify the detrimental effect that corticosteroid use during the same period has on BMD in these same patients.

Infliximab is an antagonist to TNF-α action and has been shown to be effective in the modulation of Crohn's disease activity.[22] Increased levels of circulating TNF-α have also been implicated in inhibiting bone growth[23]; in a rat model of postmenopausal osteoporosis, infusion with anti-TNF-α-binding proteins led to increased bone formation.[24] As such, we hypothesized that administration of infliximab could lead to a positive effect on BMD in patients

(A) Bisphos+

(B) Bisphos-

(C)

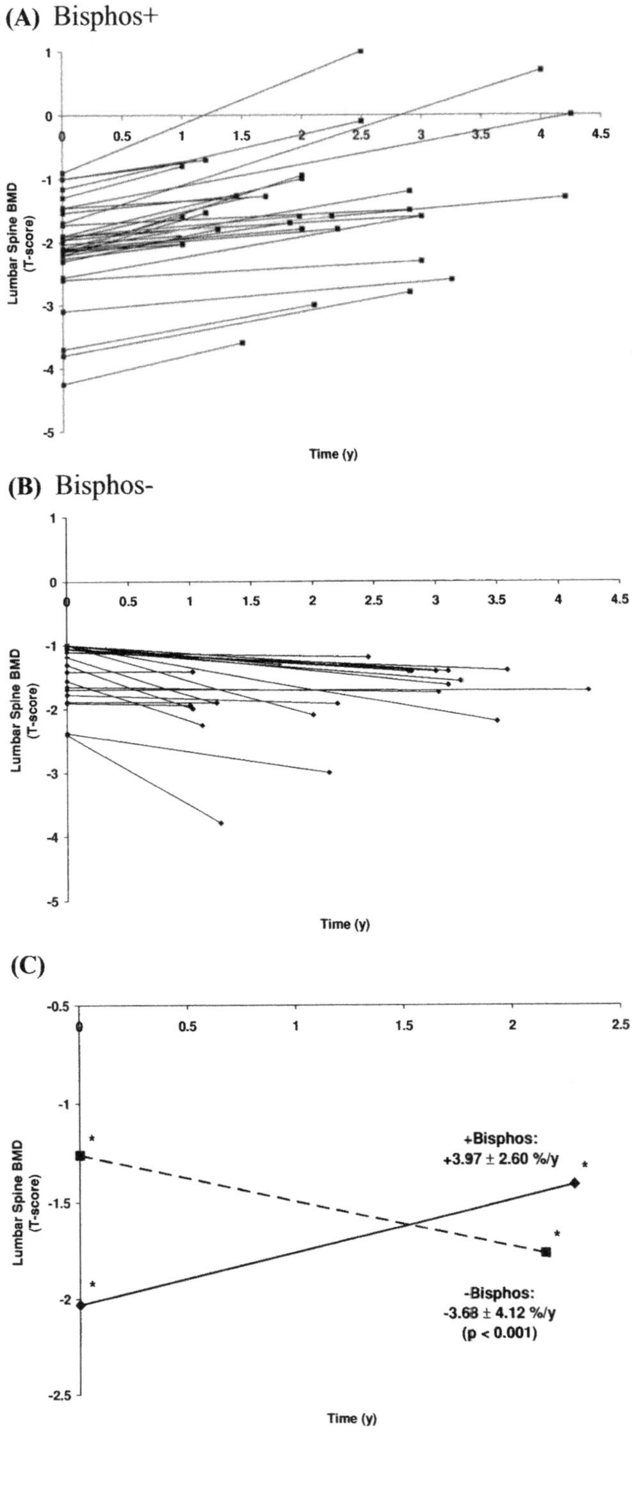

with Crohn's disease. In our analysis of the data, we found that concurrent infliximab usage provided a significant additional benefit to bisphosphonate-associated improvements in BMD over time in Crohn's patients not taking corticosteroids, even with a relatively small sample size (FIG. 1). Furthermore, although administration of infliximab alone did not have any significant effects, there was a trend for improved BMD over time in those patients strictly on infliximab, though this increase did not reach significance. It is possible that given the relatively low sample sizes, the beneficial effects of infliximab administration alone may not have been detected, which underscores the need for further studies with larger power to confirm and elaborate the effect of infliximab on bone disease in patients with Crohn's disease. In support of this argument we could add recent reports of improvement of bone turnover markers in Crohn's disease patients treated with infliximab.[8,9]

The mechanism(s) by which infliximab could reduce bone loss in the Crohn's disease population have not been investigated in humans. As mentioned above, the cytokine TNF-α itself mediates an increase in net bone loss; hence, anti-TNF-α treatment may provide direct protection against osteoclast-mediated bone destruction. Moreover, infliximab administration can also lead to decreased inflammatory disease activity over time. Hence, infliximab may provide protective effects on bone by both inhibiting the bone-destructive effect of the cytokine TNF-α as well as inhibit the proinflammatory effect of TNF-α, resulting in a decrease in the release of other inflammatory cytokines. Allali and colleagues recently published the results of a small prospective study looking at the effect of infliximab on the hip and anteroposterior (AP) BMD in 29 (6 women, 23 men) patients with spondyloarthropathy, 4 of them still receiving steroids during the study.[25] The mean increase of the lumbar spine BMD between the base line and 6 months later was 3.6% (SD ± 5.9), a statistically significant P value of 0.001. However, the clinical significance of this finding remains to be determined as measurement error could not be excluded and lateral DXA measurements were not performed. Also, BMD measurements, performed at 6-month intervals, are not generally accepted as a meaningful test.

In our study, patients with Crohn's disease and low BMD showed an increase in lumbar spine T score of almost 4% per year due to bisphosphonates (FIG. 2).

←

FIGURE 2. Change in lumbar spine bone mineral density T score over time. (**A**) Individual data for all patients in the Bisphos+ cohort. (**B**) Individual data for all patients in the Bisphos− cohort. (**C**) Summary of data in the two cohorts depicted in A and B, including the average lumbar spine T score at the beginning and end of the observational periods. The mean duration of the observational period for the Bisphos+ and Bisphos− cohorts were 2.3 and 2.2 year, respectively. *Denotes statistically significant difference between the Bisphos+ and Bisphos− cohorts at that specific data point by Mann-Whitney rank test ($P = 0.0012$).

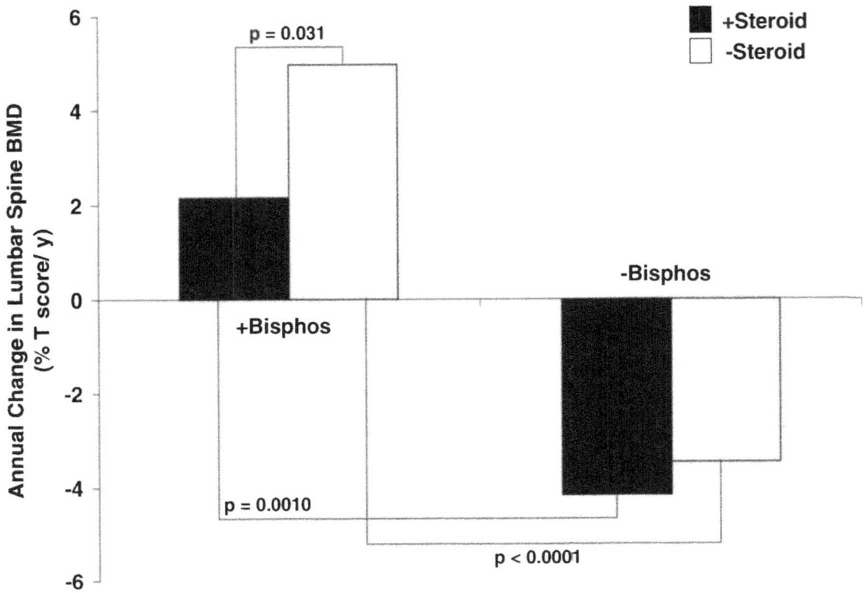

FIGURE 3. The effect of corticosteroids and bisphosphonates on mean percent change in lumbar spine T score per year. +Bisphos and –Bisphos denote those subgroups that did and did not take bisphosphonates. Solid bars represent those subgroups with documented corticosteroid use; open bars represent those without corticosteroid use. Connecting lines represent significant differences between the connected two subgroups at the *P* value listed by Mann-Whitney rank test. SD given in text.

This improvement is comparable to that obtained in the only randomized controlled trials performed in the Crohn's disease population.[16,17] In a double-blind, randomized, placebo-controlled trial by Haderslev and associates,[16] as in our case, the 32 participants were recruited from university-affiliated IBD centers. During the 12 months of their study, the mean spinal BMD increased by 4.6% (± 1.2) in patients on bisphosphonates but decreased in the placebo group by 0.9% (± 1.0).

Similar results were reported in the second randomized study.[17] Intention to treat analysis of data on 68 patients who completed 1-year observation period (up to 27 months) showed that the BMD of patients on intravenous ibandronate (1 mg every 3 months) increased by 3.5% during the first year of treatment and by a further 1.9% in the following 15 months. All patients including those in the placebo group received vitamin D_3 (1000 IU) and calcium citrate (800 mg) daily.

Reports on the rate of bone loss in patients with Crohn's disease are not conclusive. Annual bone loss in the general Crohn's population may not be as great as our study and others[16] suggest. According to a smaller ($n = 25$) prospective European cohort study recently published by De Jong and colleagues,[26] no significant change was found in the annual change in BMD among patients with a wide range of Crohn's disease activity, recruited from that university-affiliated community medical center. This is in contrast to our study where the data were derived from a patient population at the UPenn IBD Center, a tertiary care referral center. Hence, our patients may have had more severe inflammatory disease than most other patients with Crohn's disease and may not be directly comparable to the data obtained from the aforementioned studies. Further, the study by De Jong *et al.* did not perform a subgroup analysis to detail the effect of concurrent corticosteroid on BMD.

To determine the effect of bisphosphonates on spinal BMD in patients requiring any dosage of corticosteroids, we performed a subgroup analysis. Given the retrospective nature of this study, we were not able to analyze the effect of specific corticosteroids or the effect of cumulative dose on lumbar spine BMD. Patients were analyzed according to whether or not any dose of any type of corticosteroid was concomitantly taken during the observational period. Thus, any effect observed in this study associated with corticosteroid use will likely be underestimated (or overestimated), as some patients were on corticosteroids for a brief time (but not <3 months) during the observational period. Though a source of recent controversy,[27–30] it has been suggested that certain corticosteroids, such as budesonide, may not affect bone as much as others.[31] In our study, all patients were on corticosteroids in the form of prednisone and some ($n = 2$) received budesonide for only a short duration of time (several months).

Patients who required corticosteroids during the observational period exhibited significantly less benefit from bisphosphonates compared to those who did not take corticosteroids. Corticosteroids have been shown to have deleterious effects on bone mass by inhibiting osteoblast action, eliciting osteoblast apoptosis, and increasing osteoclast activity.[12,23,32,33] Bisphosphonates, on the other hand, should be effective in protecting the bone from resorption but they most likely could not prevent the slowdown on bone formation.

The need to use corticosteroids in this patient cohort may be indicative of a greater degree of inflammatory activity in Crohn's disease patients. Increased bone loss has been correlated with inflammatory disease activity; increased levels of cytokines released during the inflammatory response in Crohn's disease, including TNF-α, and IL-6, have been shown to stimulate bone degradation mainly through their action on the osteoclasts.[23] Confirming the impact of underlying inflammation on bone metabolism in Crohn's disease, Lin and colleagues have shown in a rat model of colitis that amelioration of inflammatory disease activity directly leads to steady and significant increases in bone mass; in addition, the greater the inflammatory disease activity in these models, the greater the bone loss.[3]

While the endpoint in our study was change in lumbar spine T score over time, these data serve as surrogate marker for the outcome of interest among patients with low BMD, fracture risk. We observed no fractures in the lumbar spine during the observational periods for any of the cohorts. These data were not incongruous with previous studies that estimate the rate of fracture in the IBD population to be about one fracture per 100 patient-years[21] (total number of observed patient-years in the present study was approximately 126). As the rate of fracture is so small, a retrospective single-center study to solely estimate fracture rates among different cohorts would have been impractical.

In summary, our data suggest that infliximab may work in synergy with bisphosphonates to provide additional increases in BMD in patients with Crohn's disease. We also showed that bisphosphonates confer a significant benefit in BMD in patients with Crohn's disease and osteopenia or osteoporosis. Finally, we have shown that the BMD of patients who have documented corticosteroid use do not benefit as much from bisphosphonates compared to those who do not take corticosteroids.

ACKNOWLEDGMENTS

This study was supported by the NIDDK Medical Student Research Fellowship (A.D.R., DK-07748).

REFERENCES

1. BERNSTEIN, C.N., W.D. LESLIE & M.S. LEBOFF. 2003. AGA technical review on osteoporosis in gastrointestinal diseases. Gastroenterology **124:** 795–841.
2. HYAMS, J.S., N. WYZGA, D.L. KREUTZER, *et al.* 1997. Alterations in bone metabolism in children with inflammatory bowel disease: an in vitro study. J. Pediatr. Gastroenterol. Nutr. **24:** 289–295.
3. LIN, C.L., C. MONIZ, T.J. CHAMBERS & J.W. CHOW. 1996. Colitis causes bone loss in rats through suppression of bone formation. Gastroenterology **111:** 1263–1271.
4. VAN DEVENTER, S.J. 1999. Targeting TNF alpha as a key cytokine in the inflammatory processes of Crohn's disease—the mechanisms of action of infliximab. Aliment. Pharmacol. Ther. **13**(Suppl 4): 3–8.
5. TARGAN, S.R., S.B. HANAUER, S.J. VAN DEVENTER, *et al.* 1997. A short-term study of chimeric monoclonal antibody cA2 to tumor necrosis factor alpha for Crohn's disease. Crohn's Disease cA2 Study Group. N. Engl. J. Med. **337:** 1029–1035.
6. RUTGEERTS, P., G. D'HAENS, S. TARGAN, *et al.* 1999. Efficacy and safety of retreatment with anti-tumor necrosis factor antibody (infliximab) to maintain remission in Crohn's disease. Gastroenterology **117:** 761–769.
7. REDLICH, K., S. HAYER, A. MAIER, *et al.* 2002. Tumor necrosis factor alpha-mediated joint destruction is inhibited by targeting osteoclasts with osteoprotegerin. Arthritis. Rheum. **46:** 785–792.
8. FRANCHIMONT, N., V. PUTZEYS, J. COLLETTE, *et al.* 2004. Rapid improvement of bone metabolism after infliximab treatment in Crohn's disease. Aliment. Pharmacol. Ther. **20:** 607–614.

9. RYAN, B.M., M.G. RUSSEL, L. SCHURGERS, *et al.* 2004. Effect of antitumour necrosis factor-alpha therapy on bone turnover in patients with active Crohn's disease: a prospective study. Aliment. Pharmacol. Ther. **20:** 851–857.
10. SATO, M., W. GRASSER & N. ENDO. 1991. Bisphosphonate action: alendronate localization in rat bone and effects on osteoclast ultrastructure. J. Clin. Inv. **88:** 2095–2105.
11. RESZKA, A.A. & G.A. RODAN. 2003. Bisphosphonate mechanism of action. Curr. Rheumatology. Reports **5:** 65–74.
12. AMERICAN COLLEGE OF RHEUMATOLOGY AD HOC COMMITTEE ON GLUCOCORITCOID-INDUCED OSTEOPOROSIS. 2001. Recommendations for the prevention and treatment of glucocorticoid-induced osteoporosis: 2001 update. Arthritis. Rheum. **44:** 1496–1503.
13. 2000. Osteoporosis prevention, diagnosis and therapy. National Institute of Health Consensus Development Conference Statement. March 27–29, **17:** 1–36.
14. SAAG, K.G., R. EMKEY, T.J. SCHNITZER, *et al.* 1998. Alendronate for the prevention and treatment of glucocorticoid-induced osteoporosis. Glucocorticoid-Induced Osteoporosis Intervention Study Group. N. Engl. J. Med. **339:** 292–299.
15. REID, D.M., R.A. HUGHES, R.F. LAAN, *et al.* 2000. Efficacy and safety of daily risedronate in the treatment of corticosteroid-induced osteoporosis in men and women: a randomized trial. European Corticosteroid-Induced Osteoporosis Treatment Study. J. Bone Miner. Res. **15:** 1006–1013.
16. HADERSLEV, K.V., L. TJELLESEN, H.A. SORENSEN & M. STAUN. 2000. Alendronate increases lumbar spine bone mineral density in patients with Crohn's disease. Gastroenterology **119:** 639–646.
17. VON TIRPITZ, C., J. KLAUS, M. STEINKAMP, *et al.* 2003. Therapy of osteoporosis in patients with Crohn's disease: a randomized study comparing sodium fluoride and ibandronate. Aliment. Pharmacol. Ther. **17:** 807–816.
18. 1993. Consensus development conference: diagnosis, prophylaxis, and treatment of osteoporosis. Am. J. Med. **94:** 646–650.
19. WHO STUDY GROUP. 1994. Assessment of fracture risk and its application to screening for postmenopausal osteoporosis. World Health Organ Tech. Rep. Ser. **843:** 1–129.
20. RICART, E., R. PANACCIONE, E.V. LOFTUS, *et al.* 2001. Infliximab for Crohn's disease in clinical practice at the Mayo Clinic: the first 100 patients. Am. J. Gastroenterol. **96:** 722–729.
21. COHEN, R.D., J.F. TSANG & S.B. HANAUER. 2000. Infliximab in Crohn's disease: first anniversary clinical experience. Am. J. Gastroenterol. **95:** 3469–3477.
22. SANDBORN, W.J. 2003. Strategies for targeting tumour necrosis factor in IBD. Best Practice Res. Clin. Gastroenterol. **17:** 105–117.
23. MANOLAGAS, S.C. & R.L. JILKA. 1995. Bone marrow, cytokines, and bone remodeling-emerging insights into the pathophysiology of osteoporosis. N. Engl. J. Med. **332:** 305–311.
24. PACIFICI, R. 1996. Estrogen, cytokines, and pathogenesis of postmenopausal osteoporosis. J. Bone Min. Res. **11:** 1034–1051.
25. ALLALI, F., M. BREBAN, R. PORCHER, *et al.* 2003. Increase in bone mineral density of patients with spondyloarthropathy treated with anti-tumour necrosis factor α. Ann. Rheum. Dis. **62:** 347–349.
26. DE JONG, D.J., L. MANNAERTS, L.G.M. VAN ROSSUM, *et al.* 2003. Longitudinal study of bone mineral density in patients with Crohn's disease. Dig. Dis. Sci. **48:** 1355–1359.

27. D'HAENS, G., A. VERSTRAETE, K. CHEYNS, *et al.* 1998. Bone turnover during short-term therapy with methylprednisolone or budesonide in Crohn's disease. Aliment. Pharmacol. Ther. **12:** 419–424.
28. CINO, M. & G.R. GREENBERG. 2002. Bone mineral density in Crohn's disease: a longitudinal study of budesonide, prednisone, and nonsteroid therapy. Am. J. Gastroenterol. **97:** 915–921.
29. SCHOON, E., S. BOLLANI, P. MILLS, *et al.* 2002. Budesonide versus prednisone: effect on bone mineral density in patients with ileo-cecal Crohn's disease. Am. J. Gastroenterol. **97**(Suppl 1): S272.
30. LUKERT, B.P. & L.G. RAISZ. 1990. Glucocorticoid-induced osteoporosis: pathogenesis and management. Ann. Intern. Med. **112:** 352–364.
31. LICHTENSTEIN, G.R. 2003. Evaluation of bone mineral density in inflammatory bowel disease: current safety focus. Am. J. Gastroenterol. **98:** 24–30.
32. LO-CASCIO, V., E. BONUCCI, B. IMBIMBO, *et al.* 1984. Bone loss after glucocorticoid therapy. Calcif. Tissue Int. **36:** 435–438.
33. KLAUS, J., G. ARMBRECHT, M. STEINKAMP, *et al.* 2002. High prevalence of osteoporotic vertebral fractures in patients with Crohn's disease. Gut **51:** 654–658.

Bone Remodeling in Health and Disease: Lessons From Mathematical Modeling

SVETLANA V. KOMAROVA

Faculty of Dentistry, McGill University, Montreal QC H3A 2B2, Canada

ABSTRACT: Parathyroid hormone is the single anabolic treatment for osteoporosis currently approved by U.S. Food and Drug Administration. The physiology underlying anabolic actions of parathyroid hormone remains mysterious, since only when applied intermittently, does it increase bone mass. This study employs mathematical modeling to consider multiple concomitant effects of parathyroid hormone and local regulators on bone turnover. Remarkably, due to coupling between osteoclasts and osteoblasts, anabolic effects of parathyroid hormone emerged from a single action—stimulation of osteoclastic bone resorption. This study provides a novel conceptual framework for the development of highly desirable anabolic treatments for bone diseases.

KEYWORDS: bone remodeling; power-law approximation; paracrine; autocrine; endocrine; basic multicellular unit

Bone remodeling, which involves resorption by osteoclasts followed by formation of new bone by osteoblasts, occurs in spatially and temporally discrete sites and is coordinated by multiple paracrine and autocrine factors.[1] Bone remodeling plays a major role in mineral homeostasis, by providing access to stores of calcium and phosphate. Parathyroid hormone (PTH) is secreted in response to a drop in plasma calcium levels. Acting on osteoblasts, PTH alters expression of receptor activated of nuclear factor-κB (RANKL) and osteoprotegerin (OPG), leading to a large increase in the RANKL/OPG ratio, thus stimulating osteoclastogenesis and bone resorption.[2] Most intriguingly, the overall effect of PTH on bone depends primarily on its mode of administration. Whereas a continuous increase in PTH levels decreases bone mass, intermittent PTH administration has an anabolic action on bone.[3] Mathematical modeling provides a powerful tool to predict the outcome of multiple, simultaneous actions of autocrine, paracrine, and endocrine factors on bone remodeling.[4–7]

We constructed a mathematical model describing changes in osteoblast and osteoclast populations and in bone mass at a single site of bone remodeling.[5]

Address for correspondence: Dr. Svetlana V. Komarova, Faculty of Dentistry, McGill University, 3640 University Street, Montreal, QC, Canada H3A 2B2. Voice: 514-398-4314; fax: 514-398-8900. e-mail: svetlana.komarova@mcgill.ca

Ann. N.Y. Acad. Sci. 1068: 557–559 (2006).
doi: 10.1196/annals.1346.052

Endocrine factors were assumed to affect local autocrine and paracrine effectors, which in turn govern the resident bone remodeling.[8] PTH was first modeled only to increase RANKL/OPG ratio by changing the expressions of RANKL and OPG by osteoblasts, and thus augmenting the ability of osteoblasts and their precursors to recruit osteoclasts. Since locally osteoblasts are coupled to osteoclasts through paracrine factors, PTH application resulted in the activation of both osteoclasts and osteoblasts. Continuous PTH stimulation resulted in bone loss, since resorption by more active osteoclasts always preceded bone formation. However, if PTH was promptly withdrawn, osteoclasts were rapidly removed and osteoblasts were able to rebuild bone. Importantly, at certain levels of PTH activation, osteoblasts overcompensated leading to net gain of bone mass.

In keeping with experimental data,[9] the parameters describing osteoblast (but not osteoclast) activation were found to be critical in determining whether or not remodeling will result in a gain of bone. Interestingly, different parameters were found to be most effective in initiating the changes that result in a positive remodeling balance. When PTH was assumed to act only by promoting osteoblast formation or survival, the effects on bone were negligible, because on resting surface only few osteoblasts were available to respond to PTH application. In contrast, when PTH first activated osteoclasts, the effective coupling of osteoblasts to osteoclasts resulted in prompt recruitment of osteoblasts, which were then able to benefit from stimulation by PTH.

To test the role of osteoclasts in the anabolic actions of PTH on bone, we simulated clinical studies, in which PTH was given in combination with antiresorptive agents.[10,11] In the presence of bisphosphonates, PTH applications led to smaller increases in osteoclasts, followed by recruitment of fewer osteoblasts and, as a result, the anabolic effect of PTH on bone was lost. These data are in excellent agreement with the findings that treatment of patients with the antiresorptive agent, alendronate, impairs the ability of PTH to increase bone mass.[10,11]

In summary, this study proposes a new hypothesis for the mechanism of the regulation of bone turnover by PTH. Here, PTH acts as a proresorptive agent, which stimulates osteoclasts in order to quickly restore plasma calcium levels. Locally, osteoblasts are effectively coupled to osteoclasts through paracrine factors; therefore PTH application results in the activation of both osteoclasts and osteoblasts. At certain levels of PTH activation, osteoblasts overcompensate leading to a net gain of bone mass. Thus, the model predicts that bone cells interact in such a manner that the duration of PTH application can give rise to qualitatively different outcomes—net bone loss in one case, and gain of bone mass in the other. Overall, this model provides a simple, coherent explanation for a series of otherwise surprising experimental findings, offering a conceptual framework to explore novel options for anabolic therapies for bone diseases.

REFERENCES

1. Parfitt, A.M. 1994. Osteonal and hemi-osteonal remodeling: the spatial and temporal framework for signal traffic in adult human bone. J. Cell Biochem. **55:** 273–286.
2. Suda, T. *et al.* 1999. Modulation of osteoclast differentiation and function by the new members of the tumor necrosis factor receptor and ligand families. Endocr. Rev. **20:** 345–357.
3. Rubin, M.R., & J.P. Bilezikian. 2003. New anabolic therapies in osteoporosis. Endocrinol. Metab. Clin. N. Am. **32:** 285–307.
4. Kroll, M.H. 2000. Parathyroid hormone temporal effects on bone formation and resorption. Bull. Math. Biol. **62:** 163–188.
5. Komarova, S.V. *et al.* 2003. Mathematical model predicts a critical role for osteoclast autocrine regulation in the control of bone remodeling. Bone **33:** 206–215.
6. Rattanakul, C. *et al.* 2003. Modeling of bone formation and resorption mediated by parathyroid hormone: response to estrogen/PTH therapy. Biosystems **70:** 55–72.
7. Lemaire, V. *et al.* 2004. Modeling the interactions between osteoblast and osteoclast activities in bone remodeling. J. Theor. Biol. **229:** 293–309.
8. Komarova, S.V. 2005. Mathematical model of paracrine interactions between osteoclasts and osteoblasts predicts anabolic action of parathyroid hormone on bone. Endocrinology **146:** 3589–3595.
9. Jilka, R.L. *et al.* 1999. Increased bone formation by prevention of osteoblast apoptosis with parathyroid hormone. J. Clin. Invest. **104:** 439–446.
10. Black, D.M. *et al.* 2003. The effects of parathyroid hormone and alendronate alone or in combination in postmenopausal osteoporosis. N. Engl. J. Med. **349:** 1207–1215.
11. Finkelstein, J.S. *et al.* 2003. The effects of parathyroid hormone, alendronate, or both in men with osteoporosis. N. Engl. J. Med. **349:** 1216–1226.

Modeling of Serum C-Telopeptide Levels with Daily and Monthly Oral Ibandronate in Humans

MONE ZAIDI,[a] SOLOMON EPSTEIN,[b] AND KEITH FRIEND[c]

[a]*Mount Sinai School of Medicine, New York, New York 10029, USA*

[b]*Mount Sinai School of Medicine and Doylestown Hospital, Doylestown, Pennsylvania 18901, USA*

[c]*Roche Laboratories, Nutley, New Jersey 07110, USA*

ABSTRACT: A previously validated kinetic-pharmacodynamic model was applied to changes in a marker of bone resorption (serum C-telopeptide cross-links of type I collagen [sCTX]) using data from a large-scale clinical trial of ibandronate (Monthly Oral iBandronate In LadiEs [MOBILE]). The model showed rapid and significant suppression of sCTX compared to baseline with daily and monthly ibandronate, and gradual recovery of sCTX levels between monthly doses. This modeled, post-dose oscillatory pattern of sCTX paralleled significant increases in BMD and was associated with an acceptable safety profile for patients in MOBILE, suggesting that the between-dose recovery of sCTX does not negatively affect the efficacy of once-monthly ibandronate.

KEYWORDS: postmenopausal osteoporosis; bisphosphonates; ibandronate; C-telopeptides; bone-turnover markers; kinetic-pharmacodynamic modeling

INTRODUCTION

Ibandronate is a potent nitrogen-containing bisphosphonate that exerts its antiresorptive activity on bone remodeling by inhibiting farnesyl diphosphate synthase, an enzyme in the mevalonate biosynthetic pathway.[1,2] Initial studies of the bone-binding properties of ibandronate revealed that the molecule was highly potent and it was subsequently developed as an agent offering extended between-dose (intermittent) intervals for the treatment of postmenopausal osteoporosis.

Address for correspondence: Mone Zaidi, M.D., Ph.D., Mount Sinai School of Medicine, One Gustave L. Levy Place, Box 1055, New York, NY 10029. Voice: 1-212-241-8797; fax: 1-212-426-8312.

e-mail: mone.zaidi@msnyuhealth.edu

Ann. N.Y. Acad. Sci. 1068: 560–563 (2006).
doi: 10.1196/annals.1346.058

Bone-turnover markers, such as urinary and serum C-telopeptide cross-links of type I collagen (uCTX and sCTX), offer an accepted measure of efficacy for bisphosphonate therapy. To assist in the determination of appropriate ibandronate- dosing regimens for future trials, a kinetic-pharmacodynamic (K-PD) model of uCTX profiles was developed using sample data obtained from previous studies of oral and i.v. treatment.[3–5] A K-PD model estimates dose-response characteristics, as opposed to a pharmacokinetic-pharmacodynamic (PK-PD) model that estimates dose-concentration–response characteristics. Aided by data from this K-PD model, the safety, pharmacokinetics, and pharmacodynamics of monthly ibandronate were investigated in a Phase I, dose-ranging trial, Monthly Oral Pilot Study (MOPS).[6] Subsequently, the efficacy and safety of several intermittent oral ibandronate doses were compared with the approved 2.5-mg daily dose in a large-scale Phase III clinical trial, MOBILE (Monthly Oral iBandronate In LadiEs).[7] This same K-PD model was validated for sCTX by satisfactorily predicting 1- and 2-year percentage sCTX changes based on comparisons with the empirical data available from MOBILE, but without using MOBILE sCTX data to alter the model.

METHODS

MOBILE was a multinational, randomized, Phase III, noninferiority study of 1609 women with postmenopausal osteoporosis. Several dosing regimens, including 2.5-mg daily ibandronate (FDA-approved in 2003) and 150-mg monthly ibandronate (FDA-approved in 2005) were investigated in this trial.[7] Relative change from base line at 1 year in lumbar spine bone mineral density (BMD) was the primary end point. Secondary end points included a confirmatory analysis of relative change in lumbar spine BMD at 2 years and relative and absolute changes in sCTX, concentrations of which were determined at base line and immediately before dosing at 3, 6, 12, and 24 months. The predose sCTX values represented residual suppression that occurred since the previous dose of the drug.

The parameters used in the K-PD model included ibandronate dose, equilibration rate constant, relative change in sCTX per day following placebo treatment, placebo treatment effect on sCTX (disease progression), effect of vitamin D and calcium supplementation, and sCTX concentration. One advantage of this model is that it allows assessment of sCTX suppression at any time point during the treatment, thus data are not limited to the specific sampling times set forth by the clinical trial protocol.

RESULTS AND DISCUSSION

Modeled sCTX concentrations for the 2.5-mg daily and 150-mg monthly regimens over a continuous 3-year treatment period are shown in FIGURE 1.

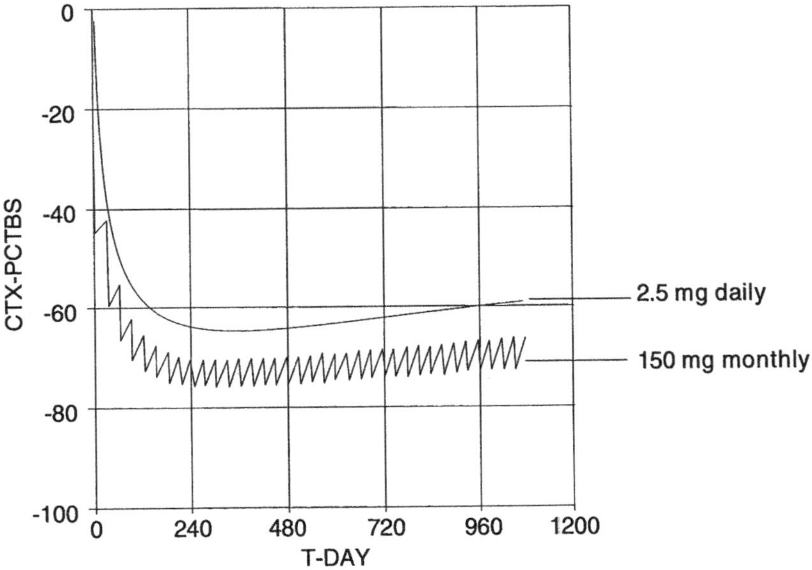

FIGURE 1. Median change in CTX from base line with oral ibandronate 2.5-mg daily versus 150-mg monthly.

Excellent correlation was found between modeled values and the empirical data available from the MOBILE trial.

According to the modeled sCTX profiles, both the daily and monthly regimens delivered rapid sCTX reduction, with the monthly 150-mg dose providing consistently greater reductions in sCTX when compared with the 2.5-mg daily dose. Both regimens maintained a level of reduction associated with significant fracture protection.[8] The post-dose concentrations with daily and monthly doses dropped rapidly into the normal premenopausal range[9] and remained there for the duration of therapy.

At steady state, a small variation in the maximum and minimum sCTX concentration is predicted with the monthly dose. This does not appear to affect efficacy, as assessed by BMD gains. For example, the relative increases in lumbar spine BMD in MOBILE were 5.0% and 6.6% for 2.5 mg daily and 150-mg monthly ibandronate, respectively, at 2 years.[10] Thus, the temporary partial recovery of sCTX levels between monthly doses does not appear to negatively affect efficacy. Physiological variations in uCTX have been observed during the menstrual cycle of healthy premenopausal women,[11] supporting the concept that oscillations in bone turnover are compatible with good skeletal health.

REFERENCES

1. Rogers, M.J. 2003. New insights into the molecular mechanisms of action of bisphosphonates. Curr. Pharm. Des. **9:** 2643–2658.

2. BARRETT, J., E. WORTH, F. BAUSS, *et al.* 2004. Ibandronate: a clinical pharmacological and pharmacokinetic update. J. Clin. Pharmacol. **44:** 951–965.
3. PILLAI, G., R. GIESCHKE, T. GOGGIN, *et al.* 2004. A semimechanistic and mechanistic population PK-PD model for biomarker response to ibandronate, a new bisphosphonate for the treatment of osteoporosis. Br. J. Clin. Pharmacol. **58:** 618–631.
4. GIESCHKE, R & N. HAYASHI. 2002. Exploring oral and intravenous ibandronate dosing regimens by clinical trial simulation [abstract P36]. Osteoporos. Int. **13** (Suppl 3): S23.
5. GIESCHKE, R. & J.-Y. REGINSTER. 2004. Successful prediction of biomarker response to oral monthly ibandronate [abstract P359SU]. Osteoporos. Int. **15** (Suppl 1) (doi: 10.1007s00198-004-1620-7): S97.
6. REGINSTER, J.Y., K.M. WILSON, E. DUMONT, *et al.* 2005. Monthly oral ibandronate is well tolerated and efficacious in postmenopausal women: results from the monthly oral pilot study. J. Clin. Endocrinol. Metab. **90:** 5018–5024.
7. MILLER, P.D., M.R. MCCLUNG, L. MACOVEI, *et al.* 2005. Monthly oral ibandronate therapy in postmenopausal osteoporosis: 1-year results from the MOBILE study. J. Bone Miner. Res. **20:** 1315–1322.
8. HOCHBERG, M.C., S. GREENSPAN, R.D. WASNICH, *et al.* 2002. Changes in bone density and turnover explain the reductions in incidence of nonvertebral fractures that occur during treatment with antiresorptive agents. J. Clin. Endocrinol. Metab. **87:** 1586–1592.
9. GARNERO, P., O. BOREL & P.D. DELMAS. 2001. Evaluation of a fully automated serum assay for C-terminal cross-linking telopeptide of type I collagen in osteoporosis. Clin. Chem. **47:** 694–702.
10. REGINSTER, J.-Y., S. ADAMI, P. LAKATOS, *et al.* 2006. Efficacy and tolerability of once-monthly oral ibandronate in postmenopausal osteoporosis: 2-year results from the MOBILE study. Ann. Rheum. Dis. Pub. **65:** 654–661.
11. GORAI, I., Y. TAGUCHI, O. CHAKI, *et al.* 1998. Serum soluble interleukin-6 receptor and biochemical markers of bone metabolism show significant variations during the menstrual cycle. J. Clin. Endocrinol. Metab. **83:** 326–332.

NFATc1: A Novel Anabolic Therapeutic Target for Osteoporosis

HYEONJU YEO, JAY M. MCDONALD, AND MAJD ZAYZAFOON

Department of Pathology, The University of Alabama at Birmingham, Birmingham, Alabama 35233, USA

ABSTRACT: Bone loss and osteoporosis are major public health problems in the elderly. With increasing life expectancy in the United States, the number of people that will develop age-related bone loss and osteoporosis is expected to rise to over 61 million by 2020. Osteoblast differentiation is a crucial aspect of bone formation and remodeling, a process severely compromised in osteoporosis. Almost all the FDA-approved treatments for building healthier bones, excluding parathyroid hormone (PTH), do not address the decrease in osteoblast differentiation seen in osteoporosis and rather are designed to target osteoclasts and bone resorption. The purpose of this study is to examine the effects of NFAT inhibition on osteoblast differentiation and to elucidate the mechanism of its action. Here we demonstrate that the inhibition of calcineurin (Cn) by using cyclosporine A (CsA) increases osteoblast differentiation, both *in vivo* and *in vitro*. Furthermore, the specific inhibition of NFATc1 by siRNA increased Fra-2 expression in osteoblasts. Taken together, our results point the way to a novel mechanism to aid in the development of anabolic treatment for osteoporosis.

KEYWORDS: osteoporosis; NFAT; AP-1; calcineurin; osteoblasts; cyclosporin

INTRODUCTION

NFAT is regulated by Ca^{2+} and the Ca^{2+}/calmodulin-dependent serine-threonine phosphatase, calcineurin (Cn). In resting cells, NFAT proteins are phosphorylated and reside in the cytoplasm. Upon activation, they are dephosphorylated by Cn, translocate to the nucleus, and become transcriptionally active. Cyclosporin A (CsA) and FK506, widely used immunosuppressive drugs, are known to elicit their responses by decreasing the expression of interleukins primarily via inhibiting the activation of the Cn/NFAT signaling pathway. It is documented that bone disease, in the form of osteopenia and

Address for correspondence: Majd Zayzafoon, The University of Alabama at Birmingham, 533 LHRB, 701 19th Street South, Birmingham, AL 35294. Voice: 205-934-5574; fax: 205-975-9927.
e-mail: mzayzafo@path.uab.edu

Ann. N.Y. Acad. Sci. 1068: 564–567 (2006). © 2006 New York Academy of Sciences.
doi: 10.1196/annals.1346.053

osteoporosis, is a serious complication of organ transplantation in hmans[1]. Although this phenomenon has been attributed to the use of CsA, the role of the underlying disease, pre- and post- transplantation treatments such as steroids, methotrexate and anticoagulants should not be ruled out as contributors to post-transplantation osteoporosis[2]. The role of NFAT in the growth and development of bone have previously focused on the effects of NFAT activation on osteoclastogenesis. The specific roles of Cn and NFAT in osteoblast differentiation are not well understood and in some instances, reports of their roles are contradictory. Here we examine the direct effects of CsA and NFATc1 on osteoblast differentiation.

METHODS

An *in vivo* mouse calvarial model was used. Briefly, Balb/c mice, 5 days old, were injected for 10 days subcutaneously over the parietal region of the skull. At the end of the study, tissues were washed with PBS, fixed with 10% formalin, and embedded in methyl methacrylate. The murine clonal osteogenic cell line, MC3T3-E1, established from newborn mouse calvaria, was used for this study. Cells were treated with CsA or transfected with an NFATc1 siRNA. At the end of the study, cells were harvested and RNA, and cytoplasmic and nuclear protein extractions were performed. Real-time polymerase chain reaction (PCR), Western blotting, and gel shifts assays were performed.

RESULTS

The inhibition of Cn/NFAT by CsA, *in vivo*, produces high-turnover osteopenia characterized by an increase in osteoclasts and osteoblast differentiation.[3] In contrast, CsA leads to decreases in osteoclast differentiation and in bone resorption in cultured isolated osteoclasts and neonatal mouse calvarial and fetal rat long bone.[4,5] To determine the effects of Cn/NFAT inhibition on osteoblast differentiation *in vivo* we used a mouse calvarial model. This method was previously used by us and by others to examine the direct effect of drugs on osteoblast differentiation. DMSO or CsA (1 μM) dissolved in PBS were injected subcutaneously for 10 days into the calvarial region of mouse pups (5 days old). Mice were then sacrificed, calvariae were harvested and tissue was processed for staining and bone histomorphometry. Treatment with CsA dramatically increased the number of osteoblasts (FIG. 1) and bone mineralization. Histomorphometric analyses of calvariae demonstrate a 31% increase in osteoblast number (N.Ob-BS) and an 18% increase in bone volume (BV/TV). These results support the role of Cn/NFAT in osteoblast formation and differentiation *in vivo*.

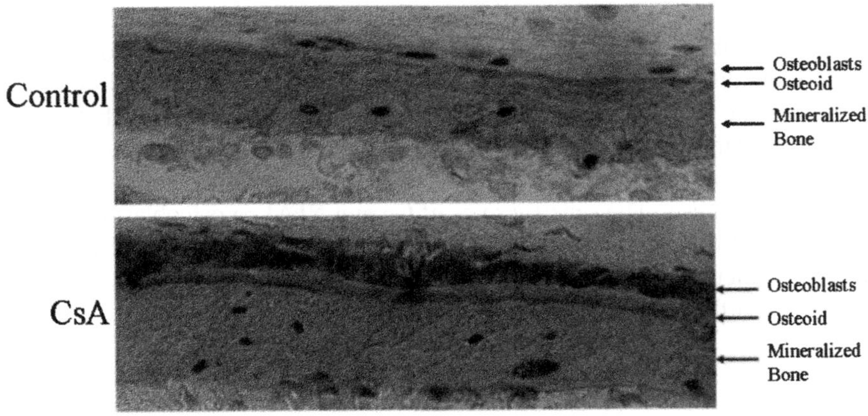

FIGURE 1. CsA increases osteoblast numbers and bone volume *in vivo*. DMSO (Control) or 1 μM CsA dissolved in PBS were injected subcutaneously for 10 days into the calvarial region of 5-day-old mice. At the end of the study, the mice were sacrificed, calvariae dissected, and modified Goldner's TriChrome was performed.

To demonstrate the role of Cn/NFAT in osteoblast differentiation *in vitro*, MC3T3-E1 cells were cultured for 14 days and 21 days and treated with different concentrations of CsA. Cells were stained for alkaline phosphatase activity and mineralization. In addition, osteoblastic gene marker expression was also examined. Our results demonstrate that the inhibition of Cn/NFAT increases osteoblast differentiation dose dependently. Furthermore, quantitative polymerase chain reaction analyses demonstrate that the inhibition of Cn/NFAT by CsA, dose dependently, increases the expression of alkaline phosphatase and osteocalcin. Taken together, these data confirm that the inhibition of Cn/NFAT by CsA increases osteoblast differentiation and bone formation both *in vivo* and *in vitro.*

To elucidate the mechanism by which NFAT inhibition increases bone formation, we examined the expression of Fra-2, a transcription factor that is known to be associated with osteoblast differentiation. Here we show that CsA increases Fra-2 gene and protein expression as well as AP-1 DNA binding activity. Furthermore, the direct inhibition of NFATc1 by siRNA increases Fra-2 protein expression in osteoblasts.

CONCLUSION

We demonstrate that NFATc1 negatively regulates osteoblast differentiation by regulating Fra-2 expression. Findings from our study will provide the first documentation of a novel molecular mechanism that describes the role of NFATc1 in osteoblast differentiation. A basis will thereby be provided for the development of a new target for drug design and therapeutic intervention to combat osteoporosis.

REFERENCES

1. SHANE, E., M. RIVAS, D.J. MCMAHON, R.B. STARON, S.J. SILVERBERG, M.J. SEIBEL, D. MANCINI, R.E. MICHLER, K. AARONSON, V. ADDESSO, & S.H. LO. 1997. Bone loss and turnover after cardiac transplantation. J. Clin. Endocrinol. Metab. **82**: 1497–1506.
2. KATZ, I.A. & S. EPSTEIN. 1992. Post transplantation bone disease. J. Bone Miner. Res. **7**: 123–126.
3. KATZ, I. *et al.* 1994. Influence of age on cyclosporin A-induced alterations in bone mineral metabolism in the rat in vivo. J. Bone Miner. Res. **9:** 59–67.
4. KLAUSHOFER, K. *et al.* 1987. Cyclosporine A inhibits bone resorption in cultured neonatal mouse calvaria. J. Pharmacol. Exp. Ther. **243:** 584–590.
5. ORCEL, P., M.A. DENNE, & M.C. DE VERNEJOUL. 1991. Cyclosporin-A in vitro decreases bone resorption, osteoclast formation, and the fusion of cells of the monocyte-macrophage lineage. Endocrinology **128:** 1638–1646.

Mechanical Loading Differentially Regulates Membrane-Bound and Soluble RANKL Availability in MC3T3-E1 Cells

DAE WON KIM, HAHN-JUN LEE, JAIME A. KARMIN, SE EUN LEE, SEONG-SIL CHANG, BEN TOLCHIN, SHIPHIN LIN, SAM K. CHO, ANTHONY KWON, JAE MOK AHN, AND FRANCIS YOUNG-IN LEE

Center for Orthopaedic Research, Department of Orthopaedic Surgery, Columbia University, New York, New York 10032, USA

ABSTRACT: To understand the biochemical response of RANKL in response to mechanical loading, MC3T3-E1 cells were biequiaxially stretched. A murine RANKL cDNA with double epitopes, pEF6 HA-RANKL-V5His, was transfected into MC3T3-E1 cells, which were then stretched. Endogenous RANKL protein expression increased in response to mechanical loading. Membrane-bound RANKL (HA-RANKL-V5His) increased in cell lysates while soluble RANKL (RANKL-V5His) decreased in the conditioned media after mechanical loading. This may have resulted from the decreased activity of TACE after mechanical loading. Increased membrane-bound RANKL may be one of the mechanisms through which osteoblasts adapt to mechanical loading by regulating osteoclastogenic activity in a region-specific manner.

KEYWORDS: RANKL; TACE; mechanical loading; osteoblast

INTRODUCTION

Osteoblasts play a pivotal role in bone remodeling and the regulation of osteoclastogenesis by activating important bone active cytokines, such as receptor activator of nuclear factor-kappa B ligand (RANKL).[1,2] The effect of mechanical loading on bone quality is confined to the region where mechanical loading is directly applied and does not extend to remote regions. Physical exercise and mechanical loading may govern bone remodeling by regulating the availability of RANKL to surrounded osteoclast precursors.

On the other hand, remote bones may not have a significant influence due to the region-specificity of regulation of membrane-bound RANKL by osteoblasts. In the present study, we sought the effects of mechanical loading on

Address for correspondence: Francis Y. Lee, M.D., Center for Orthopaedic Research, Department of Orthopaedic Surgery Columbia University, 630 W. 168th Street, BB14-1411, New York, NY 10032. Voice: 212-305-3293; fax: 212-305-8271.
e-mail: fl127@columbia.edu

Ann. N.Y. Acad. Sci. 1068: 568–572 (2006). © 2006 New York Academy of Sciences.
doi: 10.1196/annals.1346.054

RANKL protein expression in conjunction with TACE activity in MC3T3-E1 preosteoblast cells. Mechanical loading may differentially regulate membrane-bound and soluble RANKL in MC3T3-E1 cells.

MATERIALS AND METHODS

Mechanical Loading Condition

MC3T3-E1 cells purchased from American Type Culture Collection (ATCC, Manassas, MD) were cultured in 6-well Bioflex culture plates (Flexcell International, Hillsborough, NC) with propagation medium. After 24 h, the medium was replaced with osteogenic medium. Next, sinusoidal, cyclical stretching (0, 0.25, 0.5, 1, and 2.5%) was applied at a frequency of 1 Hz for 15 min/day over 7 days, using a Flexcell 4000 system (Flexcell International).

Plasmid Construction, Talon-Pull Down Assay, and Immunoblotting

A murine RANKL cDNA with double epitopes (N-terminal HA and C-terminal V5His) was generated by polymerase chain reaction (PCR), and cloned into a pEF6V5His TOPO vector (Invitrogen, La Jolla, CA) (See FIG. 1A.) Two micrograms of pEF6 HA-RANKL-V5His were transfected into human embryonic kidney (HEK) 293 cells (8×10^5) in 6-well plates with Superfect (Qiagen, Valencia, CA). After 24 h, medium containing 100 nM of phorbol-12-myristate-13-acetate (PMA) (Calbiochem, San Diego, CA) and 100 nM of TAPI II was added to test whether ectopically expressed HA-RANKL-V5 His is under the control of TACE (FIG. 1B). Identical amounts of plasmid were transfected using Fugene 6 (Roche Diagnostics, Branchburg, NJ) into MC3T3-E1 cells, which were then stretched. To detect secreted RANKL-V5His in the medium, 1 mL of conditioned medium was incubated with 10 μg of Talon (Clontech, San Diego, CA) on a rotating plate at 4°C for 12 h and centrifuged at 12,000 rpm for 20 s. The remaining pellet was suspended with 2X sample buffer. Each harvested cell was lysed with a buffer A containing Tris/Cl, pH 7.4 (10 mM), NaCl (150 mM), 1% Triton X-100, 0.25% Nonidet P-40, EDTA (2 mM), and a protease inhibitor cocktail tablet (Roche Diagnostics). EDTA-free lysis buffer B (2.5 mM Tris/Cl pH 7.3, 135 mM NaCl, 2.6 mM KCl, 1% NP-40, 10 mM 1.10-phenathroline) was used to detect the active form of TACE. RANKL (Calbiochem), V5 (Invitrogen), HA (Covance, Princeton, NJ), GAPDH (Chemicon, Temecula, CA), and TACE antibodies were used.

RESULTS

The effect of mechanical loading on endogenous RANKL protein expression was examined in MC3T3-E1 cells. Cells were loaded at strains of 0 (control),

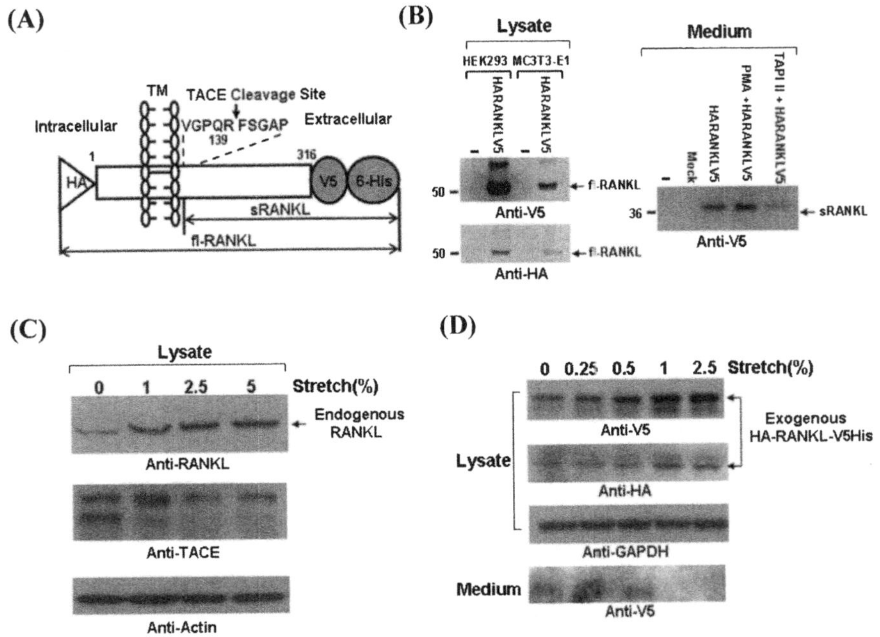

FIGURE 1. (**A**) Shown is representation of double-tagged RANKL (HA tag at its N-ter and V5 and histidine tags at its C-ter). (**B**) pEF6 HA-RANKL-V5His was transfected into HEK293 or MC3T3-E1 cells. Expression was confirmed with anti-HA and -V5 antibodies using cell lysates. A talon-pull down assay was performed with the medium from pEF6 HA-RANKL-V5His-transfected cells, which were treated with PMA (100 nM) and TAPI II (100 nM). Secreted RANKL-V5His in the medium was detected by anti-V5 antibody. (**C**) MC3T3-E1 cells were subject to sinusoidal, cyclical stretching (0, 1, 2.5, and 5%) at a frequency of 1 Hz for 15 min/day for 7 days. Actin was used as a loading control. Active form of TACE decreases while endogenous RANKL protein increased as strain increases. (**D**) pEF6 HA-RANKL-V5His was transfected into MC3T3-E1 cells, which were stretched at strains of 0, 0.25, 0.5, 1, and 2.5% at a frequency of 1 Hz for 15 min/day for 7 days. HA-RANKL-V5His increased in the cell lysate while secreted RANKL-V5His decreased in the conditioned medium as strain increased. GAPDH was used as a loading control.

1, 2.5, and 5% for 15 min/day for 7 days. Immunoblotting of cell lysates showed a strain-dependent increase in RANKL protein expression (FIG. 1C). Differential regulation of membrane-bound RANKL by mechanical loading may have resulted from altered functionality of TNF-α converting enzyme (TACE) that is responsible for ectodomain shedding.[3] There was a trend of decreasing concentrations of the active form of TACE as mechanical strains increased (FIG. 1C). The trend was reciprocal to the changes in RANKL protein concentrations in cell lysates, suggesting that TACE plays a role in mechanical load-induced RANKL expression.

Since ectodomain shedding of ectopically expressed HA-RANKL-V5His was inhibited by TAPI II, a TACE inhibitor (FIG. 1B), we used this system to further verify the aforementioned finding. MC3T3-E1 cells transfected with pEF6HA-RANKL-V5His were mechanically loaded at strains of 0 (control), 0.25, 0.5, 1, and 2.5% for 15 min/day for 7 days. Ectopically expressed HA-RANKL-V5His increased in the cell lysate while soluble RANKL-V5His decreased in the conditioned medium as strain increased (FIG. 1D). This finding also supports the conclusion that RANKL ectodomain shedding is regulated by mechanical loading in conjunction with TACE activity. It is also possible that a decreased availability of pro-TACE caused by mechanical loading alters total TACE activity. To examine this possibility, mRNAs from MC3T3-E1 cells subjected to the same mechanical loading were collected. Real-time polymerase chain reaction showed that TACE mRNA was not changed at higher strains (data not shown).

DISCUSSION

Bone is a dynamic organ that continuously adapts to mechanical loading.[4,5] The *in vivo* effect of physical exercise on RANKL production is still unknown. Mechanical adaptation may be specifically confined to the bone, which is either loaded or unloaded. Concerning the role RANKL plays in mechanical load-induced bone remodeling, several studies were restricted to determining only the changes in RANKL mRNA expression after mechanical loading. No intensive study was available focusing on RANKL protein expression in response to mechanical loading. In the present study, therefore, we examined the changes of RANKL at the protein level in response to mechanical loading, since availability of RANKL may be one of the key factors governing bone remodeling. We were able to demonstrate that membrane-bound RANKL increases and soluble RANKL decreases as strain increases with our double-tagged RANKL expression system.

TACE was also shown to be influenced by mechanical loading. The active form of TACE was decreased by mechanical loading. Collectively, mechanical loading may differentially regulate RANKL while affecting the functionality of TACE via an unknown mechanism. However, this issue requires further investigation. A thorough examination of the effects of mechanical loading on the expression of RANKL and other bone active cytokines will illuminate the molecular mechanisms by which osteoblasts adapt to mechanical loading.

REFERENCES

1. THEOLEYRE, S., Y. WITTRANT, S.K. TAT, *et al.* 2004. The molecular triad OPG/RANK/RANKL: involvement in the orchestration of pathophysiological bone remodeling. Cytokine Growth Factor Rev. **15:** 457–475.
2. GORI, F., L.C. HOFBAUER, C.R. DUNSTAN, *et al.* 2000. The expression of osteoprotegerin and RANK ligand and the support of osteoclast formation by

stromal-osteoblast lineage cells is developmentally regulated. Endocrinology **141:** 4768–4776.

3. LUM, L., B.R. WONG, R. JOSIEN, *et al.* 1999. Evidence for a role of a tumor necrosis factor-alpha (TNF-alpha)-converting enzyme-like protease in shedding of TRANCE, a TNF family member involved in osteoclastogenesis and dendritic cell survival. J. Biol. Chem. **274:** 13613–13618.
4. HUGHES-FULFORD, M. 2004. Signal transduction and mechanical stress. Sci STKE. **249:** RE12.
5. HAN, Y., S.C. COWIN, M.B. SCHAFFLER & S. WEINBAUM. 2004. Mechanotransduction and strain amplification in osteocyte cell processes. Proc. Natl. Acad. Sci. USA **101:** 16689–16894.

Index of Contributors

Printed and bound by CPI Group (UK) Ltd, Croydon, CR0 4YY

15/09/2026

14962391-0004